100 ICONIC WATCHES

GISBERT L. BRUNNER

teNeues

CONTENTS INHALT

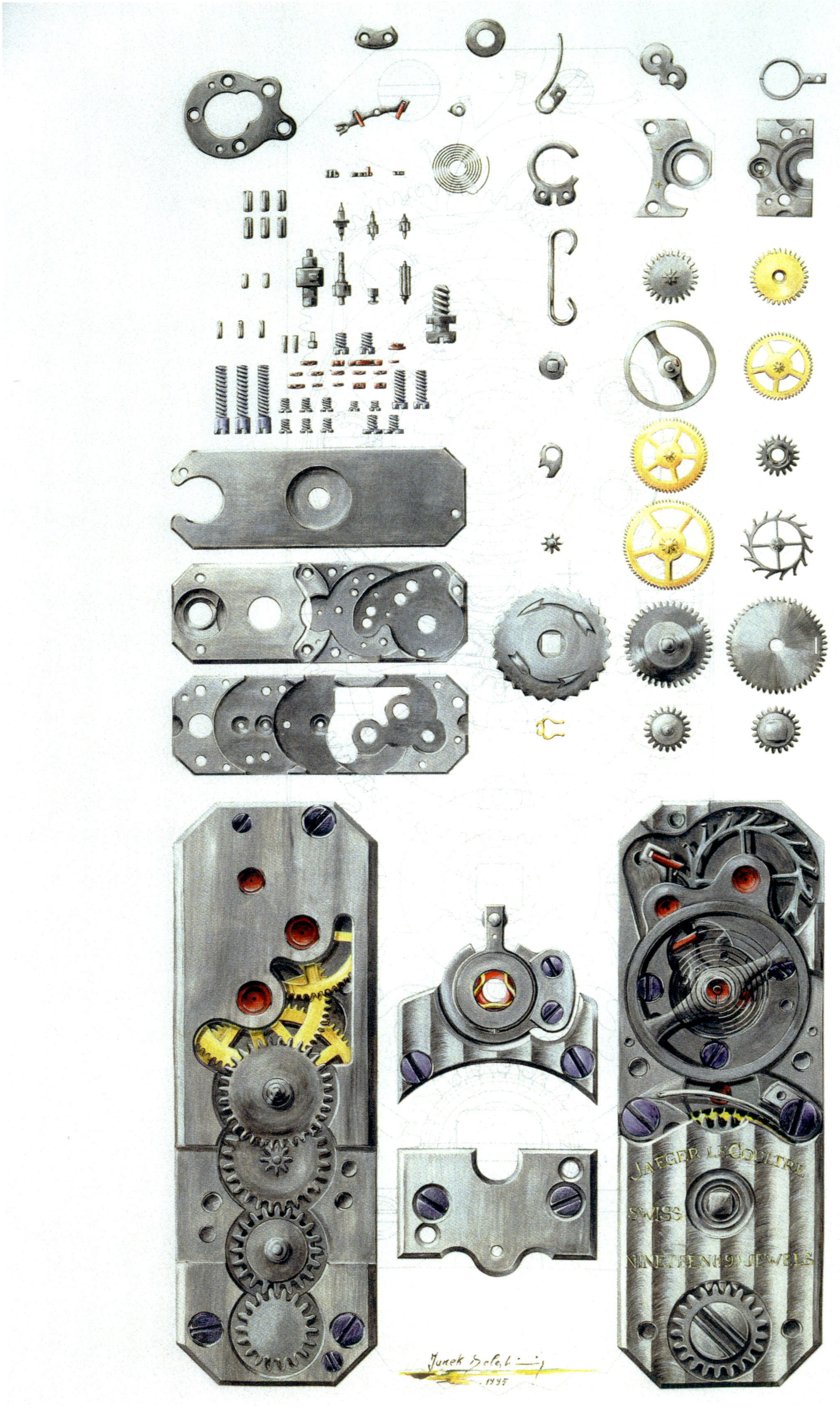
JAEGER LECOULTRE
SWISS
NINETEEN(19)JEWELS

FOREWORD VORWORT

WRIST CLASSICS AND ICONS

Success has many fathers, but failure is always an orphan. As in all areas of business life, this adage is also true of the watch industry. Top or flop? Which will it be for the many new watch models launched on the market every year? Some sink into oblivion before the year is out, others enjoy decades of success. Watches, like all other accessories, are influenced by trends and the spirit of the times. Referring to the latter, Søren Kierkegaard said that it soon made widowers of those who married it. When it comes to watches, the postulations of the Danish philosopher some 150 years ago are not so iron-clad. Although in fairness quite a few of the modern creations are pretty short-lived. Bought today, barely "in" tomorrow, and "out" the day after. Yet the princely sums often demanded call for longer-term companionship. Such a promise is kept by classic models that have held their own on the market for years or even decades. This species includes the Cartier Santos, Jaeger-LeCoultre Reverso and Audemars Piguet Royal Oak, to name but a few. By definition, classics are "exemplary products of the highest order, the products of outstanding literary, artistic or scientific achievement by creative minds whose qualities radiate sophisticated genius". Naturally, the spectrum of what can be ranged under this lofty concept is inexhaustible. Conferring of this oft-abused attribute can only be objectified to a certain extent.

Besides classics, the watch world also abounds with icons and cult images. They stand for objects that represent creative, technical or artisan values to an outstanding degree. Icons symbolise the philosophy, virtuosity or quite simply the appearance of a watch brand. Sometimes, they are even regarded as synonymous with brands, because the name of the watch model is almost inseparably linked to them.

Either way, this book presents 100 timepieces of a unique character. Distinctive elements include design, sustainability, architecture, engineering, top performance or simply the importance of a long pedigree.

Readers may ponder as to why precisely a particular watch is featured in the book and others are not. In this case, objective criteria are joined by the subjective perception of the author after 60 years of intensive engagement with watches. And, last but not least, by personal taste, for which even in the twenty-first century there is still no accounting for.

Gisbert L. Brunner

KLASSIKER & IKONEN FÜRS HANDGELENK

Erfolge haben viele Väter; der Misserfolg ist Waisenknabe. In der Uhrenindustrie gilt diese Erkenntnis genauso wie in allen anderen Bereichen des wirtschaftlichen Alltagslebens. „Top oder Flop?" heißt es bei jedem der vielen neuen Uhrenmodelle, die jedes Jahr auf den Markt strömen. Manche davon überleben kein Jahr, andere richten sich dauerhaft und erfolgreich in einer Szene ein. Wie bei allen anderen Accessoires ist sie auch bei Uhren geprägt von Trends und Zeitgeist. Letzterem bescheinigte schon Søren Kierkegaard, dass er all jene rasch zu Witwern mache, welche sich mit ihm vermählen. Ganz so apodiktisch, wie es der dänische Philosoph vor etwa 150 Jahren postulierte, liegen die Dinge bei Armbanduhren sicherlich nicht. Aber gar nicht so wenige der modernen Kreationen besitzen tatsächlich recht kurze Halbwertszeiten. Heute gekauft, morgen gerade noch tragbar und übermorgen schon wieder daneben. Dabei verlangten die oft stattlichen Preise nach längerer Lebensgemeinschaft. Eine solche versprechen Klassiker, die sich über Jahre oder gar Jahrzehnte am Markt behaupten. Zu dieser Spezies gehören, um hier nur einige Beispiele zu nennen, die seit mehr als 100 Jahren am Markt vertretene „Santos" von Cartier, die „Reverso" von Jaeger-LeCoultre oder die „Royal Oak" von Audemars Piguet. Per definitionem handelt es sich bei Klassikern um „mustergültige Produkte ersten Ranges, Resultate herausragender literarischer, künstlerischer oder wissenschaftlicher Leistungen schöpferischer Menschen, welche die Merkmale einer ausgereiften Meisterschaft in sich tragen." Naturgemäß groß ist die Bandbreite dessen, was sich unter diesen bedeutungsvollen Begriff subsumieren lässt. Außerdem ist die Zuerkennung des oft missbrauchten Attributs nur in Maßen objektivierbar.

Neben Klassikern kennt die Uhrenwelt auch noch Ikonen oder Kultbilder. Sie stehen für Objekte, welche gestalterische, technische oder handwerkliche Werte in herausragender Weise repräsentieren. Ikonen symbolisieren die Philosophie, die Kunstfertigkeit oder schlichtweg den Auftritt einer Uhrenmarke. Mitunter gelten sie sogar als Synonym, weil sich der Name eines Uhrenmodells beinahe untrennbar mit dem der dahinterstehenden Marke verknüpft.

Wie dem auch sei: Dieses Buch präsentiert 100 Zeitmesser mit ganz besonderem Charakter. Auszeichnende Elemente sind Design, Nachhaltigkeit, Konstruktion, Technik, Bestleistung oder ganz einfach die Bedeutung in der langen Historie der Armbanduhren.

Natürlich stellt sich da und dort die Frage, warum gerade diese Uhr im Buch vertreten ist und nicht jene. Zu objektiven Kriterien gesellt sich hier das subjektive Empfinden des Autors nach 60 Jahren intensiver Beschäftigung mit Armbanduhren. Nicht zu vergessen ist schließlich der persönliche Geschmack, über den sich auch im 21. Jahrhundert weiterhin nicht streiten lässt.

Gisbert L. Brunner

<< Jaeger-LeCoultre Cal. 101

ALPINA 2016

SEASTRONG DIVER HERITAGE

DIVER'S WATCH IMMERSED IN TRADITION

In 1967, the Alpina 10 made a name for itself. This diver's wristwatch has a compressor case by Ervin Piquerez SA (EPSA). The experienced case-making company had already presented its innovative patented design in 1956. An integrated spring mechanism prevented compression and potential damage to the rubber gasket when the case was being screwed down. Increasing diving depth and thus water pressure press the two parts of the case ever more firmly together. Additional features included two crowns with a characteristic grid pattern. One was used to wind and adjust the hands, the other to adjust the diving time rotating bezel located in the interior in both directions. This watch is water-resistant to ten bar.

Its comeback was celebrated in 2016 with the Alpina Seastrong Diver Heritage. The model's 1960s predecessor is recalled by the two crowns in the right case flank and the rotating bezel, which had been moved inside the case at the time for good reason. Incidentally, the 60-minute division also benefited non-divers. The arrow can be positioned so that the approaching hour hand indicates when the parking time is running out. Pasta connoisseurs will also appreciate this feature to achieve the perfect al dente spaghetti. In this case, the arrow should be positioned well ahead of the minute hand. When it reaches the arrow, it is time to remove the pasta. Thanks to its current design with a larger 42-millimetre stainless-steel case, it is water-resistant to 30 bar. Timekeeping is ensured by the automatic AL-525 movement based on a Selita SW200.

TAUCHERUHR MIT TRADITION

1967 machte bei Alpina die „Alpina 10“ von sich reden. Diese Taucherarmbanduhr verfügte über eine sogenannte „Compressor-Schale“ der Ervin Piquerez SA (EPSA). Seine neuartige und deshalb auch patentierte Konstruktion hatte der erfahrene Gehäusefabrikant schon 1956 vorgestellt. Bei ihr verhinderte ein integrierter Federmechanismus beim Verschrauben des Bodens das Zusammenquetschen und die daraus möglicherweise resultierende Beschädigung des Gummidichtungsrings. Zunehmende Tauchtiefe und damit einhergehend steigender Wasserdruck pressten die beiden Gehäuseteile immer fester zusammen. Zu den Merkmalen gehörten ferner zwei Kronen mit charakteristischem Gittermuster. Eine diente dem Aufziehen und Richten der Zeiger, die zweite dem Einstellen des innenliegenden Tauchzeit-Drehrings in beide Richtungen. Bis zu zehn bar widerstand diese Uhr dem Druck des nassen Elements.

Das Comeback erfolgte 2016 in Gestalt der „Alpina Seastrong Diver Heritage“. Ans Vorbild aus den 1960er-Jahren erinnerten zwei Bedienelemente in der rechten Gehäuseflanke und der seinerzeit aus guten Gründen ins Gehäuseinnere verlagerte Drehring. Seine 60-Minuten-Teilung kam übrigens auch Nichttauchern zugute.

Beispielsweise kann man den Merkpfeil so positionieren, dass der sich annähernde Stundenzeiger das Ablaufen der Parkzeit signalisiert. Am heimischen Herd leistet besagtes Feature wertvolle Dienste beim Al-dente-Kochen der Spaghetti. In diesem Fall positioniert man den Merkpfeil entsprechend weit vor dem Minutenzeiger. Ist er dort angekommen, muss die Pasta aus dem Wasser. Dank aktueller Gehäusekonstruktion ist das auf 42 Millimeter gewachsene Edelstahlgehäuse bis zu 30 bar wasserdicht. Für die Uhrzeit ist das auf einem Sellita SW200 basierende Automatikkaliber AL-525 zuständig.

2016

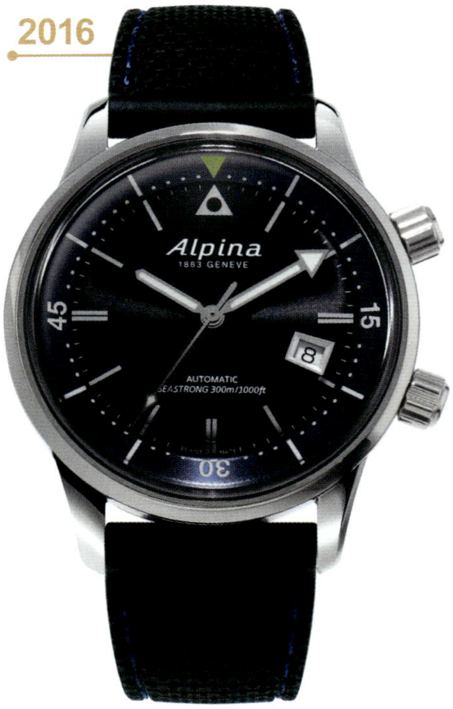

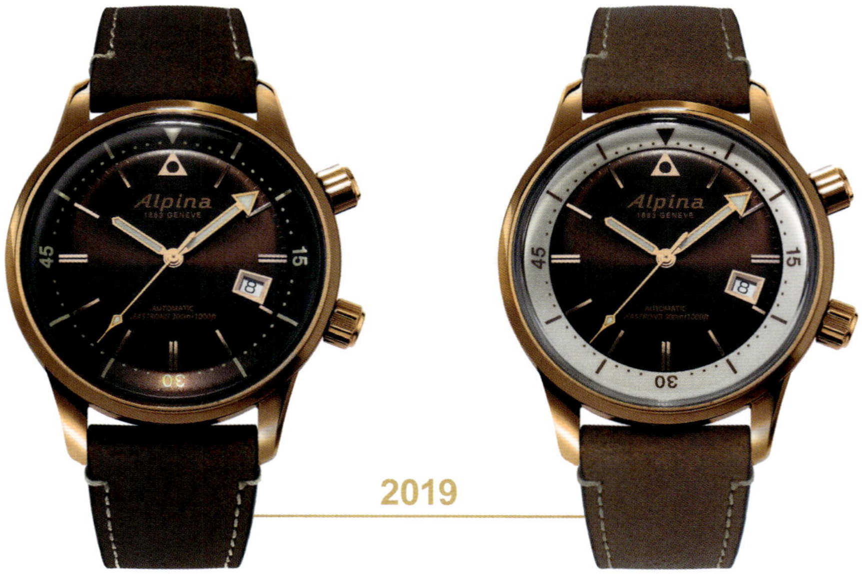

2019

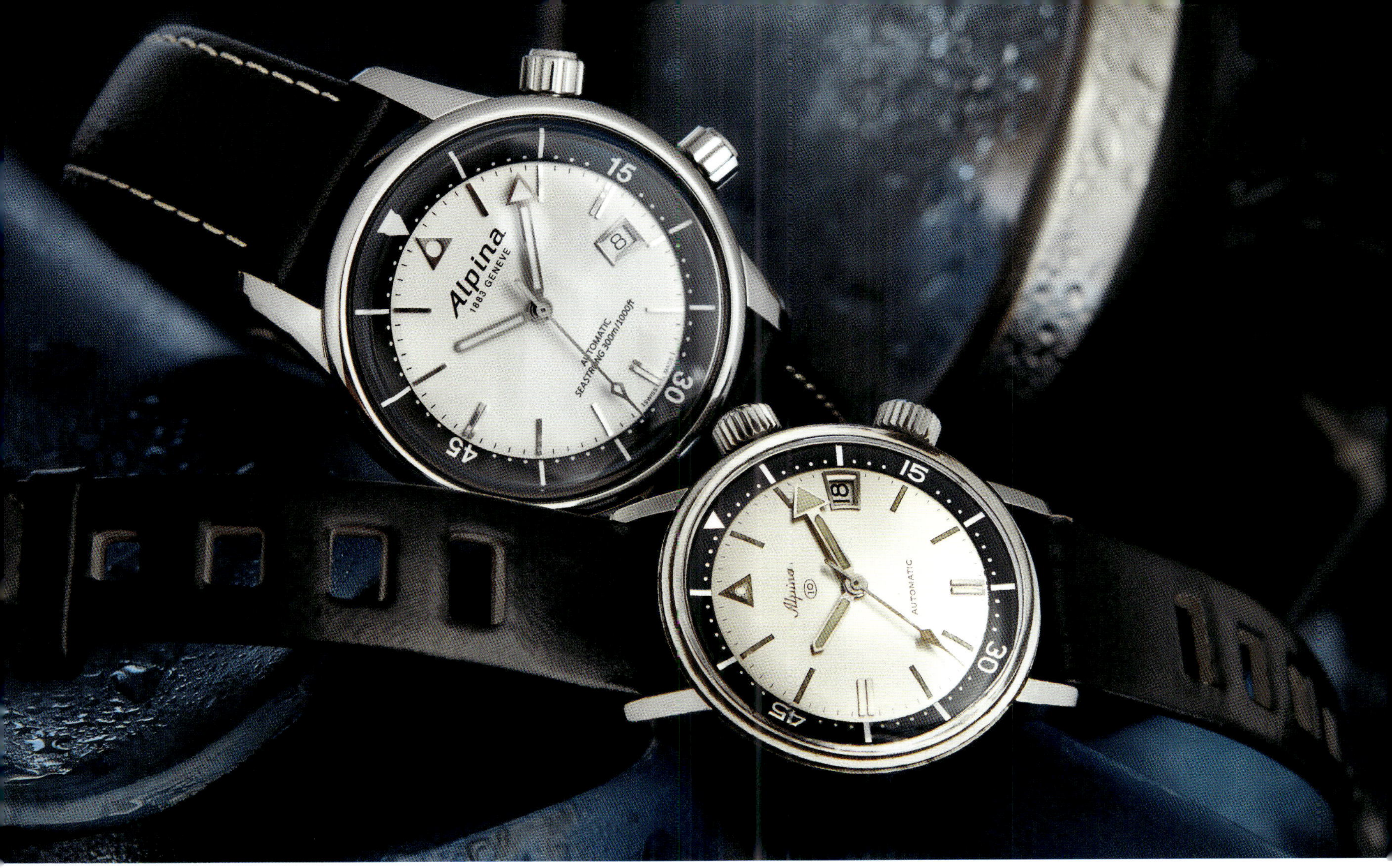
Alpina
1883 GENEVE
AUTOMATIC
SEASTRONG 300m/1000ft
Alpina
AUTOMATIC

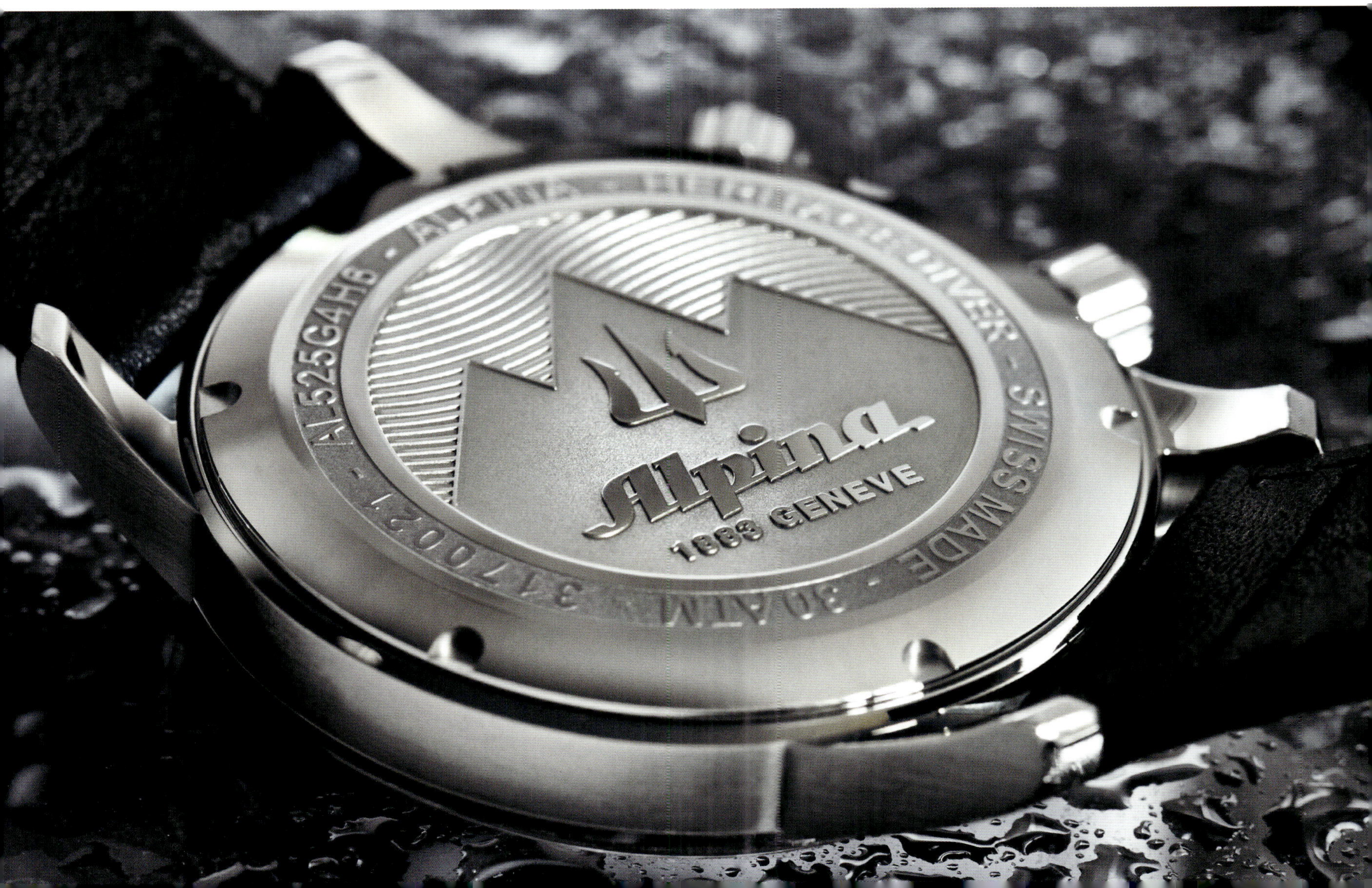
Alpina
1883 GENEVE
AL 525G4H6
SWISS MADE
30 ATM

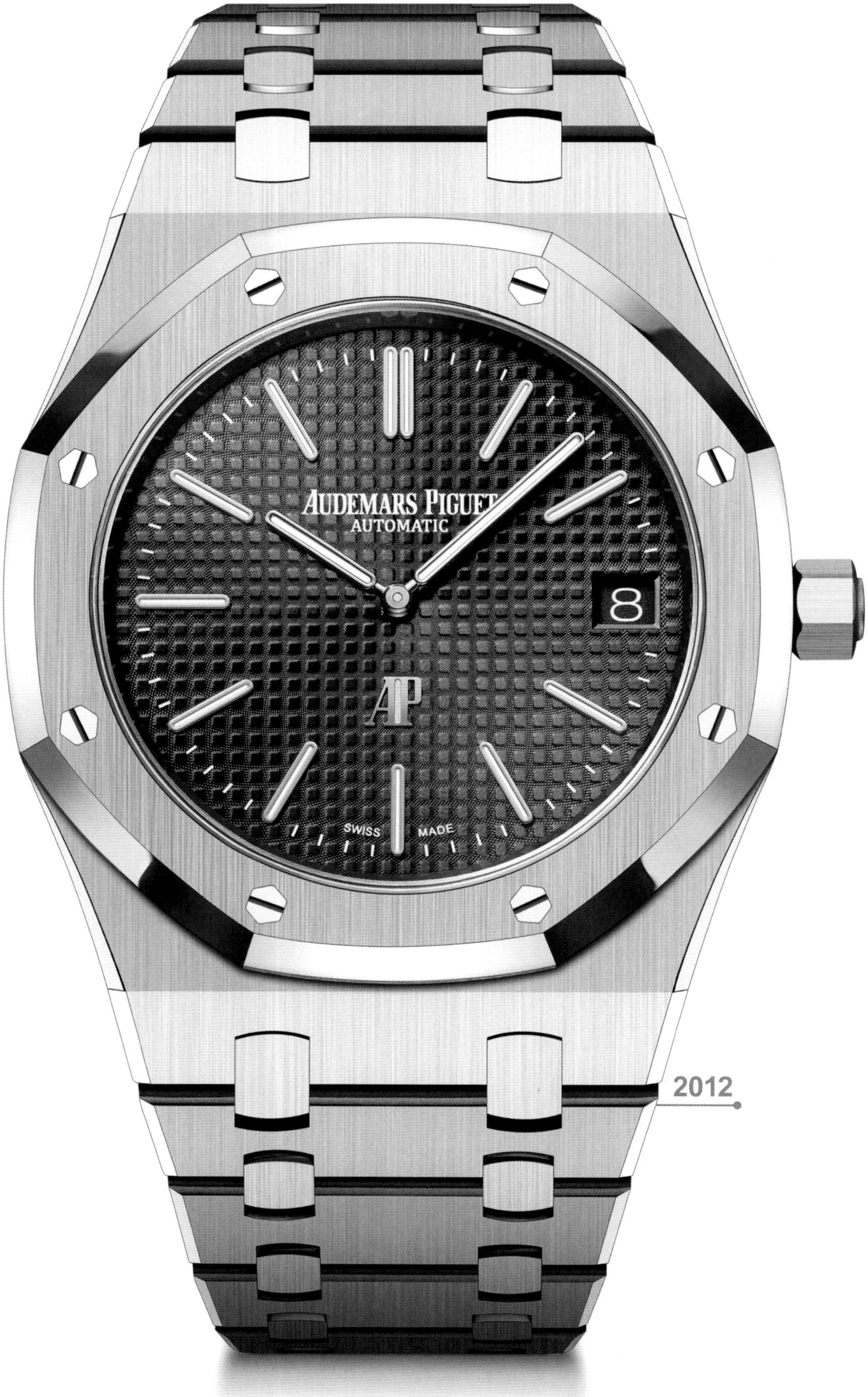

2012

1972 AUDEMARS PIGUET
ROYAL OAK

FIFTY PLUS, BUT BY NO MEANS OLD

Today, Audemars Piguet is one of the luxury-watch industry's revenue billionaires. In 1966, when Georges Golay took over the helm, the family business, founded in 1875, was financially beleaguered. Some 5,000 watches were produced annually at Le Brassus. Against this backdrop, a call from Carlo de Marchi, Charles Bauty and Charles Dorot shortly before the 1970 Basel watch fair was a stroke of providence for the traditional watchmaker. The sales agents, also known casually as the "three musketeers", wanted a universal wristwatch. One that would cut a dash behind the wheel of a sports car as well as on a sailing yacht, in a stylish club, at a gala dinner and in a concert or theatre hall. This desire for the proverbial squaring of the circle prompted the experienced manager to contact Gérald Genta. He had a tall order: a sporty elegant wristwatch the likes of which had never been seen before. And it had to be waterproof, of course.
Literally overnight the idea of a watch case inspired by a diver's helmet germinated in the designer's mind. This is reflected in the eight screws and visible connection of the lunette with the monobloc case for the movement. Initial sketches already impressed. However, it was not until 12 months later that the timepiece, named the Royal Oak, finally came into being. The first decisive moment was the choice of material for the case and bracelet. On this occasion, steel, refi-ned to jeweller's standard, was considered. This said, in the early 1970s steel was an unusual material for a premium luxury watch, with even a hint of the vulgar. In the design department, jokey allusions to stainless-steel sinks did the rounds. But Georges Golay was unabashed. Over a period of several weeks a case with a brushed and polished steel surface was created. And it found favour. Another problem was constituted by the mass production of the bracelet, because no two links are the same. By then it was obvious that the hand refinement of steel would catapult the price of the timepiece into the region of a gold watch. Audemars Piguet therefore settled for a limited edition of 1,000 for the initial Reference 5402ND/344 to mitigate potential loss. Of these, 400 were intended for the Italian market and a further 400 for the domestic market.
Due to the diameter of 38.8 millimetres and a considerable retail price the equivalent of around €1,850, reception of the model in Basel was muted. The same money would buy an Audemars Piguet timepiece in gold with the flat automatic 2121 calibre, which was also used in the Royal Oak. A Volkswagen Beetle cost the equivalent of around €3,500. >

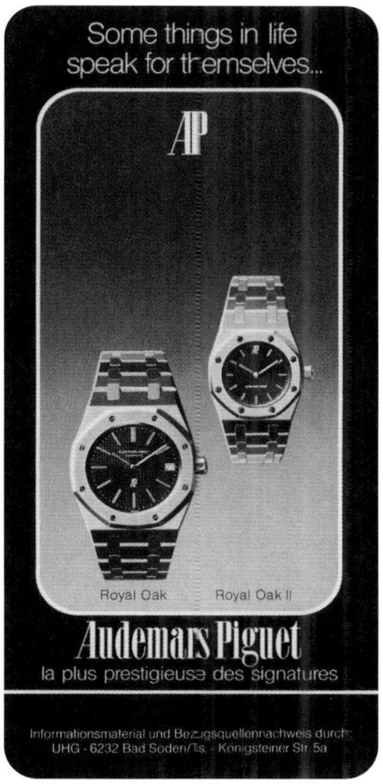

Advertisement, 1976

MEHR ALS 50, ABER KEINESWEGS ALT

Heutzutage gehört Audemars Piguet zu den Umsatzmilliardären in der Luxusuhrenbranche. 1966, als Georges Golay das Ruder übernahm, stand es um das 1875 gegründete Familienunternehmen finanziell nicht zum Besten. Jedes Jahr entstanden in Le Brassus circa 5.000 Uhren. Unter diesen Vorzeichen kam dem Vollblutkaufmann der Anruf von Carlo de Marchi, Charles Bauty und Charles Dorot kurz vor der Basler Uhrenmesse des Jahres 1970 durchaus gelegen. Die Vertriebsagenten, salopp auch „drei Musketiere" genannt, begehrten eine universell nutzbare Armbanduhr. Bella Figura sollte sie hinter dem Steuer eines Sportwagens ebenso machen wie auf der Segelyacht, im stylischen Club, beim Galadiner und im Konzert- oder Theatersaal. Das Verlangen nach der sprichwörtlichen Quadratur des Kreises veranlasste den erfahrenen Manager kurz vor Beginn der Basler Uhrenmesse zum Anruf bei Gérald Genta. Von ihm wollte er nicht weniger als eine sportlich-elegante Armbanduhr, die es so noch nie gegeben hatte. Natürlich wasserdicht. Sozusagen über Nacht gedieh die Idee, das Taucherhelmprinzip auf eine Uhrenschale zu übertragen. Gekennzeichnet durch acht Schrauben und deutlich sichtbare Verbindung der Lünette mit dem Monobloc-Container für das Uhrwerk. Bereits die ersten Zeichnungen überzeugten. Bis zur finalen Realisierung des „Royal Oak" getauften Zeitmesser zogen jedoch weitere zwölf Monate durch die Schweizer Lande. Die erste Stunde der Wahrheit schlug bei der Wahl des Werkstoffs für Gehäuse und Armband. Bei dieser Gelegenheit kam auch Stahl ins Spiel, feinbearbeitet nach Juwelierstandard. Gleichwohl war dieses Material für eine Top-Luxusuhr in den frühen 1970er-Jahren eher seltsam und sogar ein wenig vulgär. Rasch ulkte man in der Designabteilung über Küchenspülbecken aus Nirosta-Stahl. Aber Georges Golay ließ sich nicht aus dem Konzept bringen. Während vieler Wochen entstand ein Gehäuse mit gebürsteter und polierter Oberfläche. Und das gefiel. Ein weiteres Problem verkörperte die Serienfertigung des verlaufenden Armbands, da keines der Bandglieder dem anderen gleicht. Spätestens dann wurde klar, dass die handwerkliche Veredelung von Stahl den Preis des Ensembles in die Region von Gold katapultieren würde. Also beschränkte sich Audemars Piguet zur Begrenzung möglicher Verluste auf 1.000 Exemplare der Initial-Referenz 5402ST/344. Davon waren 400 für den italienischen und weitere 400 für den eidgenössischen Markt gedacht.
Angesichts des Durchmessers von 38,8 Millimetern und eines beachtlichen Publikumspreises von umgerechnet rund 1.850 Euro war die Resonanz in Basel eher verhalten. >

1972

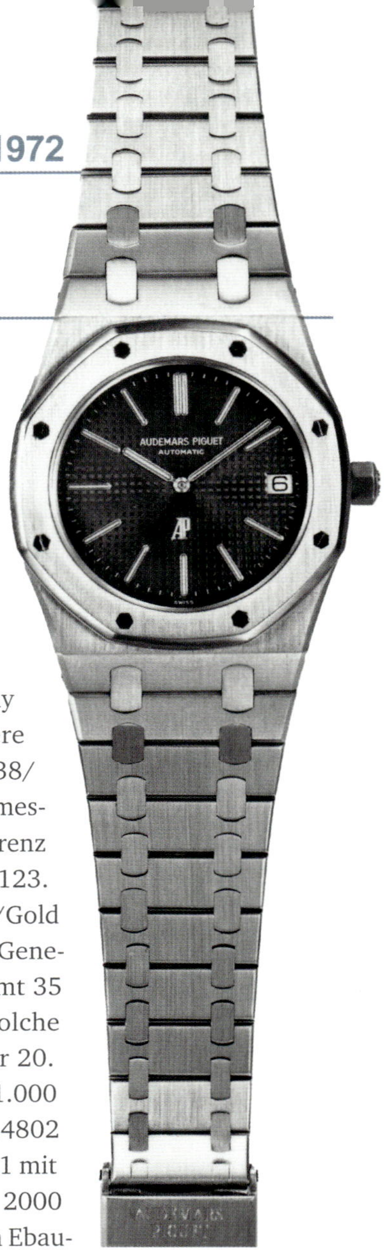

From 1976 onwards, Georges Golay placed the Royal Oak on a broader basis with iterations. Women were offered the Reference 8638/124 Automatic with a diameter of 29 millimetres. Reference 4100/477, measuring 35 millimetres, was powered by the 2123 automatic calibre. Steel/gold and solid-gold versions joined pure steel. The 1990 Audemars Piguet general catalogue included a total of 35 Royal Oak models, including some with complications and a quartz movement. In an edition of 1,000, the jubilee Reference 14802 with a see-through caseback, calibre 2121 and specially designed oscillating weight marked the model's twentieth anniversary. From 2000 onwards, the movement, the ébauche of which had been supplied by LeCoultre, became a genuine manufacture piece. Nonetheless, the venerable Royal Oak Jumbo with its diameter of 39 millimetres had to struggle for survival on repeated occasions. In 2011, there were only two countries in the world, Italy and Germany, in which the then steel Reference 15202ST sold well. The breakthrough came with the model's fortieth anniversary. Increased demand led to waiting lists and rising prices in the parallel market. The announcement of the planned end of this reference in 2021 ignited a further boom.

The model's fiftieth anniversary was celebrated by the similarly ultra-flat Reference 16202. The newly developed automatic calibre 7121 is inside. Specially designed rotors are reserved for round anniversaries. Those who deem steel, gold, platinum or titanium too profane can opt for the classic Royal Oak, including bracelet, in black, white or blue ceramic.

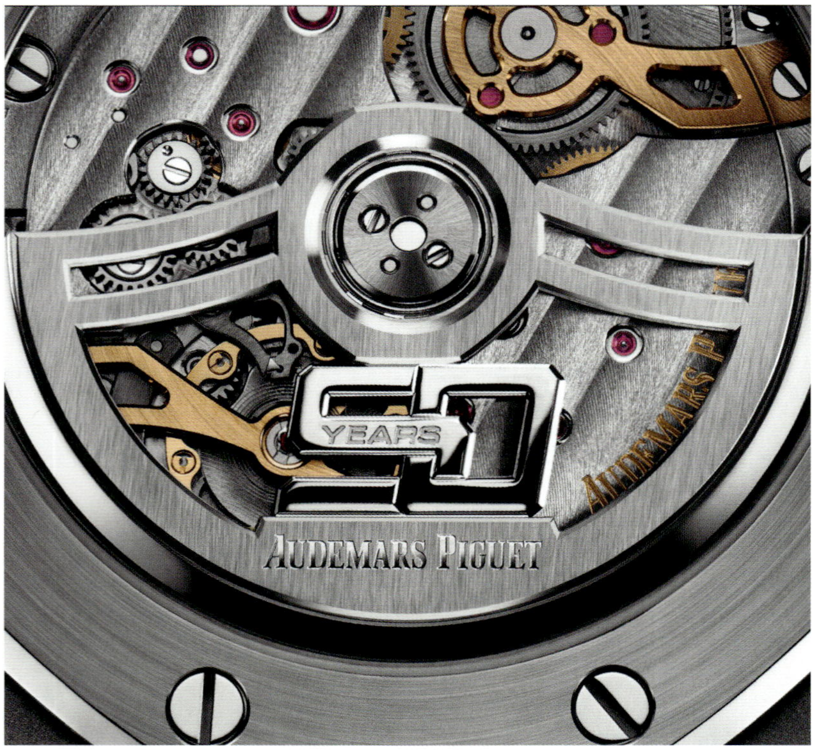

Fürs gleiche Geld gab es bei Audemars Piguet auch goldene Zeitmesser mit dem flachen Automatikkaliber 2121, welches auch in der Royal Oak Verwendung fand. Ein Volkswagen Käfer schlug mit umgerechnet ca. 3.500 Euro zu Buche.

Mit etlichen Derivaten stellte Georges Golay die „Royal Oak" ab 1976 auf eine breitere Basis. Damen bekamen die Referenz 8638/124 Automatic mit 29 Millimetern Durchmesser. In der 35 Millimetern messenden Referenz 4100/477 tickte das Automatikkaliber 2123. Zu puristischem Stahl gesellten sich Stahl/Gold und massives Gold. Im Audemars Piguet-Generalkatalog von 1990 fanden sich insgesamt 35 „Royal Oak"-Modelle, darunter auch solche mit Komplikationen und Quarzwerk. Der 20. Geburtstag im Jahr 1992 brachte die auf 1.000 Exemplare limitierte Jubiläumsreferenz 14802 mit Sichtboden und Automatikkaliber 2121 mit speziell ausgeführter Schwungmasse. Ab 2000 präsentierte sich besagtes Uhrwerk, dessen Ebauche bis dahin LeCoultre beigesteuert hatte, als echtes Manufaktur-Œuvre. Gleichwohl musste der altehrwürdige flache „Royal Oak Jumbo" mit seinen 39 Millimetern Durchmesser immer wieder ums Überleben kämpfen. 2011 gab es mit Italien und Deutschland nur noch zwei Länder auf der Welt, in der sich die damals aktuelle Stahlreferenz 15202ST gut verkaufte. Erst ab 2012 und dem 40. Geburtstag brach das Eis. Erhöhte Nachfrage führte zu Wartelisten und steigenden Preisen am Parallelmarkt. Die Ankündigung des Endes besagter Referenz mit Ablauf 2021 löste einen zusätzlichen Boom aus.

Zum 50. Geburtstag erschien die ebenfalls ultraflache Referenz 16202 auf der Bildfläche. Ausgestattet mit dem neu entwickelten Automatikkaliber 7121. Nur im Jahr des runden Geburtstags besitzt es nämlich einen speziell gestalteten Rotor. Menschen, denen Stahl, Gold, Platin oder Titan zu profan sind, erhalten die klassische „Royal Oak" samt Armband auch in schwarzer, weißer oder blauer Keramik.

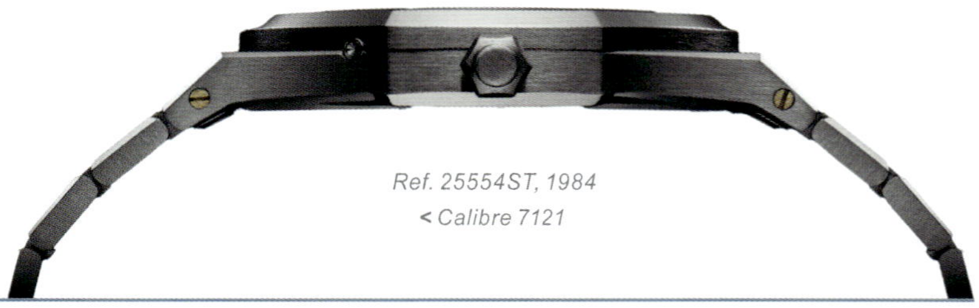

Ref. 25554ST, 1984
< Calibre 7121

1. *Technical worksheet / Technisches Arbeitsblatt, 1972*
2. *Georges Golay*
3. *Jacqueline Dimier, 1980*
4. *Launch of the first Royal Oak for women / Veranstaltung zur Einführung der ersten Royal Oak für Damen, Paris, 1976*
5. *Calibre 2121, 1985*
6. *Advertisement / Werbeanzeige, 1977*

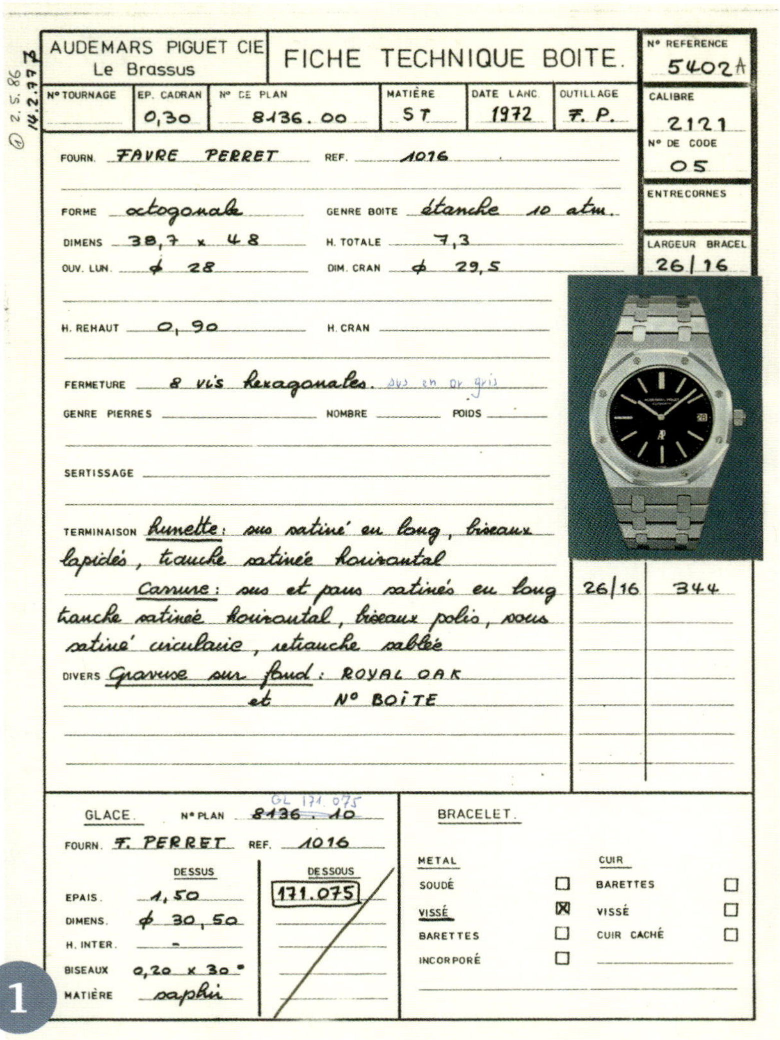

AUDEMARS PIGUET CIE Le Brassus — FICHE TECHNIQUE BOITE — N° REFERENCE 5402A

N° TOURNAGE	EP. CADRAN	N° DE PLAN	MATIÈRE	DATE LANC.	OUTILLAGE	CALIBRE
	0,30	8136.00	ST	1972	F.P.	2121

N° DE CODE 05
ENTRECORNES
LARGEUR BRACEL. 26/16

FOURN. FAVRE PERRET REF. 1016
FORME octogonale GENRE BOITE étanche 10 atm.
DIMENS. 38,7 x 48 H. TOTALE 7,3
OUV. LUN. ϕ 28 DIM. CRAN ϕ 29,5
H. REHAUT 0,90 H. CRAN
FERMETURE 8 vis hexagonales. vis en or gris
GENRE PIERRES NOMBRE POIDS
SERTISSAGE
TERMINAISON lunette: sus satiné en long, biseaux lapidés, tranche satinée horizontal
Carrure: sus et paus satinés en long tranche satinée horizontal, biseaux polis, sous satiné circulaire, retranche sablée
DIVERS Gravure sur fond: ROYAL OAK et N° BOÎTE

26/16 344

GLACE N° PLAN 8136.10 (GL 171.075)
FOURN. F. PERRET REF. 1016

	DESSUS	DESSOUS
EPAIS.	1,50	171.075
DIMENS.	ϕ 30,50	
H. INTER.	–	
BISEAUX	0,20 x 30°	
MATIÈRE	saphir	

BRACELET

METAL		CUIR	
SOUDÉ	☐	BARETTES	☐
VISSÉ	☒	VISSÉ	☐
BARETTES	☐	CUIR CACHÉ	☐
INCORPORÉ	☐		

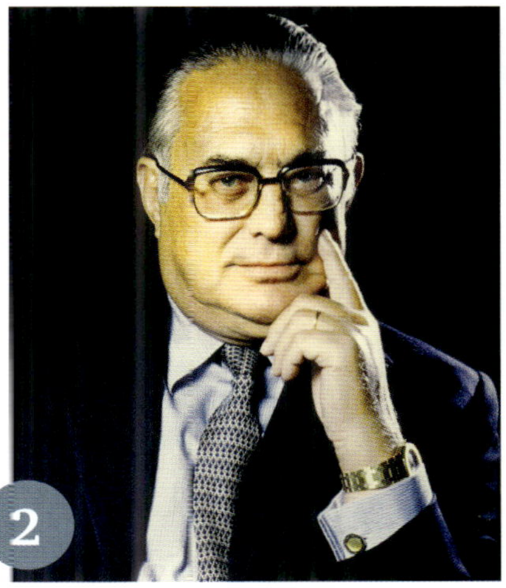

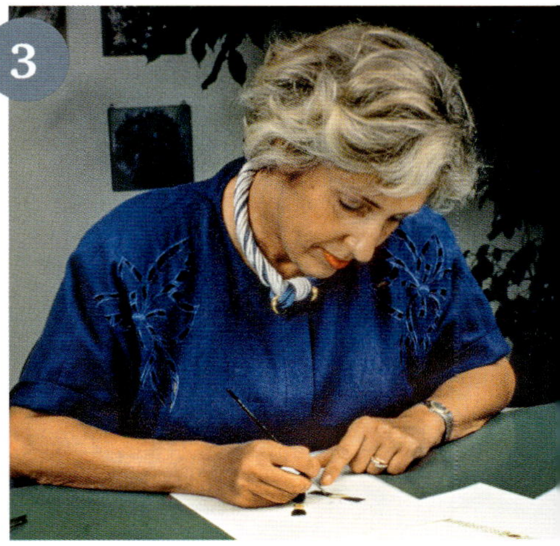

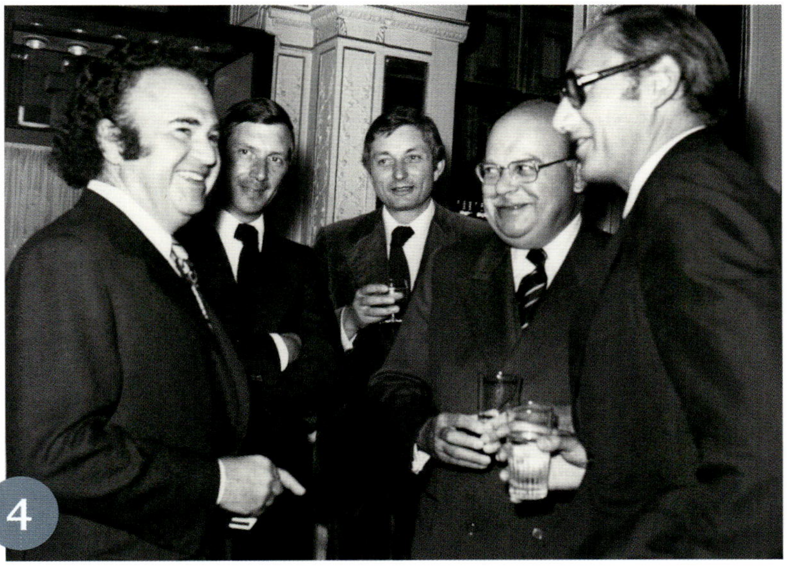

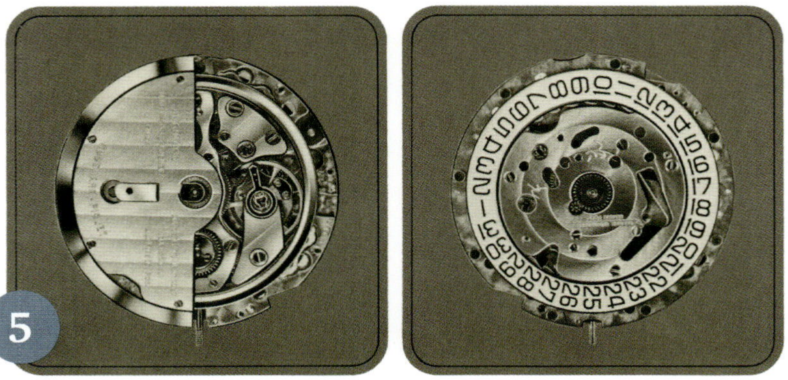

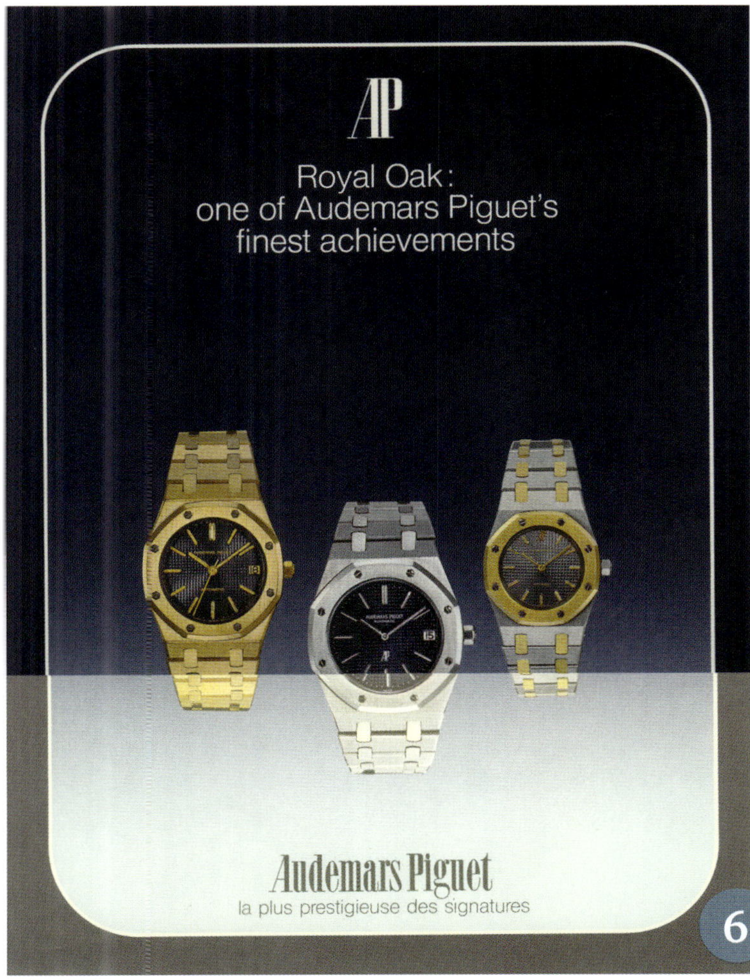

AUDEMARS PIGUET 1993

ROYAL OAK OFFSHORE

THE LE BRASSUS WALRUS

Presented in 1993, the Royal Oak Offshore was not a favourite of Gérald Genta. The creator of the slim Royal Oak therefore derisively referred to this opulent chronograph as a walrus. Audemars Piguet did not demur because this striking wristwatch soon became a roaring success. Actor Arnold Schwarzenegger played a part in this. He took a shine to the Offshore and promptly bought one for himself. As incredible as it may seem, between 2007 and 2010, the family-run watchmaker sold more Offshore than classic Royal Oak models. Considerations of withdrawing the classic flat Royal Oak from the market were therefore by no means far-fetched.

Another reason for the popularity of the Offshore is its show-stopping appearance on the wearer's wrist. The case consists of the base, middle, rubber gaskets and bezel, the whole held together by screws running all the way through. The visible gaskets afford a significant contribution to its water resistance of up to ten bar. Because Emmanuel Gueit deliberately designed this timepiece to be larger than life, it can also be worn over sport apparel. The rubber guards on the pushers and crown provide additional protection from heavy blows.

The original movement was a modular chronograph calibre 2226/2840. The front-mounted stopwatch module was supplied by Joux Valley specialist Dubois-Dépraz.

A soft iron cage protects the movement of reference 25721 from magnetic fields. Countless iterations of this wristwatch with dimensions from 27 to 48 millimetres were created over three decades. Audemars Piguet currently uses its proprietary automatic calibre 4401 in the chronograph. It consists of 381 components with a power reserve of some 70 hours.

WALROSS AUS LE BRASSUS

Gérald Genta mochte sie definitiv nicht, die 1993 vorgestellte „Royal Oak Offshore". Deswegen bezeichnete der Schöpfer der flachen „Royal Oak" diesen opulenten Chronographen spöttisch als „Walross". Audemars Piguet focht dies nicht an, denn diese markante Armbanduhr entwickelte sich rasch zu einem Erfolg. Einen Beitrag dazu leistete Schauspieler Arnold Schwarzenegger, der spontan Gefallen an der „Offshore" fand und ein Exemplar für sich erwarb. Erstaunlich aber wahr: In den Jahren 2007 bis 2010 verkaufte die Familienmanufaktur mehr „Offshore"-Modelle als klassische „Royal Oaks". Überlegungen, die klassische flache Royal Oak vom Markt zu nehmen, waren deshalb keineswegs aus der Luft gegriffen.

Ein anderer Grund für die Popularität der „Offshore" resultiert aus dem nachgerade unübersehbaren Auftritt am Handgelenk. Das Gehäuse besteht aus Boden, Mittelteil, Kautschukdichtungen und Lünette. Durchgehende Schrauben halten das Ensemble zusammen. Die sichtbaren Dichtungen leisten einen wichtigen Beitrag zur Wasserdichtigkeit bis zehn bar Druck.

Weil Emmanuel Gueit diesen Zeit-Boliden absichtlich überdimensioniert konzipiert hat, kann er auch über einem Sportdress getragen werden. Bei heftigen Stößen bieten die Kautschukhauben über Krone und Drückern zusätzlichen Schutz.

Ursprünglich kümmerte sich ein modular aufgebautes Chronographenwerk vom Kaliber 2226/2840 um die Zeit. Das vorderseitig montierte Stoppmodul steuerte der Spezialist Dubois-Dépraz aus dem Vallée de Joux bei. Übrigens schützte bei der Referenz 25721 ein Weicheisenmantel die Mechanik vor Magnetfeldern. Während dreier Jahrzehnte entstanden beinahe unzählige Varianten dieser Armbanduhr mit Durchmessern von 27 bis 48 Millimetern. Aktuell verbaut Audemars Piguet im Chronographen das hauseigene Automatikkaliber 4401. Selbiges besteht aus 381 Komponenten und besitzt rund 70 Stunden Gangautonomie.

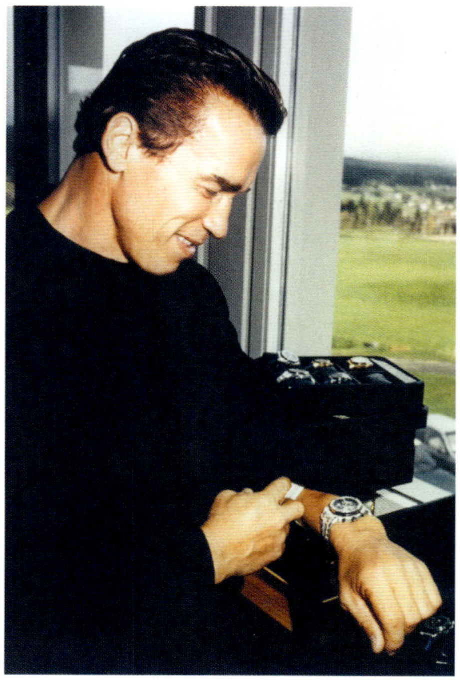

Arnold Schwarzenegger at / bei Audemars Piguet, 1997

Emmanuel Gueit & Jacqueline Dimier, c. 1990

Royal Oak Offshore 25721, No. 32, 1993

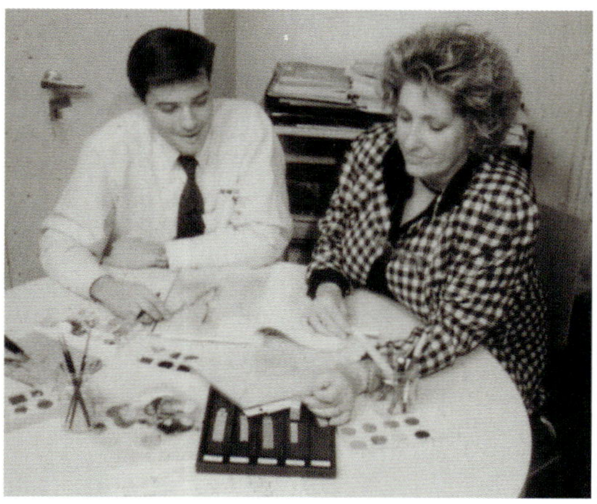

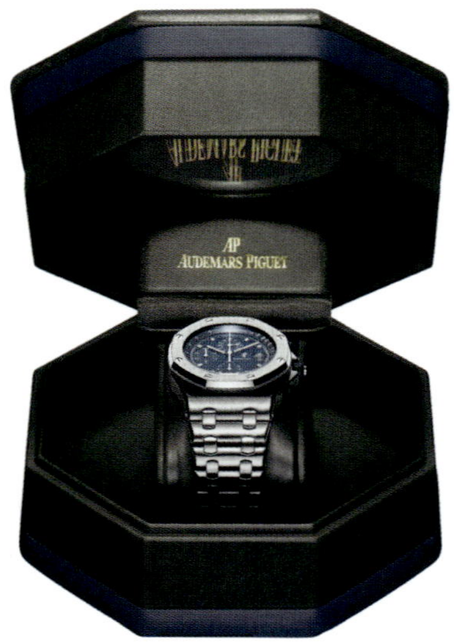

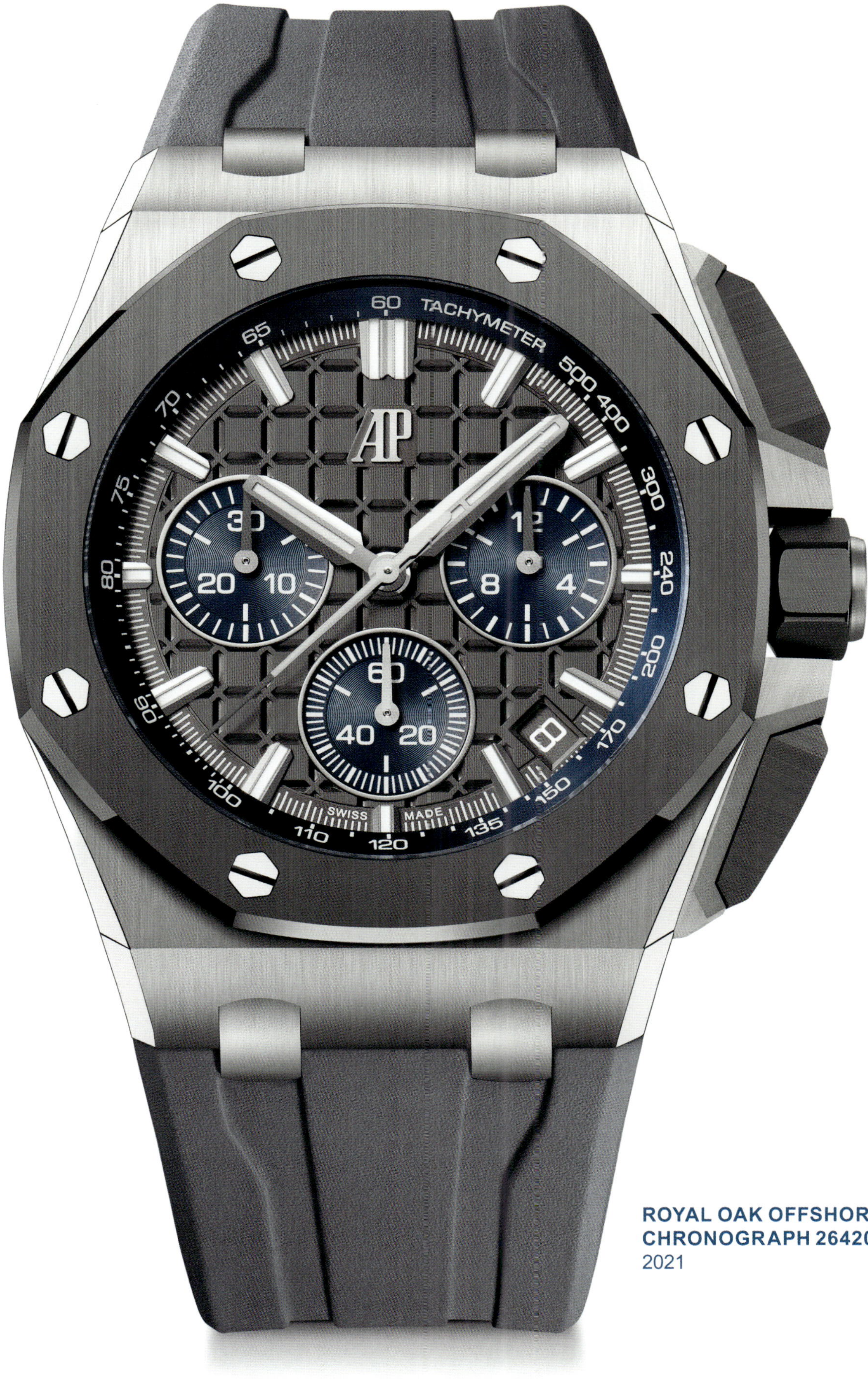

ROYAL OAK OFFSHORE CHRONOGRAPH 26420IO,
2021

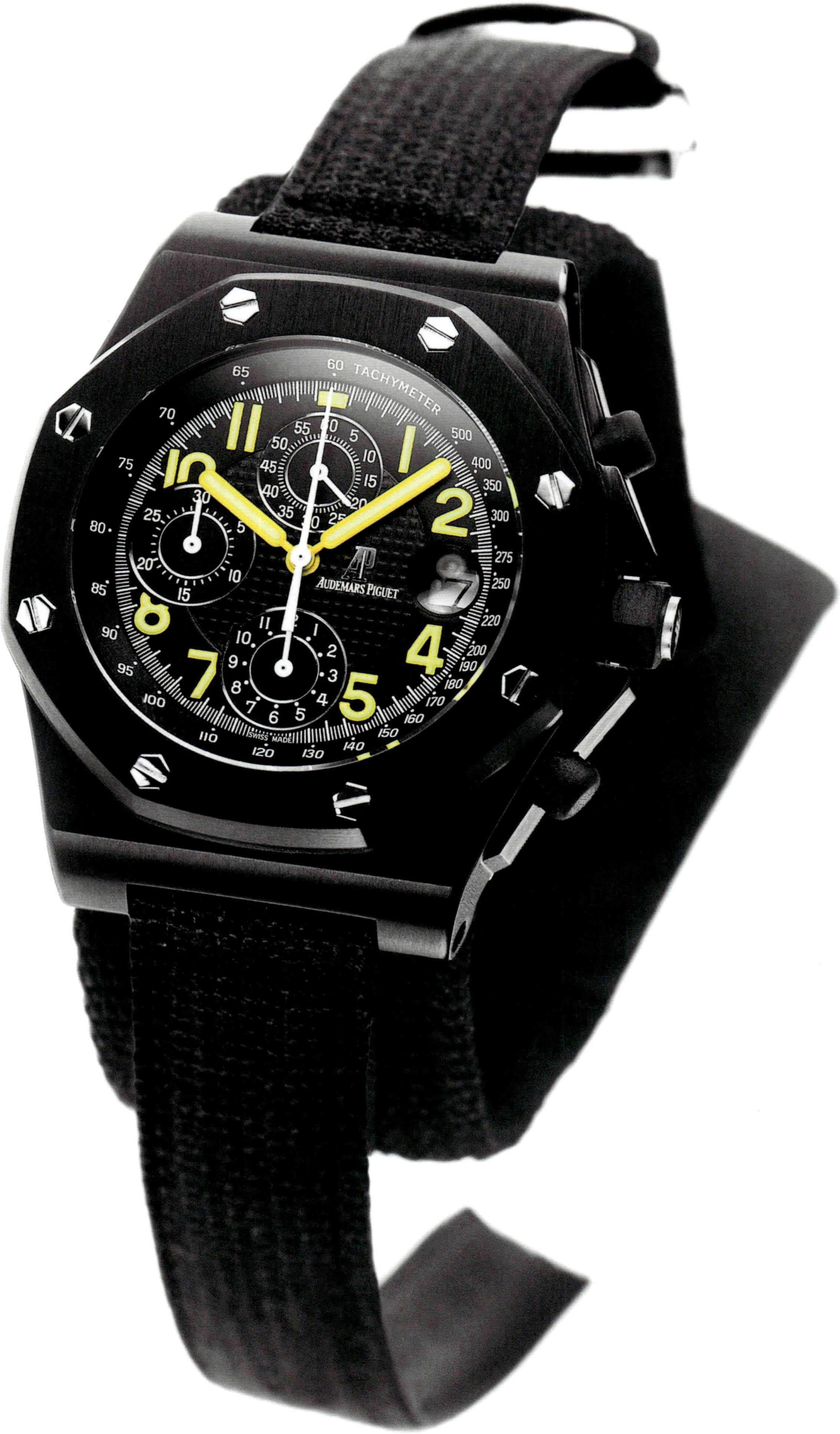
TACHYMETER
AUDEMARS PIGUET
SWISS MADE

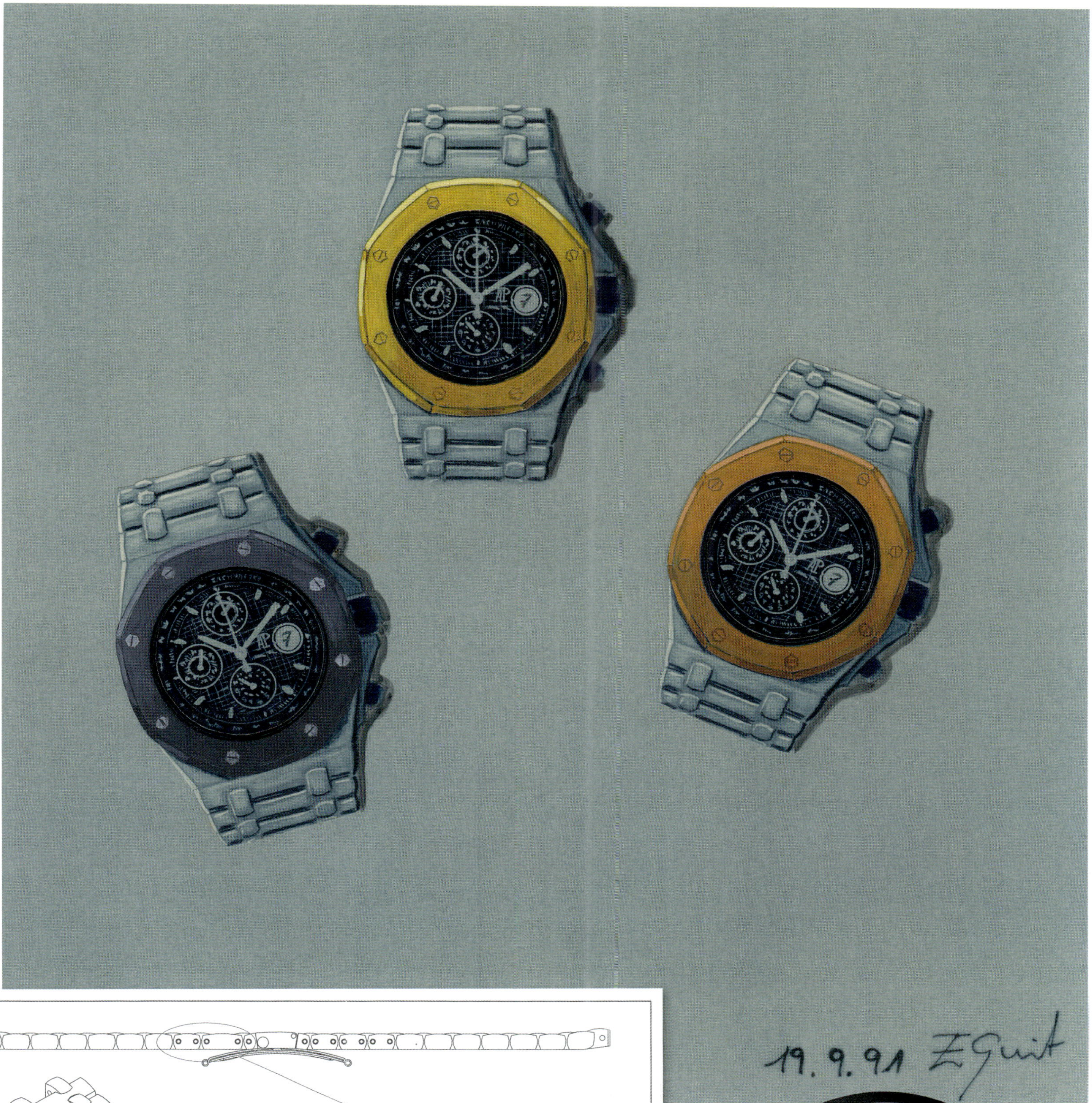

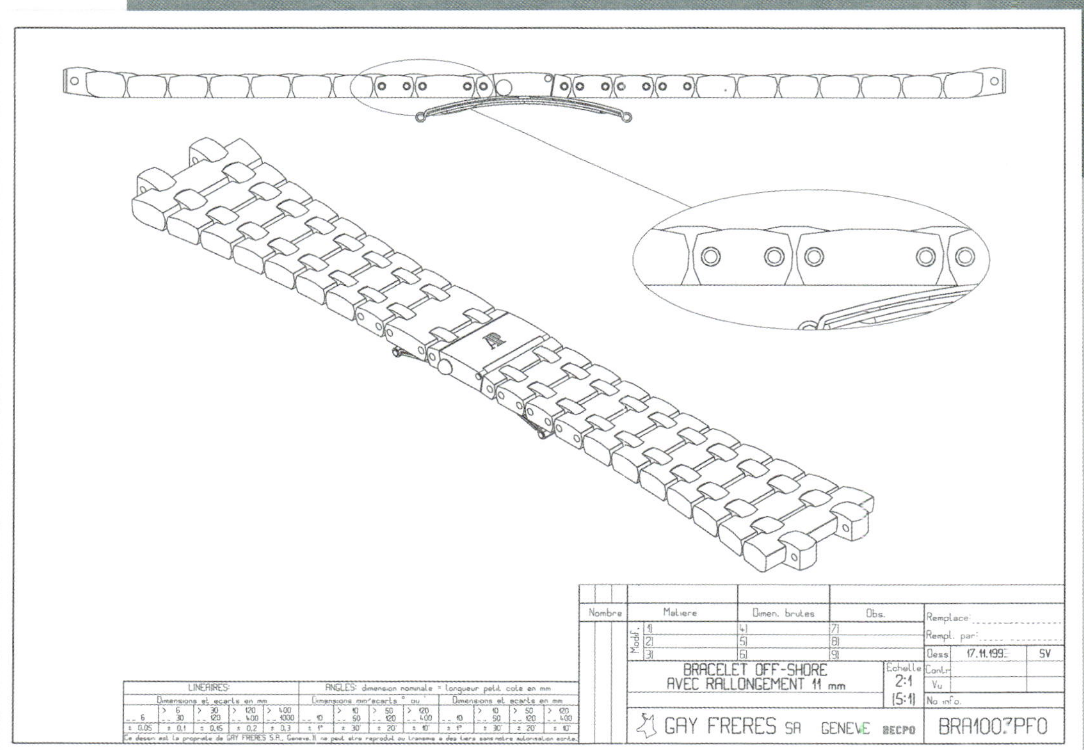

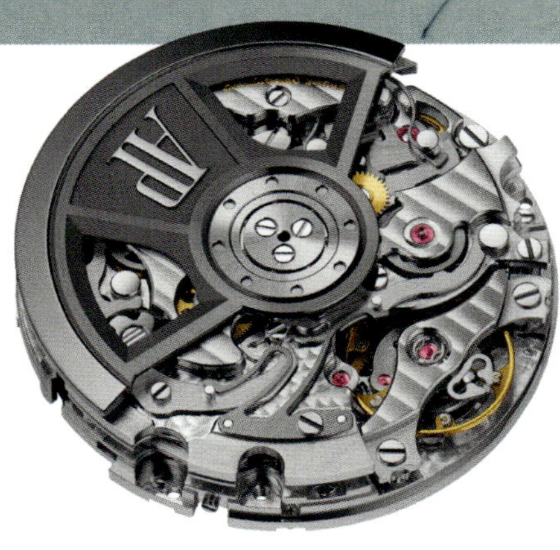

<< *Royal Oak Offshore End of Days 25770SN, 1999*

From top | Von oben: *Royal Oak Offshore sketch, 1991*

Plan of Bracelet | Zeichnung des Armbands 25721, 1993

Calibre 4401

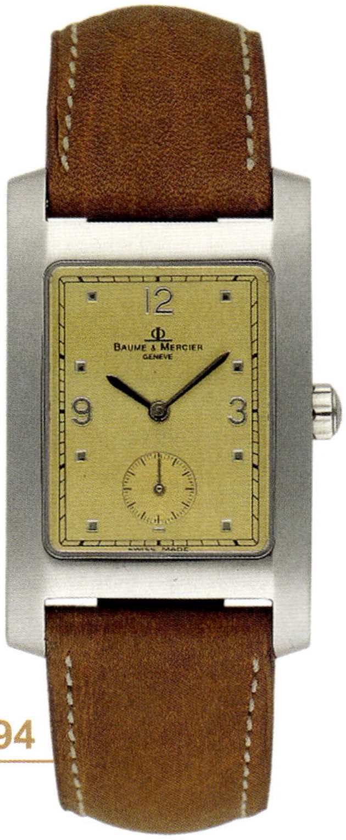

1994

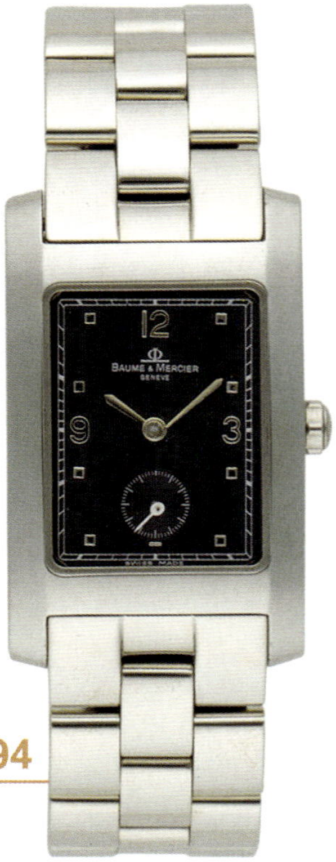

1994

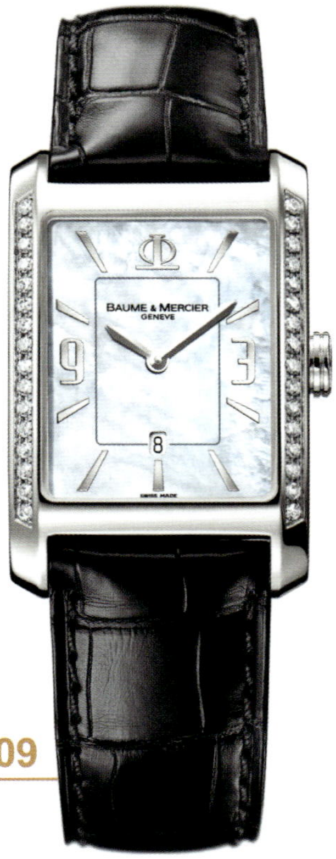

2009

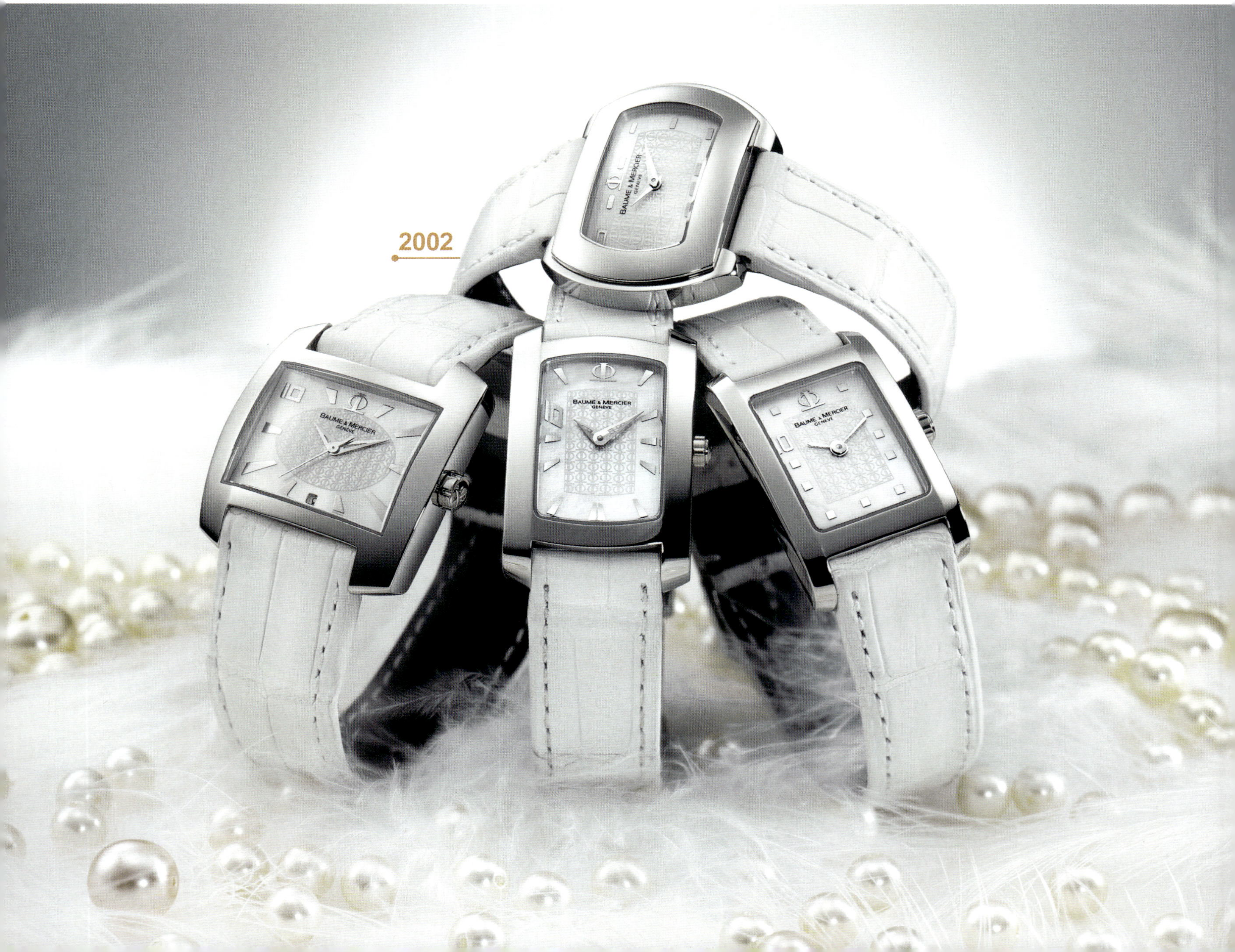

2002

BAUME & MERCIER
HAMPTON

LONG ISLAND HERE WE COME

The elegant laid-back lifestyle on Long Island is the stuff of legend. Even back in the 1930s, those who could afford it and thought they were somebody spent their weekends there. After the end of the Second World War, artists and writers such as Jackson Pollock, Willem de Kooning, Larry Rivers, Jasper Johns, Robert Rauschenberg, George Plimpton, Peter Matthiessen and John Steinbeck lent the Hamptons the flair of a colony of creative minds. Its fame attracted Andy Warhol, Roy Lichtenstein and Julian Schnabel. Pop, film and fashion stars such as Steven Spielberg, Billy Joel, Paul Simon, Calvin Klein and Donna Karan followed. For them and the moneyed elite of New York a house in the Hamptons was a must. This cachet inspired Baume & Mercier to name a 1995 model quite simply Hampton. With its sleek rectangular case, reminiscent of the 1930s, the newcomer quickly captured the market. In 2011/2012, a new Hampton united past, present and future in a single wristwatch. It did not quite achieve the success enjoyed by the first generation. In 2020, this spurred Baume & Mercier to revisit the good old days. Sleek and purist once more, the Hampton is available in various iterations. Anyone who acquired one of the 2022 edition of 102 in honour of Pierre Soulages can congratulate themselves. Its dial sports one of the French artist's characteristically abstract works of art. In 2024, Baume & Mercier launched a second tribute to Pierre Soulages with the "Hampton Polyptyque Edition – Musée Soulages 10th Anniversary", limited to 328 pieces.

AB NACH LONG ISLAND

Wahrhaft legendär ist er, der lässig-distinguierte Lebensstil auf Long Island. Wer es sich leisten konnte und etwas auf sich hielt, verbrachte dort schon in den 1930er-Jahren sein Wochenende. Nach dem Ende des Zweiten Weltkriegs bescherten Künstler und Schriftsteller wie Jackson Pollock, Willem de Kooning, Larry Rivers, Jasper Johns, Robert Rauschenberg, George Plimpton, Peter Matthiessen oder John Steinbeck den Hamptons das Flair einer Kolonie kreativer Köpfe. Der Ruf wirkte wie ein Magnet auf Andy Warhol, Roy Lichtenstein oder Julian Schnabel. Pop-, Film- und Modestars wie Steven Spielberg, Billy Joel, Paul Simon, Calvin Klein und Donna Karan folgten. Sie und der New Yorker Geldadel betrachteten ein Haus auf den Hamptons als Must-have. Unter diesen Vorzeichen taufte Baume & Mercier eine 1995 vorgestellte Armbanduhr kurz und knapp „Hampton". Mit seinem schlichten rechteckigen, an die 1930er-Jahre erinnernden Gehäuse behauptete sich der Newcomer schnell am Markt. 2011/2012 vereinigte eine neue „Hampton" Vergangenheit, Gegenwart und Zukunft in einer einzigen Armbanduhr. An den Erfolg der ersten Generation konnte sie freilich bei Weitem nicht anknüpfen. Deswegen blickte Baume & Mercier im 2020 zurück auf gute alte Zeiten. Die nun wieder schnörkellos und puristisch gestaltete „Hampton" gibt es in unterschiedlichen Ausführungen. Freuen kann sich, wer 2022 eines der 102 Exemplare zu Ehren von Pierre Soulages ergattern konnte. Ihr Zifferblatt bringt eines der charakteristisch abstrakten Kunstwerke des französischen Künstlers ans Handgelenk. 2024 lancierte Baume & Mercier mit der auf 328 Stück limitierten „Hampton Polyptyque Edition – Musée Soulages 10th Anniversary" eine weitere Hommage an Pierre Soulages Werk.

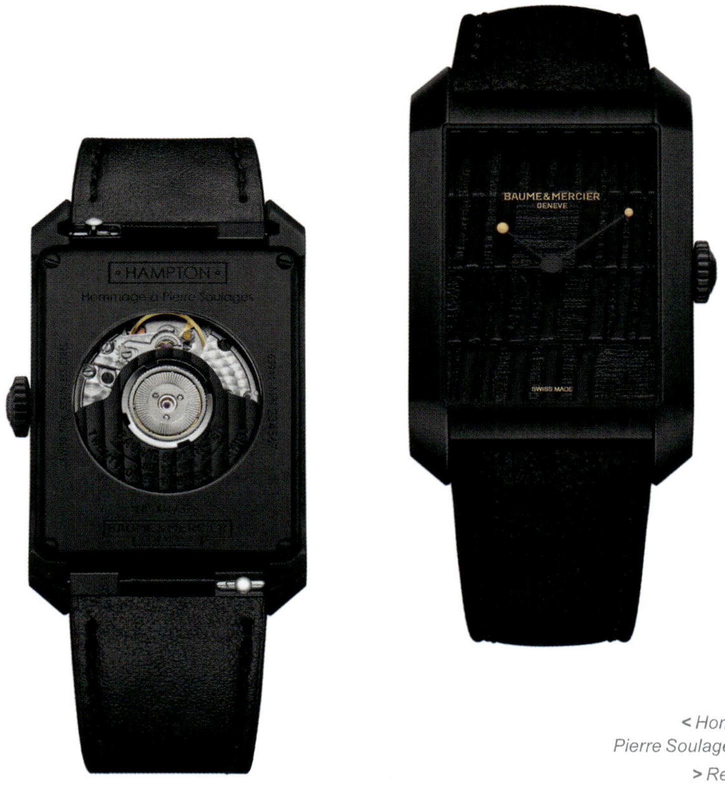

< Hommage à Pierre Soulages 10741
> Ref. 10472

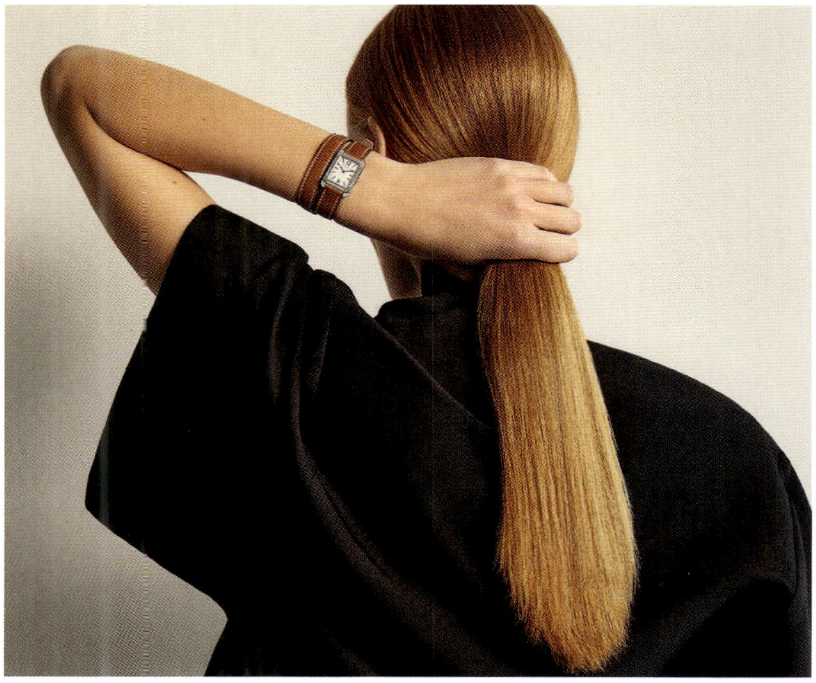

BAUME & MERCIER 1973
RIVIERA

FIVE GENERATIONS SINCE 1973

The Piaget family pursued its chronometric activities in the secluded La Côte-aux-Fées. One mainstay not being sufficient, in 1965 it acquired the majority shareholding in the traditional brand Baume & Mercier. In 1973, a huge oil crisis rocked the Western world. The Swiss watchmaking industry also bore the brunt of this event, not to mention the consequences of the quartz revolution. Labouring under these adverse circumstances, the Baume & Mercier management focused its attention on its collection. Audemars Piguet's much-discussed Royal Oak could not be ignored, nor should it have been. However, it had to be borne in mind that while their own clients also appreciated luxury, they were not nearly as affluent. The watch designed by Jean-Claude Gueit that was never produced springs to mind in this context. The designer had placed his expertise at the disposal of Audemars Piguet, Concord, Corum, Girard-Perregaux, Hermès, Omega, Patek Philippe and Vacheron Constantin. Piaget, however, rejected Gueit's idea of a sporty, elegant watch with a tiger-eye dial as not being extravagant enough. Baume & Mercier seized the opportunity. Boasting its brand signature and a tuning fork movement, the debut took place in 1973.

The model named Riviera spontaneously evoked associations with holidays and the seaside. Over the years, more than 200,000 models have graced both male and female wrists. One of the reasons for the timepiece's success was its high recognition value. The striking dodecagonal bezel stems from the fact that day and night each last twelve hours, and dials therefore usually have twelve digits or indexes. In addition, Baume & Mercier offered its clients a lot of wristwatch for their money with a traditional focus on affordable luxury.

Over the decades, the Riviera collection has not only witnessed a widening of the range, but also metamorphoses that achieved varying degrees of success. In 1979, a second generation with flatter quartz movements came onto the market. Henceforth, the dial also bore the Riviera logo. In 1985, the third generation displayed a softer design, a rounder case and a redesigned link bracelet. In its 1993 anniversary year, with more than 200 iterations to its name, Riviera accounted for around a third of the brand turnover. Among the brand ambassadors was Formula 1 driver Jacques Villeneuve. After the less successful fourth generation, Baume & Mercier relaunched the model in 2021. An exclusive mechanism, manufactured by sister company ValFleurier in the Val de Travers, distinguishes the Riviera Baumatic. The automatic calibre 13-1975A offers a power reserve of 120 hours. In addition to the chronographs, there is also a limited model with a perpetual calendar. For Baume & Mercier, revisiting the past has been well worth the effort, for this fifth generation of the fifty-plus Riviera is considered the bestseller of the collection.

FÜNF GENERATIONEN SEIT 1973

Ihren chronometrischen Aktivitäten ging die Familie Piaget im abgeschieden La Côte-aus-Fées nach. Weil es sich auf einem Bein nicht sonderlich gut steht, erwarb sie 1965 die Aktienmehrheit an der Traditionsmarke Baume & Mercier. 1973 litt die westliche Welt unter einer massiven Ölkrise. Deren Auswirkungen und dazu die Folgen der Quarz-Revolution spürte natürlich auch die eidgenössische Uhrenindustrie. Gebeutelt von der nicht unbedingt einfachen Situation, richtete das Baume & Mercier-Management sein Augenmerk auf die Kollektion. Dabei konnte und durfte die viel diskutierte „Royal Oak" von Audemars Piguet selbstverständlich nicht unberücksichtigt bleiben. Zu bedenken galt es aber auch, dass die eigene Klientel zwar ebenfalls luxusaffin, aber bei Weitem nicht so zahlungskräftig war. Unter diesen Vorzeichen erinnerte man sich an einen nicht realisierten Uhrenentwurf von Jean-Claude Gueit. Der Designer hatte sein Können unter anderem in die Dienste von Audemars Piguet, Concord, Corum, Girard-Perregaux, Hermès, Omega, Patek Philippe oder Vacheron Constantin gestellt. Piaget jedoch hatte seine Idee einer sportlich-eleganten Armbanduhr mit Tigerauge-Zifferblatt als zu wenig extravagant verworfen. Also kam Baume & Mercier zum Zuge. Mit dieser Hersteller-Signatur und einem Stimmgabel-Uhrwerk erfolgte das Debüt in besagtem Jahr 1973.

Der Modellname „Riviera" weckte spontan Assoziationen zu Urlaub und Badefreuden. Im Laufe der Jahre gelangten mehr als 200.000 Exemplare an maskuline und feminine Handgelenke. Der Grund für den Erfolg lag unter anderem im hohen Wiedererkennungswert. Das augenfällige Zwölfeck der Lünette resultierte aus der Tatsache, dass Tag und Nacht jeweils zwölf Stunden dauern und Zifferblätter deshalb zumeist zwölf Ziffern oder Indexe aufweisen. Außerdem bot Baume & Mercier dem Kundenkreis viel Armbanduhr fürs Geld.

Im Laufe der Jahrzehnte erlebte die „Riviera"-Kollektion nicht nur eine Ausweitung der Produktpalette, sondern auch mehr oder minder erfolgreiche Metamorphosen. 1979 kam eine zweite Generation mit flacheren Quarzwerken auf den Markt. Fortan trug auch das Zifferblatt den Schriftzug „Riviera". Die dritte Generation des Jahres 1985 erhielt ein weicheres Design, ein runderes Gehäuse sowie ein umgestaltetes Gliederband. Im Jubiläumsjahr 1993 steuerte die „Riviera" mit ihren mehr als 200 Varianten etwa ein Drittel zum Markenumsatz bei. Zu den Markenbotschaftern gehörte der Formel-1-Pilot Jacques Villeneuve. Nach der weniger erfolgreichen vierten Generation startete Baume & Mercier 2021 neu durch. Exklusive Mechanik, hergestellt von der Schwester ValFleurier im Val de Travers, zeichnet die „Riviera Baumatic" aus. Über 120 Stunden Gangautonomie verfügt das Automatikkaliber 13-1975A. Neben Chronographen gibt es auch ein limitiertes Modell mit ewigem Kalender. Für Baume & Mercier zahlt sich die Rückbesinnung aus, denn diese fünfte Generation der mehr als 50 Jahre alten „Riviera" gilt als Bestseller in der Kollektion.

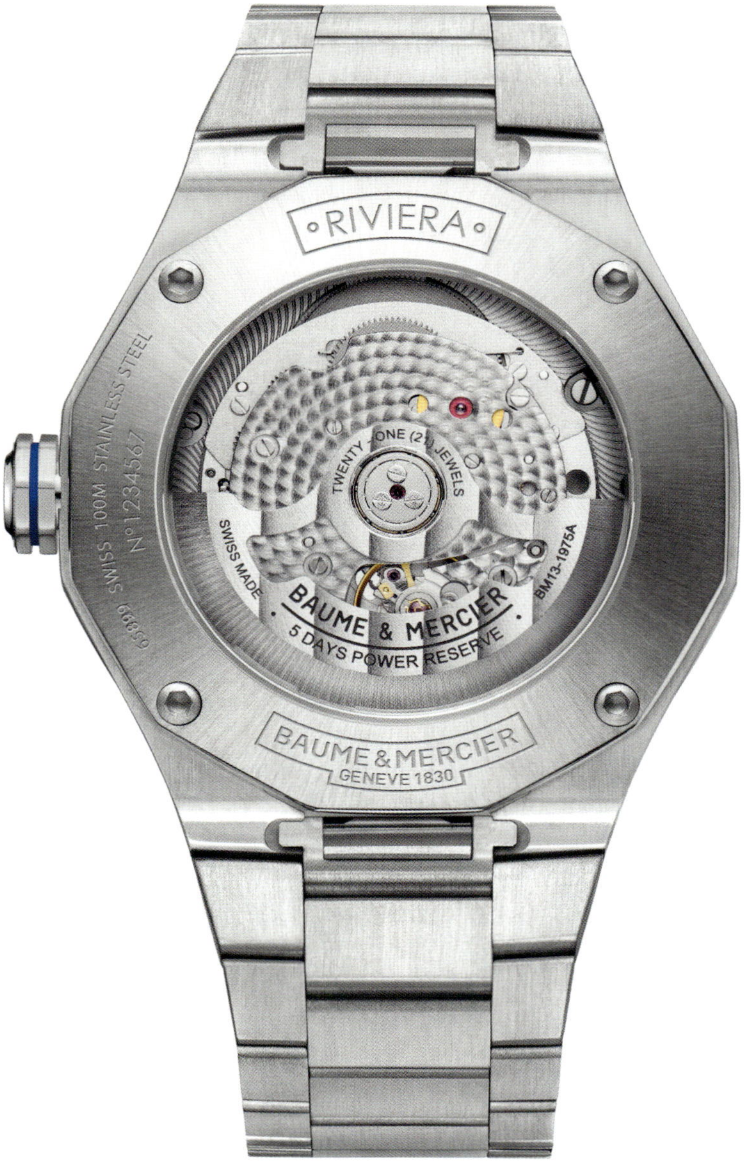
RIVIERA
BAUME & MERCIER
5 DAYS POWER RESERVE
SWISS MADE
BAUME & MERCIER
GENEVE 1830

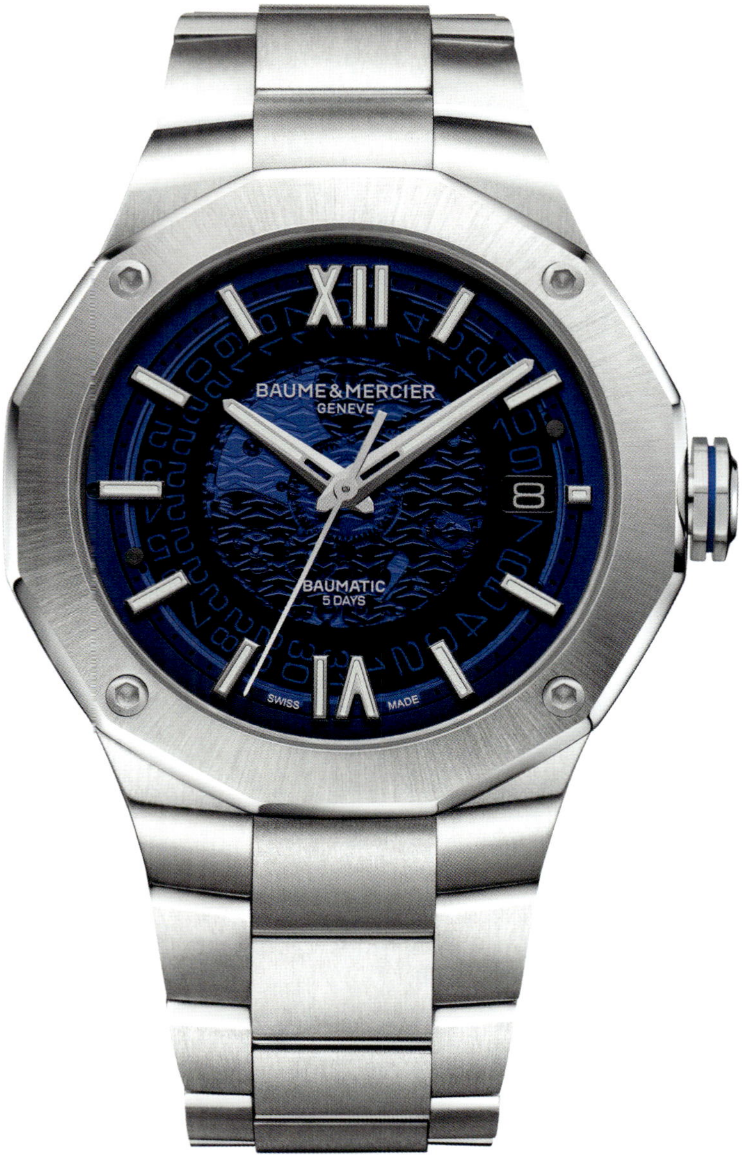
BAUME & MERCIER
GENEVE
BAUMATIC
5 DAYS
SWISS
MADE

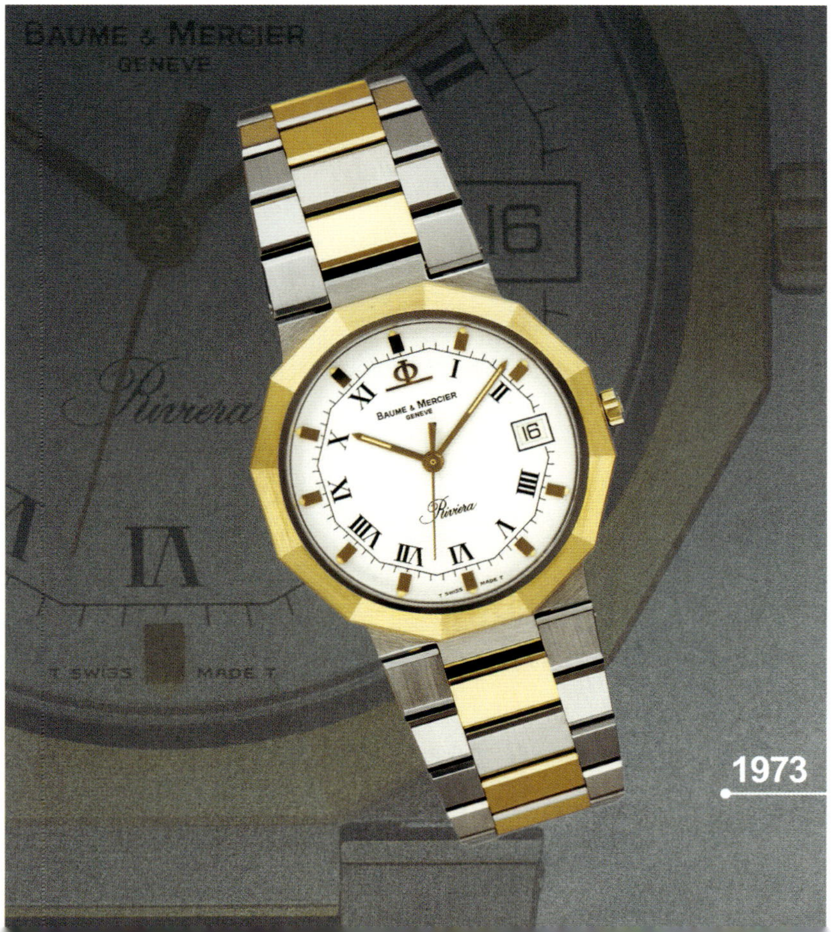
BAUME & MERCIER
GENEVE
Riviera

1973

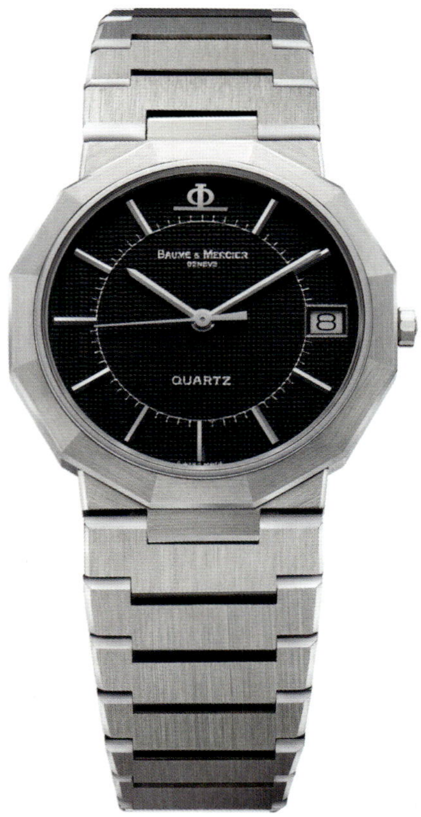
QUARTZ

BLANCPAIN FIFTY FATHOMS

1953

DIVER'S WATCH OF EPIC DEPTHS

Their assignment held huge appeal. In 1952, Commander Robert "Bob" Maloubier and Lieutenant Claude Riffaud, professional divers and officers of the French Navy, were tasked to build up an elite corps of combat divers. In addition to other equipment, wristwatches were also required that satisfied the high demands of the combat divers. But such timepieces were few and far between on the market. This inspired Maloubier and Riffaud to design one themselves. The subsequent quest for a competent watchmaker was by no means easy. Eventually, the two stumbled across the small watchmaker Rayville. In 1933, its owner Betty Fiechter had inherited the company from her boss, Frédéric-Emile Blancpain, the last member of the well-known watchmaking dynasty. She helped the two Frenchmen to put their ideas into practice: optimum readability of the dial in all light conditions, automatic movement, diving time rotating bezel and a steel case waterproof up to 50 fathoms (91.45 metres). This value was by no means coincidental. At the time, greater depths could not be reached by divers dependent on oxygen and nitrogen cylinders. The first prototypes probably date back to 1953. In 1954 or 1955, a wristwatch emerged that was to make history, not least because of its patented unidirectional bezel. It is little wonder that the Fifty Fathoms soon graced the wrists of divers in the American, German, French and Israeli navies. Jacques-Yves Cousteau and his team also used these professional timepieces in their filming of *The Silent World*, winner of a Golden Palm in Cannes in 1956.

After several years of submersion, the Fifty Fathoms resurfaced in 2003. Its design closely followed that of the original model. It was waterproof up to 300 metres or 165 fathoms. Since then, the watch series has been the flagship of the brand revived by Jean-Claude Biver, which became part of Swatch Group in 1992. Several iterations that recall the past to varying degrees are available. In 2023, the five models in the Blancpain X Swatch Scuba Fifty Fathoms anniversary edition caused a stir. Inside the sealed bioceramic case is a non-serviceable Sistem51 automatic movement.

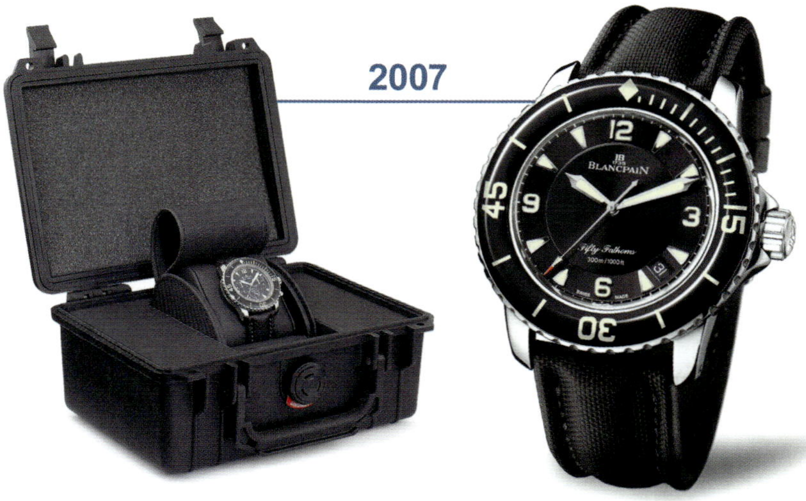

TAUCHERUHR MIT GESCHICHTE

Der Auftrag war in höchstem Maße verlockend. 1952 ging er an Commander Robert „Bob" Maloubier und Leutnant Claude Riffaud, Profitaucher und Offiziere der französischen Marine. Ihr Job: Aufbau der Eliteeinheit „Les Nageurs de Combat". Neben anderem Equipment brauchte es damals auch Armbanduhren, welche den hohen Anforderungen der Kampfschwimmer genügten. Und die waren am Markt extrem selten gesät. Also brachten Maloubier und Riffaud ihre Idealvorstellungen höchstpersönlich zu Papier. Die anschließende Suche nach einem kompetenten Fabrikanten gestaltete sich keineswegs einfach. Nach einigen Umwegen gelangte das Duo an die kleine Uhrenmanufaktur Rayville. Deren Besitzerin Betty Fiechter hatte 1933 ihren Chef Frédéric-Emile Blancpain als letztes Mitglied der bekannten Uhrmacherdynastie beerbt. Gemeinsam mit den beiden Franzosen setzte sie deren Pläne in die Praxis um: optimale Lesbarkeit des Zifferblatts bei allen Lichtverhältnissen, Automatikwerk, Tauchzeit-Drehring sowie bis zu 50 Faden (91,45 Meter) wasserdichtes Stahlgehäuse. Dieser Wert kam übrigens nicht von ungefähr. Das Gemisch aus Sauerstoff und Nitrogen gestattete Flaschentauchern damals keine größeren Tiefen. 1953 gab es vermutlich erste Prototypen. 1954 oder erst 1955 wurde eine Armbanduhr Realität, welche u. a. wegen ihrer patentierten, nur einseitig verstellbaren Drehlünette Zeitgeschichte schrieb. Kein Wunder, dass die „Fifty Fathoms" rasch an die Handgelenke von Tauchern der amerikanischen, deutschen, französischen und israelischen Kriegsmarine fand. Auch Jacques-Yves Cousteau und sein Team benutzten diese Profi-Instrumente bei ihren Dreharbeiten zum Film *Welt des Schweigens*, der 1956 in Cannes eine Goldene Palme erhielt.

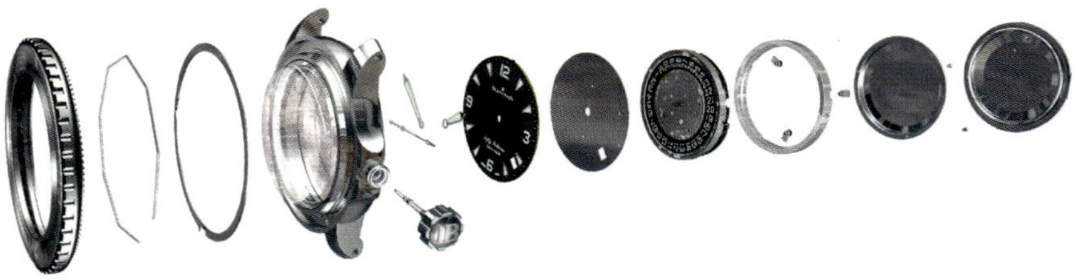

Nach mehreren Jahren in der Versenkung kehrte die „Fifty Fathoms" 2003 zurück. Und zwar in einem Design, das dem ursprünglichen weitgehend ähnelte. Wasserdicht bis 300 Meter oder 165 Faden. Seitdem ist die Uhrenlinie das Aushängeschild der von Jean-Claude Biver wiederbelebten und seit 1992 zur Swatch Group gehörenden Marke. Zur Wahl stehen zahlreiche Varianten mit mehr oder minder starkem Bezug zur Vergangenheit. Für eine Überraschung sorgte das Jahr 2023 mit fünf Ausführungen der „Blancpain X Swatch Scuba Fifty Fathoms" getauften Jubiläumsedition. Ihr fest versiegeltes „Biokeramik"-Plastikgehäuse birgt ein nicht reparierbares Sistem51 Automatikwerk.

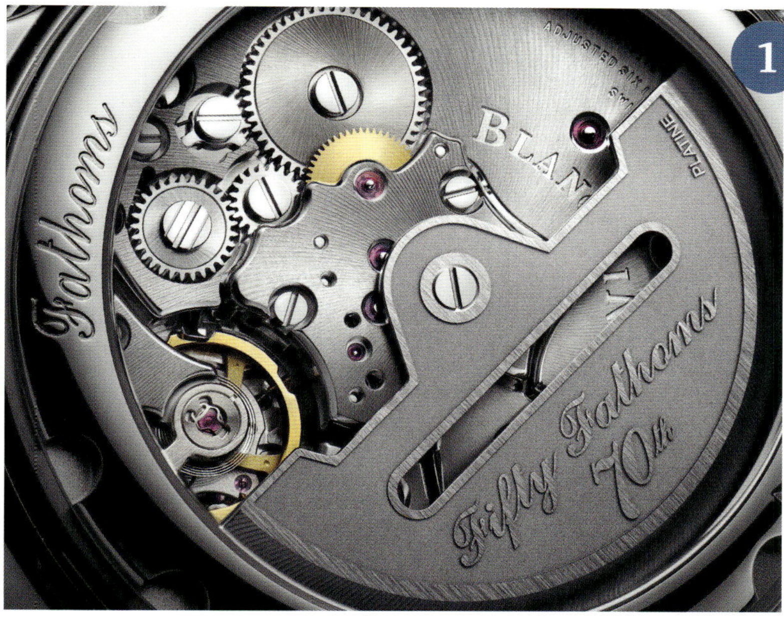

1

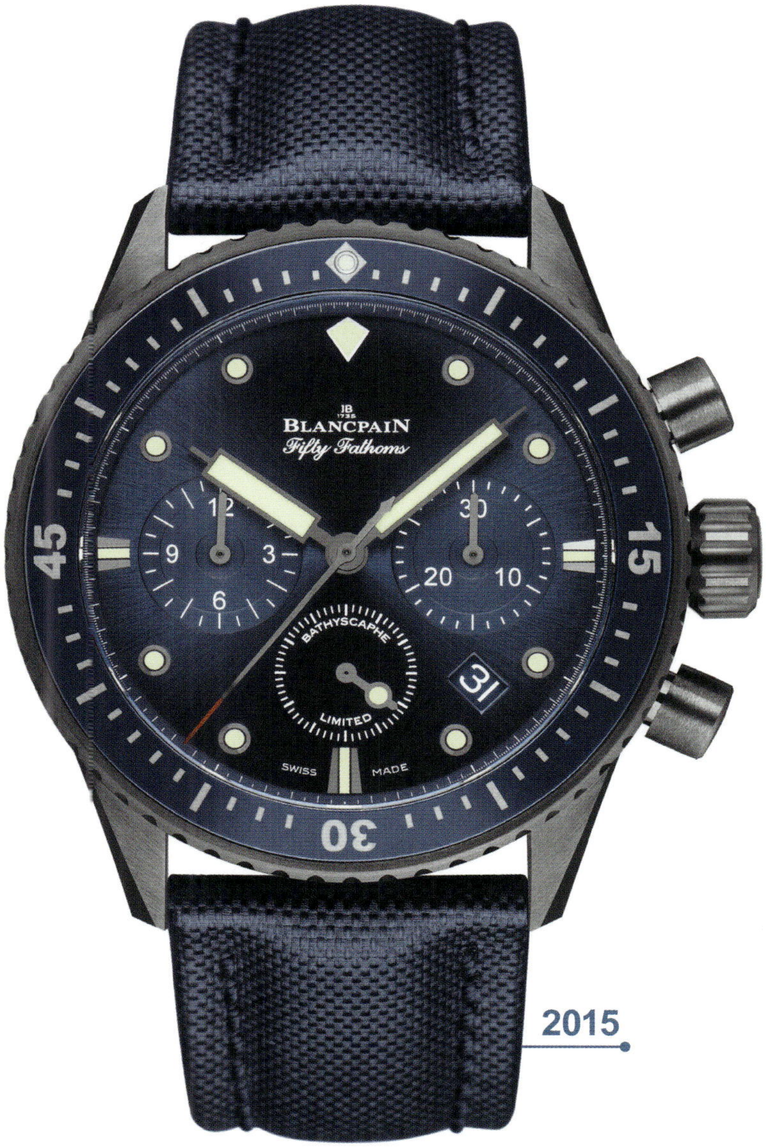

2015

1 *Movement | Werk, 2023*

2 *Fifty Fathoms No Rad, 2021*

3 *Jean-Jacques Fiechter, CEO of Blancpain from 1950 to 1980 | Geschäftsführer von 1950 bis 1980*

4 *Fifty Fathoms, 2017*

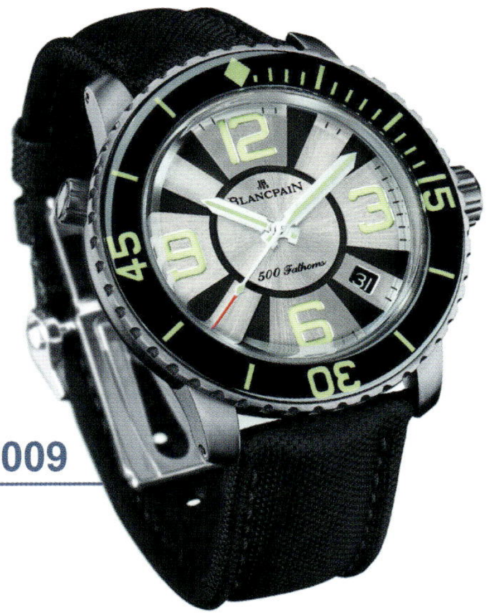

2009

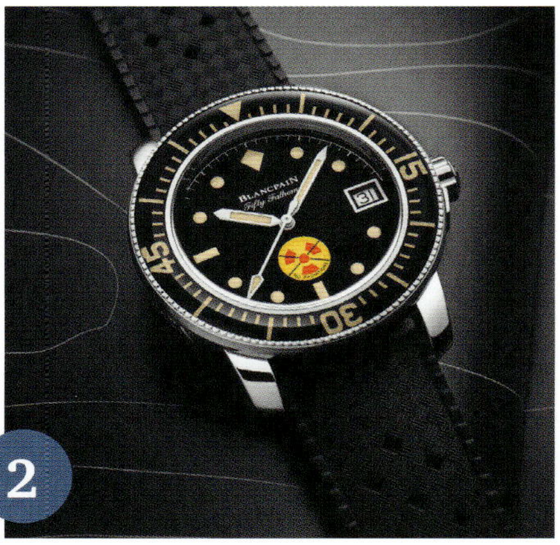

2

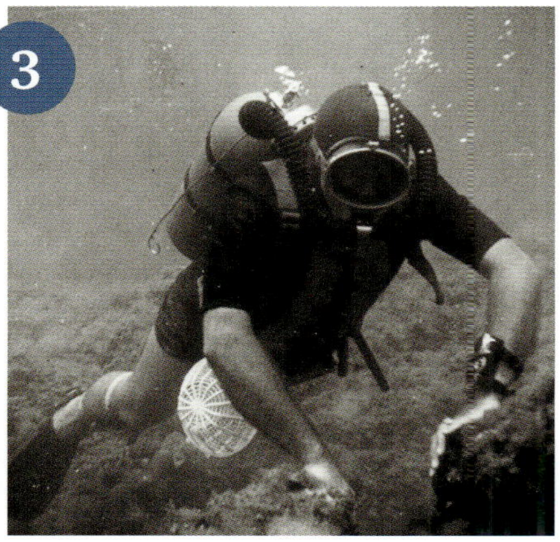

3

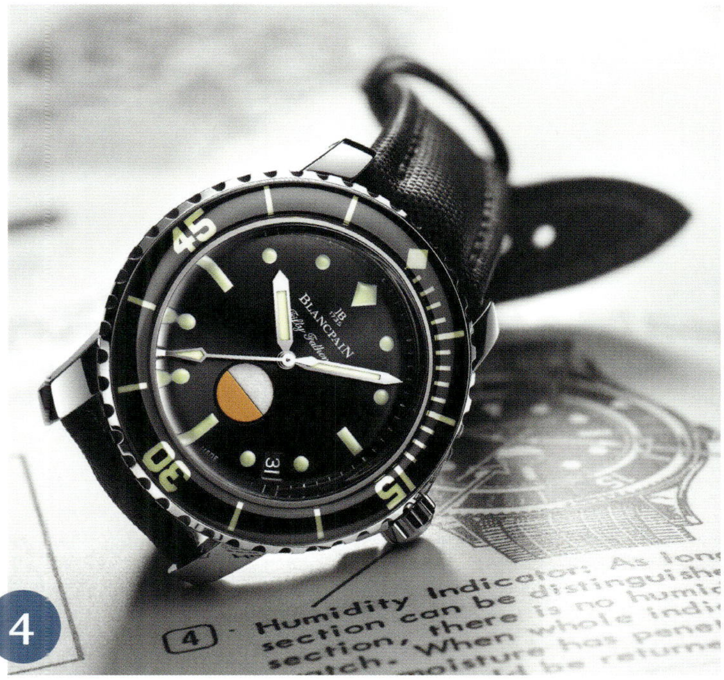

4

BOVET 2010

AMADÉO

AMADÉO FROM ROMANDY

Bovet's history dates back to the early nineteenth century. In 1822, the 25-year-old watchmaker Edouard Bovet set up his own pocket watch business aimed at the Chinese market. There, the brand soon became a synonym for exquisite Swiss horology thanks to the watchmaker's high standards of engineering, craftsmanship and aesthetics. After an eventful history in the twentieth century, Pascal Raffy launched a renaissance of the traditional manufactory in 2001. His watch line is distinguished by the winding crown at 12 o'clock and an arc-shaped lug that acts as a crown guard, referencing its distinguished pedigree. The name of the line alone, the Amadéo Fleurier, recalls the deep watchmaking roots of this picturesque region. At first glance, associations are evoked of a Lépine pocket watch. Only on closer inspection does its less discernible flexibility become apparent. An ingenious patented bracelet system enables the watch to be easily reversed or indeed to use this masterpiece as a pocket watch or table clock. Watch lovers with very differing demands will be sure to find what they are looking for in the Amadéo Fleurier portfolio. In addition to the artistic guilloché dial, the Virtuoso V model offers a five-day movement with a jumping hour and retrograde minute indications. The top-of-the-range models offer intricate tourbillons with manually skeletonised and engraved movements. Eloquent examples include the Virtuoso III and the 2013 Amadéo Amadéo Tourbillon with a seven-day power reserve. The mechanical inner workings for these gems are of course created under the manufactory's own roof. The corresponding studios are located in the villages of Tramelan and Môtiers. There, expert watchmakers hand-craft the watches from the first screw to the last, afterwards vouching with their good name for meticulous work and perfect function.

„AMADÉO" AUS DER WESTSCHWEIZ

Die Geschichte von Bovet reicht zurück bis ins frühe 19. Jahrhundert. 1822 machte sich der 25-jährige Uhrmacher Edouard Bovet mit Taschenuhren speziell für den chinesischen Markt selbständig. Durch technisch, handwerklich und optisch erstklassige Produkte entwickelte sich die Marke im „Reich der Mitte" zum Synonym für feinste Schweizer Zeitmesskunst. Nach wechselvoller Geschichte im 20. Jahrhundert startete Pascal Raffy 2001 eine Renaissance der traditionsreichen Manufaktur. Kennzeichen seiner ersten Uhrenlinie: Aufzugskrone bei „12" und darüber ein Bügel, welcher als Bandanstoß dient und gleichzeitig an die lange Vergangenheit erinnert. Allein schon wegen ihres Namens stellt die Linie „Amadéo Fleurier" einen engen Bezug zu den tief reichenden Uhr-Wurzeln in dieser malerischen Region dar. Der erste Blick weckt spontane Assoziationen zu einer Lépine-Taschenuhr. Die optisch weniger wahrnehmbare Flexibilität erschließt sich erst bei näherer Betrachtung. Ein ausgeklügeltes und deshalb auch patentiertes Armbandsystem gestattet es, die Uhr mit wenigen Handgriffen zu wenden oder das Œuvre tatsächlich auch als Taschen- oder Tischuhr zu nutzen. Im breiten „Amadéo Fleurier"-Spektrum werden Uhrenliebhaberinnen und -liebhaber mit ganz unterschiedlichen Ansprüchen fündig. Beispielsweise bietet das Modell „Virtuoso V" neben dem kunstvoll guillochierten Zifferblatt ein Fünf-Tage-Werk mit springender Stunden- und retrograder Minutenindikation. Die oberste Liga verkörpern filigran gearbeitete Tourbillons mit manuell skelettiertem und graviertem Uhrwerk. Beredte Beispiele sind hier die „Virtuoso III" oder das 2013 vorgestellte „Amadéo Amadéo"-Tourbillon, dessen Gangautonomie ganze sieben Tage beträgt. Selbstverständlich entsteht das mechanische Innenleben für diese Kleinodien unter dem eigenen Dach. Die entsprechenden Ateliers befinden sich in den Ortschaften Tramelan und Môtiers. Dort fertigen kompetente Handwerker ihre Uhren komplett von A bis Z. Hinterher bürgen sie mit ihrem guten Namen für sorgfältige Arbeit und einwandfreie Funktion.

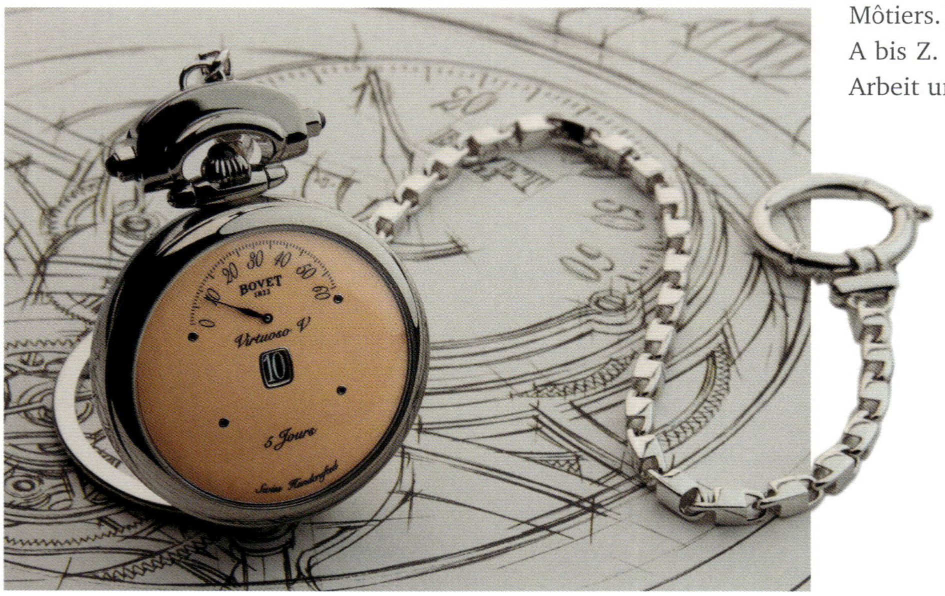

Amadéo Fleurier Virtuoso V table clock / Tischuhr, Ref. ACHS041, 2022

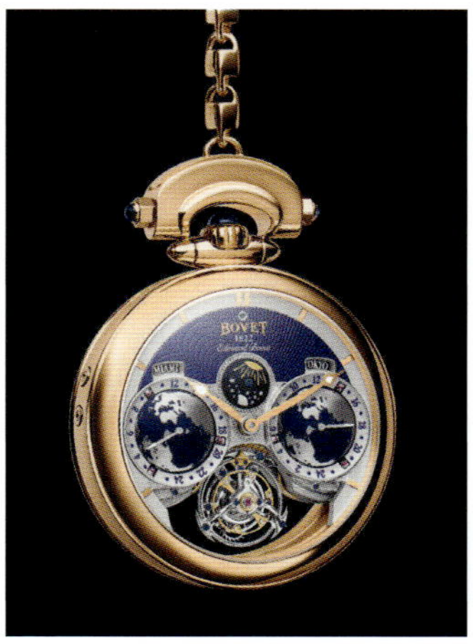

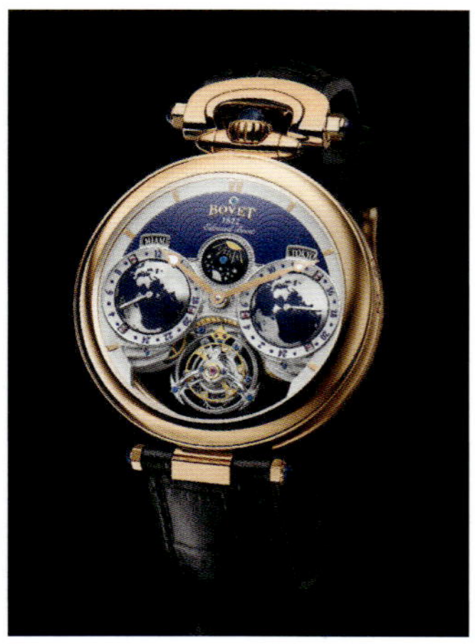

***Left:** Edouard Bovet Tourbillon, pocket watch & wristwatch / Taschenuhr & Armbanduhr*
***Right:** Amadéo Amadéo*

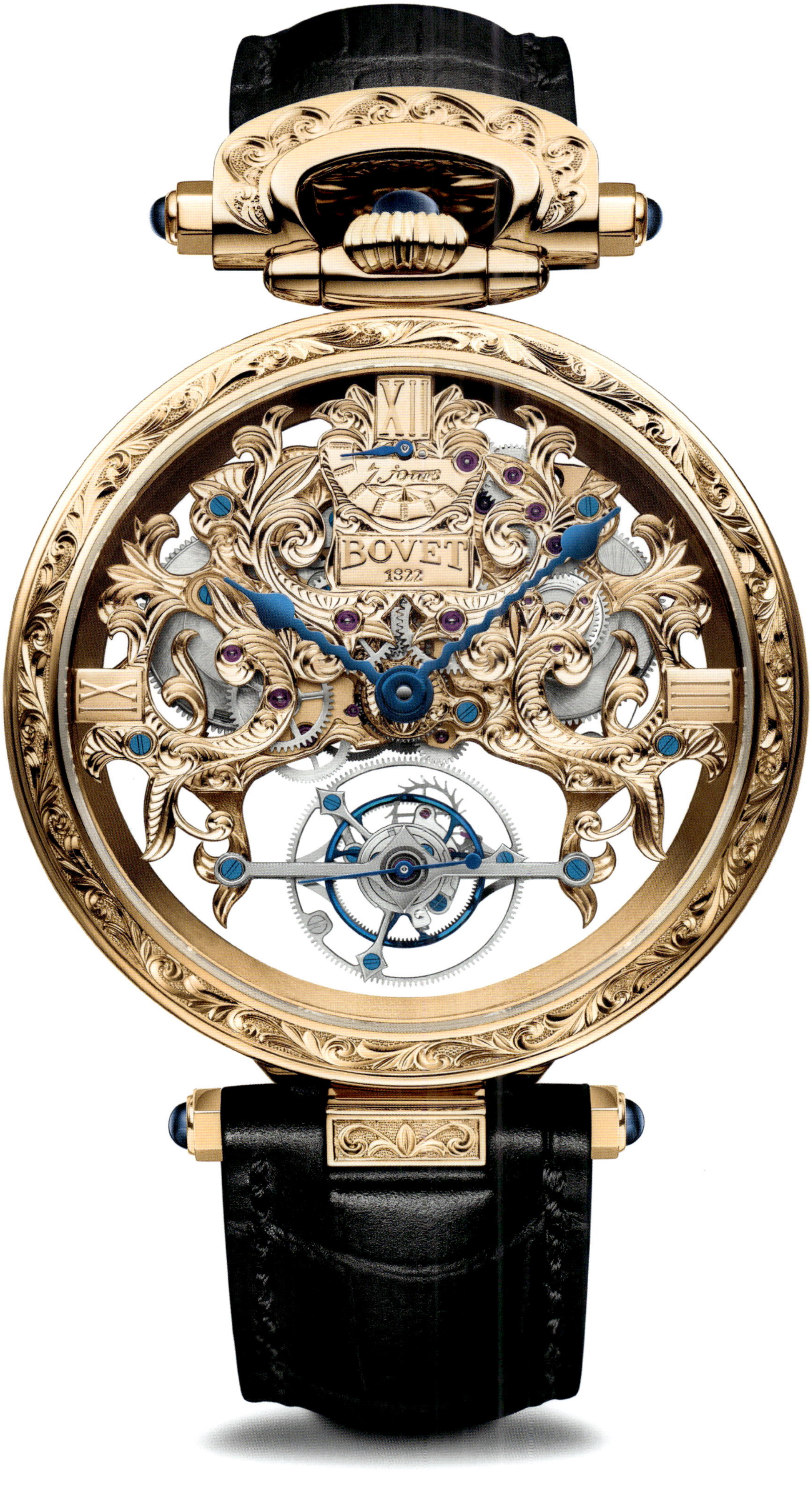

BOVET
1822

VIRTUOSO III

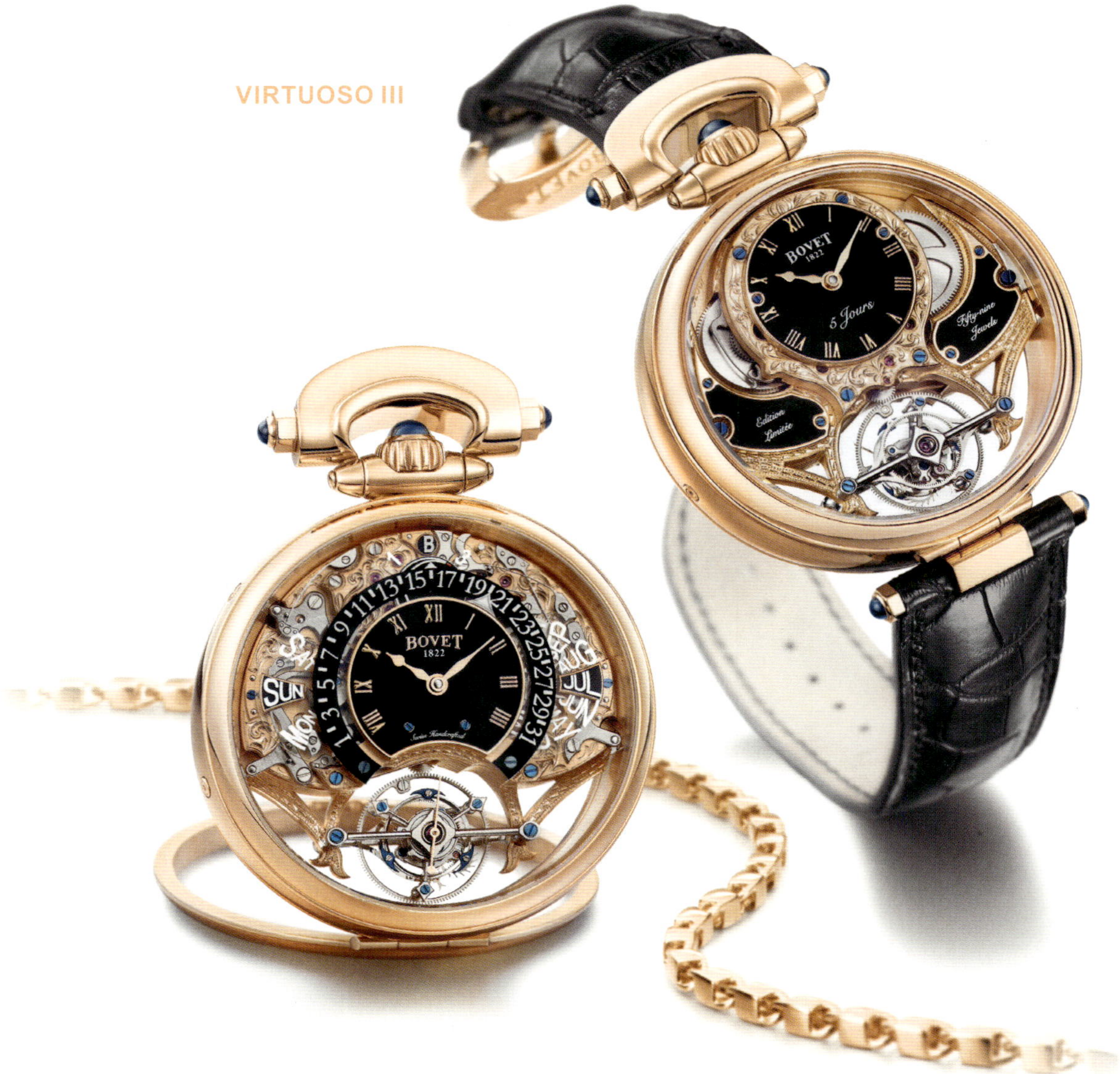

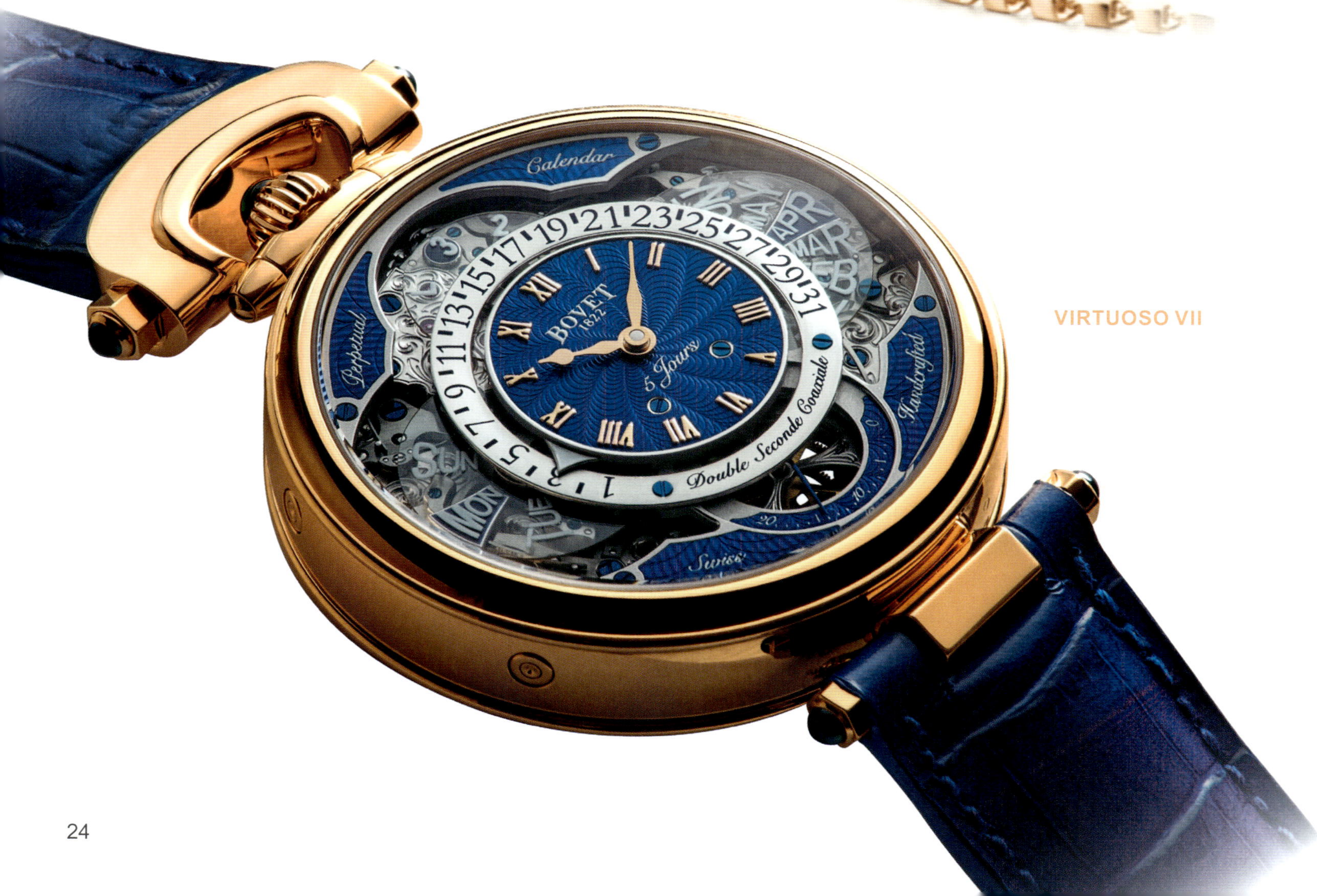

VIRTUOSO VII

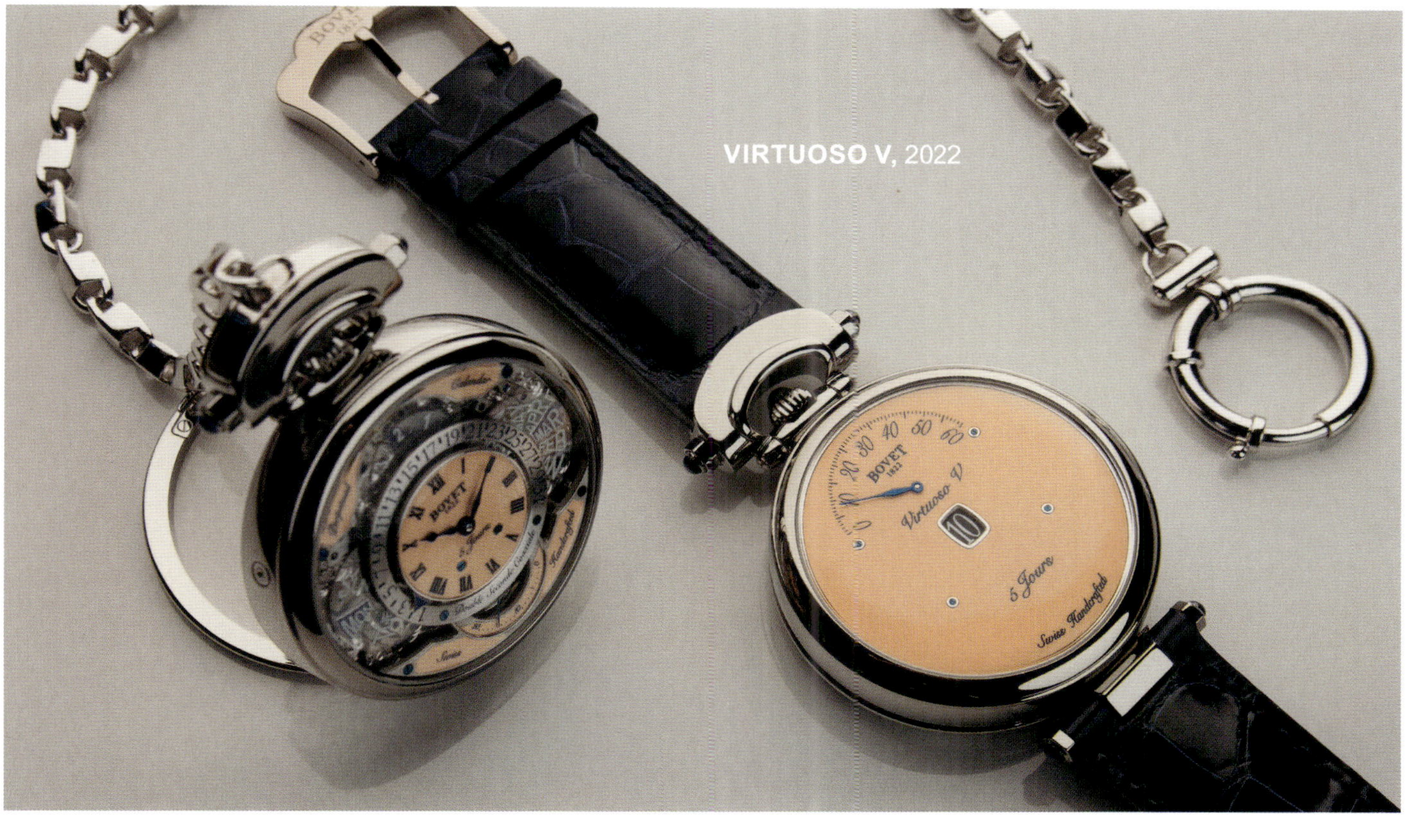

VIRTUOSO V, 2022

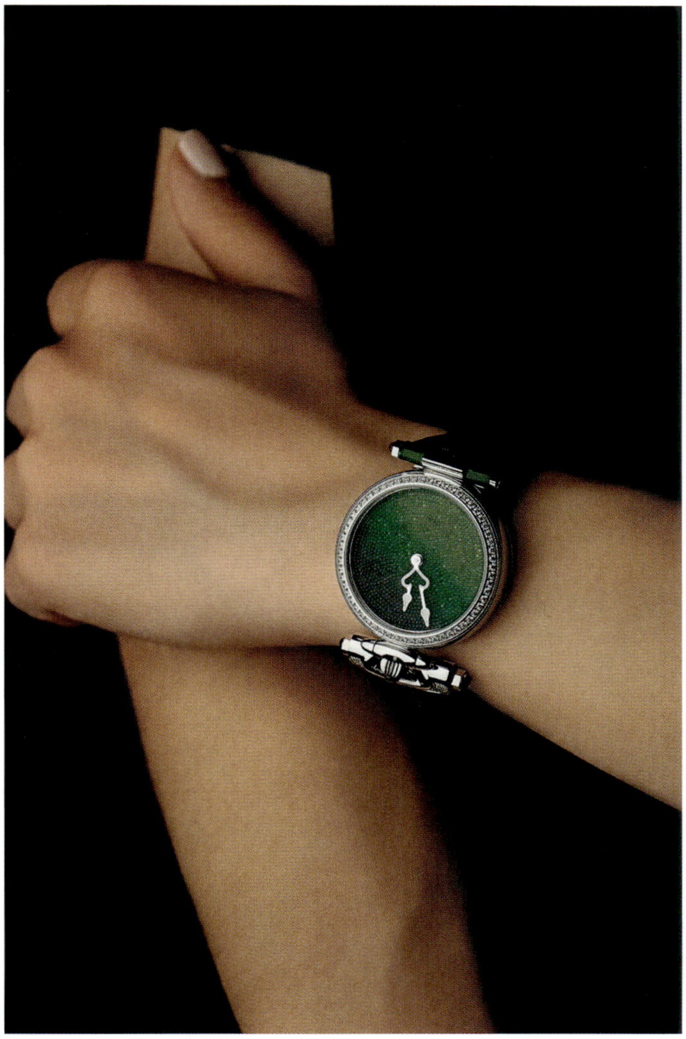

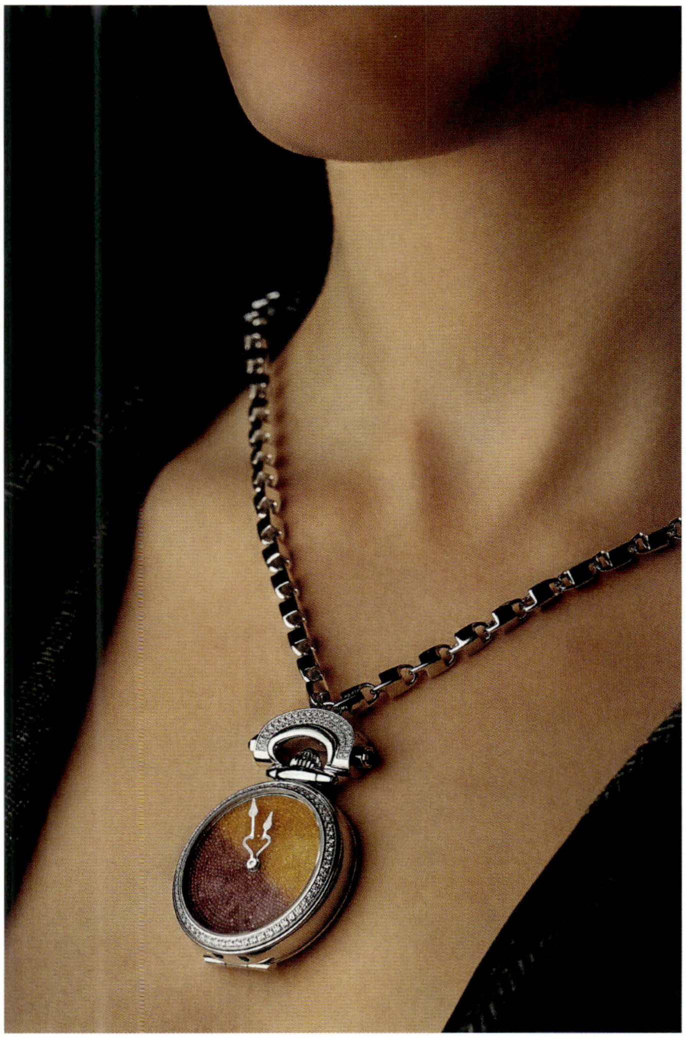

MISS AUDREY SWEET ART, 2021

BREGUET
LA TRADITION

2005

HOMAGE TO ABRAHAM-LOUIS BREGUET

The take on this Breguet could not be more modern. The Classique La Tradition 7027 ideally reflects the spirit of the French watchmaker of Swiss origin, from the distinctive case with its fluted flanks to its 33-millimetre hand-wound calibre 570 DR assembled from 215 components. To observe the inner workings of this timepiece you do not even have to take it off your wrist. Almost everything is presented under the open-work sapphire crystal. Firstly, there is the dial positioned eccentrically at 12 o'clock. The hand-engraved guilloché dial is topped by rotating blue hands with hollow moon tips that became known as Breguet hands. The power reserve indicator comes in a double format. One is located on the front and the other on the back of the hand-wound movement. After all, the entire gear train is in view for all to admire. As one would expect, the classic screw-regulated balance features a Breguet overcoil hairspring with a sharply curved outer end to ensure an accurate backward swing. The fine, and therefore fragile, pivots of the balance wheel are protected by a pare-chute or parachute shock absorber. Three more key movement figures of interest: 34 jewels, 50-hour power reserve, and to ensure relaxed observation of the balance wheel oscillations, a moderate balance frequency of just three hertz.

HOMMAGE AN ABRAHAM-LOUIS BREGUET

Mehr neuzeitliches Breguet geht fast nicht. Die „Classique La Tradition 7027" spiegelt den Spirit des französischen Uhrmachers mit Schweizer Wurzeln nachgerade idealtypisch wider. Das beginnt beim charakteristischen Gehäuse mit kannelierten Flanken und setzt sich fort im 33 Millimetern großen, aus 215 Komponenten assemblierten Handaufzugskaliber 570 DR. Zum Beobachten der funktionalen Abläufe muss man diesen Zeitmesser nicht einmal vom Handgelenk nehmen. Nahezu alles präsentiert sich unter dem vorderseitigen Saphirglas. Da ist zunächst das exzentrisch bei „12" positionierte Zifferblatt. Vor der manuell guillochierten Scheibe rotieren blaue, nach Abraham-Louis Breguet benannte Zeiger mit durchbrochener „Pomme"-Spitze. Gleich doppelt vorhanden ist die Indikation für den aktuell verbleibenden Energievorrat des Federhauses. Eine befindet sich auf der Vorder-, die andere auf der Rückseite des Handaufzugswerks. Schließlich präsentiert sich den Blicken der Betrachter auch das komplette Räderwerk. Bei der klassischen Schraubenunruh sorgt selbstverständlich eine sogenannte Breguet-spirale mit hochgebogener Endkurve für akkuraten Rückschwung. Die dünnen und daher bruchgefährdeten Zapfen der Unruhwelle schützt eine Stoßsicherung à la „parachûte". Drei weitere Zahlen zum Uhrwerk: 34 funktionale Steine, 50 Stunden Gangautonomie und, damit sich die Unruhschwingungs-Oszillationen in aller Ruhe verfolgen lassen, eine moderate Unruhfrequenz von nur drei Hertz.

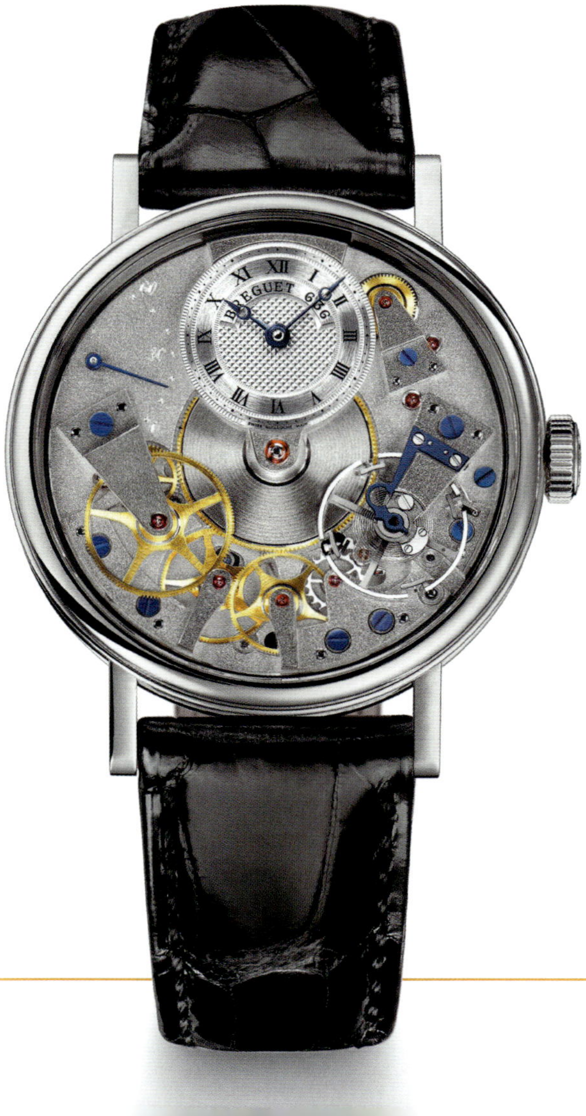

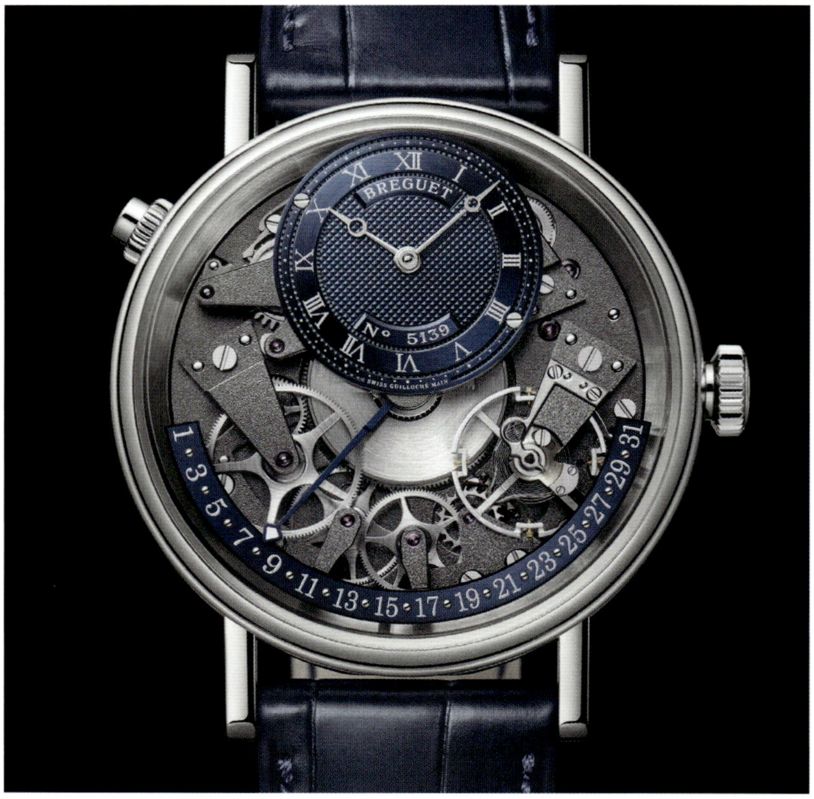

La Tradition 7037, 2009 / La Tradition 7597 / La Tradition 7027, 2009

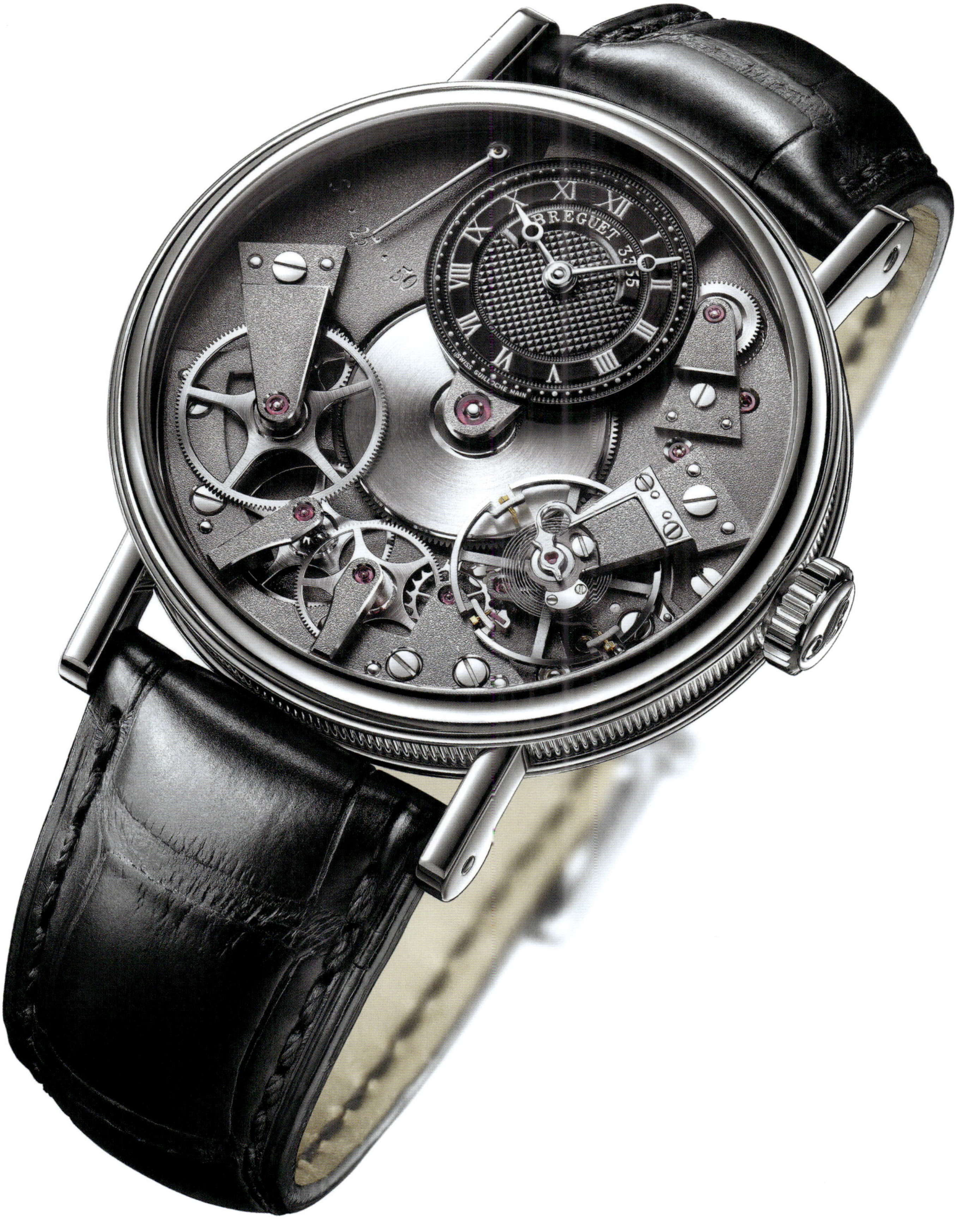
BREGUET
3755

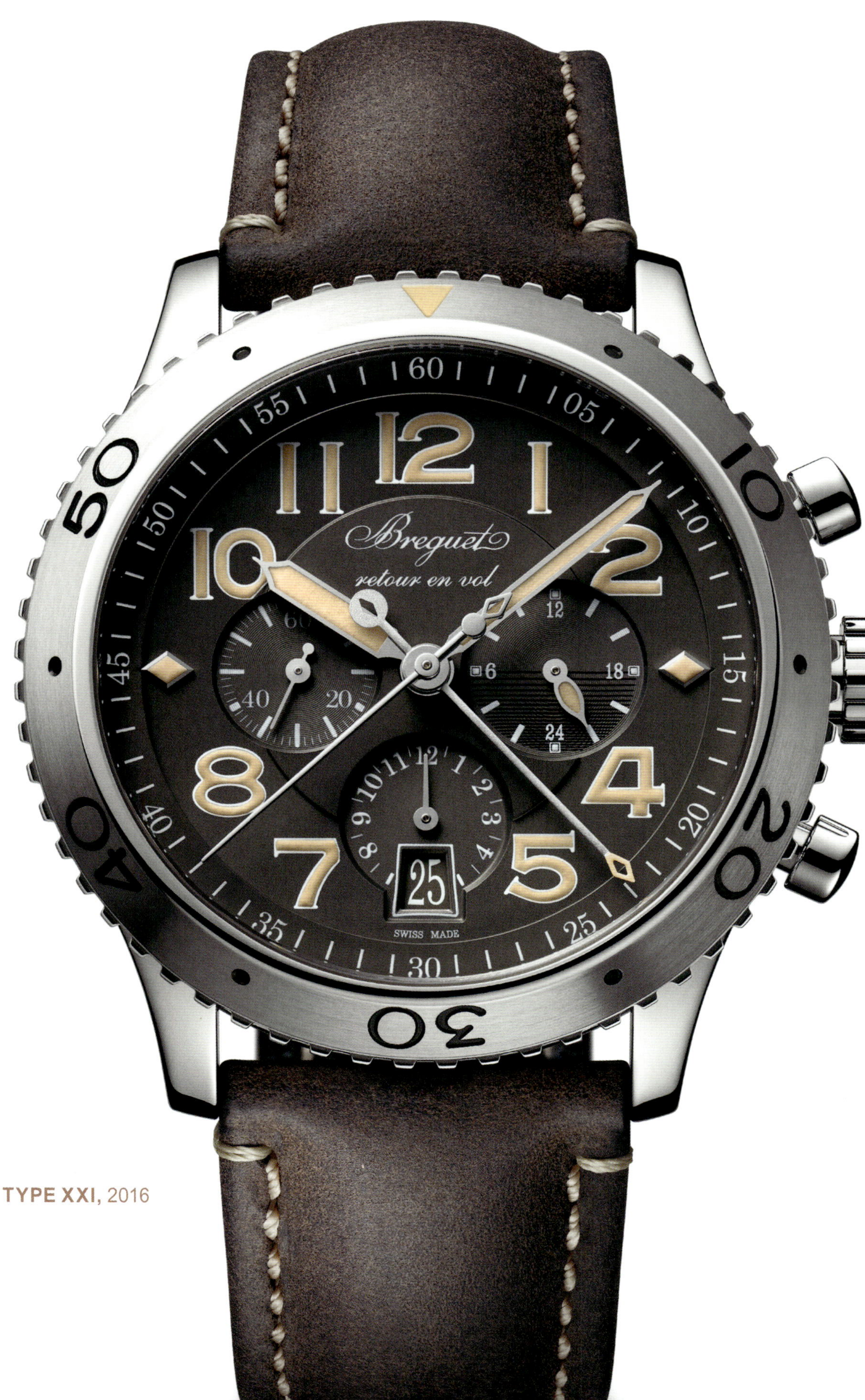

TYPE XXI, 2016

BREGUET
TYPE XX

NOT JUST FOR HIGH-FLYERS

In 1870, the Brown family acquired the horological legacy of Abraham-Louis Breguet. Under their aegis, chronographs with counters and tachymeter scales were produced in the mid-twentieth century. The target audience included pilots. In close cooperation with airlines and aircraft manufacturers, Breguet continuously developed these special watches. One client was the French Flight Test Centre (CEV) in Bretigny, which was responsible for the testing and approval of all new aircraft in France. From 1954, Breguet produced three generations of the Type XX watch chronograph. Strictly speaking, the timepiece was derived from the company's first watch chronograph launched in 1935. Characteristics include a non-slip rotating bezel and a "retour-en-vol" mechanism. Also known as a flyback, this complication is used to reset and restart the chronograph with one click, without needing to stop, then reset and start again like a standard chronograph. Breguet produced similar products for the pilots of the French Navy and French Air Force. A large ball-shaped crown permits winding even wearing gloves. For short flying manoeuvres such as waiting loops, a comparatively large and therefore easy-to-read 15-minute counter is sufficient. In the design of the dial and hands emphasis was placed on optimum readability even in adverse light conditions. For the movements, Breguet and its clients insisted on the best components available at the time. They were, for example, the 14-line calibre 22 by the reputable ébauche manufacturer Valjoux. In 1995, the company, now part of Swatch Group, presented a modern version with the automatic calibre Lémania 1377. Breguet currently offers the Type XX chronographs 2057 and 2067 with proprietary automatic calibres 7281 and 728.

NICHT NUR FÜR PILOTEN

1870 erwarb die Familie Brown die uhrmacherische Hinterlassenschaft von Abraham-Louis Breguet. Unter ihrer Ägide entstanden in der Mitte des 20. Jahrhunderts u. a. Chronographen mit Zähler und Tachymeterskala. Zur Zielgruppe gehörten beispielsweise Piloten. In enger Kooperation mit Luftfahrtgesellschaften und Flugzeugherstellern entwickelte Breguet diese Spezialuhren kontinuierlich weiter. Unter anderem für das Centre d'Essai en Vol (CEV) in Bretigny, welches in Frankreich für die Prüfung und Zulassung aller neuen Flugzeuge verantwortlich zeichnete, fertigte Breguet ab 1954 drei Generationen des Armbandchronographen „Typ XX" in unterschiedlichen Ausführungen. Genau genommen leitete sich der Stopper ab vom ersten, 1935 lancierten Armbandchronographen des Hauses. Kennzeichen sind eine griffige Drehlünette sowie ein „retour-en-vol"-Mechanismus. Mit Hilfe dieser permanenten Nullstellung, auch Temposchaltung oder Flyback genannt, lässt sich der Chronographenzeiger aus dem Lauf heraus nullstellen und ohne weiteren Knopfdruck unverzüglich neu starten. Ähnliches produzierte Breguet auch für die Piloten der französischen „Marine National" sowie für die „Armée de l'Air". Eine große, kugelförmige Krone gestattet das Aufziehen selbst mit Handschuhen. Für kurze Flugbewegungen, zum Beispiel Warteschleifen, reicht ein vergleichsweise großer und daher gut ablesbarer 15-Minuten-Zähler. Bei der Gestaltung von Zifferblatt und Zeigern stand optimale Ablesbarkeit selbst unter ungünstigen Lichtverhältnissen im Vordergrund. Bei den Uhrwerken verließen sich Breguet und die Abnehmer nur auf die besten, seinerzeit verfügbaren Komponenten. Und die stammten, beispielsweise in Gestalt des 14-linigen Kalibers 22, vom renommierten Ebauche-Fabrikanten Valjoux. 1995 präsentierte das heutige Mitglied der Swatch Group eine zeitgemäße Version mit dem Automatikkaliber Lémania 1377. Aktuell offeriert Breguet die „Type XX" Chronographen 2057 und 2067 mit den hauseigenen Automatikkalibern 7281 bzw. 728.

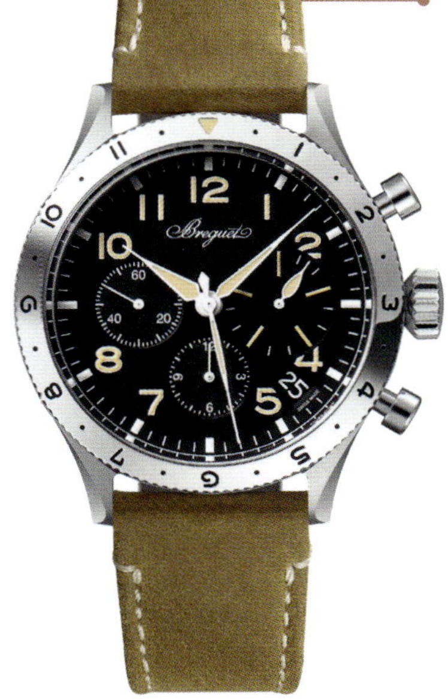

2023

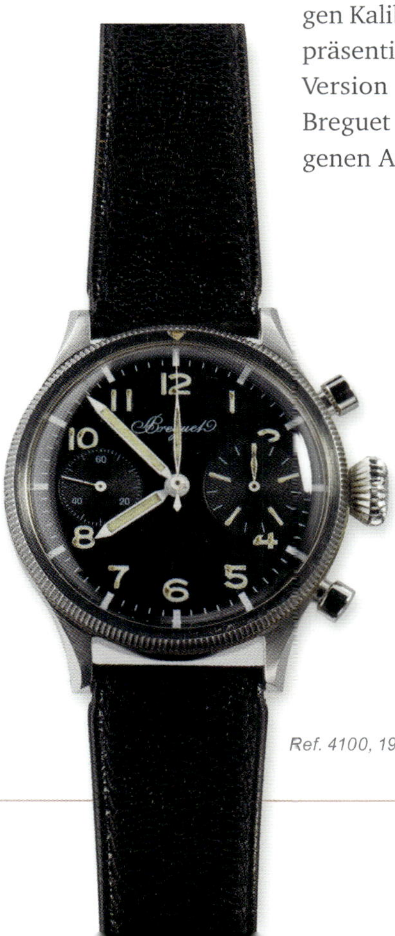

Ref. 4100, 1950s

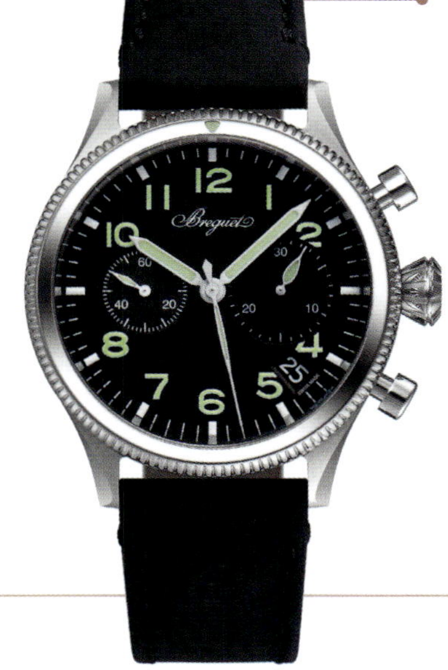

2023

BREITLING 1952

NAVITIMER

NAVIGATION AND TIMER

There is probably no other watch name that is so often mispronounced. Legally protected for Breitling in 1955, the name Navitimer has nothing to do with the navy. Developed in 1952 by Willy Breitling, the chronograph with a circular slide rule was primarily designed to make life easier for pilots in the pre-GPS age when engaged in the difficult task of coordinating time and navigation.

Two years later, the American Aircraft Owners and Pilots Association (AOPA), the world's largest pilots' organisation, chose this creation as its official timepiece. The association's wing logo was emblazoned at "12". The Navigation Timer, or Navitimer for short, was born. The distinctive instrument, which harks back to the Chronomat of the 1940s in its design, possessed a feature with significantly optimised calculation options. For this purpose, the creator and manufacturer of this multifunctional product with a hand-wound Venus 178 calibre advertised it as a "personal on-board instrument" and as an "intelligent instrument". Its logarithmic slide rule was ideal for multiplying, dividing or calculating average values or fuel consumption, average gain in altitude or distances on ascent and landing. It could also convert kilometres into sea or land miles and vice versa with ease. Thanks to this wristwatch, the laborious and sometimes quite dangerous jotting down on paper was largely eliminated. Over the decades, the classic Navitimer chronograph underwent various metamorphoses, mainly revolving around the case and the calibres. However, no one dared to question the model's high practicality with many mathematical functions, even though by the 1970s people no longer needed a slide rule.

It is this feature to which its incredibly high recognition value is largely ascribed and which accounts for the fascination that the Navitimer still exerts today. Connoisseurs can unfailingly spot this Breitling icon from a distance of 10 metres or more, regardless of whether an Eta movement or the proprietary automatic calibre B01 is inside or not. The Navitimer with split-seconds function, also known by the French horological term rattrapante, is the epitome of bar-raising. Visually, it remains true to form.

NAVIGATION UND TIMER

So oft falsch ausgesprochen wird vermutlich kein anderer Name eines Uhrenmodells. Mit „Navy", dem englischen Begriff für Marine, hat der erst 1955 für Breitling rechtlich geschützte Name „Navitimer" nämlich nichts zu tun. Vielmehr besaß der 1952 von Willy Breitling entwickelte Chronograph mit kreisförmigem Rechenschieber in erster Linie die Aufgabe, Piloten im Vor-GPS-Zeitalter den schwierigen Job der Koordination von Zeit und Navigation zu erleichtern.

Zwei Jahre später erkor die amerikanische Aircraft Owners and Pilots Association (AOPA) als weltweit größte Pilotenvereinigung diese Kreation zu ihrem offiziellen Zeitmesser. Das Flügellogo des Verbands prangte bei „12". Damit war der „Navigation Timer" oder kurz „Navitimer" geboren. Das markante, funktional aus dem „Chronomat" der 1940er-Jahre abgeleitete Instrument verfügte über deutlich optimierte Rechenmöglichkeiten. In der Werbung bezeichnete der Entwickler und Hersteller sein multifunktionales Erzeugnis mit dem Handaufzugskaliber Venus 178 als „persönliches Bordinstrument" und „intelligentes Instrument". Seine logarithmische Rechenscheibe eignete sich u. a. zum Multiplizieren, Dividieren, Errechnen von Durchschnittswerten oder Treibstoffverbrauch, durchschnittlichem Höhengewinn oder Entfernungen beim Aufsteigen und Landen. Des Weiteren gelang das Umrechnen von Kilometer in See- oder Landmeilen und umgekehrt völlig mühelos. Dank dieser Armbanduhr konnte das umständliche und mitunter auch recht gefährliche Hantieren mit Papier und Bleistift weitgehend entfallen. Im Laufe der Jahrzehnte erlebte der klassische „Navitimer"-Chronograph verschiedene Metamorphosen, die vor allem das Gehäuse und die verwendeten Kaliber betrafen. An der hohen Funktionalität mit vielfältigen Rechenfunktionen wagte jedoch niemand zu rütteln, obwohl spätestens seit den 1970er-Jahren niemand mehr einen Rechenschieber brauchte.

Aber speziell aus diesem Merkmal resultiert jener immens hohe Wiedererkennungswert, der die anhaltende Faszination des „Navitimer" bis heute ausmacht. Kenner identifizieren diese Breitling-Ikone mühelos aus zehn Metern Entfernung und mehr, egal ob im Gehäuseinneren beispielsweise ein Eta-Uhrwerk oder das hauseigene Automatikkaliber B01 tickt. Einen Komparativ zum Normalen verkörpert der „Navitimer" mit Schleppzeiger-Funktion, im Uhrmacher-Französisch Rattrapante genannt. Optisch bleibt auch er ganz auf der bewährten Linie.

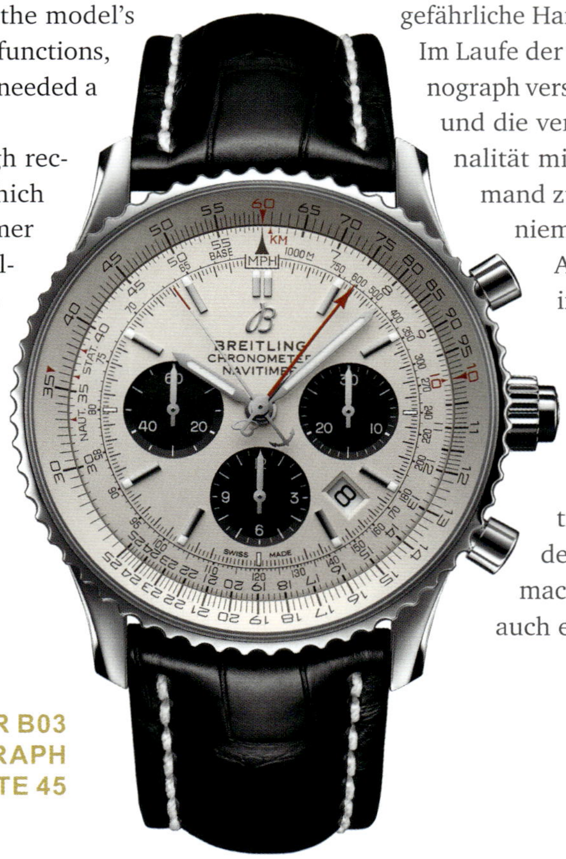

**NAVITIMER B03
CHRONOGRAPH
RATTRAPANTE 45**

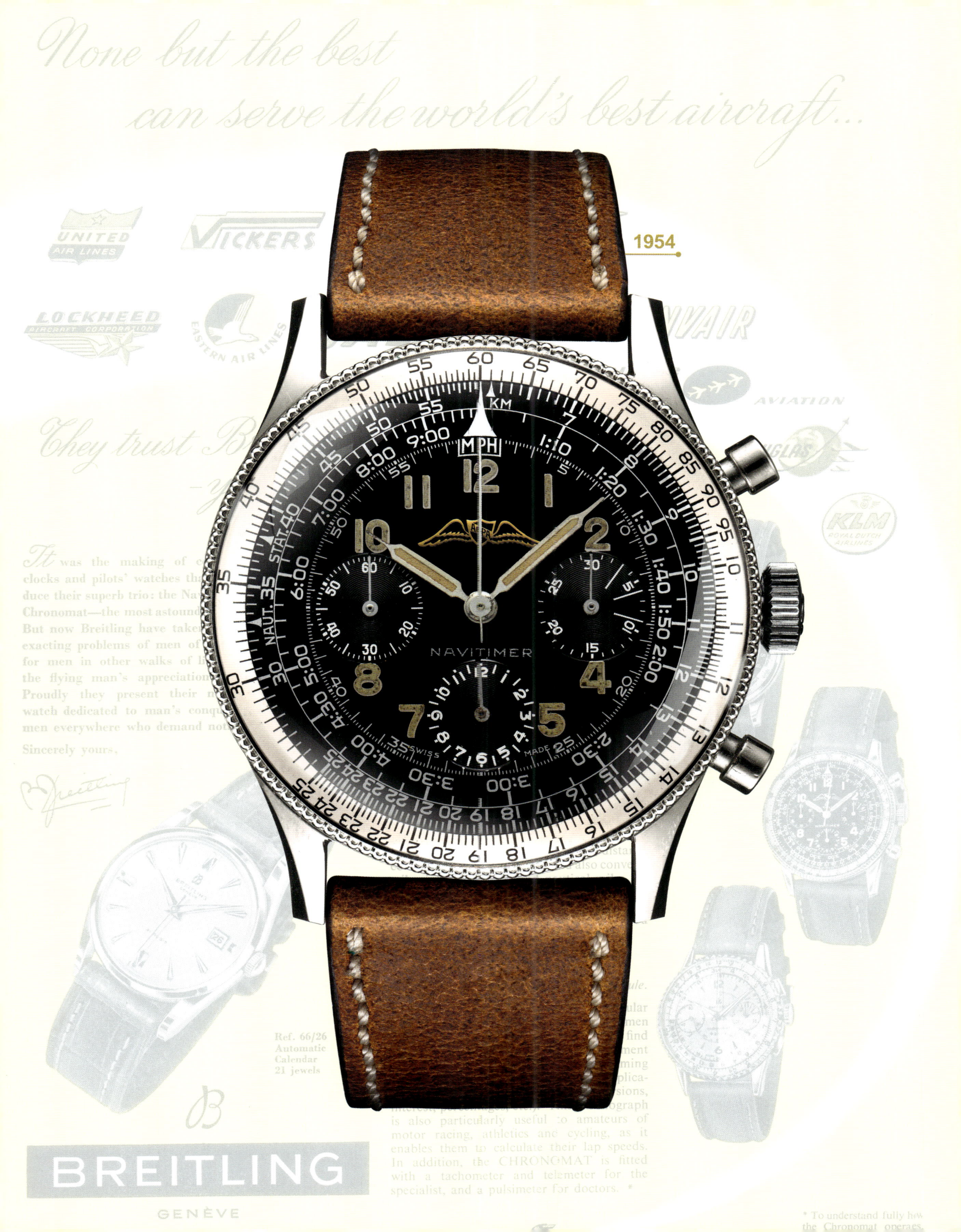
None but the best
can serve the world's best aircraft...
UNITED AIR LINES
VICKERS
LOCKHEED AIRCRAFT CORPORATION
EASTERN AIR LINES
AVIATION
KLM ROYAL DUTCH AIRLINES
NAVITIMER
SWISS MADE
Ref. 66/26
Automatic
Calendar
21 jewels
is also particularly useful to amateurs of
motor racing, athletics and cycling, as it
enables them to calculate their lap speeds.
In addition, the CHRONOMAT is fitted
with a tachometer and telemeter for the
specialist, and a pulsimeter for doctors. *
Sincerely yours,
BREITLING
GENÈVE
1954

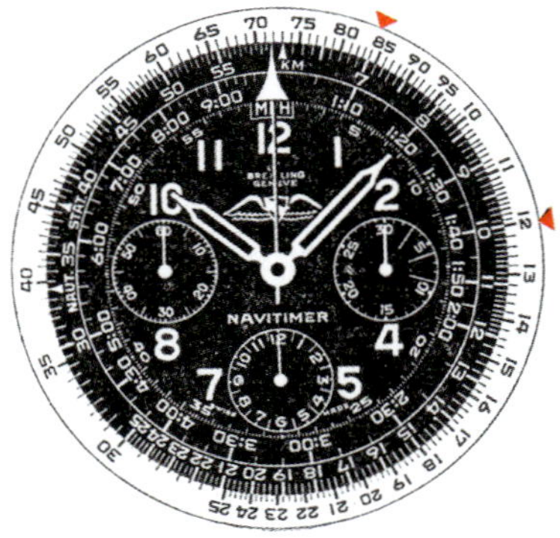

TO MULTIPLY TWO NUMBERS

To multiply 7 by 12, position the number 12 on the outer scale facing the 10 mark on the inner scale. The answer, 84, appears opposite the number 7 on the inner scale. In the same way, and without moving the bezel, the wearer can also read the answers to 5 × 12, 6 × 12, 8 × 12, etc.

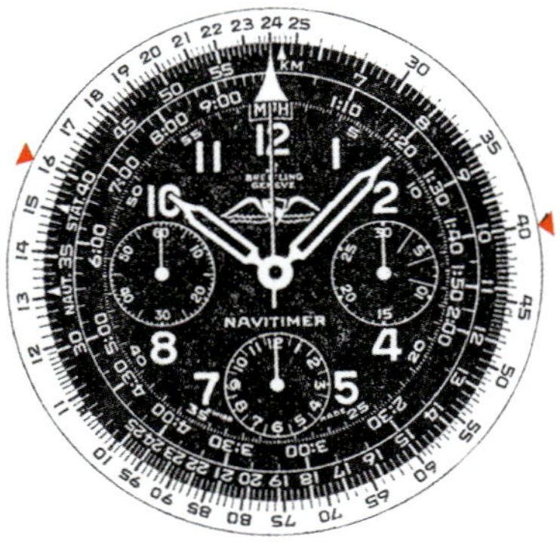

TO DIVIDE TWO NUMBERS

To divide 160 by 4, position the number 16 (=160) on the outer scale facing the number 40 (=4) on the inner scale. The answer, 40, appears opposite the number 10 on the inner scale.

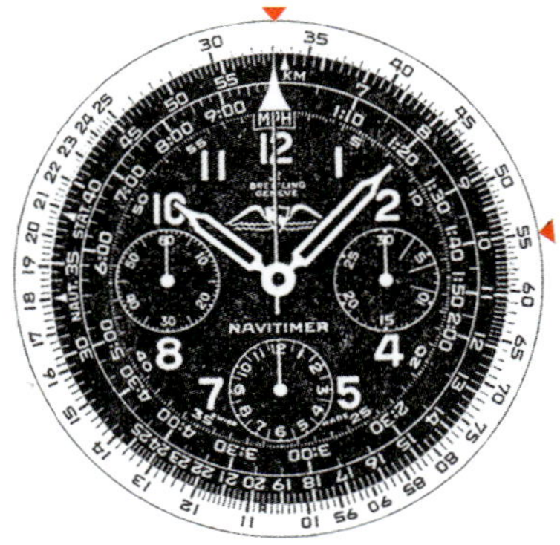

TO CALCULATE DESCENT SPEED

In this example, the pilot of a glider flying at a height of 3300 meters observes that he is descending at a speed of 1 meter/second or 60 meters/minute. What is the remaining flight time? To find the solution, position the number 33 (=3300) on the outer scale facing the number 60 on the inner scale. The answer, 55 minutes, appears opposite the number 10 on the inner scale.

NAVITIMER SLIDE RULE RECHENSCHIEBER

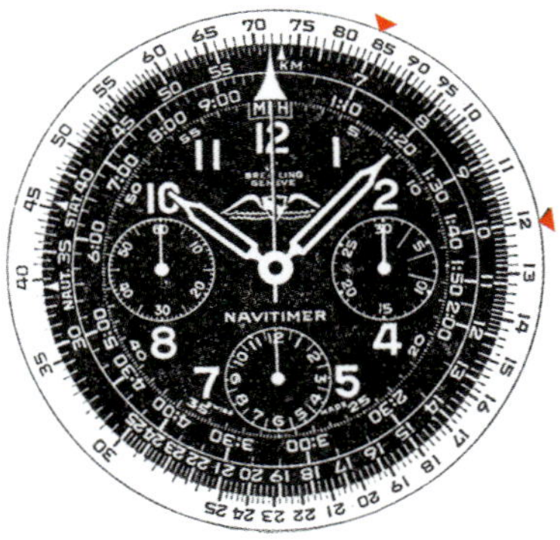

WIE MAN ZWEI ZAHLEN MULTIPLIZIERT

Um 7 mit 12 zu multiplizieren, richten Sie die Zahl 12 auf der äusseren Skala auf die 10er-Markierung der inneren Skala aus. Die Lösung, 84, erscheint gegenüber der Zahl 7 auf der inneren Skala. Auf dieselbe Weise und ohne die Lünette zu bewegen kann der Träger auch die Lösungen der Rechenoperationen 5 × 12, 6 × 12, 8 × 12 etc. ablesen.

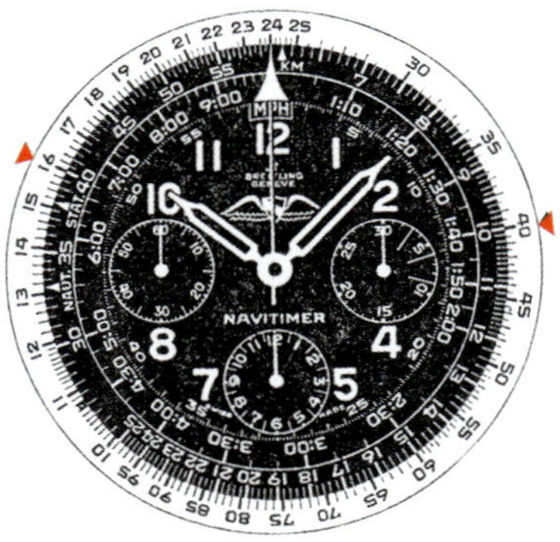

WIE MAN ZWEI ZAHLEN DIVIDIERT

Um 160 durch 4 zu dividieren, richten Sie die Zahl 16 (= 160) auf der äusseren Skala auf die Zahl 40 (= 4) auf der inneren Skala aus. Die Lösung, 40, erscheint gegenüber der Zahl 10 auf der inneren Skala.

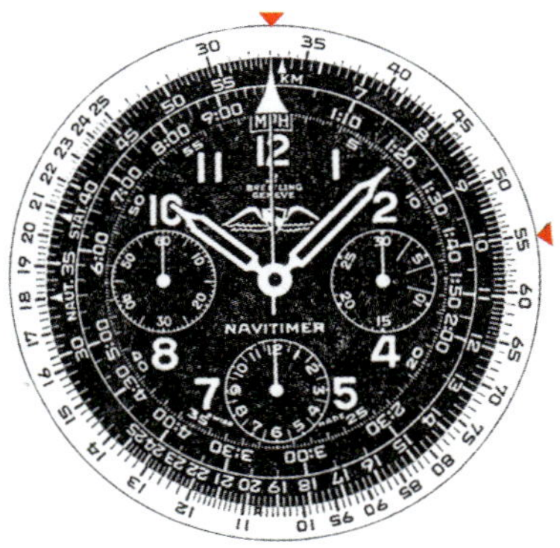

WIE MAN EINE SINKRATE BERECHNET

In diesem Beispiel stellt der Pilot eines Segelflugzeugs auf einer Höhe von 3300 Metern fest, dass er mit einer Geschwindigkeit von 1 Meter pro Sekunde oder 60 Metern pro Minute sinkt. Wie lange ist die noch verbleibende Flugzeit? Die Lösung: Richten Sie die Zahl 33 (= 3300) auf der äusseren Skala auf die Zahl 60 auf der inneren Skala aus. Die Lösung, 55 Minuten, erscheint gegenüber der Zahl 10 auf der inneren Skala.

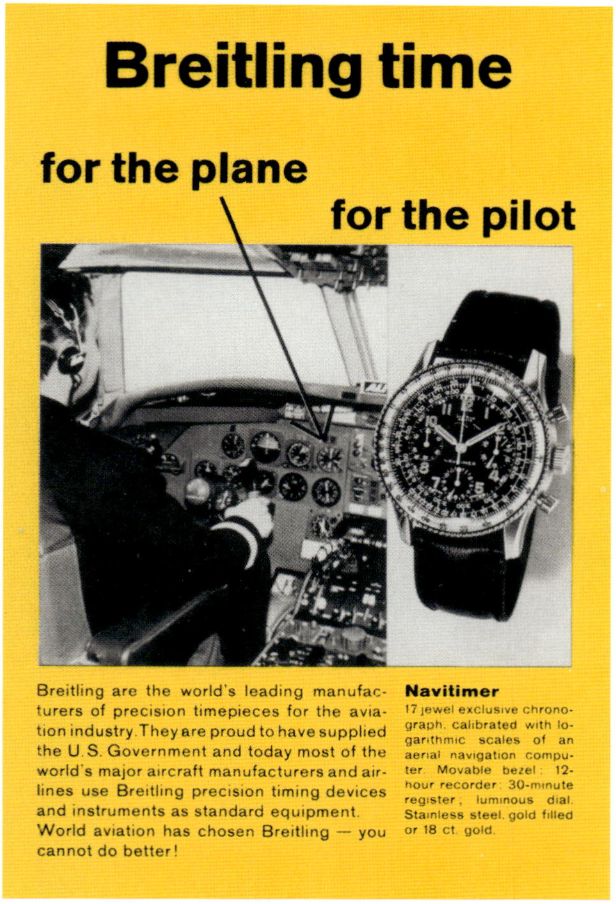

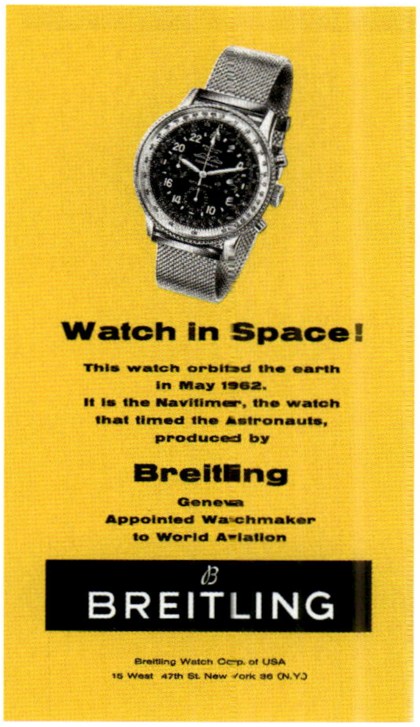

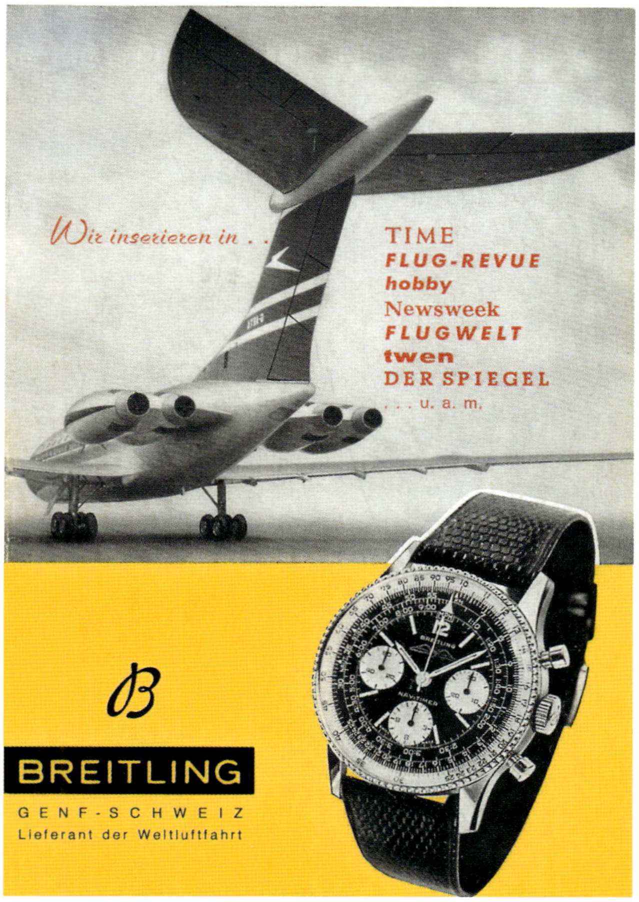

Advertisements from the 1950s and 60s / Werbeanzeigen aus den 1950er- und 60er-Jahren

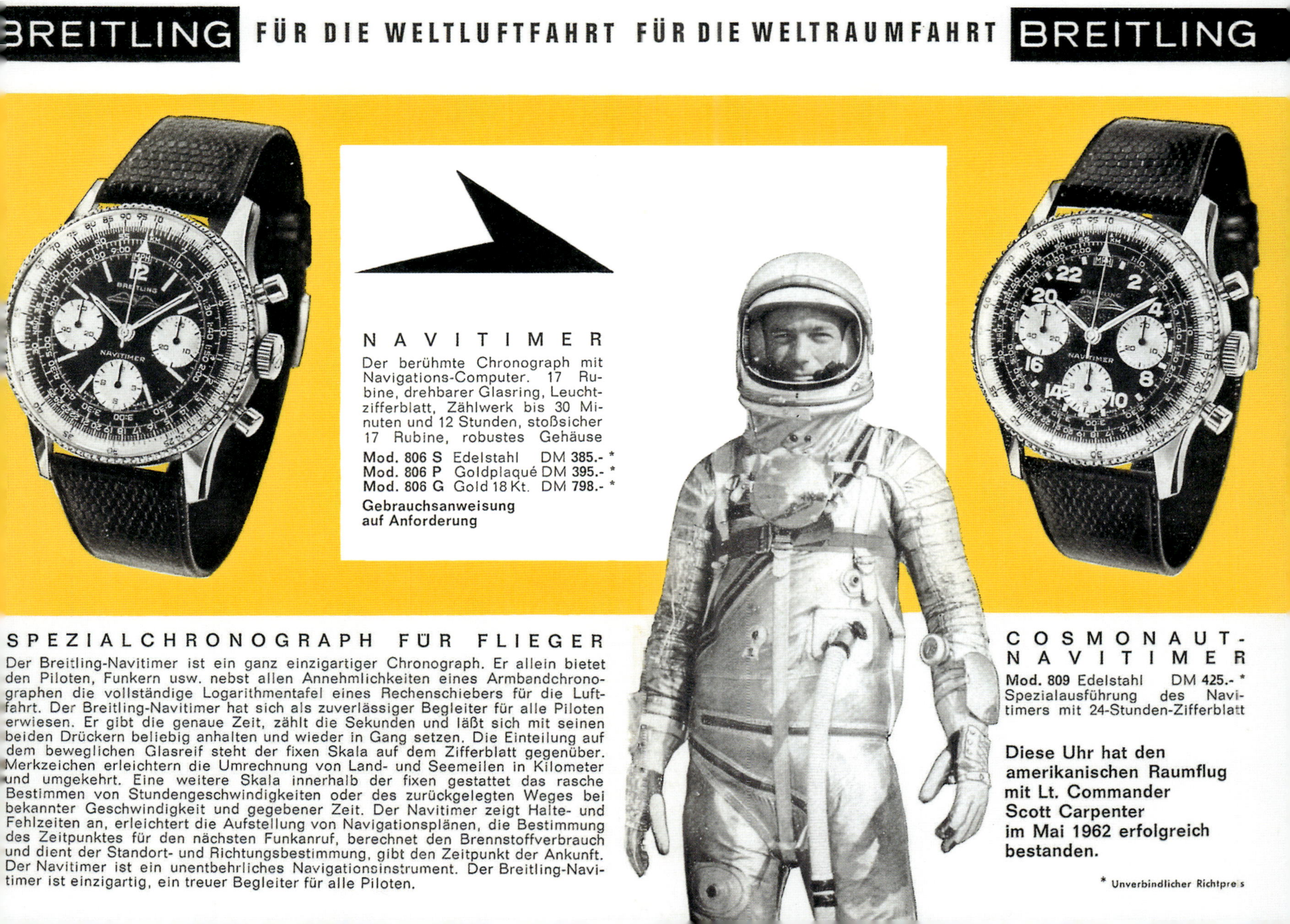

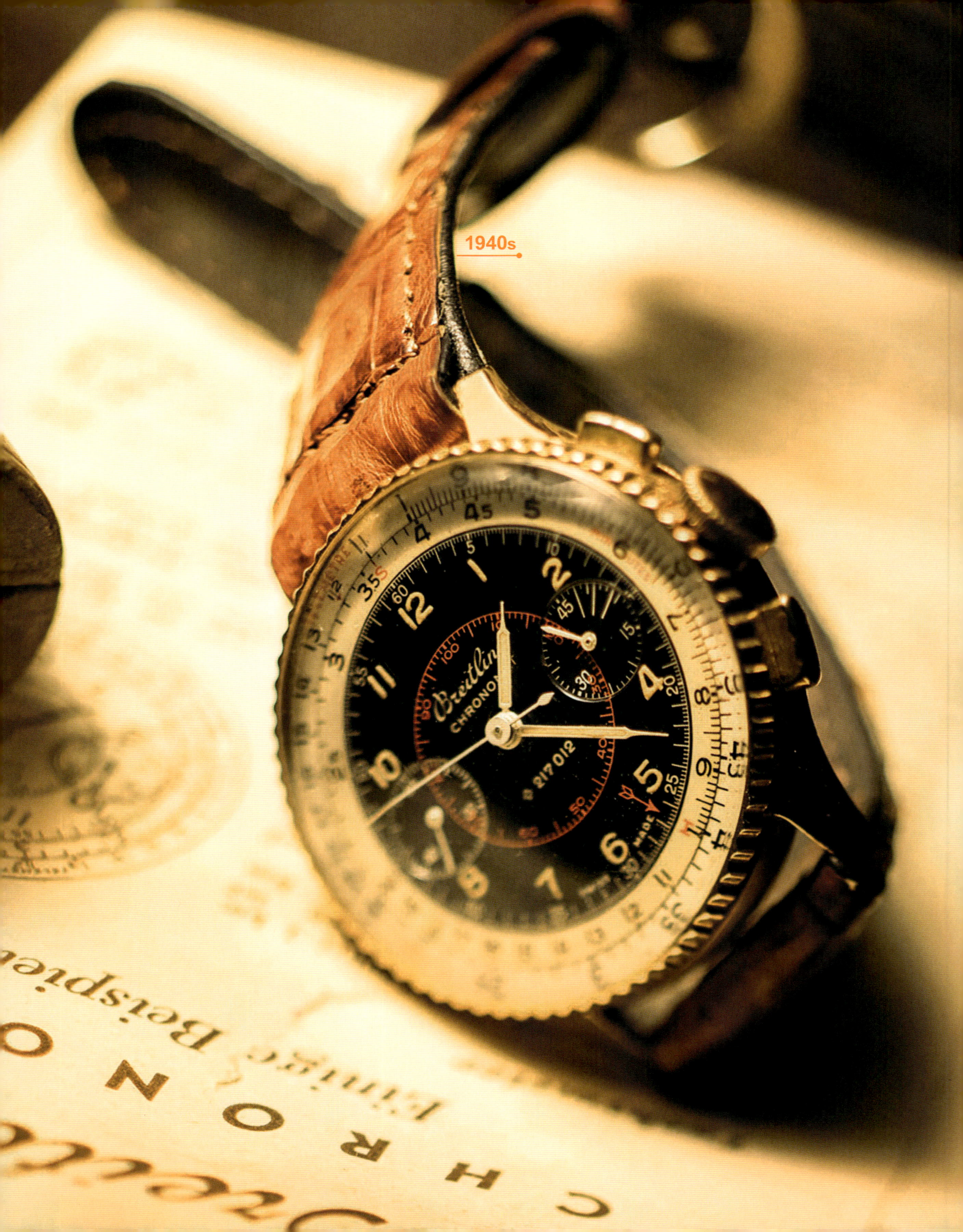

1940s

BREITLING CHRONOMAT

CHRONOMAT

The origins of the Breitling Chronomat can be traced back to 1941. The patent for the premiere model, issued in 1940, was for a tool watch with chronograph and slide rule bezel for a variety of mathematical operations. The name thus combines the abbreviations of chronograph and mathematics. In the pre-electronic age, the timepiece appealed to athletes, scientists, engineers and even the military. In hindsight, the Chronomat, with its different cases and dials, is now regarded as the forebear of another Breitling icon, the Navitimer. In 1969, when Breitling was one of the pioneers of the automatic chronograph, the Chrono-Matic model was introduced. In 1979, Ernest Schneider took over the family business, which had suffered financially as a result of the quartz crisis and whose owner was also suffering from severe health problems. In search of promising propositions, the Chronomat in question was discovered in the brimming company archive. What could be more fitting to mark the brand's centenary in 1984 than to give this name a new design character and a refreshing new lease of life. This generation no longer had anything to do with maths. Instead, the chronograph contained an automatic winding mechanism. The basis for the newcomer was a bracelet stopper developed in 1983 for the "Frecce Tricolori". For safety reasons, the distinctive bezel was now unidirectional. If necessary, the easy-grip rider tabs at 15 and 45 minutes can be swapped. Because of this and other outstanding features, the pilots of the Italian aerobatic squadron opted for this chronograph as their service watch. In 2020, the latest generation of the Chronomat was launched in various iterations. All models reflect the model's horological heritage to varying degrees. Since 2009, marking the brand's 125th anniversary, Breitling now uses its proprietary calibre B01.

CHRONOMAT

Bis ins Jahr 1941 lassen sich die Wurzeln des Breitling „Chronomat" zurückverfolgen. Das 1940 ausgefertigte Patent fürs Premierenmodell galt einer Instrumentenuhr mit Chronograph und kreisförmigem Rechenschieber für ganz unterschiedliche mathematische Operationen. So gesehen versteht sich der Name als Kürzel aus Chronograph und Mathematik. Im vorelektronischen Zeitalter sprach das Instrument Sportler, Wissenschaftler, Techniker und auch Militärs an. Rückblickend gilt der mit unterschiedlichen Gehäusen und Zifferblättern gefertigte „Chronomat" jedoch als Urahn einer weiteren Breitling-Ikone namens „Navitimer". Ab 1969, als Breitling zu den Pionieren des Automatik-Chronographen gehörte, gab es das Modell „Chrono-Matic". 1979 übernahm Ernest Schneider das infolge der Quarzkrise finanziell angeschlagene Familienunternehmen, dessen Inhaber zudem unter massiven gesundheitlichen Beeinträchtigungen litt. Auf der Suche nach Umsatzträchtigem fand sich im prall gefüllten Firmenarchiv besagter „Chronomat". Ergo lag es förmlich auf der Hand, diesen Namen zum 100. Markenjubiläum im Jahr 1984 mit neuem, gestalterisch jedoch völlig andersartigem Leben zu erfüllen. Mit Mathematik hatte diese Generation nichts mehr zu tun. Stattdessen gesellte sich zum Chronographen ein automatischer Aufzug. Grundlage für den Newcomer bildete ein 1983 für die „Frecce Tricolori" entwickelter Armbandstopper. Aus Sicherheitsgründen lässt sich der markante Glasrand nur in eine Richtung drehen. Bei Bedarf sind die griffigen Lünettenreiter mit den Zahlen 15 und 45 vertauschbar. Wegen dieser und anderer herausragender Eigenschaften erkoren die Piloten der italienischen Kunstflugstaffel diesen Chronographen zu ihrer Dienstuhr. 2020 ging die neueste Generation des „Chronomat" in verschiedenen Versionen an den Start. Alle Modelle knüpfen mehr oder minder stark an die lange zurückreichende Geschichte an. Im Gehäuseinneren verbaut Breitling jedoch das 2009 zum 125. Geburtstag eingeführte Manufakturkaliber B01.

1969

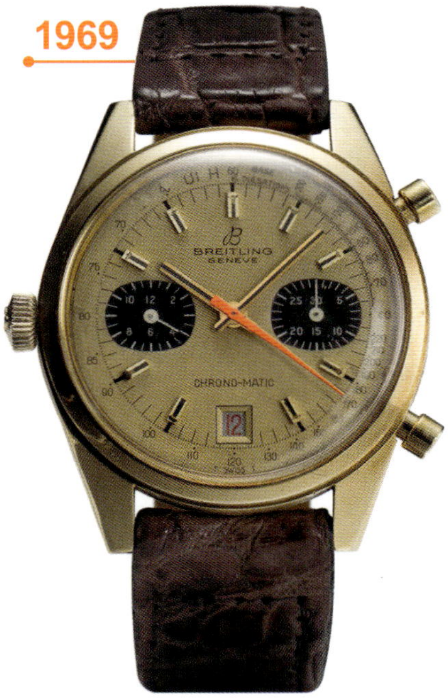

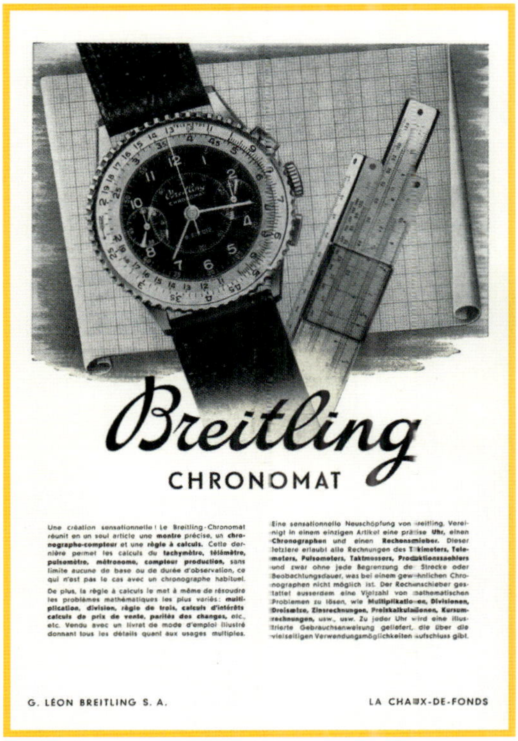

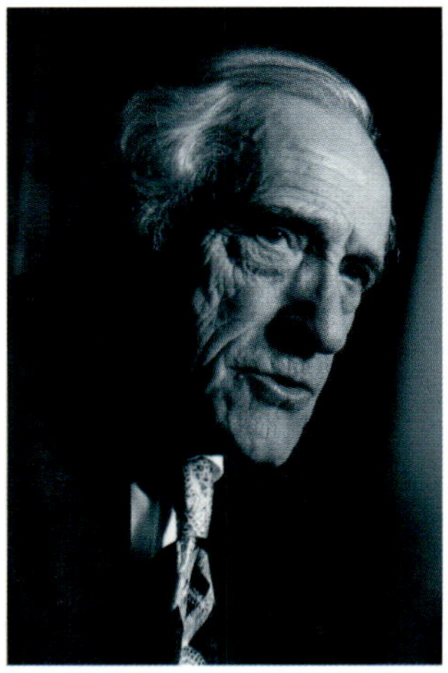

From left / Von links:
Chrono-Matic, Ref. 2116, 38mm
Advertisement, 1942
Ernest Schneider

CO-PILOT CHRONO-MATIC REF. 7651, 1969

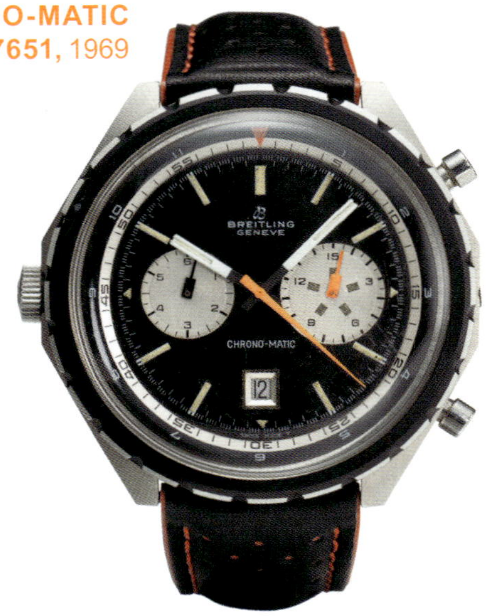

SUPEROCEAN CHRONO-MATIC REF. 2105, 1970

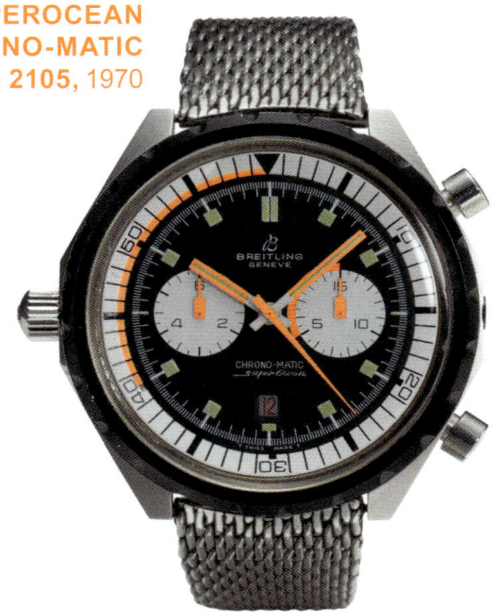

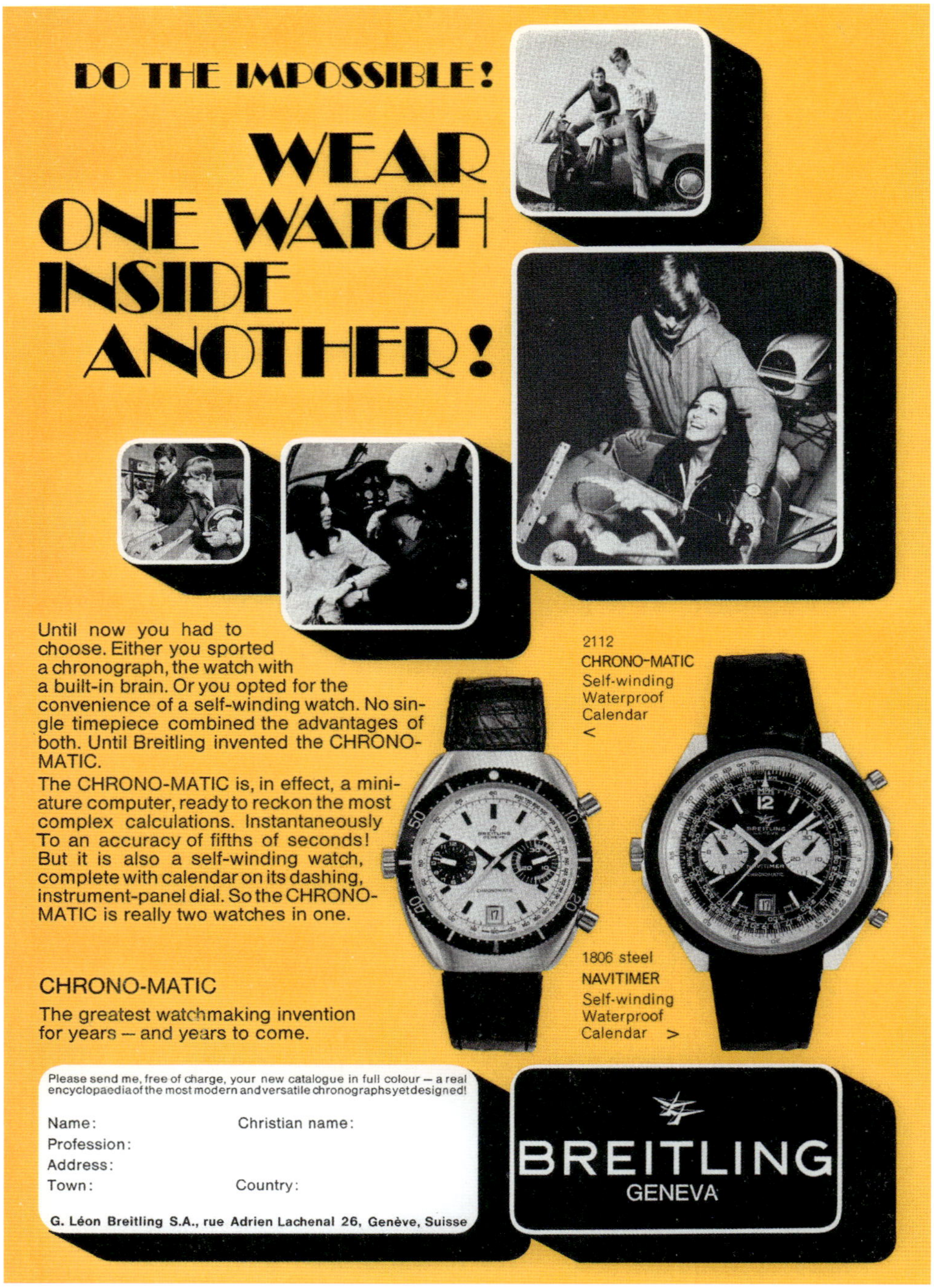

Advertisement, 1970s

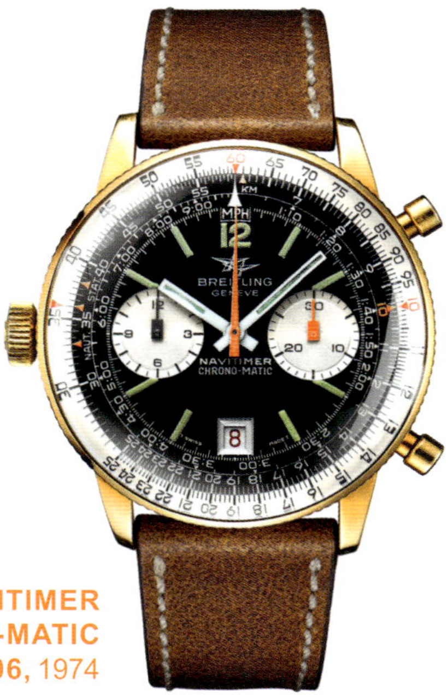

NAVITIMER CHRONO-MATIC REF. 8806, 1974

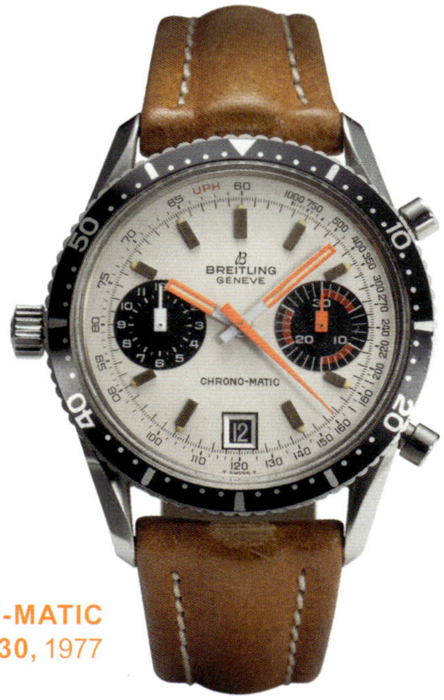

CHRONO-MATIC REF. 2130, 1977

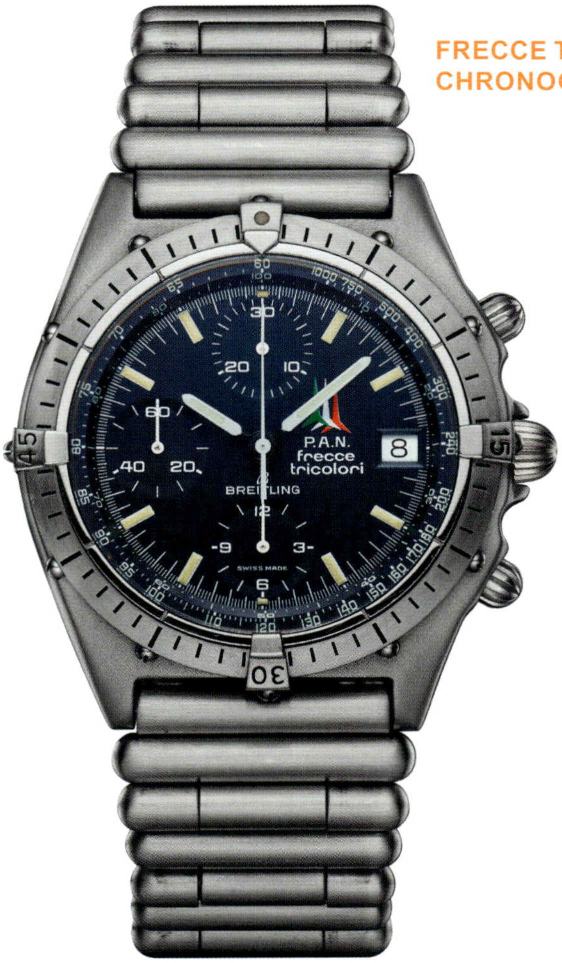

FRECCE TRICOLORI CHRONOGRAPH, 1983

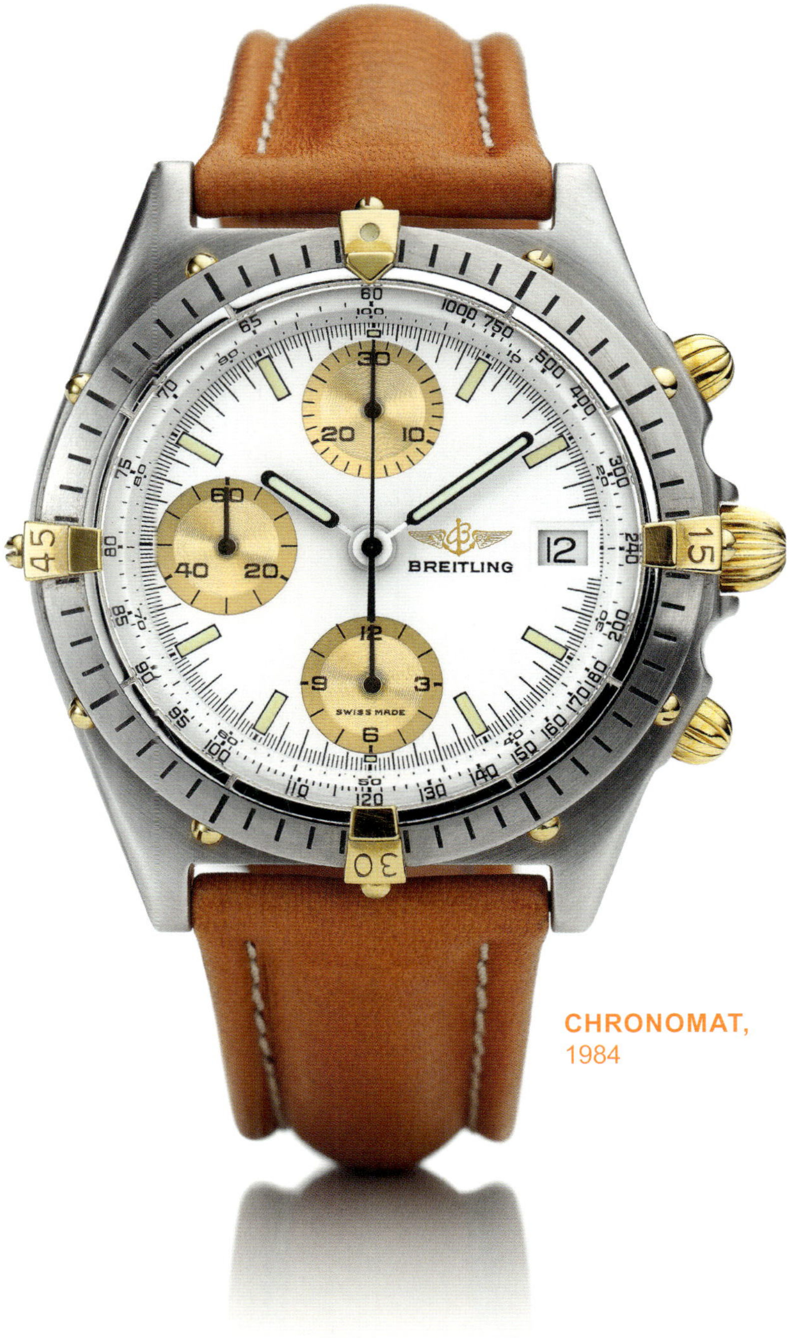

CHRONOMAT, 1984

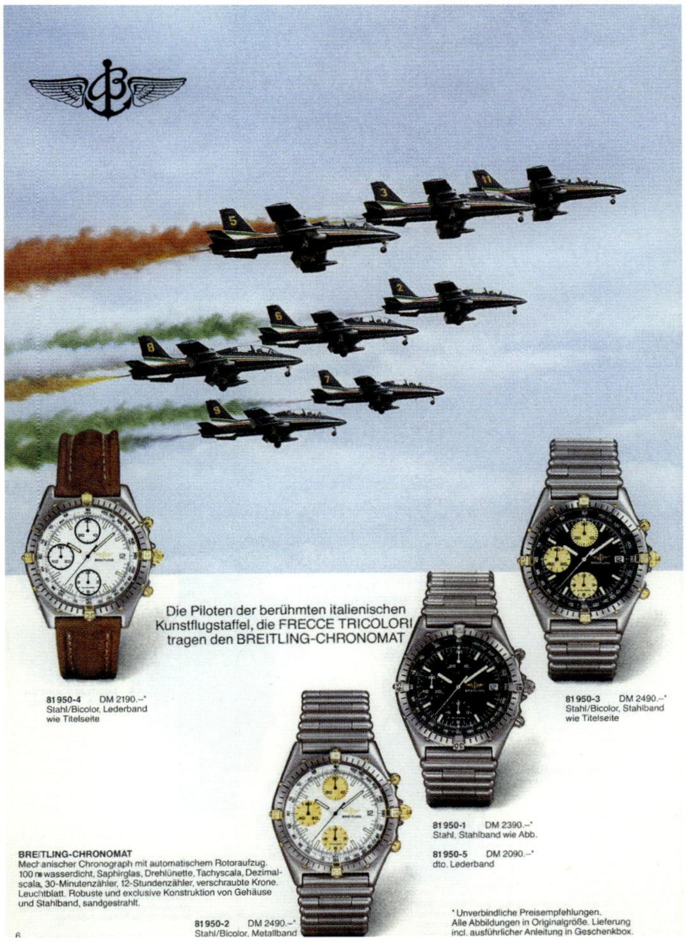

Die Piloten der berühmten italienischen Kunstflugstaffel, die FRECCE TRICOLORI tragen den BREITLING-CHRONOMAT

81950-4 DM 2190.–*
Stahl/Bicolor, Lederband
wie Titelseite

81950-3 DM 2490.–*
Stahl/Bicolor, Stahlband
wie Titelseite

81950-1 DM 2390.–*
Stahl, Stahlband wie Abb.

81950-5 DM 2090.–*
dto. Lederband

BREITLING-CHRONOMAT
Mechanischer Chronograph mit automatischem Rotoraufzug. 100 m wasserdicht, Saphirglas, Drehlünette, Tachyscala, Dezimalscala, 30-Minutenzähler, 12-Stundenzähler, verschraubte Krone. Leuchtblatt. Robuste und exclusive Konstruktion von Gehäuse und Stahlband, sandgestrahlt.

81950-2 DM 2490.–*
Stahl/Bicolor, Metallband

* Unverbindliche Preisempfehlungen. Alle Abbildungen in Originalgröße. Lieferung incl. ausführlicher Anleitung in Geschenkbox.

6

Breitling catalogue, 1987

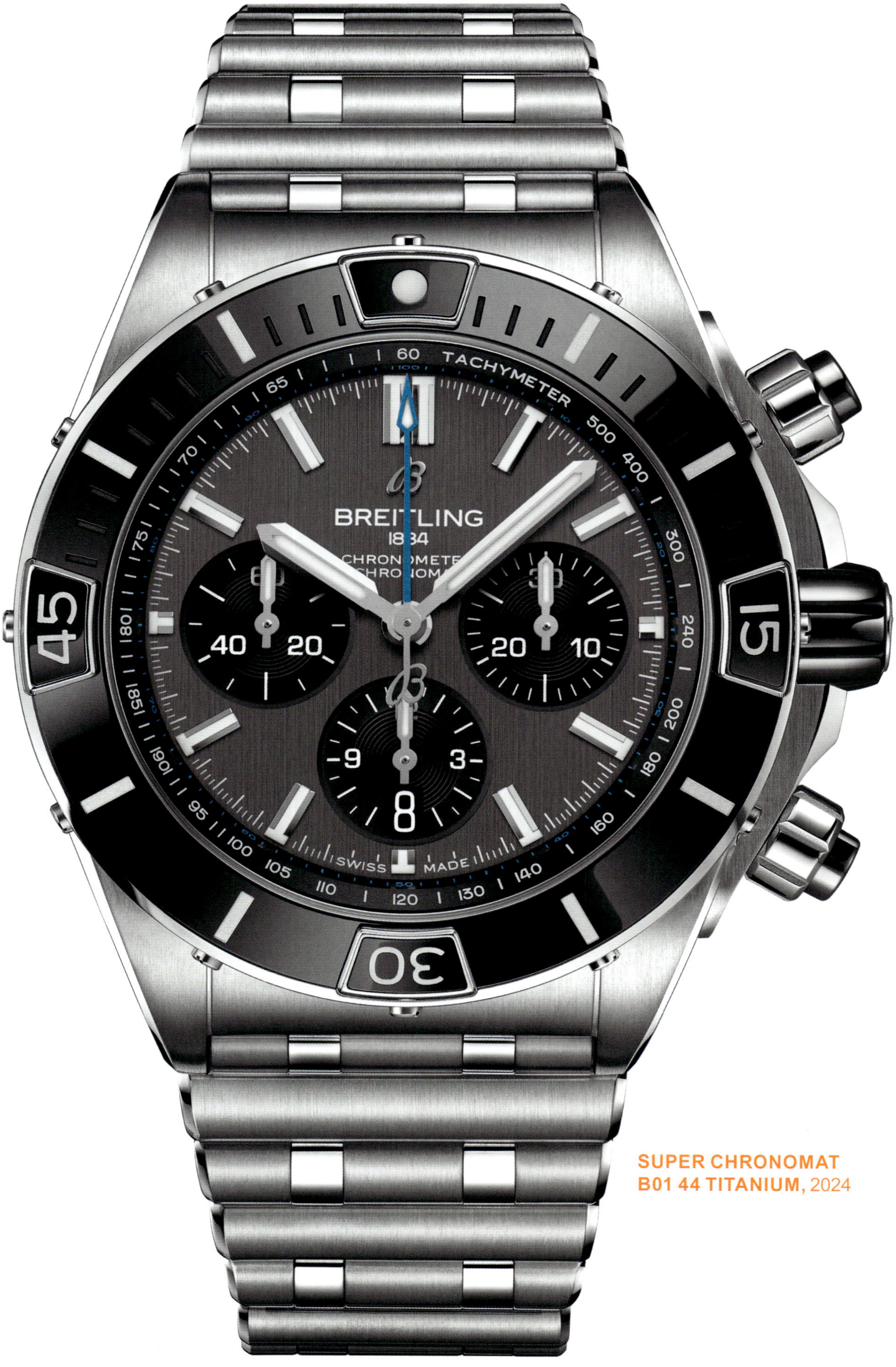

SUPER CHRONOMAT B01 44 TITANIUM, 2024

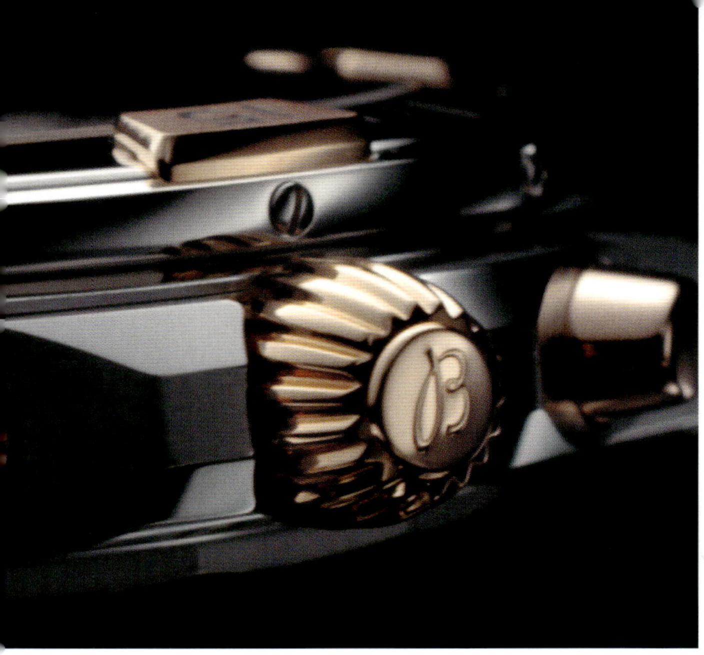

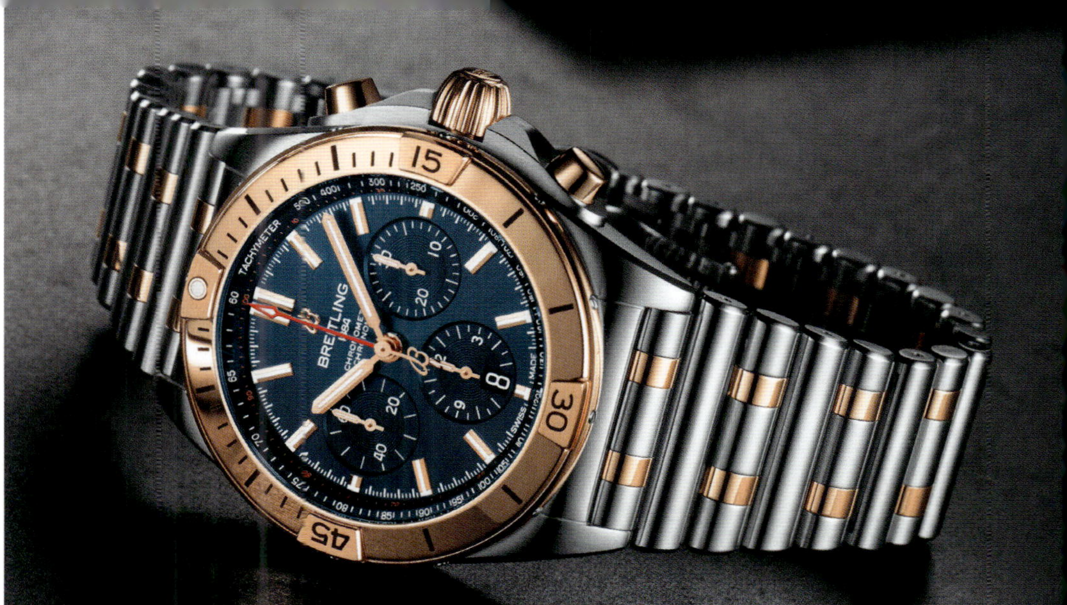

Two-tone Chronomat B01 42, 2020

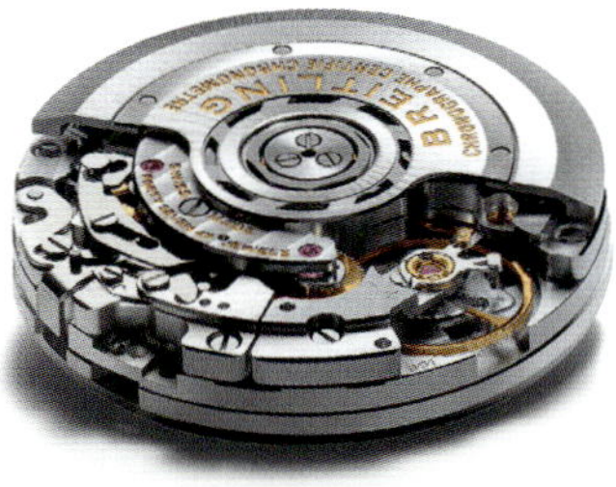

Manufacture Calibre B01, 2009

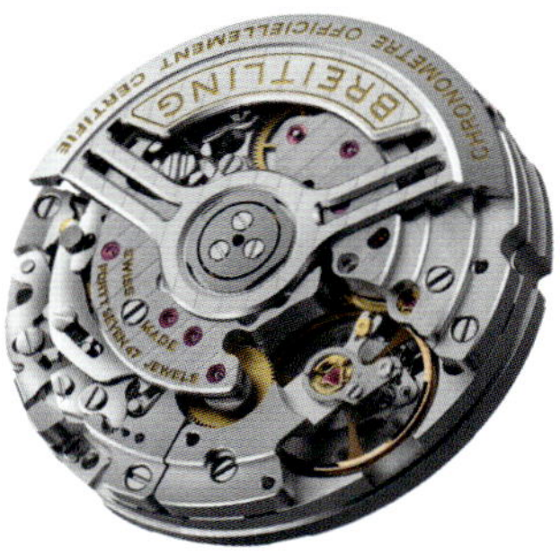

Upgraded Manufacture Calibre B01, 2022

2024

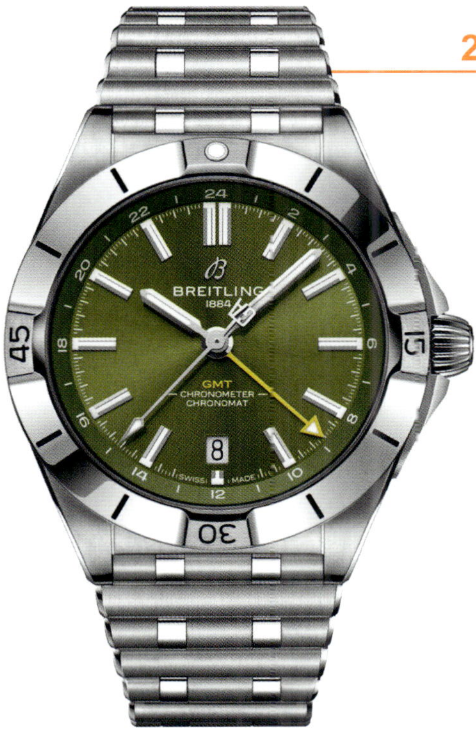

CHRONOMAT AUTOMATIC GMT 40
GIANNIS ANTETOKOUNMPO

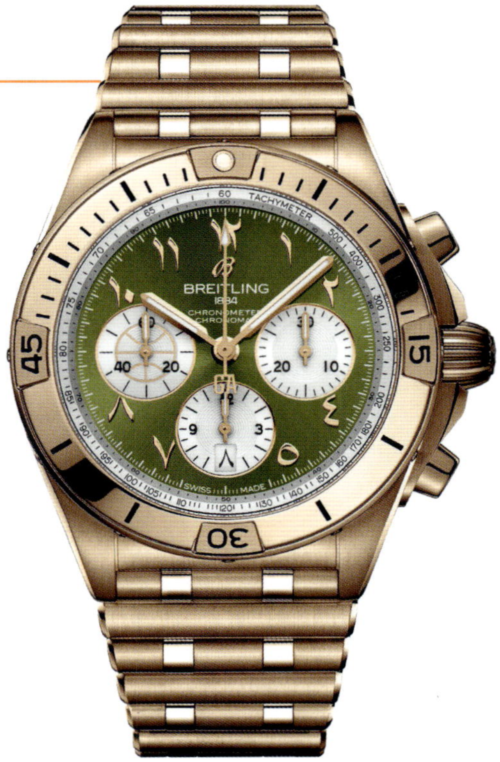

CHRONOMAT B01 42
GIANNIS ANTETOKOUNMPO

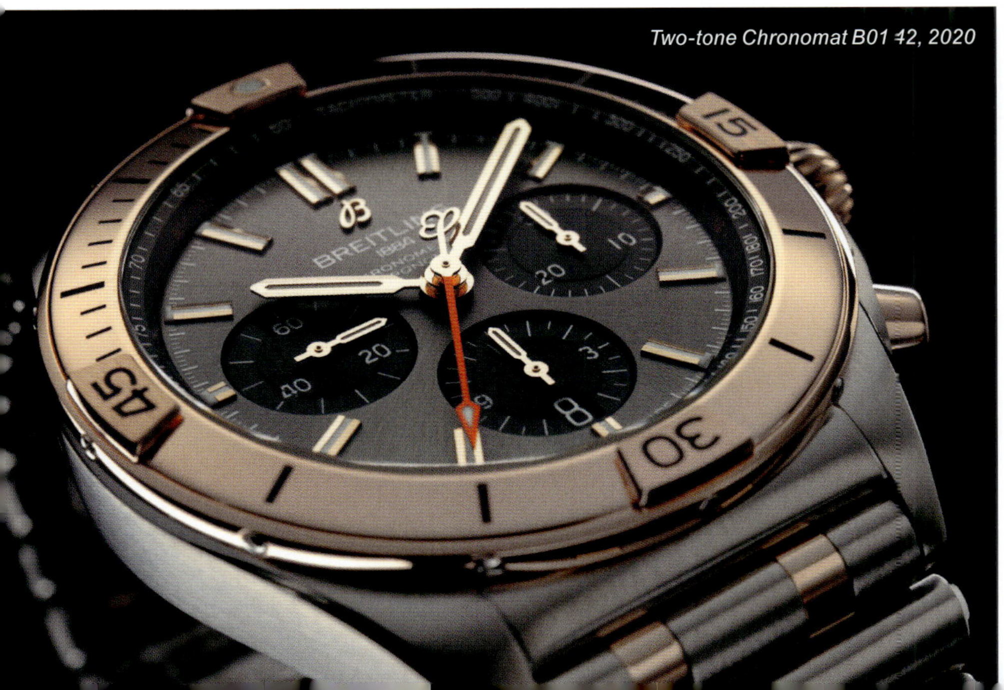

Two-tone Chronomat B01 42, 2020

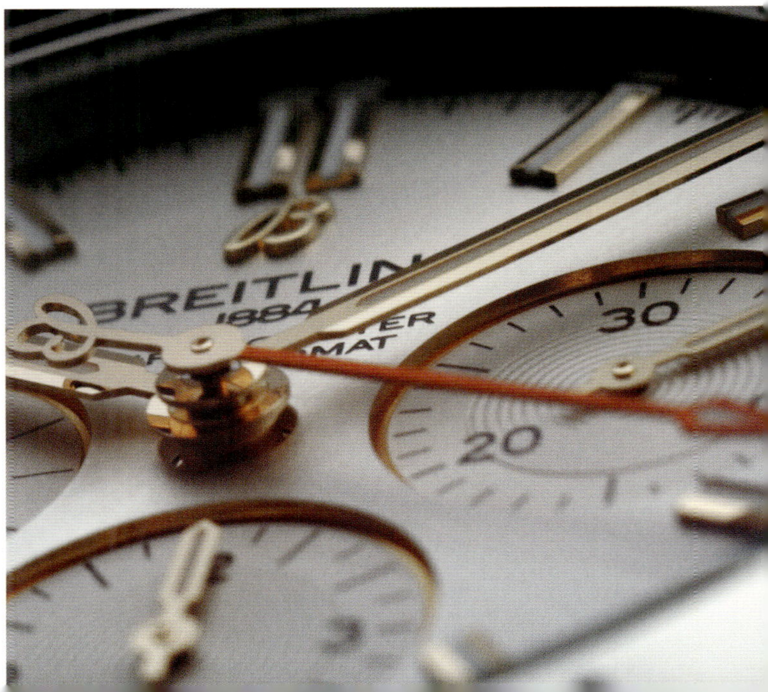

CARL F. BUCHERER 2018

MANERO TOURBILLON DOUBLE PERIPHERAL

ON THE PERIPHERY

In 2005, the watch brand Carl F. Bucherer decided to develop its own proprietary calibre. The concept of the CFB A1000 included an automatic winding system with an oscillating weight mounted on the periphery. The patented rotor is carried on DLC (diamond-like carbon)-coated rollers. It relied on spring-retained rockers for exact positioning. A dynamic shock absorption system was also added. The market launch took place in 2008.
This was the first time that Carl F. Bucherer presented an automatic movement in which a peripheral rotor winding system could be successfully realised in large quantities. The chronometer-certified A2000 automatic movement, which was first used in the Manero Peripheral in 2016, added a new facet to Carl F. Bucherer's range of manufacture movements.
Complementing the premiere model is the 2018 Manero Tourbillon Double Peripheral powered by the CFB T3000 calibre. Partnering the automatic winding system is a peripheral tourbillon. Its architecture was unlike anything ever seen before. Due to the peripheral support, the cage seems to float inside the case. It can be viewed equally well from either side of the case. This is possible thanks to the three ceramic ball bearings on the periphery. They ensure a stable connection for the cage and low-friction rotations to compensate for gravity-related errors. Made of silicon, the pallet and escape wheel operate almost friction-free due to their smooth surfaces whilst making a significant contribution to the power reserve of at least 65 hours. Evidencing its perfect function and precise regulation is a chronometer certificate for the Manero Tourbillon Double Peripheral.

AUSSENLAGER

2005 fasste die Uhrenmarke Carl F. Bucherer den Entschluss zur Entwicklung eines eigenen Manufakturkalibers. Das Konzept des CFB A1000 beinhaltete einen Selbstaufzug durch peripher drehende Schwungmasse. Die patentierte Rotorlagerung erfolgte mit Hilfe von DLC (diamond like carbon)-beschichteten Rollen. Federarme bewirkten ihre exakte Positionierung. Hinzu gesellte sich ferner eine dynamische Stoßsicherung. 2008 erfolgte die Markteinführung.
Damit präsentierte Carl F. Bucherer erstmals ein Automatikwerk, bei dem ein peripherer Rotoraufzug erfolgreich in größeren Stückzahlen realisiert werden konnte. Das als Chronometer zertifizierte Automatikwerk A2000, welches 2016 erstmals in der „Manero Peripheral“ zum Einsatz kam, erweiterte die Palette der Manufakturwerke von Carl F. Bucherer um eine neue Facette.
Als Komparativ zum Premierenmodell versteht sich die „Manero Tourbillon Double Peripheral“ von 2018. In ihr tickt das Kaliber CFB T3000. Zum Selbstaufzug gesellt sich ein wiederum peripher, also außen gelagertes Tourbillon. Ein solches hat es in der Geschichte des Drehgangs noch nicht gegeben. Durch die seitliche Lagerung scheint der Käfig im Uhrwerk zu schweben. Von beiden Gehäuseseiten ist es gleichermaßen gut wahrnehmbar. Möglich machen es drei wiederum seitlich angeordnete Keramikkugellager. Sie halten das Drehgestell sicher und bewirken reibungsarme Rotationen zum Ausgleich der schwerkraftbedingten Fehler. Anker und Ankerrad bestehen aus Silizium, arbeiten wegen der glatten Oberfläche nahezu reibungsfrei und leisten auf diese Weise einen wesentlichen Beitrag zu den mindestens 65 Stunden Gangautonomie. Beweis für einwandfreie Funktion und präzisen Gang ist ein offizielles Chronometerzeugnis für das „Manero Tourbillon Double Peripheral“.

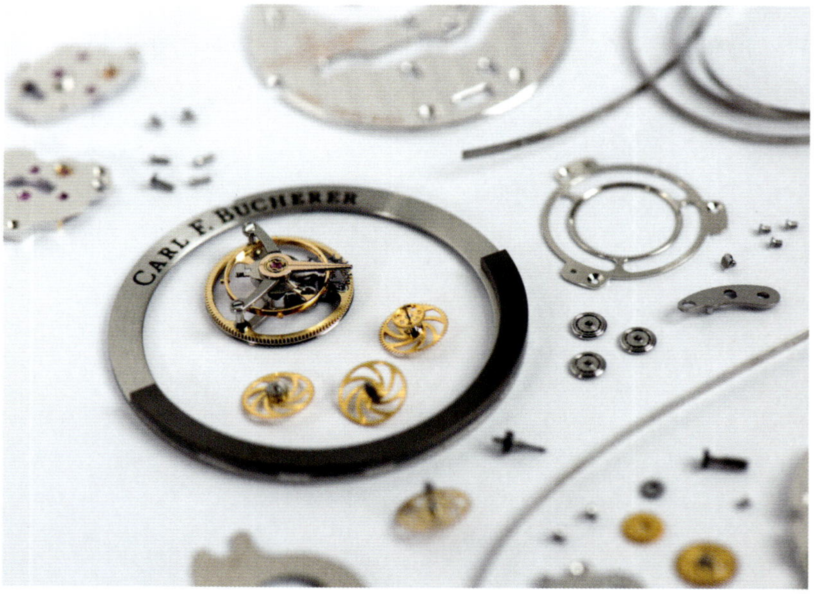

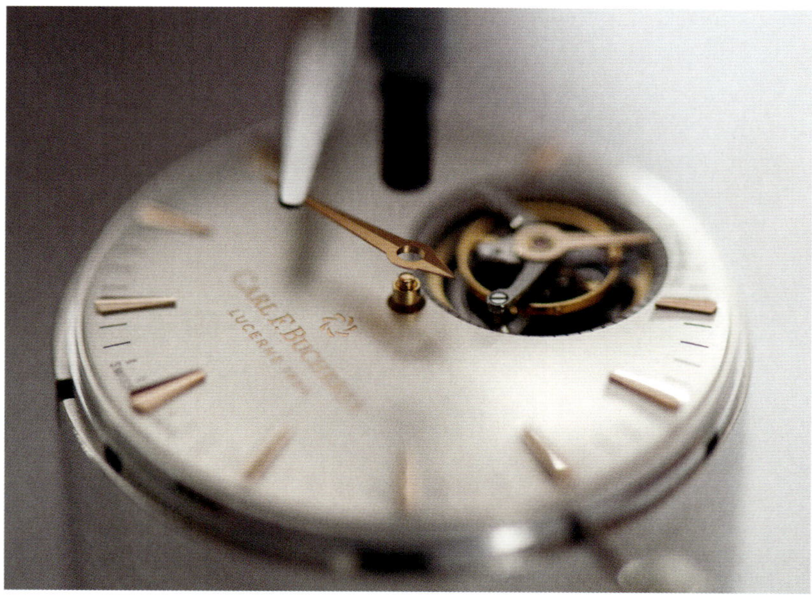

Components and assembly of Manero Tourbillon Double Peripheral | Komponenten und Montage der Manero Tourbillon Double Peripheral

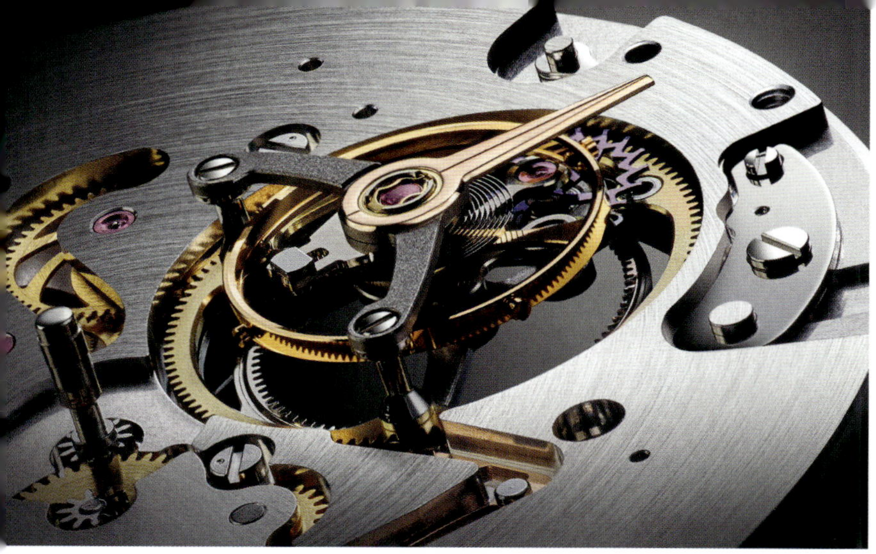

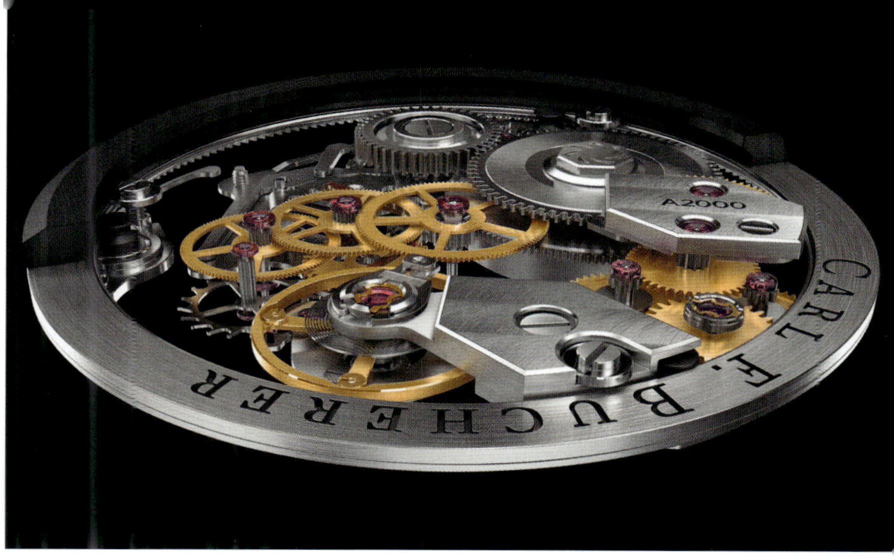

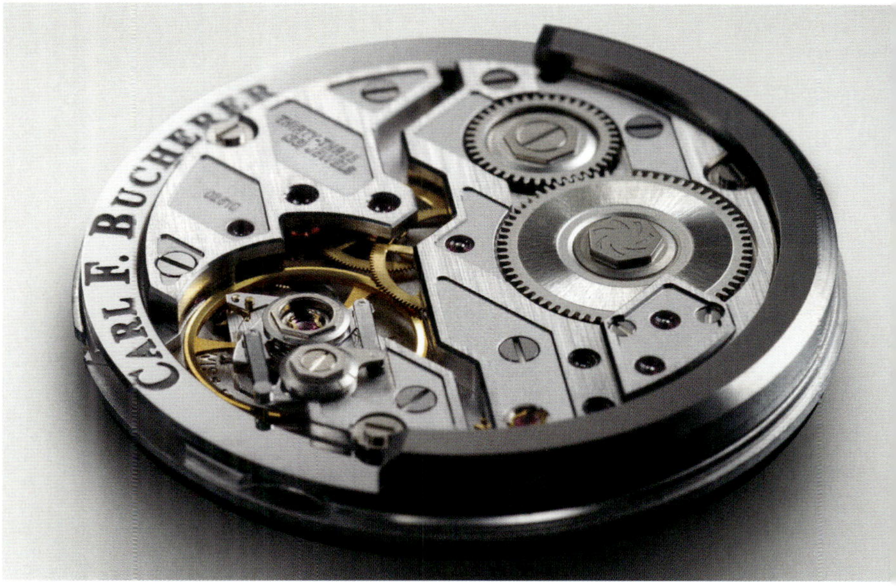

From top left, clockwise | Von oben links im Uhrzeigersinn: Calibres T3000, A2000, A1000

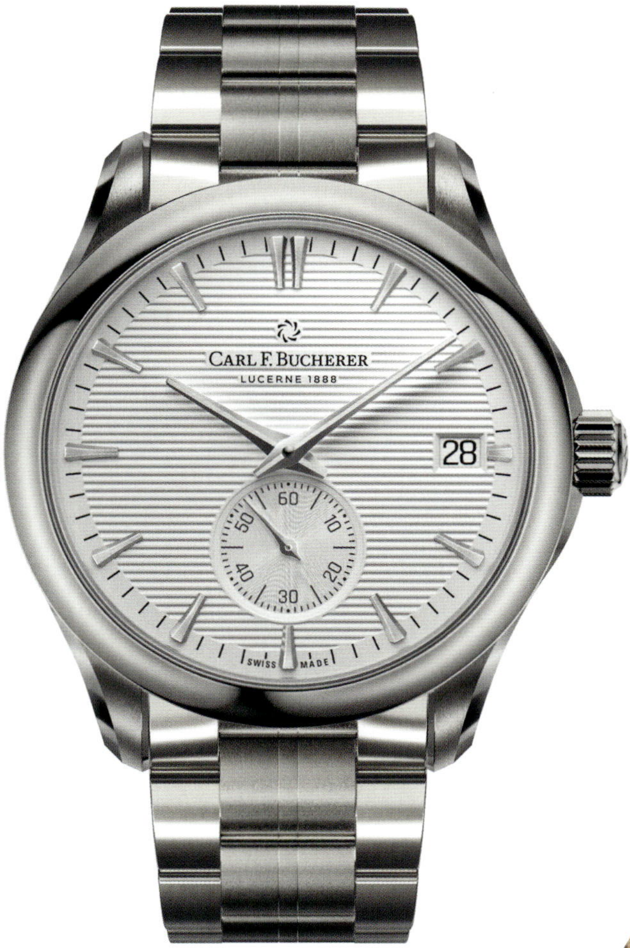

MANERO PERIPHERAL,
2024

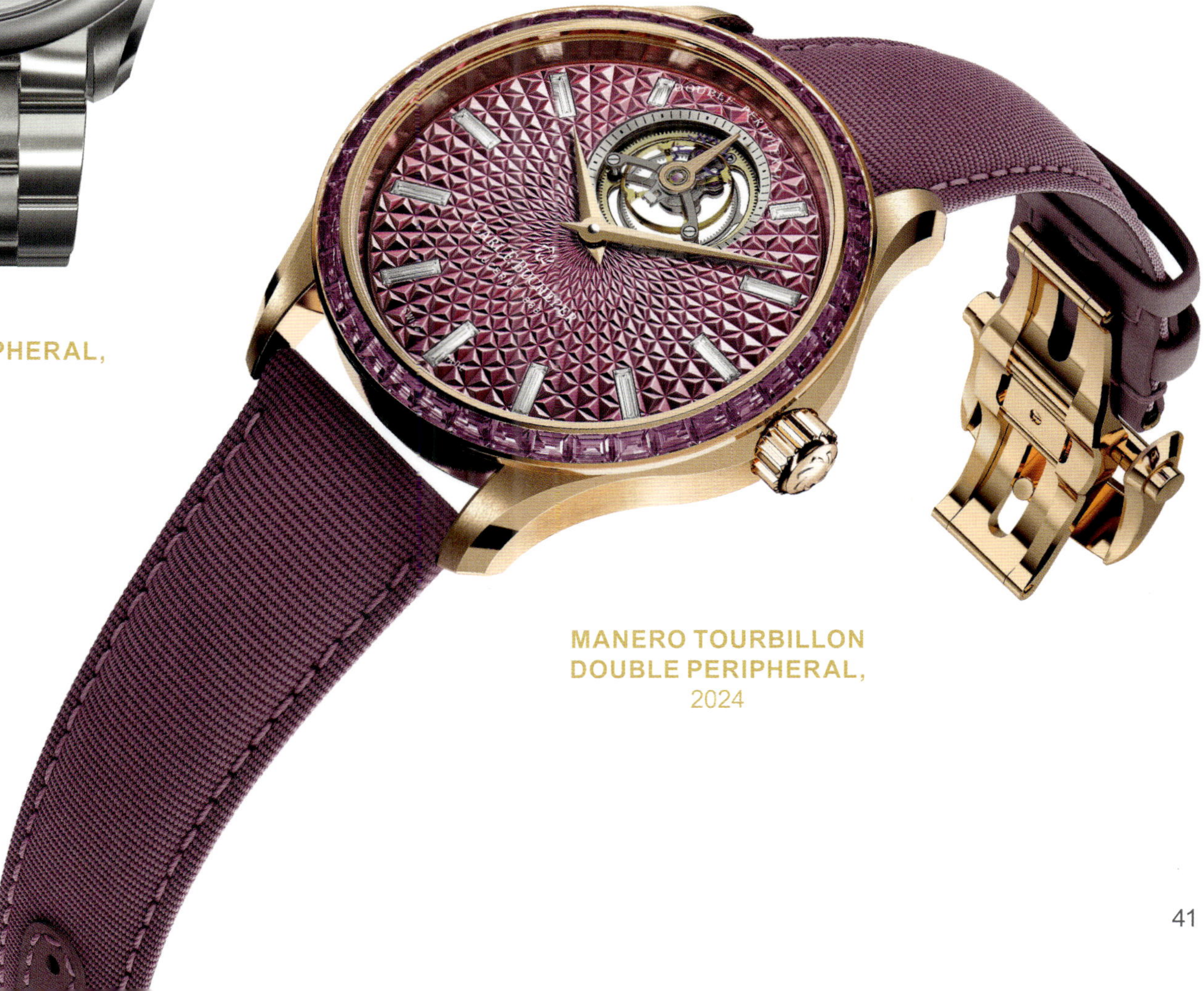

MANERO TOURBILLON DOUBLE PERIPHERAL,
2024

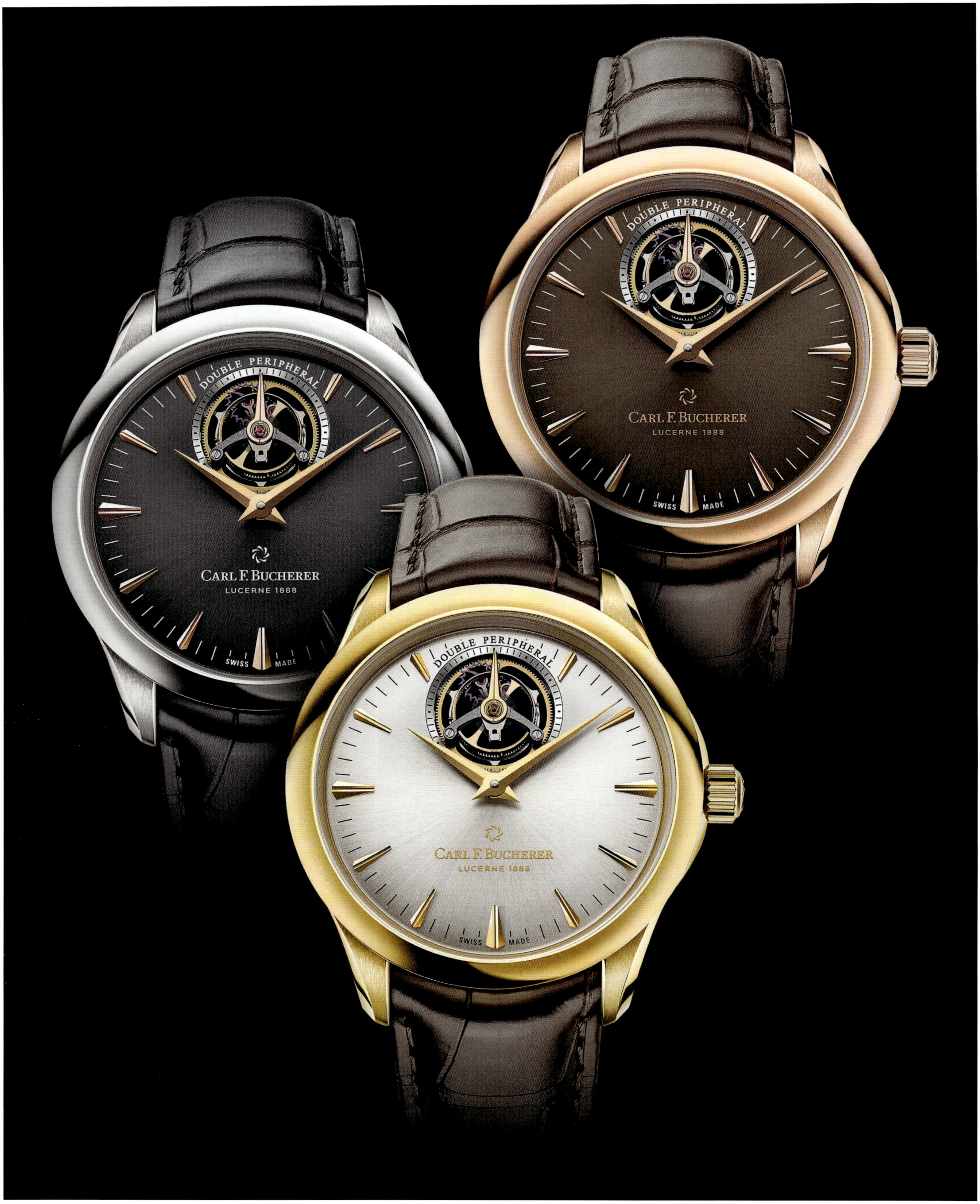
DOUBLE PERIPHERAL
CARL F. BUCHERER
LUCERNE 1888
SWISS MADE
DOUBLE PERIPHERAL
CARL F. BUCHERER
LUCERNE 1888
SWISS MADE
DOUBLE PERIPHERAL
CARL F. BUCHERER
LUCERNE 1888
SWISS MADE

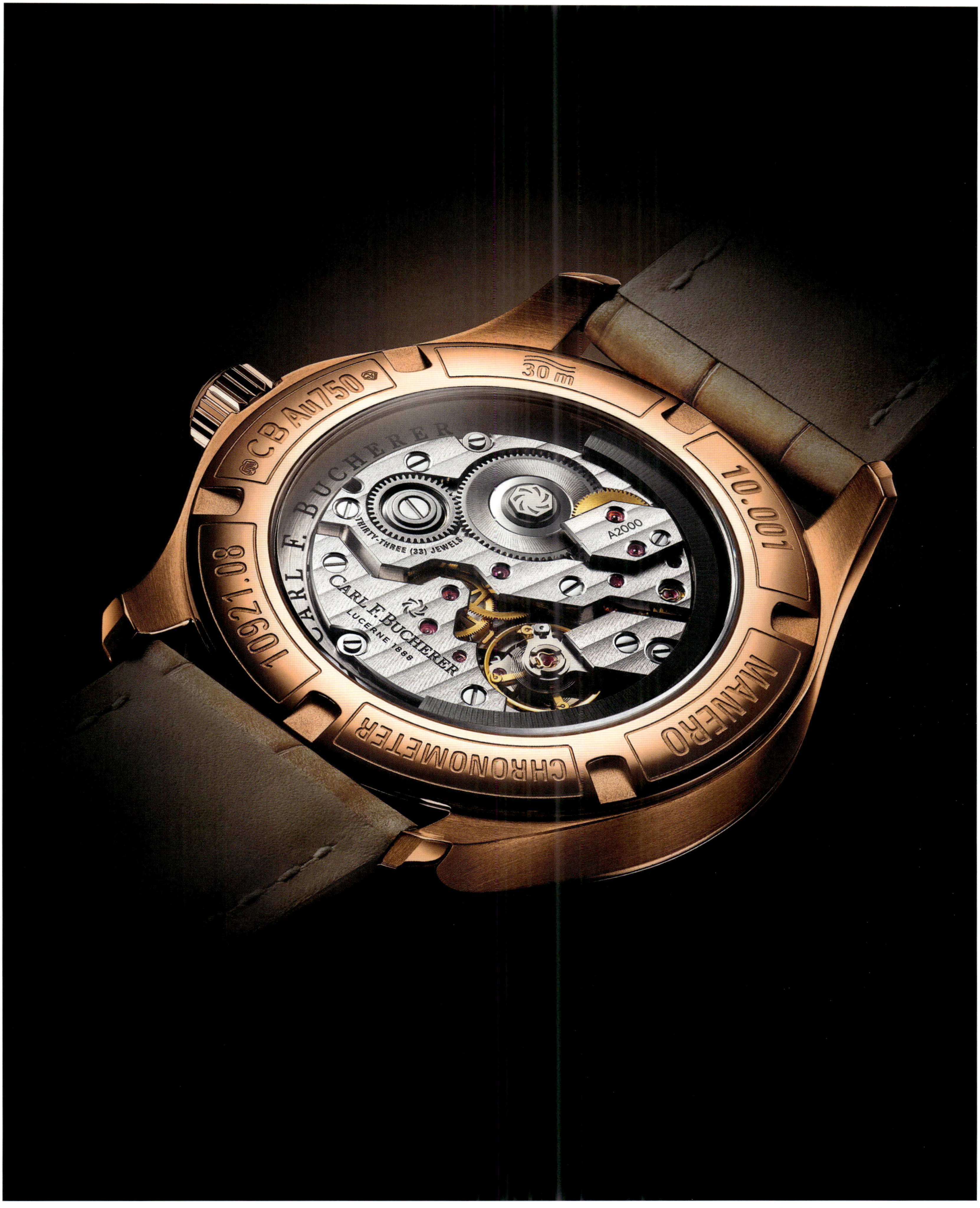

Top | Oben: Manero Peripheral, caseback

BULGARI

1977

BULGARI BULGARI

UNMISTAKABLE ORIGINS

Bulgari almost inevitably evokes associations with Bella Italia. Founded in 1884 in the capital Rome, the luxury label is now part of the French LVMH group. Today, jewellery, accessories, perfume and hotels enjoy equal status with wristwatches. The latter came onto the market in the early 1920s in a feminine design. The decisive step into the horological future was taken in 1975 and thus in the throes of the quartz revolution. The astute product designers of the Bulgari Roma had tackled the electronic challenge, combining the increasingly popular digital LCD display with a classic case design. The upshot was one of the first luxury wristwatches with quartz movement and non-analogue time display. The design of the discreet round case recalled a disc cut out of a Corinthian column. Emulating Roman emperors who demonstrated their omnipotence by engraving their names on coins, Bulgari followed suit by adding his name and the timepiece's place of origin to the wide glass bezel. Two years later, in 1977, the family-owned company brought off a grand coup with the help of Gérald Genta: the purist, consummately styled and consistently cylindrical wristwatch followed the basic design of the Roma, but otherwise made little attempt to conceal its exclusive origins. The Italians immortalised their name on the wide, flat bezel not once, but twice, and as prominently as possible.

Since its debut, the oft-imitated, but still unrivalled design icon Bulgari Bulgari has enjoyed the best of health. It is, after all, an eloquent witness to the fact that design language can hold its own in fast-moving times.

UNVERKENNBARE PROVENIENZ

Beinahe zwangsläufig erinnert Bulgari an Bella Italia. In der Hauptstadt Rom schlug 1884 die Geburtsstunde des mittlerweile zum französischen LVMH-Konzern gehörenden Luxuslabels. Auf Schmuck, Accessoires, Parfüms und Hotels versteht es sich heutzutage ebenso wie auf Armbanduhren. Letztere gelangten Anfang der 1920er-Jahre in femininer Ausführung auf den Markt. Der entscheidende Schritt in die uhrmacherische Zukunft erfolgte 1975 und damit im Zuge der Quarz-Revolution. Bei der Bulgari „Roma" hatten clevere Produktgestalter die elektronische Herausforderung angenommen und das zunehmend beliebte digitale LCD-Display mit einem klassischen Gehäusedesign kombiniert. Auf diese Weise entstand eine der ersten Luxusarmbanduhren mit Quarzwerk und nicht-analoger Zeitanzeige. Das Design der dezenten, runden Schale weckte Erinnerungen an eine Scheibe, herausgeschnitten aus einer korinthischen Säule. Ähnlich römischen Kaisern, die umfassende Macht durch ihren Namenszug auf Münzen demonstrierten, gravierte Bulgari seinen Namenszug und den Herkunftsort des Produkts unübersehbar in den breiten Glasrand. Zwei Jahre später, also 1977, landete das Familienunternehmen mit Hilfe von Gérald Genta den ganz großen Wurf: Die schnörkellos durchgestylte und konsequent zylindrisch geformte Armbanduhr folgte dem Basisdesign der „Roma", machte darüber hinaus aber noch weniger Hehl aus ihrer vornehmen Herkunft. Gleich zweimal und zudem auch größtmöglich verewigten die Italiener ihren Namenszug in der breiten, flachen Lünette.

Seit dem Debüt erfreut sich die oft imitierte, gestalterisch aber nie erreichte „Bulgari Bulgari" bester Gesundheit. Schließlich ist sie beredte Zeugin dafür, dass sich Formensprache selbst in schnelllebigen Zeiten ihre Gültigkeit bewahrt.

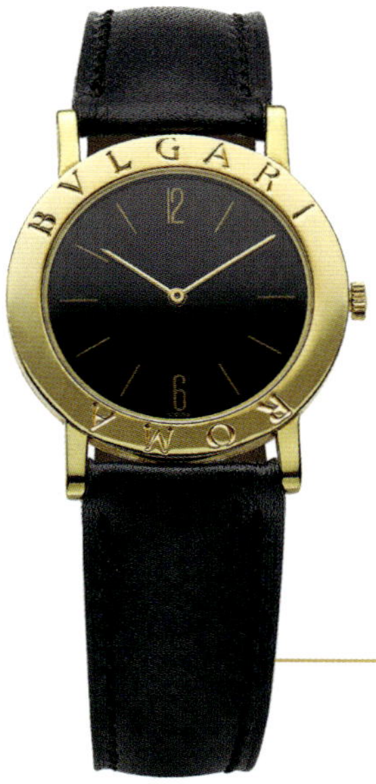

1975

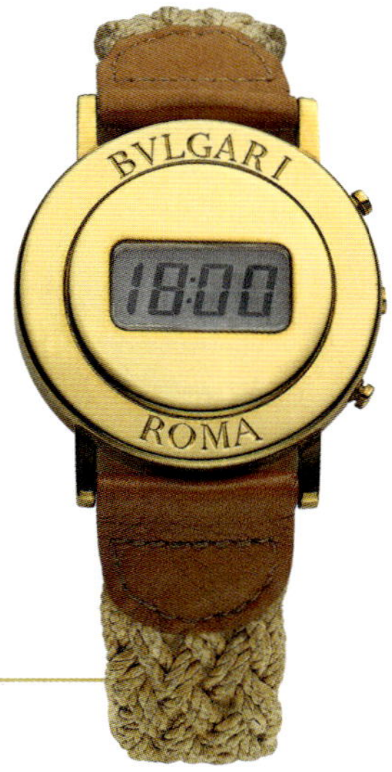

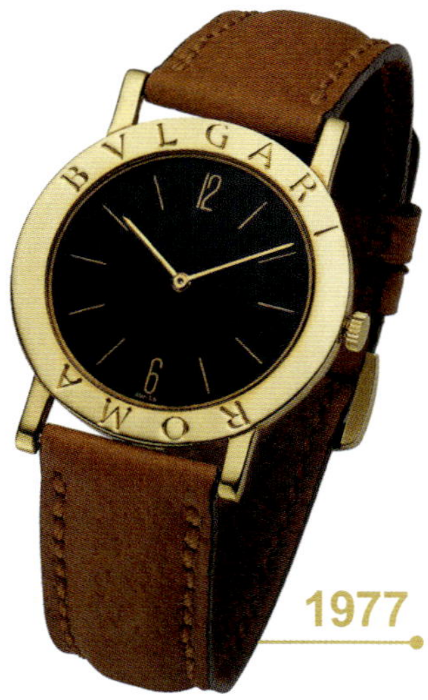

1977

1997

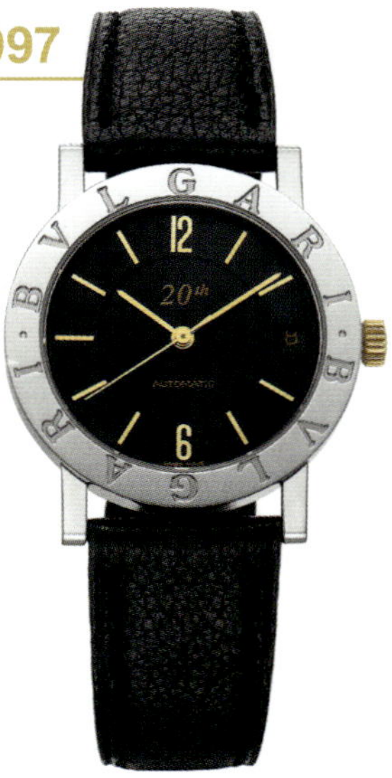

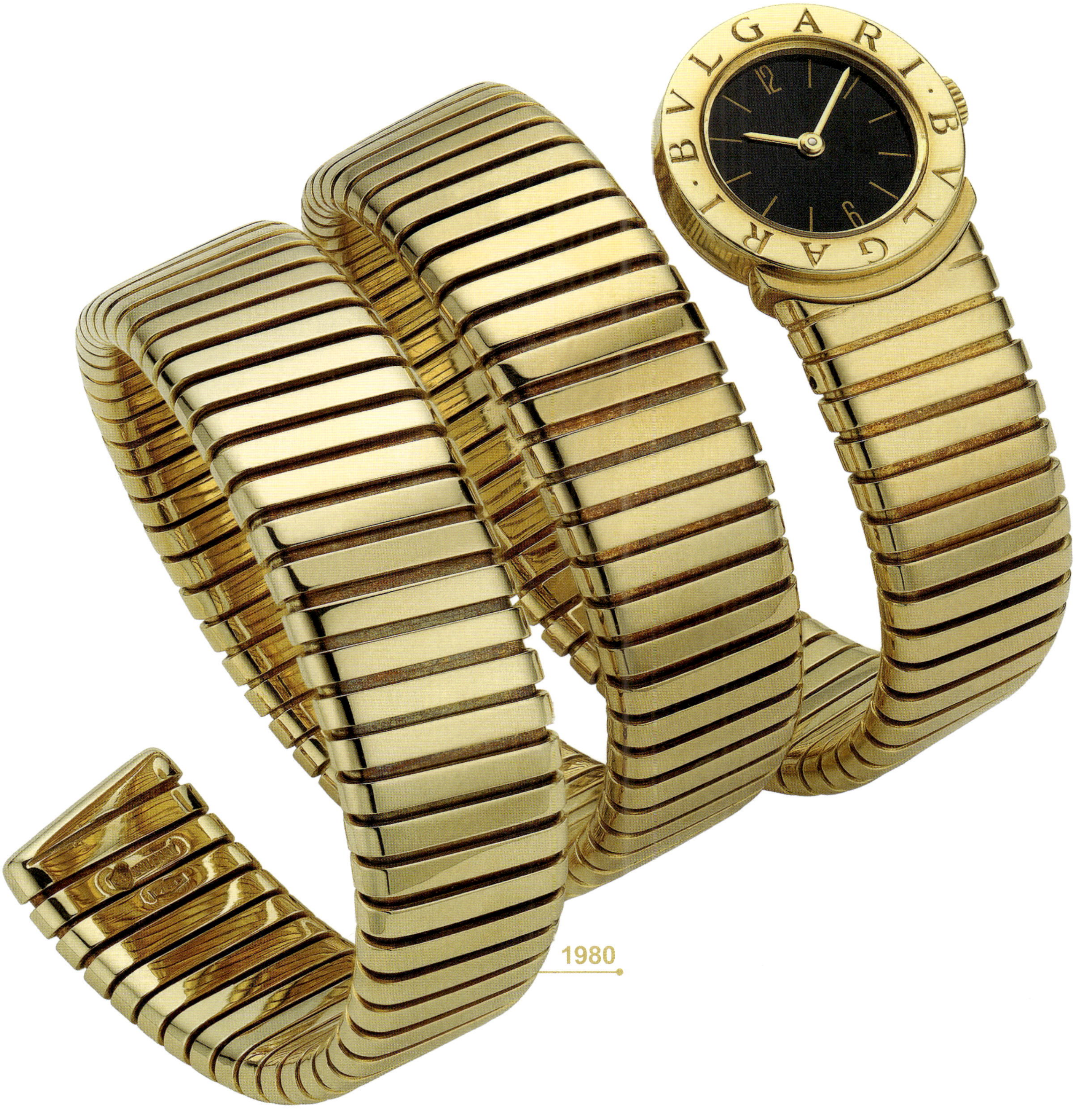

1980

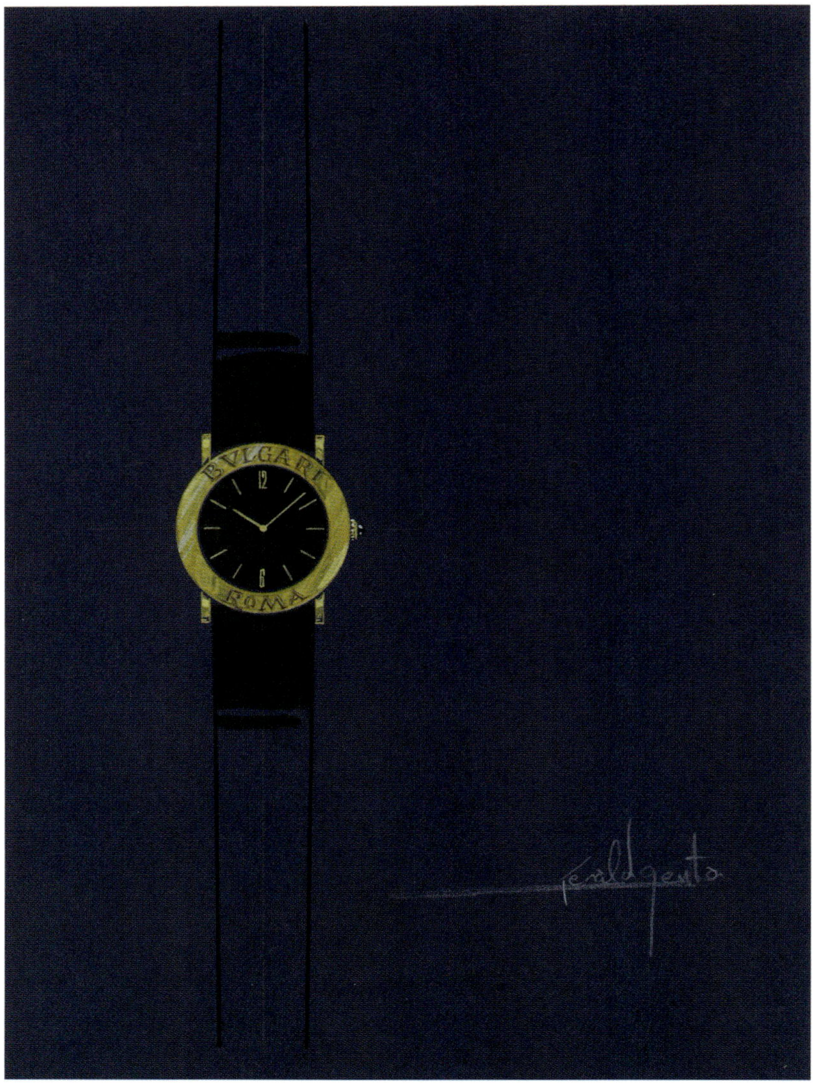

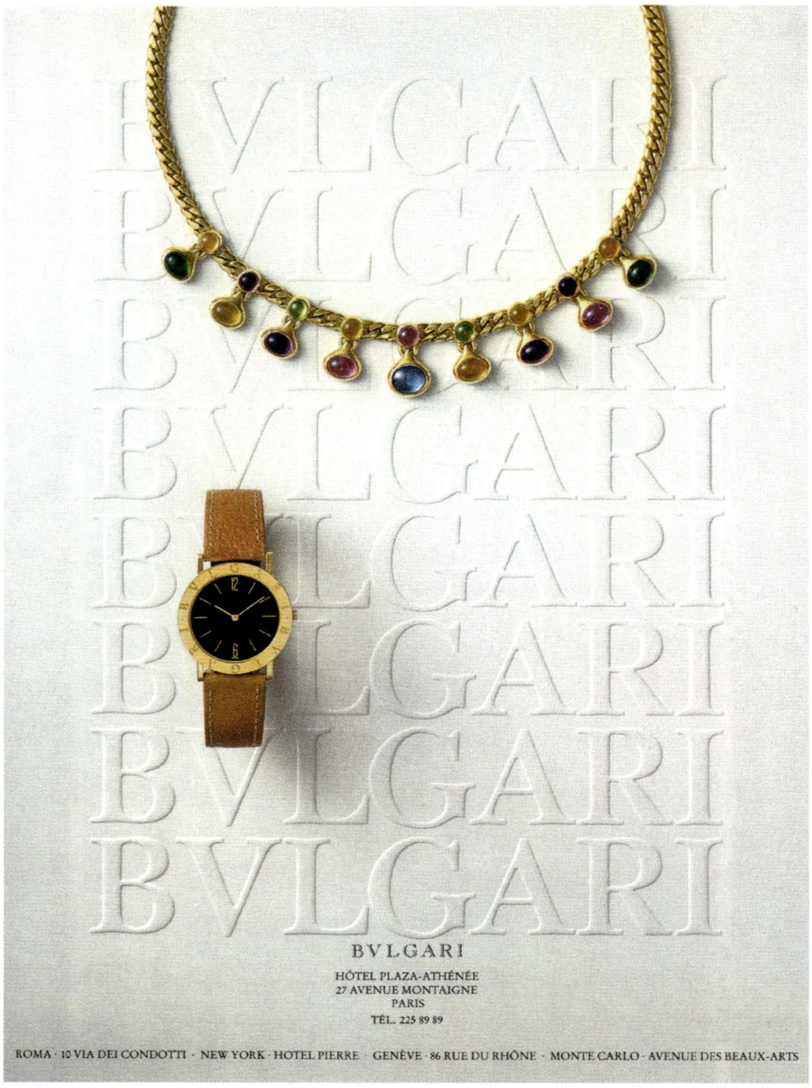

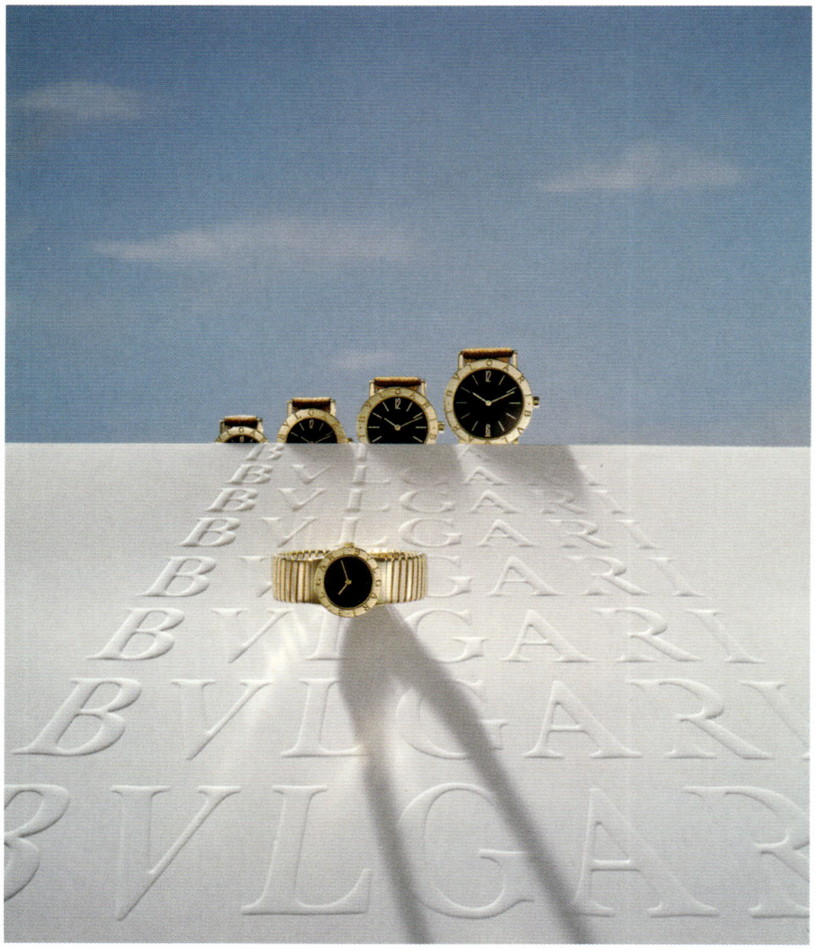

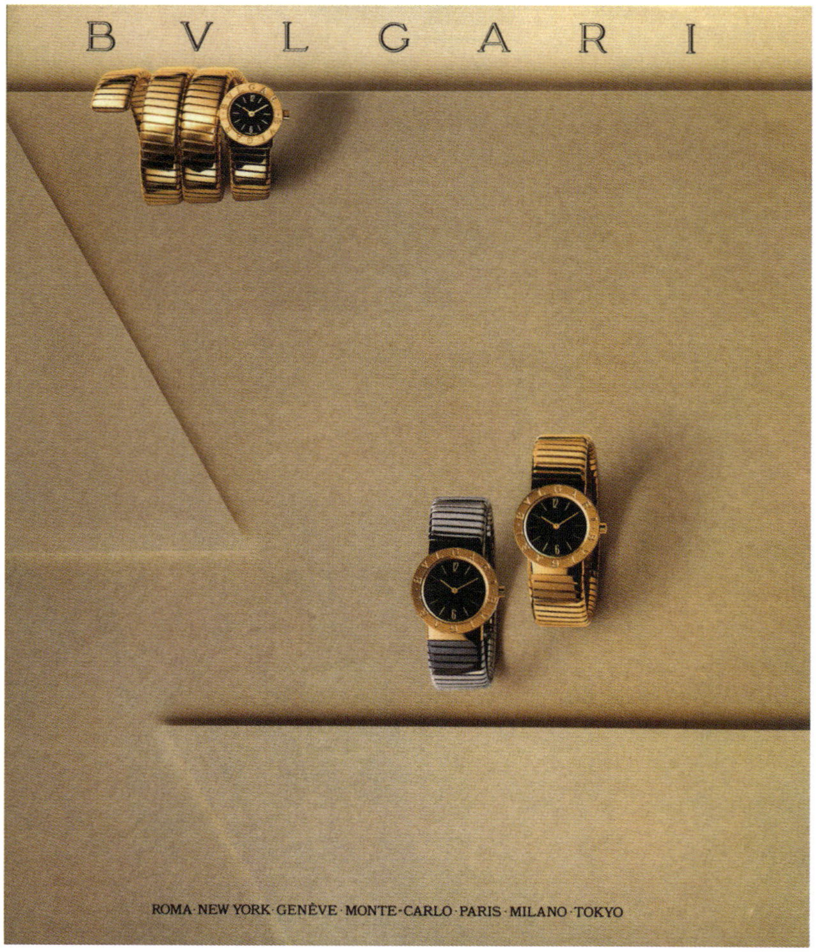

From top left, clockwise: *Original drawing Roma watch, 1975* **/** *Advertising campaign from 1978–1985* **/** *Advertising campaign from 1990–1996* **/** *Advertising campaign from 1981–1989*

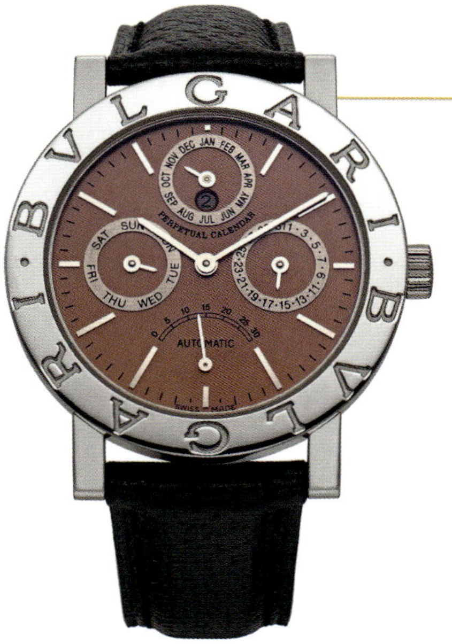

2000

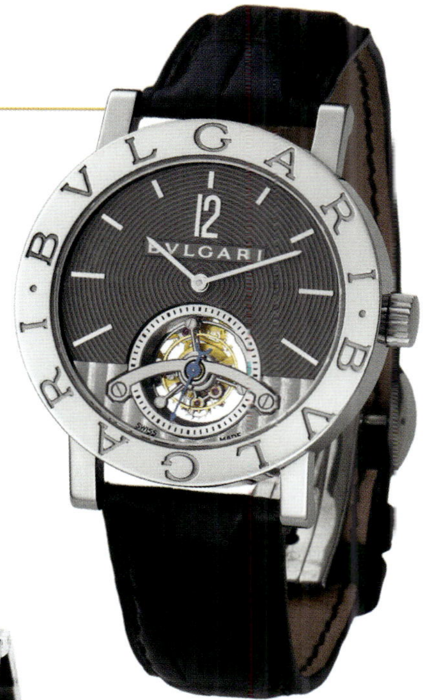

2005

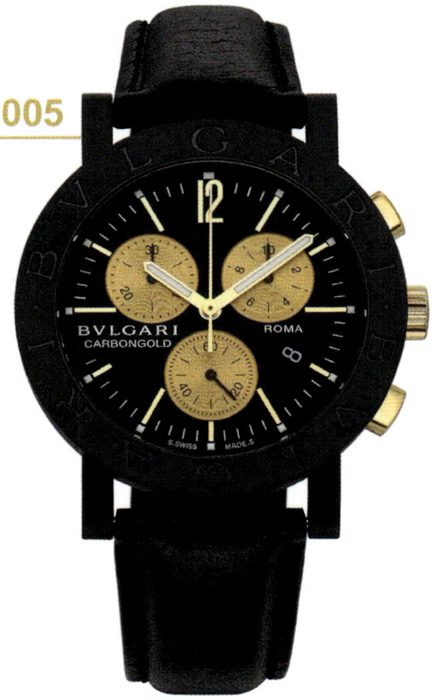

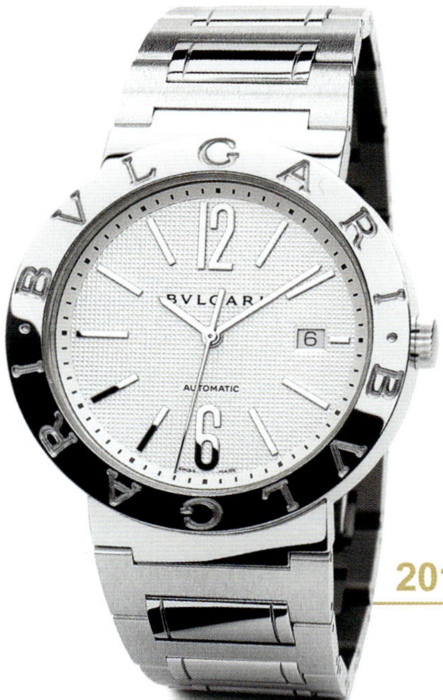

2013

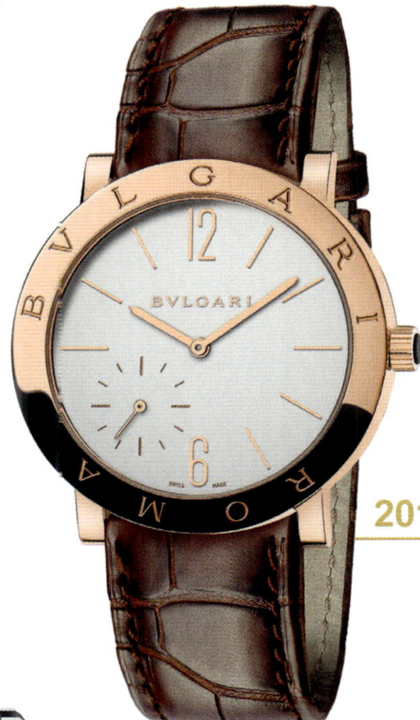

2015

2017

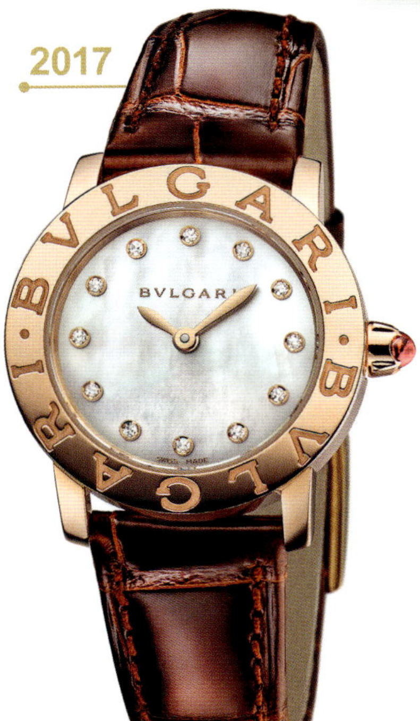

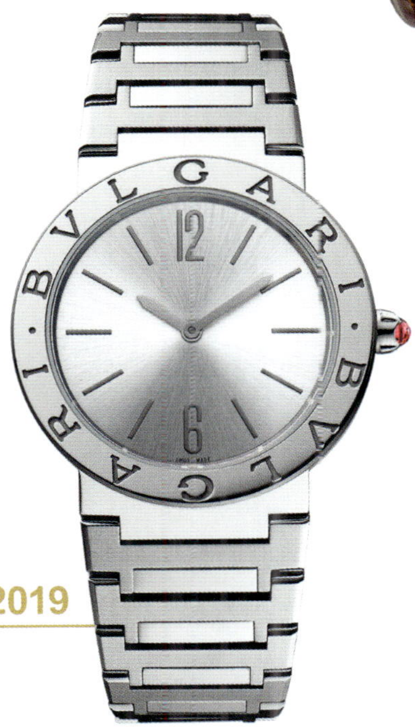

2019

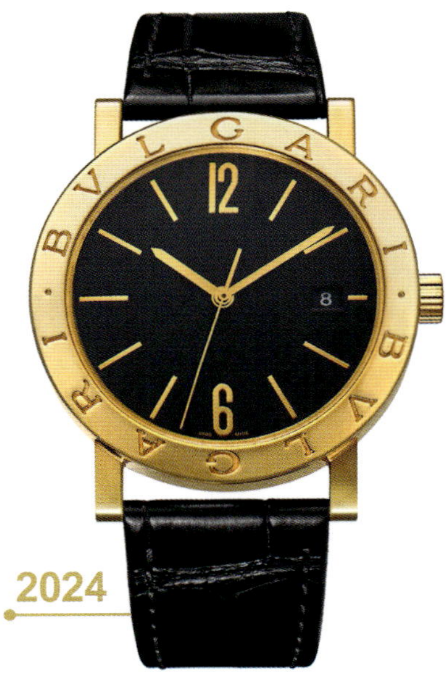

2024

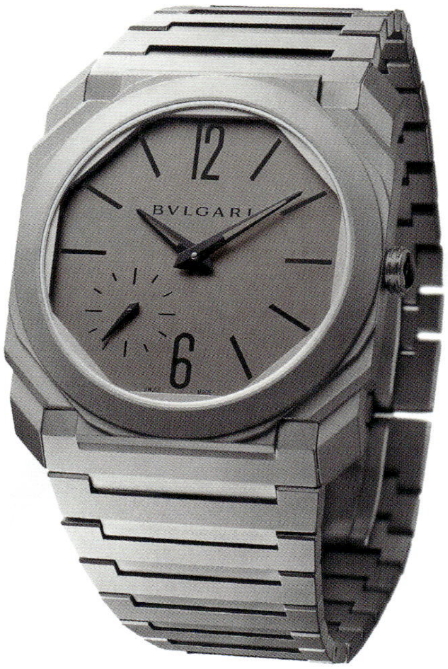
BVLGARI

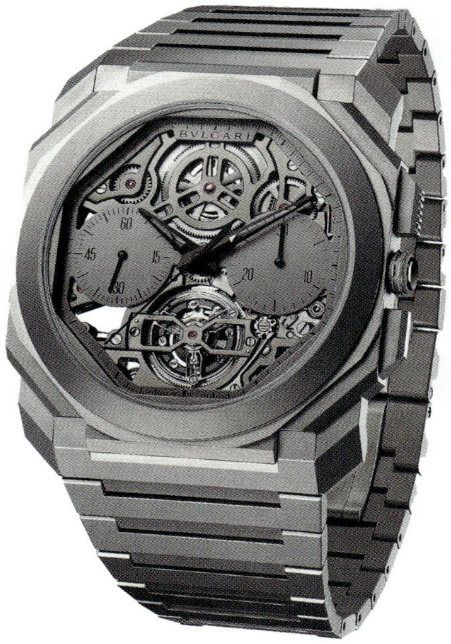
BVLGARI

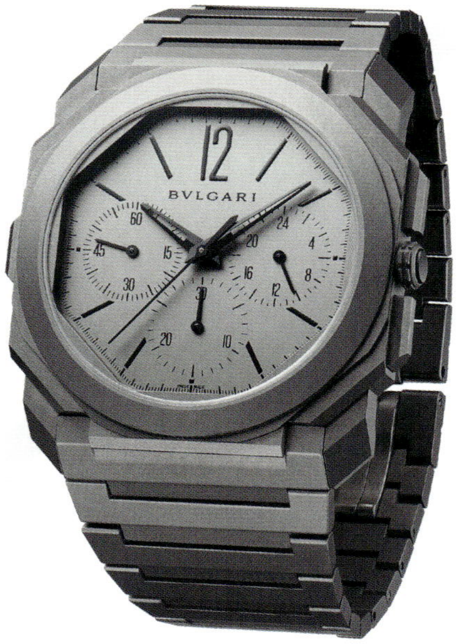
BVLGARI

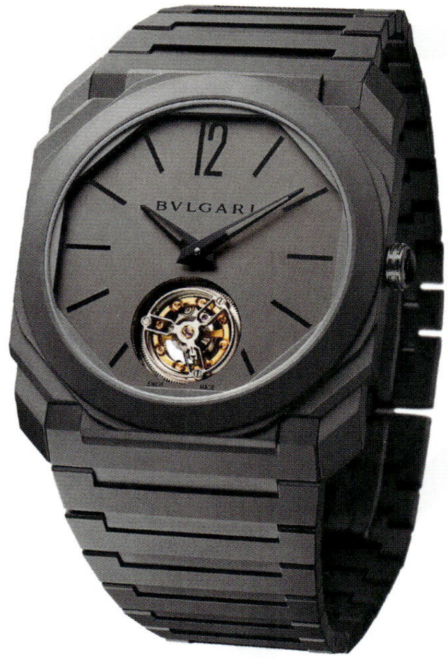
BVLGARI

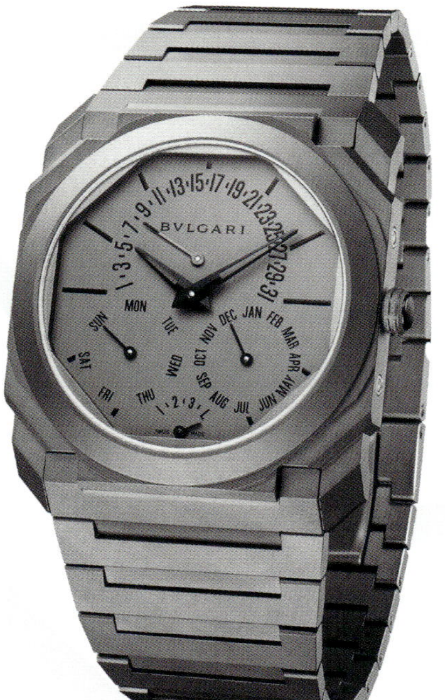
BVLGARI

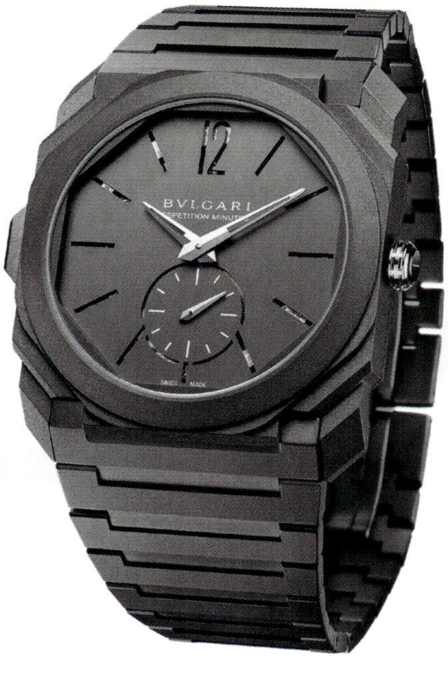
BVLGARI

BULGARI
OCTO FINISSIMO

EXQUISITE WATCHMAKING

Rapid successes in the field of watchmaking led to the founding of Bulgari Time S.A. in Neuchâtel in 1980. In 2000, Daniel Roth and Gérald Genta, two specialists for complex mechanical movements from the remote Joux Valley, joined the team. Speaking of Gérald Genta: Following the proven mantra of purist lines, in the 1980s the design guru with Italian roots created an octagonal classic, inspired by the Roman Basilica of Maxentius. The name Octo is not coincidental. As the designer, who died in 2011, remarked, "At that time, it was the thing to design watches to suit the wrist that were also comfortable to wear. My designs always have very fluid, aesthetic lines and are tailor-made for the human anatomy. Look at my octagon. It does not have straight lines; it has anatomical lines." The ultra-flat Octo Finissimo was also designed according to this principle. Following the Octo Finissimo Tourbillon and the Octo Finissimo Minute Repetition, Bulgari surprised visitors at the 2017 Basel Watch Fair with a third superlative.

The Octo Finissimo Automatique is an embodiment of the world's best micro-rotor performance. Its proprietary calibre BVL 138 impresses with a minimalist total height of just 2.23 millimetres. The 40 mm titanium case offers protection against water up to five bar. The sapphire crystal caseback is a point of honour. Hour, minute and second hands rotate in front of a titanium dial. Bulgari pairs the mechanical flyweight with a flat titanium bracelet or crocodile strap. If they wish, owners can immortalise themselves by having their name engraved on the left case flank. As an alternative to titanium, a stainless-steel version has also been available since 2020. Incidentally, for those with a penchant for complexity, the Octo Finissimo is also available with the world's thinnest automatic chronograph or as the thinnest wristwatch with a perpetual calendar. In all, the Bulgari Octo Finissimo line boasts seven world records. Launched in spring 2024, the Octo Finissimo Ultra COSC measures a mere 1.70 millimetres in height. Incredibly, the hand-wound Calibre BVL 180 inside is just 1.50 millimetres flat.

FEINSTE UHRMACHERKUNST

Rasch einsetzende Erfolge auf dem Gebiet der Uhrmacherkunst führten 1980 zur Gründung der Bulgari Time S.A. in Neuchâtel. 2000 gelangten mit Daniel Roth und Gérald Genta zwei kompetente, im abgeschiedenen Vallée de Joux beheimatete Spezialisten für komplexe mechanische Uhrwerke an Bord. Apropos Gérald Genta: Nach dem bewährten Motto klarer Linienführung kreierte der Design-Guru mit italienischen Wurzeln in den 1980er-Jahren einen achteckigen, von der römischen Maxentiusbasilika inspirierten Klassiker. Der Name „Octo" kommt also nicht von ungefähr. „In dieser Zeit", so der 2011 Verstorbene, „war es ein Phänomen, Uhren zu kreieren, die an das Handgelenk angepasst und gleichzeitig angenehm zu tragen waren. Meine Entwürfe weisen stets sehr fließende, ästhetische Linien auf und sind für die menschliche Anatomie maßgeschneidert. Mein typisches Achteck ist kein Oktogon mit geraden, sondern eines mit anatomischen Linien." Diesem Credo folgte auch die ultraflache „Octo Finissimo". Nach dem „Octo Finissimo Tourbillon" und der „Octo Finissimo Minutenrepetition" überraschte Bulgari während der Basler Uhrenmesse 2017 mit einem dritten Superlativ.

Die „Octo Finissimo Automatique" verkörpert eine Mikrorotor-Weltbestleistung. Ihr Manufakturkaliber BVL 138 besticht durch minimalistische Gesamthöhe von nur noch 2,23 Millimetern. Schutz vor Wasser bis zu fünf bar Druck bietet das 40 Millimeter messende Titangehäuse. Ehrensache ist ein Saphirglas-Sichtboden. Zeiger für Stunden, Minuten und Sekunden bewegen sich vor einem Titan-Zifferblatt. Das mechanische Fliegengewicht liefert Bulgari mit flachem Titan- oder Krokoband. Auf Wunsch können sich die Besitzerinnen und Besitzer in der linken Gehäuseflanke mit ihrem Namenszug verewigen. Alternativ zu Titan gibt es seit 2020 auch eine Version in Edelstahl. Menschen mit Faible für Kompliziertes erhalten die „Octo Finissimo" übrigens auch mit dem weltweit flachsten Automatik-Chronographen oder als derzeit flachste Armbanduhr mit ewigem Kalender. Insgesamt zeichnen sieben Weltrekorde die „Octo Finissimo"-Linie von Bulgari aus. Nur noch 1,7 Millimeter misst die im Frühjahr 2024 lancierte „Octo Finissimo Ultra COSC" in der Höhe. Dabei ist das darin verbaute Handaufzugskaliber BVL 180 fast schon unglaubliche 1,50 Millimeter flach.

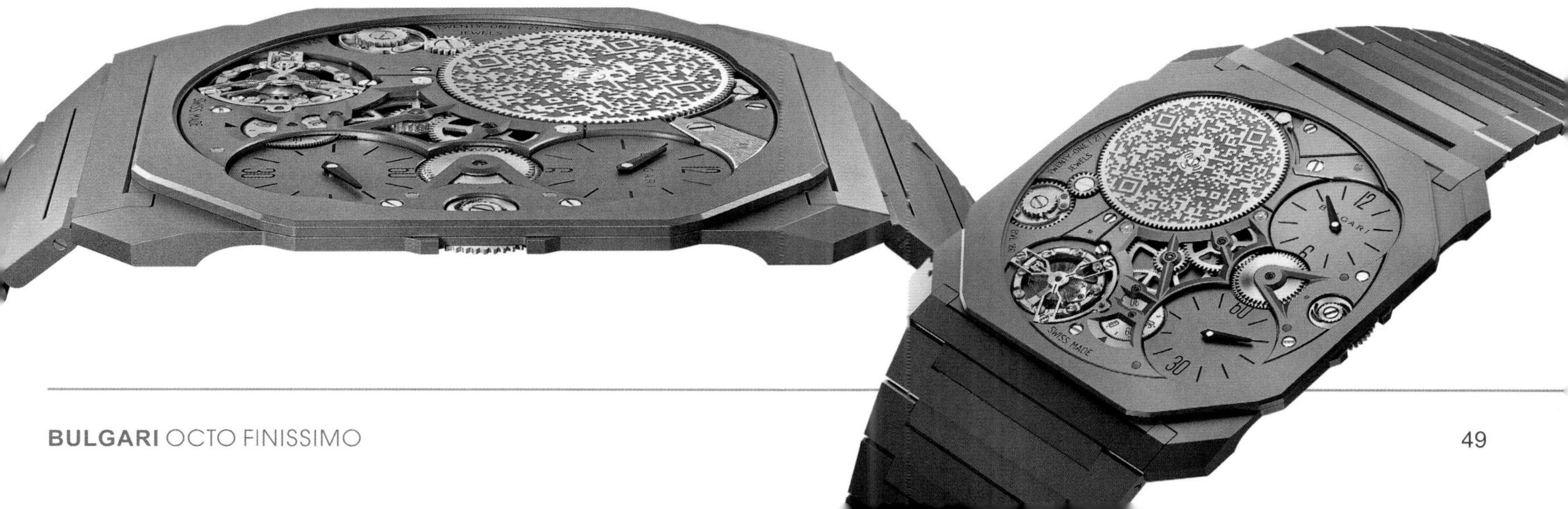

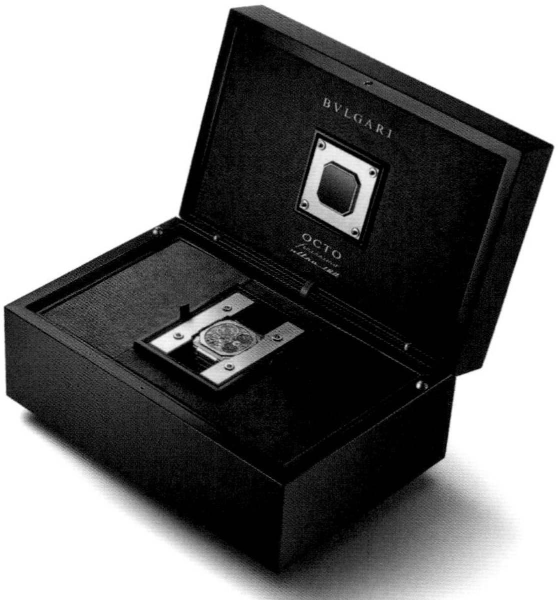

BVLGARI
OCTO

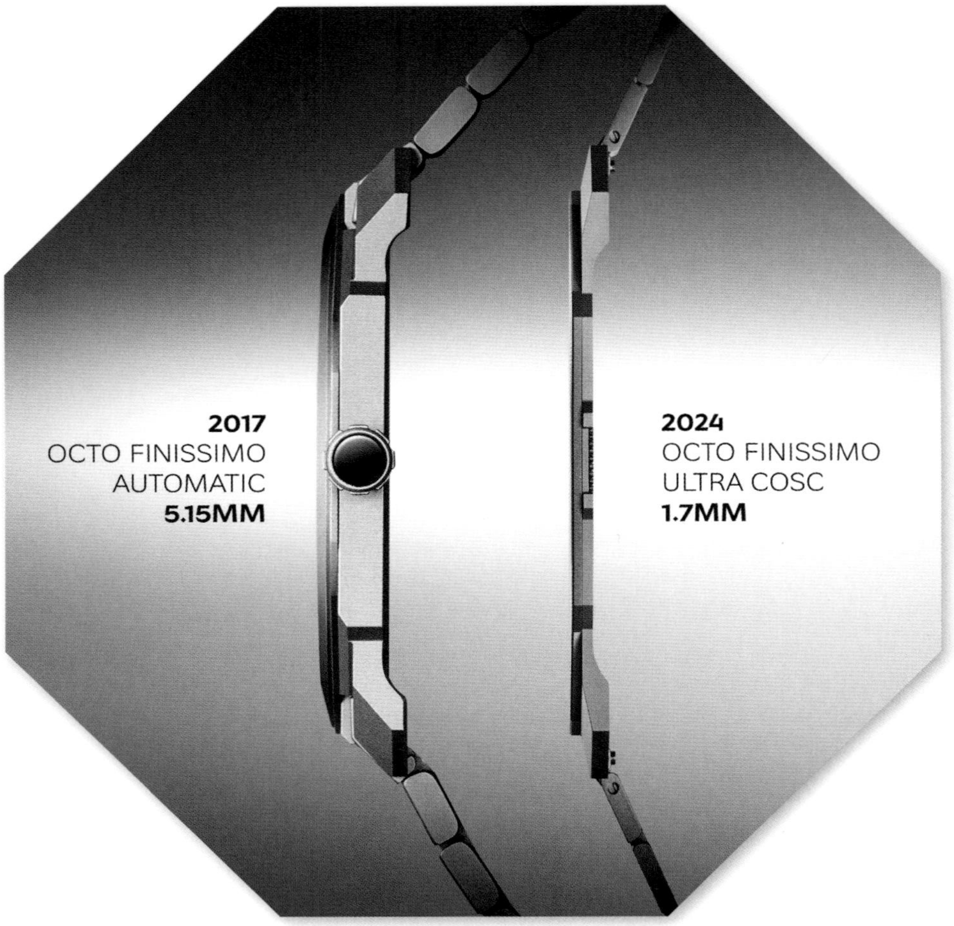

2017
OCTO FINISSIMO
AUTOMATIC
5.15MM
2024
OCTO FINISSIMO
ULTRA COSC
1.7MM

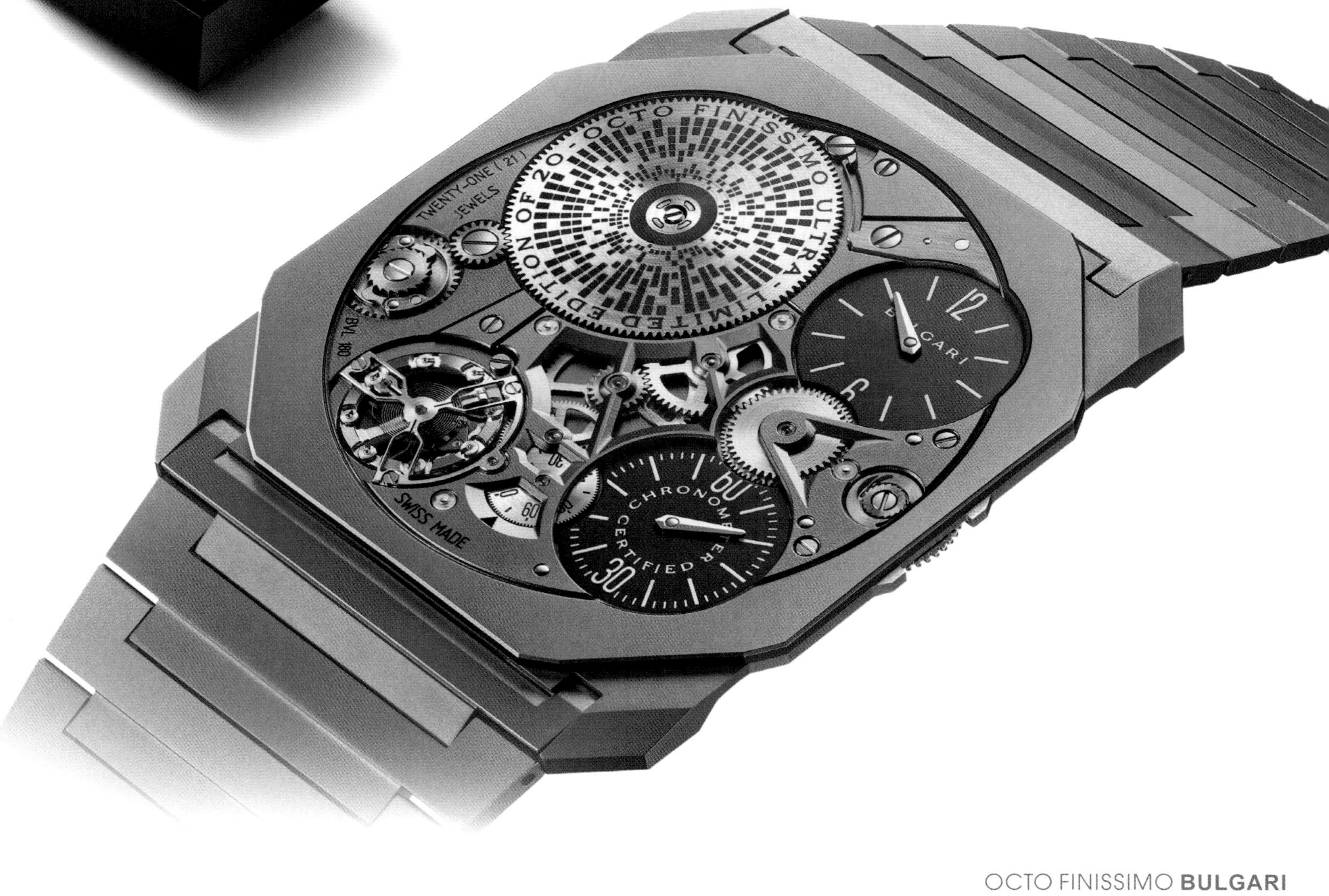

LIMITED EDITION OF 20 - OCTO FINISSIMO ULTRA
TWENTY-ONE (21)
JEWELS
BVL 180
SWISS MADE
BVLGARI
CHRONOMETER
CERTIFIED

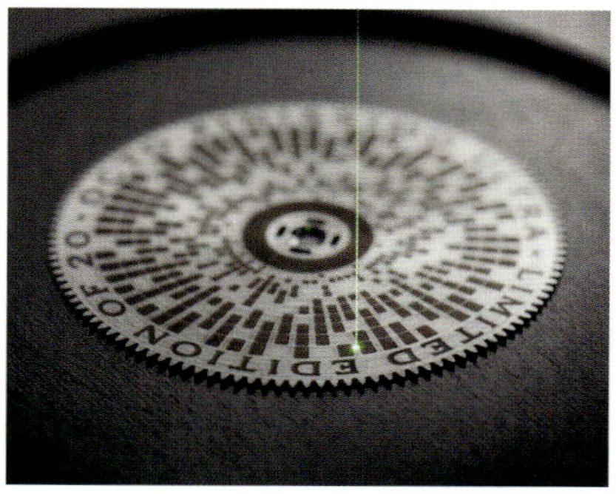

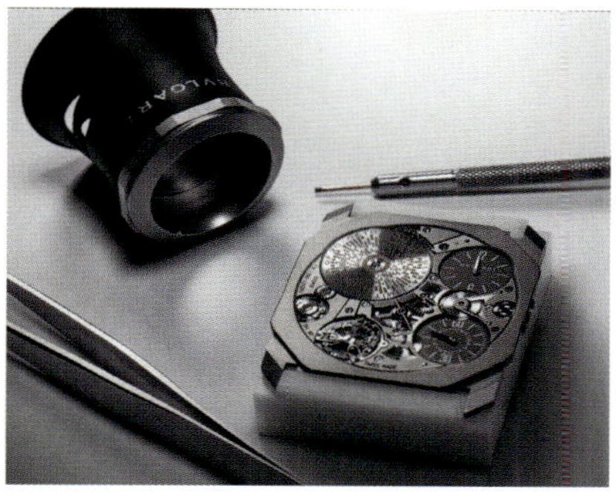

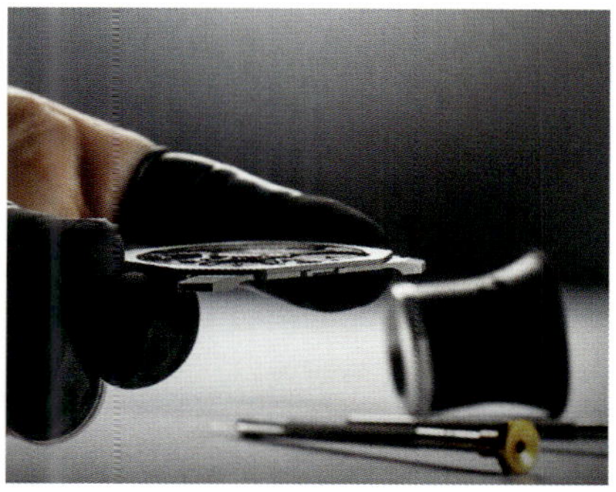

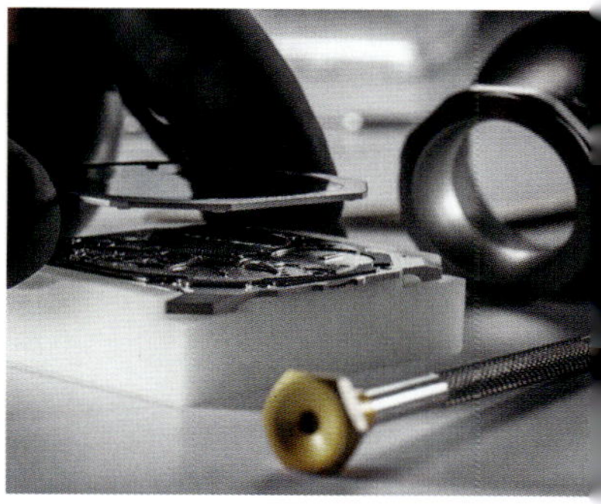

RECORD-BREAKING CRAFTSMANSHIP

The Octo Finissimo Ultra COSC is the world's thinnest watch, with just 1.70 millimetres in height. Given that the limits of movement slimness had already been reached with the Octo Finissimo Ultra in 2022, achieving an even slimmer profile for the Octo Finissimo Ultra COSC involved a rework of the case by optimising the sapphire crystal. The engineers were able to shave off that critical tenth of a millimetre – and establish a new world record.

In terms of the overall structure, the caseback serves as the mainplate, upon which the 170 components of the manually wound BVL 180 manufacture calibre are assembled. These components are selected using 1/10-micron optical measuring systems, so as to be able to select the parts that are closest to the perfect dimension. Adequate rigidity was ensured by crafting the caseback/mainplate from particularly dense, hard and ultra-resistant tungsten carbide.

To maximise the use of the watch's surface area, the hour and minute dials, the large barrel – with a 50-hour power reserve – and the escapement cleverly play with the circular geometry of the bezel opening, adjusting it ever so slightly.

HANDWERKSKUNST MIT WELTREKORD

Die Octo Finissimo Ultra COSC ist mit 1,70 Millimeter Höhe die dünnste Uhr der Welt. Da die Limiten der möglichst geringen Uhrwerkhöhe bereits 2022 bei der Octo Finissimo Ultra erreicht worden waren, musste das Gehäuse der Octo Finissimo Ultra COSC überarbeitet und das Saphirglas optimiert werden, um ein noch flacheres Profil zu erhalten. In akribischen Schritten gelang es den Ingenieuren, den entscheidenden Zehntelmillimeter abzuspecken und einen neuen Weltrekord aufzustellen.

Möglich wurde dies, weil der Gehäuseboden gleichzeitig als Uhrwerkplatine dient, auf der die 170 Komponenten des Manufakturkalibers BVL 180 mit Handaufzug montiert werden. Diese Einzelteile werden mit optischen Messsystemen im Zehntel-Mikrometer-Bereich ausgewählt, um den perfekten Abmessungen am nächsten zu kommen.

Die Herausforderung, ein für diese Dimensionen möglichst robustes Gehäuse zu entwickeln, wurde durch die Herstellung des als Werkplatine dienenden Bodens aus Wolframkarbid gelöst, einem besonders dichten, harten und extrem widerstandsfähigen Metall.

Um die Oberfläche der Uhr optimal zu nutzen, sind das Zifferblatt mit Stunden- und Minutenzeiger, das große Federhaus mit seiner Gangreserve von 50 Stunden und die Hemmung geschickt in die kreisförmige Geometrie der Lünettenöffnung eingepasst.

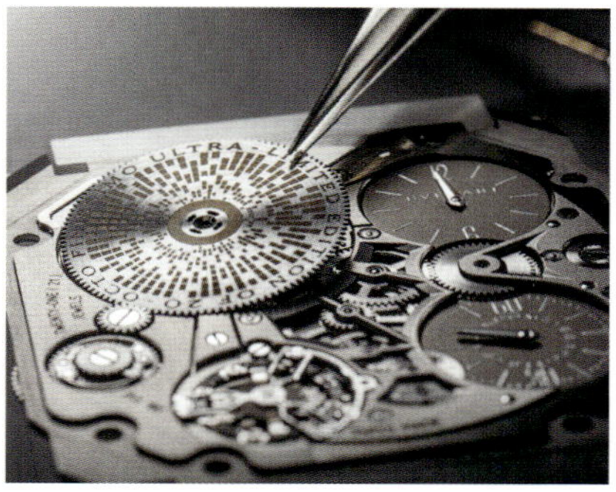

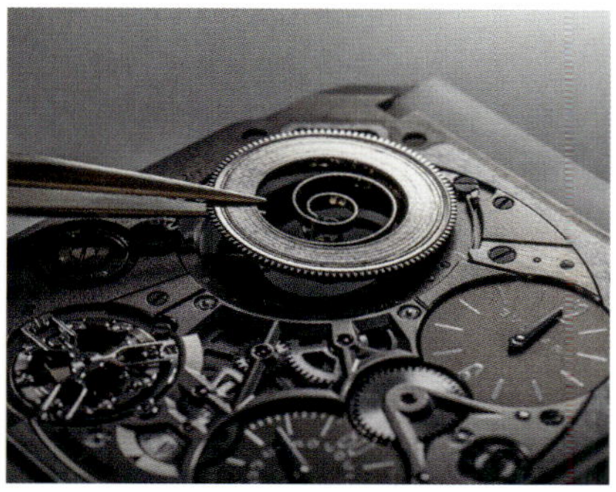

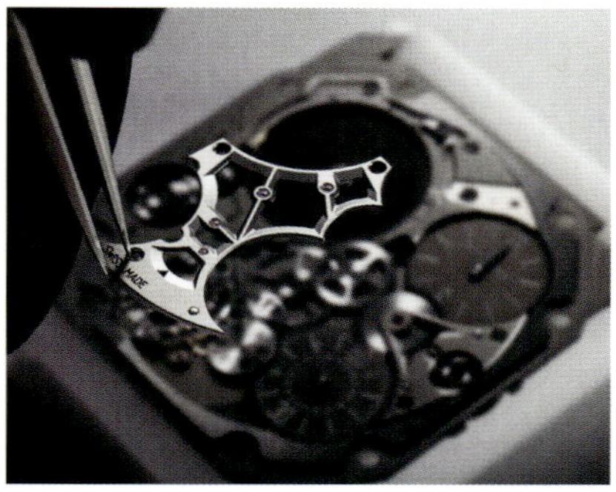

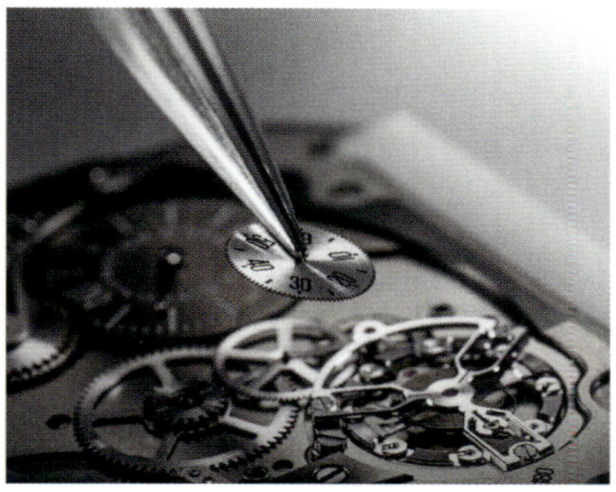

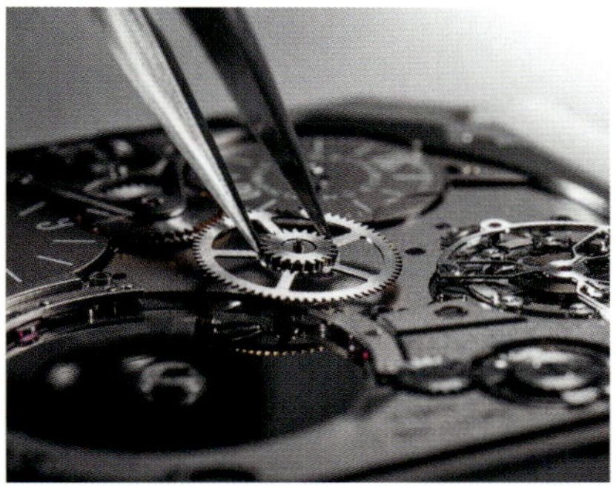

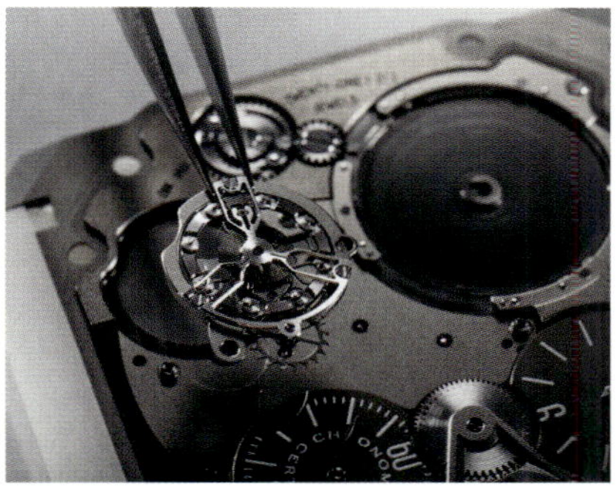

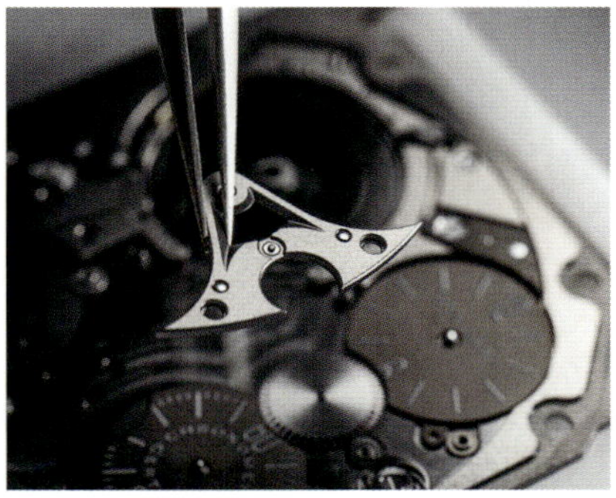

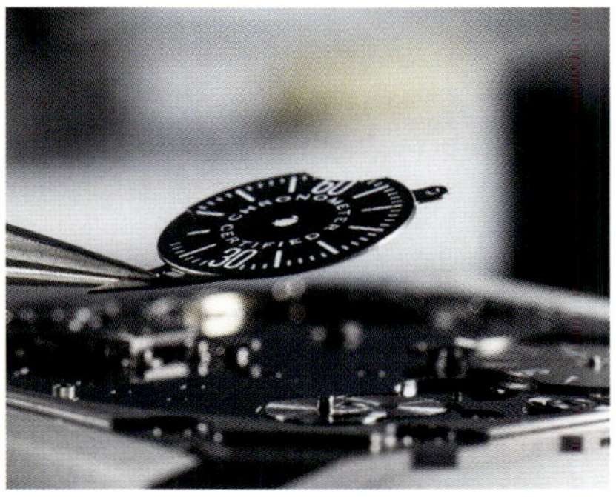

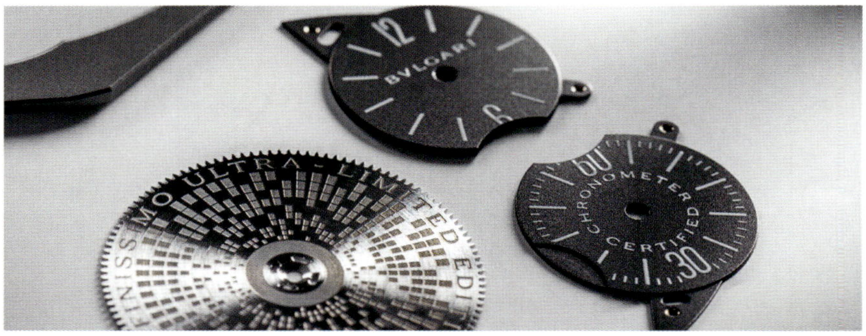

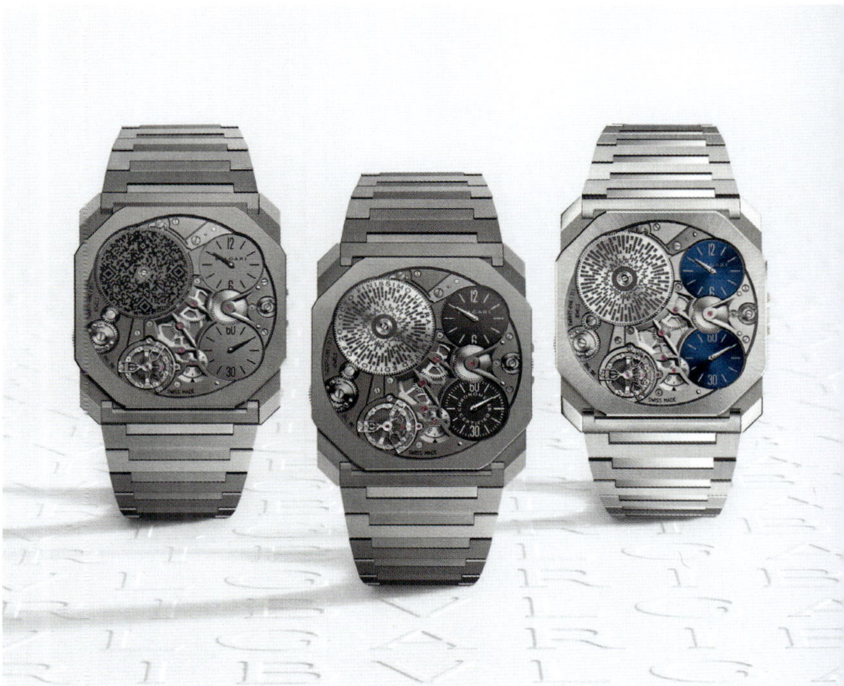

1915

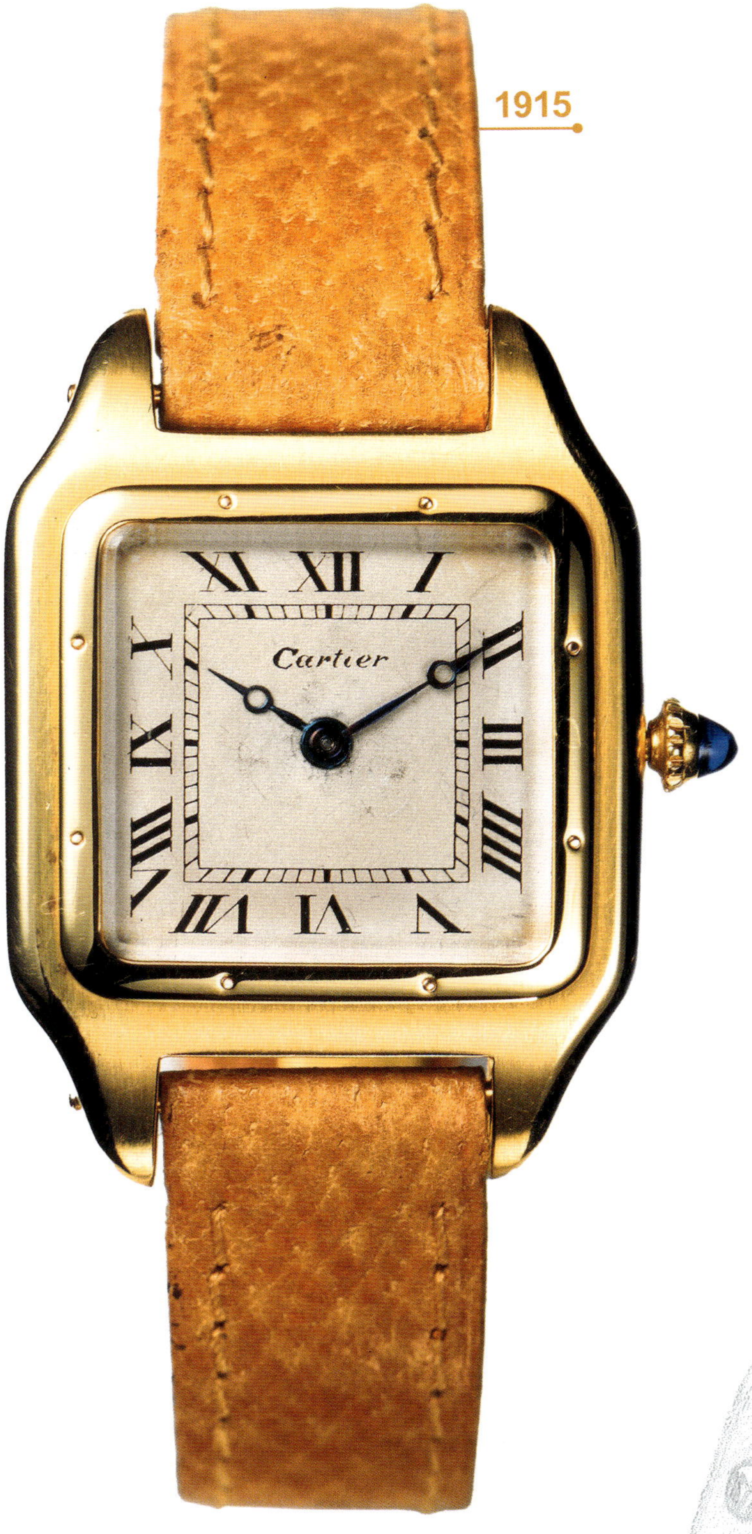

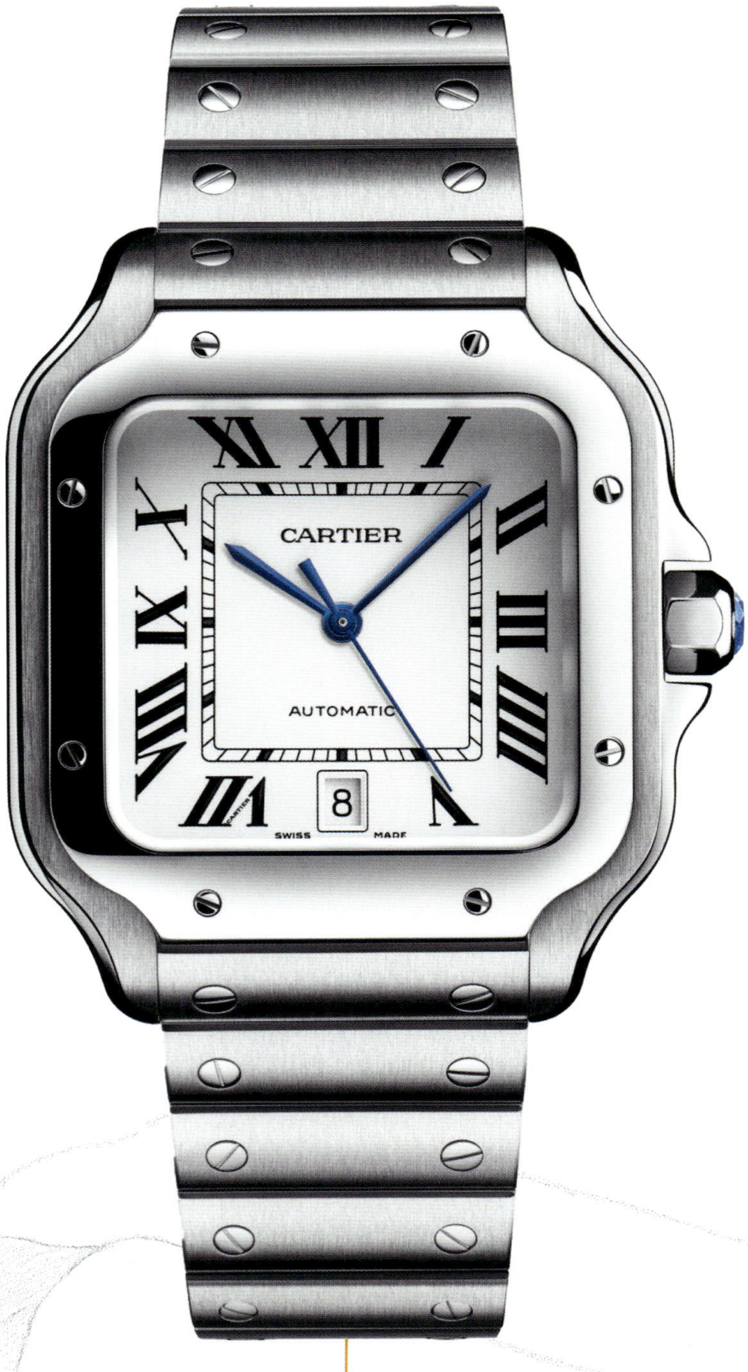

2021

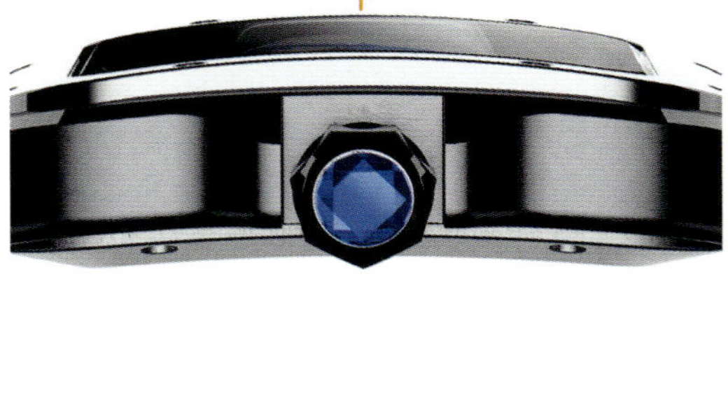

CARTIER SANTOS

THE EMANCIPATED WRISTWATCH

In their infancy, wristwatches were reduced derivatives of timepieces carried in a pocket or jacket. Round and with soldered lugs for the strap. Obviously, this fell short in terms of a permanent solution. Louis Cartier was one of the first to realise this. The descendant of Louis-François Cartier, founder of the company and "jeweller to kings", was not only a talented designer, but also a popular habitué of Parisian high-society salons. This resulted in an encounter with the Brazilian Alberto Santos-Dumont at a distinguished soirée held by Baron Deutsch de la Meurthe. The chance meeting was soon to develop into a remarkable friendship. When the avid aviation pioneer was to be honoured with the Prix de la Meurthe for his record airship flight in 1901 at the luxury restaurant Maxim's, his friend Louis was of course one of the guests. In 1904, Louis Cartier received an invitation to a select meeting in Santos-Dumont's exclusive apartment on the Champs-Élysées. The South American playboy and philanthropist did not beat about the bush. High up in the air and with both hands on the control stick, it was difficult to check the time using a pocket watch. Not a man to need a second hint, back in his studio the product designer immediately set to work. That same year, he completed his masterpiece: a square wristwatch with rounded edges, screwed glass bezel and integrated lugs. Alberto Santos-Dumont was delighted. So much so that he wore the wristwatch on 12 November 1907 during a sensational flight at an altitude of 220 metres in the new 14 bis flying machine. This major event prompted the Parisian high society to clamour for the trailblazing Santos. The jeweller to the rich and famous could not have been happier to meet the demands of his discerning clients. Thus emancipated from the design strictures of the pocket watch, from 1908 Cartier offered the wristwatch to privileged contemporaries. From 1911, it was also available as a serial product. Although design metamorphoses followed, visually the Cartier Santos remained true to itself. As it continues to do so today. This icon has never been more popular. Quite apart from that, it merits the venerable title of a superlative. No other watch can look back on a longer and, more importantly, uninterrupted history.

DIE EMANZIPIERTE ARMBANDUHR

In ihren Kinderjahren waren Armbanduhren verkleinerte Derivate der in Tasche oder Weste getragenen Zeitmesser. Rund, mit angelöteten Bügelschlaufen zur Befestigung des Bands. Auf Dauer war das natürlich zu wenig. Als einer der Ersten erkannte das Louis Cartier. Der Nachfahre von Louis-François Cartier, seines Zeichens Gründer des Unternehmens und „Juwelier der Könige", war nicht nur begnadeter Designer, sondern auch gern gesehener Gast in der Pariser Society. Als solcher lernte er im Rahmen einer vornehmen Soiree des Baron Deutsch de la Meurthe den Brasilianer Alberto Santos-Dumont kennen. Aus dem eher zufälligen Zusammentreffen entwickelte sich rasch eine bemerkenswerte Männerfreundschaft. Als der passionierte Flugpionier bald darauf im Nobelrestaurant „Maxim's" für seinen Rekordflug mit einem motorisierten Lenkballon im Jahre 1901 den Prix de la Meurthe entgegennehmen sollte, war sein Freund Louis natürlich mit von der Partie. 1904 erhielt Louis Cartier eine Einladung zum Privatissimum in Santos-Dumonts noble Wohnung auf den Pariser Champs Elysées. Dort kam der südamerikanische Lebemann und Philanthrop unverblümt zur Sache. Hoch in den Lüften und mit beiden Händen am Steuerknüppel sei die Zeit per Taschenuhr nur schwer zu kontrollieren. So etwas ließ sich Cartier natürlich nicht zweimal sagen. In seinem Atelier machte sich der Produktgestalter unverzüglich an die Arbeit. Noch im gleichen Jahr war sein Werk fertiggestellt: eine quadratische Armbanduhr mit abgerundeten Ecken, verschraubtem Glasrand und integrierten Bandanstößen. Alberto Santos-Dumont zeigte sich begeistert. Deshalb begleitete sie ihn auch am 12. November 1907 beim sensationellen 220-Meter-Hüpfer mit dem neuen Motor-Fluggerät „14 bis". Bedingt durch dieses Großereignis verlangte die Pariser Hautevolee ebenfalls nach der bahnbrechenden „Santos". Der Wunsch anspruchsvoller Kunden war dem Nobeljuwelier natürlich Befehl. Ab 1908 offerierte Cartier die erste, vom Taschenuhrdesign emanzipierte Armbanduhr in einer Mini-Edition für privilegierte Zeitgenossen. Ab 1911 war sie auch als tickendes Serienprodukt erhältlich. Gestalterische Metamorphosen folgten. Aber optisch blieb sich die Cartier „Santos" immer treu. Das gilt bis in die Gegenwart. Heute erfreut sich diese Ikone größerer Beliebtheit als je zuvor. Ganz abgesehen davon gebührt ihr das ehrenwerte Prädikat eines Superlativs: Keine andere Armbanduhr kann auf eine längere und vor allem niemals unterbrochene Geschichte zurückblicken.

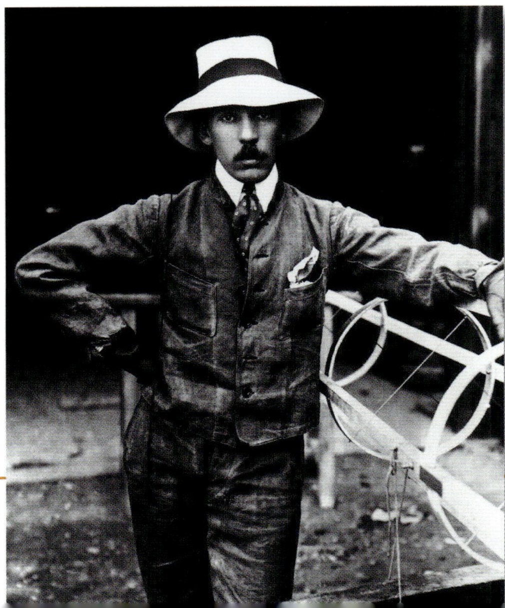

Alberto Santos-Dumont

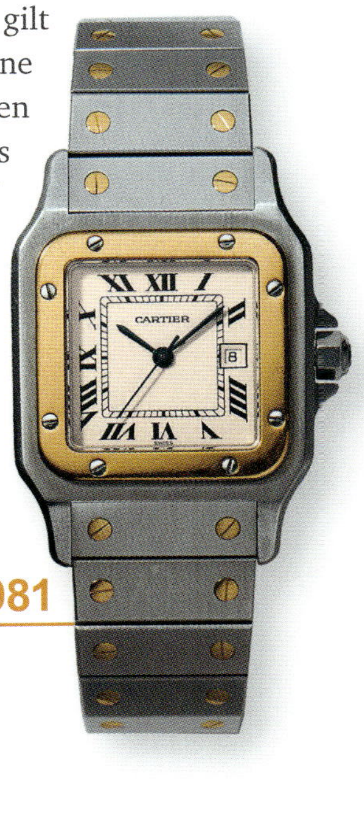

1981

CARTIER
TANK

1919

A PEACEFUL AFFAIR

"If all tanks were made by Cartier, we'd have time to live in peace." This remark by French designer Jean-Charles de Castelbajac from 1994 is more topical than ever. The classic timepiece has only a fleeting relationship with battle tanks. It owes its design and name to the military track vehicle, euphemistically called a tank, observed by Louis Cartier in 1916 as it made its way to the battlefield of the Somme. It inspired rough sketches which evolved into wristwatches. Visually, once more, they broke with conventions. Rectangular with "brancards" or vertical sides and, of course, integrated lugs. Worn on the wrist, Louis Cartier's Tank did indeed reflect the contours of its terrifying namesake. After his Santos, the French designer made a further contribution to the creative emancipation of the wristwatch from the predominantly round pocket watch design. In 1918, General John Joseph Pershing, commander-in-chief of the American troops in France, Colonel William Hayward and other senior officers of the Expeditionary Corps of the US Armed Forces were the first to sport a Cartier Tank. It was a discreetly ticking token of gratitude for the liberation of the Grande Nation from enemy troops. The recipients could only enjoy the exclusivity of their timepiece for a year. In 1919, the Cartier Tank went into serial production. Its century-old success story stems from its legendary status, wide range of models, and the undeniable fact that it is one of the longest continuously produced wristwatches ever. Mario Adorf, Pierce Brosnan, Truman Capote, Jacques Chirac, Gary Cooper, Alain Delon, Catherine Deneuve, Stewart Granger, Elton John, Calvin Klein, Madonna, Jean-Pierre Melville, George Michael, Yves Montand, Jackie Onassis, Yves Saint Laurent, Peter Sellers and Andy Warhol are just some of the celebrities to wear a Tank. >

FRIEDLICHE ANGELEGENHEIT

„Wenn alle Tanks von Cartier fabriziert würden, hätten wir Zeit zu einem Leben in Frieden." Dieses 1994 postulierte Statement des französischen Designers Jean-Charles de Castelbajac ist aktueller denn je. Mit Kampfpanzern hat der zeitbewahrende Klassiker nur am Rande zu tun. Seine Optik und den Namen erhielt er, als Louis Cartier im Jahr 1916 die martialischen, euphemistisch „Tank" genannten Kettengefährte auf dem Weg zur Sommeschlacht beobachtet hatte. Aus ersten Skizzen wurden passende Armbanduhren. Optisch einmal mehr völlig anders als das damals Übliche. Rechteckig mit bombierten Flanken und natürlich integrierten Bandanstößen. Mit diesem Auftritt am Handgelenk erinnerte Louis Cartier formal in der Tat ein wenig an die furchterregenden Panzer. Nach seiner „Santos" leistete der französische Designer einen weiteren Beitrag zur gestalterischen Emanzipation der Armbanduhr vom vorwiegend runden Taschenuhr-Design. Als erste Menschen durften sich 1918 General John Joseph Pershing, Oberbefehlshaber der amerikanischen Truppen in Frankreich, Oberst William Hayward sowie andere hohe Offiziere des Expeditionskorps der US-Streitkräfte mit einer Cartier „Tank" schmücken. Es handelte sich um einen dezent tickenden Ausdruck des Danks für die Befreiung der „Grande Nation" von den gegnerischen Truppen. Über die Exklusivität ihres Zeitmessers konnten sich die Beschenkten jedoch nur ein Jahr lang freuen. 1919 trat die Cartier „Tank" ins Stadium der Serienfertigung. Unsterblicher Mythos, breite Modellvielfalt sowie die unbestreitbare Tatsache, dass es sich um eine der am längsten ununterbrochen hergestellten Armbanduhren handelt, machen den Erfolg während mehr als 100 Jahren aus. Mario Adorf, Pierce Brosnan, Truman Capote, Jacques Chirac, Gary Cooper, Alain Delon, Catherine Deneuve, Stewart Granger, Elton John, Calvin Klein, Madonna, Jean-Pierre Melville, George Michael, Yves Montand, Jackie Onassis, Yves Saint Laurent, Peter Sellers und Andy Warhol sind nur einige der Prominenten mit einer „Tank" am Handgelenk. >

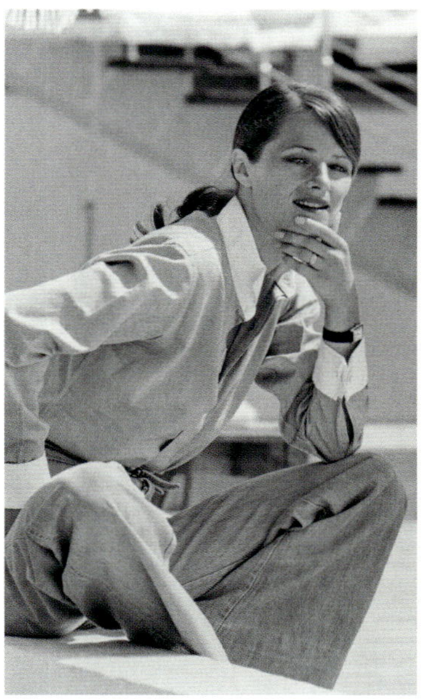

1922

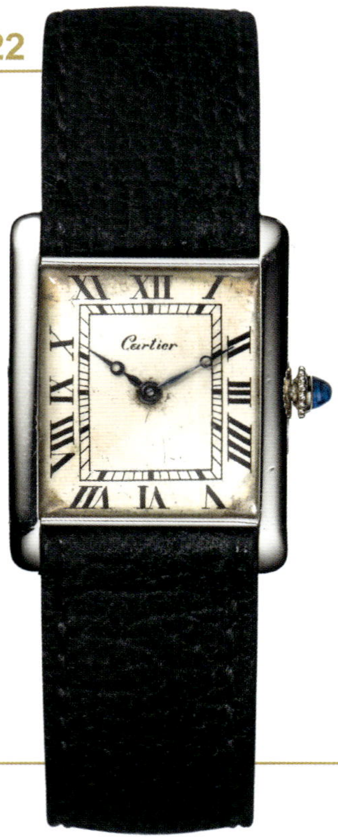

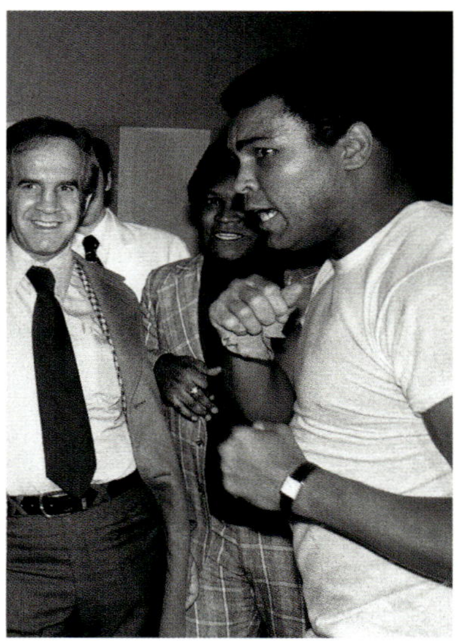

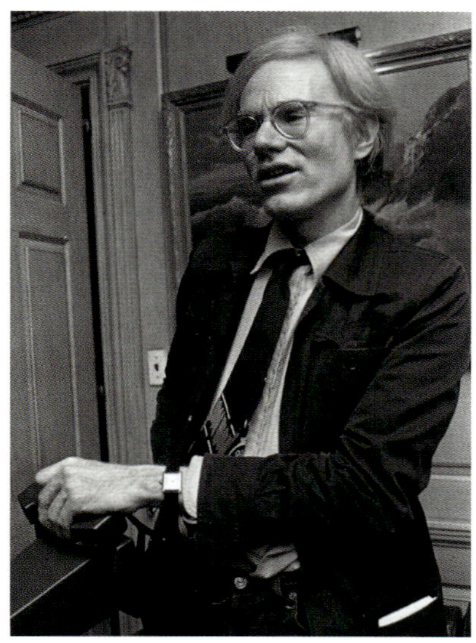

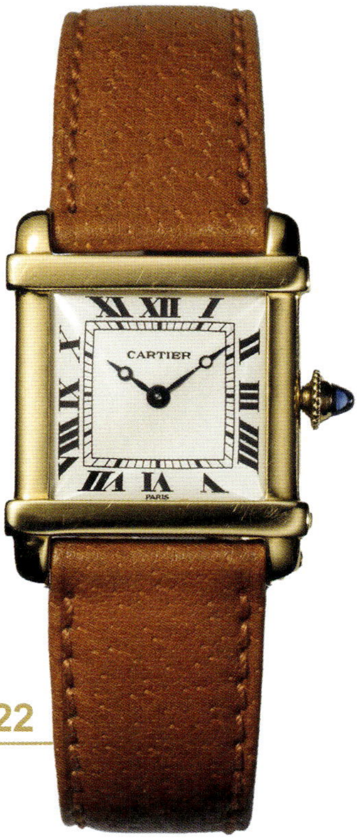

1922

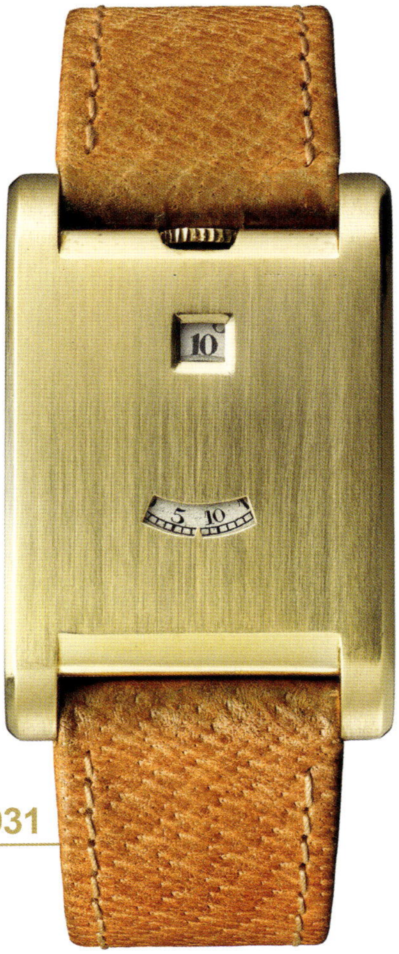

1931

1936

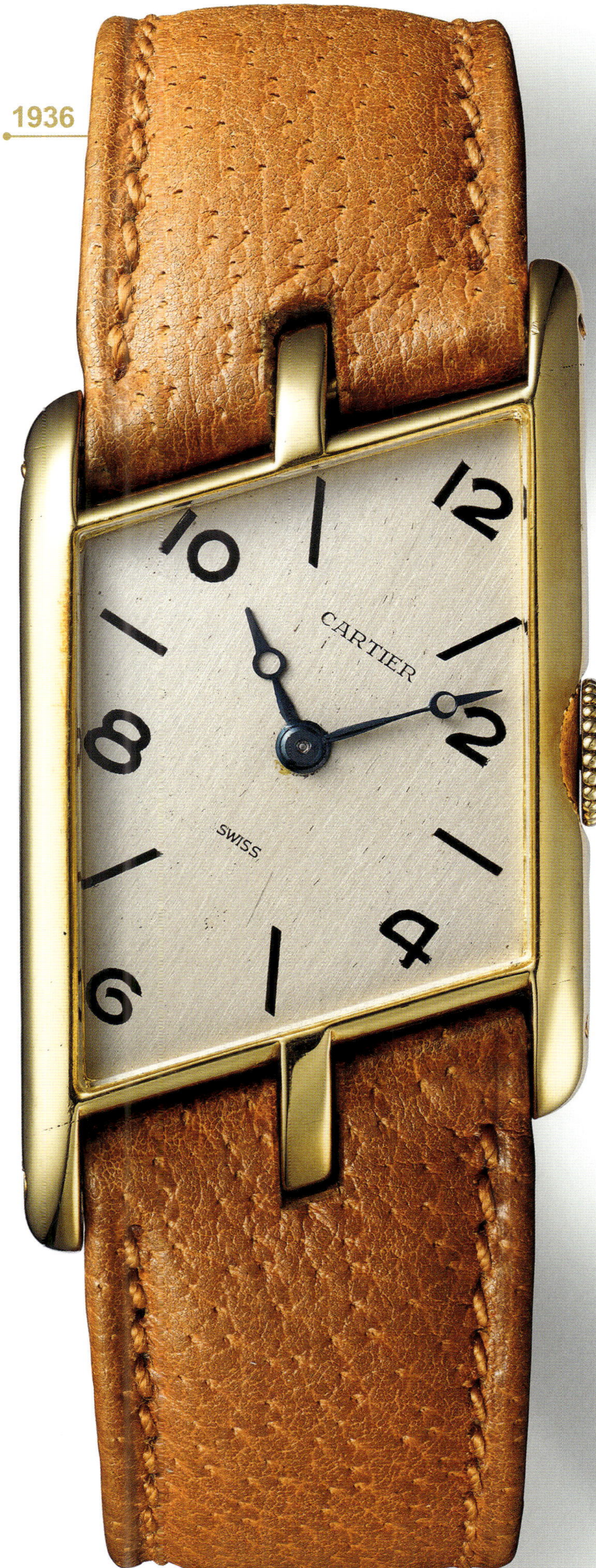

Left page | Linke Seite:
Charlotte Rampling | Muhammad Ali | Andy Warhol

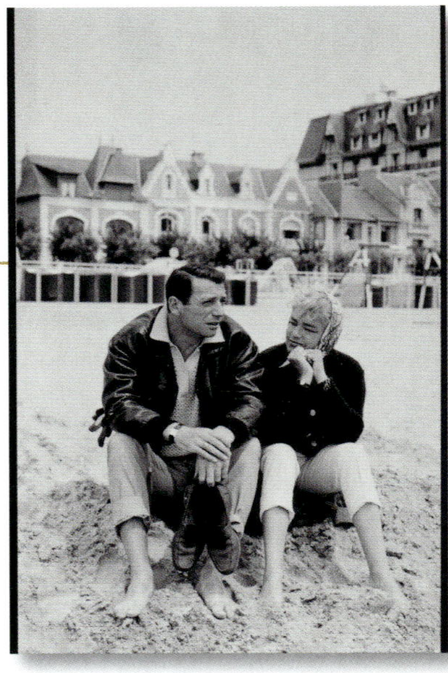

The Maharaja of Kapurthala, who already owned some 250 Cartier watches and who employed a servant with the sole task of winding them up, even commissioned an exclusive custom-made model. In 1924, the wealthy Indian received the forerunner of the Tank Cintrée. Rudolf Valentino was so attached to his watch that the Hollywood star insisted on wearing it during the filming of *The Son of Sheikh* in 1926, although it was a complete anachronism.

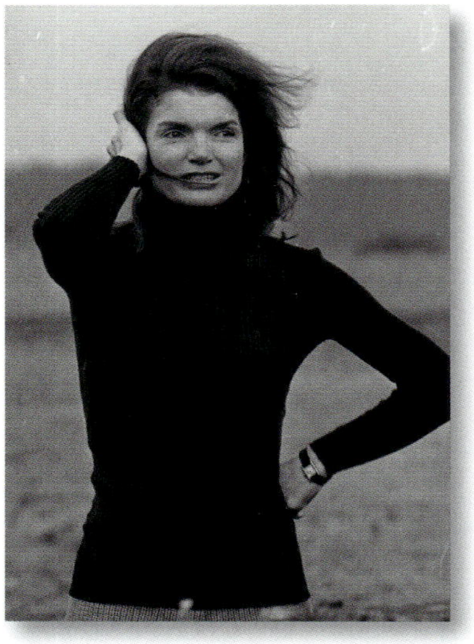

Cartier fulfilled the wish of the Pasha of Marrakech for a waterproof version in 1931 with the Tank Étanche. Other milestones include the Tank Savonnette with a protective hinged lid, the Petite Tank Rectangle and the elongated Tank Allongée, the Tank à Guichets with digital time display, as well as the Tank Asymétrique with a parallelogram-shaped case. The latter strikingly illustrates the versatility of this watch.

In the 1970s, contemporaries of more slender means could delight in acquiring a Tank from the Les Must de Cartier collection. In 1989, the vaulted Tank Américaine was presented and in 1996, a reinterpretation of the emblematic original, the Tank Française. The latter is available for the first time with a series metal link bracelet.

Der Maharadscha von Kapurthala, dem bereits rund 250 Cartier-Uhren gehörten und der einen Diener speziell zum Aufziehen der Werke beschäftigte, wünschte sich sogar eine exklusive Sonderanfertigung. 1924 erhielt der wohlhabende Inder den Vorläufer der „Tank Cintrée". Rudolfo Valentino wuchs sein Exemplar so sehr ans Handgelenk, dass der Hollywood-Star 1926 bei den Dreharbeiten zu *Der Sohn des Scheichs* auf seiner geliebten, im Film indes gänzlich unpassenden Armbanduhr bestand.

Den Wunsch des Paschas von Marrakesch nach einer wasserdichten Version erfüllte Cartier 1931 in Gestalt der „Tank Étanche". Als weitere Meilensteine gelten die „Tank Savonnette" mit schützendem Klappdeckel, die „Petite Tank Rectangle" und die gestreckte „Tank Allongée", die „Tank à Guichets" mit digitaler Zeitanzeige sowie die „Tank Asymétrique" mit Parallelogramm-Schale. Letztere zeugte ausdrucksstark von der Wandlungsfähigkeit dieser Armbanduhr.

Zeitgenossen mit begrenztem Budget erfreuten sich in den 1970er-Jahren an einer „Tank" aus der Kollektion „Les Must de Cartier". Das Jahr 1989 brachte die gewölbte „Tank Américaine" und 1996 eine gestalterische Neuinterpretation des emblematischen Originals, die „Tank Française". Letzte gibt es erstmals serienmäßig mit metallenem Gliederband.

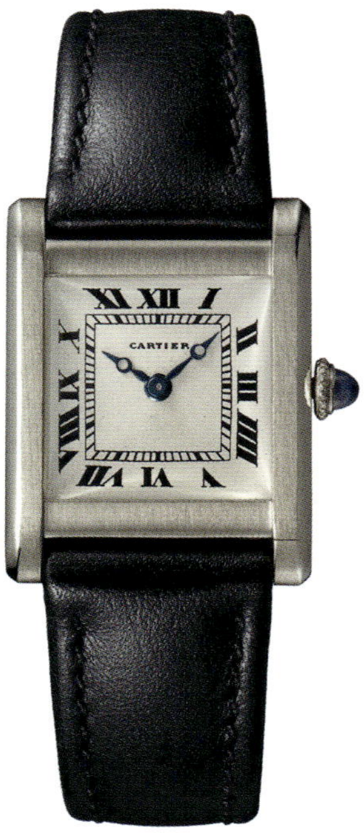

TANK, 1919

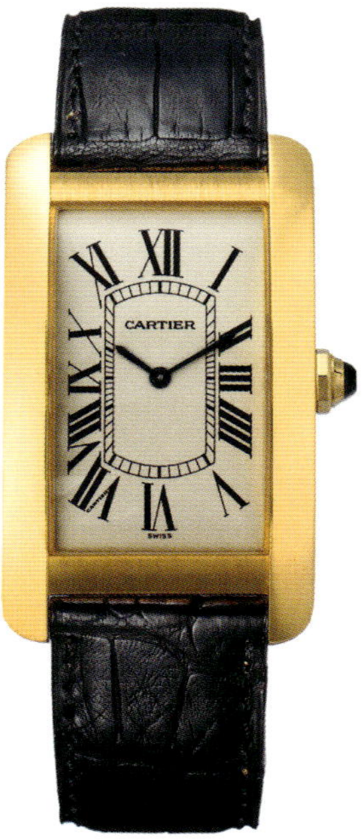

AMÉRICAINE, 1989

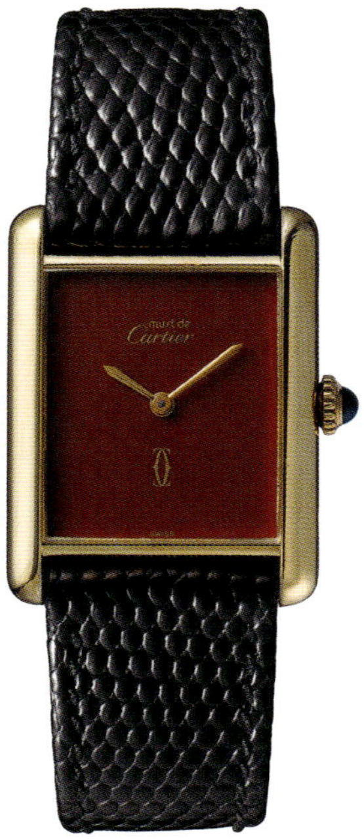

MUST, 1977

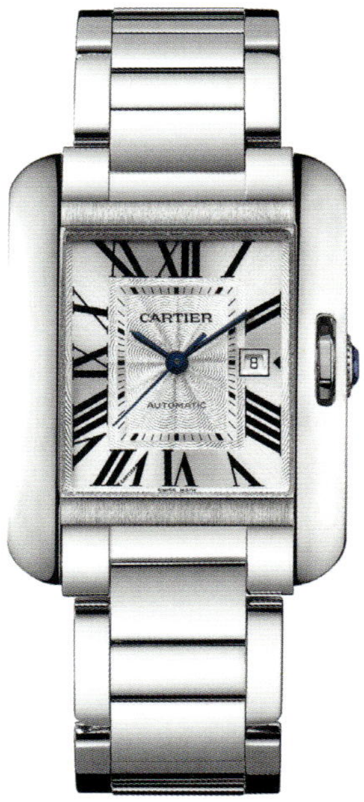

ANGLAISE, 2012

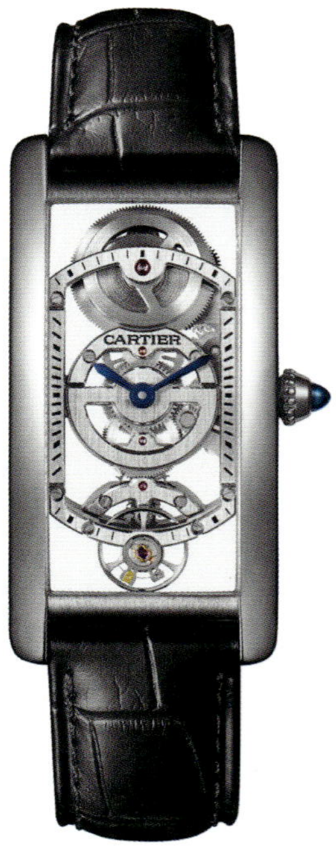

CINTRÉE SKELETON, 2017

FRANÇAISE, 1996

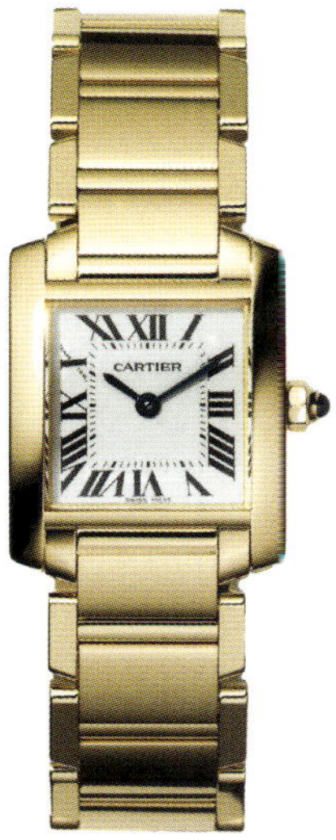

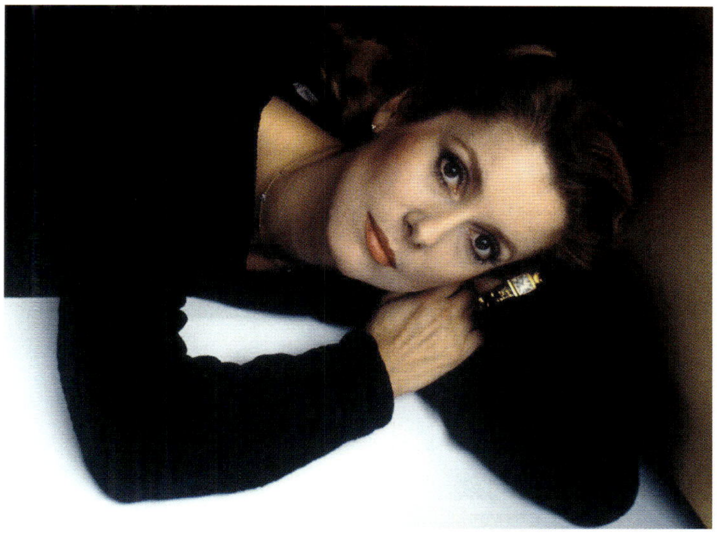

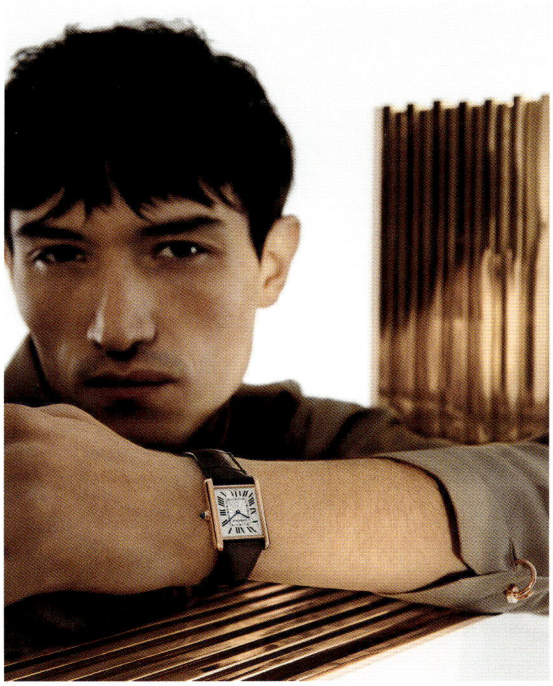

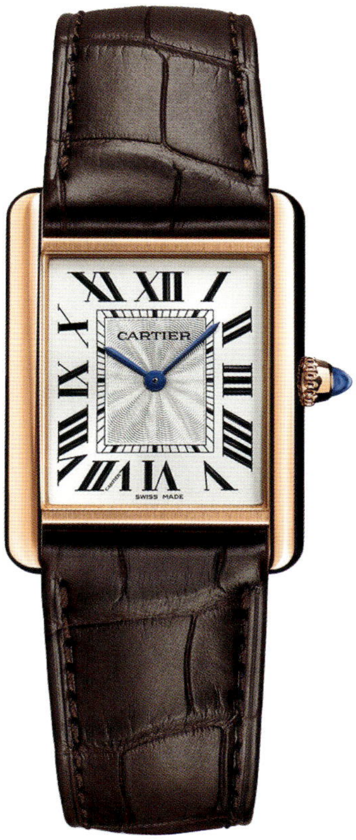

LOUIS CARTIER, 2023

Left page | Linke Seite:
Jackie Onassis | Yves Montand & Simone Signoret
Right page, top right | Rechte Seite, oben rechts:
Catherine Deneuve
Bottom | Unten:
Advertisement for Tank Américaine |
Werbeanzeige für die Tank Américaine, 1993

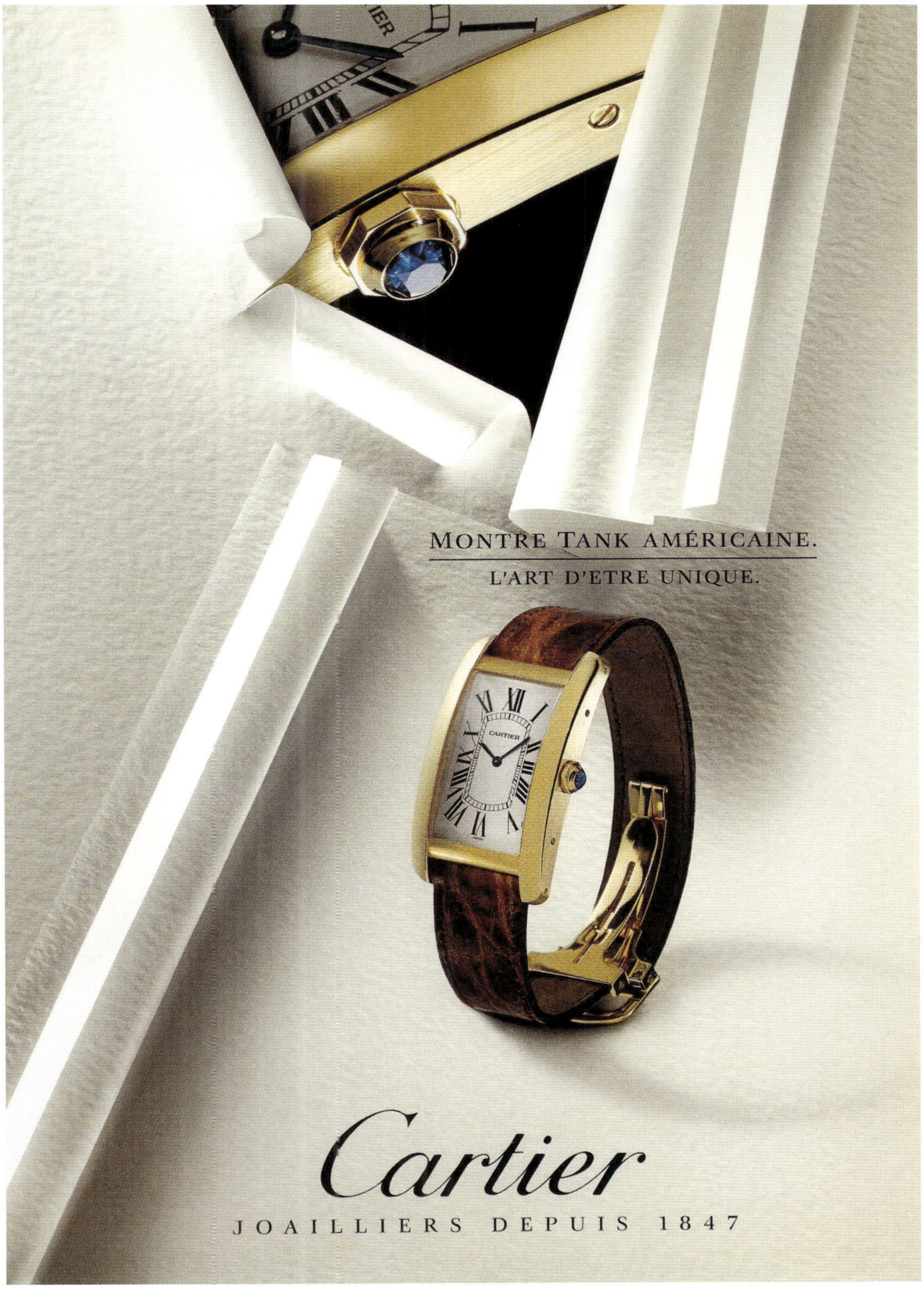

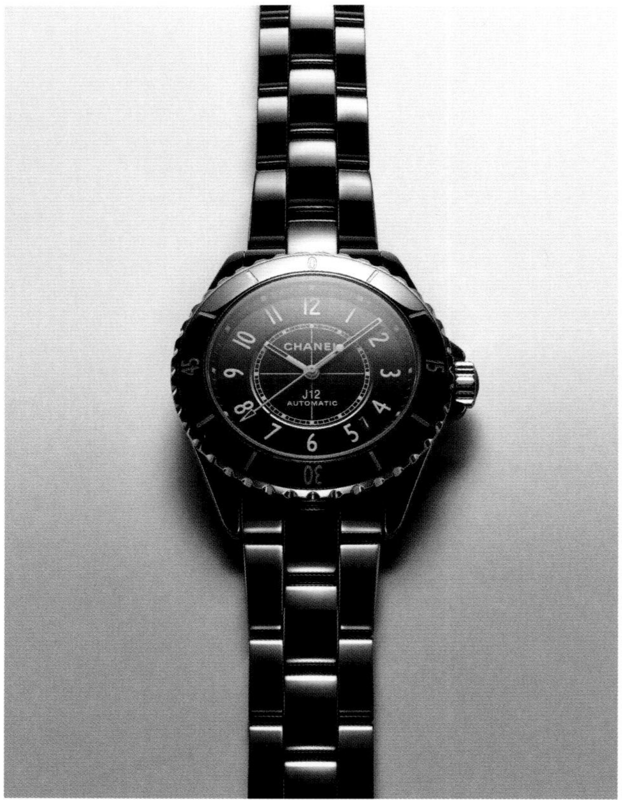
CHANEL
J12
AUTOMATIC

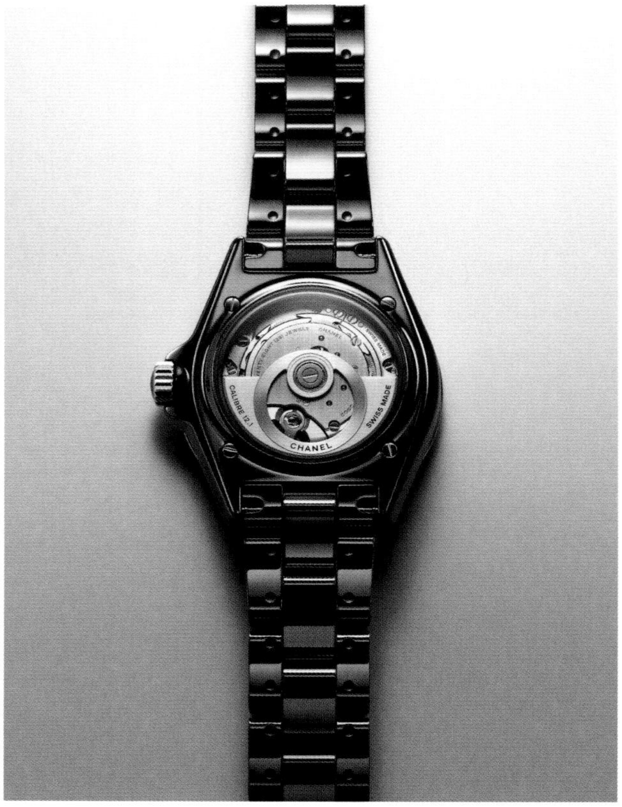
CHANEL
SWISS MADE

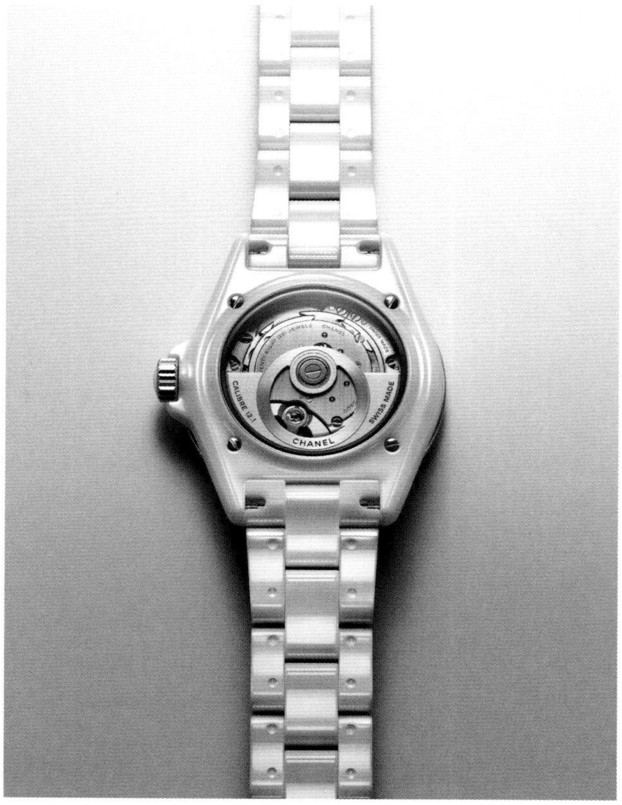
CHANEL
SWISS MADE

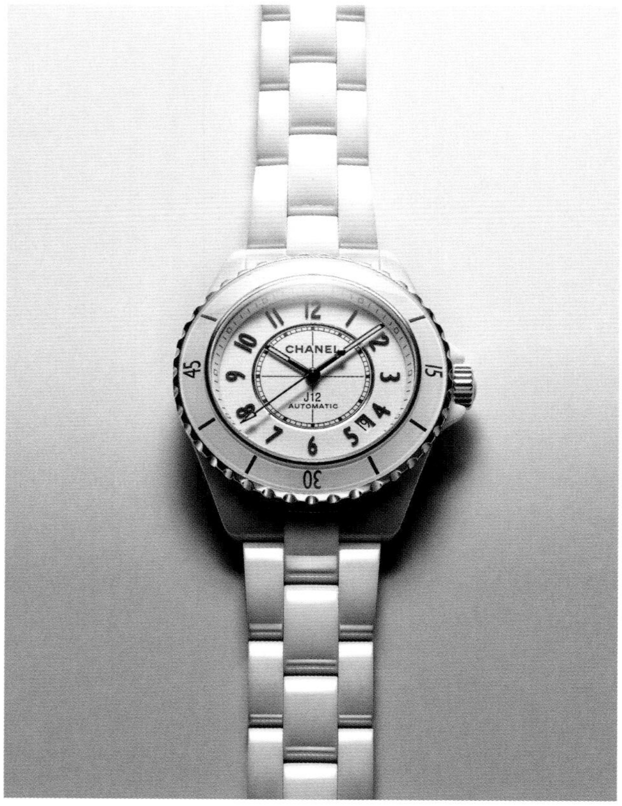
CHANEL
J12
AUTOMATIC

CHANEL
J12

ENDURING BEAUTY

For a comparatively brief 20 years, the Chanel J12, launched in 2000, became a fixture on the international watch scene. When Jacques Helleu created the eye-catching timepiece, he initially had himself in mind. The creative director of the fragrance and beauty as well as watches and jewellery divisions of the fashion label, founded in 1910 by Gabrielle "Coco" Chanel, was inspired by racing cars and the sleek IACC yachts of the America's Cup. One of the racing boats, the *J12*, was to become the namesake for the sporty newcomer, dressed all in black. Mademoiselle Chanel held this colour in high esteem. Because Helleu not only had an eye for every detail, but also for enduring beauty, he selected scratch-resistant ceramic for the case and bracelet. In 2003, it was joined by a white version. Although designed by a man primarily for a male audience, the J12, initially equipped with an ETA automatic movement, tended to appeal more to the fairer sex. Increasingly, women of fashion sought larger, more prominent means of expression on their wrist. The ensuing, almost endless stream of iterations therefore comes as little surprise.

The present-day J12 bears the signature of Arnaud Chastaingt. He never met Jacques Helleu, who died in 2007. The designer subjected the fully fledged icon to a radical remake. Around 70 per cent of the visible elements were modified. What immediately catches the eye is the more finely notched rim of the bezel, the steeper domed bezel insert and the inner minute track. The greatest step into the future, however, is revealed by the mechanical inner workings of the classic model. They hail from the Kenissi manufactory in Le Locle, in which Chanel holds a 20 per cent stake. Due to the slightly higher automatic calibre 12.1 and the desire to maintain the ergonomics of the case, Chanel opted for a monocoque structure. Its see-through floor showcases the unusually designed rotor.

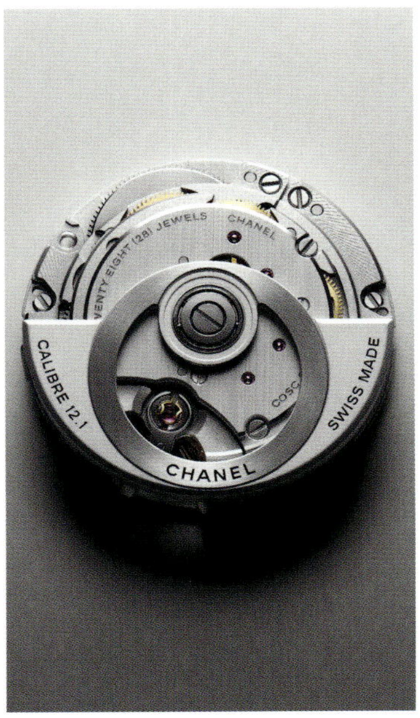

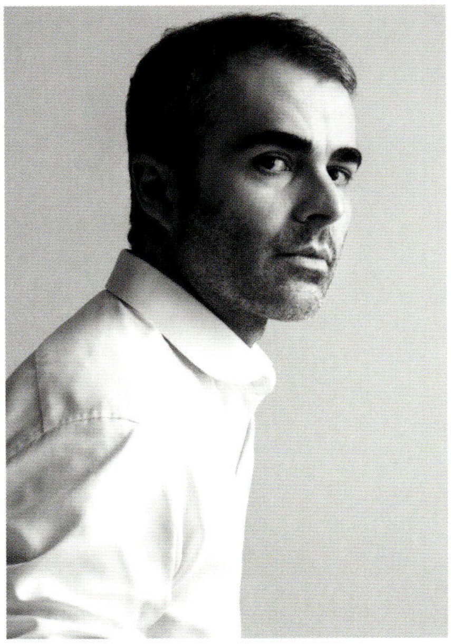

EWIGE SCHÖNHEIT

Während vergleichsweise kurzer 20 Jahre entwickelte sich die 2000 lancierte Chanel „J12" zu einem festen Bestandteil der internationalen Uhrenszene. Als Jacques Helleu den augenfälligen Zeitmesser kreierte, dachte er zunächst einmal an sich selbst. Beim Design ließ sich der Kreativdirektor der Sparten Parfum und Beauty sowie ab 1987 auch Uhren und Schmuck des 1910 von Gabrielle „Coco" Chanel gegründeten Modelabels von Rennwagen und schnittigen IACC-Yachten des „America's Cup" leiten. Eines der Rennboote, die J12, diente als Namenspate für den sportiven, ganz in Schwarz ausgeführten Newcomer. In der Gunst von Mademoiselle Chanel stand diese Farbe ganz oben. Weil Helleu nicht nur jedes gestalterische Detail berücksichtigte, sondern auch ewige Schönheit ins Auge fasste, wählte er für Gehäuse und Armband kratzfeste Keramik. 2003 gesellte sich ein weißes Pendant hinzu. Obwohl von einem Mann in erster Linie an Männer adressiert, punktete die anfangs mit einem Eta-Automatikwerk ausgestattete „J12" vom Start weg primär beim zarten Geschlecht. Mehr und mehr Damen von Welt verlangten nach Größerem und Ausdruckstarkem für ihr Handgelenk. Die anschließende Deklination in beinahe zahllose Varianten verwundert also nicht. Die Gegenwart der „J12" steht im Zeichen von Arnaud Chastaingt, der den 2007 verstorbenen Jacques Helleu nicht persönlich kennenlernen konnte. Der Designer unterzog die volljährige Ikone einer gründlichen Überarbeitung. Rund 70 Prozent des Sichtbaren erfuhr dabei eine Modifikation. Ins Auge stechen beispielsweise der feiner gerändelte Glasrand, das stärker bombierte Lünetten-Inlay und auch die Eisenbahn-Minuterie im Zifferblattzentrum. Der größte Schritt in die Zukunft zeigt sich indessen beim mechanischen Innenleben des klassischen Modells. Es stammt von der in Le Locle angesiedelten Manufaktur Kenissi, an der Chanel eine 20-prozentige Beteiligung hält. Bedingt durch das etwas höher bauende Automatikkaliber 12.1 und den Wunsch, die Gehäuse-Ergonomie zu erhalten, entschied sich Chanel für eine Monocoque-Konstruktion. Damit einher geht ein Sichtboden, durch den sich der außergewöhnlich gestaltete Rotor präsentiert.

Left / Links: *Calibre 12.1*
Right / Rechts. *Arnaud Chastaingt*

CHOPARD 1976/1993

HAPPY DIAMONDS / HAPPY SPORT

HAPPY WITH CHOPARD

In 1976, 13 years after the acquisition of Chopard, Karl and Karin Scheufele presented a wristwatch that was soon to make history. Believe it or not, the premiere model of the Happy Diamonds was a men's dress watch with an 18-carat white-gold case and bracelet. It was inspired by a waterfall in the Black Forest and created as a showpiece for the Golden Rose of Baden-Baden design award. Not even the owners, with their passion for jewellery, had dared dream of the success story that unfolded. After the jury had elected this model as the most interesting design of 1976, Chopard launched its serial production, which was powered by the 6¾-line hand-wound calibre ETA 2442. In 1993, Caroline Scheufele was inspired by this model. The current co-president increased its sporty flair and paired it with steel and diamonds, an unusual alliance in its day.

As with Happy Diamonds, the sparkling gemstones moved freely between two sapphire crystals. Given the circumstances, it was little wonder that Happy Sport quickly became an icon of the Geneva family company. In the years after 1993, more than a thousand iterations including Happy Fish, Happy Ocean and Happy Snowflakes were created. In 2016, interchangeable bracelets provided even greater freedom. To mark the collection's twenty-fifth anniversary, a woman's watch was presented powered by the 20.4-millimetre proprietary calibre 09.01 C with self-winding rotor.

HAPPY MIT CHOPARD

1976, im Jahr 13 nach Übernahme des Hauses Chopard, präsentierten Karl und Karin Scheufele eine Armbanduhr, die rasch Geschichte schreiben sollte. Beim Premierenmodell der „Happy Diamonds" handelte es sich, man mag es fast nicht glauben, um eine Herrenuhr mit 18-karätigem Weißgoldgehäuse und -band. Inspiriert von einem Wasserfall im Schwarzwald und kreiert als Schaustück für den Design-Award „Goldene Rose von Baden-Baden". Von der anschließenden Erfolgsstory wagten nicht einmal die schmuckaffinen Eigentümer zu träumen. Nachdem die Jury dieses Modell jedoch zum interessantesten Entwurf des Jahres 1976 gekürt hatte, startete Chopard mit der Serienproduktion, ausgestattet mit dem 6¾-linigen Handaufzugskaliber ETA 2442. An dieses Modell knüpfte Caroline Scheufele 1993 an. Die heutige Co-Präsidentin entwickelte eine sportlicher auftretende und damals absolut ungewöhnliche Allianz aus Stahl und Diamanten.

Wie bei der „Happy Diamonds" bewegten sich die funkelnden Edelsteine frei zwischen zwei Saphirgläsern. Dass die „Happy Sport" sehr schnell zu einer Ikone der Genfer Familienmanufaktur avancierte, verwunderte unter diesen Umständen nicht. In den Jahren nach 1993 entstanden mehr als tausend Varianten wie beispielsweise „Happy Fish", „Happy Ocean" oder „Happy Snowflakes". Noch mehr Freiheit brachten 2016 die leicht auswechselbaren Armbänder. Und zum 25. Jubiläum erhielten Frauen auch noch das 20,4 Millimeter kleine Manufakturkaliber 09.01 C mit Rotor-Selbstaufzug.

HAPPY FISH

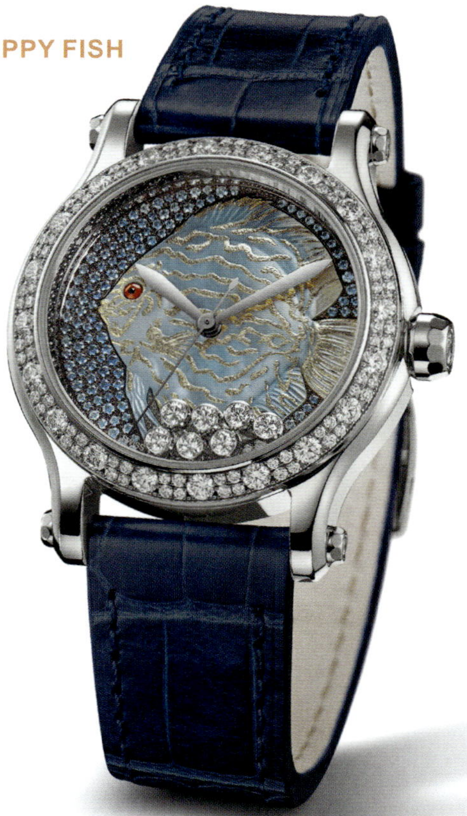

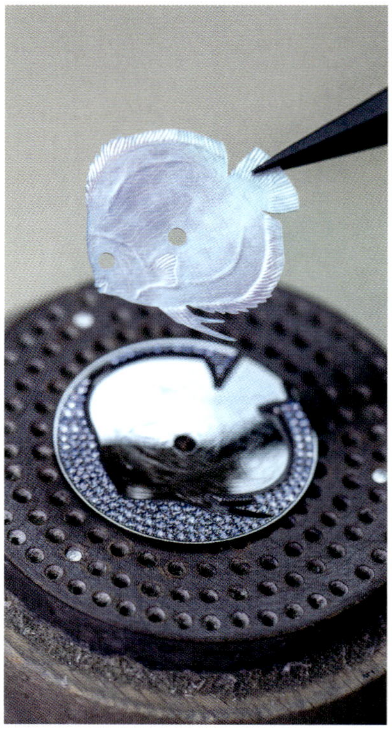

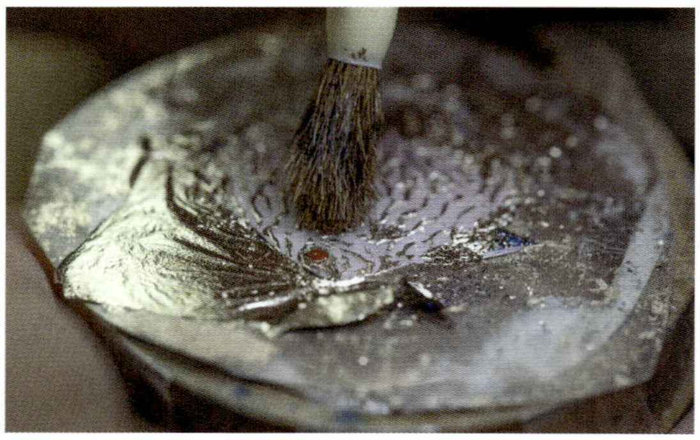

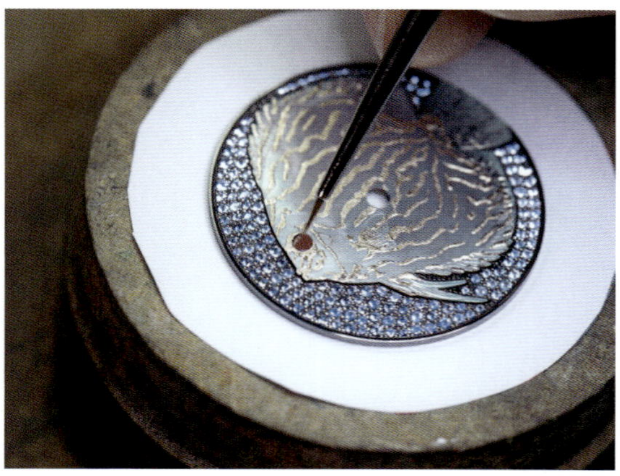

Making of Happy Fish | *Herstellung der Happy Fish*
Right page | **Rechte Seite:** *Happy Sport*

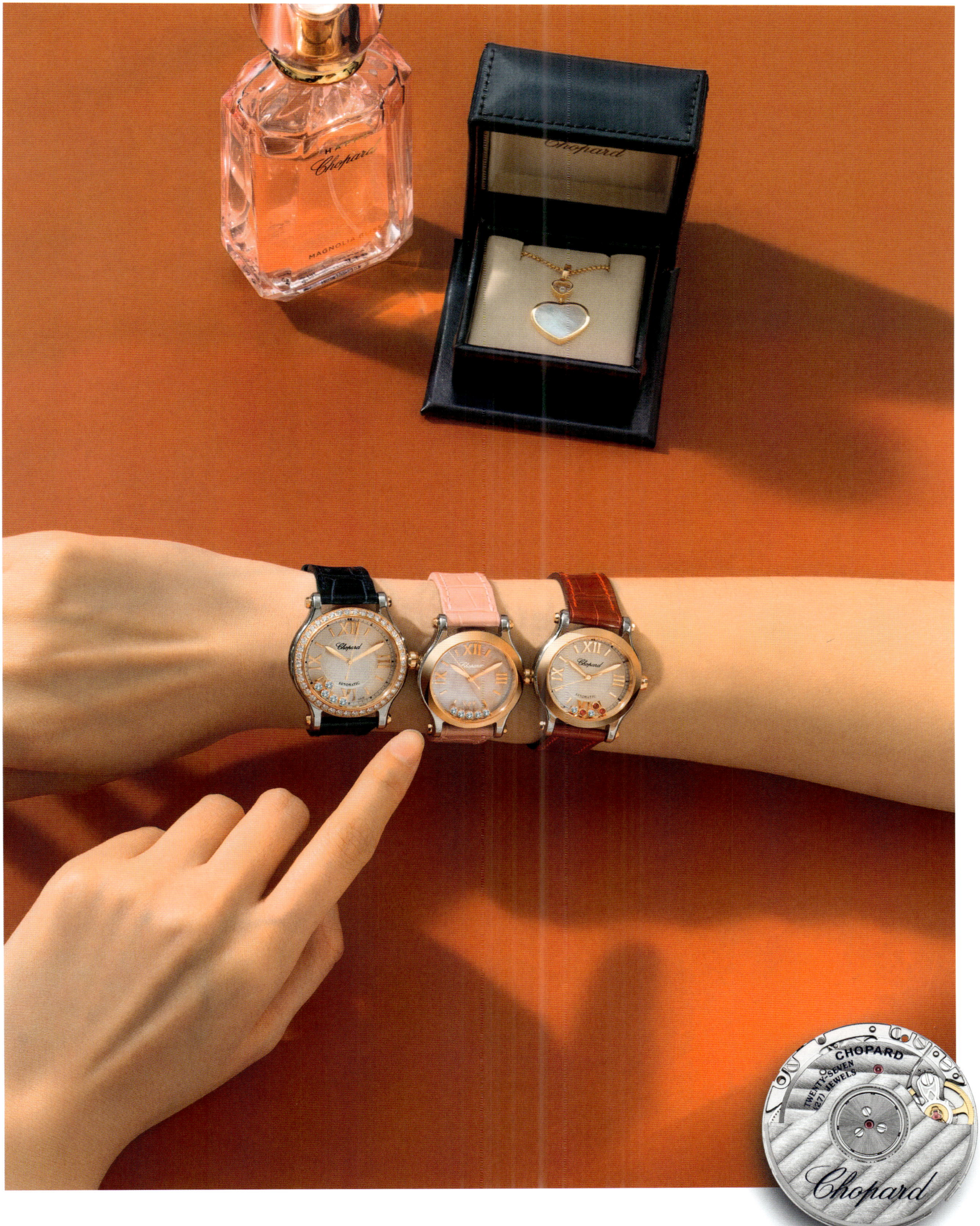
Chopard
Chopard
CHOPARD
TWENTY-SEVEN
(27) JEWELS
Chopard

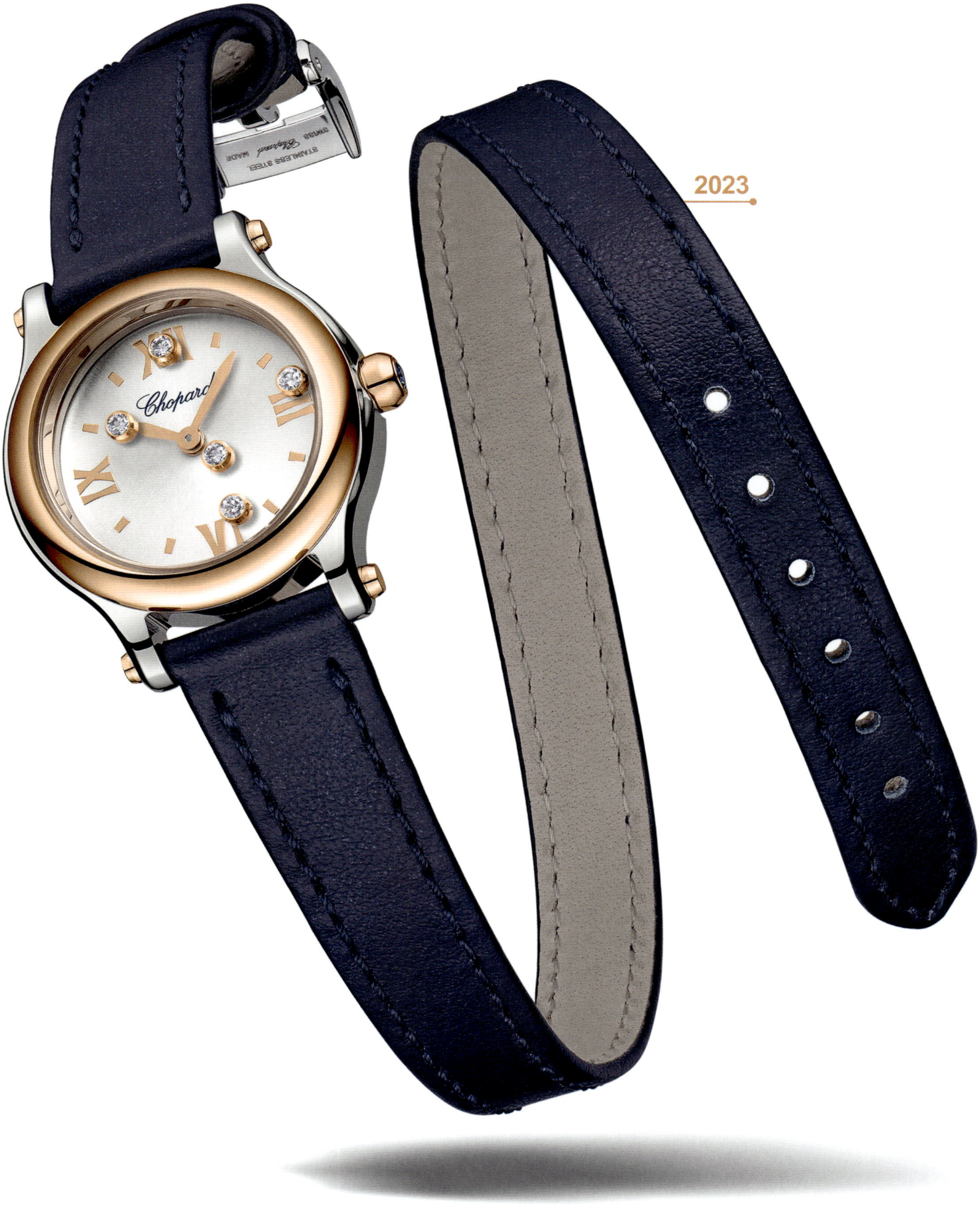
2023
Chopard

1)

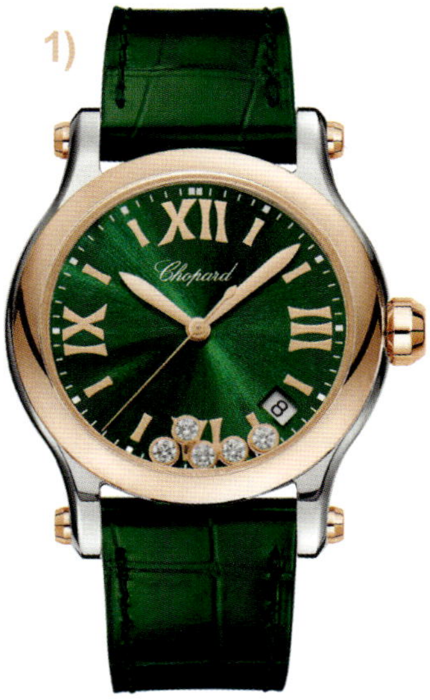

2)

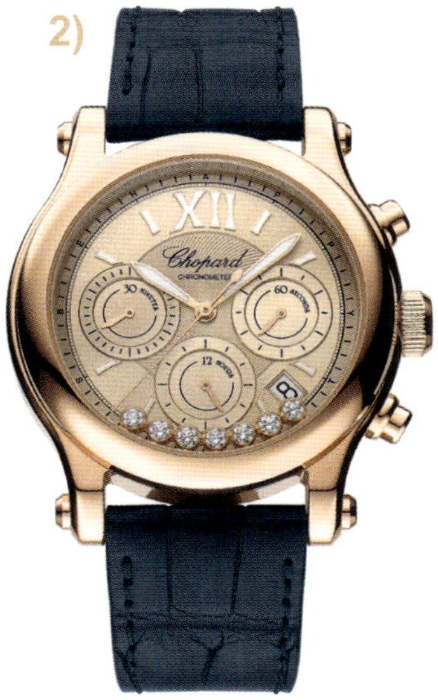

3)

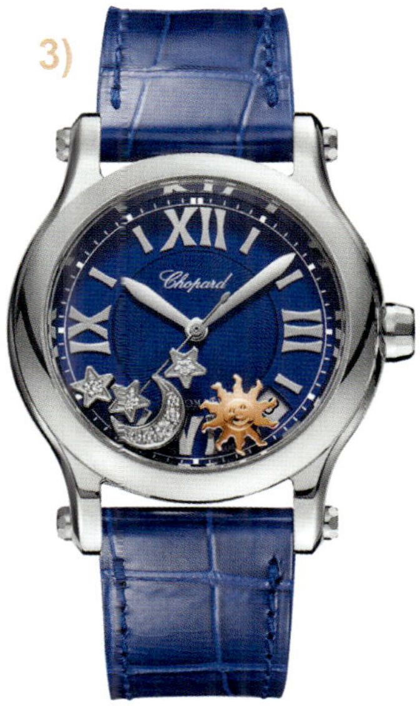

4)

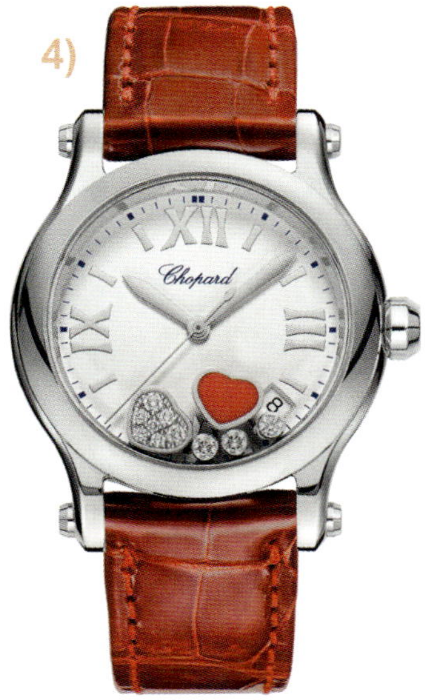

5)

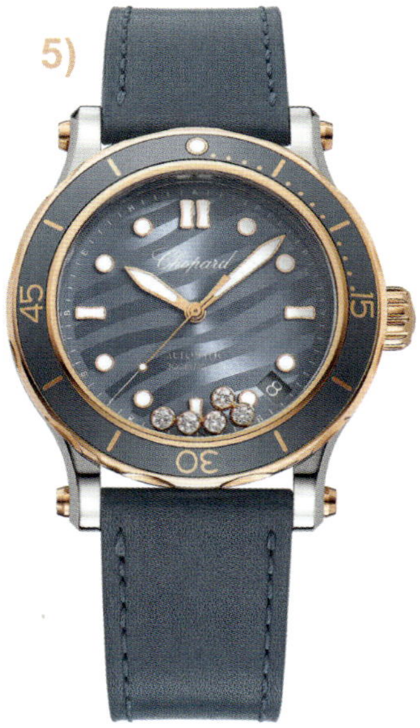

6)

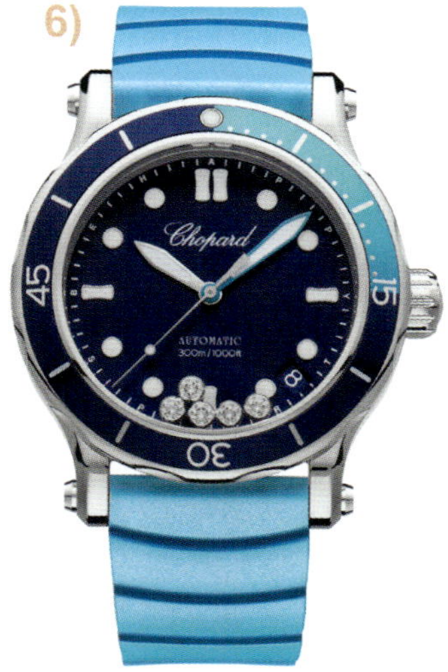

7)

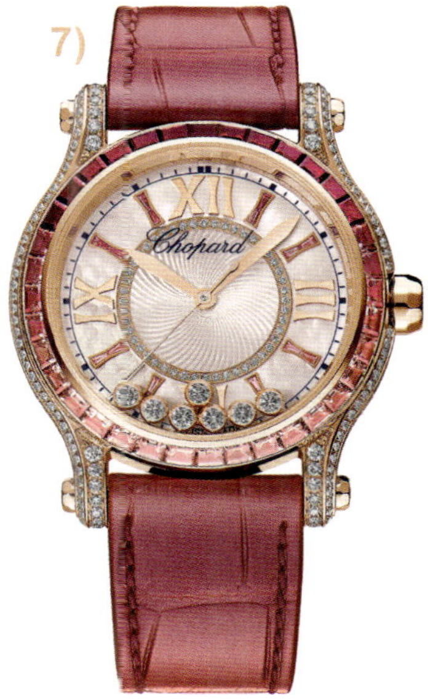

8)

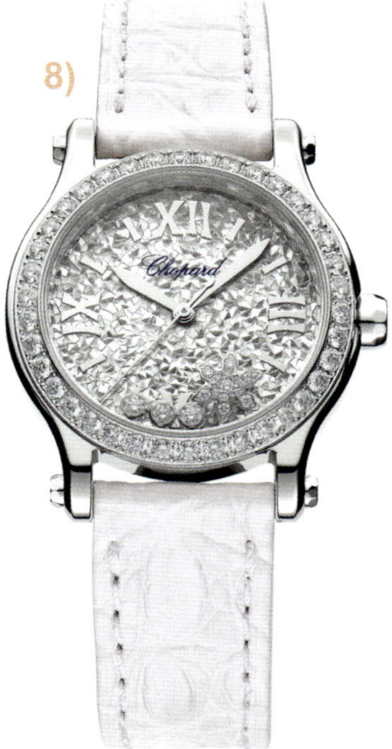

9)

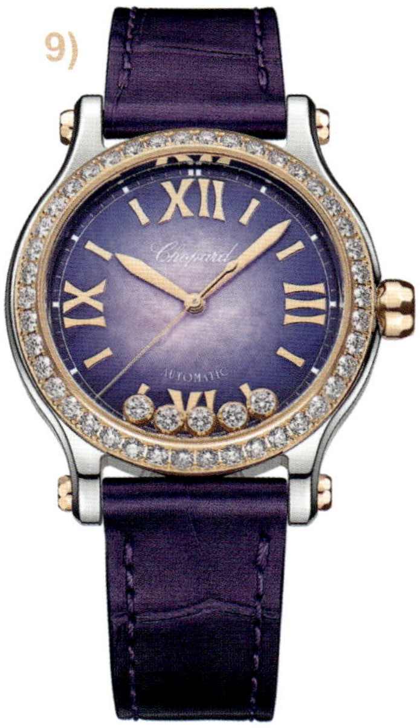

10)

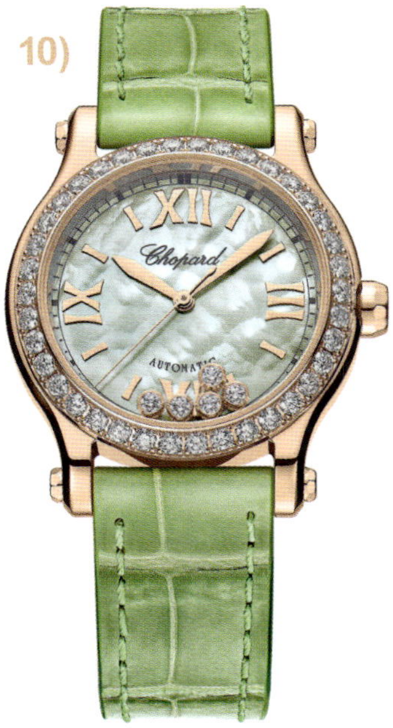

11)

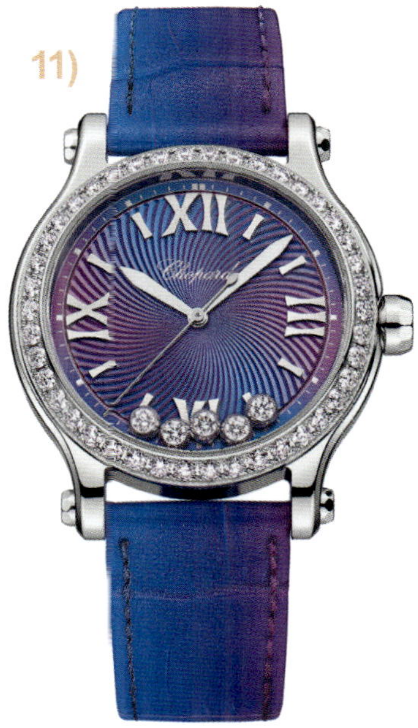

1) Happy Sport Green, 2019
2) Happy Sport Chrono, 2022
3) Happy Sport Sun, Moon and Stars, 2018
4) Happy Hearts, 2019
5) Happy Ocean, 2021
6) Happy Ocean, 2017
7) Happy Sport Rose de Caroline, 2024
8) Happy Snowflakes, 2021
9) Happy Sport, 2023
10) Happy Sport, 2023
11) Happy Sport, 2023

1

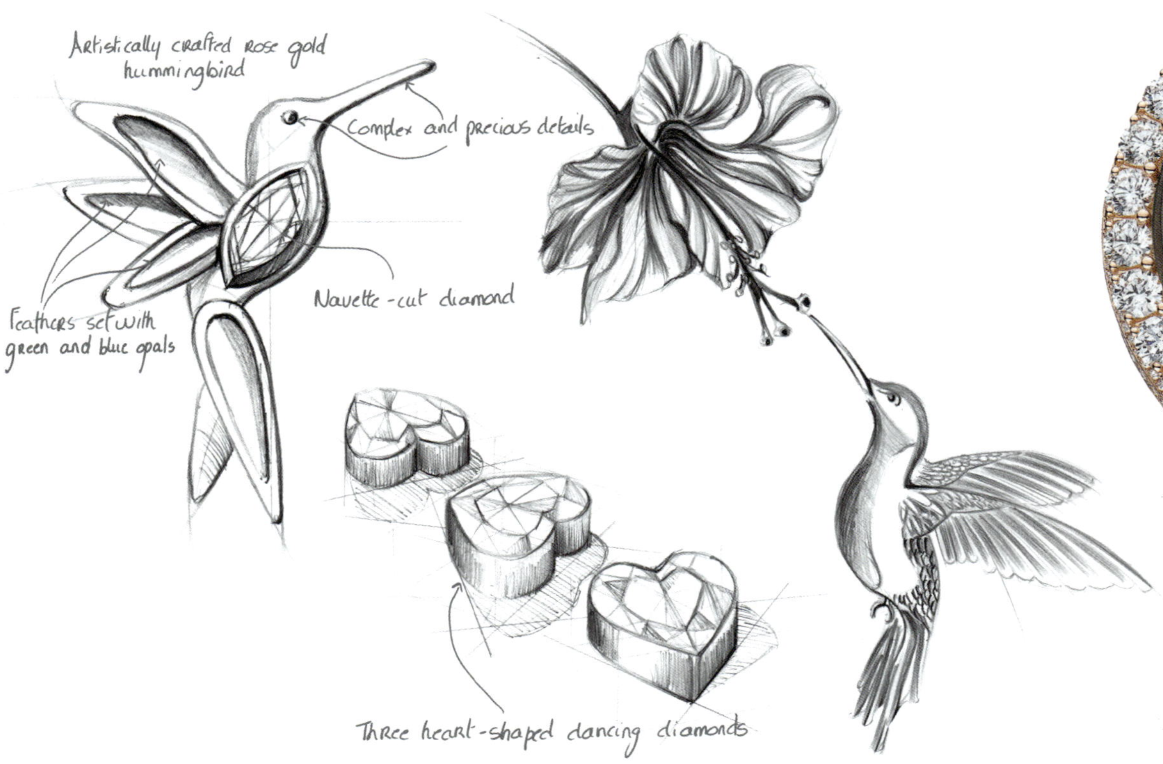
Artistically crafted rose gold hummingbird
Complex and precious details
Feathers set with green and blue opals
Navette-cut diamond
Three heart-shaped dancing diamonds

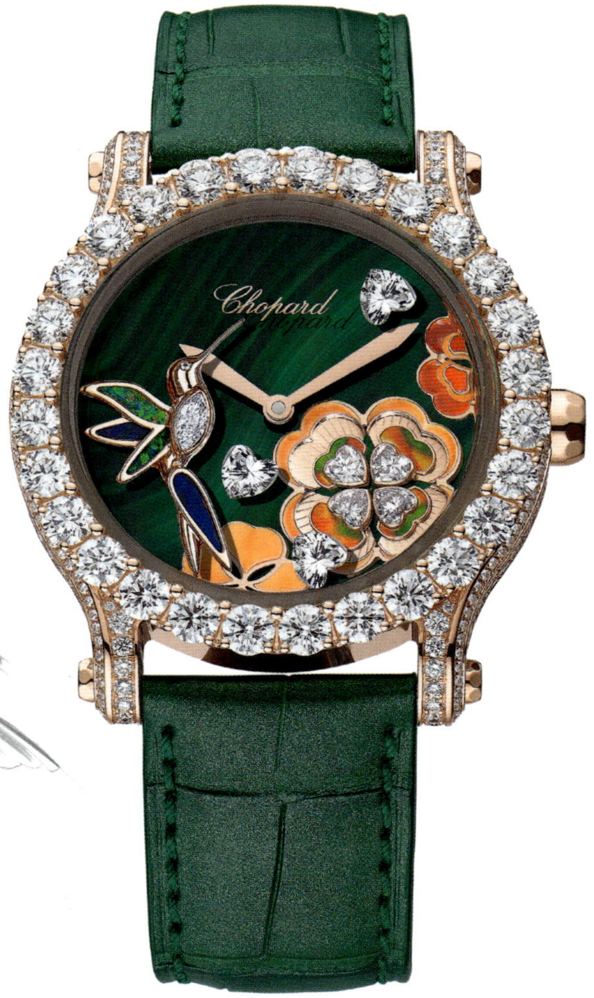
Chopard

2

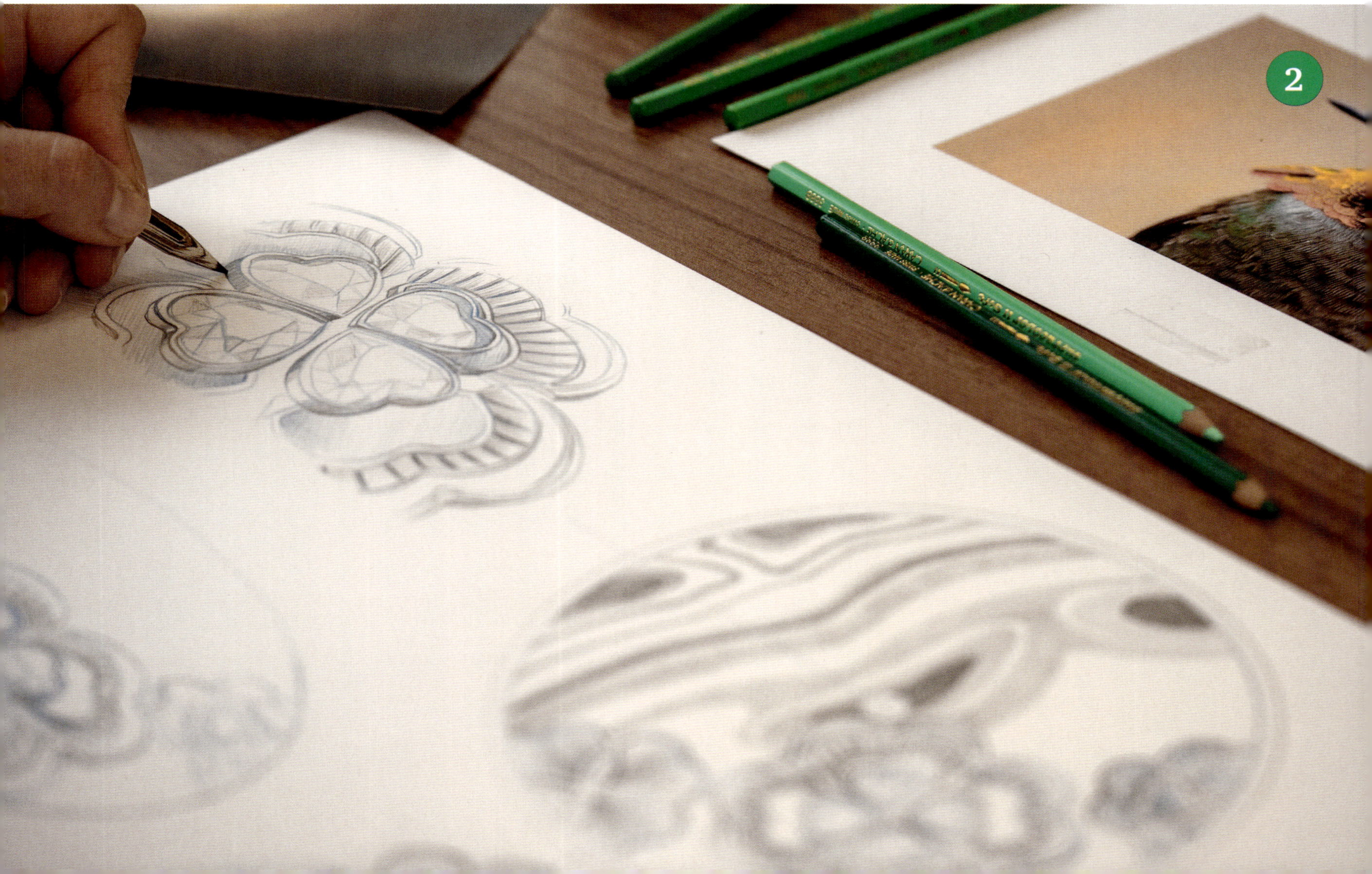

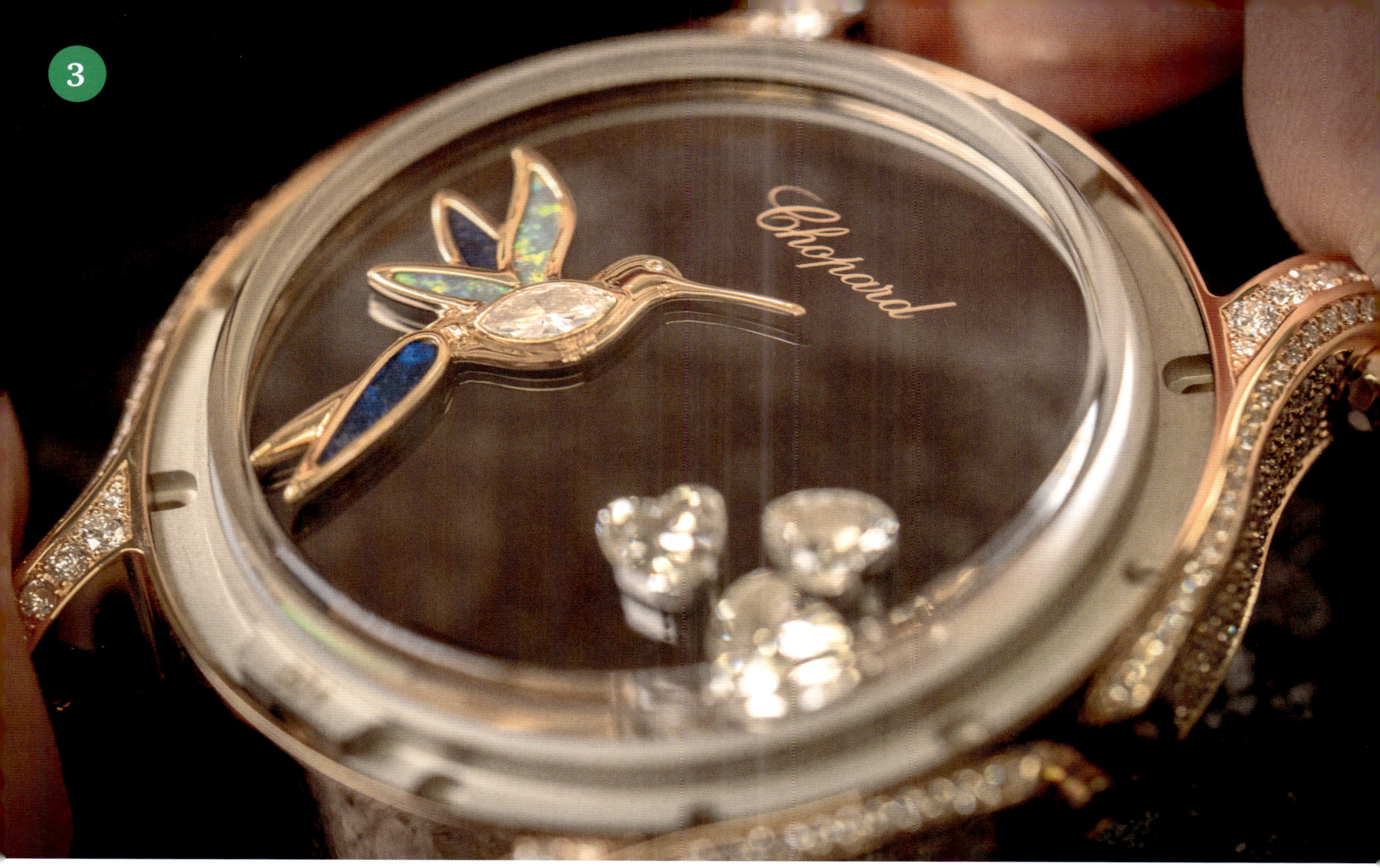

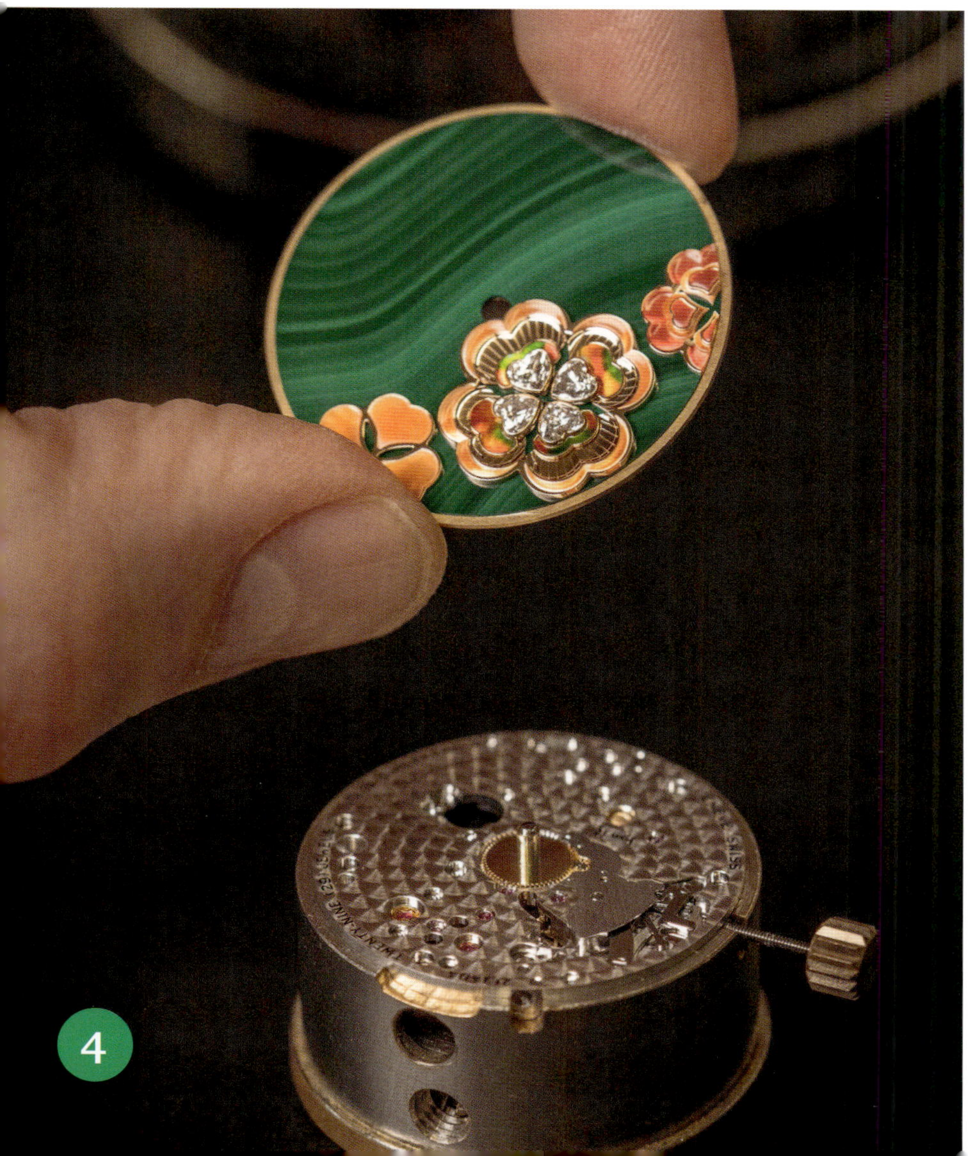

MAKING OF HAPPINESS

1. *Happy Sport Métiers d'Art, limited to eight pieces | limitiert auf acht Stück*

2. *Artisan drawing the fascinating miniature ecosystem | Ein Kunsthandwerker zeichnet das faszinierende Miniatur-Ökosystem*

3. *The case is crafted from ethical gold, magnificently hand-decorated and entirely set with sparkling gems. | Das Gehäuse ist aus ethischem Gold gefertigt, prächtig von Hand verziert und vollständig mit Edelsteinen besetzt.*

4. *The impression of depth exuded by the mineral dial graced with multi-dimensional marquetry inlays makes this watch a veritable work of art. | Die Tiefenwirkung ihres mit drei-dimensionalen Intarsien gefertigten Mineralzifferblatts macht die Uhr zu einem wahren Kunstwerk.*

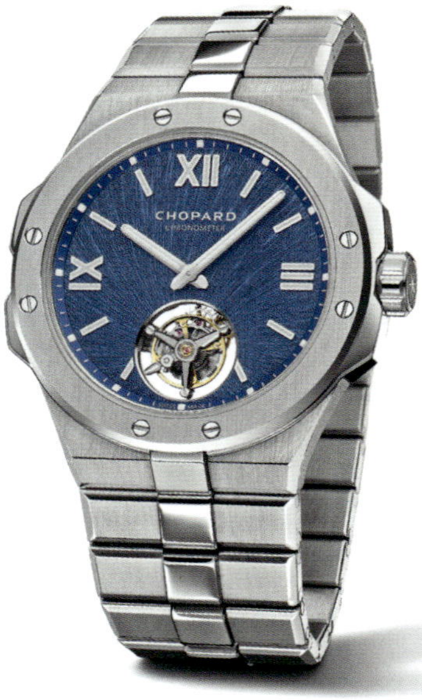

FLYING TOURBILLON,
2022

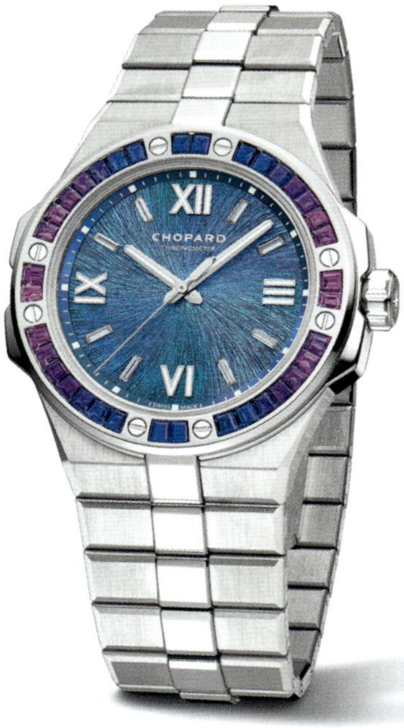

SUMMIT, 2023

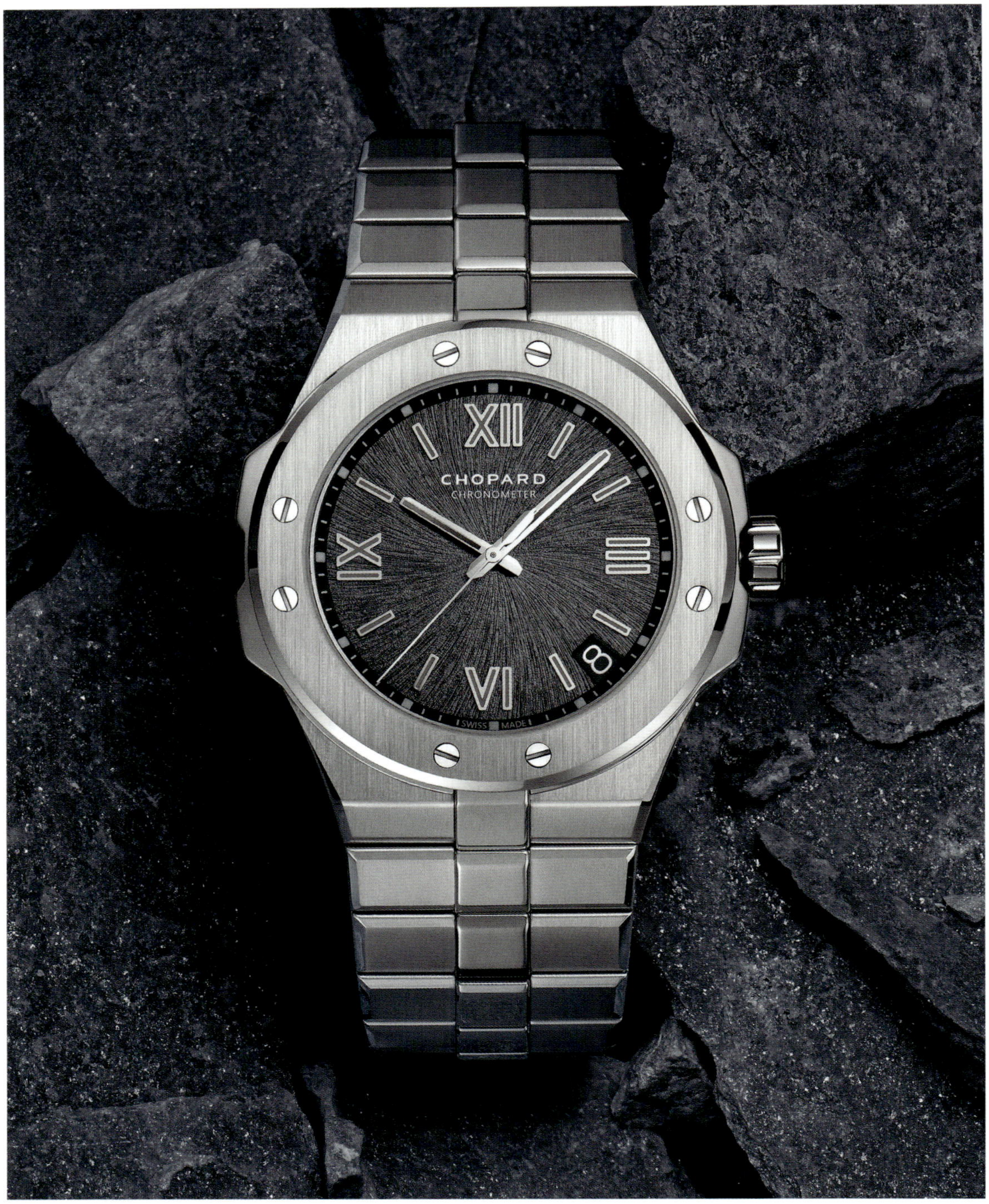

Alpine Eagle, 2019

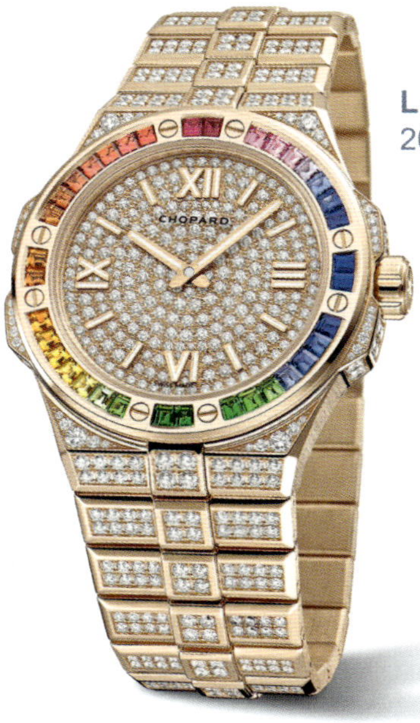

LARGE FROZEN,
2022

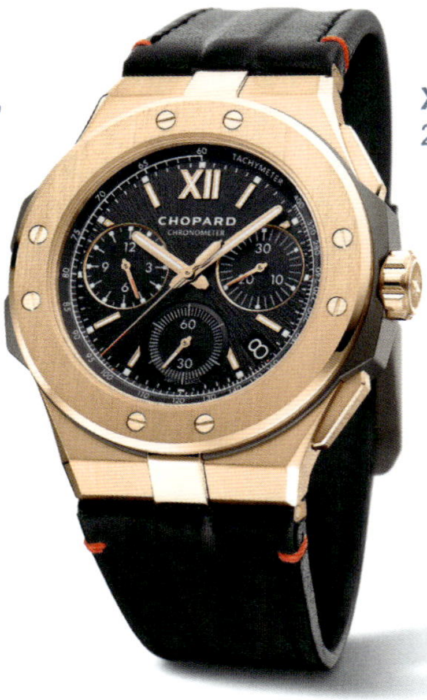

XL CHRONO,
2021

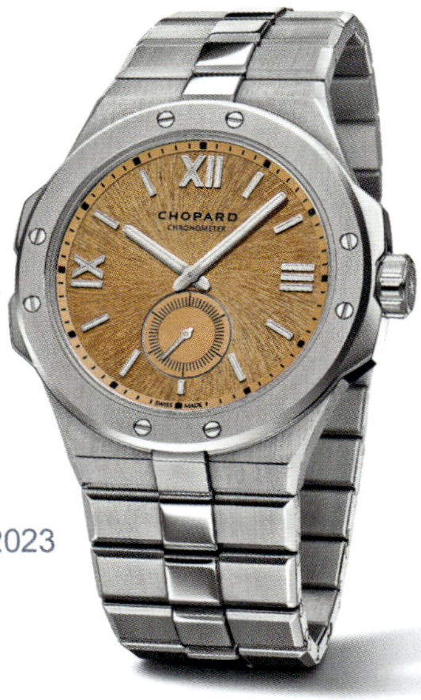

XPS, 2023

CHOPARD
ALPINE EAGLE

CHOPARD AND THE EAGLE'S EYE

For many years, the St. Moritz, designed by Karl-Friedrich Scheufele in 1980, epitomised Chopard's talent for combining sportiness with elegance. However, time inevitably took its toll. The creative spirit behind its successor was Karl-Fritz Scheufele. The son of the current co-president had discovered the St. Moritz in the archive and together with his grandfather worked on its rebirth. In 2019, the Alpine Eagle was ready for production.

The eight screws securing the bezel pay tribute to the past. The bezel however now displays a classic roundness. The spirit of the St. Moritz also lives on in the case flanks. The closest connection between past and present is evident in the purist design of the link bracelet. Available in several sizes, the premiere model shared its debut with Lucent Steel A223 by Voest-Alpine. Its development and the technology underpinning its industrial production took no less than four years. Thanks to its hypoallergenic alloy, the material used for the case and bracelet possesses the qualities of skin-friendly surgical steel. With a hardness of 223 Vickers degrees, the material is 50 per cent more resistant to abrasion than conventional steel. Logically, this reduces the risk of unsightly scratches. Its homogeneous crystal structure and purity evoke parallels with platinum or white gold.

The successful Alpine Eagle collection has grown continuously. Steel is now joined by gold, the link bracelet by a rubber strap. Lovers of complications can choose between a chronograph or tourbillon.

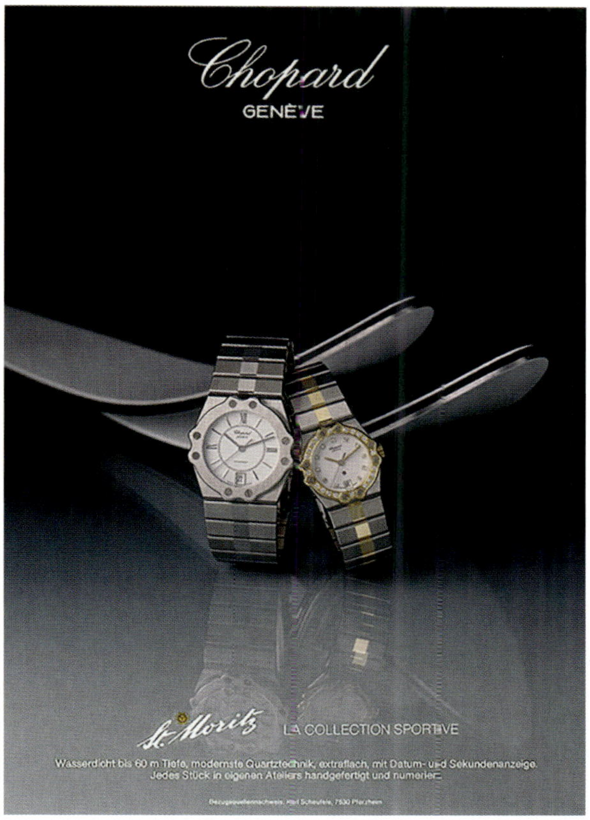

CHOPARD LÄSST DEN ADLER LANDEN

Über viele Jahre hinweg repräsentierte die 1980 von Karl-Friedrich Scheufele gestaltete „St. Moritz" die sportliche und gleichzeitig auch elegante Seite von Chopard. Irgendwann nagte der Zahn der Zeit an diesem Modell. Die Idee zu einem adäquaten Nachfolger ist Karl-Fritz Scheufele zu verdanken. Der Sohn des aktuellen Co-Präsidenten hatte besagte „St. Moritz" im Archiv entdeckt und zusammen mit seinem Großvater entsprechende Pläne entwickelt. 2019 erlangte die „Alpine Eagle" ihre Serienreife.

Bezüge zur Vergangenheit demonstrieren beispielsweise die acht Schrauben zur Befestigung des Glasrands. Der tritt nun jedoch klassisch rund in Erscheinung. Auch die Gehäuseflanken lassen das Modell „St. Moritz" andeutungsweise weiterleben. Den engsten Schulterschluss zwischen einst und jetzt gibt das schnörkellose Gliederband zu erkennen. Mit dem in mehreren Größen erhältlichen Premierenmodell debütierte bei Chopard auch „Lucent Steel A223" von Voest-Alpine. Nicht weniger als vier Jahre nahmen die Entwicklung und Verfahrenstechnologie zur industriellen Fabrikation in Anspruch. Dank seiner hypoallergenen Legierung besitzt der für Gehäuse und Band verwendete Werkstoff die Qualitäten hautfreundlichen Chirurgenstahls. Mit 223 Vickersgraden Härte ist das Material 50 Prozent abriebfester als herkömmliche Stähle. Das mindert logischerweise die Gefahr unschöner Kratzer. Bedingt durch die homogene Kristallstruktur und die damit einhergehende Reinheit werden Erinnerungen an Platin oder Weißgold wach. Mittlerweile ist die erfolgreiche „Alpine Eagle"-Kollektion kontinuierlich gewachsen. Zu Stahl gesellt sich Gold, zum Glieder- ein Kautschukband. Chronograph oder Tourbillon sprechen Menschen mit Passion für uhrmacherische Zusatzfunktionen an.

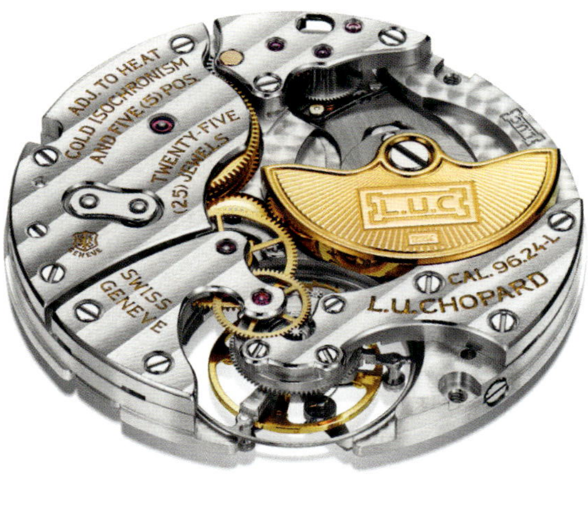

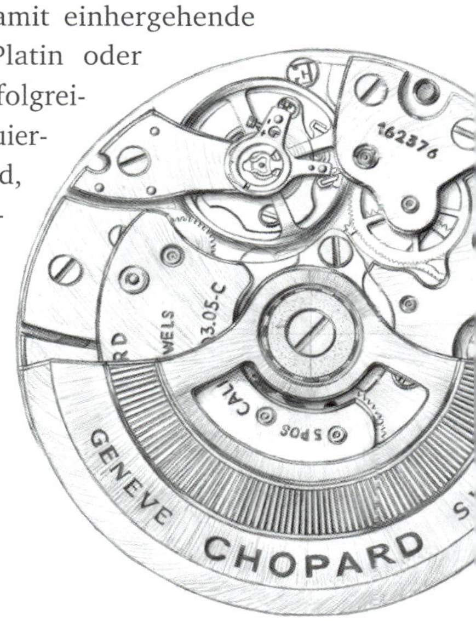

Top / Oben: Advertisement for the St. Moritz, 1980s / Werbeanzeige für das Modell St. Moritz, 1980er-Jahre
Right / Rechts: Cal. 03.05-C (sketch)
Left / Links: Cal. 96.24-L

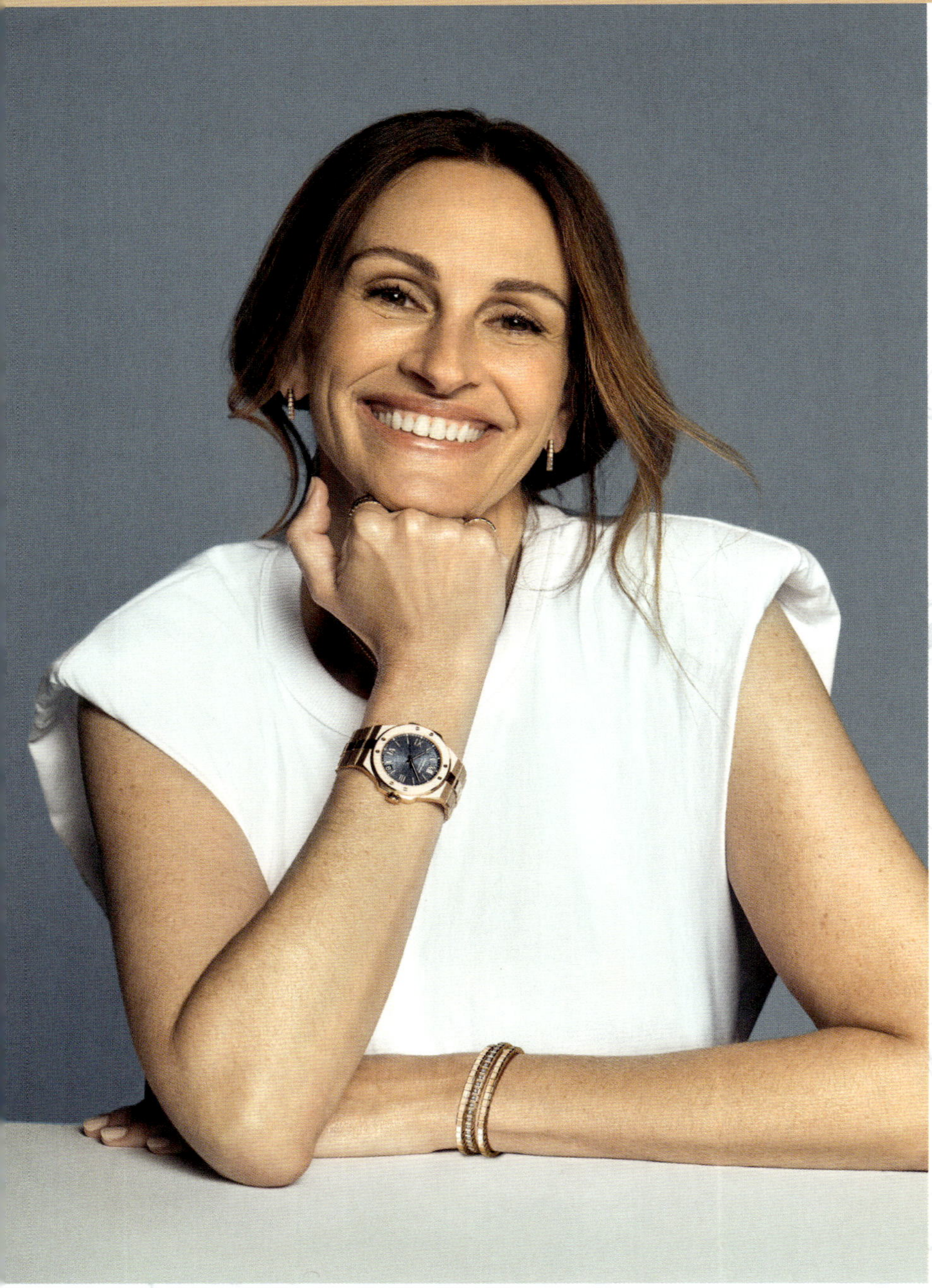

Actress Julia Roberts is the global ambassador for Chopard since 2023. | Schauspielerin Julia Roberts ist seit 2023 Chopards Markenbotschafterin.

CHOPARD LOVES CINEMA: ALPINE EAGLE COLLECTION, 2023

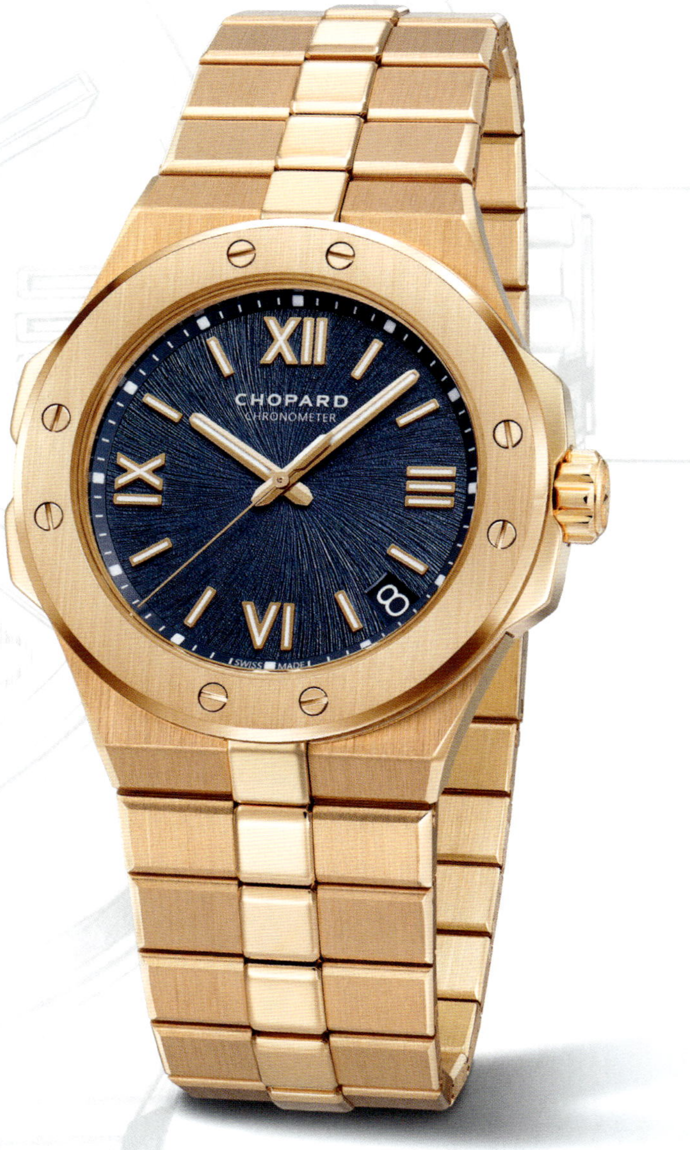

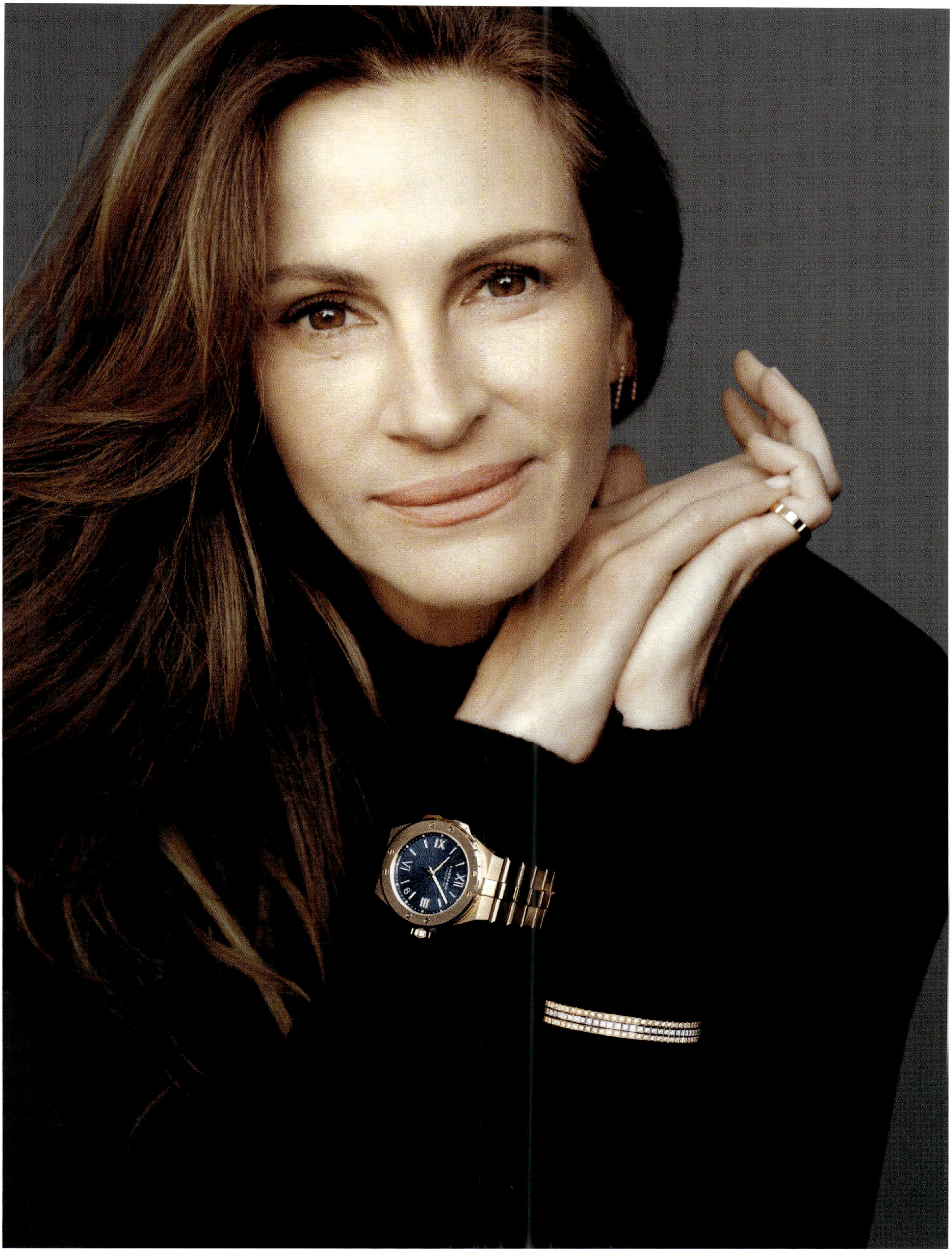

CHOPARD 1988

MILLE MIGLIA

WATCH FOR THE MILLE MIGLIA

The nickname "la corsa più bella del mondo" of the legendary Mille Miglia car race is not without justification. Starting in the time-honoured city of Brescia, "the most beautiful race in the world" covers some 1,600 kilometres of Italy's scenic country roads and picturesque towns to Rome and back again. From 1927 to 1957, the race really was a do-or-die affair in the truest sense. Daring manoeuvres and neck-breaking speeds were daily fare. They led to some serious accidents and consequently to the race's demise. Its revival was staged in 1988 as an elite vintage car rally. Only vehicles that participated in the original race are eligible to take part. Nevertheless, the list of aspirants is regularly longer than that of the lucky participants. Nowadays, speed takes a back seat. The focus is on precision and reliability. Something that the Scheufele family as vintage car enthusiasts know well. In 1963, they bought the Chopard watch brand, which has sponsored the racing event since its renaissance. They also regularly take part in the rally themselves with classic cars.

UHR ZUR MILLE MIGLIA

Keineswegs zu Unrecht trägt das legendäre Autorennen „Mille Miglia" den Beinamen „la corsa più bella del mondo". Nach dem Start in der altehrwürdigen Stadt Brescia führt „die schönste Strecke der Welt" ungefähr 1.600 Kilometer weit durch malerische Ort- und Landschaften Italiens bis nach Rom und wieder zurück. Von 1927 bis 1957 ging es bei besagter Rallye im wahrsten Wortsinn ums Ganze. Waghalsige Manöver und Höchstgeschwindigkeiten waren deshalb an der Tagesordnung. Sie führten zu teils schweren Unfällen und in der Folge zum Aus der Veranstaltung. Das Revival erfolgte 1988 als elitäre Oldtimer-Ausfahrt. Zugelassen sind ausschließlich früher partizipierende Fahrzeugtypen. Dennoch ist die Liste der Bewerber regelmäßig länger als jene der auserkorenen Teilnehmer. Tempo spielt heutzutage eine untergeordnete Rolle. Im Vordergrund stehen Präzision und Zuverlässigkeit. Dessen ist sich auch die oldtimerbegeisterte Familie Scheufele bewusst. 1963 kaufte sie bekanntlich die Uhrenmarke Chopard, welche die Rennveranstaltung seit ihrer Renaissance sponsert. Überdies nimmt sie regelmäßig selbst mit automobilen Klassikern teil.

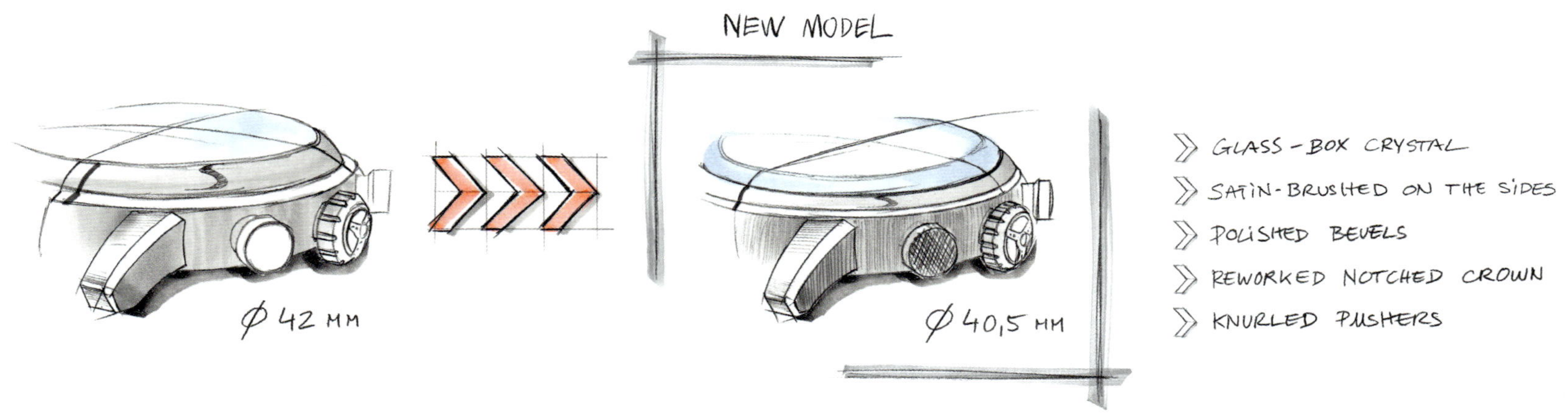

And of course, for discerning Mille Miglia participants an exclusive watch is a must. The drivers do not have to exert themselves, because they are presented with the new annual edition of the Mille Miglia line for free. Ordinary mortals can of course also enjoy this fascination. The race models, limited to the number of the respective year, can be purchased as well as several other wristwatches from the wide-ranging collection. Past models have long since acquired cult and collector's status.

Wer bei der „Mille Miglia" etwas auf sich hält, trägt natürlich eine Armbanduhr nobler Provenienz. Die Fahrerinnen und Fahrer tun sich dabei leicht, denn sie erhalten gratis ein Exemplar der jährlich neu gestalteten „Mille-Miglia"-Linie. Normalsterbliche können diese Faszination natürlich auch erleben. Die auf die jeweilige Jahreszahl limitierten Rallye-Modelle sind ebenso käuflich zu erwerben wie zahlreiche andere Armbanduhren der inzwischen breit gefächerten Kollektion. Exemplare aus der Vergangenheit besitzen in diesem Zusammenhang längst schon Kult- und Sammlerstatus.

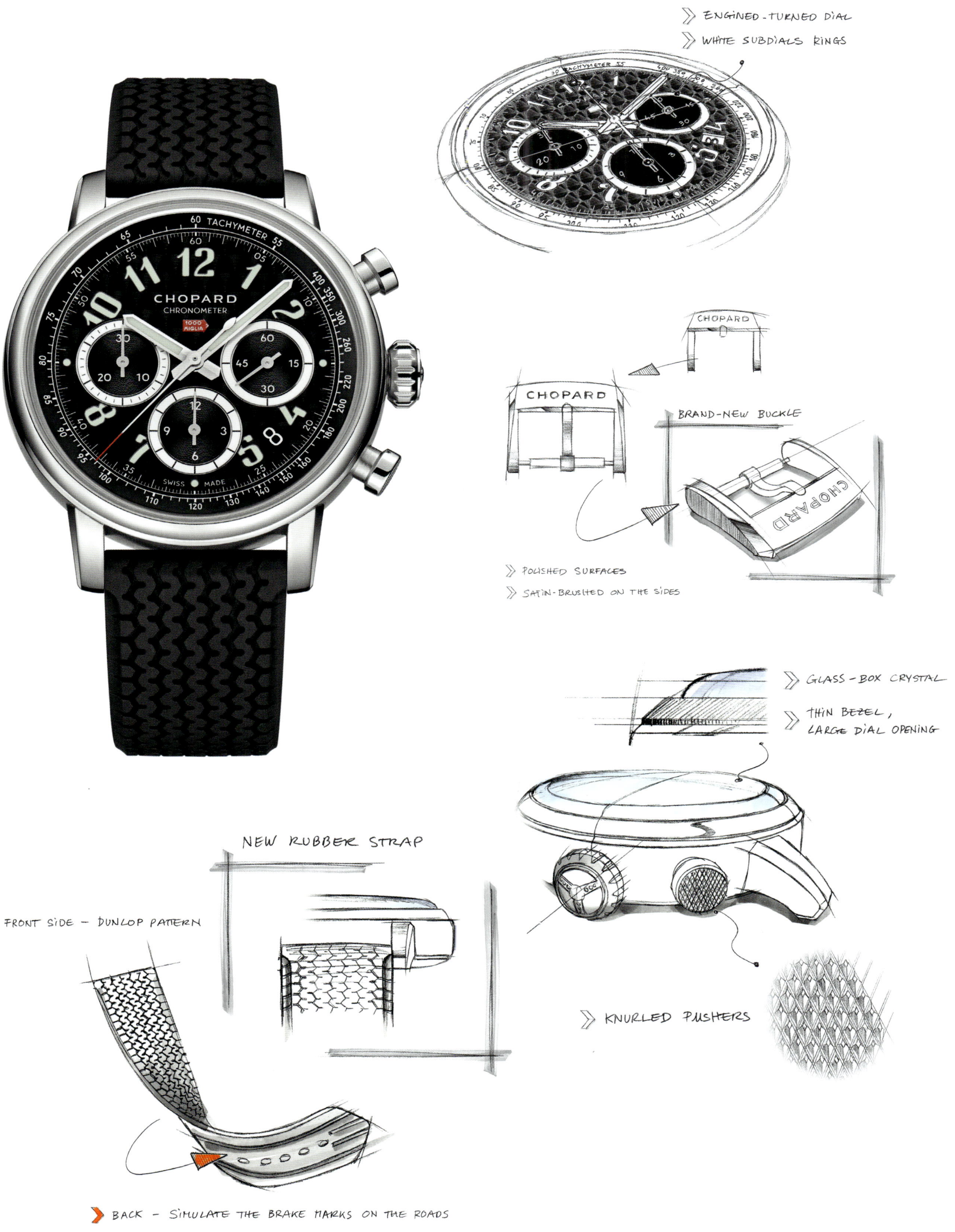
ENGINED-TURNED DIAL
WHITE SUBDIALS RINGS
CHOPARD
BRAND-NEW BUCKLE
POLISHED SURFACES
SATIN-BRUSHED ON THE SIDES
GLASS-BOX CRYSTAL
THIN BEZEL, LARGE DIAL OPENING
NEW RUBBER STRAP
FRONT SIDE – DUNLOP PATTERN
KNURLED PUSHERS
BACK – SIMULATE THE BRAKE MARKS ON THE ROADS

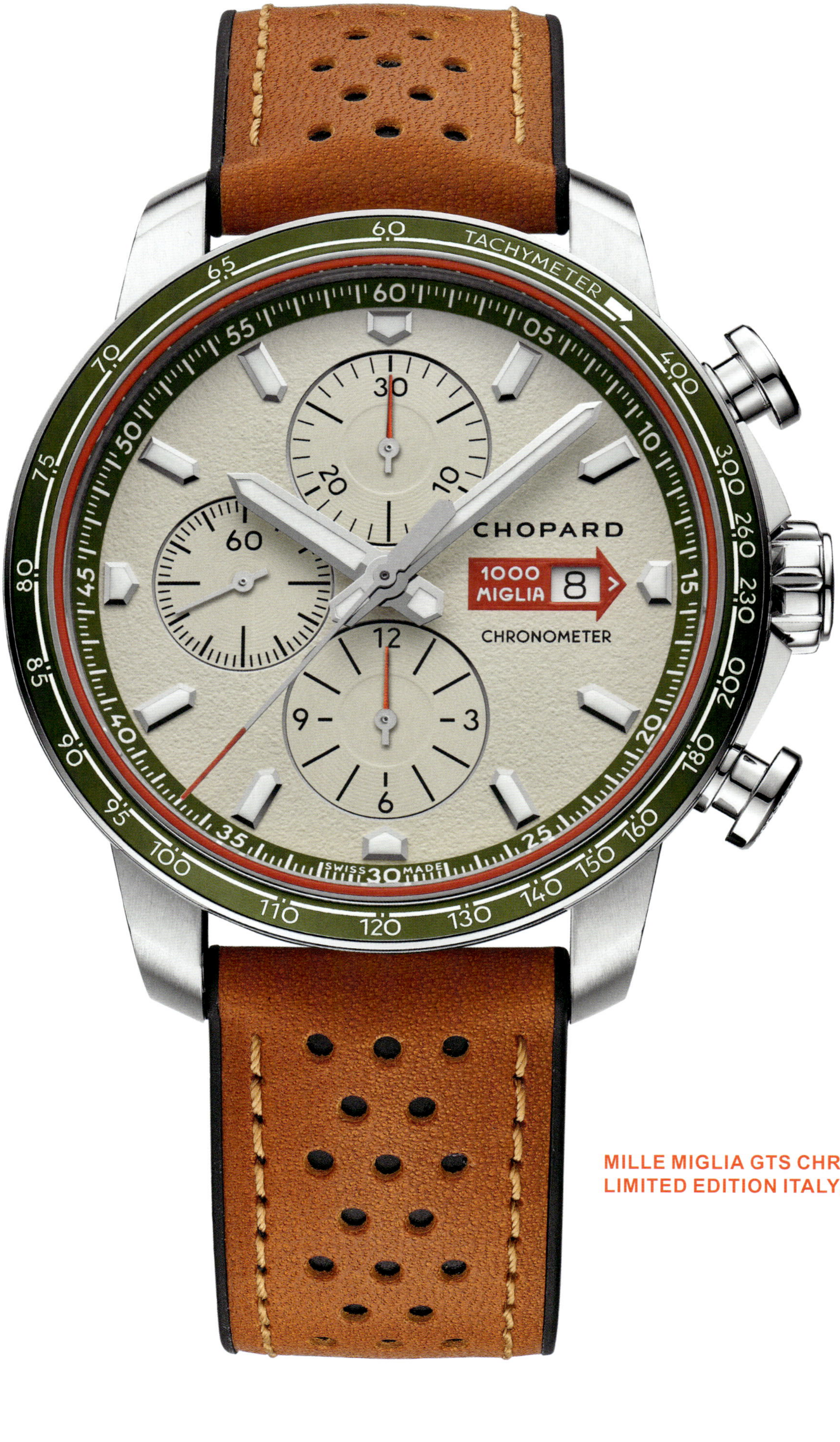

MILLE MIGLIA GTS CHRONO LIMITED EDITION ITALY, 2023

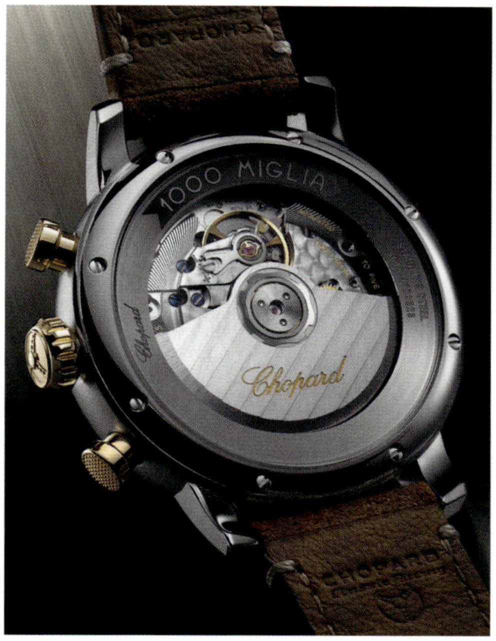

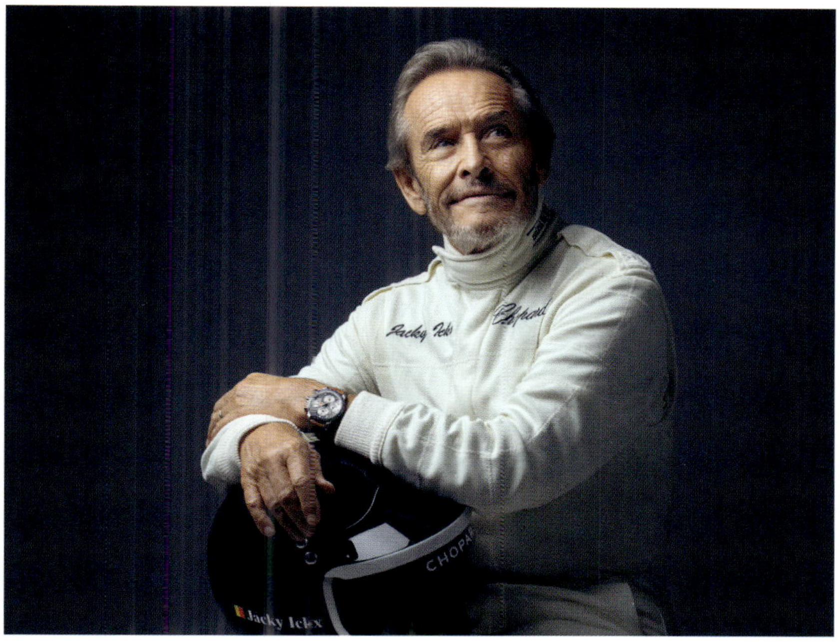

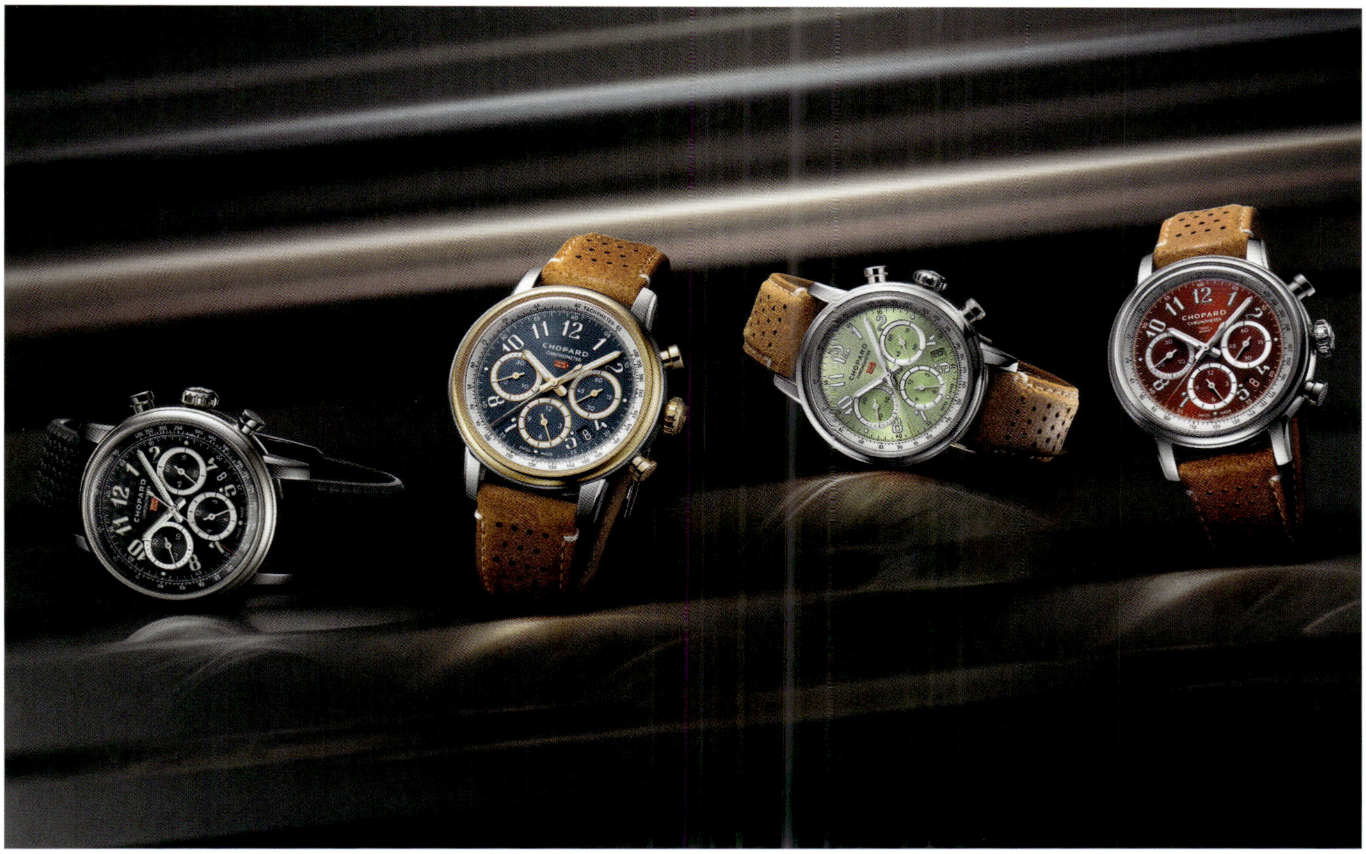

From top left | Von oben links: *Mille Miglia Classic Chronograph, caseback | Racing driver Jacky Ickx | Classic Chronograph collection 2023*

CHRONOMÈTRE FB 2RE

THE PAST IS ALIVE

In recent times, the long-forgotten and more than 250-year-old name of Ferdinand Berthoud made a comeback with a fresh chronometric face. Arbiter of the renaissance was Karl-Friedrich Scheufele. In Fleurier, the co-president of Chopard birthed the new, exquisite Chronométrie Ferdinand Berthoud, which was launched in 2015. Five kilometres away from the birthplace of the great watchmaker, who enjoyed a meteoric career in Paris, a small team employs time-honoured methods to develop, craft and market a highly exclusive, very limited edition of watches. Based at the same address, the Chopard manufactory ably assists. It produces components with the help of a formidable array of high-precision machines.

Historical aspects played a significant role in the development of the new watches and their hand-wound movements. In 1770, Ferdinand Berthoud had acquired the title of master watchmaker and horologist-mechanic by appointment to the King and the Navy. One of his outstanding creations for sea navigation was the marine chronometer No. 6, launched in 1777. Chronométrie Berthoud used it as the model for the FB 2 chronometer. Its horological attractions include a traditional fusée-and-chain system to ensure constant force transmission. In contrast to earlier designs, the barrel and the fusée are suspended. Berthoud takes the constant force system to new heights with a remontoir d'égalité, which delivers precisely defined amounts of energy in one-second cycles. The 0.01-gram second hand thus executes accurate jumps. In their design of the movement, the engineers also followed in the footsteps of Ferdinand Berthoud with a power reserve of some 50 hours. Ticking at a classical rate of 2.5 hertz, the timepiece consists of 26 bridges held by ten steel pillars. This design is typical of the marine chronometers manufactured by Ferdinand Berthoud in the eighteenth century. The crown is crafted to ease hand-winding. As soon as the mainspring is fully wound, a safety device prevents overwinding. The three hands rotate on a hand-finished two-piece enamel dial. Another touch that keeps the brand's glorious past alive. An official chronometer certificate from the COSC Testing Institute attests to the precision of this fine timepiece.

DIE VERGANGENHEIT LEBT

Sehr lange ist es noch nicht her, dass der mehr als 250 Jahre alte, zwischenzeitlich jedoch weitgehend in Vergessenheit geratene Name Ferdinand Berthoud wieder ein chronometrisches Gesicht bekam. Zu verdanken ist die Renaissance Karl-Friedrich Scheufele. In Fleurier rief der Co-Präsident des Hauses Chopard die neue und sehr feine Chronométrie Ferdinand Berthoud ins Leben, 2015 ging sie an den Start. Fünf Kilometer entfernt vom Geburtsort des großen Uhrmachers, der in Paris eine beispiellose Karriere machte, beschäftigt sich erneut eine kleine Equipe nach alter Väter Sitte mit der Entwicklung, Herstellung und dem Vertrieb streng limitierter und daher sehr exklusiver Uhr-Editionen. Tatkräftige Unterstützung leistet die im gleichen Haus residierende Chopard-Manufaktur. Dort entstehen Rohteile mit Hilfe eines beachtlichen Parks an Präzisionsmaschinen.

Bei der Entwicklung neuer Uhrenmodelle samt der zugehörigen Handaufzugswerke spielten historische Aspekte eine wichtige Rolle. 1770 hatte Ferdinand Berthoud den Titel „Meisteruhrmacher und -mechaniker des Königs und der Marine" erlangt. Zu seinen herausragenden Kreationen für die Navigation auf hoher See gehörte der 1777 fertiggestellte Marinechronometer Nr. 6. Dieser diente der Chronométrie Berthoud als Vorbild für den Chronometer „FB 2".

Zur Palette seiner uhrmacherischen Merkmale gehört ein überliefertes Kette-Schnecke-System für konstante Antriebskraft. Im Gegensatz zu früheren Konstruktionen sind das Federhaus und die Schnecke nur einseitig, also fliegend gelagert. Durch einen Gleichmäßigkeitsaufzug, der Impulse im Sekundentakt liefert, treibt Berthoud das Konstantkraftsystem noch auf die Spitze. Folglich vollzieht der 0,01 Gramm leichte Sekundenzeiger akkurate Sprünge. Auf den Spuren Ferdinand Berthouds wandern die Techniker schließlich auch bei der Architektur des Handaufzugswerks mit rund 50 Stunden Gangautonomie. Das Gestell des mit klassischen 2,5 Hertz tickenden Mikrokosmos besteht aus insgesamt 26 Brücken. Zehn stählerne Pfeiler stellen die Verbindung her. Diese Bauweise ist charakteristisch für die von Ferdinand Berthoud im 18. Jahrhundert fabrizierten Marinechronometer. Eine griffige Krone erleichtert das manuelle Aufziehen. Sobald die Zugfeder voll gespannt ist, unterbindet ein Drehmomentbegrenzer weitere Kraftzufuhr. Die drei Zeiger drehen vor einem handgefertigten zweiteiligen Emailzifferblatt. Auch dieses Merkmal hält die großartige Vergangenheit wach. Ein amtliches Chronometerzertifikat der Prüfbehörde COSC bescheinigt die Präzision des feinen Zeitmessers.

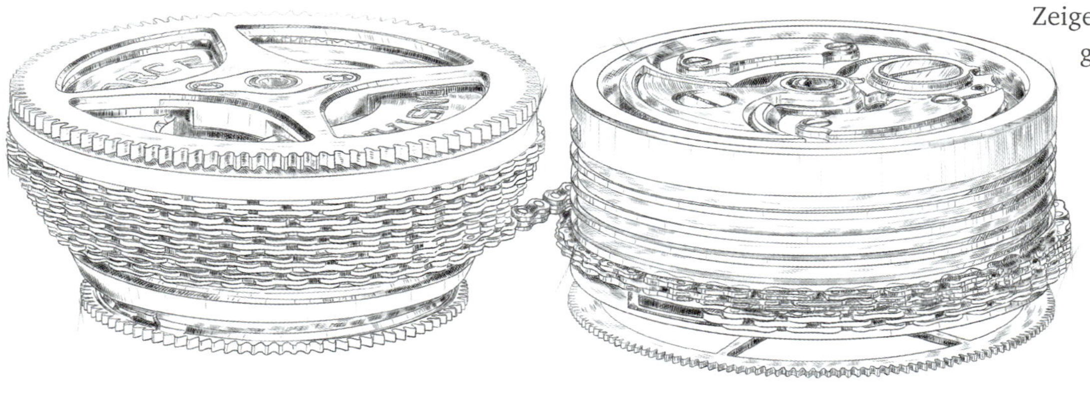

Left: *Fusee-and-chain mechanism* | ***Links:*** *Kette-Schnecke-Mechanismus*

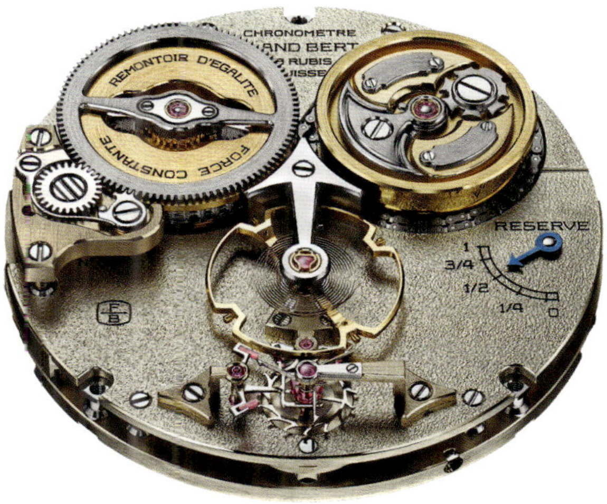

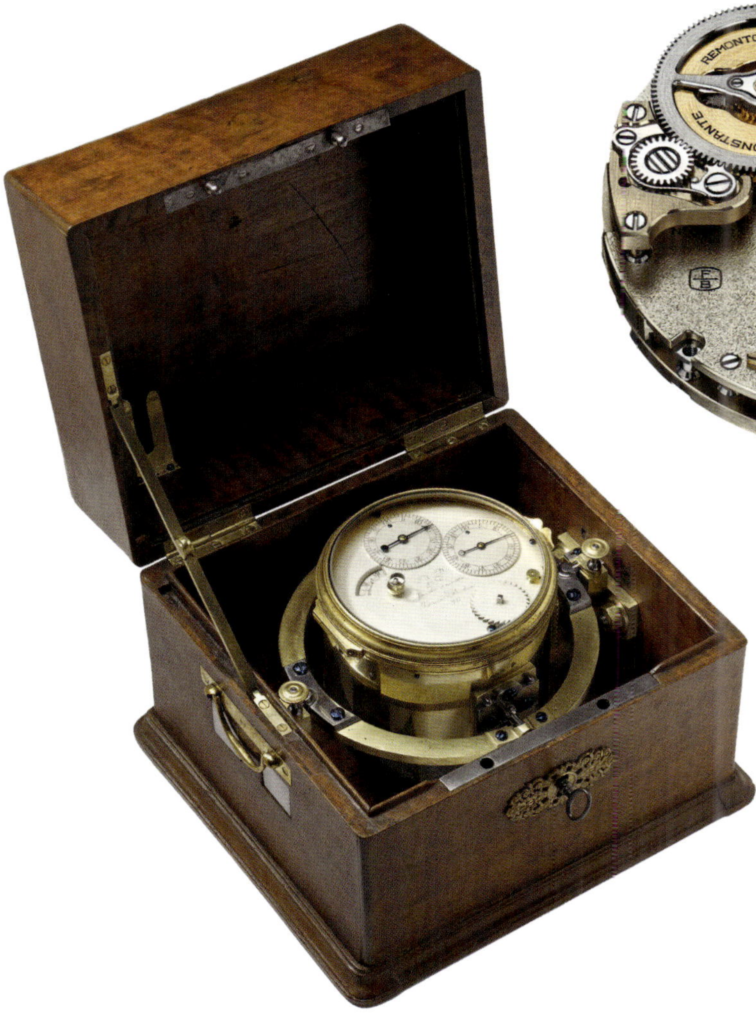

Top | Oben: Calibre FB-RE.FC, 2020
Left: Marine Clock M.M No. 6, Paris 1777, L.U.CEUM collection, Fleurier |
Links: Marinechronometer Nr. 6, Paris 1777, L.U.CEUM-Kollektion, Fleurier

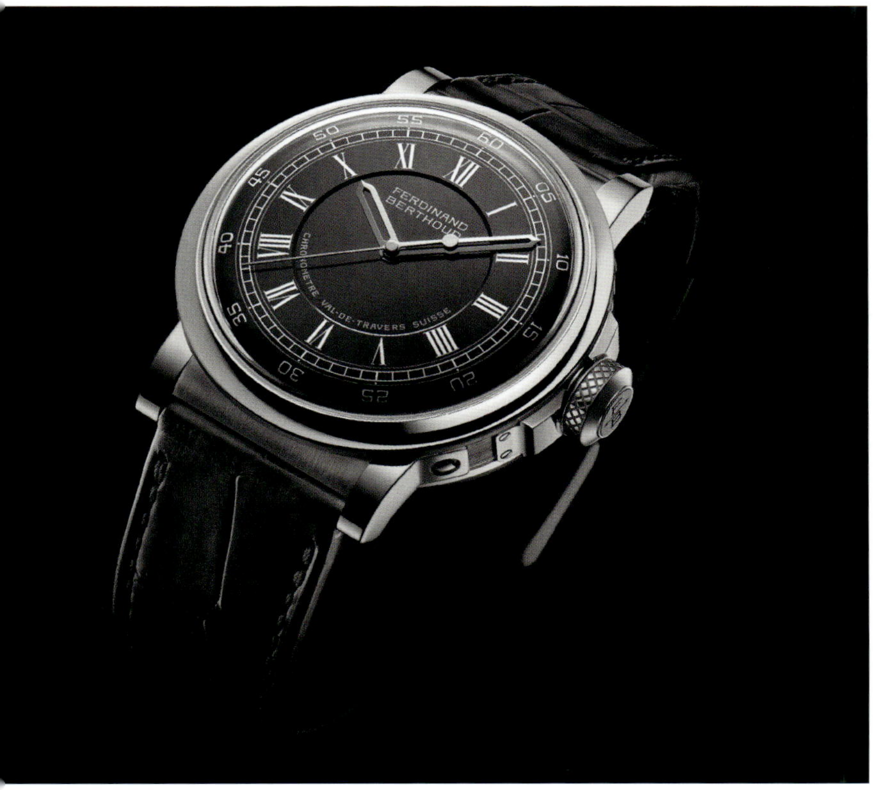

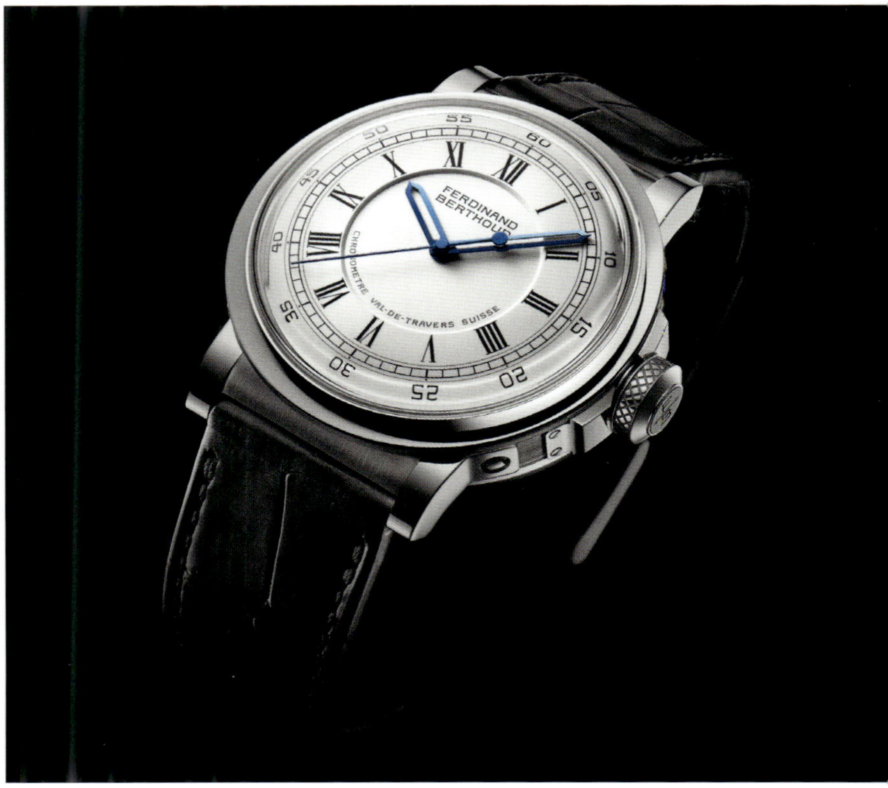

Chronomètre FB 2RE.1, 2020

CHRONOSWISS *RÉGULATEUR*

1988

INSPIRED BY PENDULUM CLOCKS

In 1988, the adage that nothing is more powerful than an idea whose time has come rang true for Munich watchmaker Chronoswiss. The idea in question was a watch by the name of Régulateur. Its special dial was inspired by high-quality precision pendulum clocks, which watchmakers used to regulate their movements. The limited first edition of this watch, available in steel, gold and bronze, was sold by founder Gerd-Rüdiger Lang in the space of a few months. In 1990, due to the lack of hand-wound movements of the discontinued Unitas 6376 calibre, the second generation of the Chronoswiss icon was launched as the Régulateur Automatique. Appropriately modified models of the Enicar calibre 122 henceforth beat inside the characteristic round cases. Later, further limited editions caused a stir: the Régulateur Rectangulaire (1993), the Grand Régulateur (1994) and, marking the model's twentieth anniversary, the Régulateur 24. In 2013, Chronoswiss, now owned by a Swiss family, celebrated its thirtieth anniversary. With a suitable watch to celebrate the occasion: the Régulateur 30 with an automatic movement, and for the first time, a jumping digital hour display. The creative potential at Chronoswiss was by no means exhausted.

The year 2016 was characterised by the launch of a new version of the Regulator. Very much in the style of our era, this evolutionary stage impresses with a three-dimensional dial. The current Regulator Manufacture with a calibre C.122 upholds tradition. The Regulator Blue Steel makes the perfect blue-hour companion, while the Open Gear ReSec offers a retrograde second hand. Evidently, the traditional Régulateur dial still holds a lot of potential even in the twenty-first century.

ABGESCHAUT VON PENDELUHREN

1988 bewahrheitete sich bei der Münchner Uhrenfabrikation Chronoswiss einmal mehr die altbekannte Spruchweisheit, dass nichts mächtiger ist als eine Idee, deren Zeit gekommen ist. Konkret ging es um eine Armbanduhr namens „Régulateur". Ihr spezielles Zifferblatt leitete sich ab von hochwertigen Präzisionspendeluhren, mit deren Hilfe Uhrmacher ihre Werke zu regulieren pflegten. Die limitierte Erst-Edition dieser Armbanduhr, erhältlich in Stahl, Gold und Bronze, verkaufte Gründer Gerd-Rüdiger Lang quasi innerhalb weniger Monate. Mangels weiterer Handaufzugswerke vom nicht mehr produzierten Kaliber Unitas 6376 folgte 1990 mit dem „Régulateur Automatique" die zweite Generation der Chronoswiss-Ikone schlechthin. Nun tickten in den charakteristischen runden Gehäusen entsprechend modifizierte Exemplare des Kalibers 122 von Enicar. Anschließend machten weitere limitierte Serien von sich reden. Zum Beispiel die „Régulateur Rectangulaire" (1993), die „Grand Régulateur" (1994) oder zum 20. Modellgeburtstag die „Régulateur 24". 2013 feierte Chronoswiss, nun im Eigentum einer Schweizer Familie, sein 30. Firmenjubiläum. Die dazu passende Uhr: „Régulateur 30" mit Automatikwerk und erstmals einer springenden, digitalen Stundenanzeige. Damit war das Kreativpotenzial im Hause Chronoswiss aber keinesfalls erschöpft.

Das Jahr 2016 stand im Zeichen einer neuen Version des „Regulator". Diese Evolutionsstufe besticht, ganz im Stil unserer Epoche, durch ein dreidimensionales Zifferblatt. An die Tradition knüpft der aktuelle „Regulator Manufacture" mit dem Kaliber C.122 an. Für Blaue Stunden eignet sich der „Regulator Blue Steel". Und über einen rückspringenden Sekundenzeiger verfügt der „Open Gear ReSec". Man sieht: Das überlieferte „Régulateur"-Zifferblatt besitzt auch im 21. Jahrhundert noch jede Menge Potenzial.

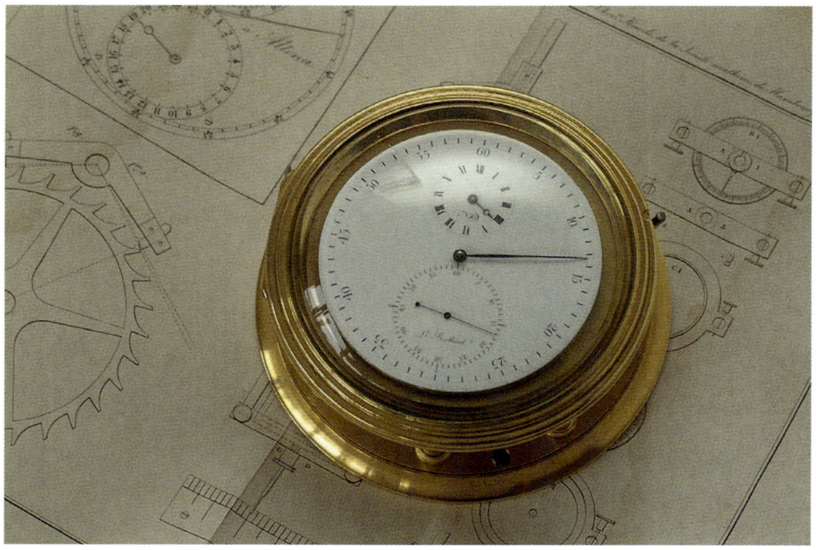

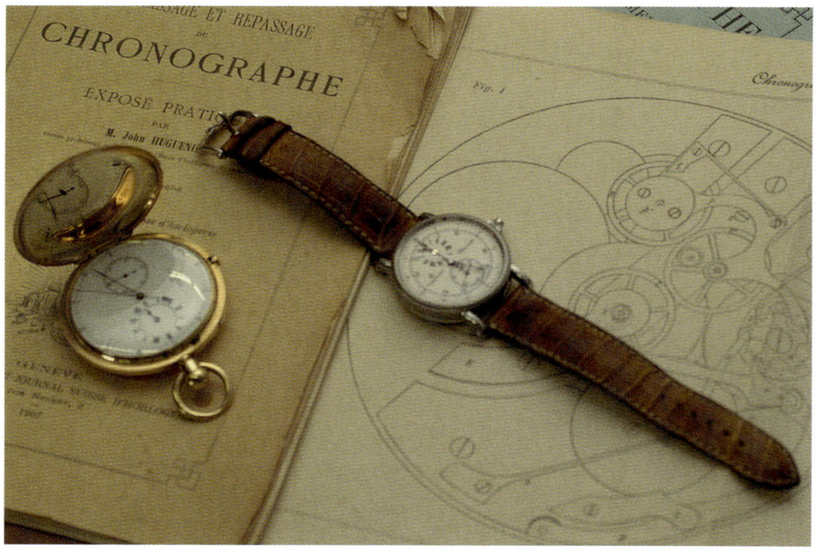

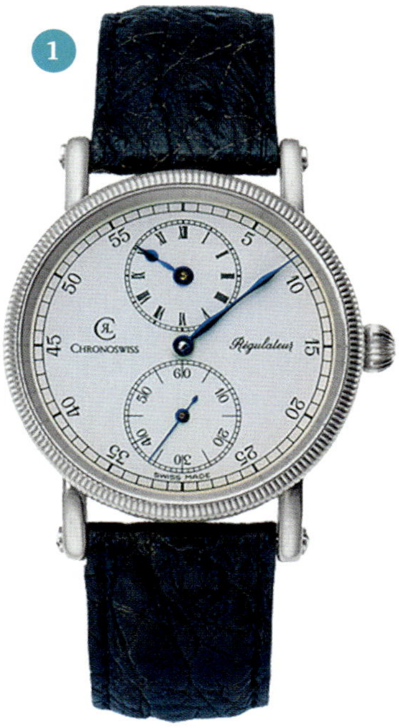

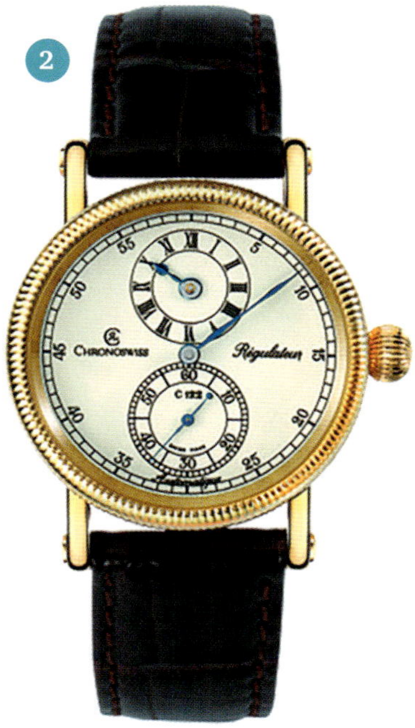

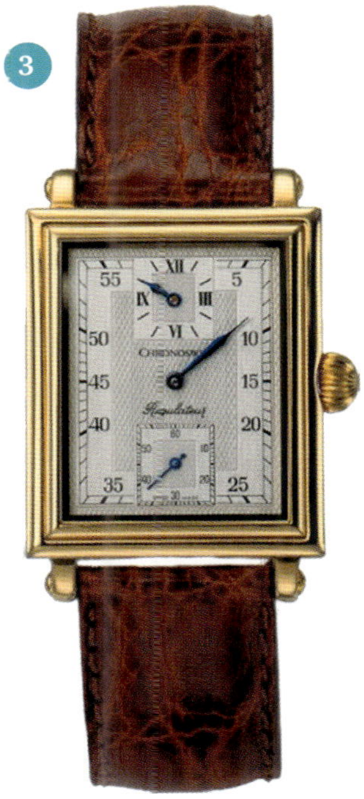

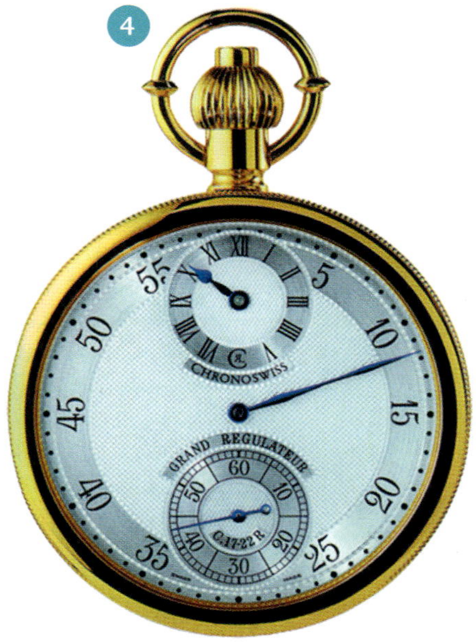

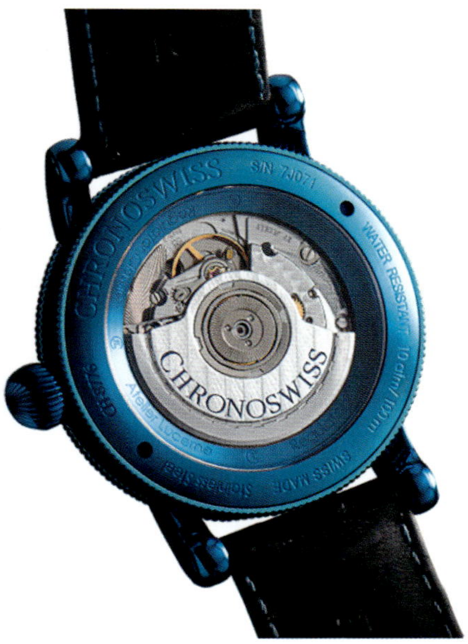

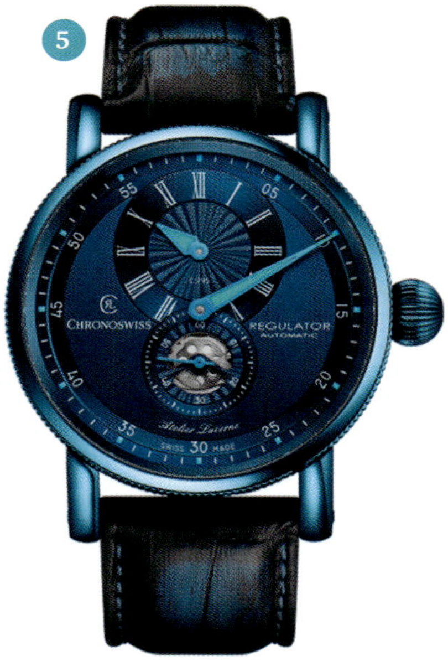

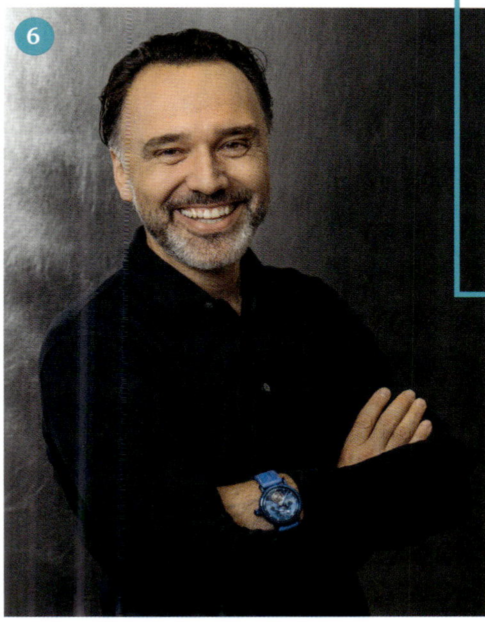

- 1 *Régulateur I, 1988*
- 2 *Régulateur II, 1990*
- 3 *Régulateur Rectangulaire, 1993*
- 4 *Régulateur pocket watch, 1994*
- 5 *Regulator Classic Blue Steel, 2020*
- 6 *Oliver Ebstein, owner and director of Chronoswiss*

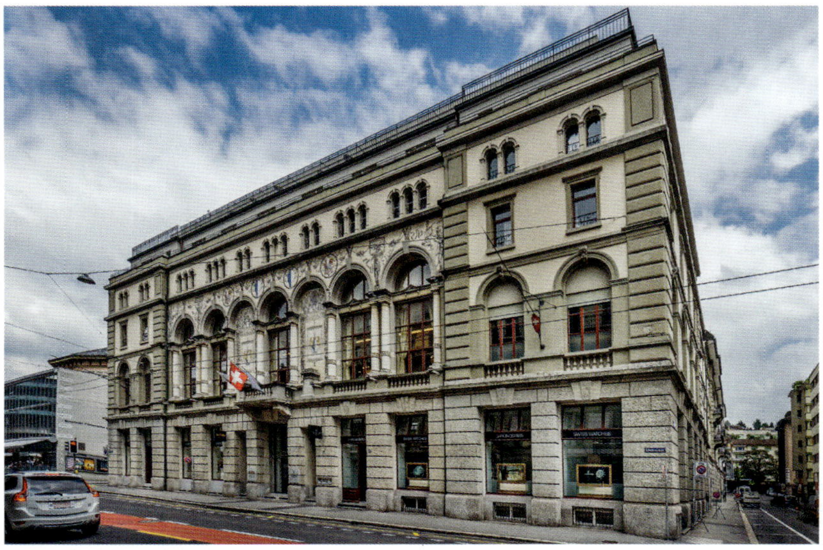

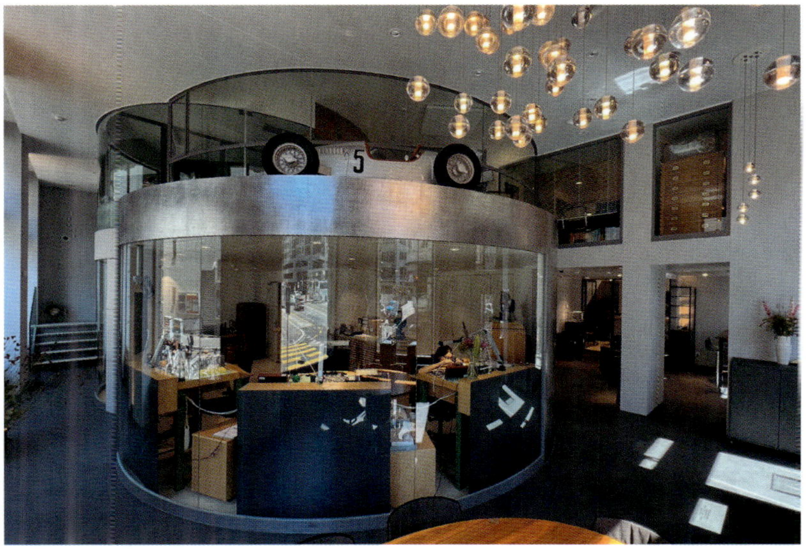

House of Chronoswiss, Lucerne

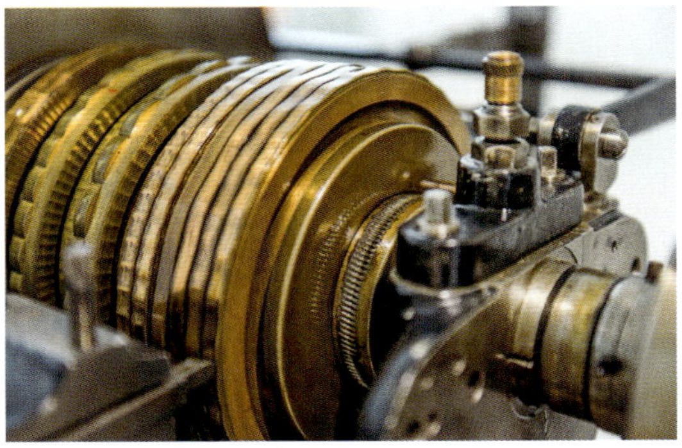
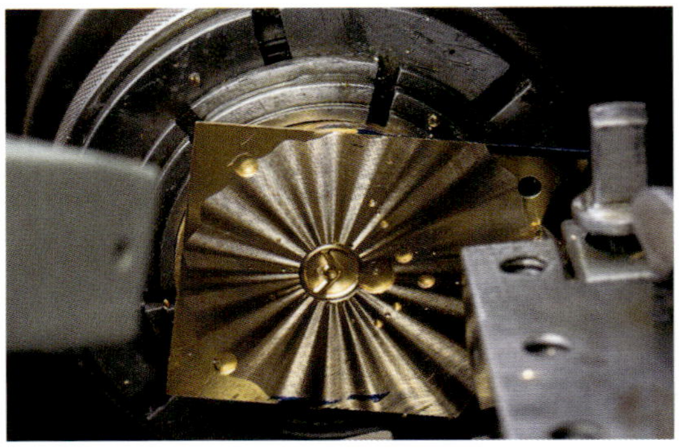
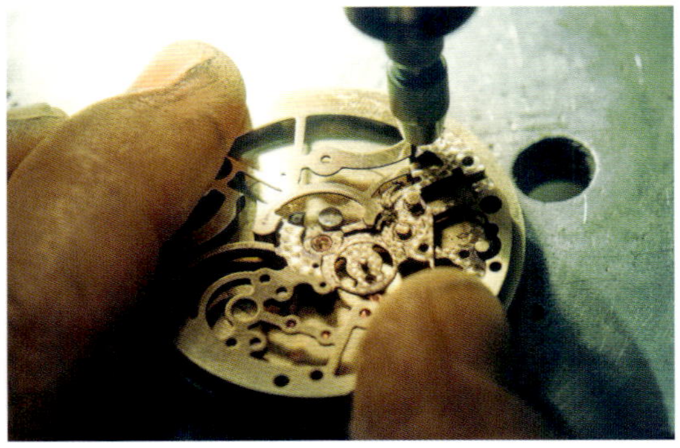
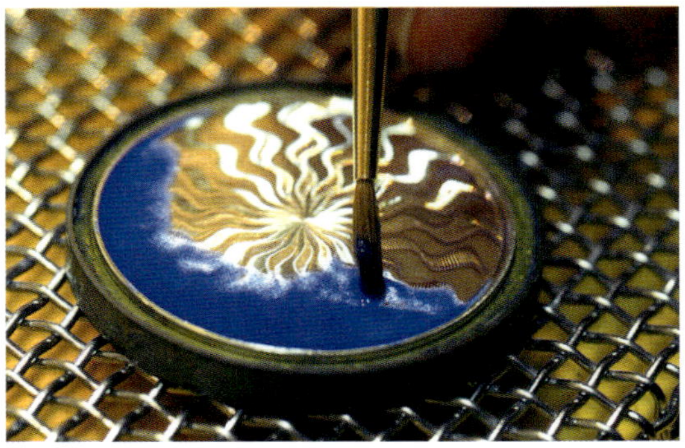
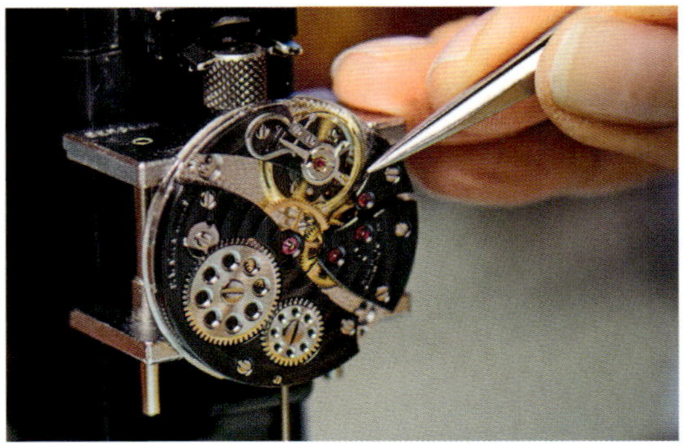
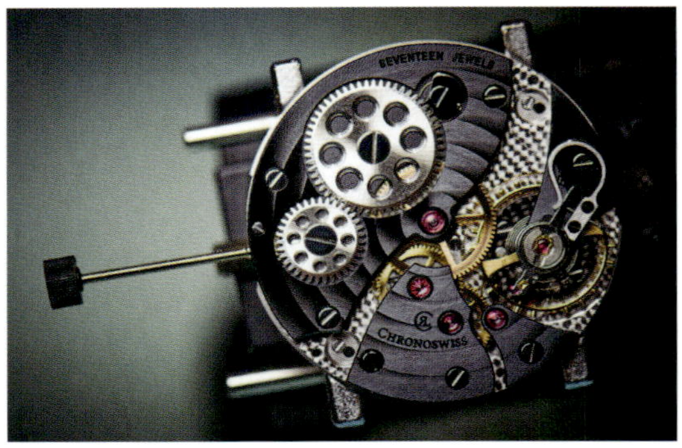

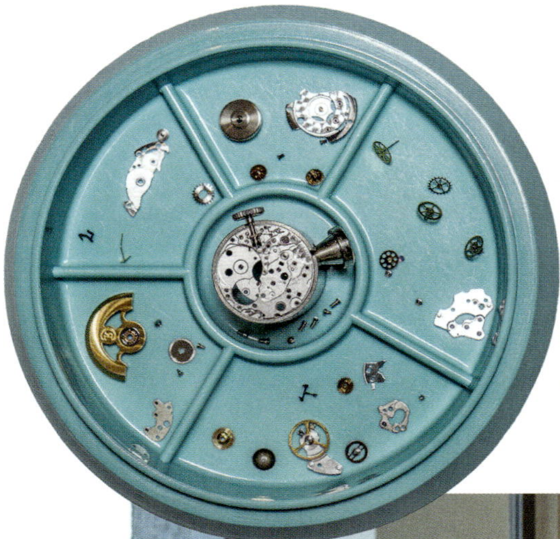
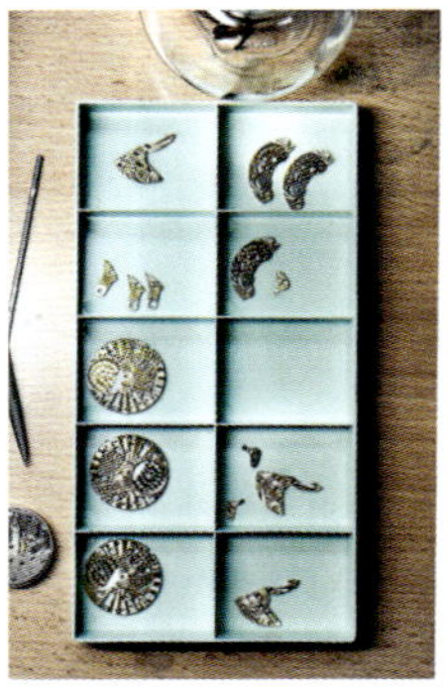
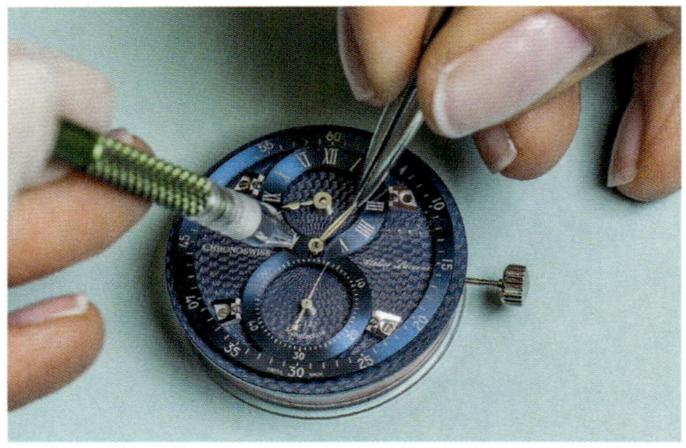

At the workshop | Im Atelier

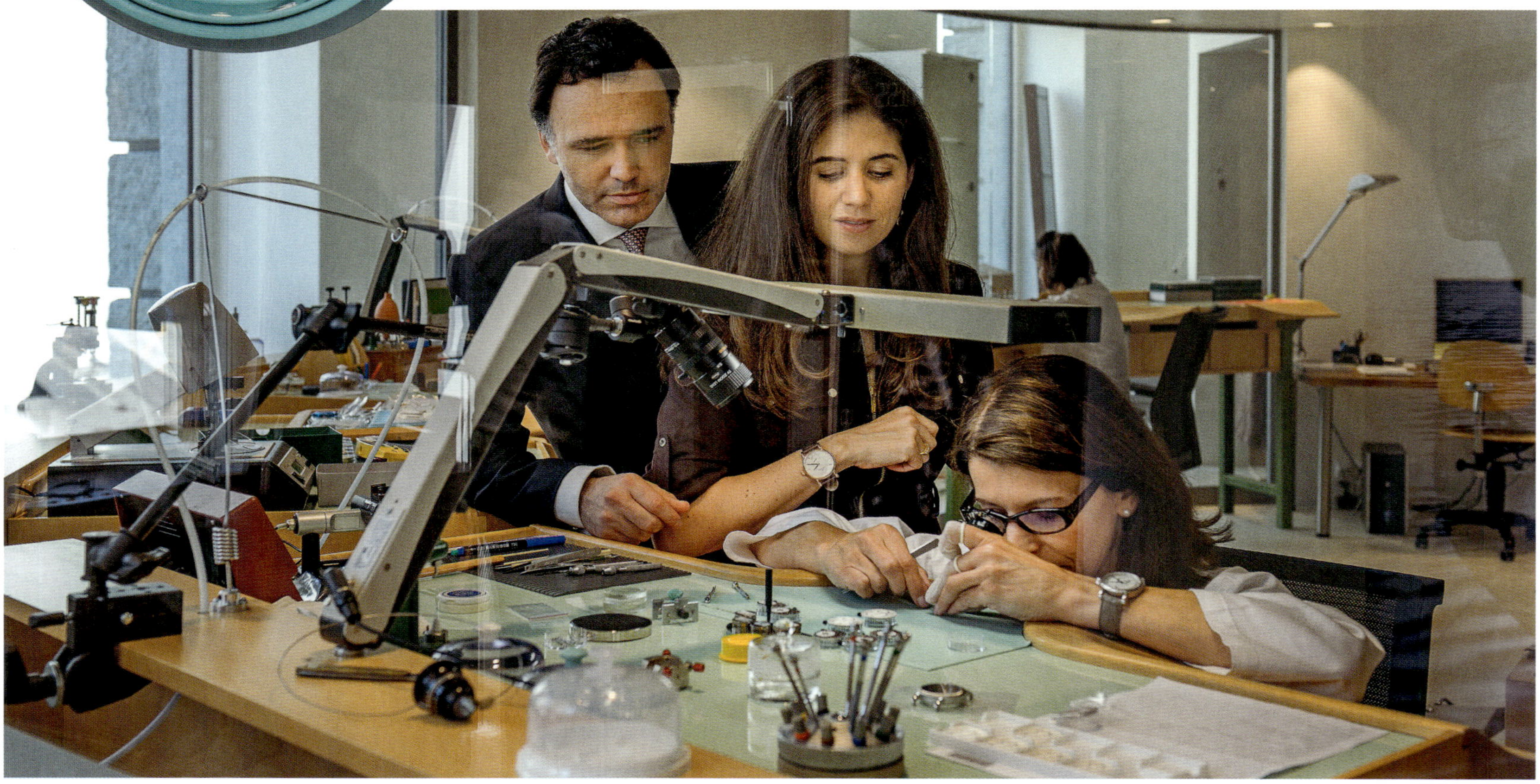

GRAND REGULATOR

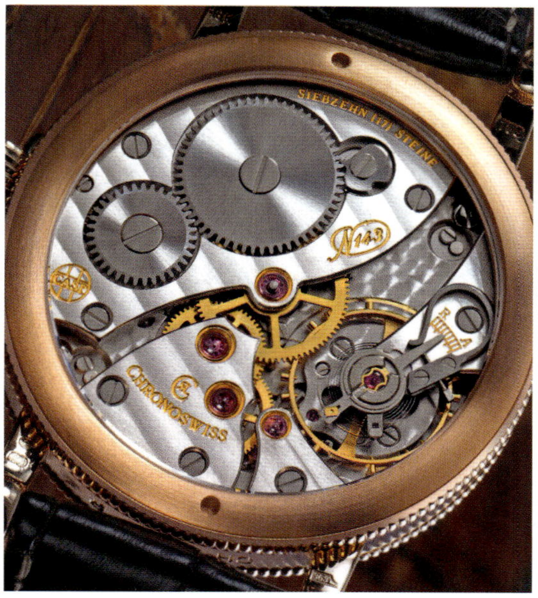

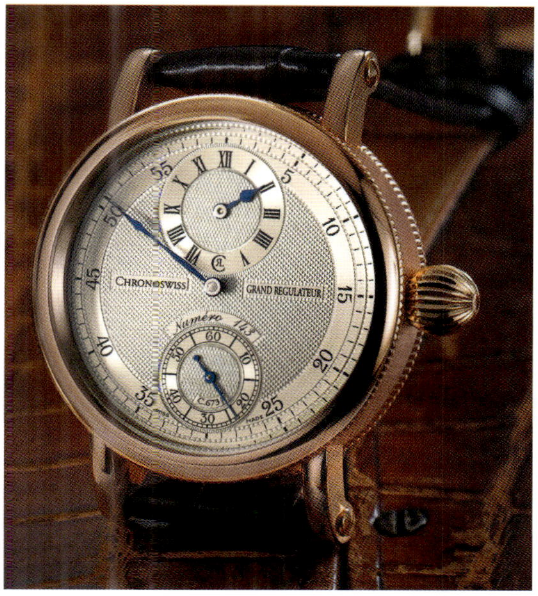

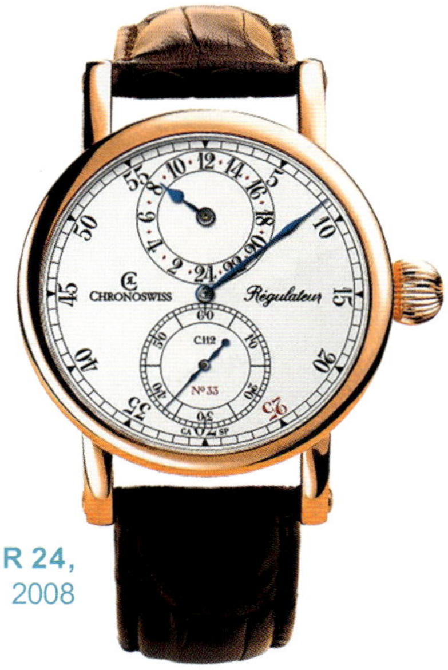

RÉGULATEUR 24, 2008

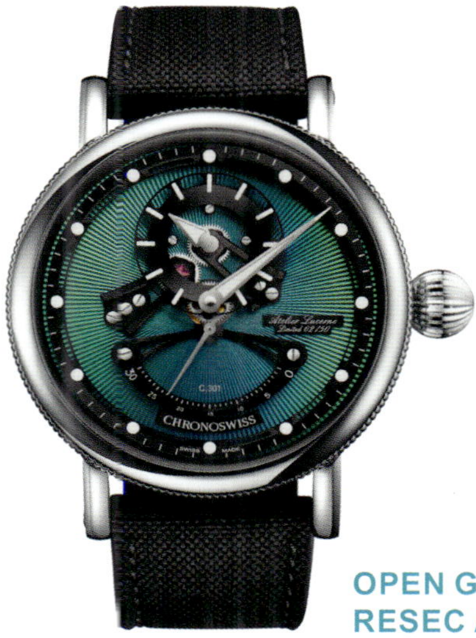

OPEN GEAR RESEC AURORA, 2023

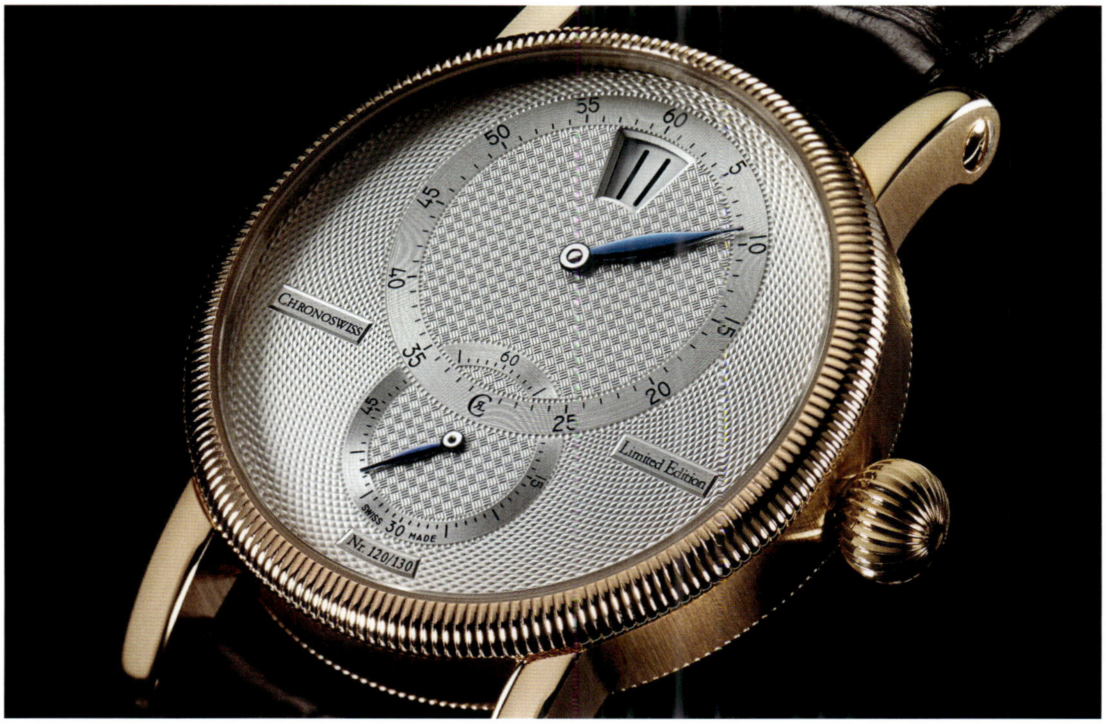

RÉGULATEUR 30, 2013

CORUM 1980

GOLDEN BRIDGE

GOLDEN BRIDGE

One day in 1977, Vincent Calabrese went for an interview at Corum. He did so on the recommendation of André Curtit, then director of the International Watch Museum in La Chaux de Fonds. The young watchmaker did not come empty-handed. In his pocket, he had a movement with a linear gear train, yet to be perfected. Jean René Bannwart, son of company founder René Bannwart, was impressed and bought the idea. Then, of course, the real work began. Implementing the idea in a watch effectively required a complete redesign. This delicate and painstaking task fell to Serge Steudler. Corum's technical director was responsible for reducing the size of the prototype. He also had to redesign the winding mechanism and hand position to ensure reliable operation for daily wear. By 1979, the 30-millimetre-long in-line baguette movement with three golden bridges one atop the other was ready for production. The name Golden Bridge was by no means outlandish. Alone the making of the movement using traditional production methods consisted of a formidable 85 work steps. That was why it was impossible to produce more than two series, each with 100 movements, per year. Even in the first models of the Golden Bridge, the dial was positioned beneath the movement, so that the filigree assembly with its fine wheels was always visible. The dream of a completely transparent wristwatch with this delicate movement was realised by Jean Claude Schwab. In Seitz S.A., the young artistic director found a committed partner who developed a faceted sapphire crystal case for Corum. Probably as a gesture of gratitude to André Curtit, the first serial-produced Golden Bridge was delivered to the International Watch Museum on 25 September 1980.

Despite the company's eventful biography, becoming part of Severin Wunderman in 1999 and then acquired by the Chinese Citychamp in 2013, the Golden Bridge remains an undisputed icon. Titanium bridges, automatic winding, rectangular, tonneau-shape or round cases: the possible variations are manifold. And of course, even after 40 years, the in-line baguette movement is still a sight to behold.

GOLDENE BRÜCKE

Eines Tages im Jahr 1977 sprach Vincent Calabrese bei Corum vor. Er tat dies auf Empfehlung von André Curtit, dem damaligen Direktor des Internationalen Uhrenmuseums in La Chaux de Fonds. Im Gepäck hatte der junge Uhrmacher ein noch unausgereiftes Uhrwerk, dessen Rädersatz streng in einer Linie angeordnet war. Jean René Bannwart, Sohn des Firmengründers René Bannwart, zeigte sich begeistert und erwarb die Idee. Doch damit begann erst die eigentliche Arbeit. Die Verwendung des Konstrukts in einer Armbanduhr erforderte faktisch eine Neukonstruktion. Diese delikate und aufwändige Arbeit kam Serge Steudler zu. Dem technischen Direktor von Corum oblag es, den Prototypen zu verkleinern. Außerdem musste er Aufzug und Zeigerstellung für einwandfreie Funktion bei täglichem Tragen umgestalten. 1979 war das 30 Millimeter lange Stabwerk mit drei übereinanderliegenden goldenen Brücken zur Serienreife gediehen. Der Name „Golden Bridge" war also keineswegs an den Haaren herbeigezogen. Alleine die Herstellung des Rohwerks nach traditionellen Fertigungsmethoden erforderte nicht weniger als 85 Arbeitsgänge. Allein deshalb konnten jährlich nicht mehr als zwei Serien zu je 100 Rohwerken aufgelegt werden. Schon bei den ersten Modellen der „Golden Bridge" lag das Zifferblatt unter dem Werk, damit man den filigranen Stab mit seinen feinen Rädern jederzeit betrachten konnte. Den Traum von einer absolut transparenten Armbanduhr mit diesem zierlichen Werk realisierte Jean Claude Schwab. In der Seitz S.A. fand der junge künstlerische Direktor einen engagierten Partner, welcher für Corum ein facettiertes Saphirglas Gehäuse entwickelte. Wohl auch als Dank an den bereits erwähnten André Curtit gelangte am 25. September 1980 die erste serienmäßig hergestellte „Golden Bridge" ins Internationale Uhrenmuseum.

Trotz der bewegten Firmenbiographie, welche Corum 1999 unter das Dach von Severin Wunderman und 2013 ins Eigentum der chinesischen Citychamp führte, ist die „Golden Bridge" unangefochtene Ikone. Titanbrücken, automatischer Aufzug, rechteckiges, tonneauförmiges oder rundes Gehäuse: Variationsmöglichkeiten gibt es viele. Dabei macht das Stabwerk auch mehr als 40 Jahre nach seiner Kreation immer wieder eine gute Figur.

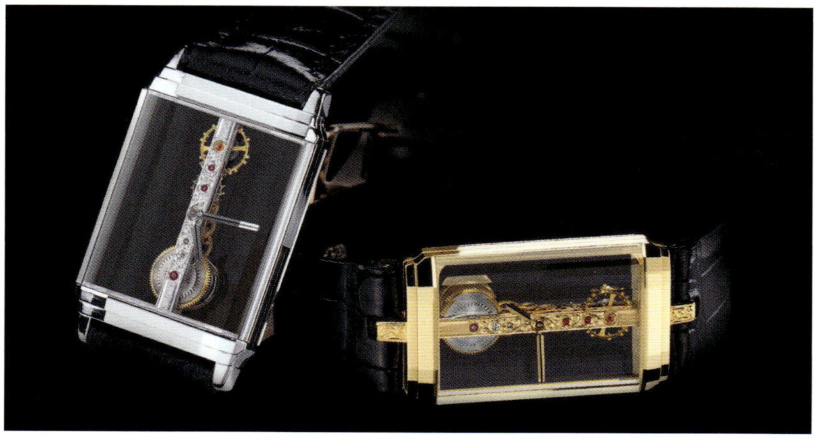

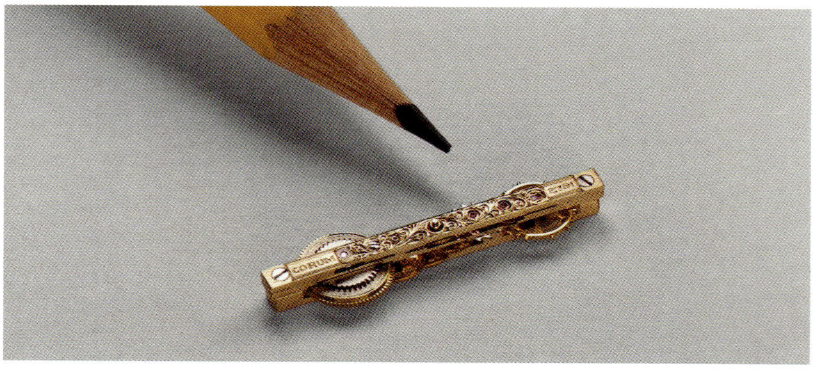

Golden Bridge models and movement from 1980 / Golden Bridge-Modelle und Uhrwerk von 1980

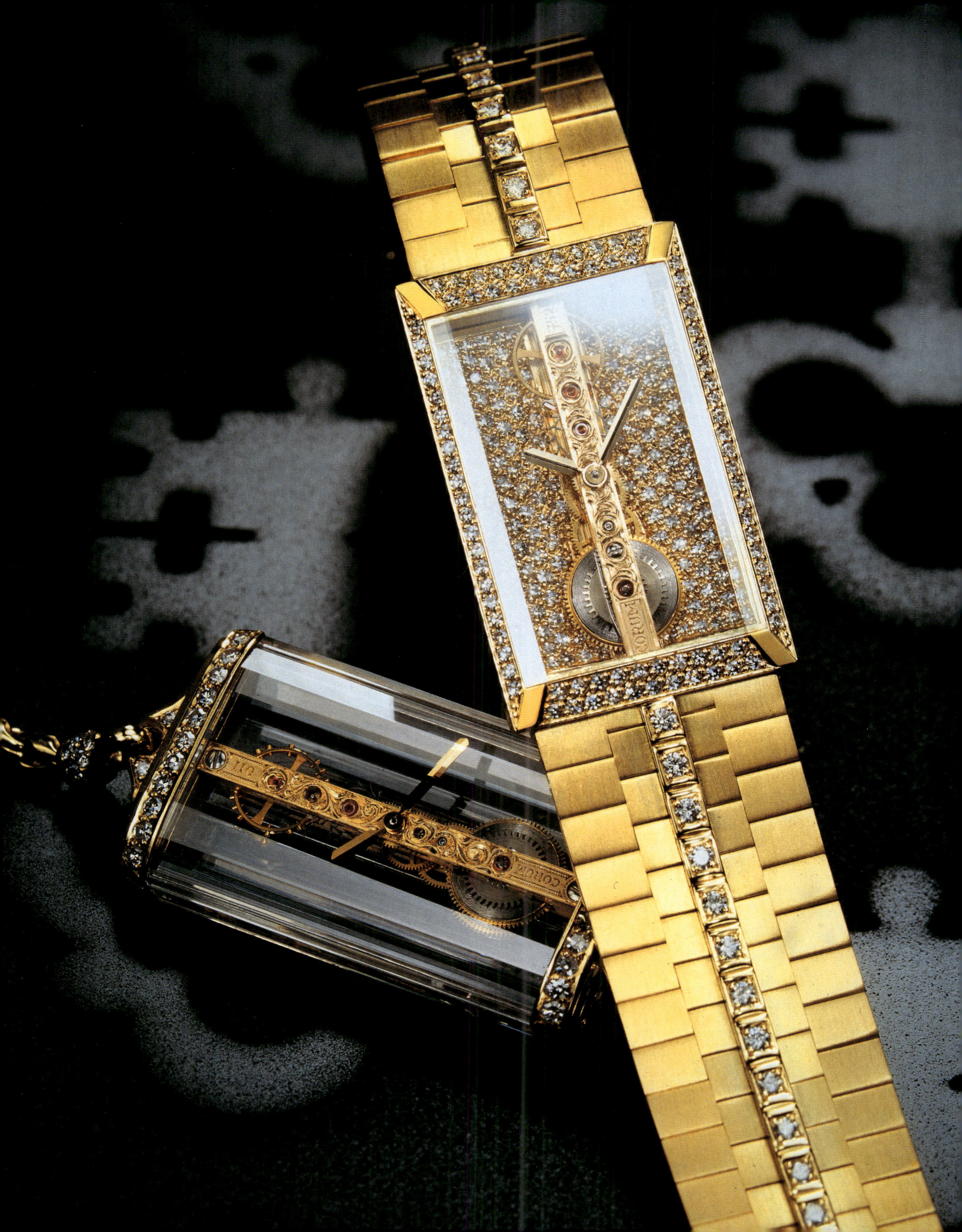
CORUM

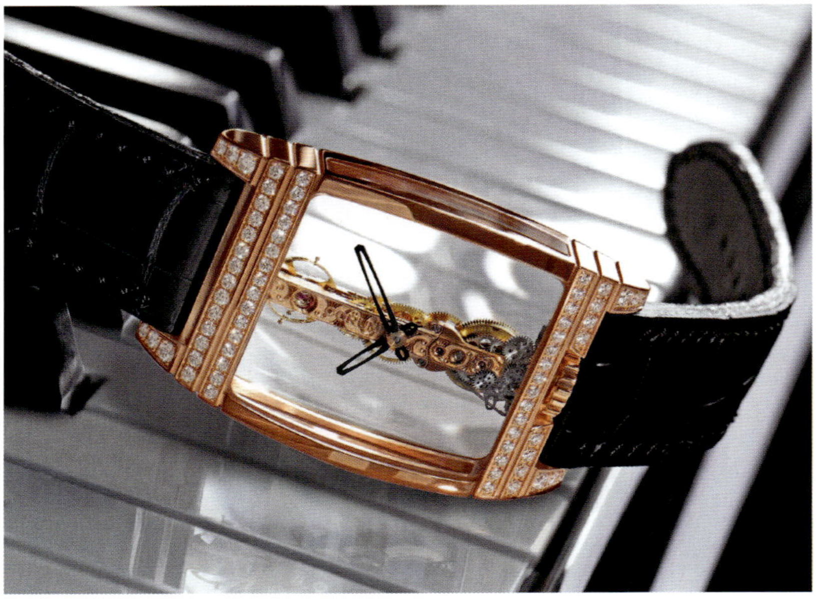

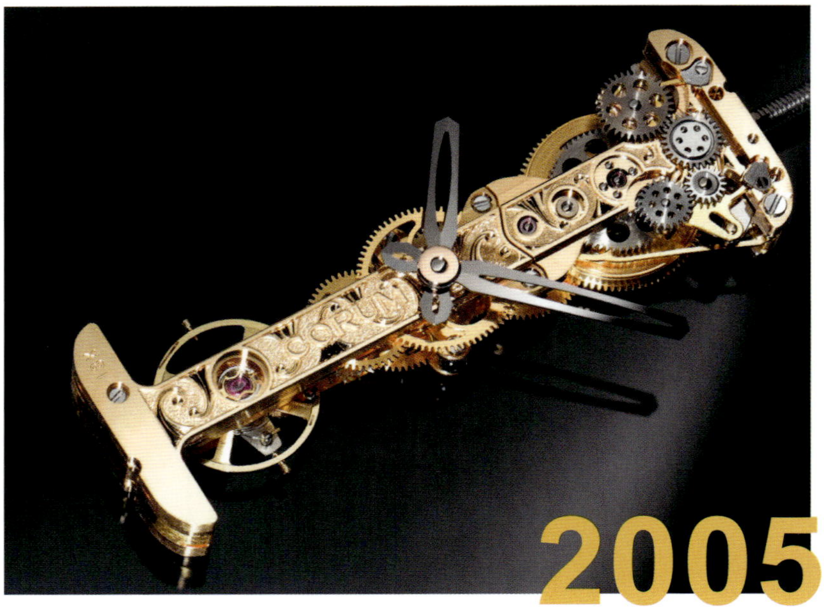

2005

2005 *Golden Bridge Classic*

2010 *Golden Bridge Tourbillon*

2010 *Golden Bridge Miss >>*

2011 *Golden Bridge Automatic*

2016 *Golden Bridge 43*

2017 *Golden Bridge Stream*

2019 *Golden Bridge Classic*

2021 *Golden Bridge Avant-Garde*

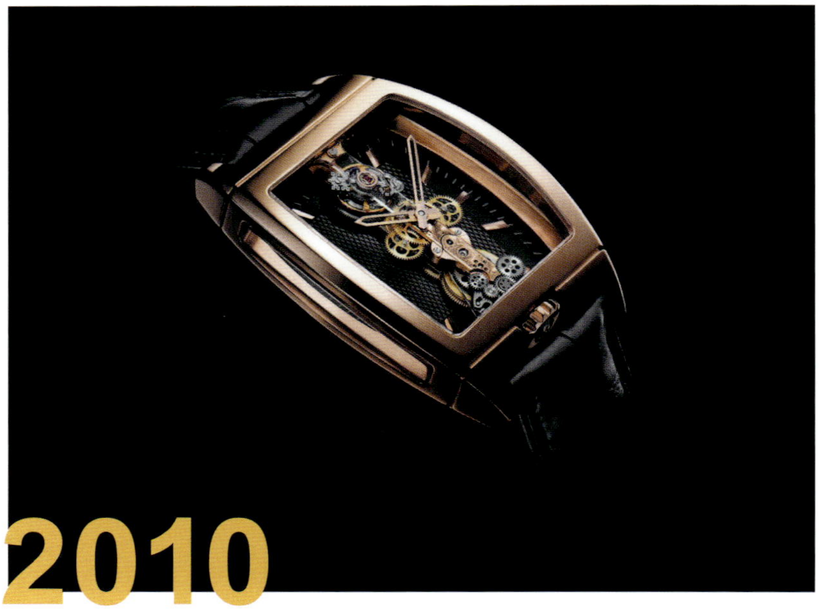

2010

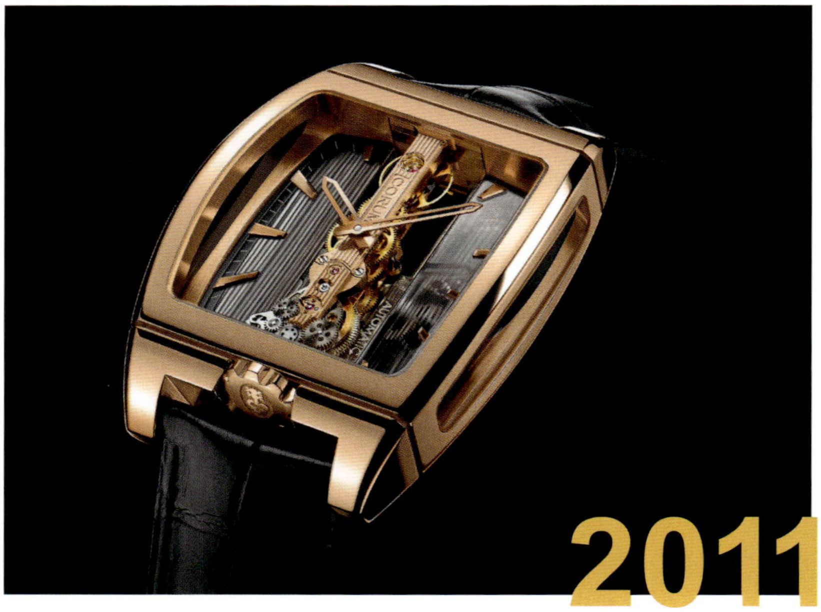

2011

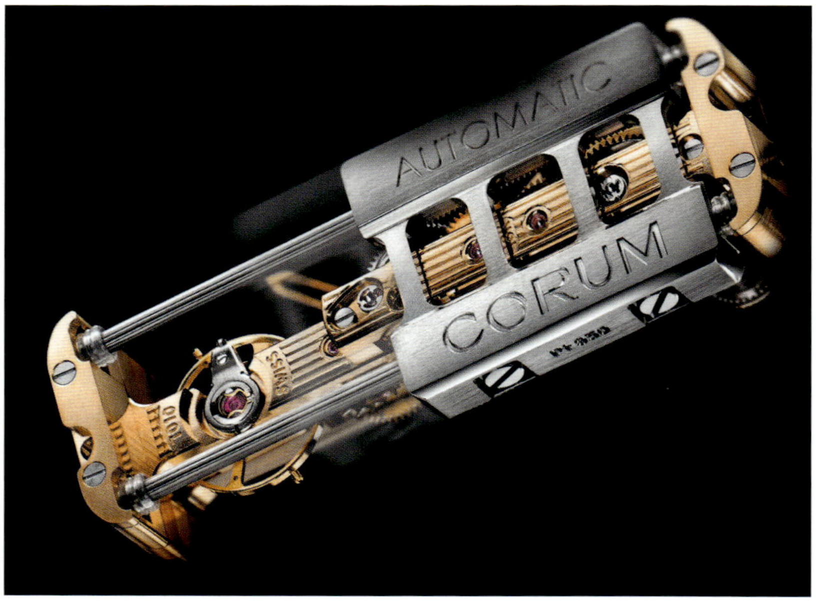

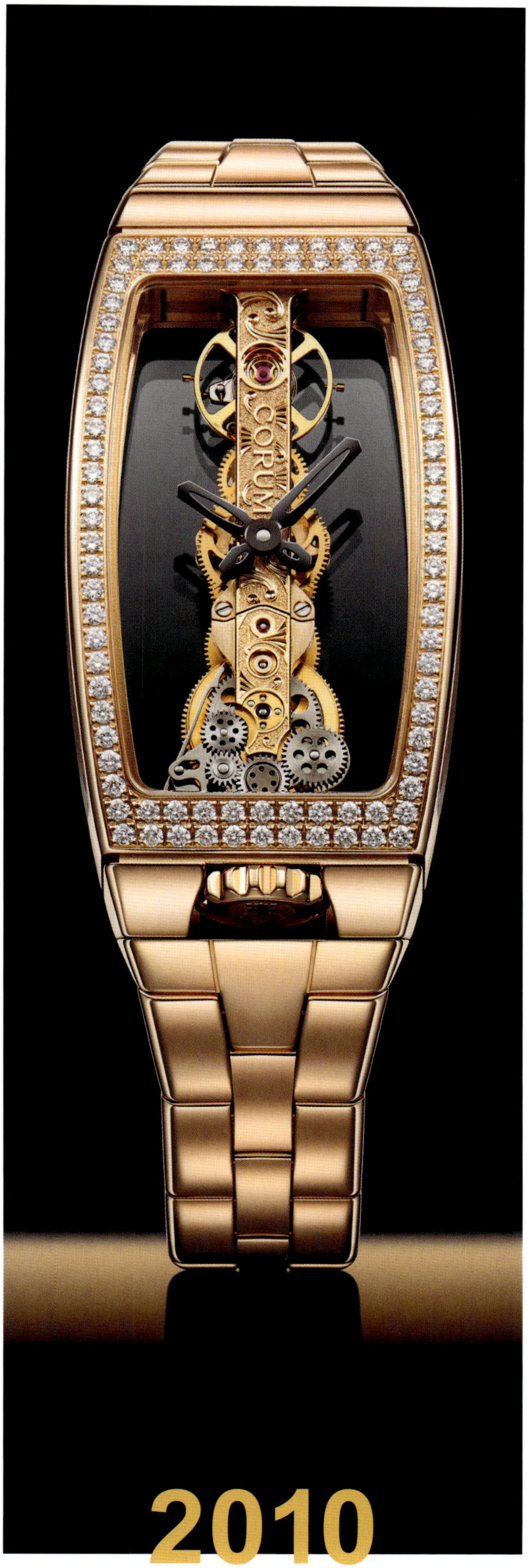

2010

2016

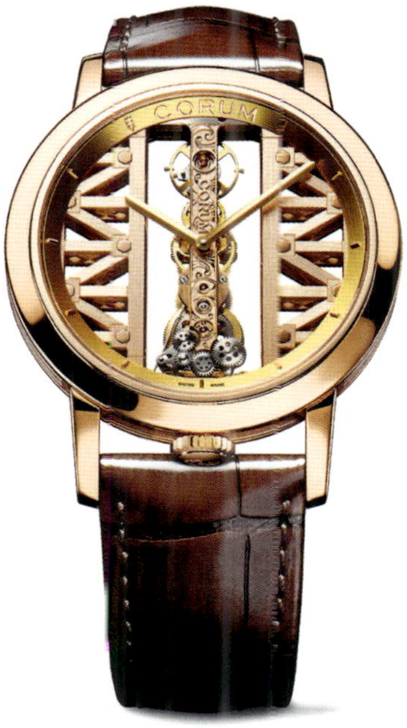

2017

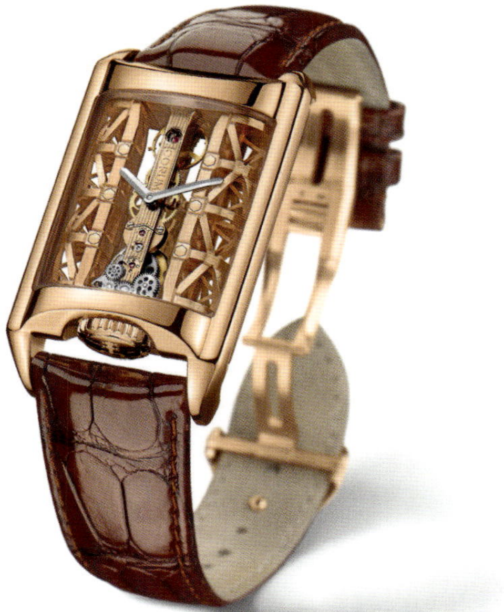

2019

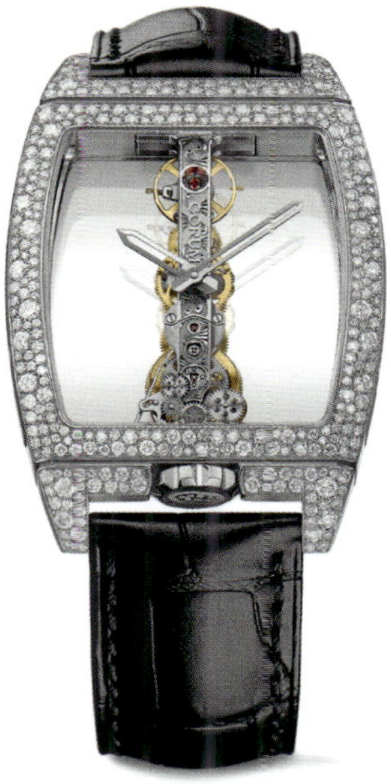

2021

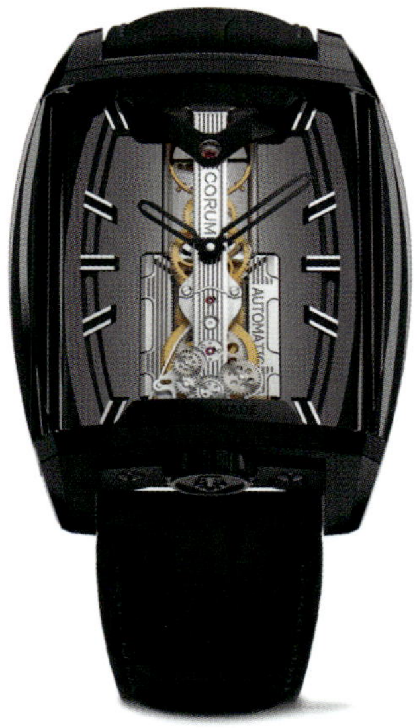

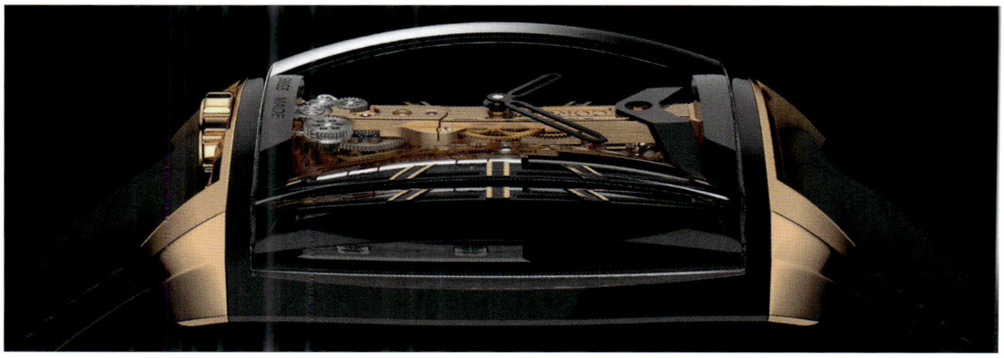

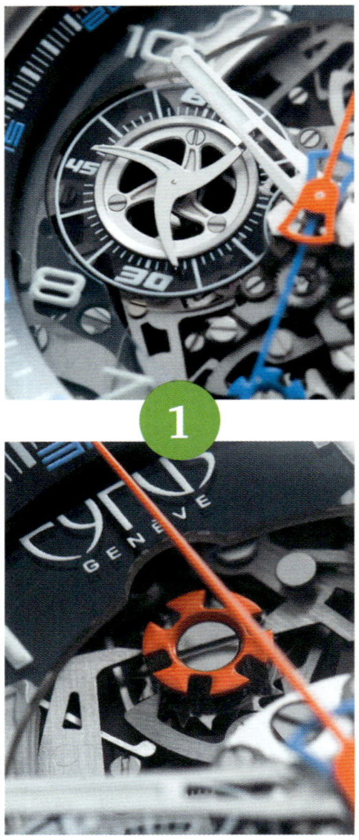

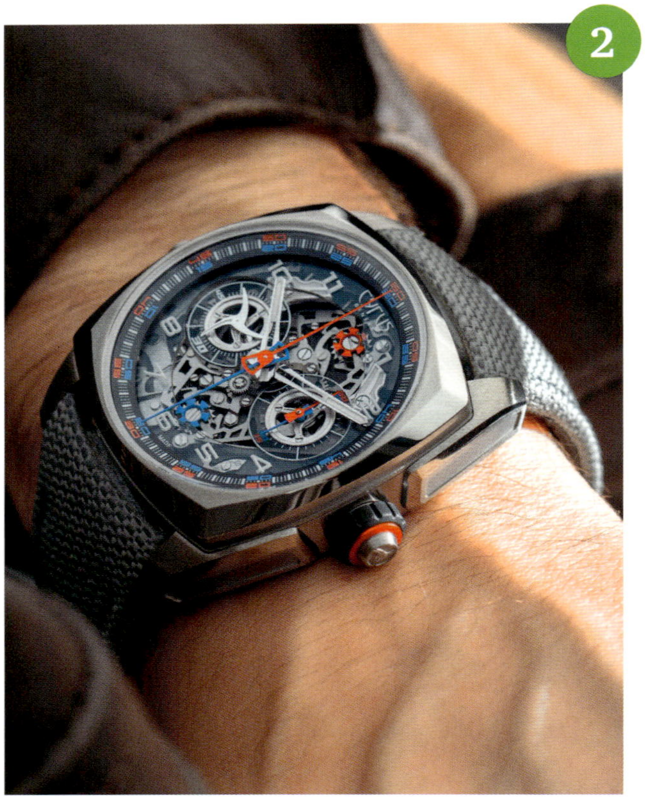

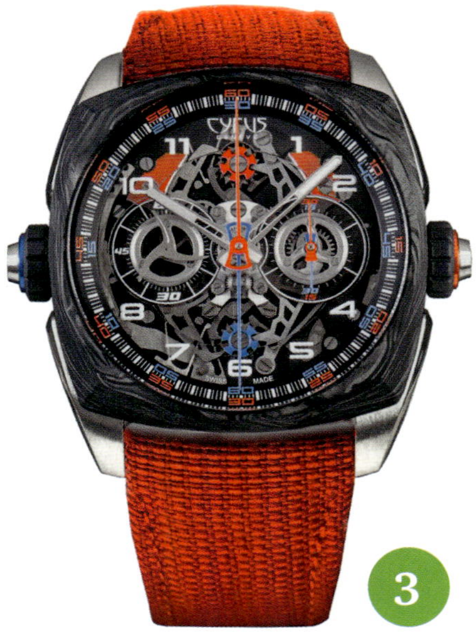

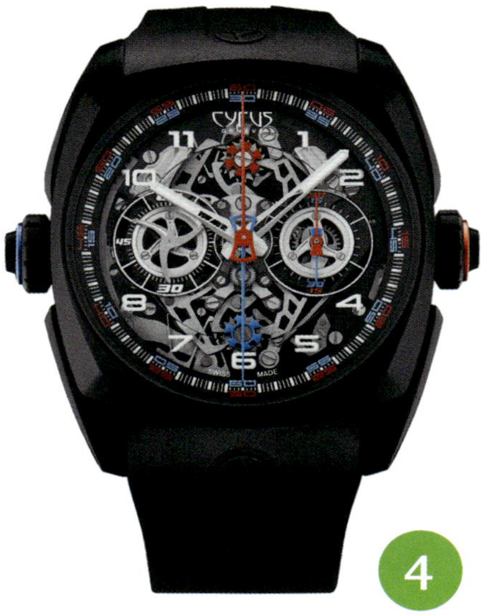

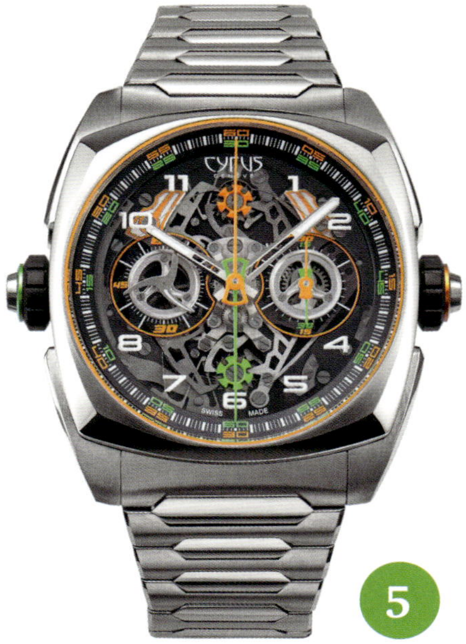

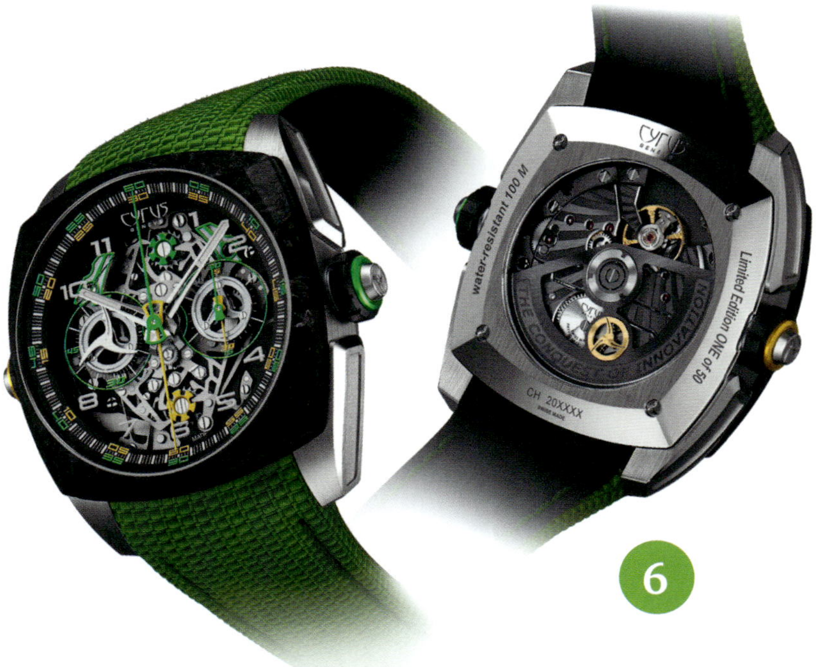

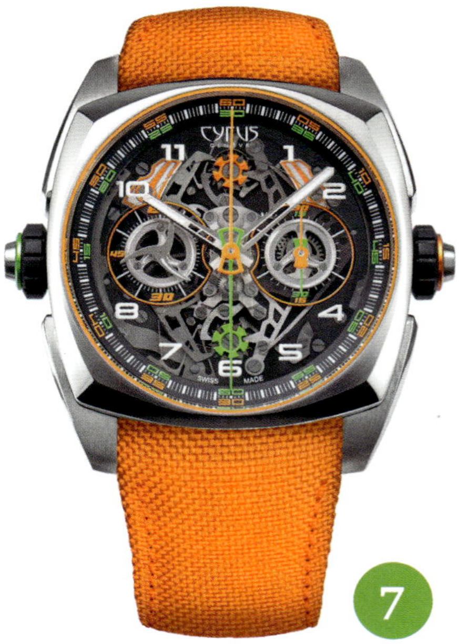

1 *Dial closeup | Zifferblatt im Detail*

2 *Klepcys DICE Titanium*

3 *Klepcys DICE Racing*

4 *Klepcys DICE Black DLC Titanium*

5 *Klepcys DICE Saffron*

6 *Klepcys DICE Lime Carbon*

7 *Klepcys DICE Saffron*

CYRUS
KLEPCYS DICE

TWO CHRONOGRAPHS IN ONE WATCH

As incredible as it may seem, the world's first wristwatch with two independently operated chronographs did not arrive on the scene until 2021. The Cyrus watch brand, which was only founded in 2010, presented its Klepcys DICE. Responsible for the development of the visible module powered by an automatic movement with separate stopwatch mechanisms is Chronode S.A. in Le Locle. Cyrus group is a majority shareholder; technical director is Jean-François Mojon. Incidentally, the four letters of "DICE" stand for "Double Independent Chronograph Evolution". To better distinguish between the two chronographs, the column wheels for the start, stop and zeroing functions are colour-coded. Corresponding with this duo are the two concentrically rotating chronograph hands in the centre of the dial and the similarly layered 30-minute totalisers at 3 o'clock. At the opposite side of the dial, the scales for the starting points are also colour-coded. Ergonomically arranged pushers and a horizontal clutch are used to operate the two chronographs. Cyrus has dispensed with a conventional dial to enable an in-depth appreciation of the complex mechanical actions. Transparent sapphire elements ensure optimum visibility. Positioned at 9 o'clock, a three-armed element modelled on the Cyrus logo acts as function control. The 36-millimetre proprietary calibre CYR 718 is 11.65 millimetres high. Each watch consists of 443 components. Of these, 218 are needed for the front chronograph module alone. The automatic movement with ball-bearing-rotor is visible through the transparent back of the 42-millimetre titanium case, which is waterproof to ten bar. It offers a power reserve of 55 hours.

ZWEI CHRONOGRAPHEN IN EINER UHR

Erstaunlich, aber wahr: Die weltweit erste Armbanduhr mit zwei unabhängig voneinander bedienbaren Chronographen ließ bis 2021 auf sich warten. Da stellte die erst 2010 gegründete Uhrenmarke Cyrus ihre „Klepcys DICE“ vor. Für die Entwicklung des sichtbaren, von einem Automatikwerk angesteuerten Moduls mit getrennten Stoppmechanismen zeichnen die mehrheitlich zur Cyrus-Gruppe gehörende Chronode S.A. in Le Locle und deren technischer Direktor Jean-François Mojon verantwortlich. Die vier Buchstaben „DICE“ stehen übrigens für „Double Independent Chronograph Evolution“. Zur besseren Unterscheidung verfügt das Schaltwerk über zwei farblich klar voneinander abgesetzte Schalträder für die Funktionen Start, Stopp und Nullstellung. Mit diesem Duo korrespondieren die beiden konzentrisch im Zifferblattzentrum drehenden Chronographenzeiger und die ebenso übereinander angeordneten 30-Minuten-Totalisatoren bei „3“. Am Zifferblatt gegenüber liegen die Startpunkte der gleichfalls farblich abgestimmten Skalen. Ergonomisch angeordneten Drückern obliegt die Bedienung der beiden Stopper mit horizontaler Räderkupplung. Damit sich das komplexe mechanische Geschehen bis in alle Details nachvollziehen lässt, hat Cyrus auf ein konventionelles Zifferblatt verzichtet. Transparente Saphirelemente gestatten optimale Sicht. Zur Funktionskontrolle rotiert bei „9“ ein dreiarmiges, dem Cyrus-Logo nachempfundenes Gebilde. 11,65 Millimeter hoch baut das 36 Millimeter messende Manufakturkaliber CYR 718. Für ein Exemplar benötigen die Uhrmacher 443 Komponenten. Davon braucht es allein 218 für das vorderseitige Chronographenmodul. Durch den Sichtboden im 42 Millimeter großen Titangehäuse, übrigens bis zehn bar wasserdicht, zeigt sich das Automatikwerk mit Kugellagerrotor. Es besitzt rund 55 Stunden Gangautonomie.

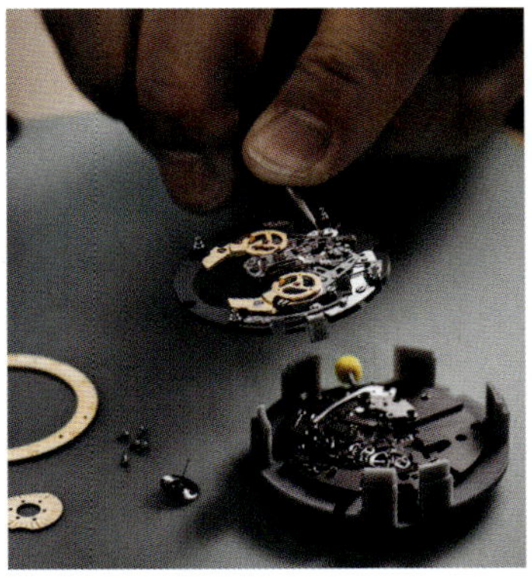

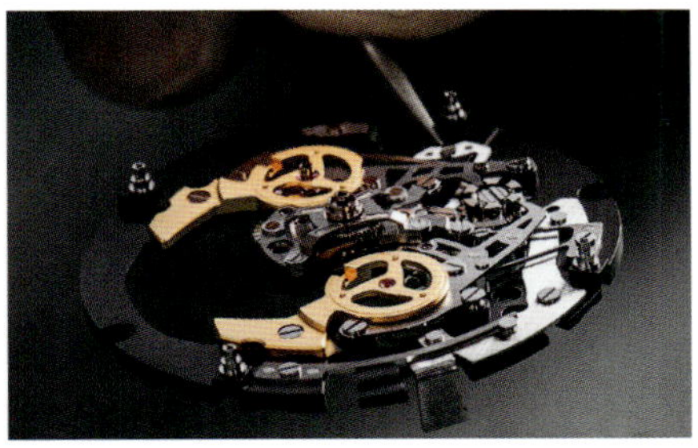

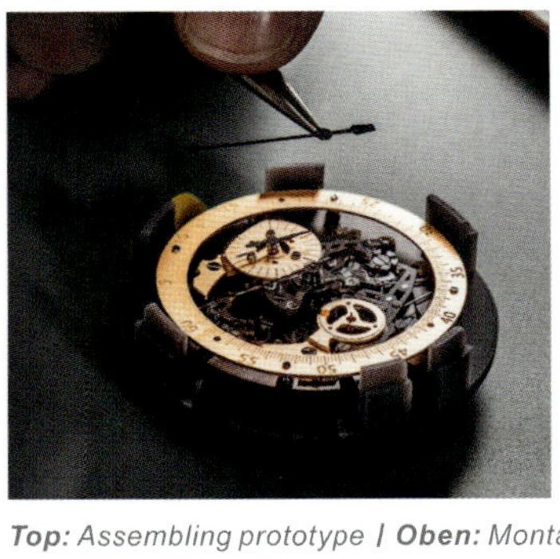

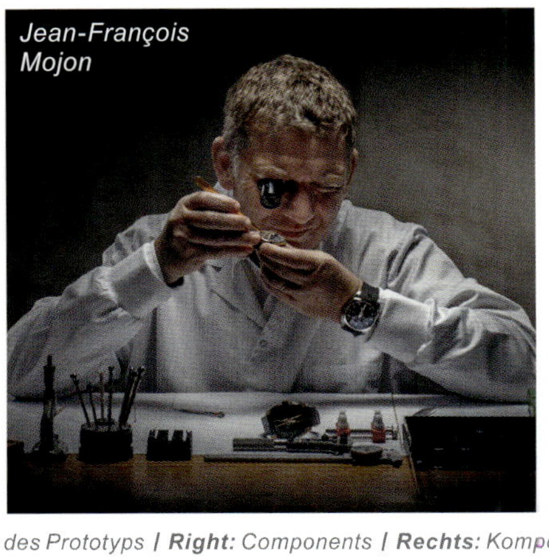

Jean-François Mojon

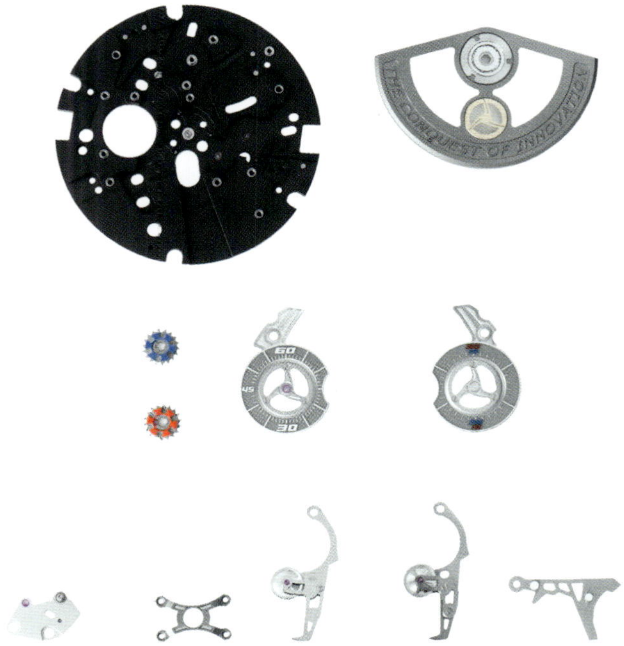

Top: Assembling prototype / *Oben:* Montage des Prototyps / *Right:* Components / *Rechts:* Komponenten

DOXA SUB

1967

DIVING WITH DOXA

The history of Doxa reaches back to 1889. But it was not until the Basel Watch Fair in 1967 that this traditional Swiss watch brand launched its diver's watches. Withstanding pressures of up to 30 ATM, the SUB 300 Professional case featured an automatic movement and diving time rotating bezel. The Doxa SUB 200 appealed to the fairer sex, while chronograph fans got their money's worth with the SUB 200 T-Graph. This chapter of the illustrious company biography was penned back in 1964 by Urs Eschle. The ebullient product developer foresaw the increasing importance of civilian underwater timepieces for ambitious recreational divers and of their availability at affordable prices. In 1964, an ingenious team headed by professional diver Claude Wesly settled down to develop them. No aspect escaped their scrutiny. One was the important combination of dial and hands given the preciousness of time spent underwater. Many dives and colour tests later, the team opted for bright signal orange. A sumptuously sized minute hand rotated in front of the brilliantly coloured dial, since diving hinges on minutes, not hours. The pressure-tight case was made from a solid steel block. It had a unidirectional rotating bezel with a minute scale and depth table. All these features characterised the Doxa SUB launched in 1967. In 1968, the Doxa SUB 300T Conquistador offered a patented helium valve. The importance of the SUB watches also stems from the fact that the elite combat diving unit of the Swiss army chose the SUB 300T Professional as their official underwater timepiece in 1969.

Because these models enjoy a resounding reputation in collectors' circles, the current Doxa 1967 Sub 300 Professional watch line sees itself as a contemporary reinterpretation. Orange is obligatory, but of course, the family business also offers other colours.

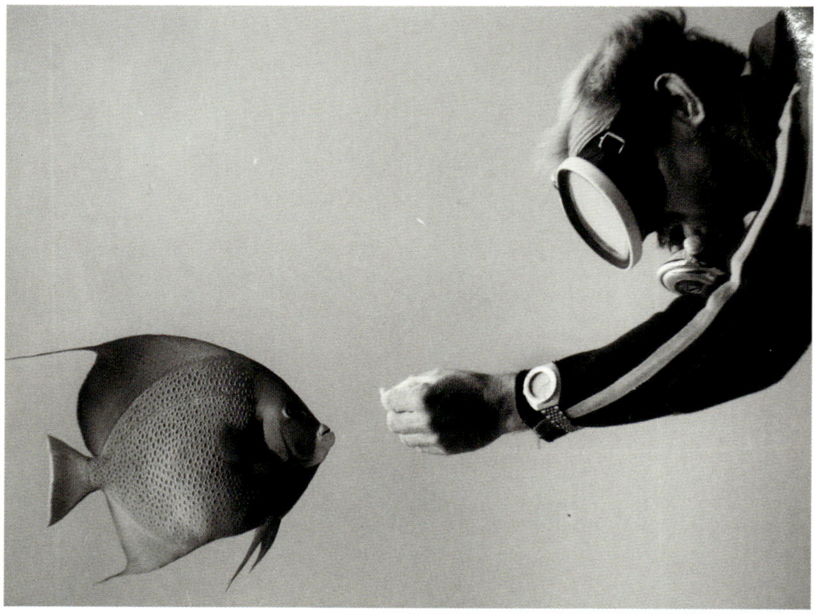

ABGETAUCHT MIT DOXA

Bis 1889 reicht die Geschichte der Firma Doxa zurück. Aber erst zur Basler Uhrenmesse 1967 gingen Taucheruhren dieser traditionsreichen Schweizer Uhrenmarke an den Start. Bis zu 30 Atmosphären reichte die Druckfestigkeit des Gehäuses der „SUB 300 Professional" mit Automatikwerk und Tauchzeit-Drehlünette. An der Doxa „SUB 200" erfreuten sich Vertreterinnen des zarten Geschlechts. Chronographenfans kamen beim „SUB 200 T-Graph" auf ihre Kosten. Seinen Anfang nahm dieses Kapitel der illustren Firmenbiographie bereits 1964, initiiert von Urs Eschle. Der quirlige Produktentwickler ahnte die zunehmende Bedeutung ziviler Unterwasserzeitmesser für ambitionierte Sporttaucher. Und zwar erhältlich zu erschwinglichen Preisen. 1964 machte sich ein mit allen Wassern gewaschenes Team um den Profitaucher Claude Wesly an die Entwicklung. Dabei mussten alle Aspekte auf den Prüfstand. Einer galt der wichtigen Kombination aus Zifferblatt und Zeiger. Schließlich ist die Zeit unter Wasser besonders kostbar. Nach zahlreichen Tauchgängen und Farbtests in verschiedenen Tiefen fiel die Entscheidung für leuchtendes Signal-Orange. Vor dem solcherart gefärbten Zifferblatt rotierte ein opulent dimensionierter Minutenzeiger, denn beim Tauchen kommt es weniger auf die Stunden als auf die Minuten an. Die Herstellung der druckdichten Schale erfolgte aus einem massiven Stahlblock. Sie verfügte über eine nur unidirektional verdrehbare Lünette mit Minutenskala und Tiefentabelle. All das kennzeichnete die 1967 lancierte Doxa „SUB". 1968 brachte die Doxa „SUB 300T Conquistador" mit patentiertem Heliumventil. 1969 kürte die Elite-Kampfschwimmereinheit der Schweizer Armee die „SUB 300T Professional" zum offiziellen Unterwasserzeitmesser.

Nachdem diese Modelle in Sammlerkreisen einen Ruf wie Donnerhall genießen, versteht sich die aktuelle Doxa-Uhrenlinie „1967 Sub 300 Professional" als zeitgemäße Reinterpretation. Modelle in Orange sind Pflicht. Aber natürlich offeriert das Familienunternehmen auch andere Farben.

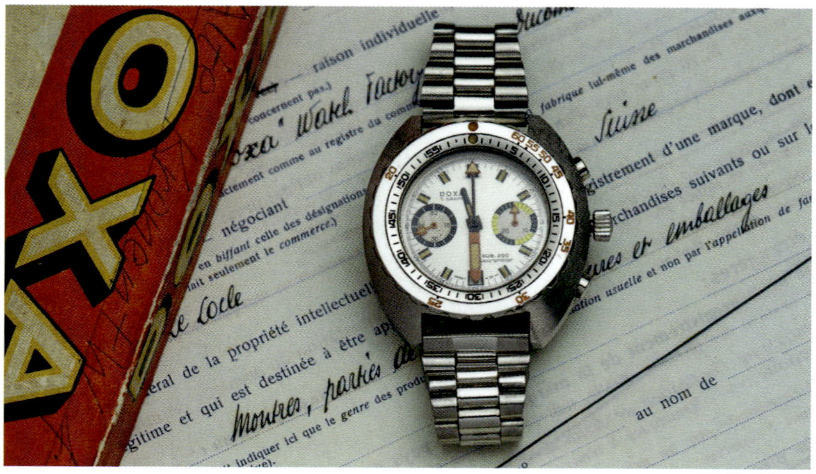

SUB 200 T-Graph

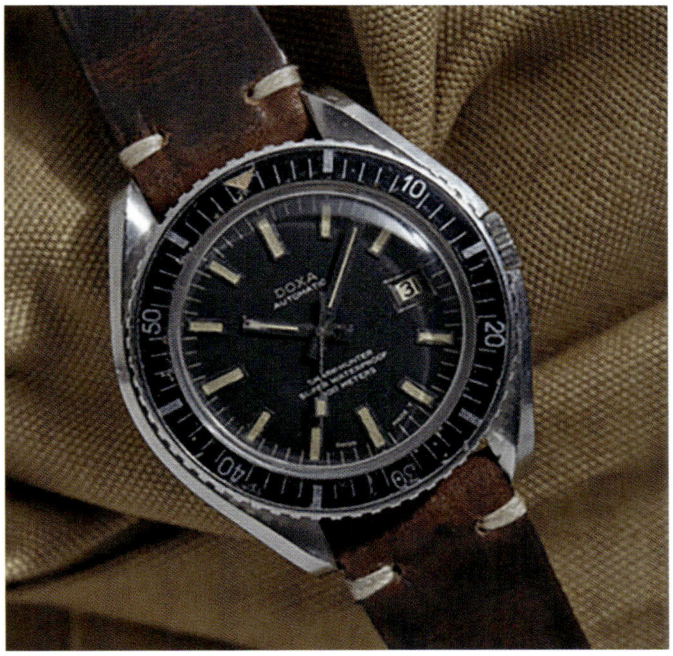

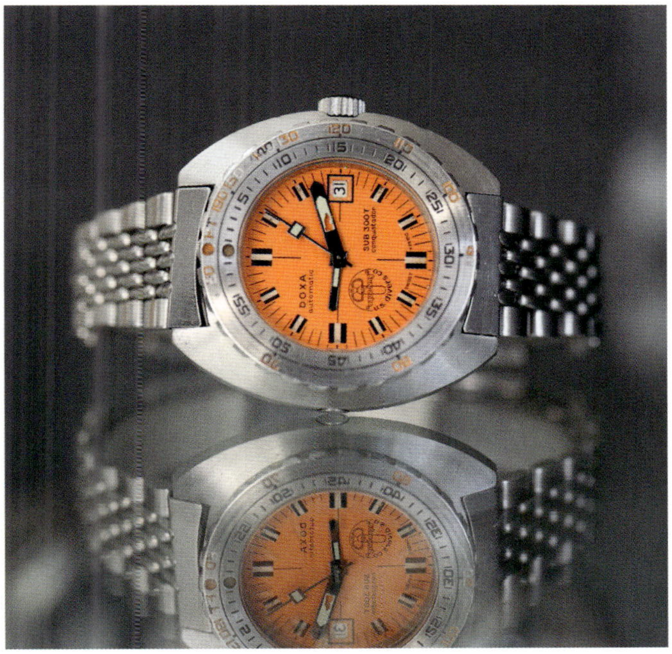

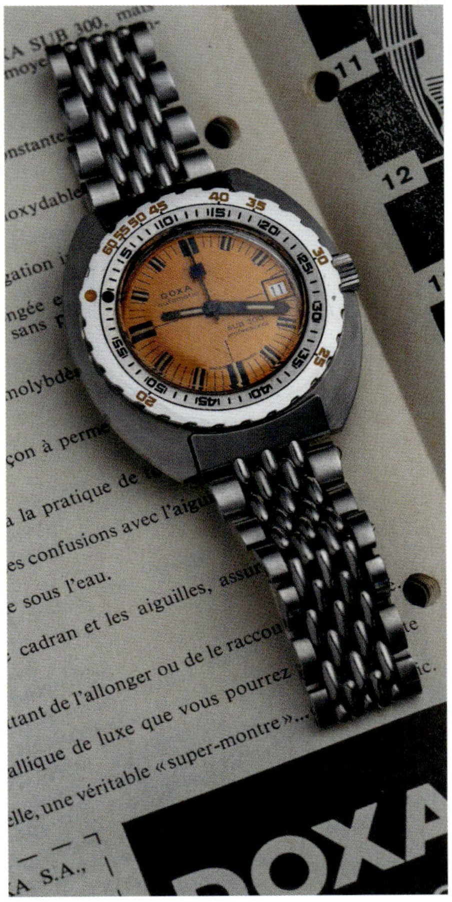

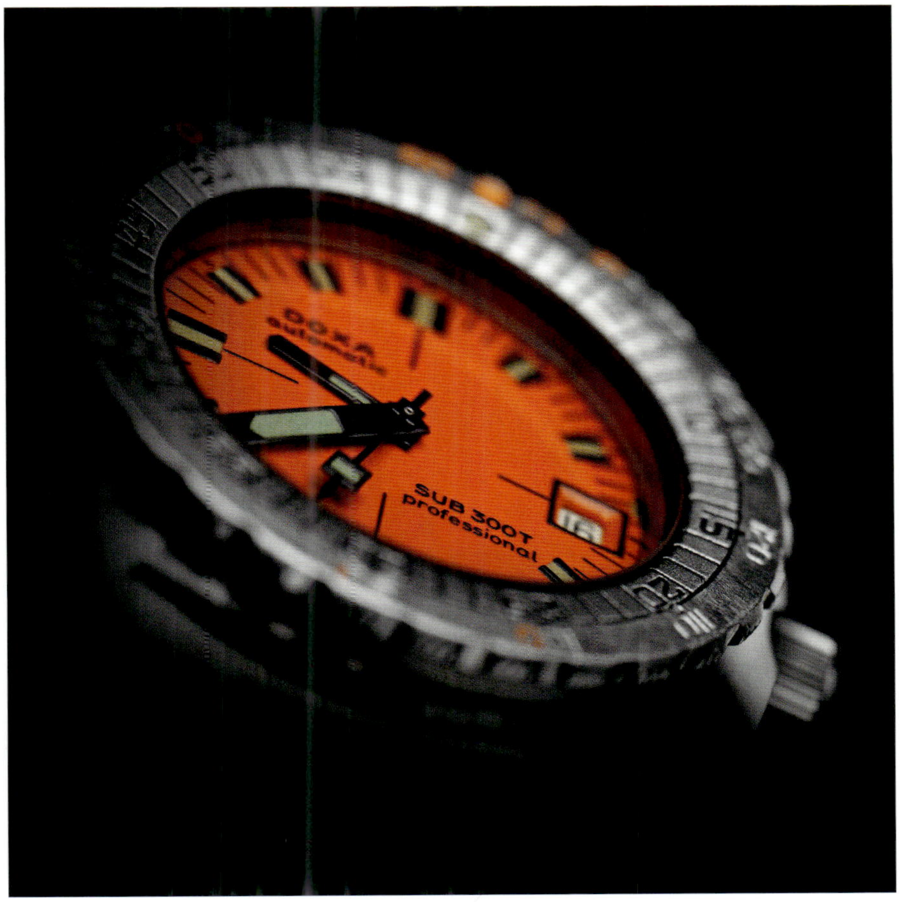

***From top left, clockwise* | *Von oben links im Uhrzeigersinr* :** *SUB 200 | SUE 300T Conquistador | Original SUB, 1967 | SUB 300 Professional*

ROGER DUBUIS EXCALIBUR

2006

TIME IN A SPIDER'S WEB

Other than their name, the current products of the brand that is a member of the Richemont empire have little in common with the timepieces designed by Monsieur Roger Dubuis from 1995 onwards. One of his first creations was the retro-style Hommage line. Roger Dubuis offered various editions, always limited to 28, including the Chronograph Bi-Retrograde with retrograde day and date indicators. In 2006, the brand image changed with the launch of the distinctive Excalibur flagship line. The Spider in particular appeals to lovers of horological extravagance, both in terms of appearance and engineering. There is of course a reason why it was dubbed the Spider. The story unfolds in 2007 as Roger Dubuis first caused a stir with his skeletonised movements. Their eye-catching appearance bore a distinct resemblance to that of filigree spider's webs. In contrast to manual skeletonising practised since 2005, the openwork Spider design emerged in the course of the development of manufacture calibres. Here, the engineers focus on optimum visibility, achieved by the congruence of the supporting components in whose bearings the gear train rotates. At the same time, the overall stability of this filigree structure must not be compromised in any way. Experienced designers ensure that the Excalibur cases match the style of these special movements. They are both protective and decorative. Because almost all Roger Dubuis products are ennobled by the Geneva seal, Excalibur Spider watches do not brook any compromise, whether with or without complications.

ZEIT IM SPINNENNETZ

Mit jenen Uhren, welche Monsieur Roger Dubuis ab 1995 fürs Handgelenk konzipierte, haben die aktuellen Erzeugnisse der zum Richemont-Imperium gehörenden Marke außer dem Namen nur noch herzlich wenig zu tun. Zu seinen Erstlingswerken gehörte beispielsweise die im Retrostil gestaltete Linie „Hommage". Roger Dubuis offerierte verschiedene, jeweils auf 28 Stück limitierte Editionen, darunter der „Chronograph Bi-Retrograde" mit rückspringenden Anzeigen für Wochentag und Datum. 2006 änderte sich das Bild der Marke mit dem Lancement der markanten Leader-Linie „Excalibur".

Speziell die „Spider" genannte Ausführung wirkt auf Menschen mit Hang zur uhrmacherischen Extravaganz sowohl optisch wie auch technisch und handwerklich besonders anziehend. Mit dem Beinamen „Spider" hat es selbstverständlich seine Bewandtnis. Die Geschichte begann 2007, als Roger Dubuis erstmals mit skelettierten Uhrwerkskonstruktionen für Aufmerksamkeit sorgte.

Ganz spontan weckt deren augenfällige Optik Assoziationen zu filigranen Spinnennetzen. Im Gegensatz zu der seit 2005 gepflegten manuellen Skelettierung entsteht die durchbrochene „Spider"-Struktur bereits im Zuge der Entwicklung von Manufakturkalibern. Bei deren Konstruktion achten die Techniker strikt auf optimalen Durchblick, ermöglicht durch die Kongruenz der tragenden Teile, in deren Lagern sich das Räderwerk bewegt. Dabei darf die lichte Struktur der Stabilität des Ganzen keinen Abbruch tun. Passend zu den solcherart gestalteten Uhrwerken kümmern sich erfahrene Designer um „Excalibur"-Gehäuse. Diese sind Schutz und Zierde zugleich. Weil Roger Dubuis fast alle Produkte durch das Genfer Siegel adelt, verbieten sich bei den „Excalibur Spider" getauften Armbanduhren Kompromisse jeglicher Art. Egal ob mit oder ohne uhrmacherische Komplikation.

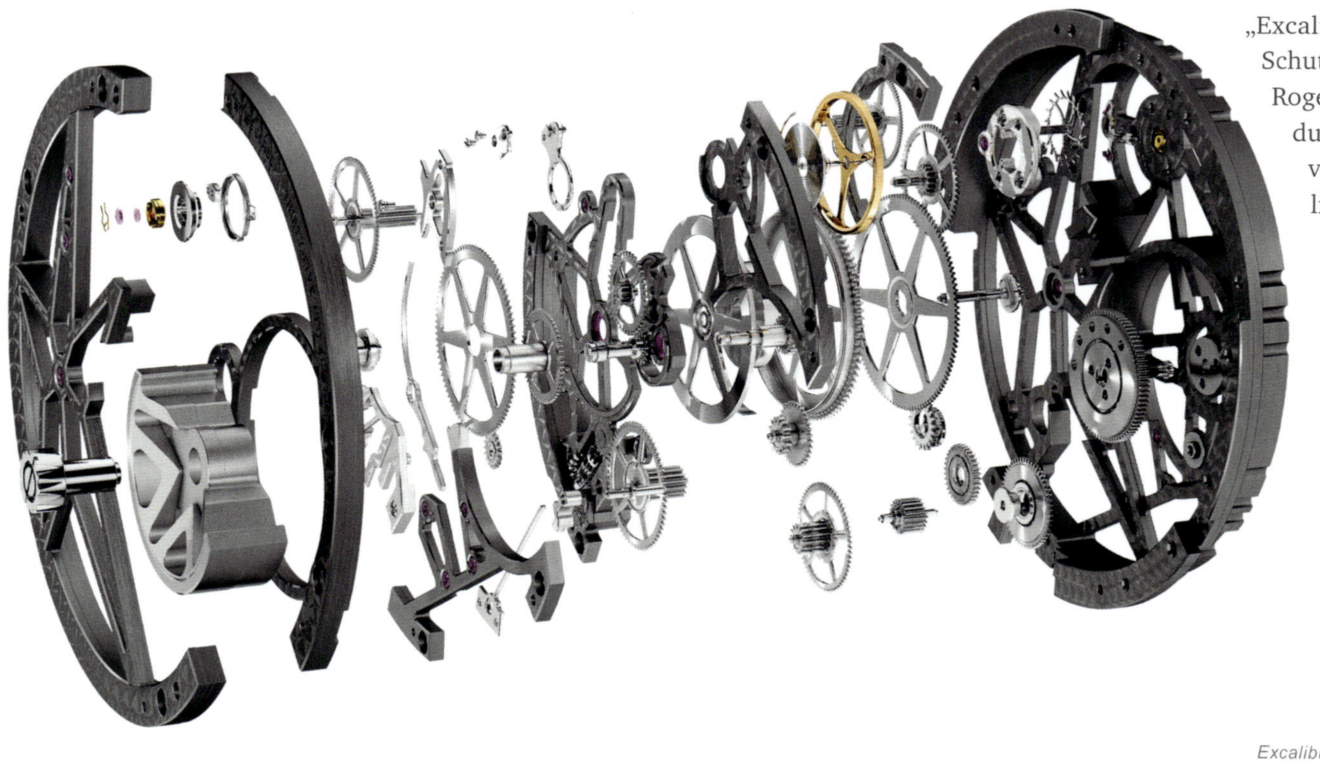

Excalibur Spider calibre, exploded view, 2015

EXCALIBUR TITANIUM MONOTOURBILLON, 2024

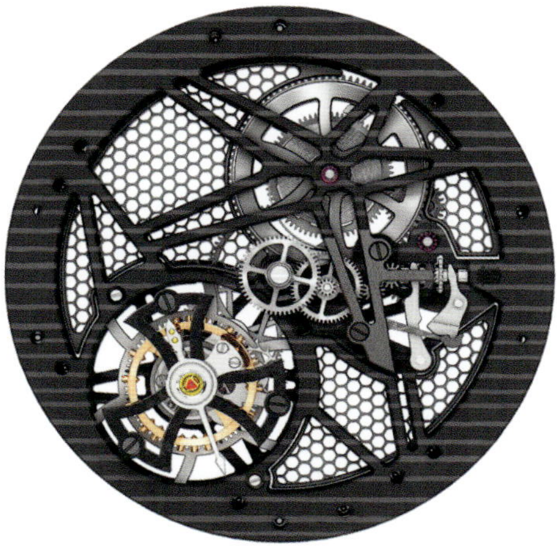

Excalibur Spider Calibre RD509SQ, 2015

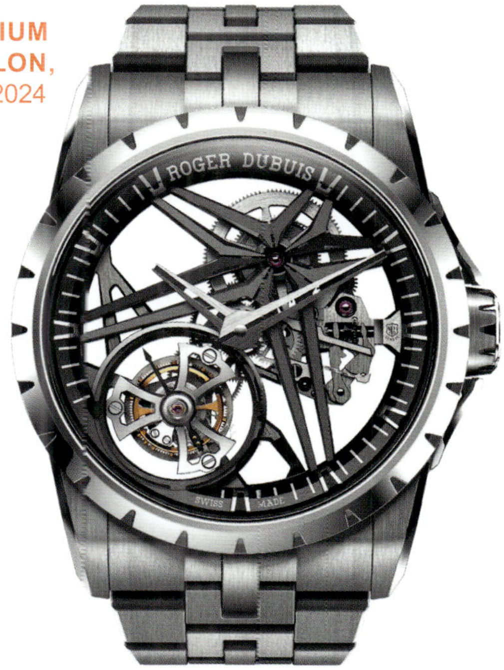

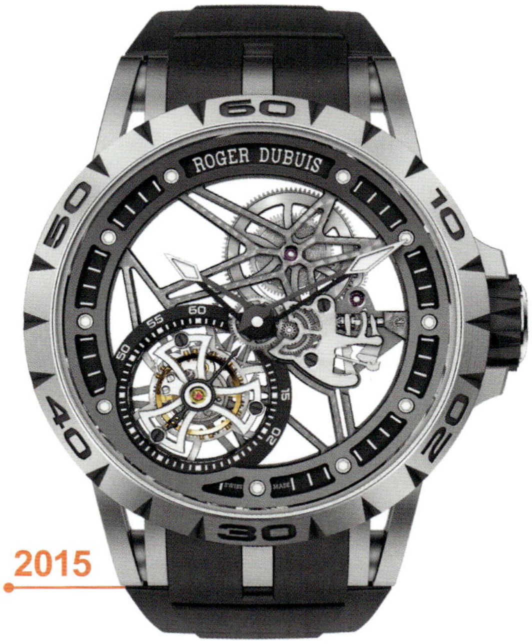

2015

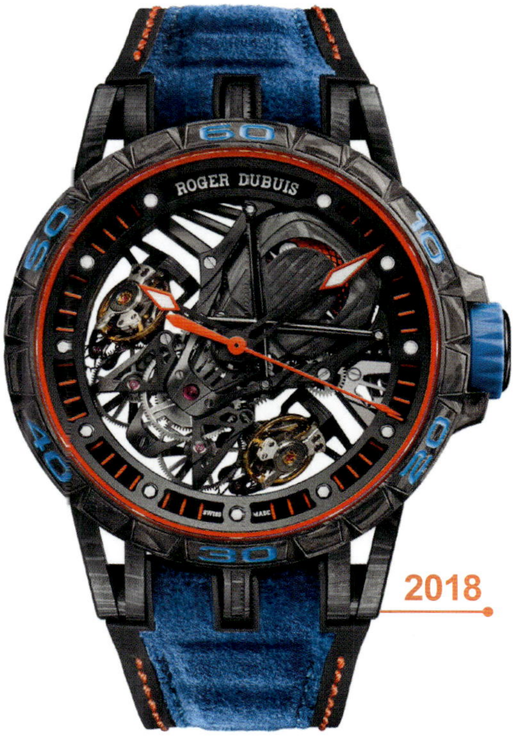

2018

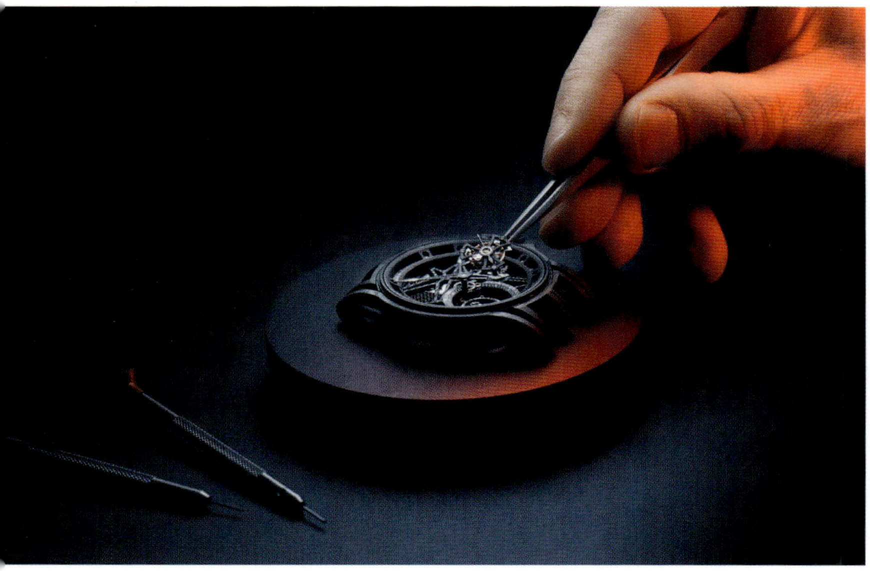

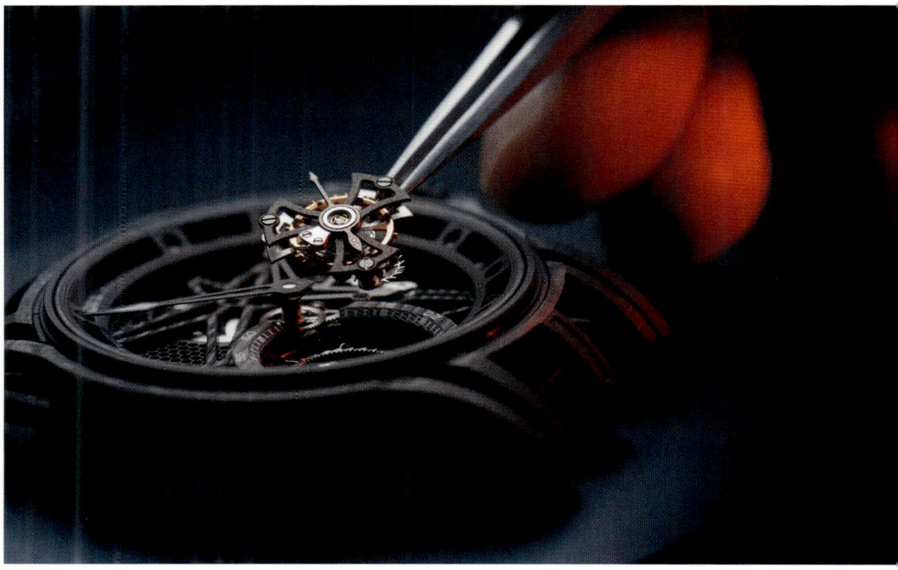

Excalibur Spider Carbon, 2019

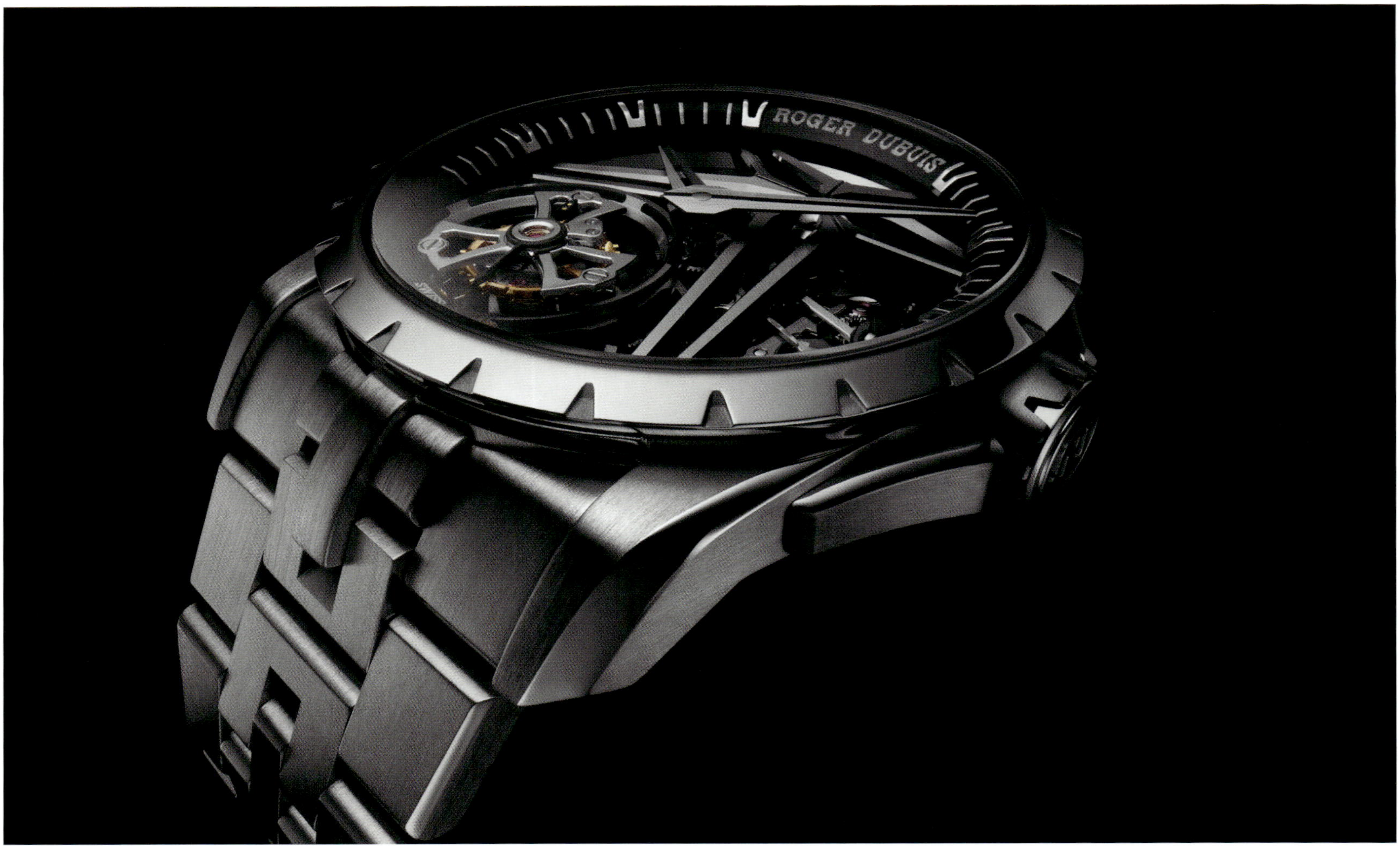

Excalibur Titanium Monotourbillon, 2024

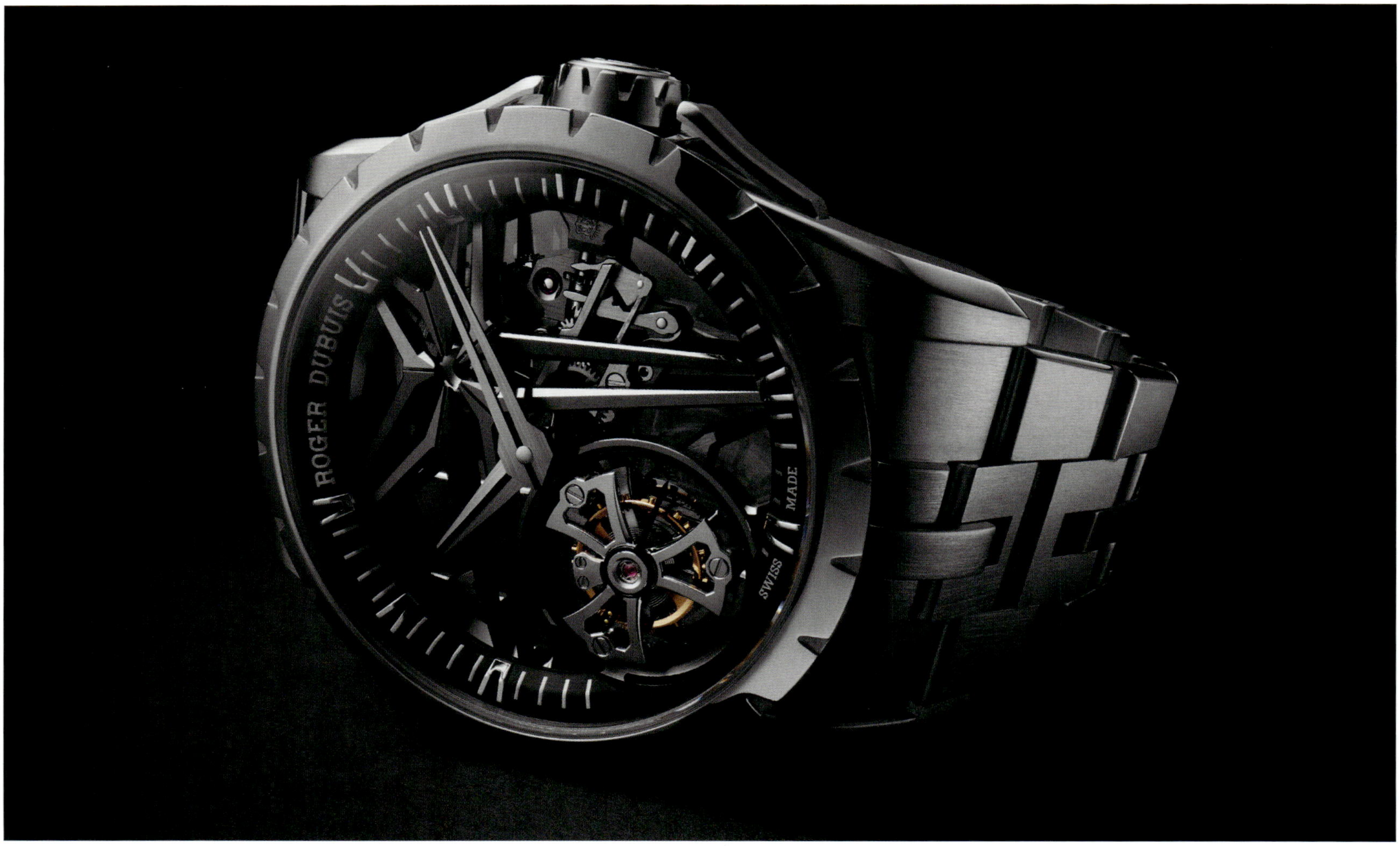

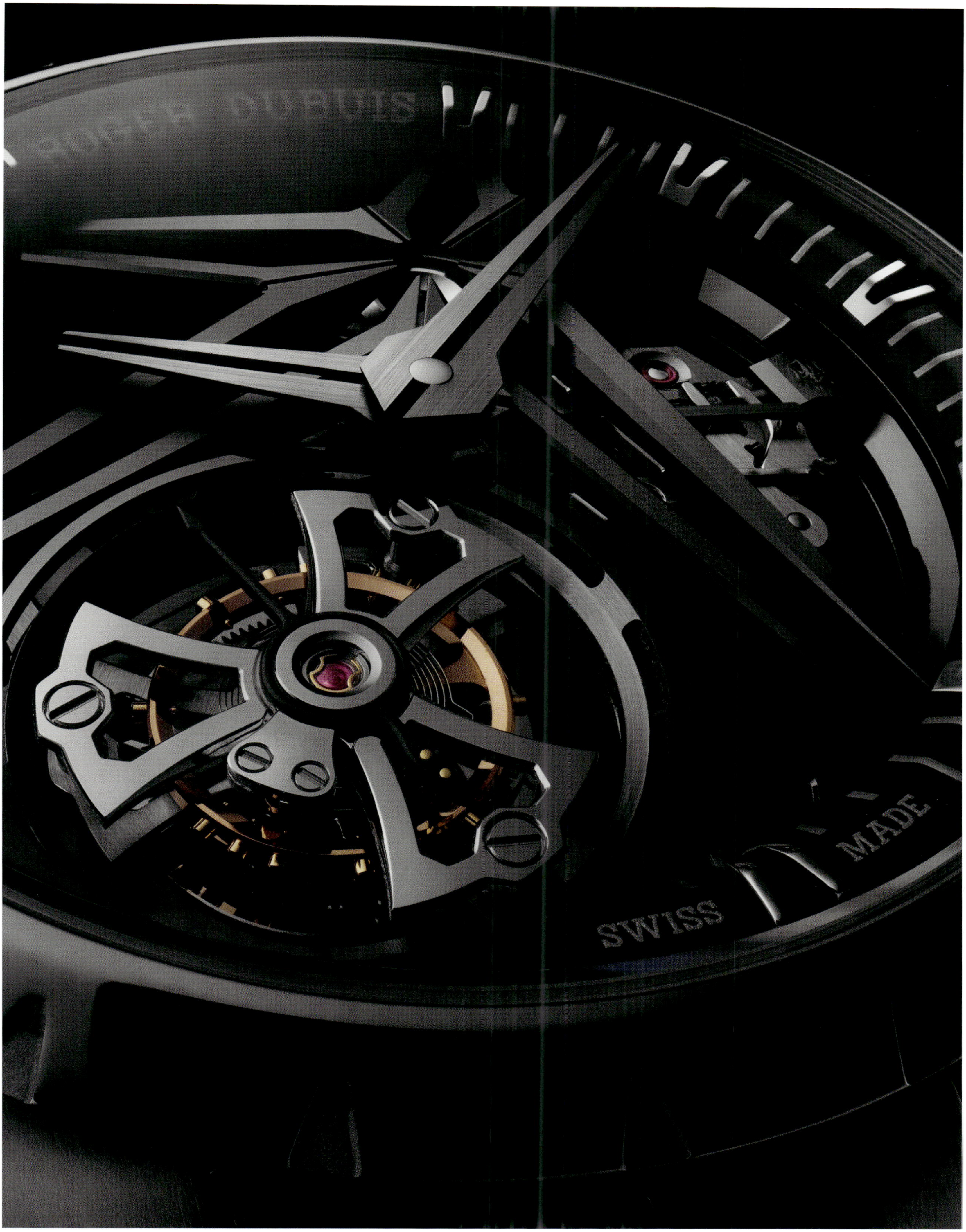
ROGER DUBUIS
SWISS
MADE

PHILIPPE DUFOUR 1996

DUALITY

A DOUBLE BALANCING ACT

At first glance, the unassuming white house differs little from the others in this remote area. Two storeys, pointed gable, windows all around. On crossing the threshold of the narrow wooden entrance door, the observant visitor will notice a small plaque with the inscription "Collège 1870". Once upon a time, the children of the tiny Jura Le Soliat community learned the three Rs here. This also goes for the three daughters of Philippe Dufour, who set up his studio in the light-filled former classroom. What he does for a living is revealed by the golden door plate with the inscription "*Horlogerie compliquée*", French for complicated watchmaking. A superficial look at Dufour's Duality does not give away its secret. Especially if you only focus on the discreet dial and hands. Closer examination of the hand-wound movement shows that appearances are deceptive. In 1996, the master watchmaker revisited the resonance principle of balances oscillating in opposite directions. Each oscillating object transmits vibrations to its surroundings. Mechanical feedback causes the two balance wheels to swing back and forth by dint of their natural oscillation frequencies. Together, the synchronised regulators achieve a higher inertia. External forces such as bumps and sudden movements are absorbed, as they inversely influence the oscillation rates of the two balances. If one accelerates, the other decelerates. The two balance wheels then revert back to their common frequency. Consequently, abrupt changes in position have no or only minor influence on the accuracy of watches with this complication. Things can be this easy when an outstanding watchmaker has a hand in the sophisticated engineering. Philippe Dufour's seemingly simple yet super-precise Duality is a shining example of such ingenious simplicity.

DOPPELSCHWUNG AM HANDGELENK

Auf den ersten Blick unterscheidet sich das schlichte weiße Haus kaum von den anderen in dieser abgeschiedenen Gegend. Zwei Stockwerke, spitzer Giebel, Fenster rundherum. Beim Eintreten durch die schmale hölzerne Eingangstür bemerkt der aufmerksame Besucher eine kleine Plakette mit der Inschrift „Collège 1870". Einstmals lernten hier die Kinder der winzigen Jura-Gemeinde Le Solliat das Lesen und Schreiben. Das gilt auch für die drei Töchter von Philippe Dufour, der sein Atelier im lichtdurchfluteten ehemaligen Klassenzimmer etabliert hat. Seine Profession verkündet ein goldenes Türschild mit der Aufschrift „Horlogerie compliquée", zu Deutsch: Fertigung komplizierter Uhren. Wirklich kompliziert erscheint Dufours „Duality" auf den ersten Blick nicht. Vor allem dann, wenn man nur das dezente Zifferblatt und die Zeiger betrachtet. Dass der Schein trügt, zeigt sich beim Studium des Handaufzugswerks. 1996 besann sich der Altmeister auf das Resonanzprinzip gegenläufig agierender Schwingsysteme. Ihm zufolge überträgt jedes oszillierende Objekt Vibrationen an seine Umgebung. Bedingt durch mechanische Gegenkopplung bewegen sich zwei Unruhreifen mit dem ständigen Mittel ihrer natürlichen Schwingfrequenzen hin und her. Gemeinsam erreichen die synchronisierten Gangregler eine höhere Trägheit. Äußere Krafteinwirkungen wie Schläge und plötzliche Bewegungen werden absorbiert, da sie die Schwingungsrate beider Unruhn gegensätzlich beeinflussen. Wird eine beschleunigt, bremst die andere entsprechend ab. Anschließend schwingen sich die Unruhn wieder auf ihre gemeinsame Frequenz ein. Folglich üben abrupte Lageänderungen keinen oder nur geringen Einfluss auf die Ganggenauigkeit der damit ausgestatteten Armbanduhren aus. So einfach können die Dinge sein, wenn sich ein überragender Uhrmacher darauf versteht, sie auf erlesene Art und Weise zu realisieren. So geschehen bei der vermeintlich einfachen und doch so präzisen „Duality" von Philippe Dufour.

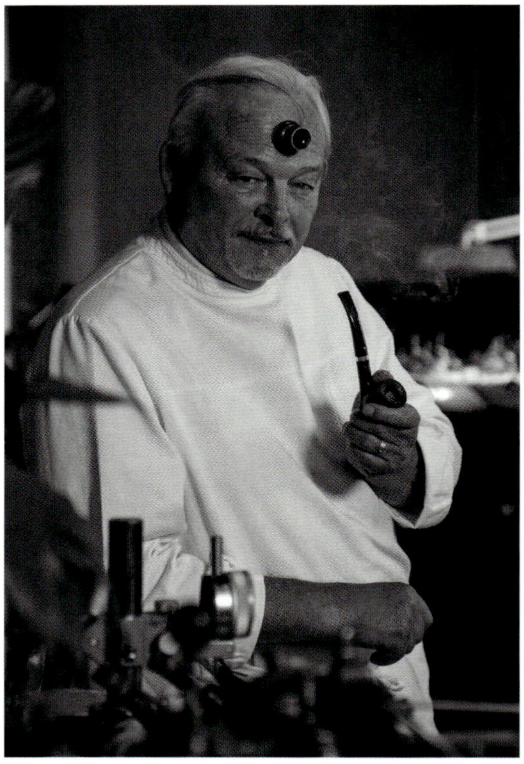

Philippe Dufour

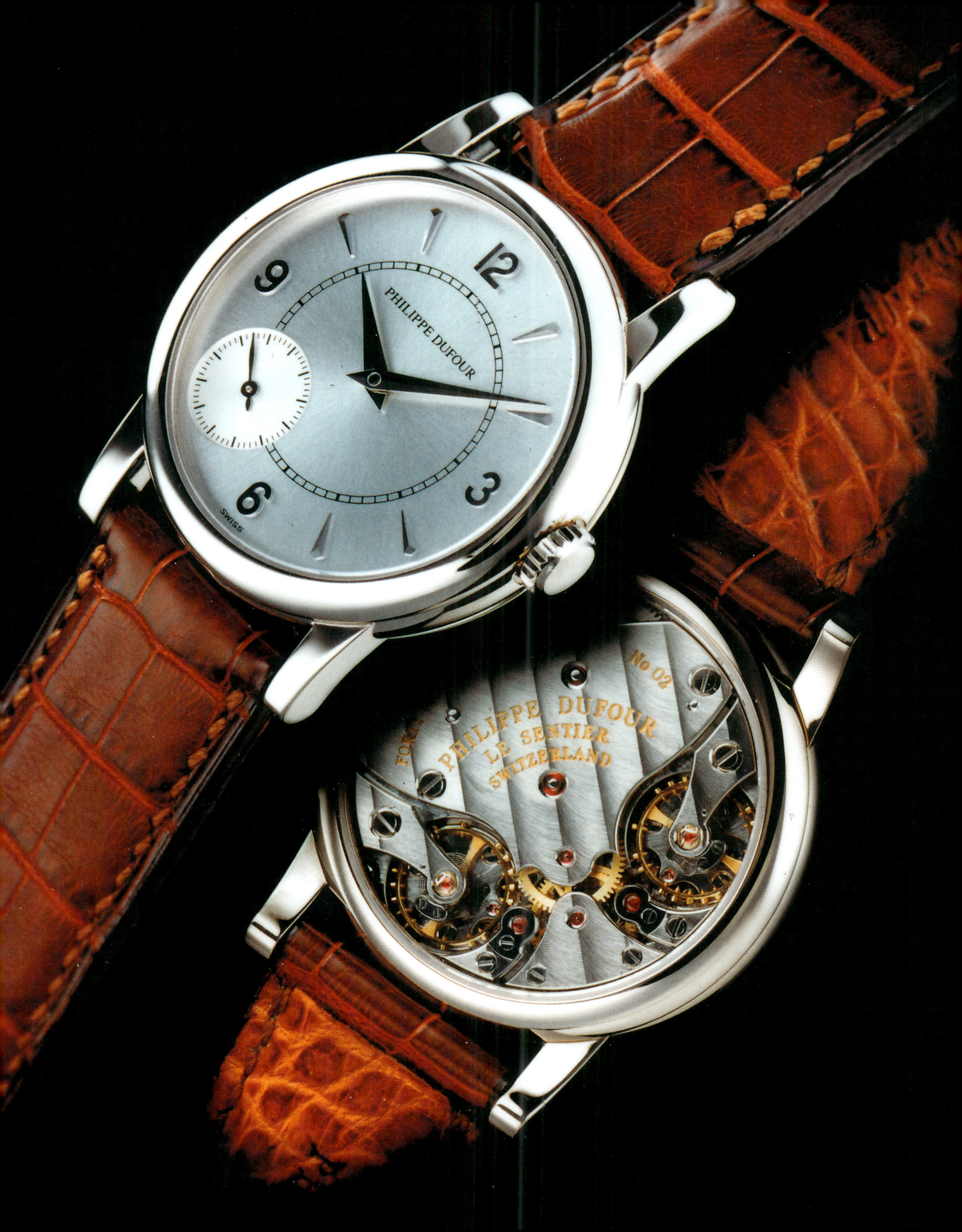
PHILIPPE DUFOUR
SWISS
No 02
PHILIPPE DUFOUR
LE SENTIER
SWITZERLAND

SPORT CLASSIC

BY EDDY SCHÖPFER FOR EBEL

The success of the watch model presented in 1977 came as something of a surprise. Ebel owner Pierre-Alain Blum had entrusted Eddy Schöpfer with the design. He, in turn, created a product that was by no means in step with the spirit of the time. Various design elements of the Sport Classic as it was named appeared much too avant-garde to secure commercial success. Yet the son of the company founder trusted his gut feeling and identified a potential winner. His intuition proved to be correct. In retrospect, the newcomer established itself as a highly respected trendsetter. Its characteristic features henceforth informed the design signature of the self-appointed "architect of time". Its distinctive look is largely owed to the hexagonal case with its soft and undeniably luscious curves. The product designer drew his inspiration from the sculptor's art. Its finely structured wave motif lent the Sport Classic a look that the changing trends of the following decades could not hope to rival. In an era which tended to favour edgy lines, the sculptural monocoque case was an exception that breathed originality.

Looking back, the combination of the sporty, fluid geometry of the hexagon with rounded contours and bevelled edges can well be described as visionary. Five visible bezel screws emphasise the dynamic element. To this day, the hand-flattering qualities of the timepiece's surfaces remain undisputed. Their texture recalls a pebble smoothed over time by water. Without a flat quartz movement, the aesthetic appearance in the late 1970s would not have been possible. Many models in the Sport Classic collection still favour this electronic means of timekeeping. Self-winding mechanical movements are of course also available.

VON EDDY SCHÖPFER FÜR EBEL

Selbstverständlich vorgezeichnet war der Erfolg des 1977 vorgestellten Uhrenmodells keineswegs. Ebel-Eigentümer Pierre-Alain Blum hatte Eddy Schöpfer mit dem Design betraut. Der wiederum hob ein Produkt aus der Taufe, das dem damaligen Zeitgeist herzlich wenig entsprach. Verschiedene Gestaltungselemente der „Sport Classic" getauften Armbanduhr wirkten für einen kommerziellen Erfolg viel zu avantgardistisch. Doch der Sohn des Firmengründers glaubte an sein Gespür und witterte eine echte Chance. Tatsächlich sollte er recht behalten. Retrospektiv betrachtet erwies sich der Newcomer als viel beachteter Trendsetter. Seine charakteristischen Gestaltungsmerkmale prägten fortan die Designhandschrift der selbst ernannten „Architekten der Zeit". Zum hohen Wiedererkennungswert trug die sechseckige Schale mit sanften, aber durchaus sinnlichen Rundungen maßgeblich bei. Bei seinem Schaffen hatte sich der Produktgestalter von der Bildhauerkunst inspirieren lassen. Ihr fein strukturiertes Wellenmotiv verschaffte der „Sport Classic" einen Look, dem die wechselnden Trends der anschließenden Jahrzehnte kaum etwas anhaben konnten. In einer Epoche, als eher kantige Linienführungen bestachen, war das skulptural geprägte Monocoque-Gehäuse eine originelle Ausnahmeerscheinung.

Rückblickend kann die Kombination aus der sportlich fließenden Geometrie des Sechsecks mit runden Linien und abgeschrägten Kanten durchaus als visionär gelten. Fünf sichtbare Lünettenschrauben unterstreichen das dynamische Element. Unbestritten waren und sind bis heute die handschmeichlerischen Qualtäten der Oberflächen. Sie erinnern an einen von viel Wasser geglätteten Kieselstein. Ohne ein flaches Quarzwerk wäre der ästhetische Auftritt in den späten 1970er-Jahren nicht möglich gewesen. Zeitbewahrende Elektronik findet sich auch heute noch in vielen Modellen der „Sport Classic"-Kollektion. Aber natürlich gibt es auch mechanisches Innenleben mit Selbstaufzug.

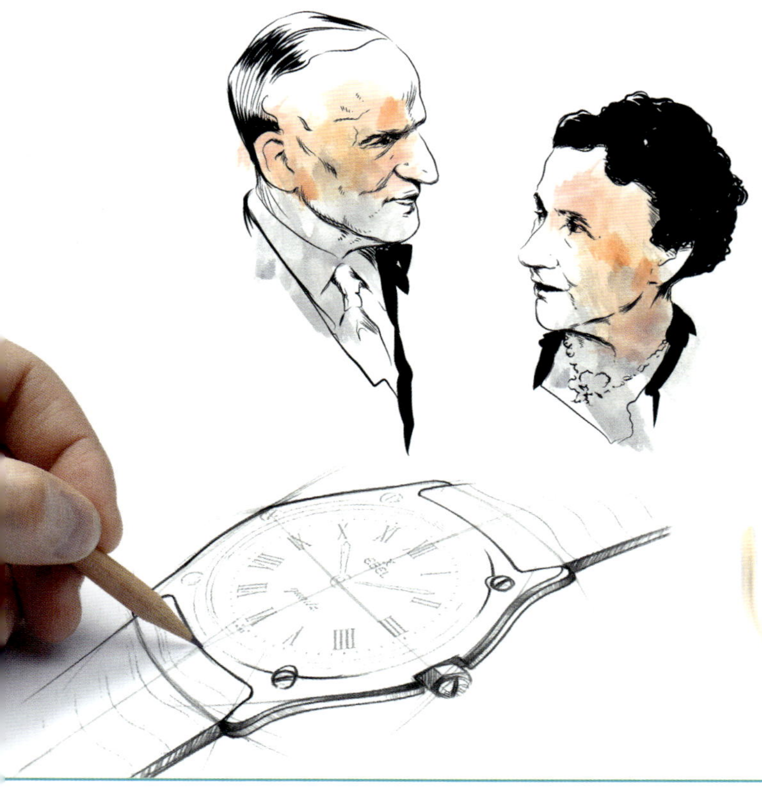

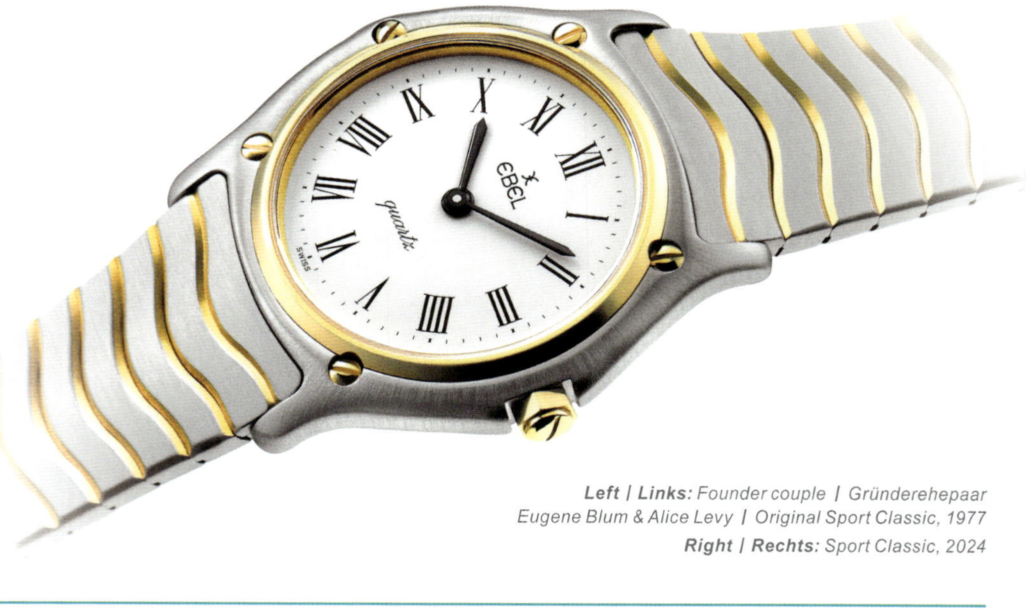

Left | Links: *Founder couple | Gründerehepaar Eugene Blum & Alice Levy | Original Sport Classic, 1977*
Right | Rechts: *Sport Classic, 2024*

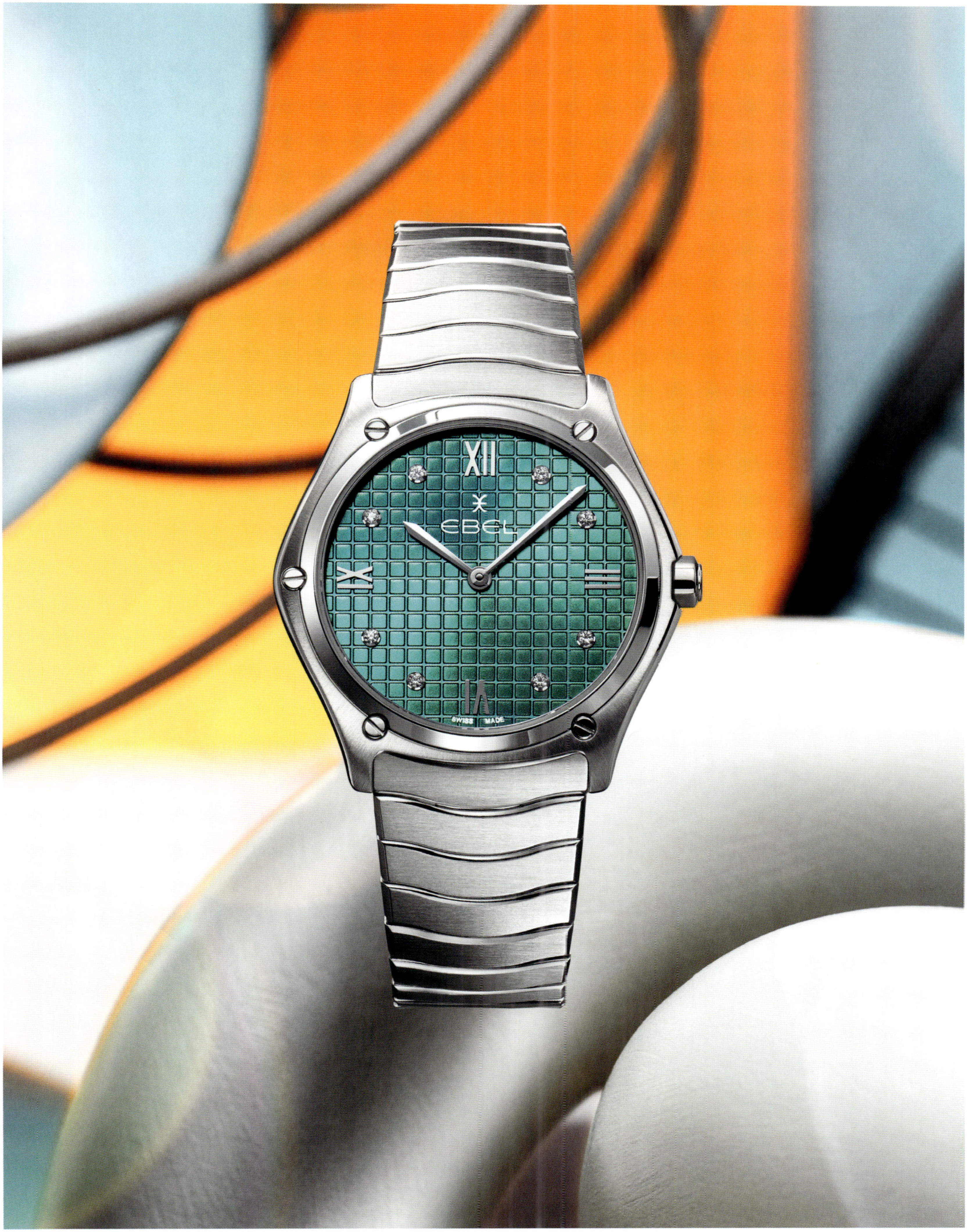
EBEL
SWISS MADE

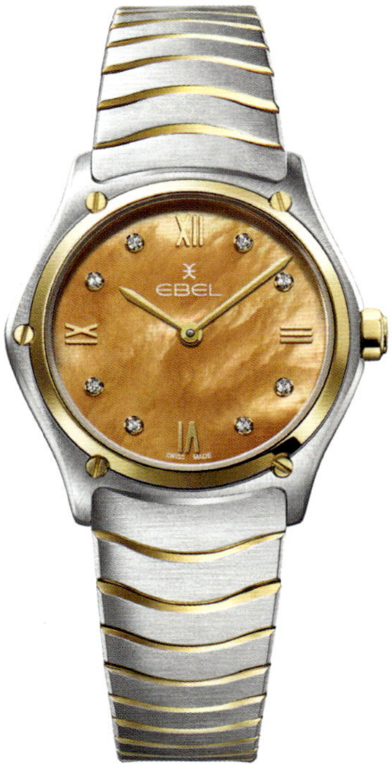

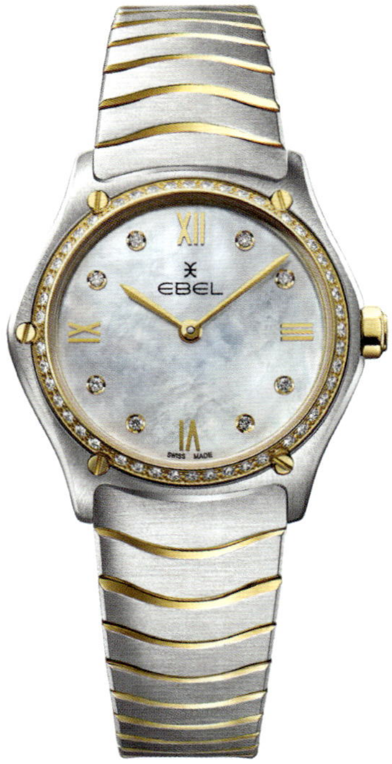

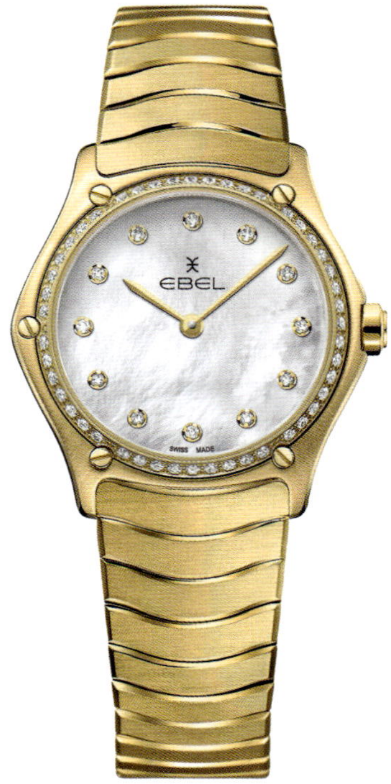

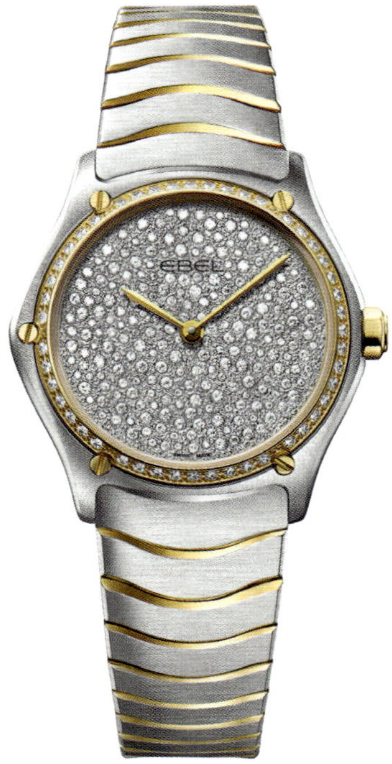

1) *Ref. 1216595*

2) *Ref. 1216390A*

3) *Ref. 1216392*

4) *Ref. 1216563*

5) *Ref. 1216527*

6) *Ref. 1216432M*

7) *Ref. 1216614*

8) *Ref. 1216420A*

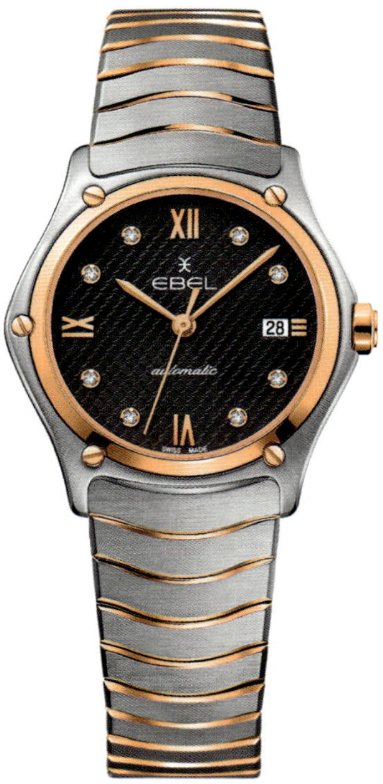

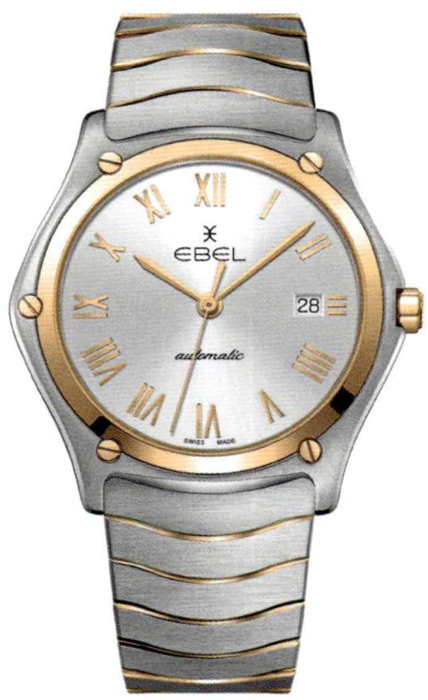

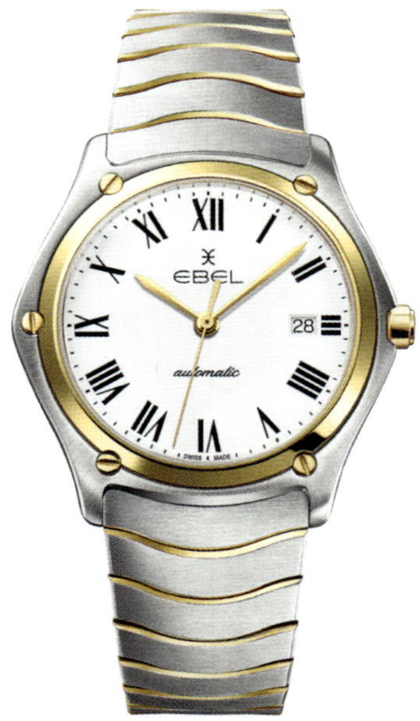

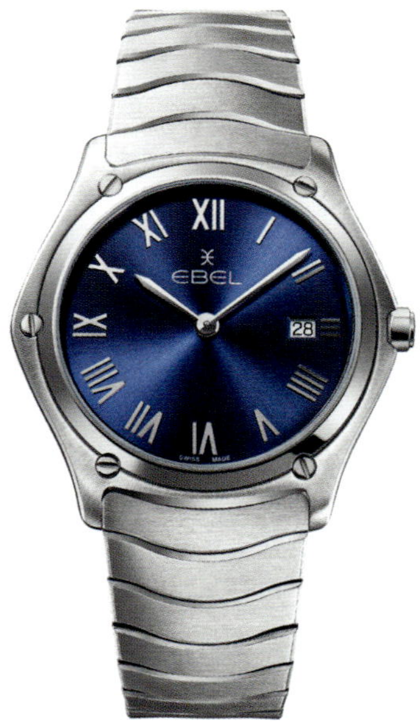

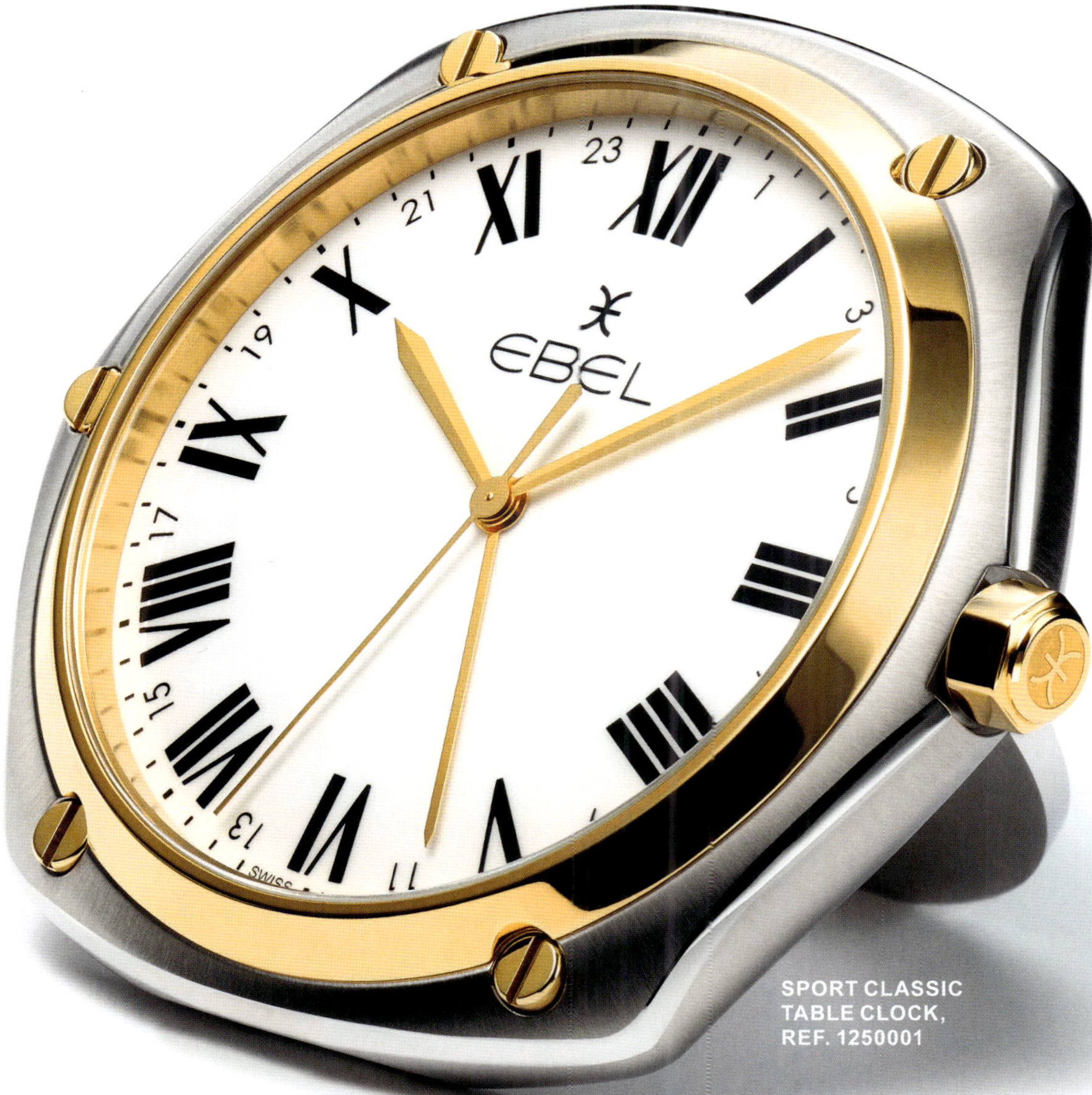

SPORT CLASSIC TABLE CLOCK, REF. 1250001

FREDERIQUE CONSTANT HIGHLIFE 1999

HIGH LIFE AT WRIST LEVEL

The origins of Frederique Constant's Highlife collection date back to 1999. It ranged from the low-end model with a quartz movement to iterations with a simple full calendar, automatic chronograph or tourbillon. The latter was paired with one of four Progress calibres. The movements, designed according to the préciplus system, were aimed at socialising tourbillon watches. Frederique Constant opted for calibre 6361.151 and renamed it FC-800. The first Highlife tourbillon was thus launched on the market, featuring a platinum case with sapphire crystal back and front, and a leather strap. There is a date indication at 12 o'clock. Following the credo that constant change is part of the essence of time, in 2020, Frederique Constant returned the Highlife collection to the test stand, subjected it to a full reworking and then relaunched it.
The newcomers also adhere to the brand's loyally cultivated principle of affordable luxury. For its bracelets, the company, now part of Citizen, follows the popular trend which allows them to be replaced without the wearer damaging their fingernails in the process. Watch lovers who appreciate purism and officially certified accuracy need look no further than the Highlife Chronometer with automatic calibre FC-303 to satisfy their passion. The date indication of the Highlife Perpetual Calendar Manufacture does not require manual correction until 2100. In this case, the manufacture calibre is the FC-775. The undisputed premium model is the Highlife Tourbillon Perpetual Calendar Manufacture, also available in solid gold. Inside is the manufacture calibre FC-975 with tourbillon and perpetual calendar.

HIGHLIFE AM HANDGELENK

Bis ins Jahr 1999 reichen sie zurück, die Wurzeln der „Highlife"-Kollektion von Frederique Constant. Das Produktspektrum reichte vom Einsteigermodell mit Quarzwerk über Versionen mit einfachem Vollkalendarium oder Automatik-Chronograph bis hin zum Tourbillon. Letzteres nutzte eines von vier Progress-Kalibern. Die nach dem System „préciplus" konzipierten Uhrwerke sollten zur Sozialisierung der Drehganguhren beitragen. Frederique Constant hatte sich für das Kaliber 6361.151 entschieden und taufte es in FC-800 um. Auf den Markt gelangte so das erste „Highlife"-Tourbillon mit beidseitig verglastem Platingehäuse und Lederband. Am Zifferblatt bei „12" fand sich ein Zeigerdatum. Nach dem Motto, dass zum Wesen der Zeit stetige Veränderung gehört, hat Frederique Constant die „Highlife"-Kollektion auf den Prüfstand gestellt und ihr 2020 nach kompletter Überarbeitung einen Relaunch beschert.
Auch die Newcomer gehorchen dem seit Anbeginn gepflegten Prinzip eines erschwinglichen Luxus. Bei den Armbändern folgt das mittlerweile zu Citizen gehörende Unternehmen dem verbreiteten Trend, Armbänder selbst ohne Beschädigung der Fingernägel auswechseln zu können. Uhrenliebhaber, die puristischen Auftritt und amtlich geprüfte Ganggenauigkeit schätzen, kommen beim „Highlife-Chronometer" mit dem Automatikkaliber FC-303 auf ihre Kosten. Erst im Jahr 2100 verlangt die Datumsanzeige des „Highlife Perpetual Calendar Manufacture" nach manueller Korrektur. Hier nennt sich das Manufakturkaliber FC-775. Unangefochtenes Spitzenprodukt ist das auch in Massivgold erhältliche „Highlife Tourbillon Perpetual Calendar Manufacture". In diesem Fall kommt das mit Drehgang und ewigem Kalender ausgestattete Manufakturkaliber FC-975 zum Einsatz.

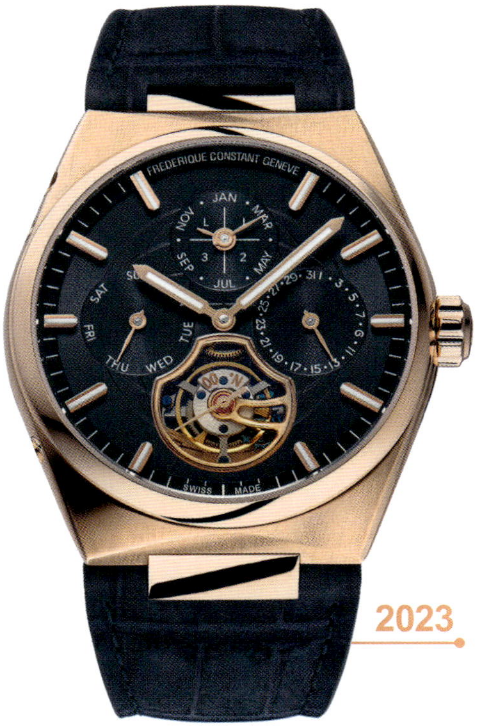

2023

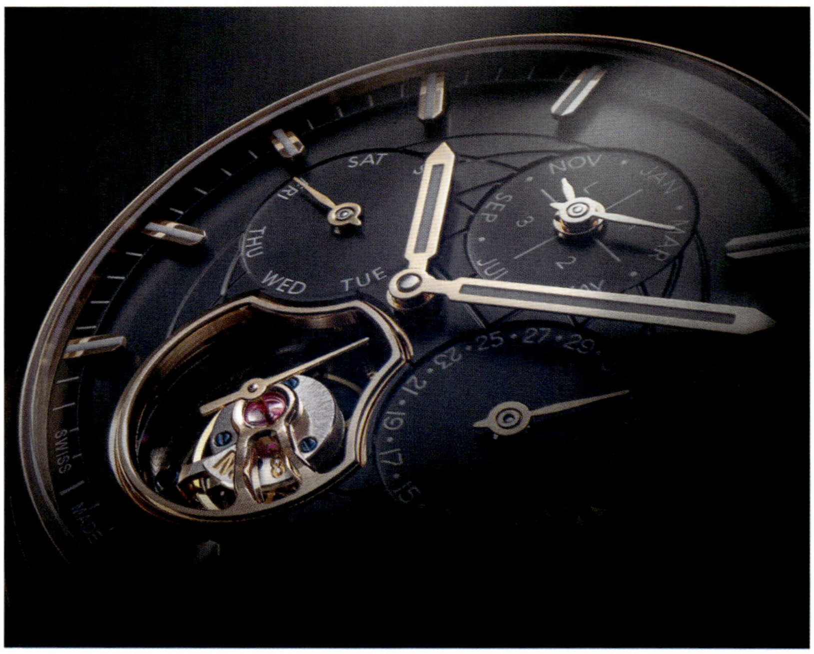

Highlife Tourbillon Perpetual Calendar Manufacture, 2023

2023

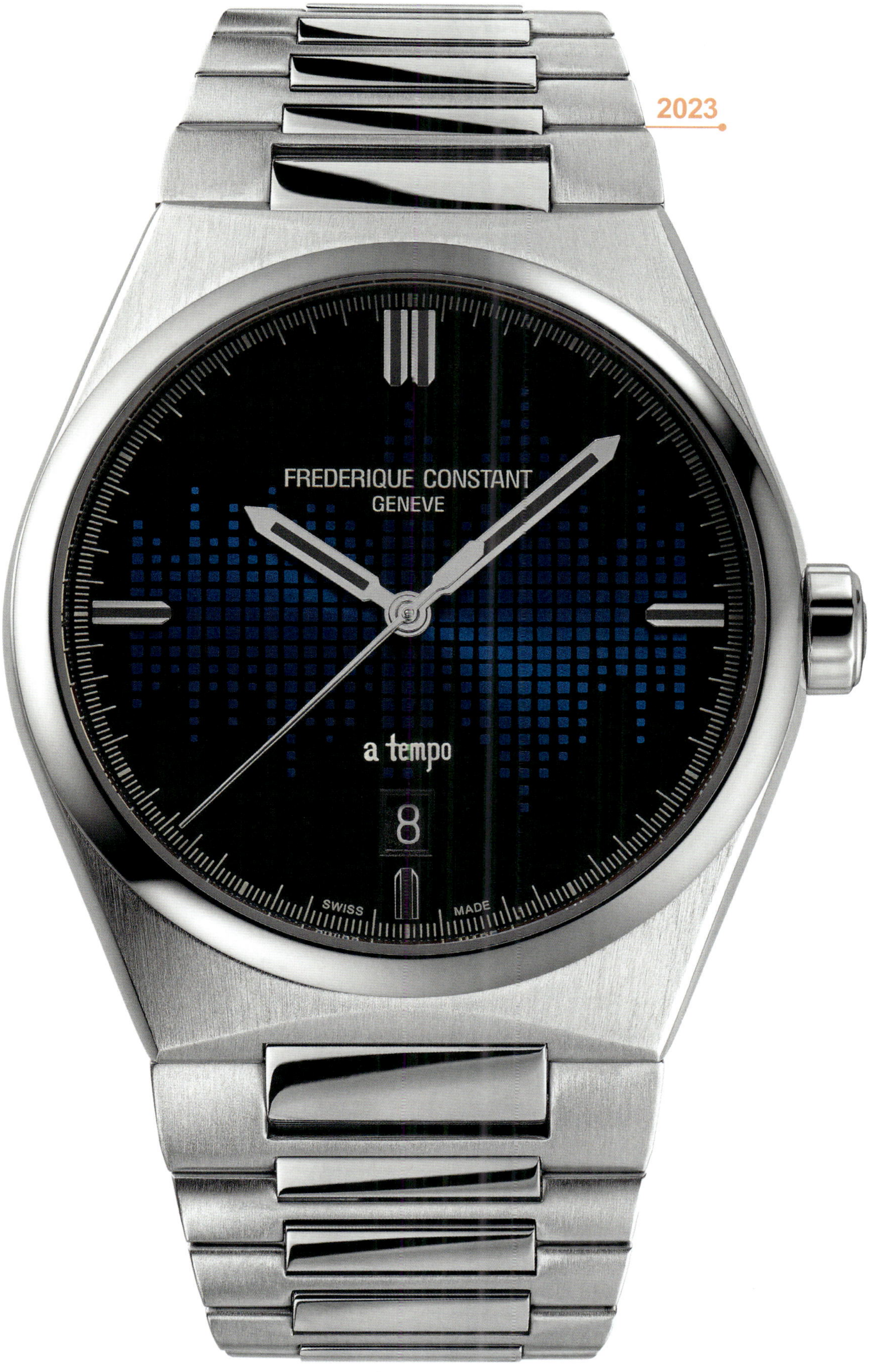

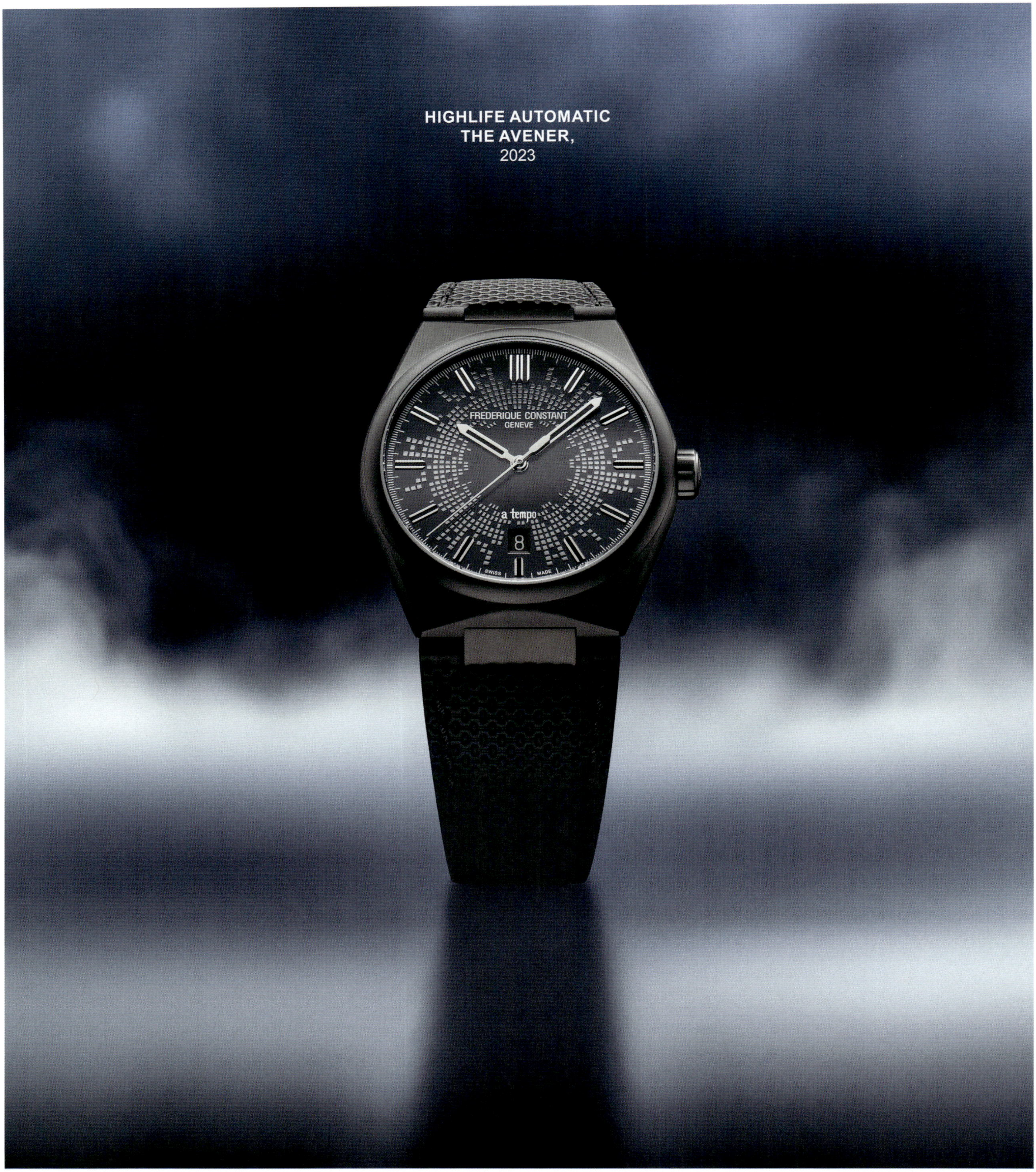

HIGHLIFE AUTOMATIC
THE AVENER,
2023

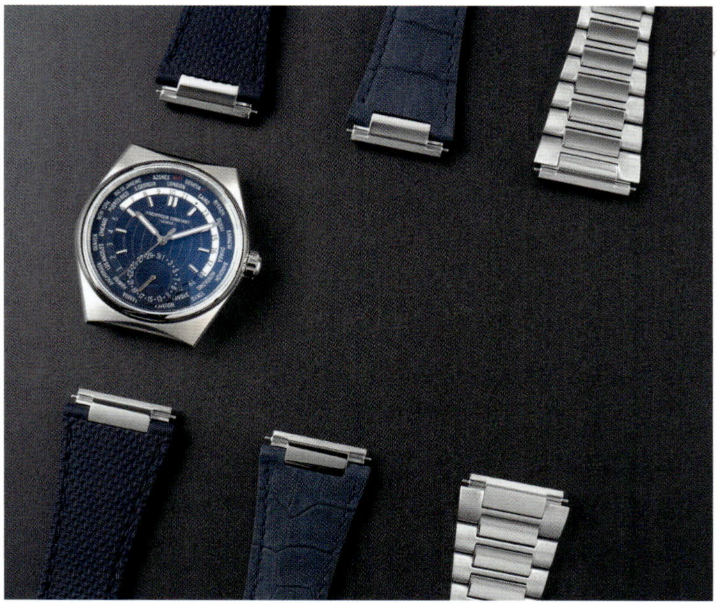

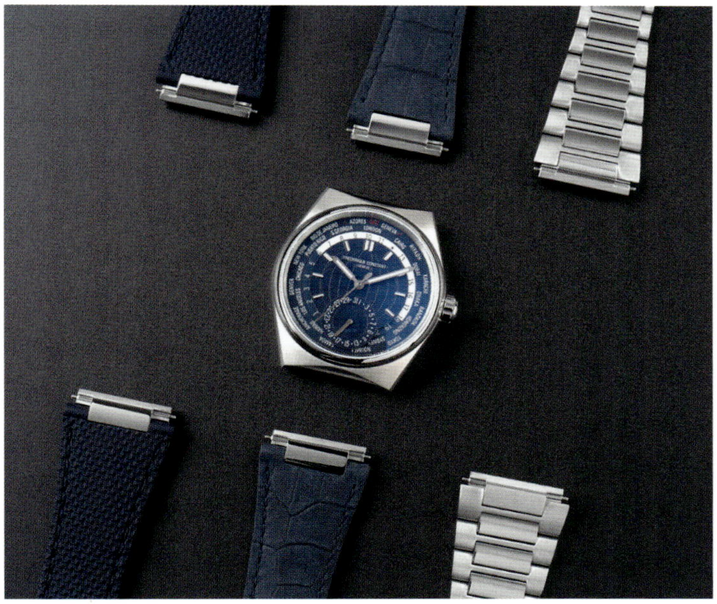

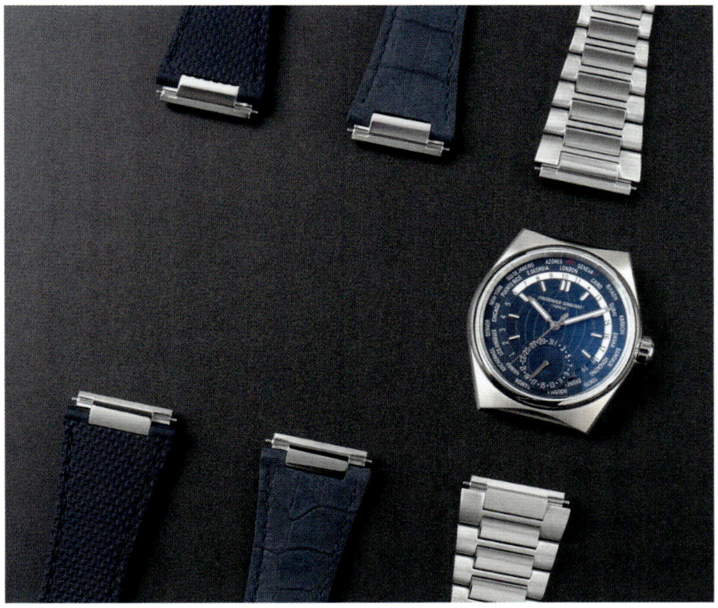

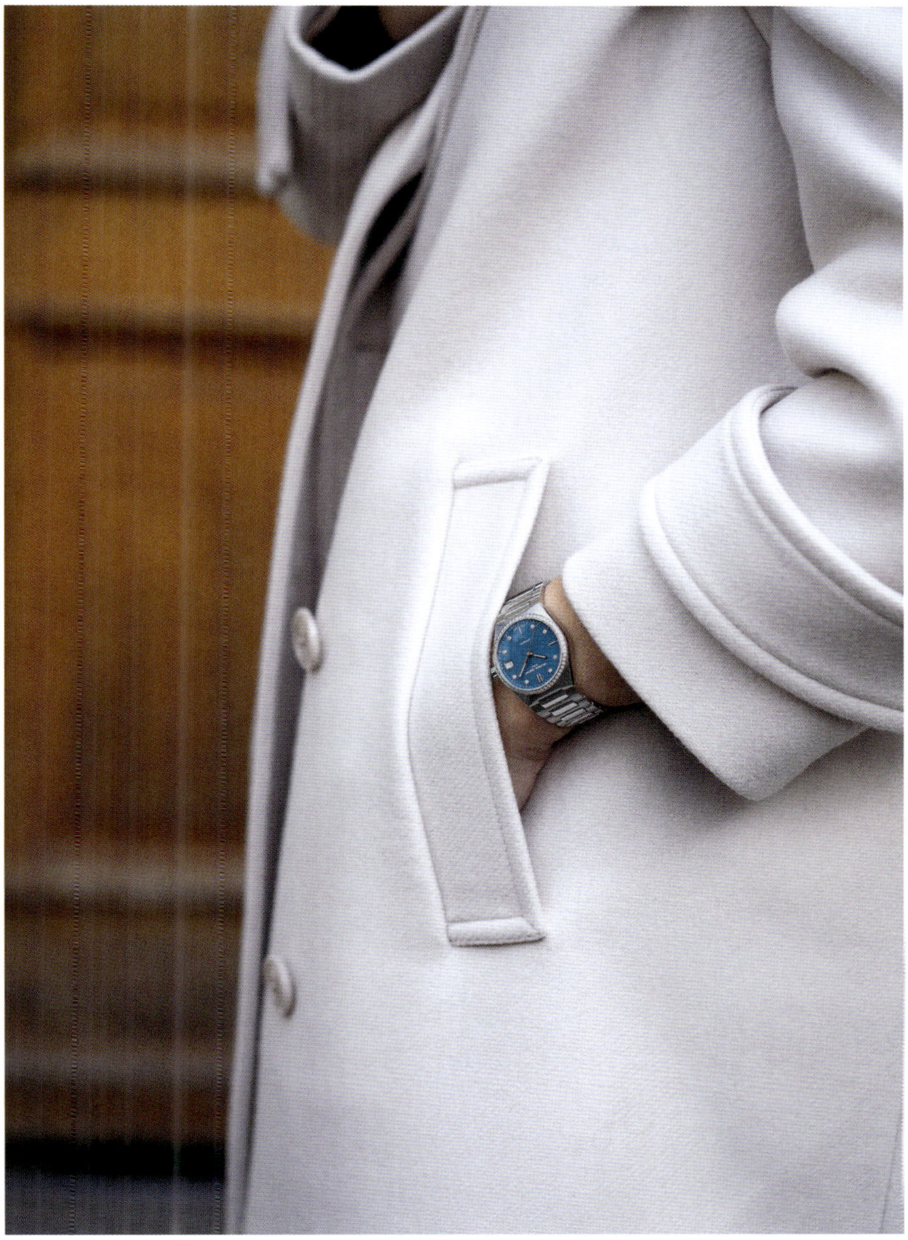

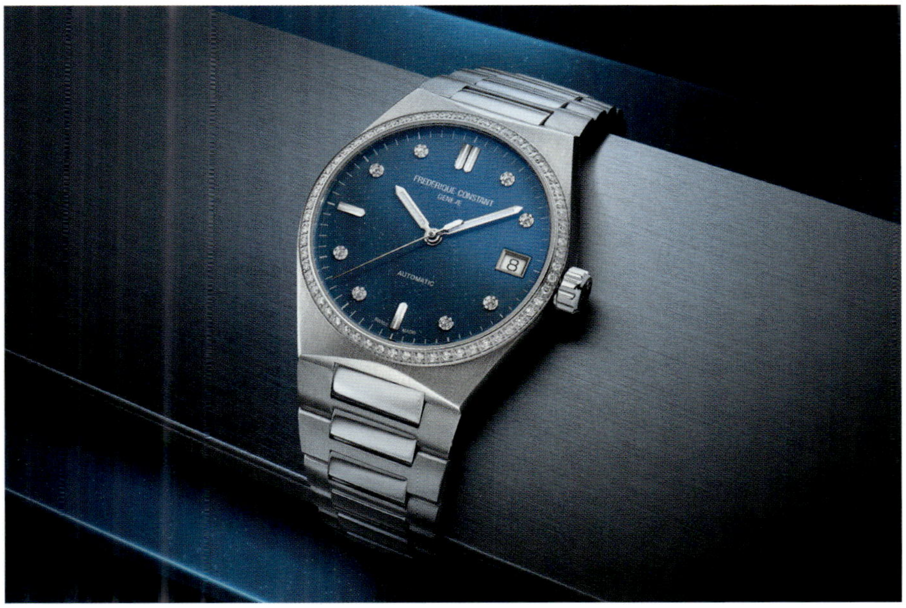

Left | Links: Highlife Worldtimer Manufacture , 2021, with three straps | mit drei Armbändern | **Right | Rechts:** Highlife Ladies Automatic, 2023

GIRARD-PERREGAUX 1991

TOURBILLON 3 BRIDGES

TOURBILLON WITH THREE BRIDGES

In 1854, Constant Girard and Marie Perregaux entered into holy matrimony. The marriage of the watchmaker and the daughter of a renowned chronometer maker led the Bautte company founded in 1791 to change its name to the present Girard-Perregaux. The bridegroom and experienced watchmaker was particularly attracted to the tourbillon as a way of offsetting the negative effects of gravity on the accuracy of vertically carried pocket watches. He therefore keenly developed and produced his own tourbillon movements. The master watchmaker mounted his innovative oscillating and escapement system in a delicate cage. His initial designs were still based on a conventional plate construction. But between 1867 and 1890, some 20 microcosms of an exceptional nature were created. Their technical novelty consisted of three prominent bridges. On 25 March 1884, Girard-Perregaux received a patent for their parallel arrangement on the mainplate. In this connection, a solid-gold pocket watch with a decorative hunter case deserves special mention. This mechanical gem, along with the mighty Eiffel Tower, was one of the highlights of the 1889 World Exhibition in Paris. Not least for this reason, the jury awarded it a prize. Its hand-wound movement with minute tourbillon had a compensation balance with 16 screws, a gold escape wheel and a chronometer escapement with rocker. Since this marvel was presented in the luxurious Mexican specialist store of Messrs Hauser and Zivy, the magnificent object was named La Esmeralda. One of its proud owners was Porfirio Díaz, the Mexican president from 1884 to 1911. In the 1960s, Girard-Perregaux succeeded in repurchasing the masterpiece. Its reacquisition inspired the comeback of the legendary three-bridge tourbillon. In 1981, the watch world noted the limited edition of a pocket watch with this complication. Only 20 models were produced. Marking its bicentenary in 1991, the traditional company presented wristwatch versions for the first time.
Since then, the bridge tourbillon, and many iterations of it, has become a fully fledged member of the Girard-Perregaux watch collection.

TOURBILLON MIT DREI BRÜCKEN

1854 schlossen Constant Girard und Marie Perregaux den heiligen Bund der Ehe. Durch die Hochzeit zwischen dem Uhrmacher und der Tochter eines angesehenen Chronometermachers erhielt die 1791 von Jean-François Bautte gegründete Manufaktur ihren aktuellen Namen Girard-Perregaux. Dem erfahrenen Ehemann hatte es das Tourbillon zur Kompensation negativer Schwerkrafteinflüsse auf die Ganggenauigkeit senkrecht getragener Taschenuhren besonders angetan. Tatkräftig packte er deshalb die Entwicklung und Fertigung eigener Drehgangwerke an. In einen filigranen Käfig montierte der Meister sein neuartiges Schwing- und Hemmungssystem. Die ersten Konstruktionen erfolgten noch in konventioneller Platinenbauweise. Aber zwischen 1867 und 1890 entstanden etwa 20 Mikrokosmen außergewöhnlicher Art. Ihre technische Besonderheit bestand in drei markanten Brücken. Für deren parallele Anordnung auf der Werkplatte konnte Girard-Perregaux am 25. März 1884 ein Patent entgegennehmen. Besondere Erwähnung verdient in diesem Zusammenhang eine massivgoldene Taschenuhr mit kunstvoll dekoriertem Sprungdeckelgehäuse. Dieses mechanische Kleinod gehörte neben dem mächtigen Eiffelturm zu den Highlights der 1889 in Paris abgehaltenen Weltausstellung. Nicht zuletzt deswegen gestand ihm die Jury einen Ehrenpreis zu. Sein Handaufzugswerk mit Minutentourbillon verfügte über eine Kompensationsunruh mit 16 Schrauben, ein goldenes Hemmrad sowie eine Chronometerhemmung mit Wippe. Da man das Wunderwerk in einem luxuriösen mexikanischen Fachgeschäft der Herren Hauser und Zivy präsentierte, erhielt das prunkvolle Objekt den Namen „La Esmeralda“. Zu den stolzen Eigentümern gehörte Porfirio Díaz, der mexikanische Präsident von 1884 bis 1911. In den 1960er-Jahren gelang es Girard-Perregaux, dieses Meisterstück zurückzukaufen. Diese Aktion begründete ein Comeback des legendären Drei-Brücken-Tourbillons. 1981 sprach die Uhrenwelt über eine limitierte Taschenuhr mit dieser Komplikation. Erhältlich waren lediglich 20 Exemplare. Und zu ihrem 200. Geburtstag im Jahr 1991 wartete die Traditionsmanufaktur erstmals mit entsprechenden Armbanduhren auf.
Seitdem gehört das mittlerweile unterschiedlich deklinierte Brückentourbillon zum festen Bestand der Uhrenkollektion von Girard-Perregaux.

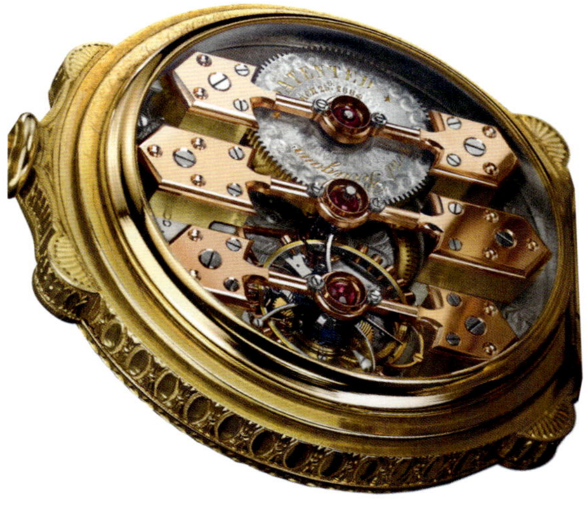

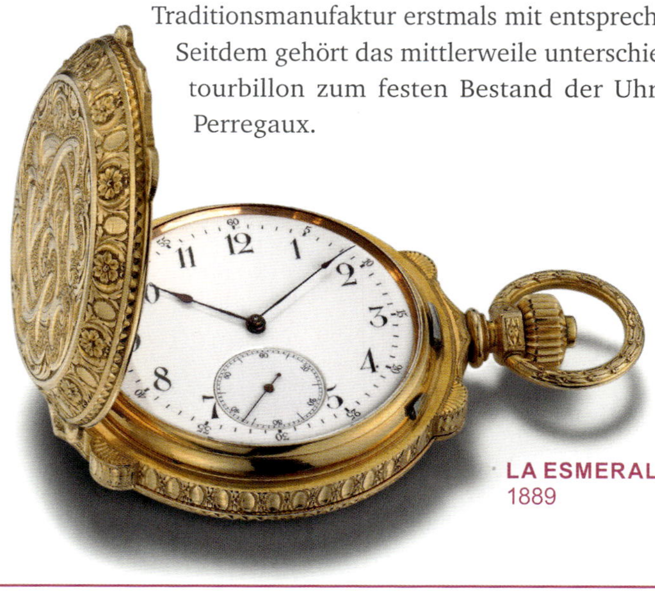

LA ESMERALDA,
1889

2016

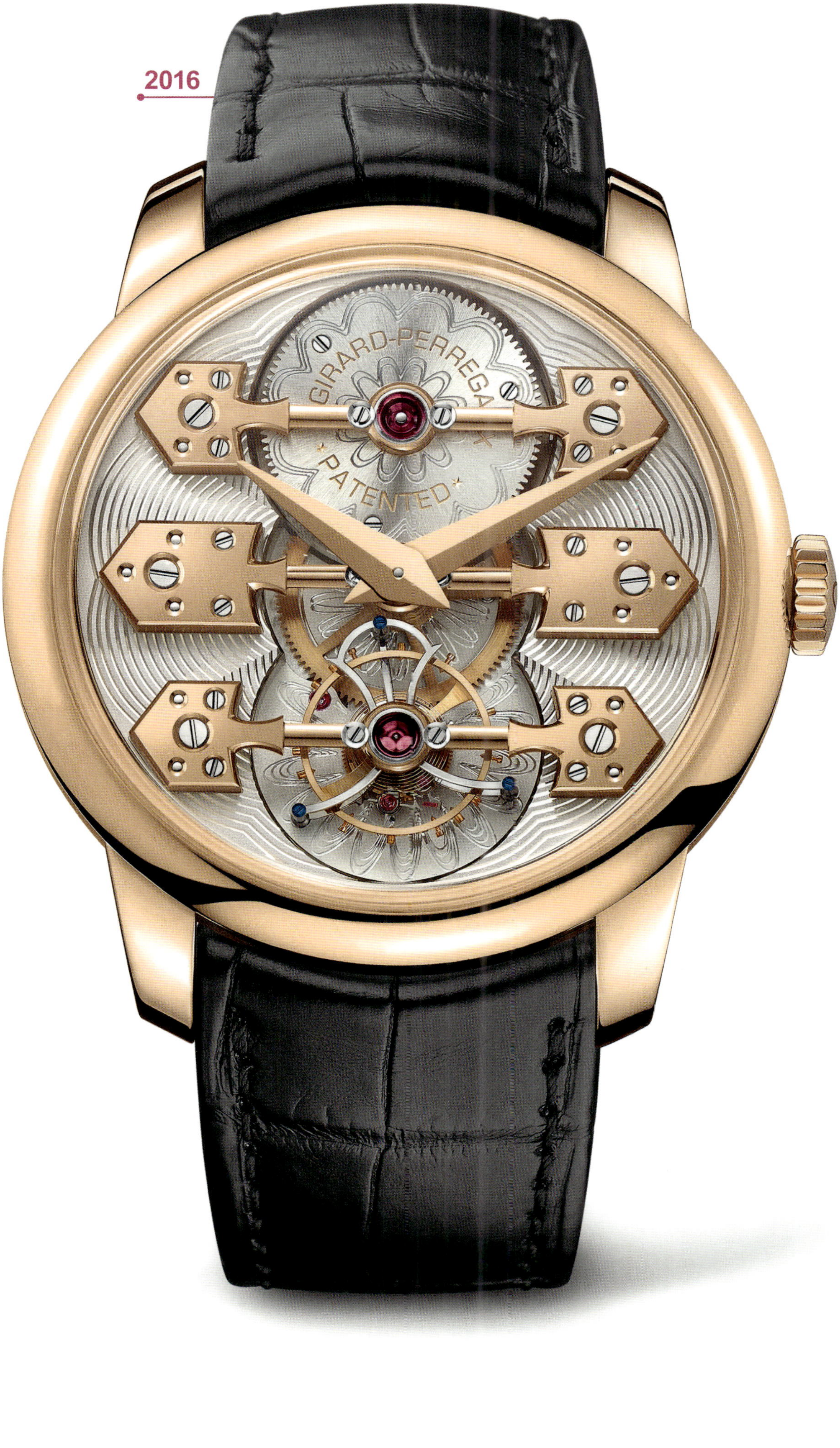

2018

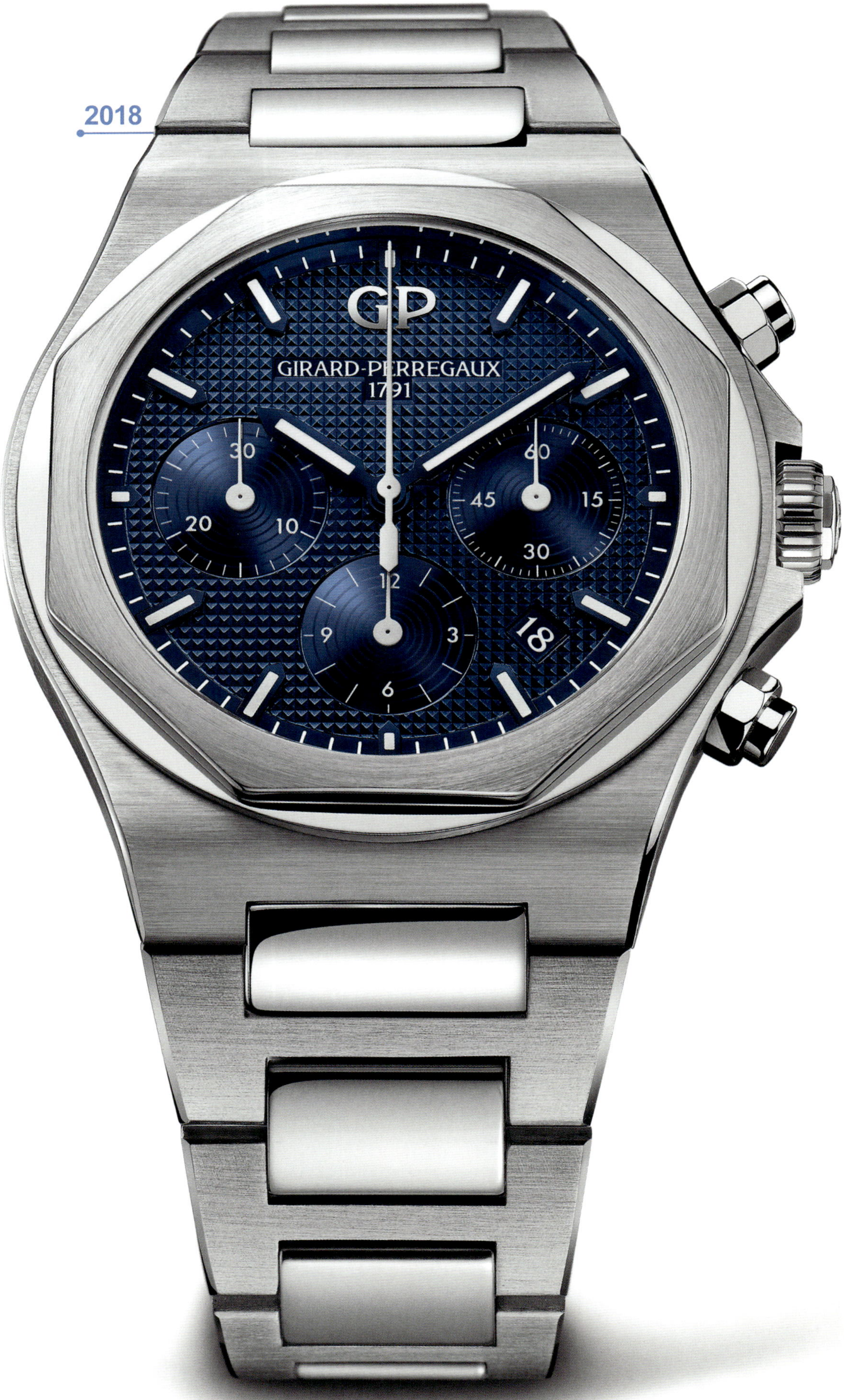

1975

GIRARD-PERREGAUX
LAUREATO

NOT RESERVED FOR LAUREATES

How did this Girard-Perregaux watch collection get its name? The answer lies in the 1970s and in the Italian importer of the venerable brand, who advertised the model – originally called Quartz Chronometer – as the "graduate ("Laureato" in Italian) of the school of Girard-Perregaux". In 1975, the first wristwatches of this name were launched on the market. Nine years later, in 1984, the time was ripe for its first makeover. Girard-Perregaux used this opportunity to present various movements with additional astronomical indications. A second design development followed a decade later. In 1995, Laureato wristwatches came on the scene. In 2003, the third stage in the collection's evolution incorporated prominent visual modifications. What immediately catches the eye are the gentler cuff-friendly lines, which characterise the Laureato to the present day. Models are available in a range of case sizes and choice of materials including stainless steel, ceramic and of course solid gold. Lovers of this watch line who browse the wide-ranging collection will be sure to find something that takes their fancy.

NICHT NUR FÜR LAUREATEN

Mit dem Namen dieser Uhrenlinie von Girard-Perregaux hat es natürlich seine Bewandtnis. Konkret geht er zurück auf die 1970er-Jahre, den italienischen Importeur der Zeitmesser dieser altehrwürdigen Marke, denn dort wurde der ursprüngliche „Quarz Chronometer" als „die Diplomierte von Girard-Perregaux" beworben – auf Italienisch nichts anderes als „Laureato". 1975 gelangten erste Armbanduhren dieses Namens auf den Markt. Neun Jahre später, also 1984, stand die erste optische Überarbeitung an. Im Zuge dessen präsentierte Girard-Perregaux auch unterschiedliche Uhrwerke mit astronomischen Zusatzindikationen. Weitere zehn Jahre ließ danach die zweite gestalterische Fortentwicklung auf sich warten. 1995 standen entsprechende „Laureato"-Armbanduhren zur Verfügung. Mit deutlich erkennbaren optischen Modifikationen zündete 2003 die dritte Evolutionsstufe. Auffallend ist eine sanftere, den Manschetten zuträgliche Linienführung, welche die „Laureato" bis in die Gegenwart auszeichnet. Wahlmöglichkeit besteht zwischen unterschiedlichen Modellen in mehreren Gehäusegrößen, erhältlich unter anderem in Edelstahl, Keramik oder natürlich auch Massivgold. Wer diese Uhrenlinie mag und sich mit der breit gefächerten Kollektion beschäftigt, wird mit Sicherheit fündig.

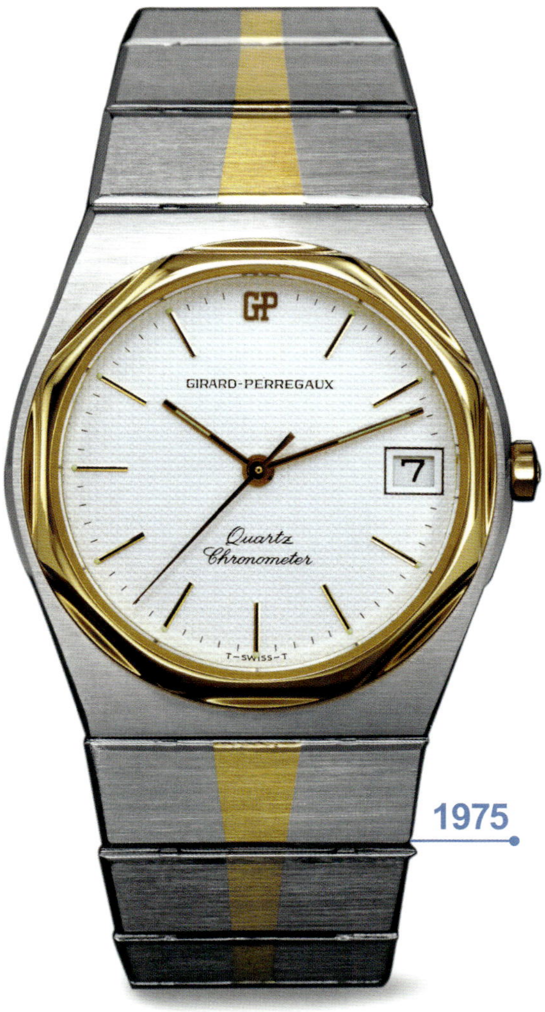

1975

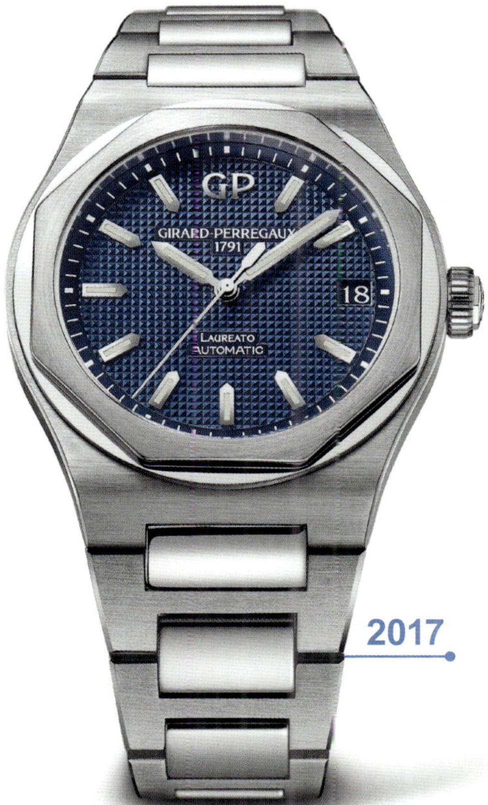

2017

2023

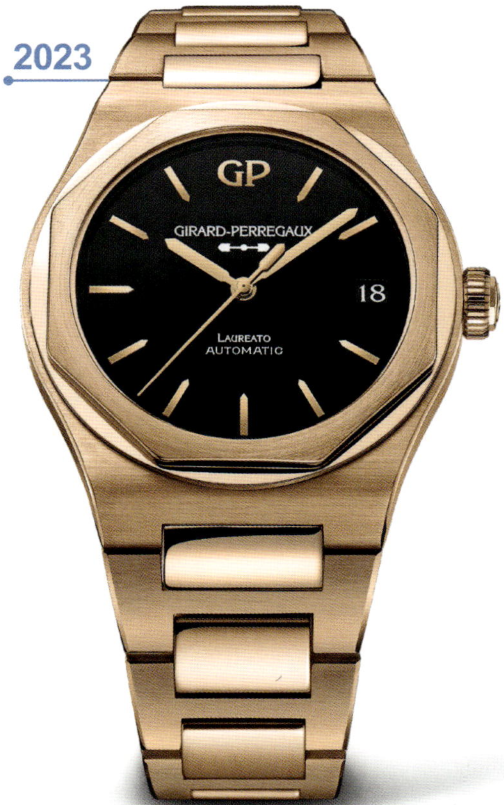

GLASHÜTTE ORIGINAL 2007

SIXTIES

GLASHÜTTE EVOKES THE SWINGING SIXTIES

In the eventful 1960s, when the Beatles, the Rolling Stones, Woodstock, Flower Power and miniskirts shaped an entire generation, Glashütte Original's automatic wristwatches bore the name Specimatik. Models of horological luxury these East German products were not. The timepieces, equipped with proprietary self-winding calibres 06-25 and 06-26, came onto the market in 1964. In their design, the engineers had focused on ease of servicing and longevity. A bidirectional rotor in heavy metal was used to wind the mainspring. Eighteen years after the fall of the Wall dividing Germany, in 2007, the German member of the Swiss Swatch Group reconnected with this stirring era of its past. Based in the secluded Müglitztal valley, the designers aimed to revive the pulsating feel of the Swinging Sixties with their retro-style watch models, paired of course with contemporary engineering. The visuals therefore reference the past, while the technology inside is consistent with the demands of the twenty-first century. This applies to both the case and the inner workings. No matter whether purist or with a large date indication or chronograph, all automatic movements are produced in-house at the Glashütte Original manufactory. The sleek round cases are available in steel or solid gold. The domed dials come in a choice of colours.

GLASHÜTTER ERINNERUNGEN AN DIE SIXTIES

In den bewegten 1960er-Jahren, als die Beatles, die Rolling Stones, als Woodstock, Flower Power und Miniröcke eine ganze Generation prägten, trugen die Automatik-Armbanduhren von Glashütte Original den Namen „Spezimatik". Von uhrmacherischem Luxus waren diese ostdeutschen Produkte weit entfernt. Ans Licht der Öffentlichkeit traten die mit hauseigenen Selbstaufzugskalibern 06-25 und 06-26 ausgestatteten Zeitmesser im Jahr 1964. Konstruktiv hatten die Techniker großen Wert gelegt auf Servicefreundlichkeit und Langlebigkeit. Stets bewirkte ein beidseitig wirkender Schwermetallrotor das Spannen der Zugfeder. 18 Jahre nach dem Fall der Deutschland trennenden Mauer, sprich 2007, knüpfte das deutsche Mitglied der schweizerischen Swatch Group an diese bewegende Epoche der eigenen Vergangenheit an. Bei den im Retrostil gehaltenen Uhrenmodellen ging es den Designern im abgeschiedenen Müglitztal um die Rückkehr des Lebensgefühls eines „swingenden" Jahrzehnts vor natürlich zeitgemäßem Hintergrund.
In diesem Sinne knüpft die Optik an Vergangenes, während die technische Seite in jeder Hinsicht den Anforderungen des 21. Jahrhunderts genügt. Das betrifft die Gehäuse ebenso wie die darin verbaute Zeitmechanik. Egal ob puristisch, ob ausgestattet mit Großdatum oder mit Chronograph: Alle Automatikwerke entstehen in der gläsernen Manufaktur von Glashütte Original. Die schlichten runden Schalen bestehen entweder aus Stahl oder massivem Gold. Wahlmöglichkeit besteht bei den Farben der bombierten Zifferblätter.

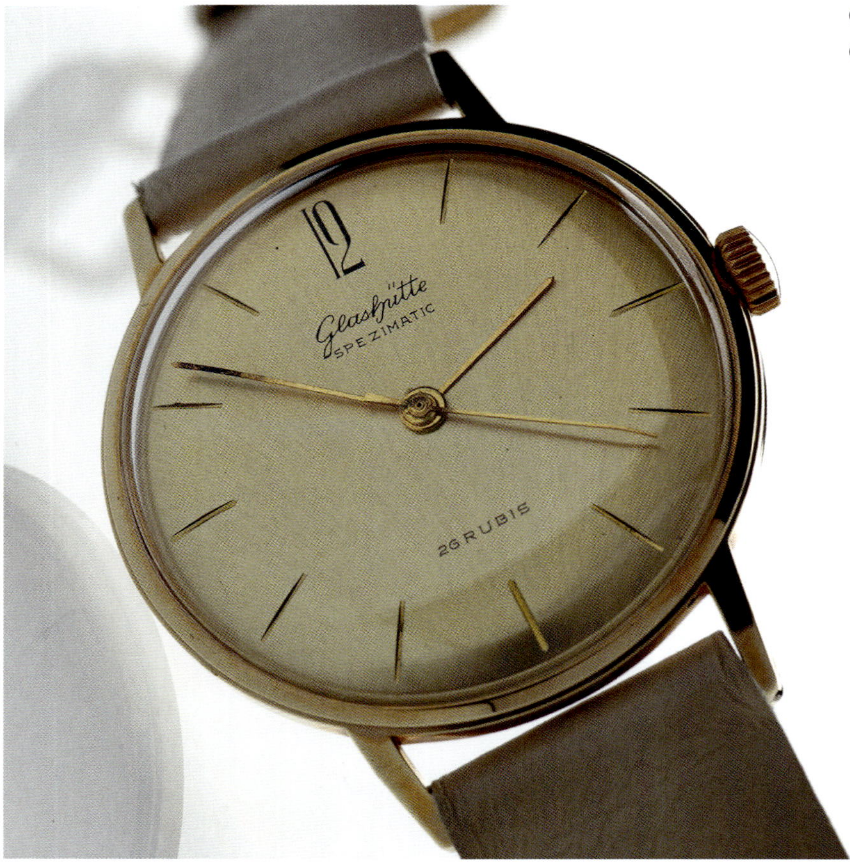

Spezimatic Ivory, 1964

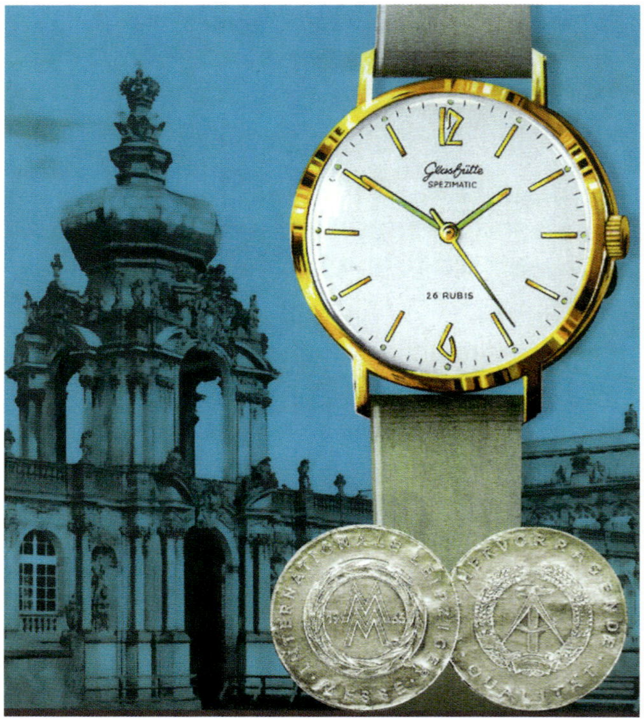

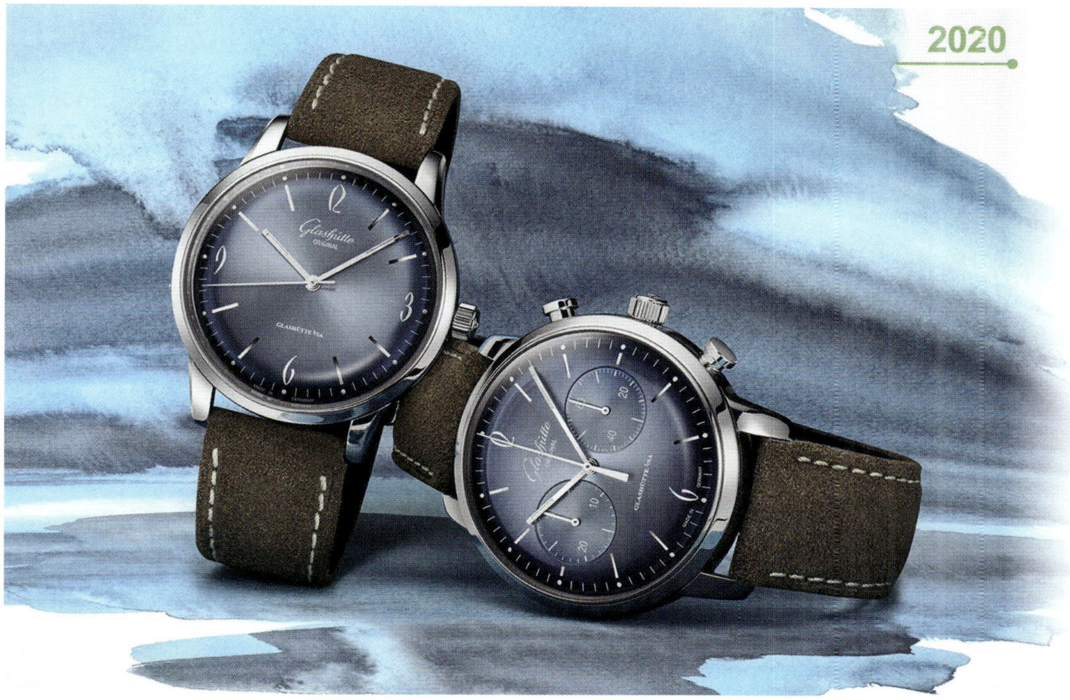

2020

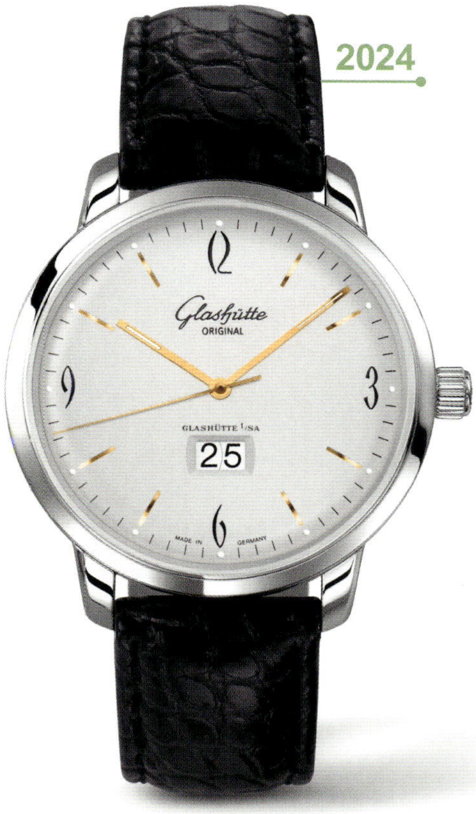

2024

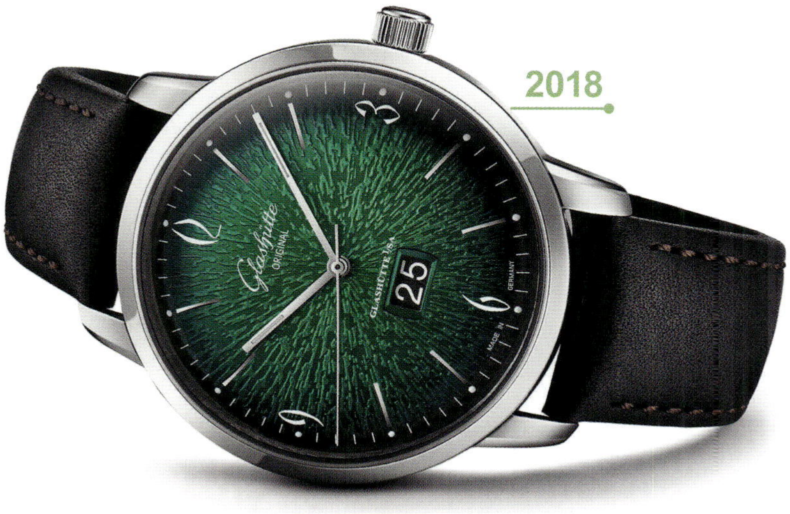

2018

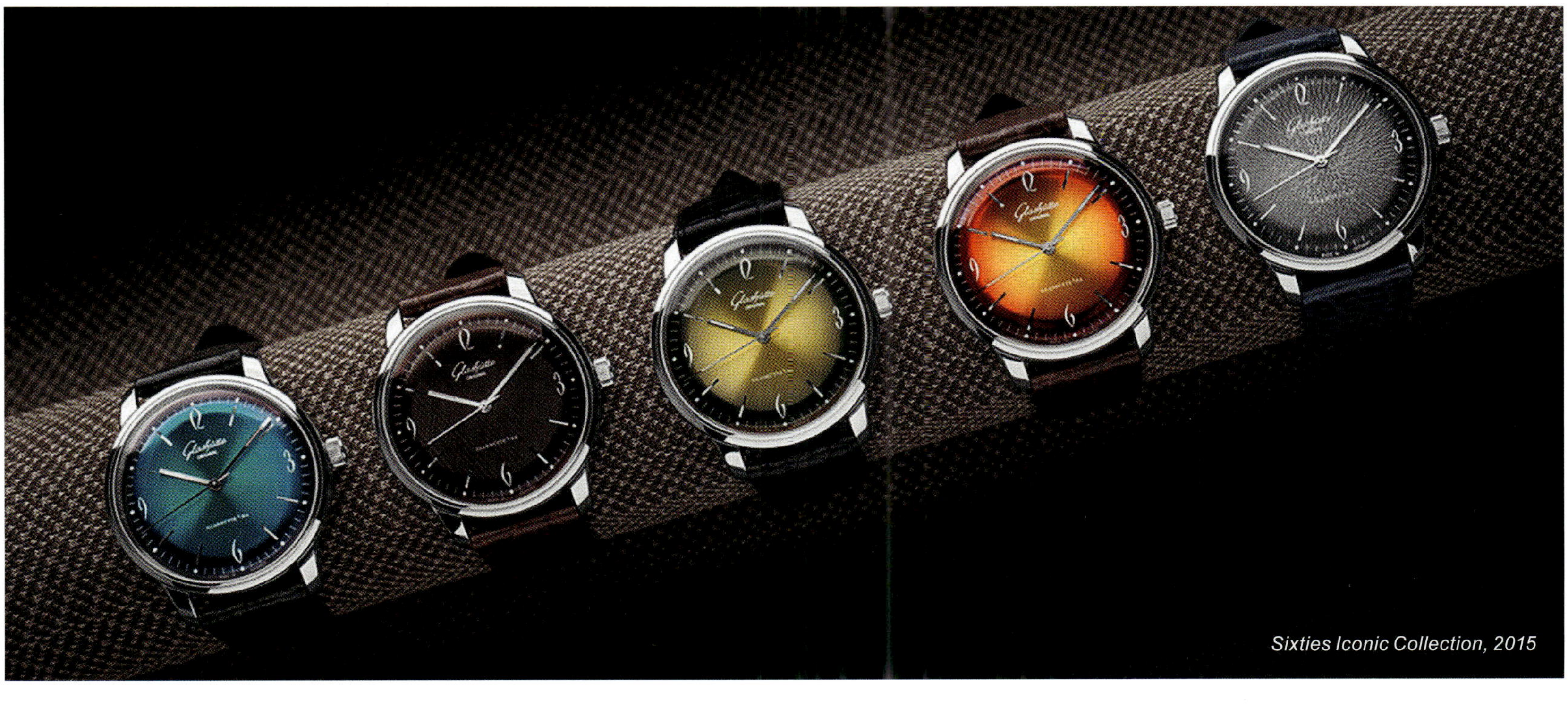

Sixties Iconic Collection, 2015

GRAND SEIKO

CAL. 9SA5

2020

EXCLUSIVE LEVER ESCAPEMENT

Marking its sixtieth anniversary in 2020, Grand Seiko presented the new Calibre 9SA5. It was developed by experienced engineers and watchmakers in Morioka, a town in the north of Japan. The biography of the 9S calibre family dates back to 1998. The most significant feature of the newcomer is its innovative oscillation and escapement system. A regulator mechanism to influence the active length of the proprietary balance spring can therefore be dispensed with. Even more remarkable is the Dual Impulse Escapement. The exclusive design represents a departure from the traditional piston-tooth lever escapement. It consists of three pallets for a continuous alternation between direct and indirect impulses. With direct impulses, the tip of a tooth on the star-shaped escape wheel has a direct influence on the pallet in the roller and thus on the balance oscillations. This is followed by an indirect impulse via the escape wheel, entry pallet, pallet and fork on the jewel, which is mounted in the roller as usual. The third element used by Grand Seiko is an exit pallet with a purely blocking function. Made of a nickel material, the escape wheel and pallet are crafted by means of photolithography technology. The overcoil with end curve is the product of thousands of simulations. Despite its energy-draining high frequency, the Calibre 9SA5 offers a power reserve of some 80 hours. This remarkable feat is not least ascribable to the 20 per cent higher efficiency of the Dual Impulse Escapement. The exacting in-house standards at Grand Seiko ensure a maximum daily deviation within a narrow range of minus three to plus five seconds. Accuracy certificates document the performance achieved by the watchmaking studio for each of the finished watches. The high technical standards aspired to by the automatic Calibre 9SA5 are rivalled by the visual components of the Reference SLGH005 White Birch. The Evolution 9 Collection 40-millimetre stainless-steel watch comes with a silver-coloured dial whose texture recalls Japanese birchwood and is inspired by the white birch forest surrounding the Shizikuishi Watch Studio.

EXKLUSIVE ANKERHEMMUNG

Zum 60. Jubiläum im Jahr 2020 präsentierte Grand Seiko das neue Kaliber 9SA5. Dessen Entwicklung oblag erfahrenen Technikern und Uhrmachern in Morioka, einer Stadt im Norden Japans. Zurück bis ins Jahr 1998 reicht die Biographie der Kaliberfamilie 9S. Die größte Besonderheit des Newcomers ist im innovativen Schwing- und Hemmungssystem zu sehen. Ein Rücker-Mechanismus, welcher die aktive Länge der selbst gefertigten Unruhspirale beeinflussen würde, ist dadurch entbehrlich. Noch bemerkenswerter ist die Doppelimpuls-Hemmung. Die exklusive Konstruktion bringt eine Abkehr von der traditionellen freien Kolbenzahn-Hemmung. Sie besitzt insgesamt drei Paletten für beständigen Wechsel zwischen direktem und indirektem Antriebsimpuls. Beim direkten Kraftstoß wirkt die Spitze eines Zahns des sternförmig gestalteten Hemmrads unmittelbar auf die in der Hebelscheibe befestigte Palette und damit auf die Unruhschwingungen ein. Anschließend erfolgt ein indirekter Impuls via Hemmrad, Eingangspalette, Ankerkörper und -gabel auf den wie üblich in der Hebelscheibe montierten Hebelstein. Als Drittes verwendet Grand Seiko eine Ausgangspalette mit ausschließlich blockierender Funktion. Ankerrad und Ankerkörper entstehen auf fotolithographischem Weg aus Nickel-Werkstoff. Resultat tausender Simulationen ist die hochgebogene Endkurve der Unruhspirale. Trotz energiezehrender Hochfrequenz läuft das Kaliber 9SA5 etwa 80 Stunden am Stück. Einen wichtigen Beitrag zu dieser beachtlichen Gangautonomie leistet der um 20 Prozent höhere Wirkungsgrad der Doppelimpuls-Hemmung. Die strengen internen Grand Seiko-Regularien gewährleisten eine maximale tägliche Gangabweichung im engen Bereich von minus drei bis plus fünf Sekunden. Gangzeugnisse dokumentieren die im Uhrmacherstudio erreichte Performance jeder der fertiggestellten Uhren. Dem hohen technischen Anspruch des Automatikkalibers 9SA5 trägt die optische Komponente der Referenz „SLGH005 White Birch“ Rechnung. Die Struktur des silberfarbenen Zifferblatts der 40 Millimeter großen Edelstahlarmbanduhr aus der „Evolution 9 Collection“ erinnert an japanisches Birkenholz und ist inspiriert vom Blick in den Weißbirkenwald rund um das Shizikuishi Uhrenstudio.

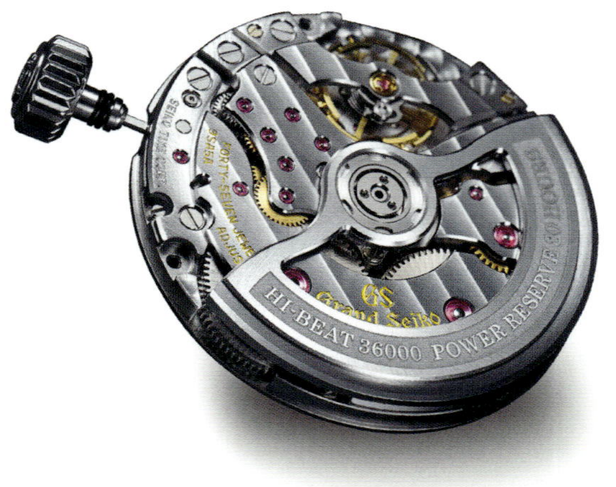

Cal. 9SA5, 2020

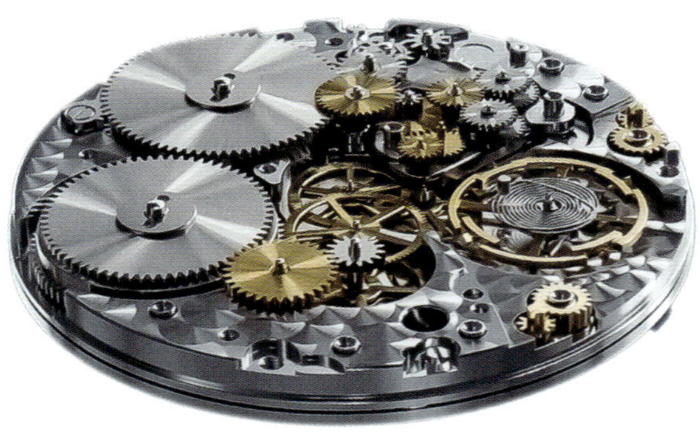

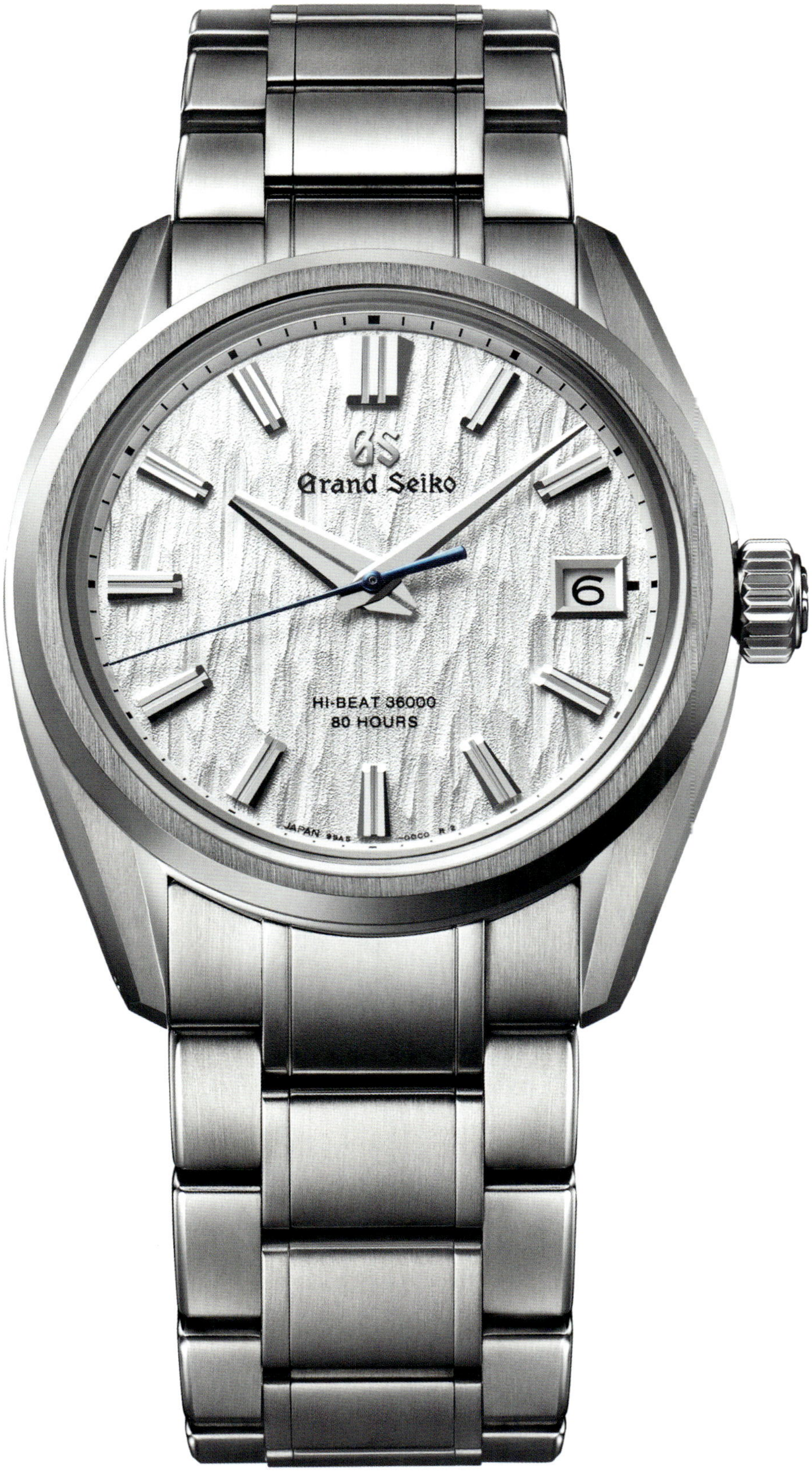

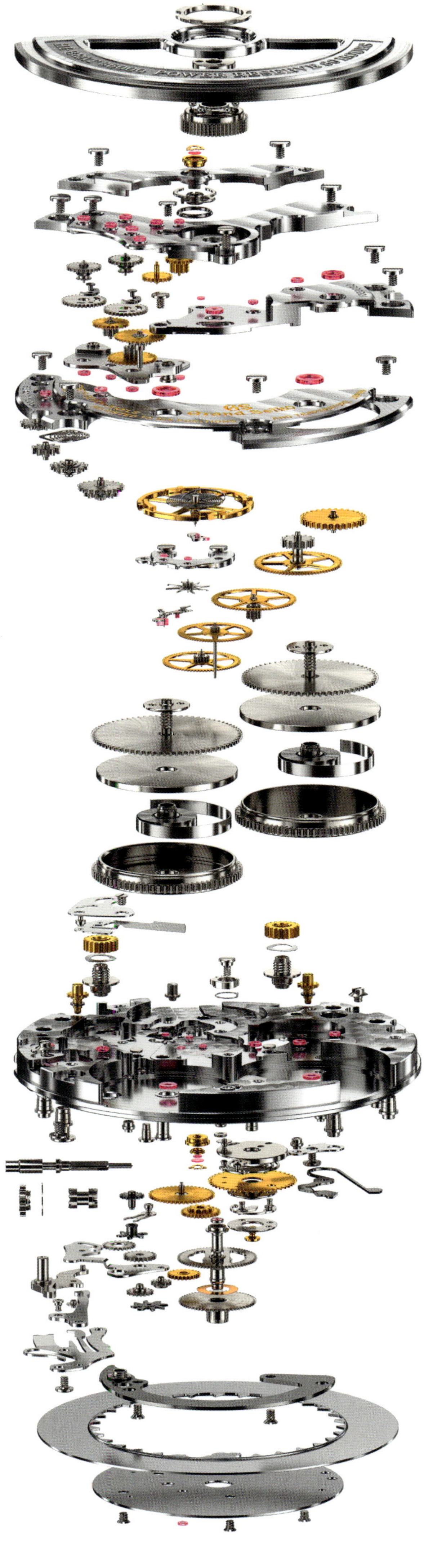

Left | Links: *Ref. SLGH005 White Birch, 2020*
Right | Rechts: *Cal. 9SA5, 2020, exploded view | Explosionsdarstellung*

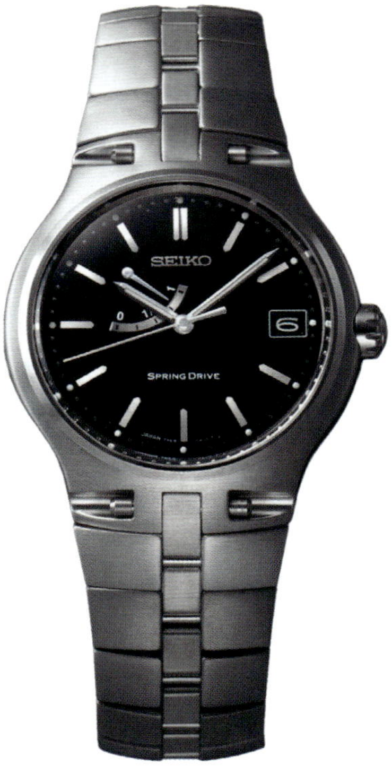

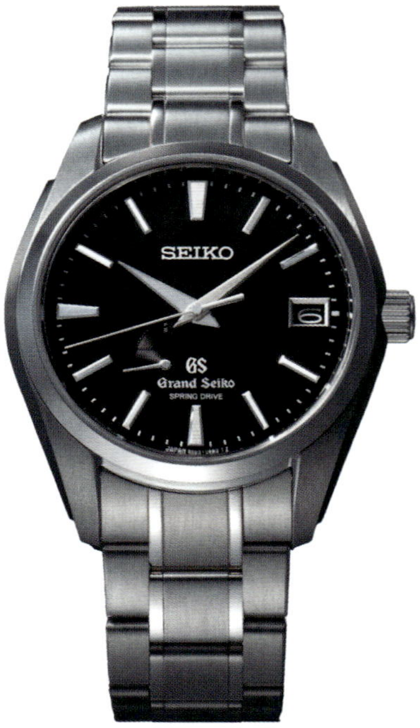

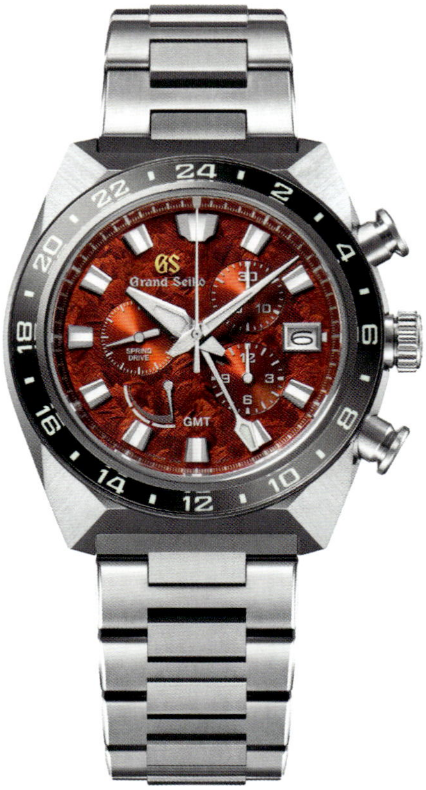

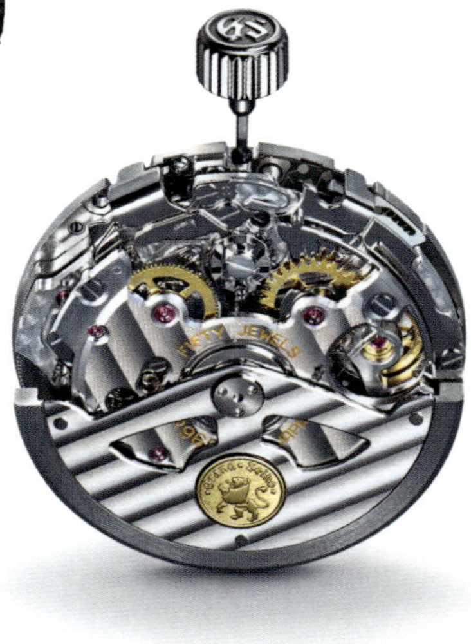

Cal. 9R96

From left: *First Spring Drive watch, 1999* **|** *Spring Drive watch, Cal. 9R65, 2004* **|** *Ref. SBGC275, 2024* **|**
Von links: *erste Uhr mit Spring Drive, 1999* **|** *Spring Drive-Uhr, Kal. 9R65, 2004* **|** *Ref. SBGC275, 2024*

Bottom: *Grand Seiko watch with Cal. 9S, 1989* **|** *Ref. SBGE307, 2024*
Unten: *Grand Seiko Uhr mit Kal. 9S, 1989* **|** *Ref. SBGE307, 2024*

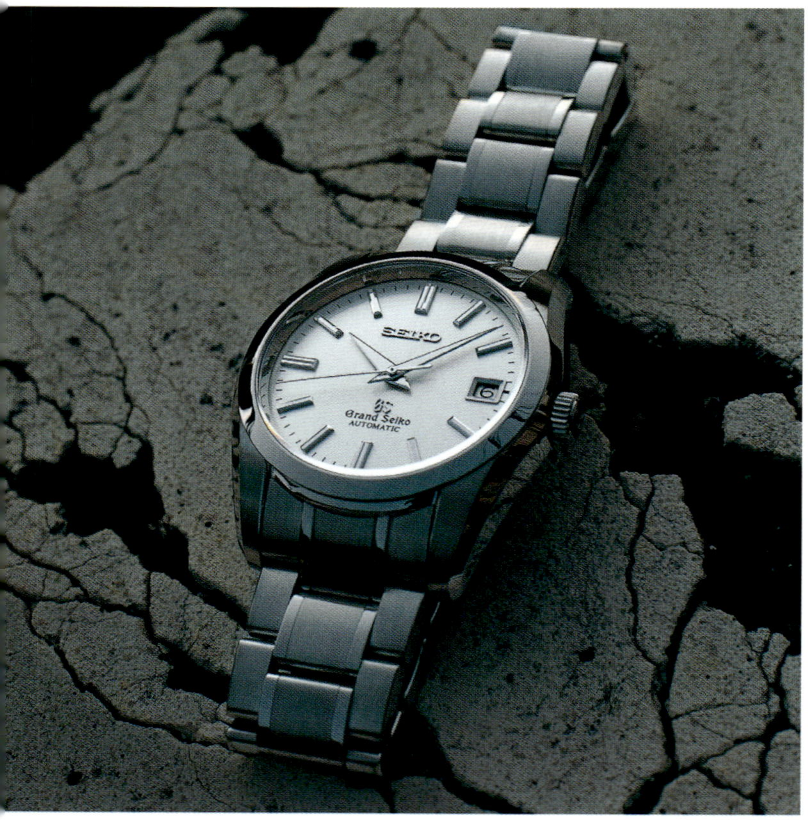

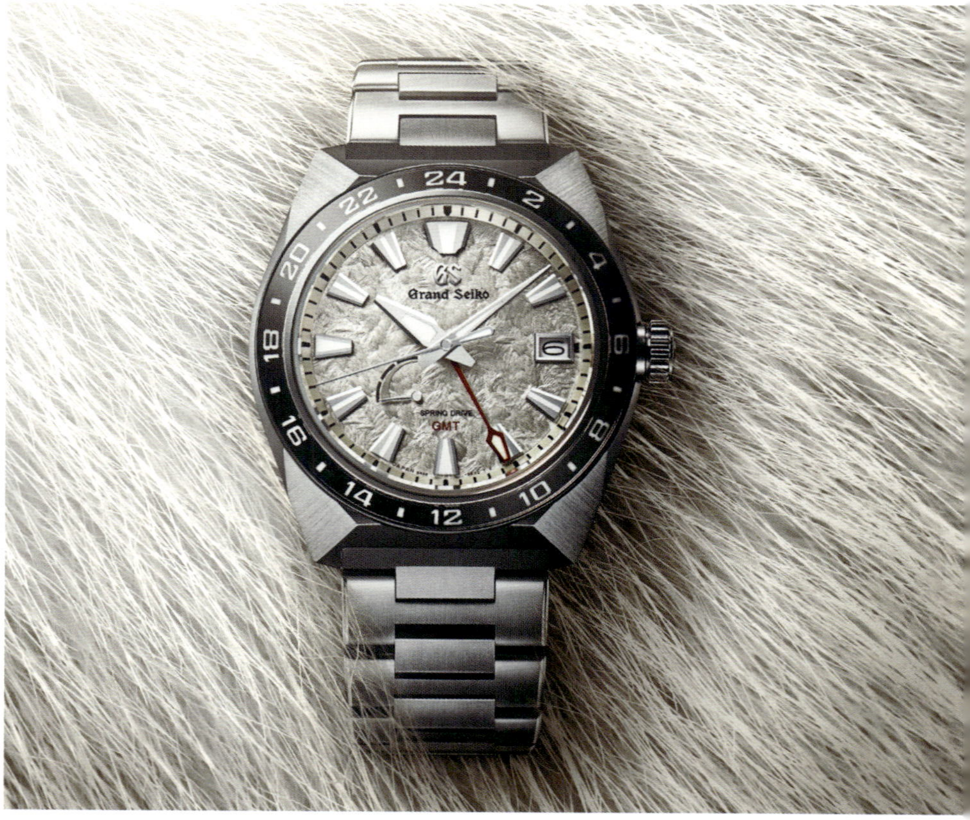

GRAND SEIKO SPRING DRIVE

GREEN TIME ENERGY

Developed from 1977 by Seiko, Spring Drive technology embodies the successful combination of conventional mechanical horology and advanced electronics. The sparking idea was a quartz watch without a battery. Because it turned out to be significantly tougher to achieve than anticipated by Japanese engineer Yoshikazu Akahane, it was 1982 before the first prototype was born. And it stopped after four hours. In 1993, the close cooperation with high-tech specialists resulted in a breakthrough and further prototypes. Six years later, the hand-wound version of a Spring Drive calibre came onto the market. Since the latter only appealed to a small target audience, a more convenient automatic iteration was a must. Seiko announced the start of mass production in its Mastery Studio, specially built for the purpose, in 2004. In technical terms, the Spring Drive represents an innovative symbiosis of traditional engineering and modern electronics. With the exception of the rate-regulating unit, everything works the same way as traditional automatic movements. Grand Seiko's in-house mainspring provides a power reserve of 72 hours. The gear train transforms the slow barrel rotations into faster, but less powerful rotations. The transmission transfers motion from the bi-directional oscillating weight to the barrel. The regulator alone heralded a new era in timekeeping. Broadly speaking, Seiko-Epson's intelligent Tri-synchro regulator resembles a tram with an eddy brake. Driven by the mainspring, a small glide wheel and magnetic rotor move constantly in a single direction. They generate the energy for the quartz-controlled electromagnetic brake. The latter ensures the outstandingly smooth action of the entire gear train. Initially, a precision of ±15 seconds per month was achieved. Grand Seiko's current Spring Drive automatic Calibre 9RA5 now deviates from the astronomical norm by just ±10 seconds over 30 days. In addition, the power reserve of the latest generation's movement is a reassuring five days. The hands rotate uniformly to indicate the time on the dial.

After this detailed description it will come as no surprise that watches with Spring Drive inner workings, now exclusively available from Grand Seiko, command the same prices as purely mechanical counterparts. The unique technology is available in various Grand Seiko watch lines, in styles ranging from elegant to sporty.

ÖKOLOGISCHER ZEIT-HYBRID

Einen Königsweg zwischen konventioneller Zeitmechanik und progressiver Elektronik verkörpert die von Seiko ab 1977 entwickelte „Spring Drive"-Technologie. Am Anfang stand die Idee einer Quarzuhr ohne Batterie. Weil sich deren Realisierung deutlich schwieriger gestaltete, als vom japanischen Ingenieur Yoshikazu Akahane ursprünglich gedacht, ließ ein erster Prototyp bis 1982 auf sich warten. Aber der lief nur vier Stunden am Stück. 1993 brachte die enge Kooperation mit Hightech-Spezialisten den Durchbruch und weitere Prototypen. Weitere sechs Jahre später stand die Handaufzugsversion eines „Spring Drive"-Kalibers zur Verfügung. Nachdem selbige nur eine kleine Zielgruppe erreichte, führte an einer komfortableren Automatikvariante kein Weg vorbei. Die Aufnahme der Serienproduktion in eigens eingerichteten „Mastery Studios" konnte Seiko 2004 vermelden.

Technisch verkörpert „Spring Drive" eine innovative Symbiose aus traditioneller Mechanik und moderner Elektronik. Bis auf die gangregulierende Baugruppe funktioniert alles wie bei überlieferten Automatikwerken. Eine Grand Seiko-eigene Zugfeder speichert Kraft für 72 Stunden Gangautonomie. Eine Kette aus Zahnrädern und -trieben formt die langsamen Federhausdrehungen in schnellere, aber kraftärmere Rotationen um. Die beidseitig wirkende Schwungmasse wirkt via Automatikgetriebe auf das Federhaus ein. Allein beim Gangregler ist eine neue Zeitrechnung angebrochen. Im weitesten Sinn ähnelt Seiko-Epsons intelligenter „Tri-Synchro-Regulator" einer Straßenbahn mit Wirbelstrombremse. Angetrieben vom Energiespeicher bewegt sich ein kleines Gleitrad samt Magnetrotor konstant in einer Richtung. Dabei generiert das Duo unmittelbar Energie für die quarzgesteuerte elektromagnetische Bremse. Letztere bewirkt einen überaus gleichmäßigen Ablauf des gesamten Räderwerks. Anfangs lag die Präzision bei plus/minus 15 Sekunden monatlich. Nur noch plus/minus zehn Sekunden weicht das aktuelle, nun von Grand Seiko gefertigte „Spring Drive"-Automatikkaliber 9RA5 während 30 Tagen von der astronomischen Norm ab. Dazu beträgt die Gangautonomie des Uhrwerks der neuesten Generation beruhigende fünf Tage. Gleichförmig drehende Zeiger stellen die Zeit auf dem Zifferblatt dar.

Nach diesen Erläuterungen sollte verständlich sein, dass inzwischen nur noch bei Grand Seiko erhältliche Armbanduhren mit „Spring Drive"-Innenleben das gleiche Preisniveau erreichen wie rein mechanische Modelle. Verfügbar ist die einzigartige Technologie in verschiedenen Grand Seiko-Uhrenlinien, angefangen bei elegant bis hin zu sportlich.

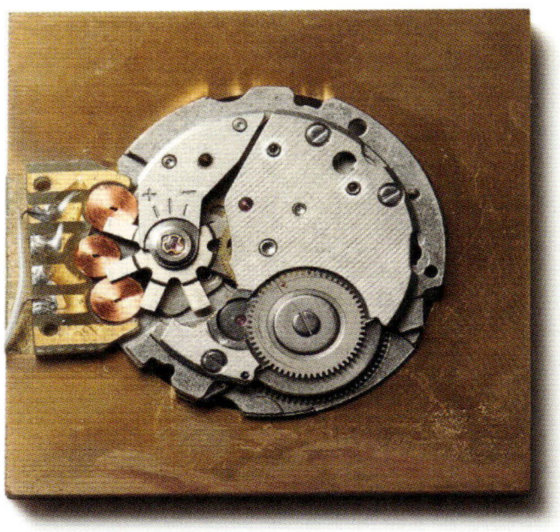

Spring Drive prototype, 1982

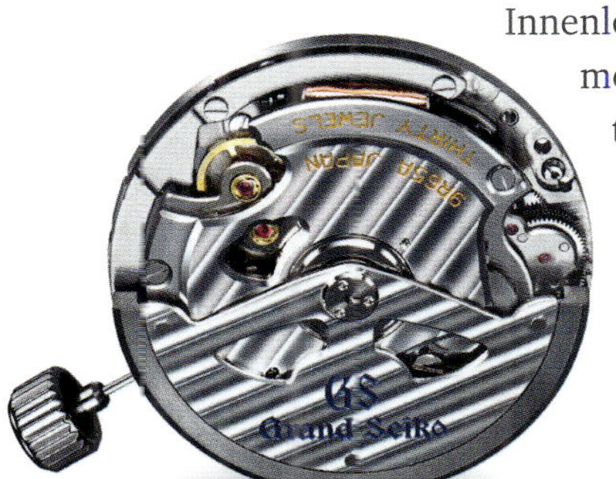

Cal. 9R65

GREUBEL FORSEY 2004

DOUBLE TOURBILLON 30°

TOURBILLON FOR THE THIRD MILLENNIUM

After the turn of the millennium, Greubel Forsey devoted four years to the design of an innovative tourbillon. It increases the performance of mechanical wristwatches significantly by compensating for the daily fluctuations in rate due to the effects of gravity caused by the ever-changing position of the watch in daily use. Two cages are mounted at an angle of 30 degrees to one another, which is why this patented timepiece was named the Double Tourbillon 30°. In a nutshell, the basic technology reveals the following: inside a large cage with a diameter of some 15 millimetres and a rotation time of four minutes, a smaller cage, inclined at an angle of 30 degrees, revolves around its axis once every 60 seconds. It holds the balance wheel with variable torque and accompanying Breguet spiral or overcoil. In sum, the double tourbillon's 128 components weigh only 1.17 grams. The key features of the Greubel Forsey invention, the 30-degree inclination of the rotational plane of the smaller cage along with the ingenious rotational speeds of the two cages, enable the balance wheel to oscillate in different positions. This means that the compensation of the rate fluctuations caused by gravity is no longer limited to the vertical plane.

The driving force behind the Double Tourbillon 30° is provided by two fast-rotating, serially switched barrels. The sliding bridle at the end of one of the barrels prevents the overwinding and breakage of the mainspring during winding. The power reserve is at least 72 hours. Understandably, the architecture of the 36.4-millimetre self-winding movement is oriented towards the innovative tourbillon mechanism. The gear train is located between two plates. With Greubel Forsey, it is superfluous to mention that the craftsmanship also meets the highest standards. In 2011, the Double Tourbillon 30° Technique scored a stunning 915 points out of 1,000 at the Concours International de Chronométrie.

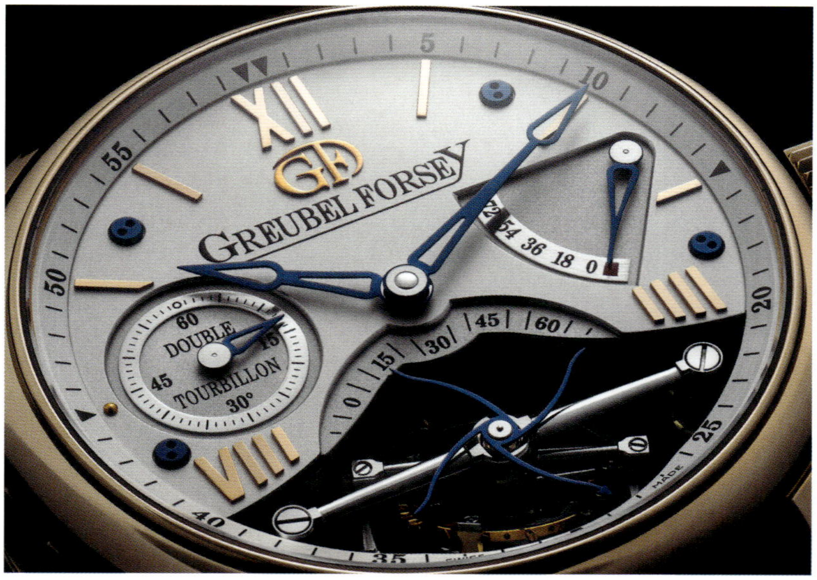

TOURBILLON DES DRITTEN JAHRTAUSENDS

Vier Jahre nahm bei Greubel Forsey nach der Jahrtausendwende eine neuartige Tourbillon-Konstruktion in Anspruch. Selbige vermag die Leistungsfähigkeit mechanischer Armbanduhren spürbar zu steigern, indem sie die von der Erdanziehungskraft verursachten Gangabweichungen in all jenen Positionen ausgleicht, welche im Alltagsleben eingenommen werden. Zwei Käfige stehen im Winkel von 30 Grand zueinander, weshalb dieser patentierte Zeitmesser auch den Namen „Doppeltourbillon 30 Grad“ erhielt. Zur grundlegenden Technik in aller Kürze nur so viel: Im Innern eines großen Drehgestells mit ungefähr 15 Millimeter Durchmesser sowie vier Minuten Rotationszeit bewegt sich ein kleineres, um 30 Grad abgewinkeltes Pendant alle 60 Sekunden einmal um seine Achse. Es beherbergt die Unruh mit variablem Trägheitsmoment und die zugehörige Breguetspirale. Summa summarum wiegen die insgesamt 128 Komponenten der Tourbillon-Duos lediglich 1,17 Gramm. Die entscheidenden Eigenschaften der Greubel Forsey'schen Erfindung, sprich die 30-Grad-Neigung der Rotationsebene des kleineren Käfigs zusammen mit den durchdachten Drehgeschwindigkeiten beider Drehgestelle, lassen die Unruh in verschiedenen Lagen schwingen. Ergo beschränkt sich die Kompensation der schwerkraftbedingten Gangabweichungen nicht mehr nur auf die Senkrechte.

Die Antriebskraft für das „Double Tourbillon 30°“ (so die originale Bezeichnung) liefern zwei schnell drehende und seriell geschaltete Federhäuser. Der Gleitzaum am Ende einer der Energiespeicher verhindert das Überspannen und Abreißen der Zugfeder beim Aufziehen. Mindestens 72 Stunden beträgt die Gangautonomie. Verständlicherweise orientiert sich die Architektur des 36,4 Millimeter messenden Handaufzugswerks ganz bewusst am innovativen Tourbillon-Mechanismus. Sein Räderwerk befindet sich zwischen zwei Platinen. Von selbst versteht sich bei Greubel Forsey, dass die handwerkliche Ausarbeitung gleichfalls allerhöchsten Standards entspricht. 2011 glänzte besagtes Doppeltourbillon „30° Technique“ beim Concours International de Chronométrie mit 915 von 1000 Punkten.

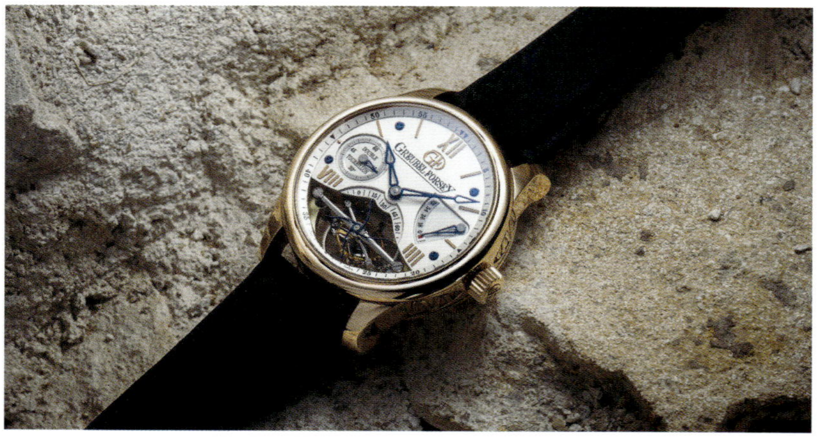

Double Tourbillon 30°, 2004

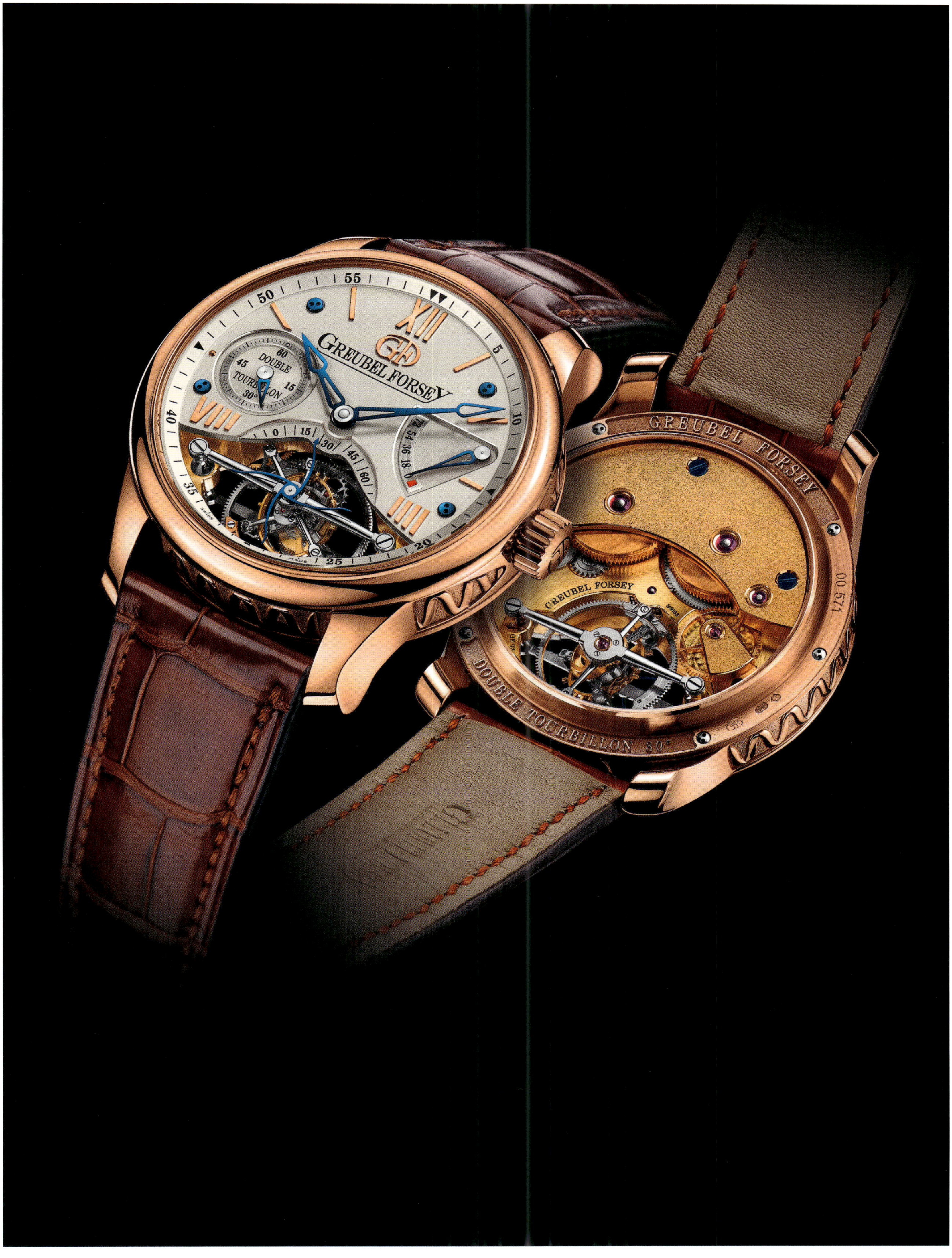
GREUBEL FORSEY
DOUBLE TOURBILLON 30°
GREUBEL FORSEY
GREUBEL FORSEY
DOUBLE TOURBILLON 30°

GROSSMANN 2013
BENU

A HAIRY PROPOSITION

Hair in the soup is a no-go for chefs. The same applies to watchmakers. A single hair can bring a gear train to a halt. Moritz Grossmann turned the problem into a virtue. A closer examination of the 38.4-millimetre manufacture Calibre 103.0 from 2013 reveals a bunch of hair held together by a small ring and some shellac. A brush of sorts, it stops the hand-wound movement to enable the wearer to set the hand to the nearest second. When the crown is pulled, the unconventional patent-pending component and associated mechanism swivel against the regulator. Gentle pressure halts the hand-wound movement. A tap of the small pusher below the winding adjusting crown sets it off again. This hairy proposition is just one of the timepiece's assets. Another is the flying tourbillon. The cage of the 700-milligram tourbillon measures a mighty 16 millimetres, the balance inside 14.2 millimetres. The cage takes three minutes to complete a rotation. The energy-conserving deceleration pays into the power reserve. A large barrel generates the energy for the power reserve of some three days. Grossmann mills the gears, which are made of a hard, amagnetic ARCAP alloy. Naturally, the lower rotational speed of the tourbillon requires a separate, and in this case indirectly driven second hand. Positioned at 9 o'clock, the friction element, a small bronze spring, ensures uniform rotations by gently pressing against a ring made of hard, resinous guaiacum wood. The hour hand rotates on the right side of the dial. An eye-catching arc-shaped segment connects the two secondary dials and eases reading of the minutes. For precisely this reason, the central minute hand has a long and a short end. Since the tourbillon, prominently positioned at 6 o'clock, interrupts the external track between 25 and 35, the short end of the hand indicates the intervening 10 minutes on the track segment below the signature. Another unconventional solution, but it works.

HAARIGE ANGELEGENHEIT

Haare in der Suppe sind jedem Koch ein Gräuel. Gleiches gilt auch für Uhrmacher. In deren Metier kann schon ein Haar zum Stillstand des Räderwerks führen. Moritz Grossmann hat aus dieser Not eine Tugend gemacht. Im 38,4 Millimeter messenden Manufakturkaliber 103.0 von 2013 findet sich gleich ein ganzes Bündel Haare, zusammengehalten von einem kleinen Ring und etwas Schellack. Der solcherart geformte Pinsel bewirkt das Anhalten des Handaufzugswerks zum Zweck sekundengenauer Zeigerstellung. Beim Ziehen der Krone schwenkt das ungewöhnliche und deshalb samt dem zugehörigen Mechanismus zum Patent angemeldete Bauteil gegen den Gangregler. Behutsamer Druck hält das Handaufzugswerk an. Betätigung des kleinen Drückers unterhalb der Aufzugs- und Zeigerstellkrone bewirkt die Reaktivierung. Diese haarige Angelegenheit ist nur ein Aktivposten. Ein weiterer besteht im fliegend gelagerten Tourbillon der sogenannten „BENU"-Uhrenlinie. Der Käfig des 700 Milligramm leichten Drehgangs misst stolze 16, die darin untergebrachte Unruh 14,2 Millimeter. Für einen vollständigen Umlauf benötigt der Käfig ganze drei Minuten. Die kräftesparende Entschleunigung kommt der Gangautonomie zugute. Als Energiequelle für rund drei Tage Gangautonomie dient ein großes Federhaus. Die Zahnräder fräst Grossmann aus einer harten, amagnetischen ARCAP-Legierung. Naturgemäß verlangt die reduzierte Umlaufgeschwindigkeit des Tourbillons nach einem separaten, in diesem Fall indirekt angetriebenen Sekundenzeiger. Das für gleichförmige Rotationen bei „9" sorgende Friktionselement besteht aus einer kleinen Bronzefeder, welche sanft gegen einen Ring aus hartem, harzreichem Pockholz drückt. Rechts am Zifferblatt dreht der Stundenzeiger. Ins Auge sticht schließlich noch ein bogenförmiges Kreissegment. Dieses verbindet die beiden Nebenzifferblätter und unterstützt zeitweise das genaue Ablesen der Minuten. Zu diesem Zweck besitzt der zentrale Minutenzeiger ein langes und ein kurzes Ende. Nachdem das prominent bei „6" positionierte Tourbillon die äußere Indexierung zwischen 25 und 35 unterbricht, weist der Zeigerstummel zehn Minuten lang auf die Hilfsskala unterhalb der Signatur. Eine ebenfalls ungewöhnliche Lösung, aber sie funktioniert.

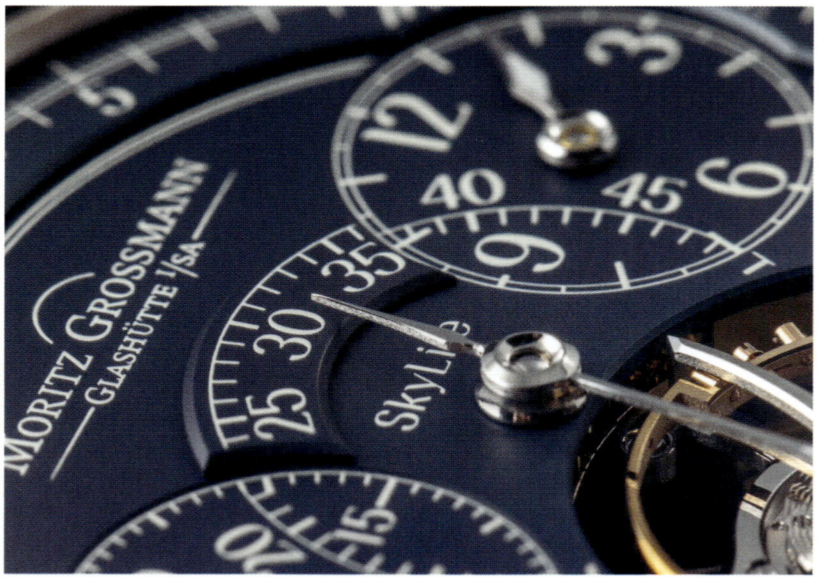

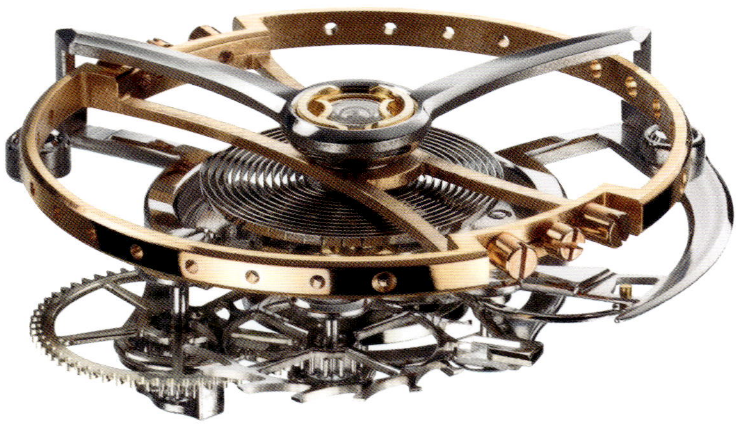

From left / Von links: *Tourbillon Skylife / Cal. 103.0, cage / Käfig*

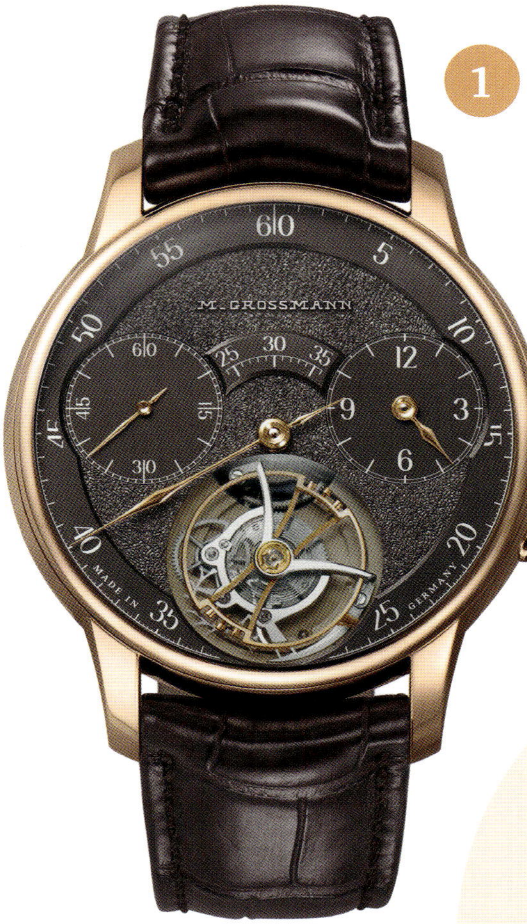

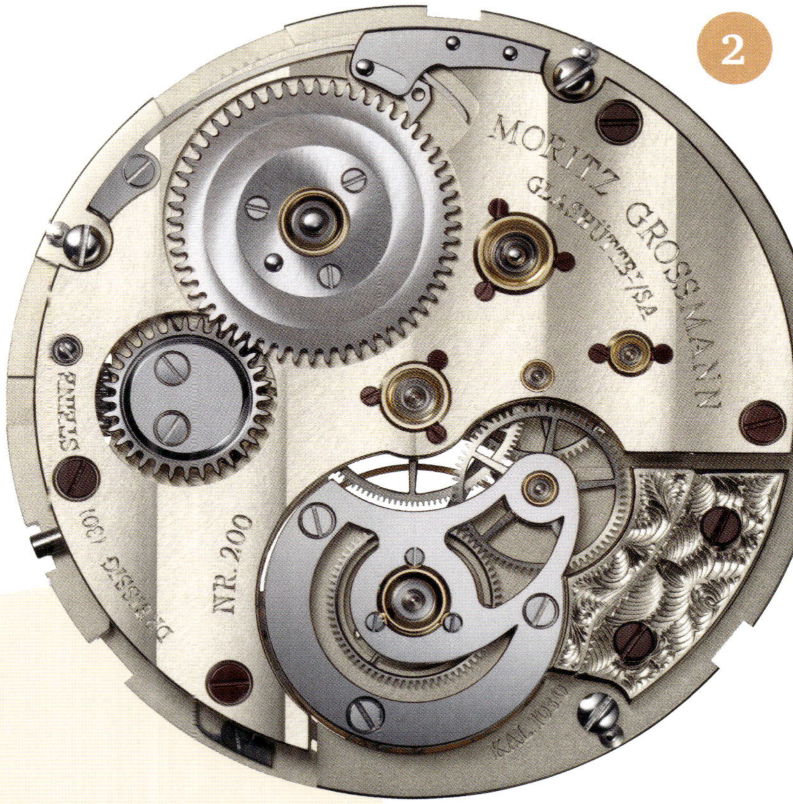

1 *Tourbillon Tremblage, 2024*

2 *Cal. MG 103.0, 2013*

3 *Tourbillon Skylife*

4 *Tourbillon, 2013*

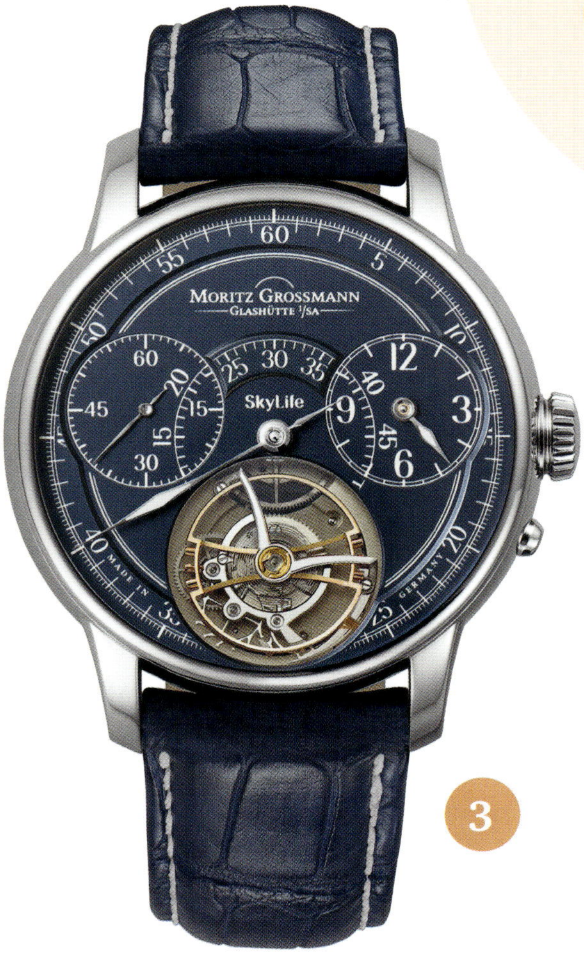

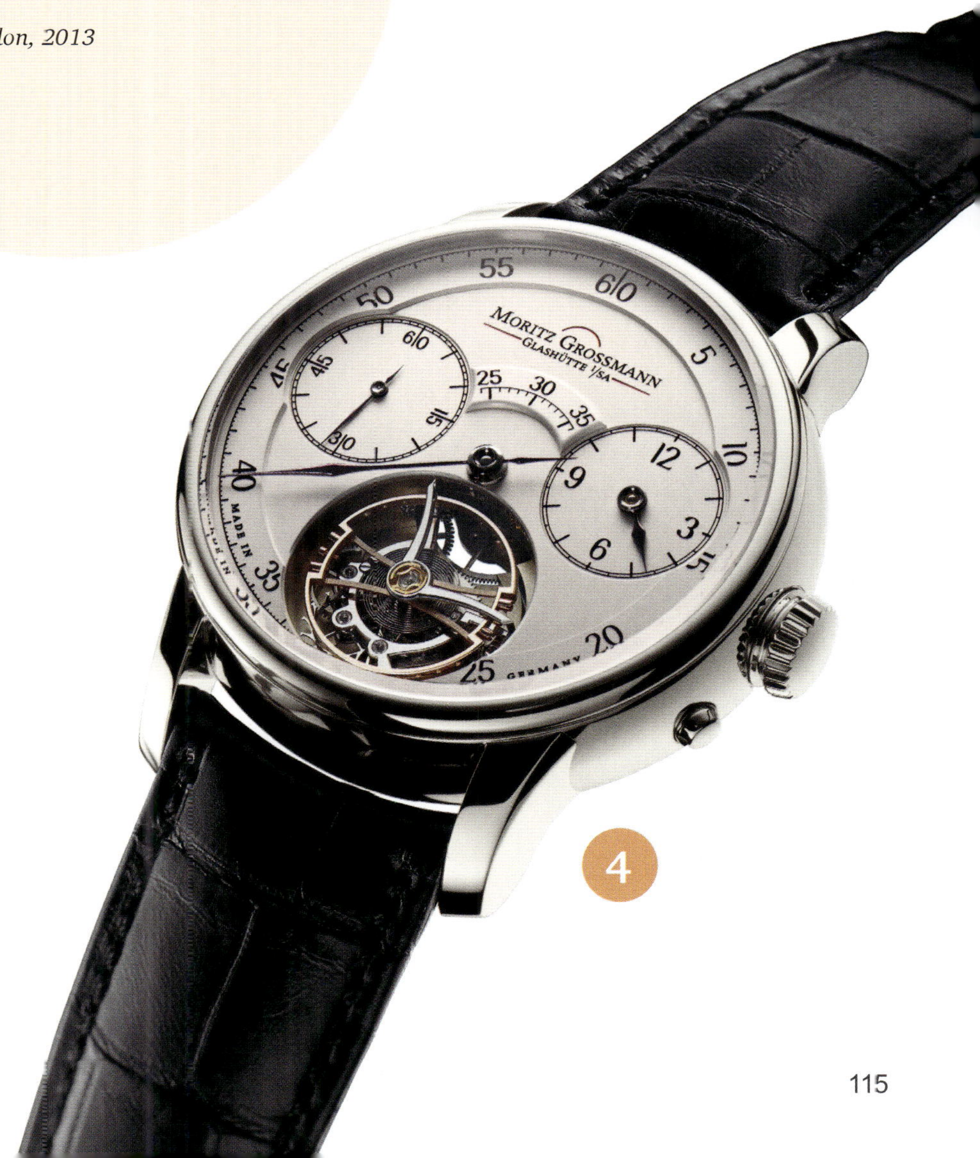

HANHART 1939

417 ES

STAINLESS STEEL RULES

Among a host of other items, the German Bundeswehr, founded in 1955, needed functional timepieces for its pilots. A chance for Black Forest-based watchmaker Hanhart to step into the breach again. The company's history goes back to 1882. Imprinted in the memories of collectors is the 1939 aviator chronograph, equipped with proprietary hand-wound calibres 40, 41 and 42. The first of these has only one pusher for the start, stop and zeroing functions. Boasting two controls, the 41 model also masters addition. The premium 42 version comes with a flyback function. This eased the pilot's life on high. Because the accidental use of the zeroing pusher could have easily resulted in disastrous consequences, the worried wife of a military pilot unceremoniously painted it with red nail polish. An action that was to give birth to a legend. Another icon that enjoys the limelight is the Reference 417. Hanhart offered the chronograph as 417 E with a matte chrome-plated brass case and from 1957 also as 417 ES Flyback with a stainless-steel case and the flyback calibre 42. The movements were all finely adjusted in six positions. The case is distinguished by its fluted rotating bezel and red marking. For six years, Hanhart produced these wristwatches for military purposes. This is why it is known as the Bundeswehr chronograph among connoisseurs. The most famous fan of the 417 ES was Steve McQueen. How the American superstar stumbled across this chronograph no one can say with certainty today. We do know though that he used it in motorcycle races. In 1962, Hanhart, now world market leader in stopwatches, temporarily discontinued the manufacture of its wristwatches. On the new 417 ES, the two letters again stand for *Edelstahl*, the German for high-grade steel. However, the diameter of the statement case with infinitely rotatable bezel has been increased from the original 39 to a stately 42 millimetres. The inner workings, be they self or hand-winding, are no longer produced in-house. Purchased from Swiss specialist Sellita, Hanhart keeps a close check on the quality of their engineering.

ES HEISST EDELSTAHL

Neben vielen anderen Materialien benötigte die 1955 ins Leben gerufene Deutsche Bundeswehr auch funktionale Zeitmesser für ihre Piloten. Hier nun schlug einmal mehr die Stunde der im Schwarzwald beheimateten Uhrenmanufaktur Hanhart. Bis ins Jahr 1882 reicht deren Geschichte zurück. Ins Sammlergedächtnis eingeprägt hat sich der Fliegerchronograph von 1939, ausgestattet mit den eigenen Handaufzugskalibern 40, 41 und 42. Ersteres besaß nur einen Drücker für die Funktionen Start, Stopp und Nullstellung. Ausgestattet mit zwei Bedienelementen gestattete das 41 auch Additionsstoppungen. Die Top-Version 42 verfügte über eine so genannte Flyback-Funktion. Diese erleichterte den Piloten das Leben hoch in den Lüften. Weil die versehentliche Benutzung des Nullstelldrückers durchaus gravierende Folgen haben konnte, färbte ihn die besorgte Gattin eines Militärpiloten kurzerhand mit rotem Nagellack ein. Und damit war flugs eine zeitschreibende Legende geboren.

Als weitere Ikone stand die Referenz 417 im Rampenlicht. Hanhart offerierte den Chronographen als „417 E" mit mattverchromtem Messinggehäuse und ab 1957 auch als „417 ES Flyback" mit Edelstahlgehäuse und besagtem Flyback-Kaliber 42. Stets waren die verbauten Werke in sechs Lagen feinreguliert. Das Gehäuse kennzeichnete der geriffelte Drehring mit Merk-Markierung. Sechs Jahre lang produzierte Hanhart diese Armbanduhren für militärische Zwecke. Daher fällt in diesem Zusammenhang regelmäßig auch der Name „Bundeswehr-Chronograph". Prominentester Fan der „417 ES" war Steve McQueen. Wie der amerikanische Superstar an diesen Chronographen kam, lässt sich heute nicht mehr nachvollziehen. Auf jeden Fall nutzte er ihn unter anderem bei Motorradrennen. 1962 stellte Hanhart, inzwischen der Weltmarktführer bei Stoppuhren, die Armbanduhrenfabrikation vorübergehend ganz ein.

Beim neuen „417 ES" stehen die beiden Buchstaben wiederum für edlen Stahl. Allerdings ist der Durchmesser des ausdruckstarken Gehäuses mit stufenlos verstellbarer Lünette von ursprünglich 39 auf 42 Millimeter gewachsen. Die zeitschreibende Mechanik, egal ob mit automatischem oder manuellem Aufzug, stammt allerdings nicht mehr aus eigener Fertigung. Hanhart bezieht sie vom Schweizer Spezialisten Sellita, achtet dabei jedoch auf sorgfältige Ausführung.

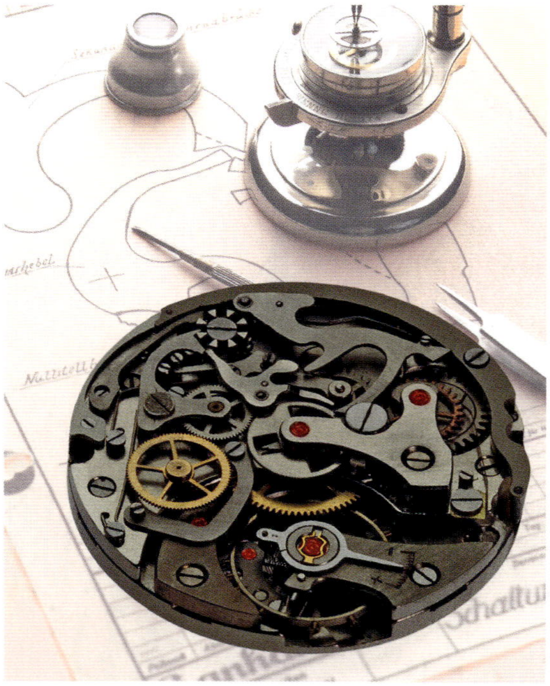

Cal. 41, 1939

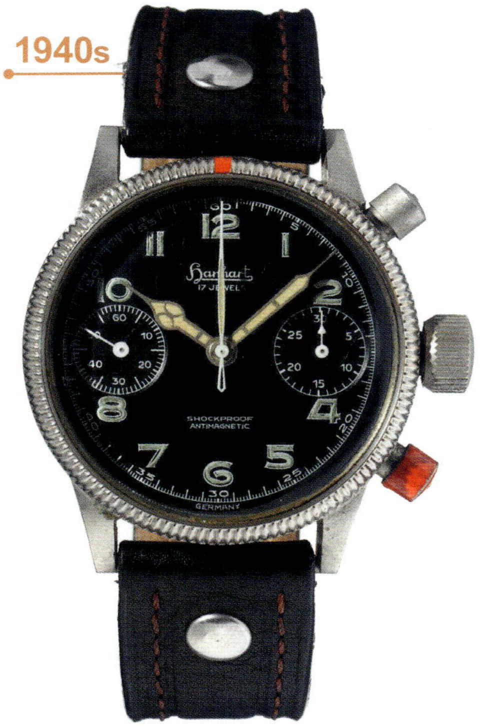
1940s

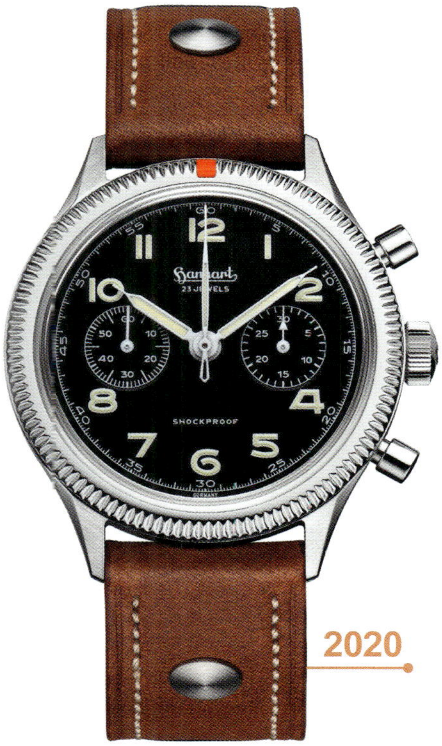
2020

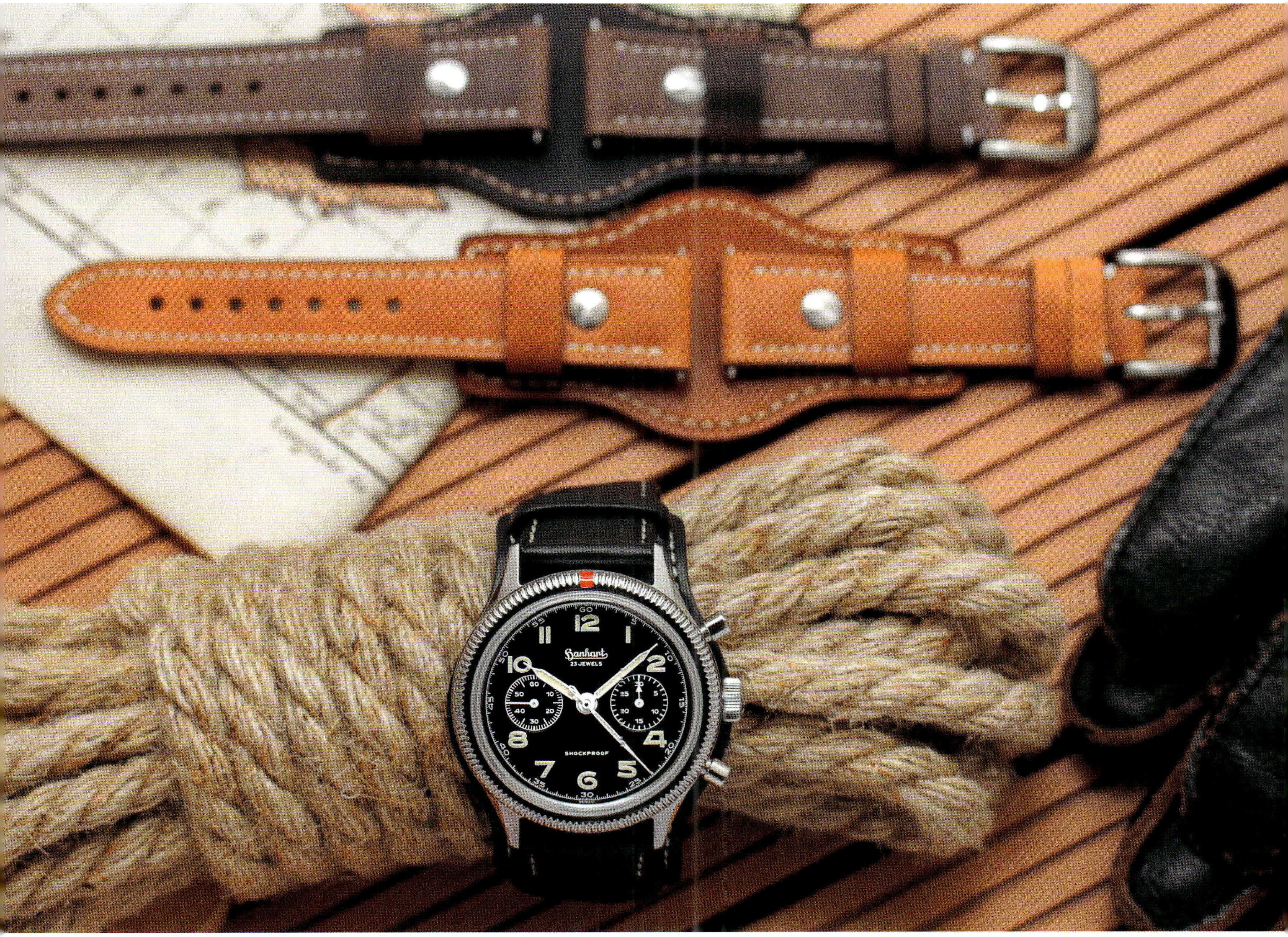

HEUER 1962
AUTAVIA

AUTO + AVIATION = AUTAVIA

When the Autavia appeared on the Heuer scene in 1933, no one had a wristwatch in mind. The new timepieces were board instruments for cars and aeroplanes, i.e. the automotive and aviation sectors. Charles-Edouard and Hubert-Bernard Heuer, the heads of the company, merged the two words to create the name Autavia. At the beginning of the 1960s, the memorable coinage found favour with Jack W. Heuer, giving rise to a new line of chronograph watches. The steely timepieces made their début in 1962. References 2446 (30-minute and 12-hour counters) and 3646 (30-minute totaliser) are powered by the exquisite hand-wound Valjoux 72 calibre with column wheel control and horizontal clutch. The two gave birth to the biography of a chronographic myth.

In the view of many collectors, the Autavia outshines other classics bearing the Heuer signature. One reason for this being the propensity of international racing driver greats to sport these chronographs on their wrists. Take Jochen Rindt, for example. He relied on the third iteration of Reference 2446, which adds to the cachet of what is known as the Rindt model. Mario Andretti preferred the 3646, also in the third generation. Over the years, both these original references have undergone several design transformations. In 1968, the Reference 2446 GMT was launched. Its hand-wound Valjoux 724 Calibre offers an additional 24-hour hand. At the end of the 1960s, Jack W. Heuer was known to be a member of the working group that developed the world's first chronograph wristwatch with micro-rotor winding system. It comes as no surprise that the Calibre 11, launched in 1969, also powered the Autavia. First in Reference 1163, which was worn by Jo Siffert, Clay Regazzoni, Derek Bell and Graham Hill. Despite the racing buzz, sales of these watches were sluggish due to the quartz revolution and its impact on the mechanical watch market. Because of the reticence displayed by traditional retailers, Jack W. Heuer had no choice but to resort to very unconventional marketing methods. For a period of some months, the US company Reynolds Tobacco inserted a voucher into every packet of its Viceroy cigarettes. It entitled the holder to purchase a modified Autavia chronograph for a mere $88. The promotion more than satisfied expectations. By the end of the campaign, Heuer had sold a very handsome 16,000 watches to customers. Racing driver Gilles Villeneuve also deserves a mention.

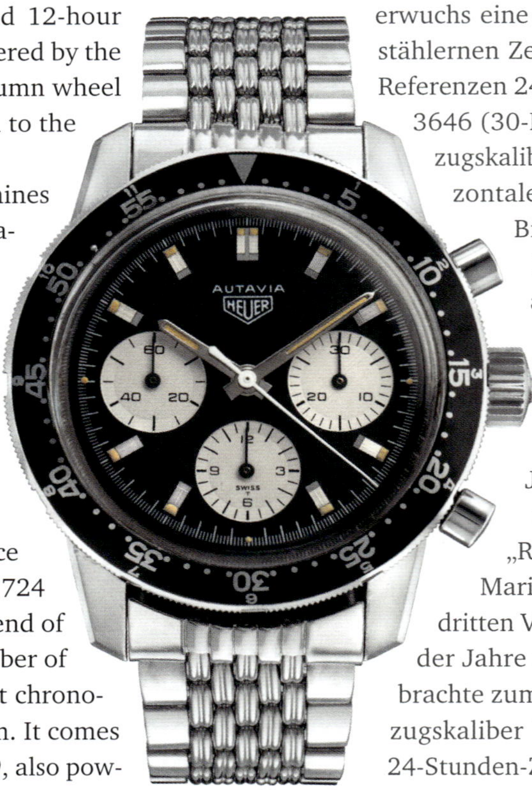

Top / Oben: Autavia Ref. 2446H "Jochen Rindt", 1969

AUTO + AVIATIK = AUTAVIA

Als die „Autavia" 1933 auf der Heuer'schen Bildfläche erschien, dachte noch niemand an einen Zeitmesser fürs Handgelenk. Vielmehr handelte es sich bei den neuen Uhren um Bordinstrumente für Automobile und die Aviatik, also die Luftfahrt. Aus beiden Wörtern generierten Charles-Edouard und Hubert-Bernard Heuer, die seinerzeit das Unternehmen lenkten, kurzerhand „Autavia". Zu Beginn der 1960er-Jahre begeisterte sich Jack W. Heuer für das einprägsame Kürzel. Daraus erwuchs eine neue Linie von Armband-Chronographen. Die stählernen Zeitschreiber gaben 1962 ihren Einstand. In den Referenzen 2446 (30-Minuten- und 12-Stunden-Zähler) und 3646 (30-Minuten-Totalisator) tickte das feine Handaufzugskaliber Valjoux 72 mit Schaltradsteuerung und horizontaler Räderkupplung. Dieses Duo begründete die Biographie eines chronographischen Mythos.

Bei zahlreichen Sammlern stellt die „Autavia" andere Klassiker mit der Signatur Heuer deutlich in den Schatten. Ein Grund ist darin zu sehen, dass internationale Größen im Automobilsport ihre Handgelenke mit diesen Stoppern schmückten. Zu ihnen gehörte beispielsweise Jochen Rindt. Er blickte auf die dritte Ausführung der Referenz 2446, was das so genannte „Rindt-Modell" besonders erstrebenswert macht. Mario Andretti setzte auf die 3646, ebenfalls in der dritten Version. Beide Startreferenzen erfuhren im Laufe der Jahre mehrere gestalterische Metamorphosen. 1968 brachte zum Beispiel die Referenz 2446 GMT. Ihr Handaufzugskaliber Valjoux 724 verfügte über einen zusätzlichen 24-Stunden-Zeiger. Ende der 1960er-Jahre hatte sich Jack W. Heuer bekanntlich an der Arbeitsgemeinschaft zur Entwicklung des weltweit ersten Armband-Chronographen mit Mikrorotor-Aufzug beteiligt. Logischerweise fand das 1969 lancierte Kaliber 11 auch seinen Einzug in die „Autavia". Zunächst in die Referenz 1163, welche Jo Siffert, Clay Regazzoni, Derek Bell und Graham Hill am Handgelenk trugen. Ungeachtet dessen gestaltete sich der übliche Verkauf dieser Armbanduhren infolge der Quarzrevolution und ihrer Auswirkungen auf den Mechanikmarkt eher schleppend. Wegen der zögerlichen Haltung des tradierten Fachhandels blieb Jack W. Heuer keine andere Wahl, als sehr unkonventionelle Vertriebsmethoden anzuwenden. In diesem Sinne legte Reynolds Tobacco in den USA einige Monate lang jeder Stange seiner Zigarettenmarke Viceroy einen Gutschein bei. Dieser autorisierte den Besitzer zum Erwerb eines modifizierten „Autavia"-Chronographen für lediglich 88 Dollar. Das Resultat konnte sich sehen lassen. Am Ende der Aktion hatte Heuer immerhin 16.000 Exemplare an den Mann gebracht. Nicht zu vergessen der Rennfahrer Gilles Villeneuve.

In 1964, he purchased an Autavia Reference 73663 with the hand-wound Valjoux 7736 Calibre. The first Autavia renaissance dawned in 2003. In his then role as honorary president of TAG Heuer, Jack W. Heuer was involved in this homage model with the Calibre 11, a combination of an Eta automatic movement and a Dubois-Dépraz module. Due to the modular design of the inner workings, as for Reference 1163, the crown, rarely needed on account of the automatic movement, was placed on the left case flank.

In 2017, the Autavia made its comeback with a manufacture automatic movement. The increase in diameter from 38 to 42 millimetres called for ingenuity on the part of the design team. But the effort was worthwhile. Jack W. Heuer was presented with a special edition to mark his 85th birthday. Its differences to the normal model were personally specified by the octogenarian. They included dark fields for the permanent second and the 30-minute and 12-hour counters on the bright sunray-cut dial. The colour of the Super-LumiNova luminous material was also changed. Since then, TAG Heuer's oldest collection has enjoyed unabated popularity.

Ab 1964 nannte er eine „Autavia"-Referenz 73663 mit dem Handaufzugskaliber Valjoux 7736 sein Eigen. Eine erste „Autavia"-Renaissance brachte das Jahr 2003. An diesem Hommage-Modell mit dem „Calibre 11", einem Mix aus Eta-Automatik und Dubois-Dépraz-Modul, hatte Jack W. Heuer in seiner damaligen Rolle als Ehrenpräsident von TAG Heuer mitgewirkt. Durch die modulare Bauweise des Innenlebens wanderte die wegen des Selbstaufzugs eher selten benötigte Krone analog zur Referenz 1163 auf die linke Gehäuseseite.

2017 kehrte die „Autavia" mit Manufaktur-Automatik zurück in die Uhrenszene. Das Anwachsen des Durchmessers von seinerzeit 38 auf 42 Millimeter verlangte dem Designerteam einiges ab. Aber der Aufwand hat sich gelohnt. Zu Ehren seines 85. Geburtstags bekam Jack W. Heuer eine Sonderedition. Die Unterschiede zum normalen Modell legte er höchstpersönlich fest. Sie bestanden unter anderem in dunklen Feldern für die Permanentsekunde sowie die 30-Minuten- und 12-Stunden-Zähler auf hellem Sonnenschliff-Zifferblatt. Änderung erfuhr auch die Farbe der Super-LumiNova-Leuchtmasse. Seitdem lebt die älteste Kollektion bei TAG Heuer so munter wie ein Fisch im Wasser.

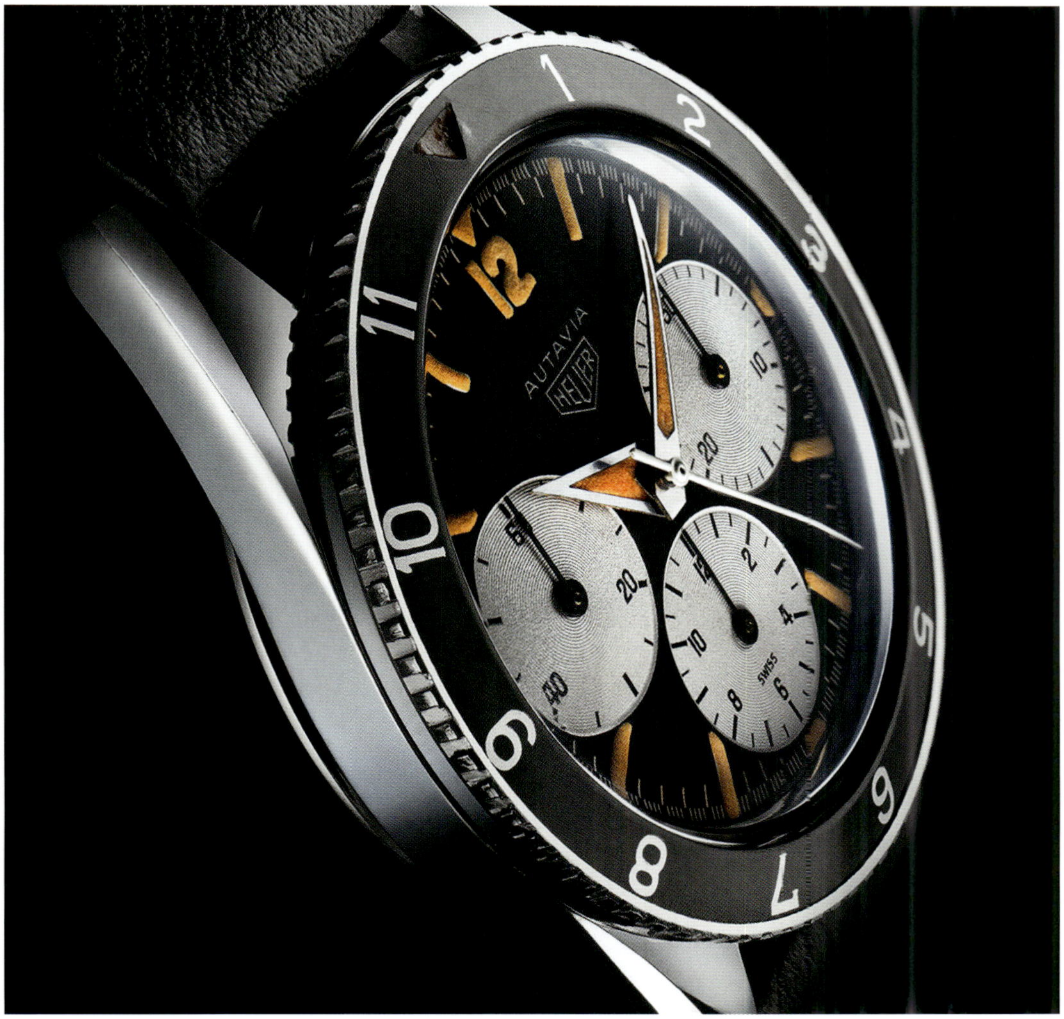

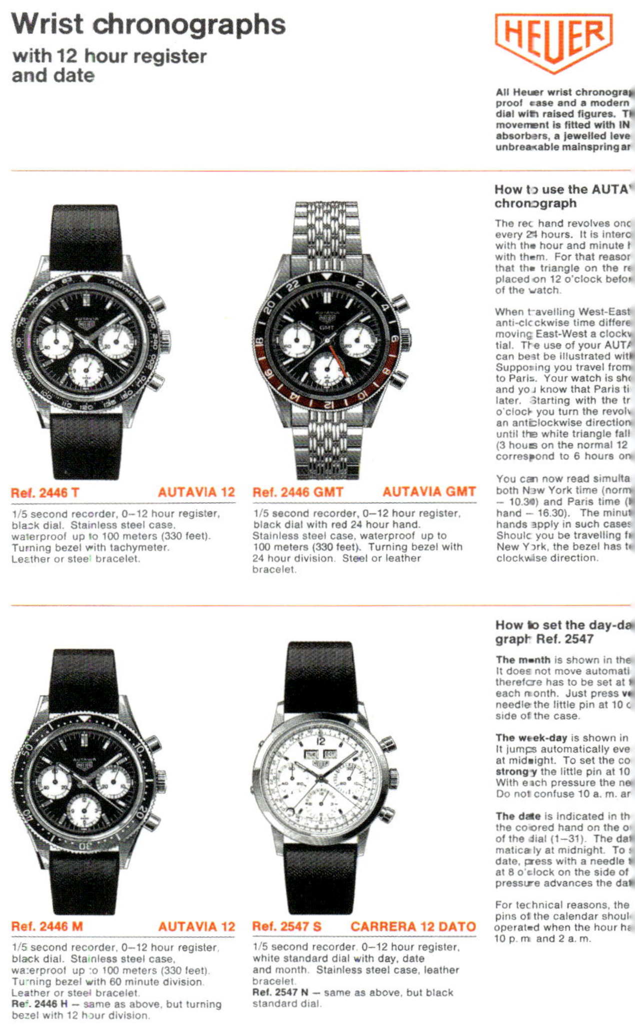

Wrist chronographs

with 12 hour register and date

All Heuer wrist chronogra proof case and a modern dial with raised figures. T movement is fitted with IN absorbers, a jewelled leve unbreakable mainspring ar

Ref. 2446 T — AUTAVIA 12

1/5 second recorder, 0–12 hour register, black dial. Stainless steel case, waterproof up to 100 meters (330 feet). Turning bezel with tachymeter. Leather or steel bracelet.

Ref. 2446 GMT — AUTAVIA GMT

1/5 second recorder, 0–12 hour register, black dial with red 24 hour hand. Stainless steel case, waterproof up to 100 meters (330 feet). Turning bezel with 24 hour division. Steel or leather bracelet.

How to use the AUTA chronograph

The red hand revolves onc every 24 hours. It is interc with the hour and minute h with them. For that reason that the triangle on the re placed on 12 o'clock befor of the watch.

When travelling West-East anti-clockwise time differe moving East-West a clockw tial. The use of your AUTA can best be illustrated wit Supposing you travel from to Paris. Your watch is sh and you know that Paris ti later. Starting with the tr o'clock you turn the revolv an anticlockwise direction until the white triangle fall (3 hours on the normal 12 correspond to 6 hours on

You can now read simulta both New York time (norm – 10.30) and Paris time (hand – 16.30). The minut hands apply in such cases Should you be travelling f New York, the bezel has t clockwise direction.

Ref. 2446 M — AUTAVIA 12

1/5 second recorder, 0–12 hour register, black dial. Stainless steel case, waterproof up to 100 meters (330 feet). Turning bezel with 60 minute division. Leather or steel bracelet.
Ref. 2446 H – same as above, but turning bezel with 12 hour division.

Ref. 2547 S — CARRERA 12 DATO

1/5 second recorder, 0–12 hour register, white standard dial with day, date and month. Stainless steel case, leather bracelet.
Ref. 2547 N – same as above, but black standard dial.

How to set the day-da graph Ref. 2547

The month is shown in the It does not move automati therefore has to be set at each month. Just press ve needle the little pin at 10 o side of the case.

The week-day is shown in It jumps automatically eve at midnight. To set the co **strongly** the little pin at 10 With each pressure the ne Do not confuse 10 a. m. ar

The date is indicated in th the colored hand on the o of the dial (1–31). The da matically at midnight. To date, press with a needle at 8 o'clock on the side of pressure advances the da

For technical reasons, the pins of the calendar shoul operated when the hour ha 10 p. m. and 2 a. m.

***From left | Von links:** First Autavia, 1962 | Catalogue 1968–1969*

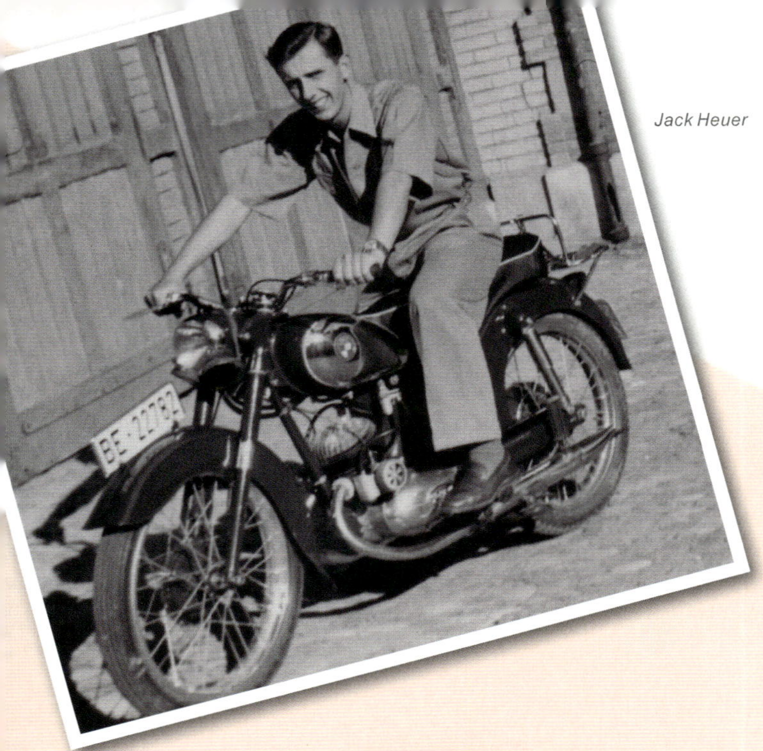

Jack Heuer

1910s

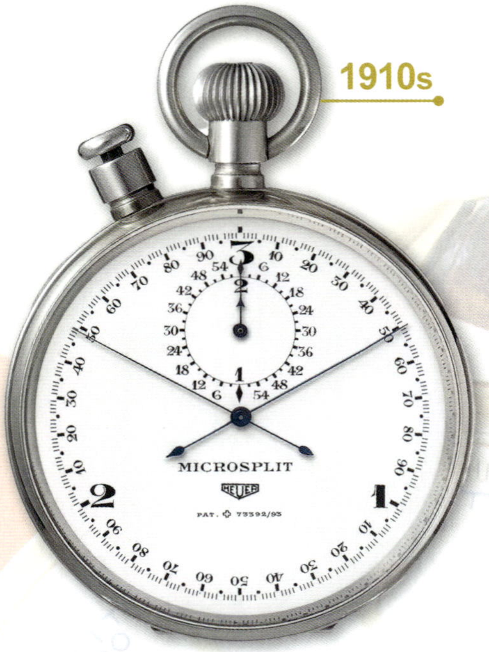

1963

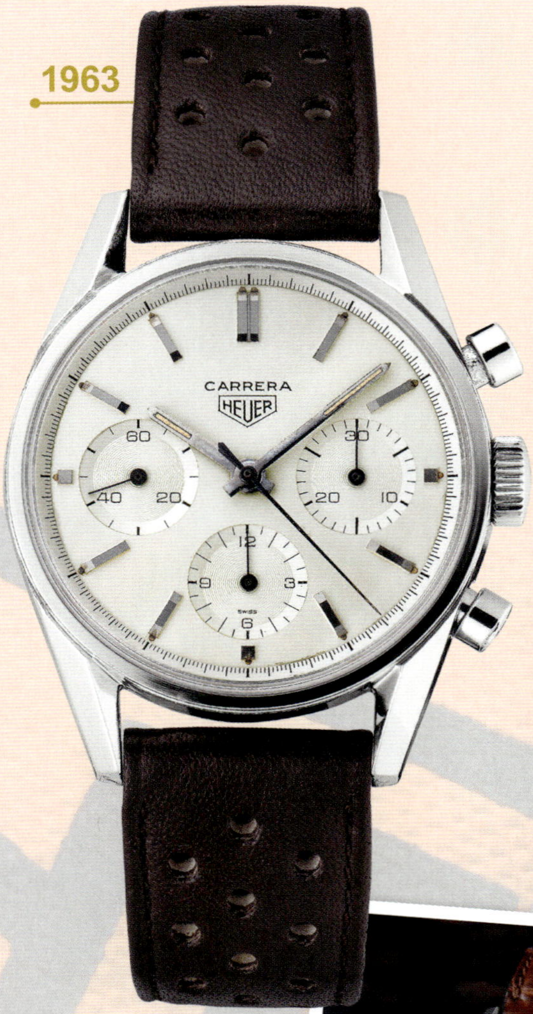

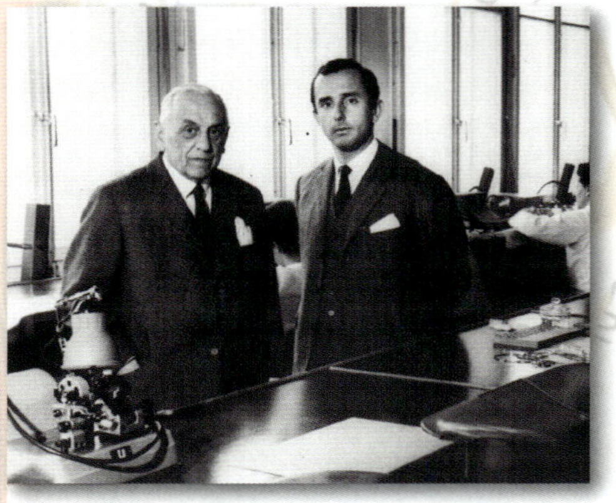

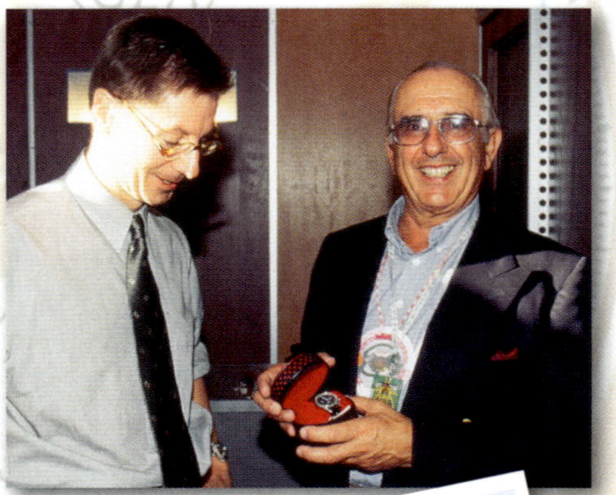

Charles-Edouard and Jack Heuer | Jack Heuer in Monza, 1996

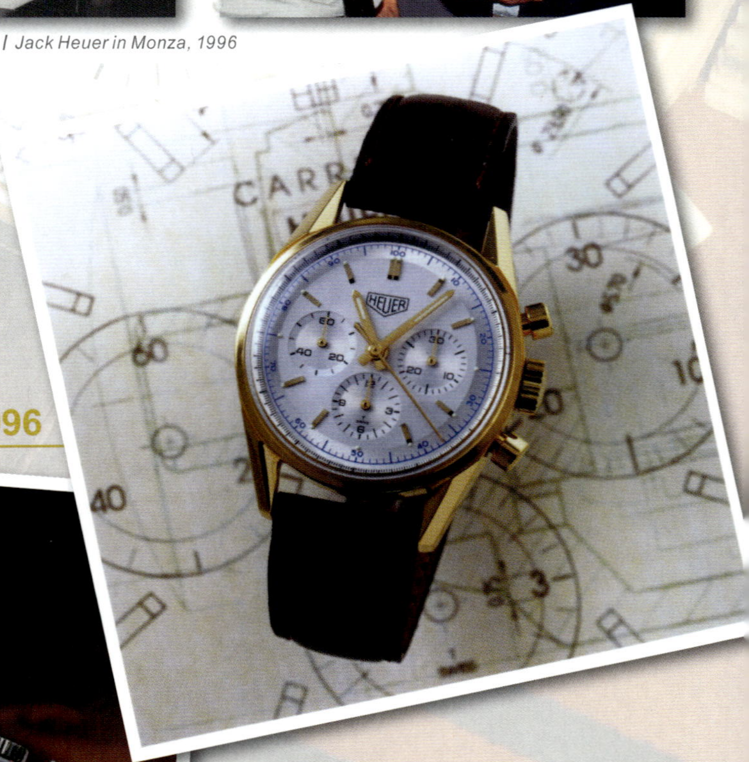

1996

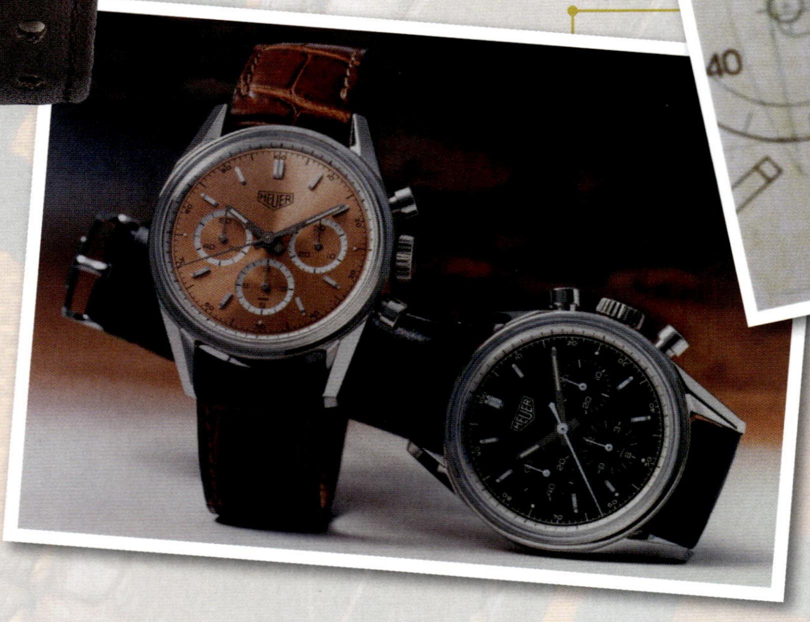

MEMORIES OF MEXICO

Presented in 1963, the Carrera undoubtedly ranks among the most recognised style icons in watch history. Its biography can be traced back to the early 1960s. Jack W. Heuer, the timepiece's intellectual father, had a keen nose for what matters when it comes to chronographs. Even more important than the movement is the design of the case and dial. Back then, opulent dimensions were not a thing. The case diameter mainly depended on the inner workings. Wide movement retaining rings and space around the calibre were frowned upon, a height of 38 millimetres deemed lofty, if not unbecomingly bulky. Along with ease of operation, legibility topped the priority list. Therefore, the owner of the family company Heuer not only set his sights on an understated design, but within the bounds of propriety, also a large dial. After intensive considerations, the qualified industrial engineer achieved a remarkable feat of ingenuity. Like his competitors, to optimise water resistance Jack W. Heuer used reinforced plexiglass with a metal tension ring. His inspired idea revolved around the first-time use of the oblique surface for the printing of the chronograph's fifth-second track. This innovation increased the size of the dial by almost two millimetres. However, this was only one aspect of the work of art envisioned by its creator. At a slightly lower depth, the secondary dials for the permanent second and totaliser(s) lend the timepiece a unique three-dimensional appearance. After completion of the comparatively modest, yet expressive prototypes, all that was needed was an appropriate name. Chance then took a hand. At the famous 12 Hours of Sebring, Florida's famous motorsport endurance race, the enterprising head of the company personally assumed responsibility for the official timekeeping. Chatting with the daring racing driver Ricardo Rodriguez, the latter waxed lyrical about the Carrera Panamericana. In the 1950s, this breakneck car race spanned Mexico from border to border. His enthusiasm was obviously contagious because on his return to the Old World, Jack W. Heuer succinctly christened his new timepiece Carrera. Associating the wristwatch with motorsport had a great and astonishingly speedy effect on sales. Even the first, still relatively modest collection gained a successful foothold in the fraught chronograph market. The classic model, the Carrera 12, was a sporty and elegant chronograph with a 30-minute and 12-hour counter, powered by the finely crafted hand-wound Valjoux 72 calibre at a rate of 2.5 hertz. >

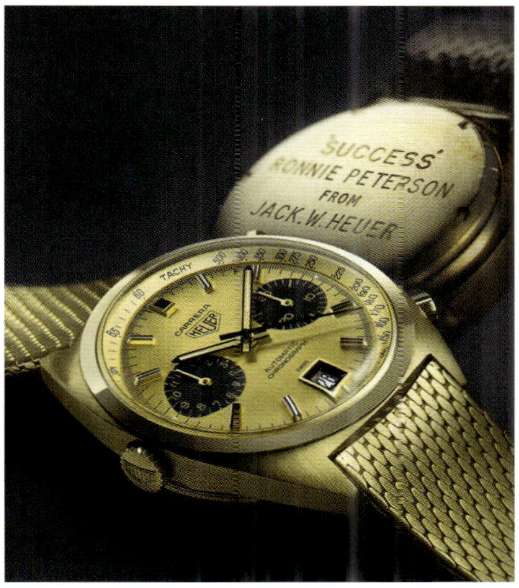

Top | Oben: *Ref. 1158 CHN "Ronnie Peterson"*

ERINNERUNG AN MEXIKO

Zu den anerkannten Stilikonen der Uhrengeschichte gehört zweifellos die 1963 vorgestellte „Carrera". Deren Biographie lässt sich bis in die frühen 1960er-Jahre zurückverfolgen. Jack W. Heuer, der geistige Vater, wusste sehr genau, worauf es bei funktionsdominierten Chronographen ankommt. Wichtiger noch als das Uhrwerk ist das Design von Schale und Zifferblatt. In jener Epoche waren opulente Dimensionen noch kein Thema. Der Gehäusedurchmesser orientierte sich primär am tickenden Innenleben. Breite Werkhalteringe und Luft rund ums verbaute Kaliber galten als unschicklich, eine Höhe von 38 Millimetern als groß, wenn nicht gar klotzig. Neben der Bedien-Ergonomie rangierte die Ablesbarkeit in der Prioritätenliste ganz oben. Daher strebte der Eigentümer des Familienunternehmens Heuer nicht nur ein sachlich-reduziert gestaltetes, sondern im Rahmen der Möglichkeiten auch großes Zifferblatt an. Nach intensiven Überlegungen gelang dem studierten Betriebsingenieur ein bemerkenswerter Kunstgriff. Zum Zweck optimierter Wasserdichte nutzte Jack W. Heuer wie auch die Mitbewerber ein armiertes Plexiglas mit metallenem Spannring. Seine nachgerade geniale Idee zielte auf die erstmalige Nutzung der schrägen, den Augen zugewandten Fläche zum Bedrucken mit der Fünftelsekunden-Indexierung für den Chronographen. Diese Eingebung brachte eine Vergrößerung des Zifferblatts um beinahe zwei Millimeter. Hierbei handelte es sich allerdings nur um einen Aspekt des angestrebten Kunstwerks. Leicht abgesenkte Nebenzifferblätter für Permanentsekunde und Totalisator(en) bewirkten eine einzigartige dreidimensionale Optik. Nach Fertigstellung der vergleichsweise schlichten, aber doch ausdruckstarken Prototypen brauchte es nur noch einen passenden Namen. Und hier half Genosse Zufall. Bei den berühmten „12 Stunden von Sebring", Floridas berühmtem Sportwagen-Langstreckenrennen, kümmerte sich der rührige Firmenchef höchstpersönlich um die offizielle Zeitnahme. Während eines Plauschs berichtete Ricardo Rodriguez, ein verwegener Fahrer, geradezu euphorisch über die „Carrera Panamericana". In den 1950er-Jahren führte dieses halsbrecherische Autorennen quer durch Mexiko. Die Begeisterung schwappte über auf Jack W. Heuer, der seinen zeitschreibenden Newcomer nach der Rückkehr in die Alte Welt kurz und bündig „Carrera" taufte. Erstaunlich schnell entfaltete die Verknüpfung von Armbanduhr und Motorsport ihre verkaufsfördernde Wirkung. Bereits die erste, noch relativ bescheidene Kollektion behauptete sich erfolgreich am angespannten Chronographenmarkt. Beim Klassiker, dem „Carrera 12", handelte es sich um einen sportlich-eleganten Stopper mit 30-Minuten- und 12-Stunden-Zähler, in dem das feine Handaufzugskaliber Valjoux 72 mit 2,5 Hertz tickte. >

2023

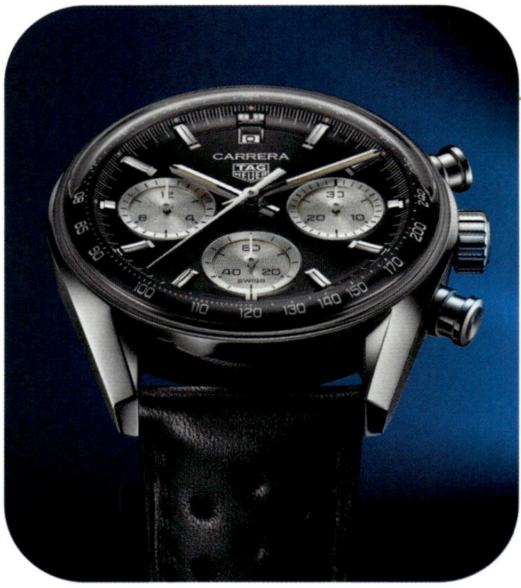

1996

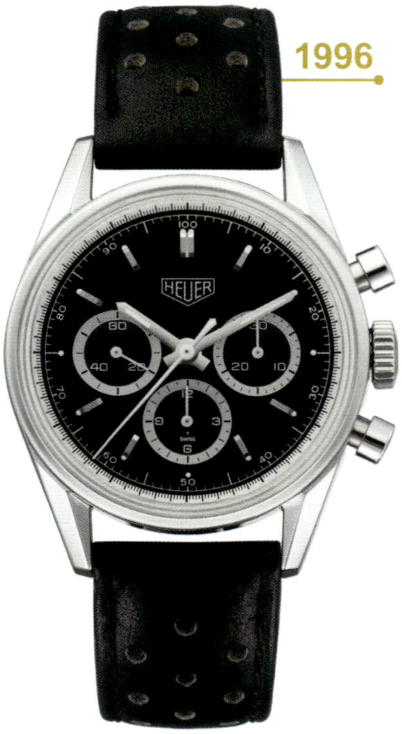

2020

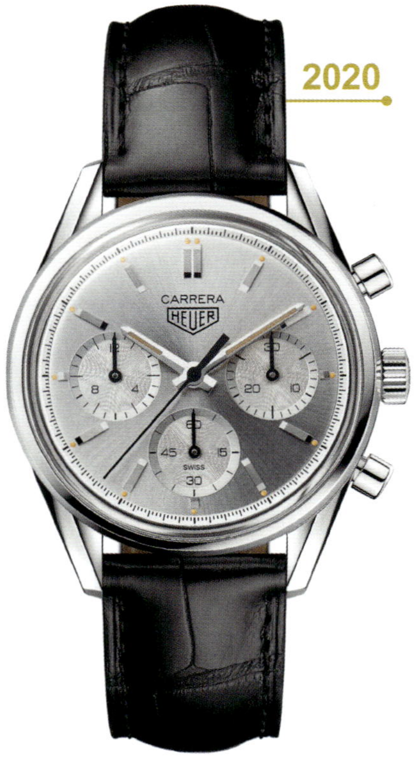

Five generations of the model preceded a temporary hiatus in the Carrera's history in 1986. Exactly ten years later, the company, now renamed TAG Heuer, celebrated the timepiece's sorely needed renaissance. Initially, legal reasons prevented the use of the memorable name. Once these were resolved, the Carrera line quickly developed into one of the mainstays of the collection. The variety of different models is truly impressive. Top of the range in technical terms are different iterations with a proprietary chronograph movement and tourbillon.

Bis 1986, als die Geschichte der „Carrera" vorübergehend endete, gab es fünf Generationen. Exakt zehn Jahre später, 1996, zelebrierte das zwischenzeitlich als TAG Heuer firmierende Unternehmen die dringend gebotene Renaissance. Zunächst war der einprägsame Name aus rechtlichen Gründen tabu. Nach Beseitigung der Namenshürden hat sich die „Carrera"-Linie rasch zu einer der tragenden Kollektionssäulen entwickelt. Wahrhaft beeindruckend ist die Vielfalt unterschiedlicher Modelle. Technisch ganz oben rangieren unterschiedliche Versionen mit hauseigenem Chronographenwerk und Tourbillon.

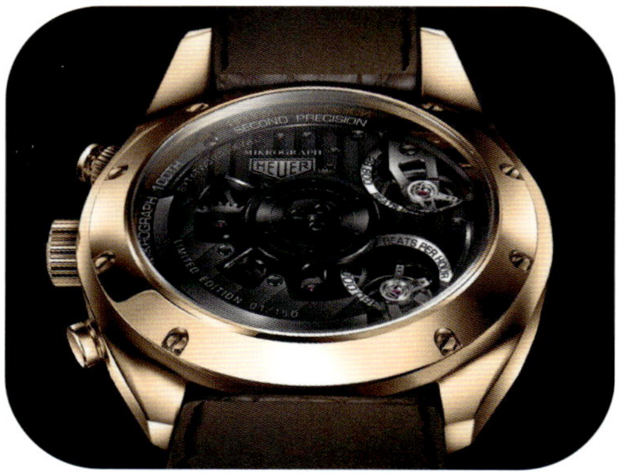

2011

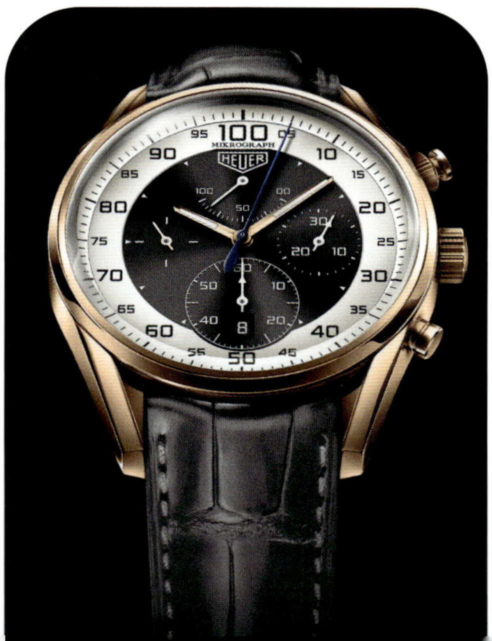

2022

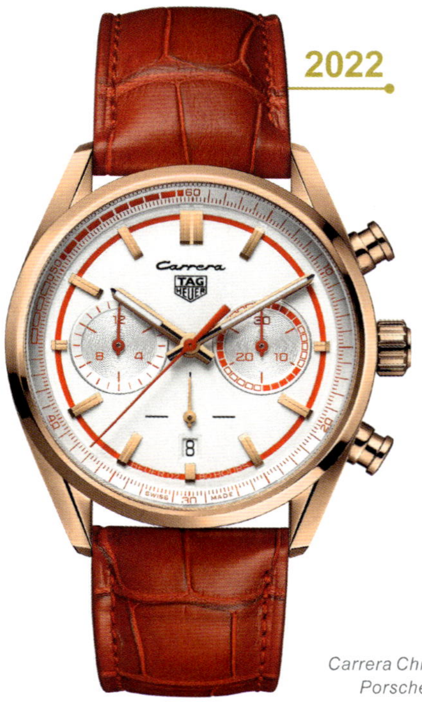

2022

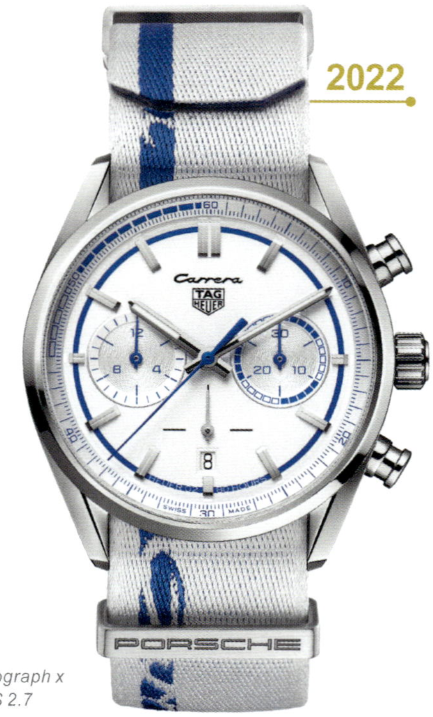

Carrera Chronograph x Porsche RS 2.7

< Mikrograph Anthracite CAR5041.FC8178

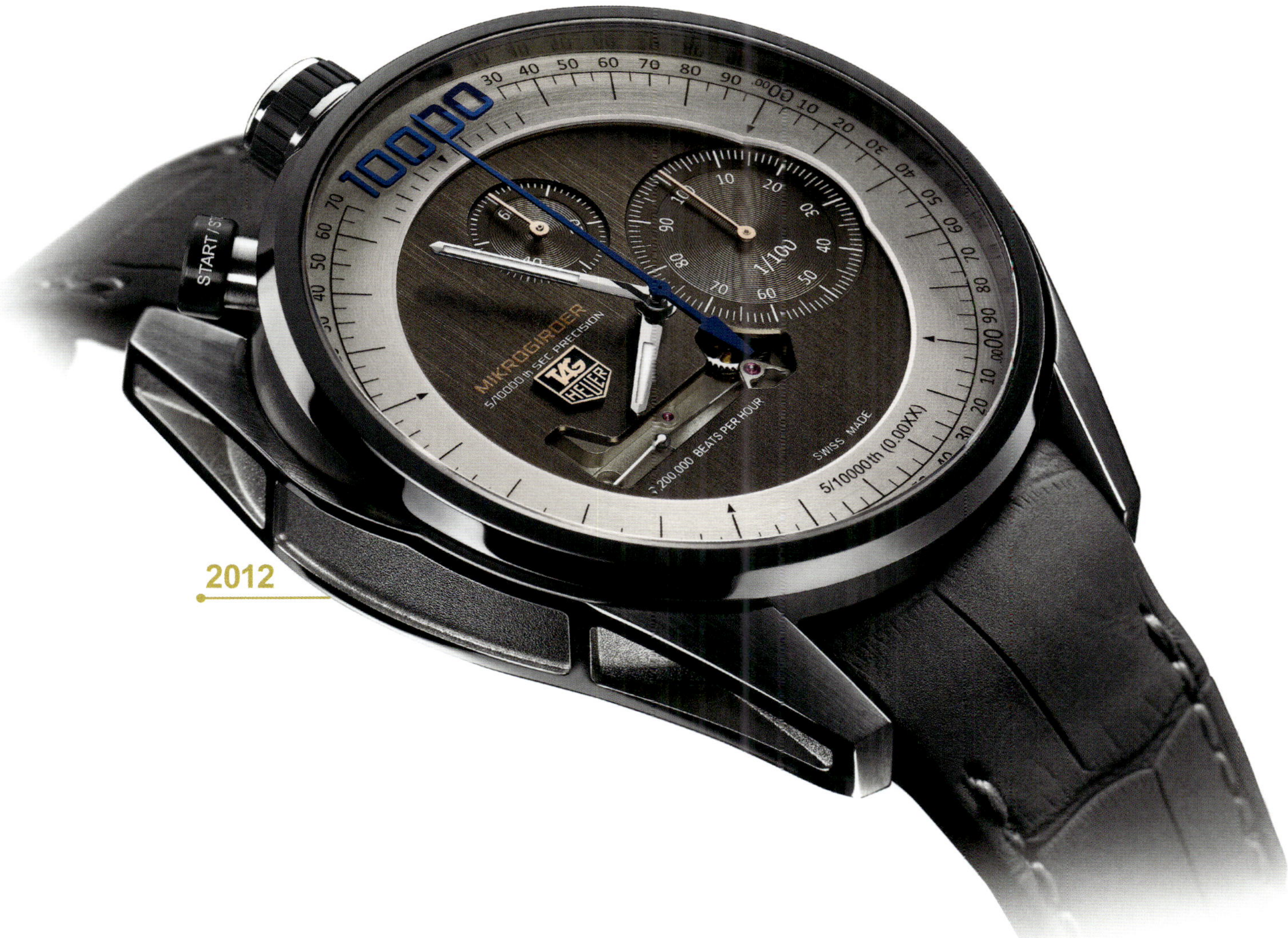

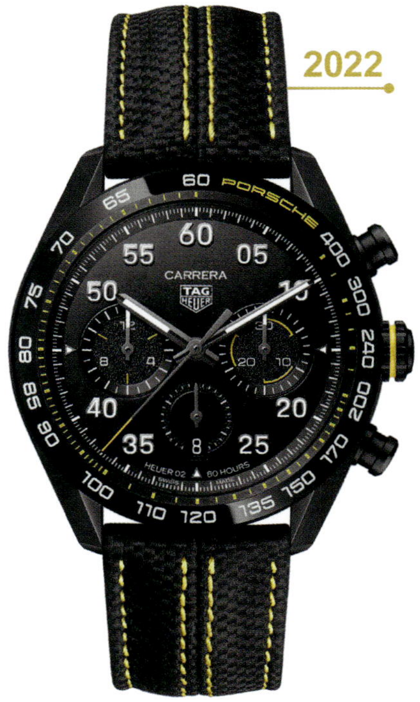

Chronograph x Porsche
Yellow Racing

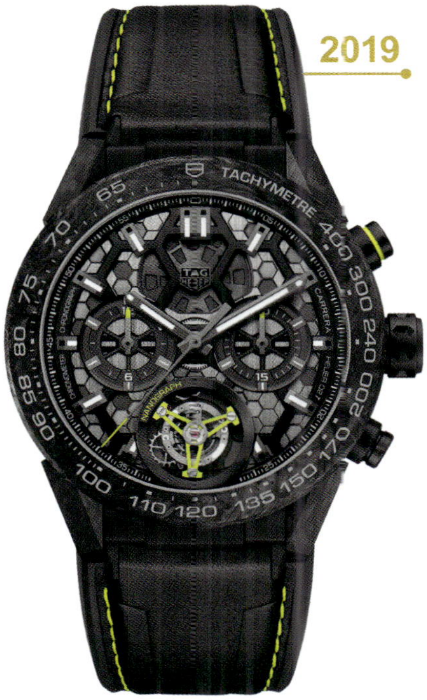

Calibre Heuer02T
Tourbillon Nanograph

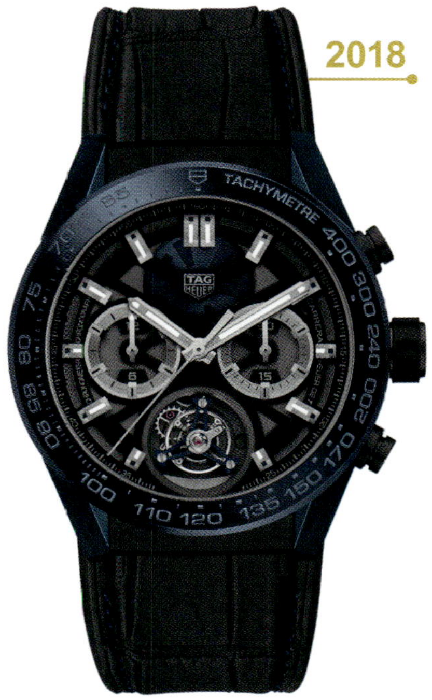

Tourbillon Chronograph
"Tête de Vipère" Chronometer

HEUER MONACO

1969

SQUARE ICON

In 1969, Jack W. Heuer, who headed the family business in the fourth generation, presented the eagerly awaited micro-rotor self-winding chronographs. In the mid-1960s, finding himself unable to finance the costly project alone, the company boss had entered a cooperation with competitor Willy Breitling. They in turn engaged two specialists to develop the innovative inner workings: Büren provided a flat automatic movement with a micro-rotor, while Dubois-Dépraz supplied the chronograph module. At Heuer, the modular Calibre 11 made its début in the well-known Autavia and Carrera lines. The third model in the chronograph collection was the Monaco, a world première with a waterproof case. It was developed by expert casemaker Ervin Piquerez SA (EPSA). The square meteor owes its current cult status to an international star and the creativity of Jack W. Heuer. "The huge amount of effort poured into its development all but exhausted our financial resources. Therefore, we had to seek smart, less expensive ways to promote sales." On the golf course, Jack W. Heuer learned that Joseph "Jo" Siffert was looking for sponsors. The offer of an annual sponsorship worth CHF 25,000 convinced the Swiss Formula 1 driver. The new partner strapped a Monaco to his wrist, placed the Heuer logo on his racing overall and attached another one to his Porsche 908 Spyder. The deal paid off in 1970, when Lee H. Katzin and Steve McQueen produced the film *Le Mans*. For the filming, Jack Heuer had supplied the set designer Don Nunley with plenty of stickers, board clocks and chronographs. When choosing his racing outfit, the Hollywood star chose the overall of Jo Siffert standing next to him. He also chose a Monaco as his watch. In July 2012, it was sold at auction for the lordly sum of $650,000. The Monaco never achieved real success, in part on account of its unusual appearance. The first chapter thus ended around 1975. >

QUADRATISCHE IKONE

1969 präsentierte Jack W. Heuer, der das Familienunternehmen in vierter Generation leitete, die sehnsüchtig erwarteten Chronographen mit Selbstaufzug durch Mikrorotor. Nachdem er sich Mitte der 1960er-Jahre außerstande gesehen hatte, das kostspielige Projekt finanziell alleine zu stemmen, war der Firmenchef eine Kooperation mit dem Konkurrenten Willy Breitling eingegangen. Zur Entwicklung der neuartigen Mechanik verpflichtete das Duo zwei Spezialisten: Büren steuerte ein flaches Automatikwerk mit Mikrorotor bei, Dubois-Dépraz lieferte das Modul mit Stoppfunktion. Bei Heuer debütierte das Modulkaliber 11 in den bekannten Linien „Autavia" und „Carrera". Dritte im Chrono-Bunde war die „Monaco", eine Weltpremiere mit wasserdichtem Gehäuse. Dessen Entwicklung stammte von der einschlägig erfahrenen Ervin Piquerez SA (EPSA). Seinen heutigen Kultstatus verdankt der quadratische Zeit-Bolide einem Weltstar und der Kreativität von Jack W. Heuer. „Nach dem gewaltigen Entwicklungs-Kraftakt waren unsere finanziellen Mittel beinahe erschöpft. Daher mussten wir intelligente, weniger kostspielige Methoden der Verkaufsförderung suchen." Auf dem Golfplatz erfuhr Jack W. Heuer, dass Joseph „Jo" Siffert auf der Suche nach Sponsoren sei. Die Offerte, den eidgenössischen Formel-1-Piloten mit jährlich 25.000 Schweizer Franken zu unterstützen, überzeugte. Der neue Partner schnallte sich eine „Monaco" ans Handgelenk, platzierte das Heuer-Logo auf seinem Renn-Overall und klebte ein weiteres auf seinen Porsche 908 Spyder. Der Deal zahlte sich bereits 1970 aus, als Lee H. Katzin mit Steve McQueen den Film *Le Mans* produzierte. Für die Dreharbeiten hatte Jack Heuer den Filmausstatter Don Nunley mit reichlich Stickern, Borduhren und Chronographen ausgestattet. Bei der Wahl seines Renn-Outfits entschied sich der Hollywoodstar für den Dress des neben ihm stehenden Jo Siffert. Als Armbanduhr wählte er ebenfalls eine „Monaco". Im Juli 2012 erlöste diese bei einer Auktion stolze 650.000 US-Dollar. Richtig erfolgreich wurde die „Monaco" trotzdem nicht, was auch an der ungewöhnlichen Optik lag. Somit endete das erste Kapitel gegen 1975. Vergeblich hatte Jack Heuer die „Monaco" damals als Spielgefährtin angepriesen und in Anzeigen erläutert, „Wie Männer richtig mit ihr spielen." >

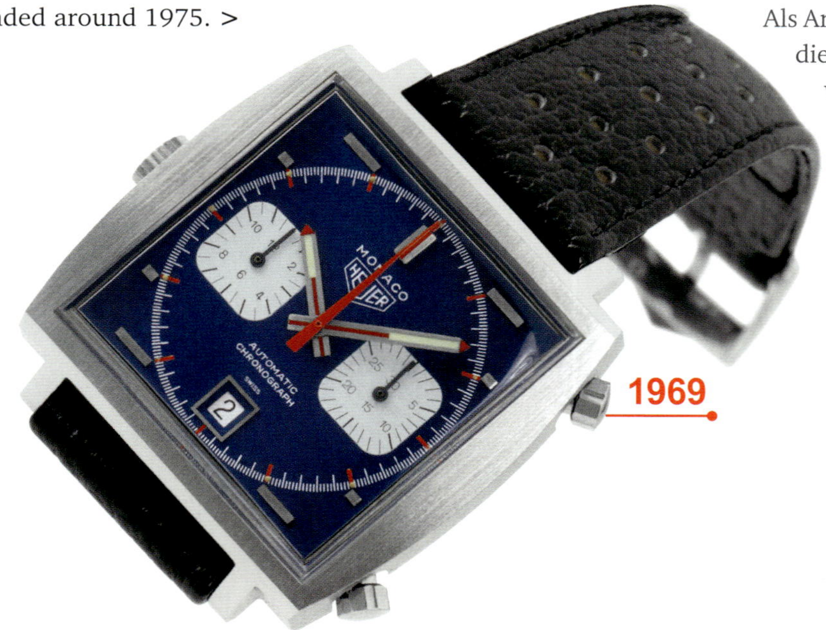

1969

Right page, from top left | Rechte Seite, von oben links:
Steve McQueen, Le Mans, 1970 | Monaco "Steve McQueen", Ref. 1133B, in the box & detail

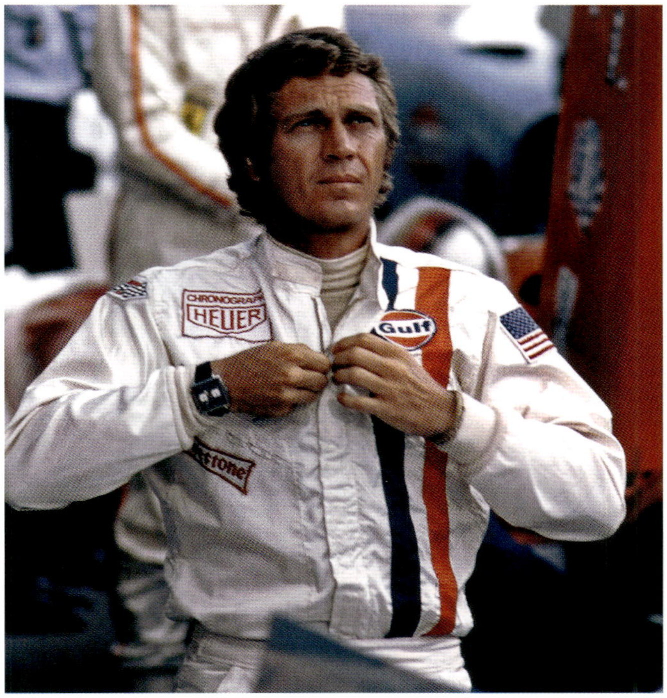
CHRONOGRAPH
HEUER
Gulf

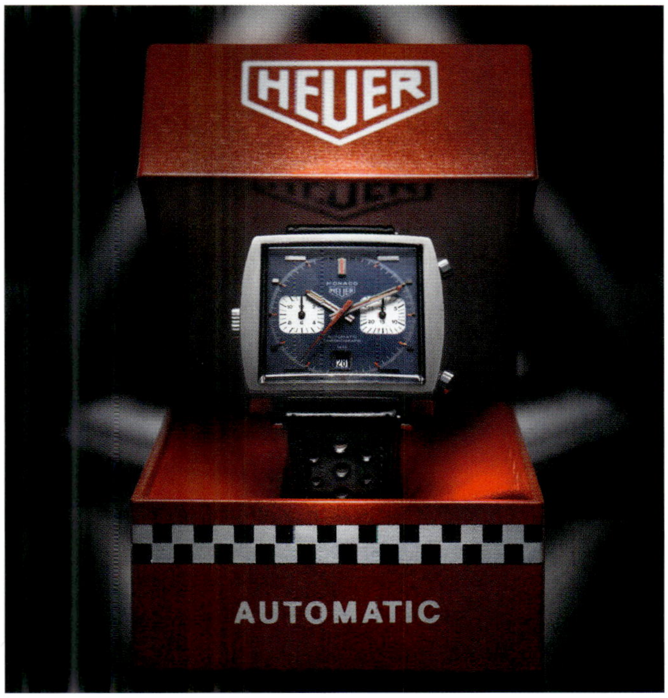
HEUER
AUTOMATIC

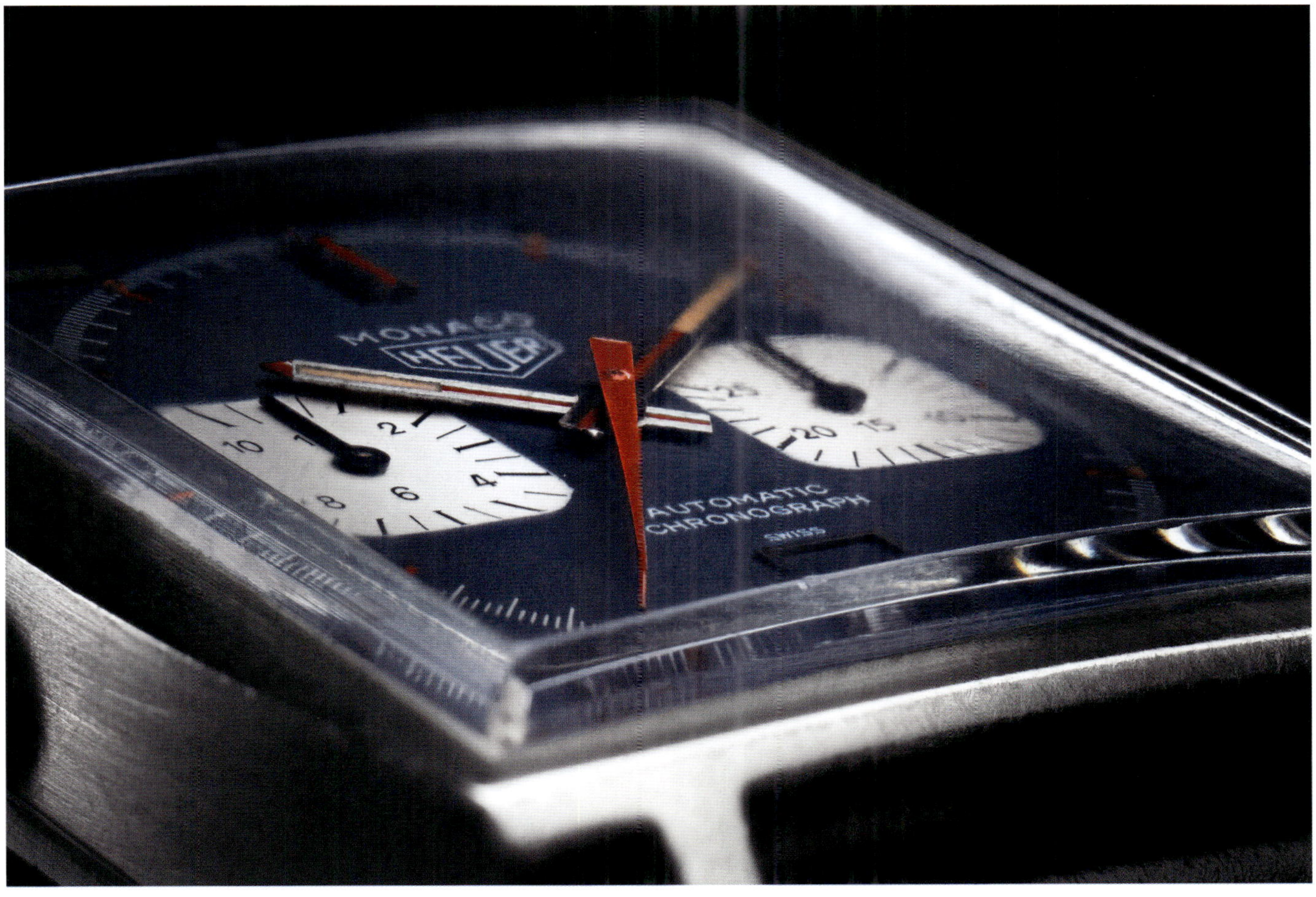
HEUER
AUTOMATIC
CHRONOGRAPH
SWISS

With little success, Jack Heuer had touted the Monaco as a female playmate; the advertising ditty ran "Have fun with Heuer's two-timing beauty."
The comeback took place in 1998 in a limited edition of 5,000. Since then, the chronographic icon has become an integral part of the TAG Heuer collection. A 2009 re-edition with a blue dial recalls the star, who died in 1980. In 2018, TAG Heuer and Gulf Oil continued their motorsport history with the Porsche 917K, Steve McQueen, the 24-hour Le Mans race and, of course, the Monaco chronograph. Blue and orange stripes prominently identify Gulf. Only 50 models commemorated the Gulf Racing team's first win at the Le Mans 24 Hours in 1968. To mark the Monaco's half-century in 2019, TAG Heuer brought out five limited, by now expensive watch editions as well as a magnificent monograph on this cult watch. In 2024, TAG Heuer moved into the premium luxury watch bracket with its bespoke Monaco Split-Seconds Chronograph. The specialist Vaucher supplies the movement of the automatic Calibre TH81-00.

Das Comeback erfolgte 1998 in einer limitierten Edition von 5.000 Exemplaren. Seitdem ist die chronographische Ikone aus der TAG Heuer-Kollektion nicht mehr wegzudenken. Unter anderem erinnerte 2009 eine Re-Edition mit blauem Zifferblatt an den 1980 verstorbenen Star. 2018 knüpften TAG Heuer und Gulf Oil an die Motorsport-Geschichte mit dem Porsche 917K, Steve McQueen, dem 24-Stunden-Rennen von Le Mans und natürlich dem „Monaco"-Chronographen an. Unübersehbare Gulf-Erkennungszeichen sind blaue und orangefarbene Streifen. Nur 50 Exemplare erinnerten an den ersten Sieg des Team Gulf Racing bei den 24 Stunden von Le Mans im Jahr 1968. Zum 50. „Monaco"-Jubiläum im Jahr 2019 brachte TAG Heuer fünf limitierte, inzwischen teuer bezahlte Uhr-Editionen sowie eine prachtvolle Monographie über diese Kult-Uhr auf den Markt. 2024 begibt sich TAG Heuer mit dem individualisierbaren „Monaco Split-Seconds Chronograph" in die Top-Luxusklasse der Armbanduhren. Die Basis des Automatikkalibers TH81-00 liefert der Spezialist Vaucher zu.

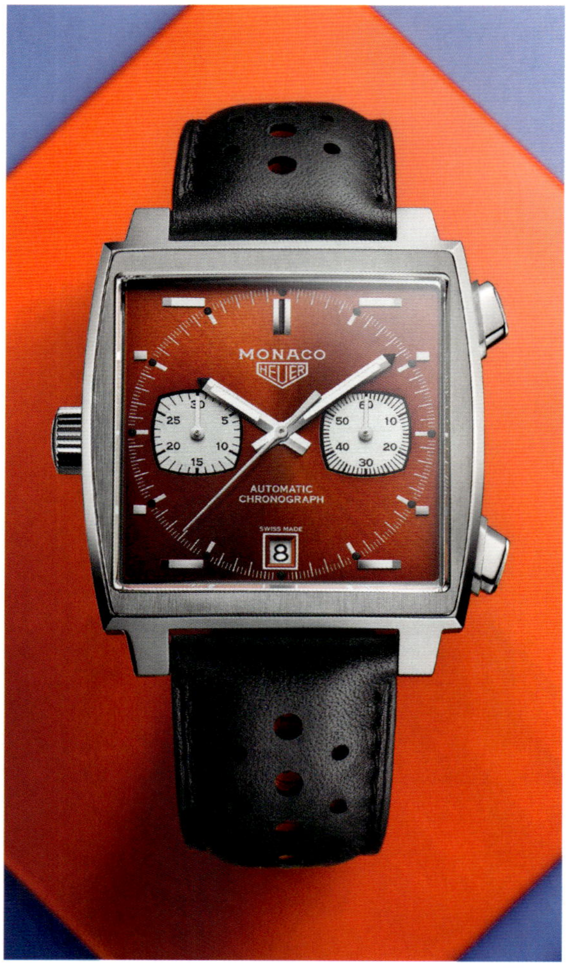

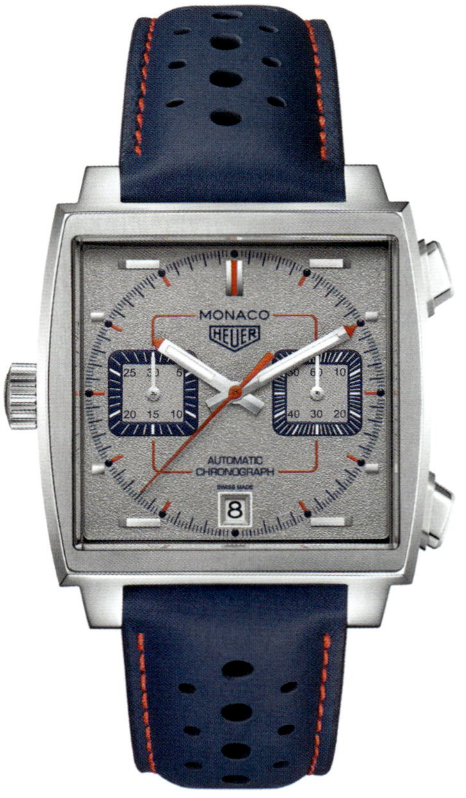

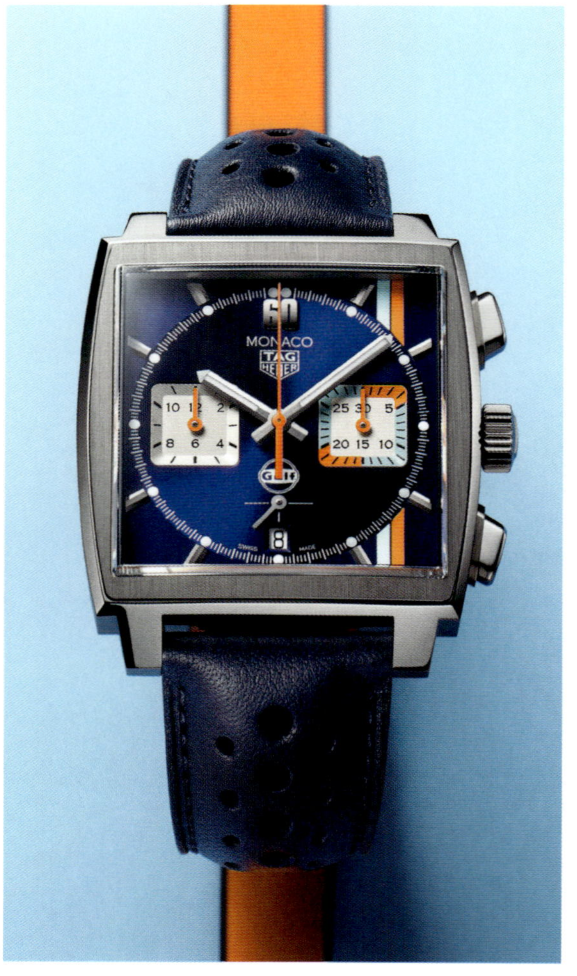

From left / Von links:
Monaco Through Time 80s, 2019 /
Monaco Through Time 90s, 2019 /
Monaco Chronograph x Gulf, 2022

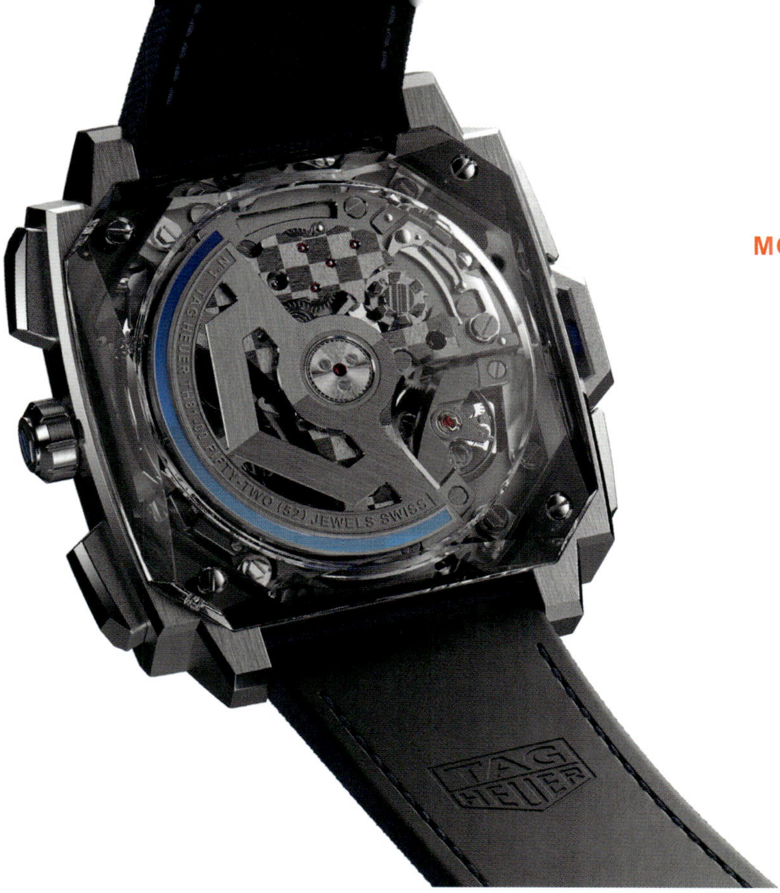

TAG HEUER
MONACO SPLIT-SECONDS
CHRONOGRAPH, 2024

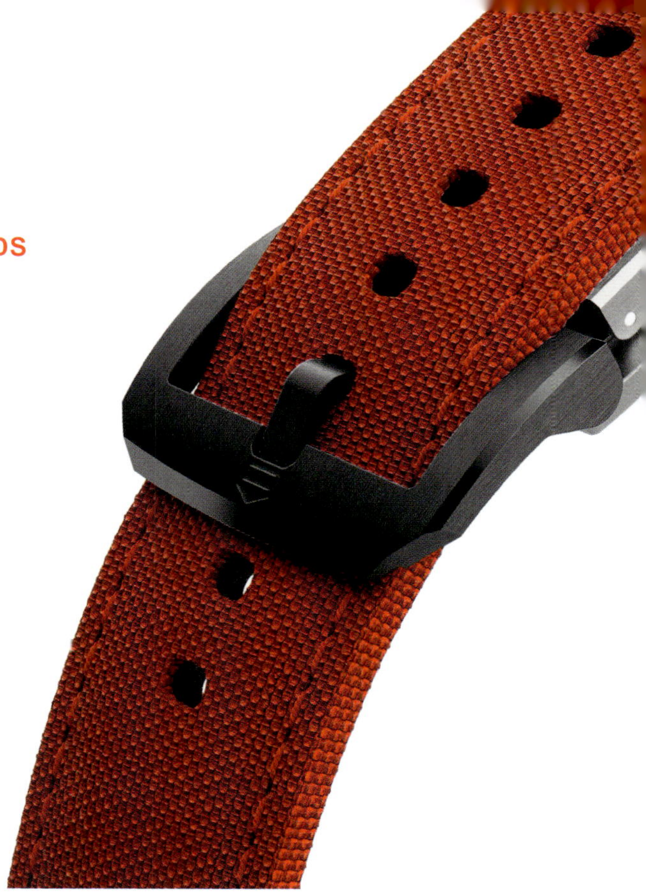

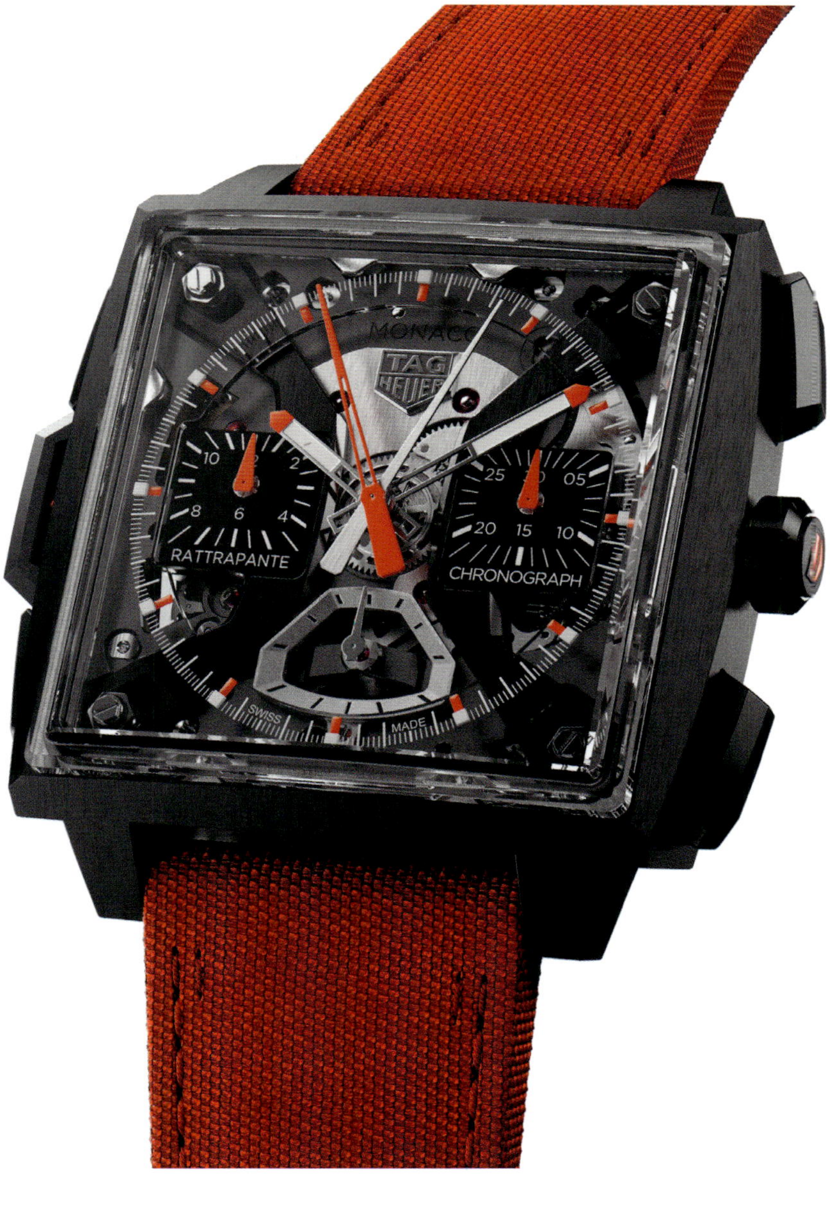

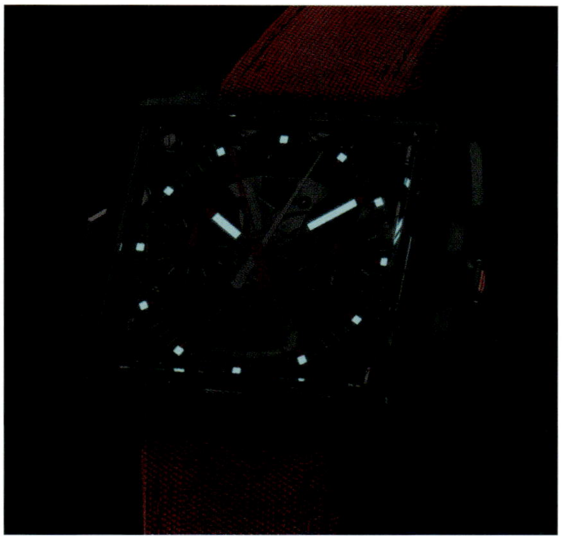

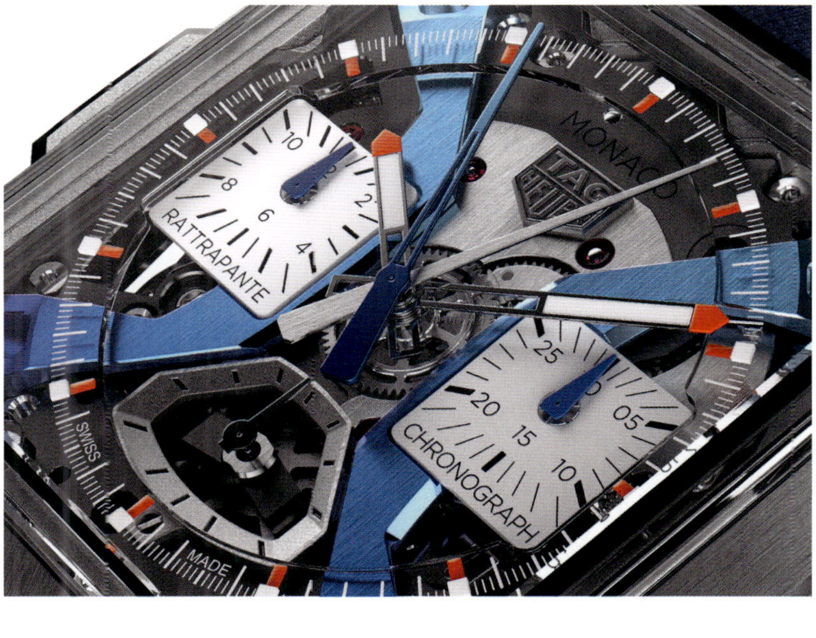

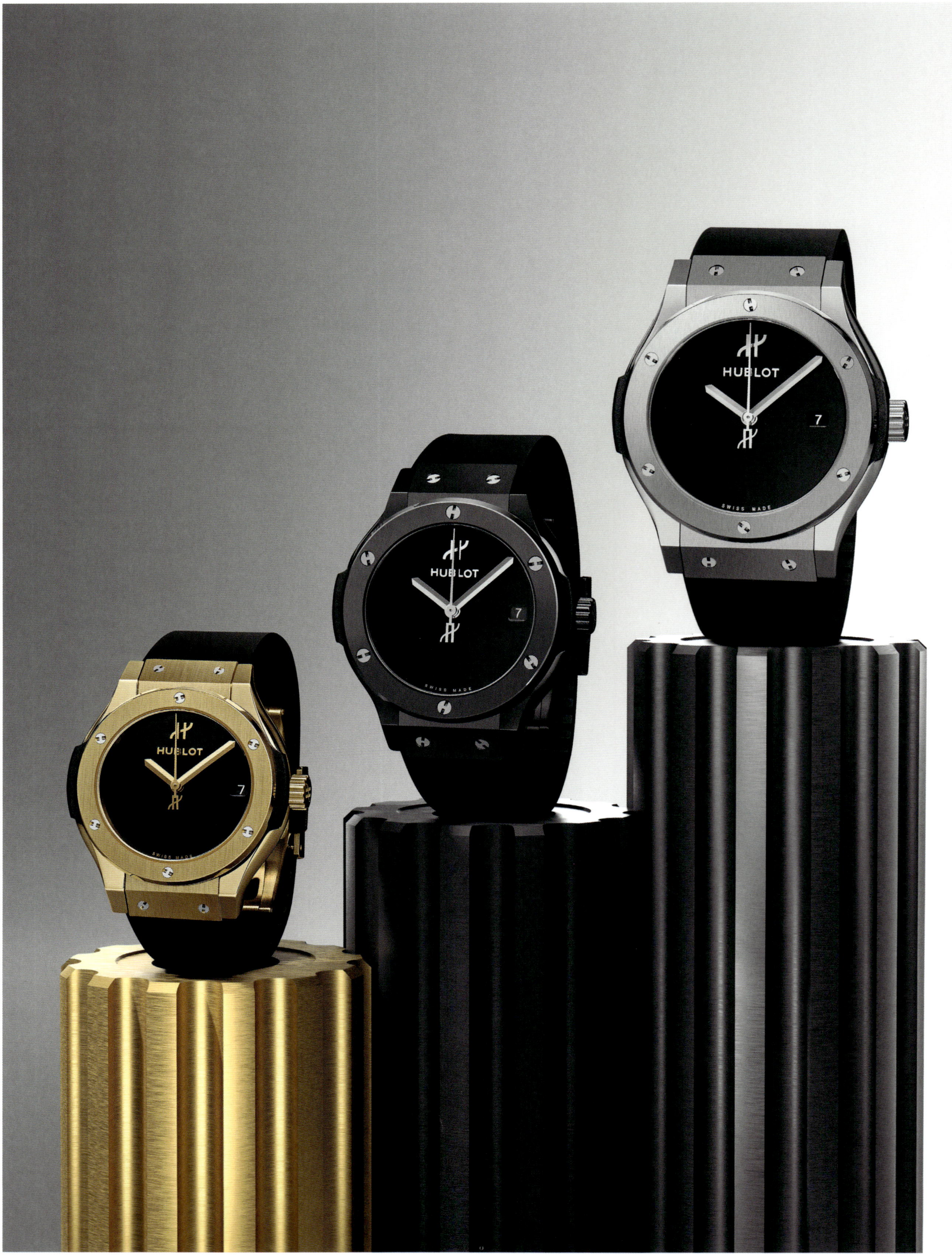
HUBLOT
7
SWISS MADE
HUBLOT
7
SWISS MADE
HUBLOT
7

HUBLOT
CLASSIC FUSION

CLASSIC FUSION

Undeterred by the quartz crisis rocking the watch industry, in 1976 Carlo Crocco founded MDM Genève. The three letters chosen by the 32-year-old scion of a wealthy Milanese watch dynasty was of course symbolic. Two referred to the first names of his wife Marie Danielle. The third stood for *montres*, French for watches. It was only later that the logogram came to signify Montre des Montres, or watch among watches. The Italian saw it as a sporty, elegant timepiece with a rubber strap, the likes of which had not existed before. In particular, the development of the bracelet delayed the launch of the first models until the Basel Watch Fair in 1980. Woven steel fibres ensure high tear resistance and an unmistakeable whiff of vanilla in the nostrils. The concept of breaking with convention is mirrored in the case. Because of its resemblance to a porthole, the French word for this type of window, *hublot*, was proposed. Soon after, the Spanish King Juan Carlos and other members of the international aristocracy developed a taste for these newcomers. At the turn of the new millennium the philanthropist company was made painfully aware of the fact that nothing is forever. Due to a lack of innovation, MDM and Hublot increasingly lost their allure. Jean-Claude Biver tackled the drop in sales and deficits with a radical shake-up and a new beginning heralded by the fulminant launch of the Big Bang. But the past was not forgotten. It lives on in the shape of the Classic Fusion Original.

CLASSIC FUSION

Ungeachtet der quarzbedingten Uhrenkrise rief Carlo Crocco 1976 die Firma MDM Genève ins Leben. Die drei Buchstaben hatte der 32-jährige Spross einer wohlhabenden Mailänder Uhrendynastie natürlich ganz bewusst gewählt. Zwei bezogen sich auf die Vornamen seiner Ehefrau Marie Danielle. Der dritte stand für *montres*, also „Uhren". Erst später mutierte das Kürzel zu *Montre des Montres*, also „Uhr unter den Uhren". Darunter verstand der Italiener sportlich-elegante Zeitmesser mit Kautschukband, welche es bis dahin noch nicht gegeben hatte. Speziell die Entwicklung des Armbands verzögerte das Lancement erster Modelle bis zur Basler Uhrenmesse 1980. Eingewebte Stahlfasern bewirkten hohe Reißfestigkeit und in die Nase stieg ausgeprägter Vanillegeruch. Auch die Schale stand im Zeichen der Idee, zeitbewahrende Dinge schlichtweg anders zu machen. Wegen der Nähe ihrer Optik zu einem Bullauge stand eines Tages das entsprechende französische Wort *hublot* im Raum. Schon bald darauf fanden der spanische König Juan Carlos und andere Mitglieder des internationalen Hochadels Gefallen an diesen Newcomern. Dass nichts für die Ewigkeit ist, diese Erfahrung musste der philanthropisch gesinnte Firmengründer Anfang des dritten Jahrtausends machen. Mangels Innovation gerieten MDM und Hublot zunehmend ins Abseits. Umsatzrückgängen und Defiziten begegnete Jean-Claude Biver durch gründliche Sanierung und fulminanten Neubeginn mit der „Big Bang". Aber die Vergangenheit geriet nicht in Vergessenheit. Sie lebt in Gestalt der „Classic Fusion Original".

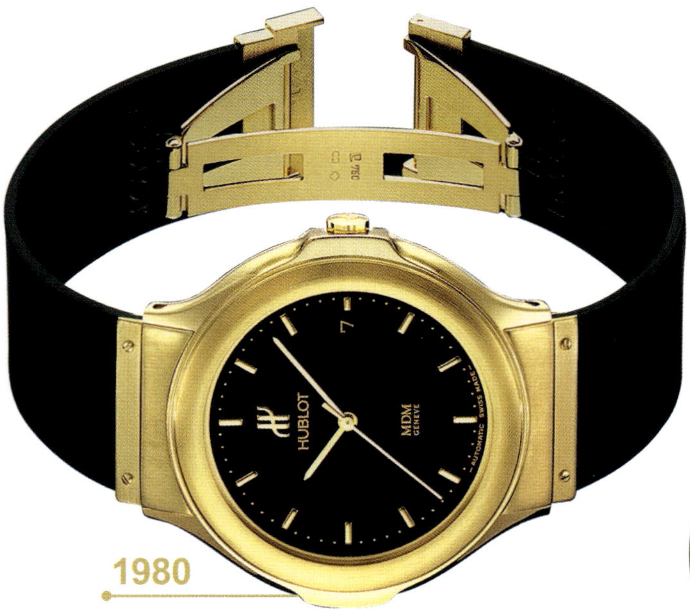

1980

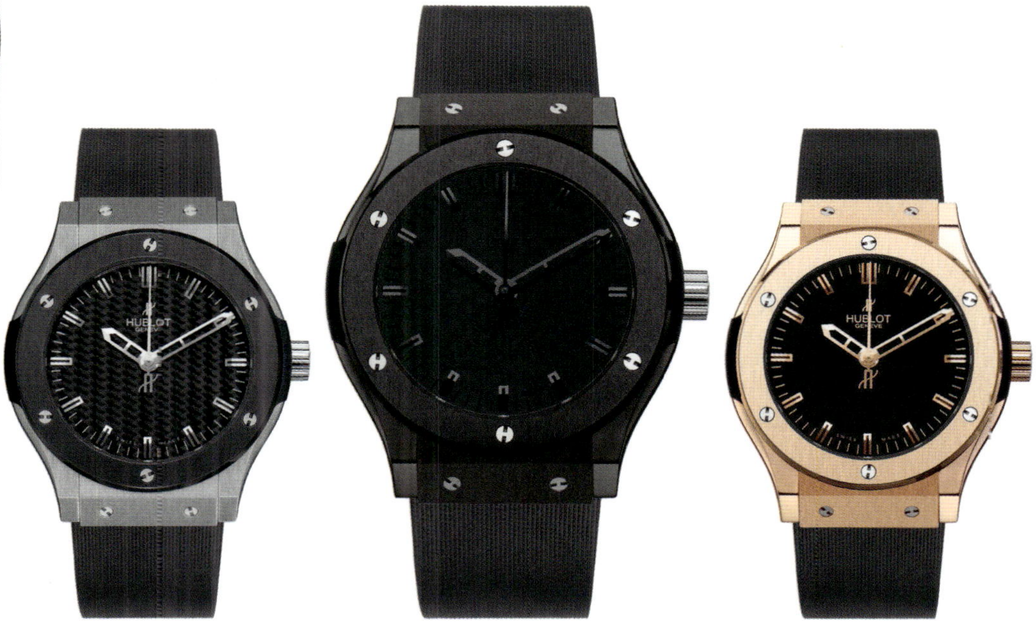

Left page | Linke Seite: *Classic Fusion Original Collection, 2023*
Right | Rechts: *Classic Fusion collection, Ref. 501.ZM.1670.RX | 501.CM.1110.RX | 501.PX.1680.RX, 2008*

HUBLOT
BIG BANG

2005

BIG BANG TICKS ON

In 2004, Jean-Claude Biver joined Hublot, taking over at the helm. Based in Nyon, the watch brand was not faring well by any means. The new captain had to invest all his navigational skills to set the floundering ship back on an even keel. Along with a future-focused brand philosophy, a sensational product to blow fresh wind into the company's sails was needed. Luckily, Biver was not at a loss for either. In the autumn, he had already taken the first steps into a new watch age with just a few strokes of his pen. Biver's sketch visualised his clear concept of fusion: "Hublot watches and the sea belong together, just as humankind and the earth belong together. Hublot fuses the sea, earth and humans together and not only invests in the project, but also in the concept of fusing these three." At this point, the second arbiter of the company's unabated success entered the stage: Mijat, a renowned Swiss designer, gave a distinctive face to a product that was soon to make history. He also played a part in the design of the case, focusing on versatility. Back then, neither had dreamed of Everybody's Darling. A watch that was meant to polarise and provoke. In 2004, Jean-Claude Biver paid a visit to his old friend Jean Todt. He did not go empty-handed. He showed his friend sketches of very different wristwatches with prominent Ferrari branding on the case, strap and winding rotor. But neither the Scuderia Ferrari team boss nor the management evinced any enthusiasm for the new Hublot creations. As a result, the art of fusion idea, aimed at blending tradition and innovation, had to seek its fortune without the envisaged partner.
In 2005 the Big Bang was completed. Of the approximately 70 components of the eye-catching case, a good dozen reveal fusionary traits: the glass bezel, gasket, top and bottom plates, case body, separate flanks left and right, the rear bezel, crown and pushers, the two trims for the lugs and of course the strap itself. Further scope is offered by the 14 screws, which keep the whole securely battened down and watertight.
The original Big Bang was all about titanium. This material is used for the central container into which the watchmakers insert the movement. It is joined by synthetic resin in any colour imaginable for the elements between the layers. The success story of the following years is spiked with imaginative ideas. The ingenious Big Bang concept enables almost any combination of traditional and completely new, sometimes also patented materials to complete the outfit. The chronographs aside, Hublot has also proved its rich ingenuity in its movements. Marking its fifteenth anniversary in 2020, the Big Bang Integrated was launched. It is defined by an integrated link bracelet. As the entire collection impressively illustrates, the concept still holds a wealth of potential. Newcomers include the Big Bang MP-11 Water Blue Sapphire with a transparent sapphire case, vertical barrels and a 14-day power reserve.

TICKENDER UHR-KNALL

2004 gelangte Jean-Claude Biver bei Hublot an Bord und zugleich auch ans Ruder. Der in Nyon beheimateten Uhrenmarke ging es nicht gut. Daher musste der neue Kapitän das schlingernde Schiff wieder auf stabilen Kurs bringen. Neben einer zukunftsweisenden Markenphilosophie brauchte es auch ein aufsehenerregendes Produkt, das frischen Wind bringen würde. Zum Glück war Biver weder um das eine noch um das andere verlegen. Erste Schritte in ein neues Uhr-Zeitalter hatte er bereits zu Papier gebracht. Die Skizze zeigte Bivers klare Vorstellungen von Fusion: „Hublot-Uhren und das Meer gehören zusammen, ebenso wie der Mensch und die Erde zusammengehören. Hublot lässt Meer, Erde und Mensch miteinander verschmelzen und investiert nicht nur in das Projekt, sondern auch in das Konzept der Verschmelzung dieser drei."
An dieser Stelle gelangte auch der zweite Vater anhaltenden Erfolgs ins Spiel: Mijat, ein anerkannter Schweizer Designer, verlieh dem, was innerhalb kurzer Zeit Geschichte schrieb, das markante Gesicht. Überdies wirkte er an der auf Vielseitigkeit angelegten Gehäusekonstruktion mit. An „Everybody's Darling" dachte damals niemand von beiden. Die Uhr sollte polarisieren und provozieren. Schon 2004 war Jean-Claude Biver zu seinem alten Freund Jean Todt gereist. Mit im Gepäck: Zeichnungen ganz unterschiedlicher Armbanduhren mit unübersehbarem Ferrari-Branding auf Gehäuse, Band und Aufzugsrotor. Aber weder der damalige Scuderia Ferrari-Teamchef noch das Management konnten sich für die neuen Hublot-Schöpfungen begeistern. Also musste besagte Fusions-Idee, die auf eine Verschmelzung von Tradition und Innovation zielte, auch ohne den intendierten Partner greifen.
2005 war die „Big Bang" fertig. Von den rund 70 Komponenten des markanten Gehäuses eignete sich ein gutes Dutzend für fusionierende Aspekte: Glasrand, Dichtring, vordere und hintere Deckplatten, Gehäusekorpus, separate Flanken links und rechts, die rückwärtige Lünette, Krone und Drücker, die beiden Verblendungen für die Bandanstöße und natürlich auch das Armband selbst. Zusätzlichen Spielraum boten jene 14 Schrauben, welche das Ganze gleichermaßen fest wie wasserdicht zusammenhielten. Bei der Ur-„Big Bang" drehte sich alles um Titan. Aus diesem Werkstoff bestand der zentrale Container, in dem die Uhrmacher das Werk fixierten. Hinzu gesellte sich beliebig färbbares Kunstharz für die Elemente zwischen den Schichten. Die Erfolgsgeschichte während der anschließenden Jahre verknüpfte sich mit reichlich Phantasie. Das geniale „Big Bang"-Konzept gestattete nahezu jede Kombination klassischer und völlig neuer, teilweise auch patentierter Werkstoffe für das Outfit. Auch bei den verbauten Uhrwerken hat Hublot über die Chronographen hinaus jede Menge Einfallsreichtum bewiesen.
Zum 15. Geburtstag im Jahr 2020 ging die „Big Bang Integrated" an den Start. Ihr Kennzeichen ist ein integriertes Gliederband. Damit ist das Potenzial noch lange nicht ausgeschöpft. Zu den neuesten Mitgliedern gehört die „Big Bang MP-11 Water Blue Sapphire" mit transparentem Saphirgehäuse, stehend angeordneten Federhäusern und 14 Tagen Gangautonomie.

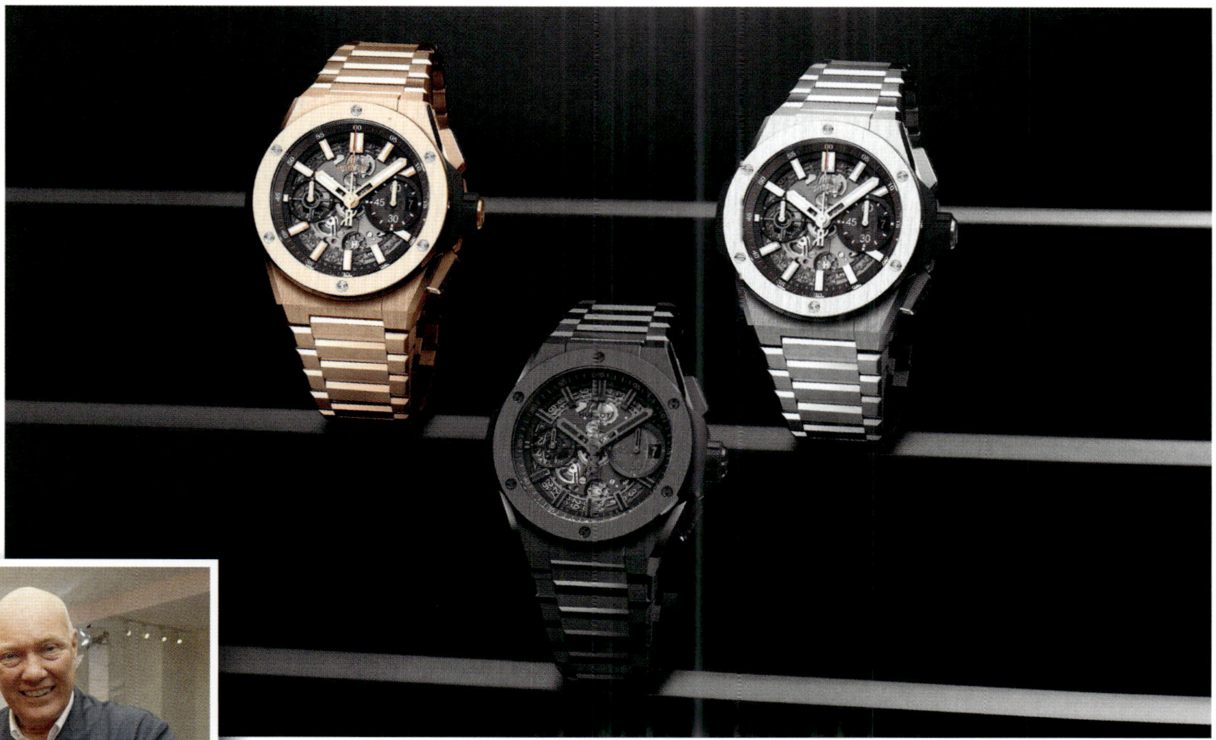

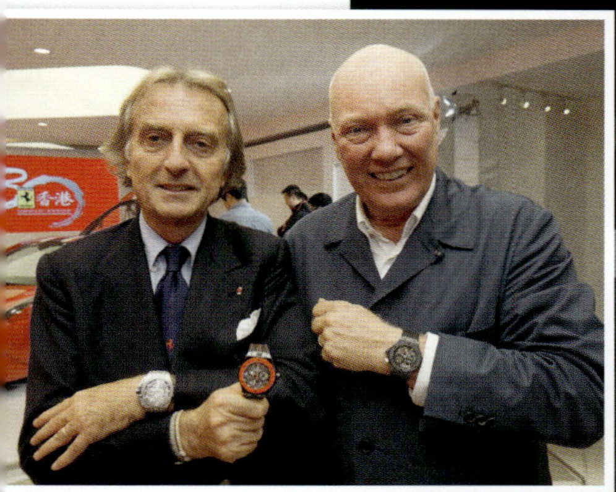

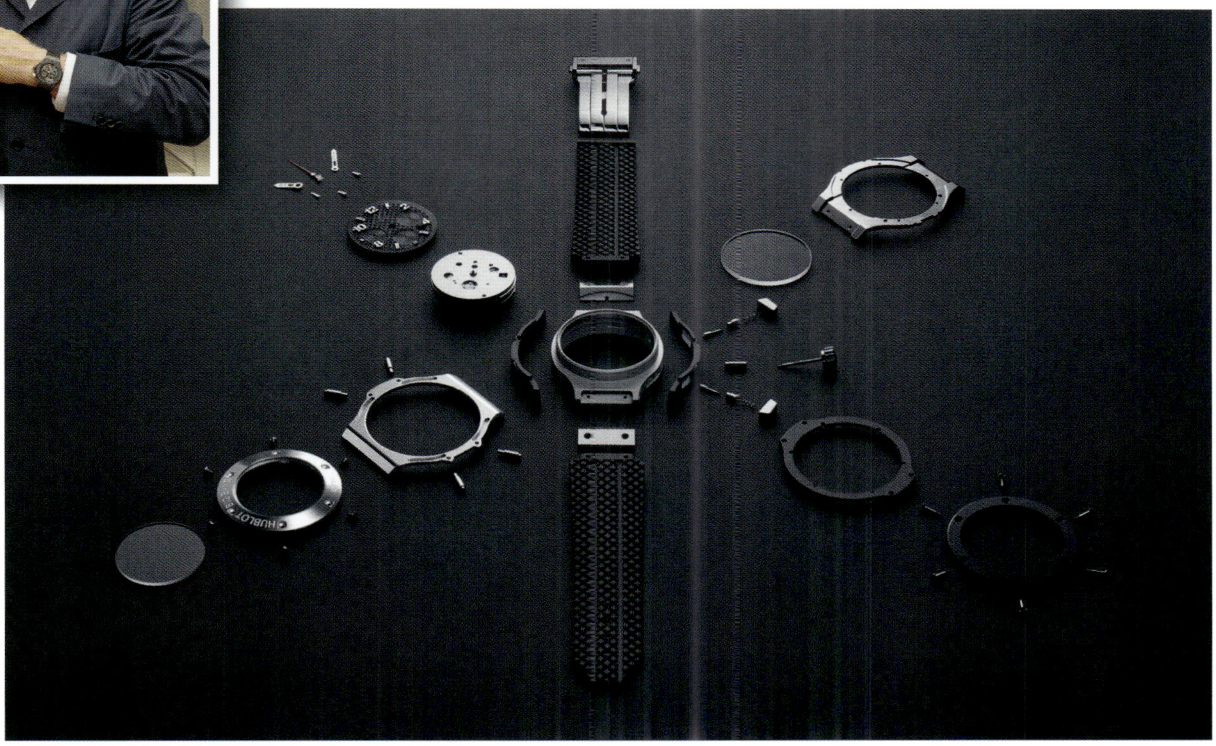

From top* | *Von oben: *Big Bang Integrated King Gold, All Black, Titanium, 2020* | *Luca Cordero di Montezemolo & Jean-Claude Biver* | *Components of* | *Komponenten der Big Bang Original Steel Ceramic*

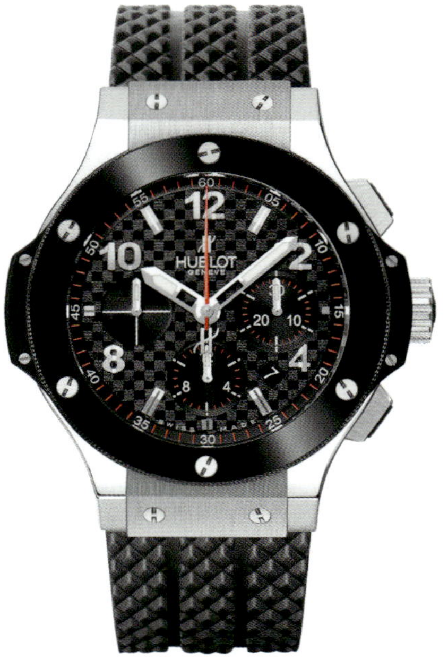

2005

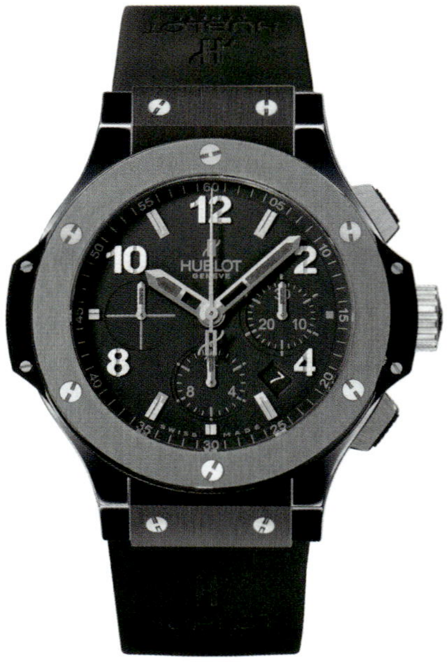

2006

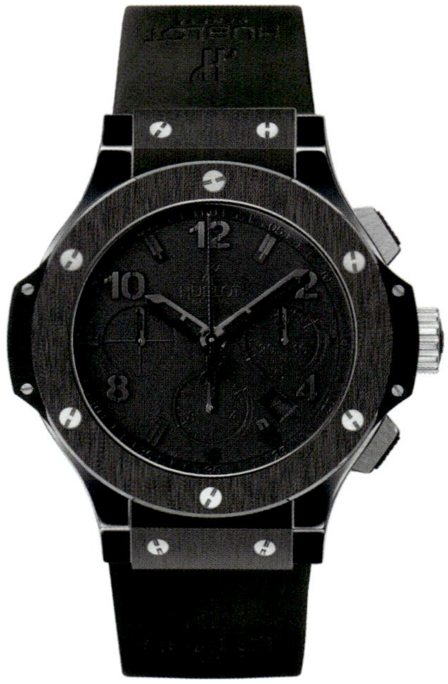

2006

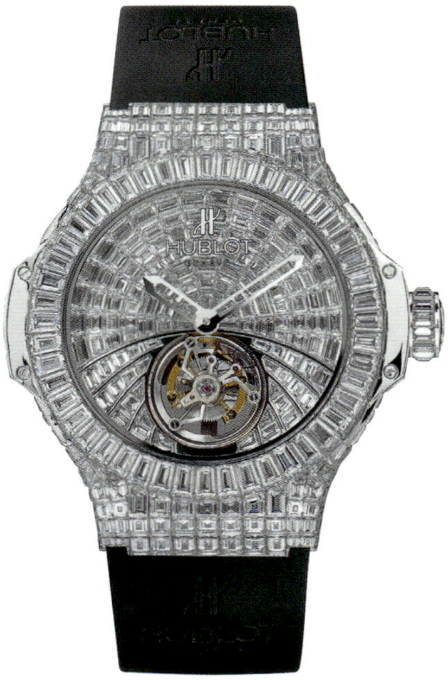

2007

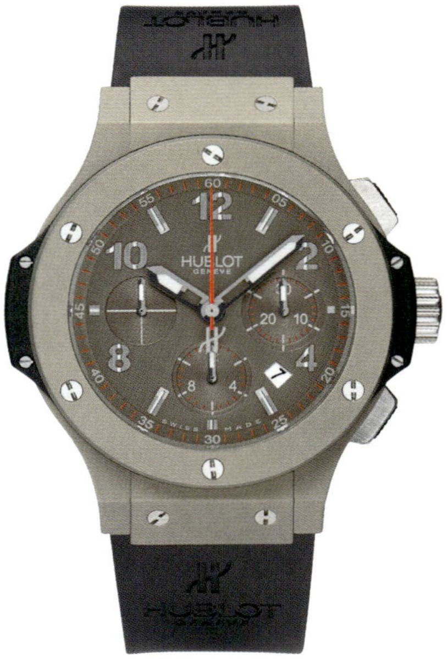

2008

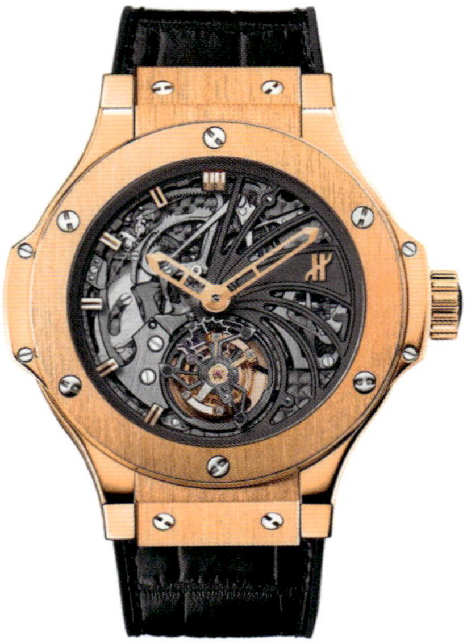

2009

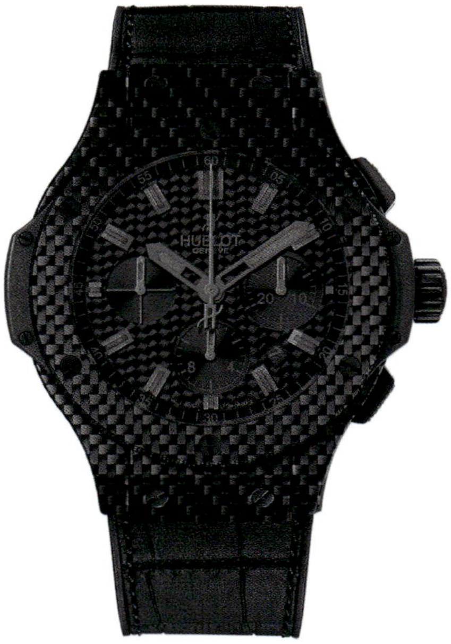

2011

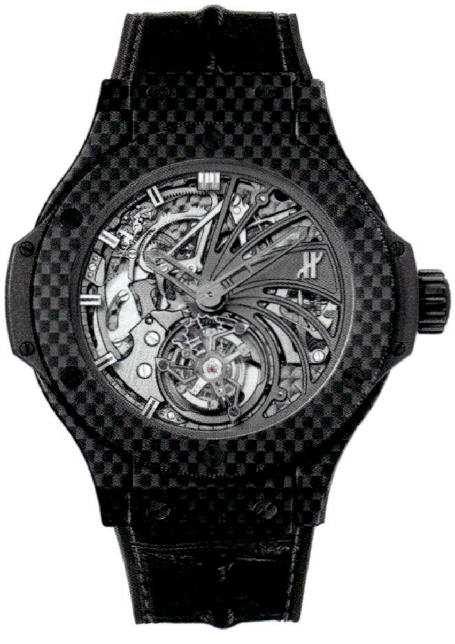

2012

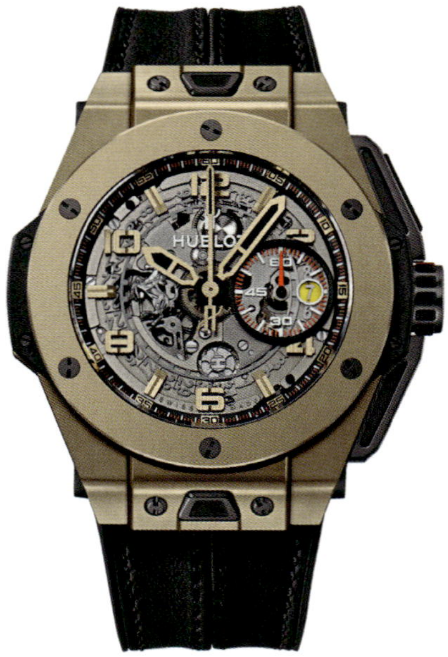

2012

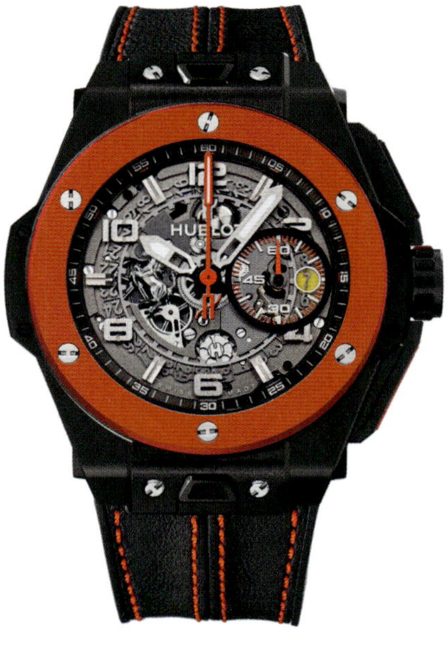

2013

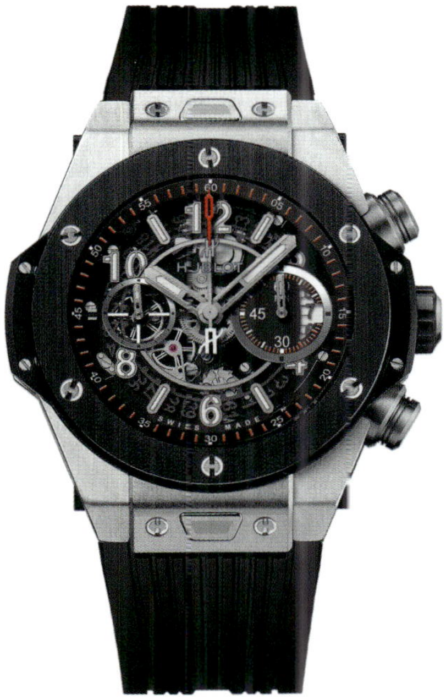

2013

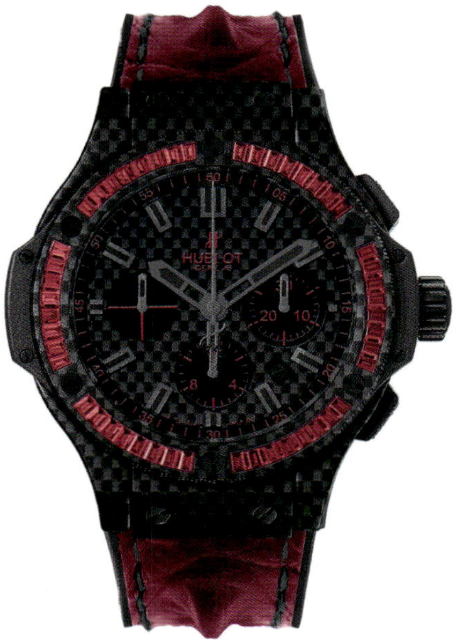

2013

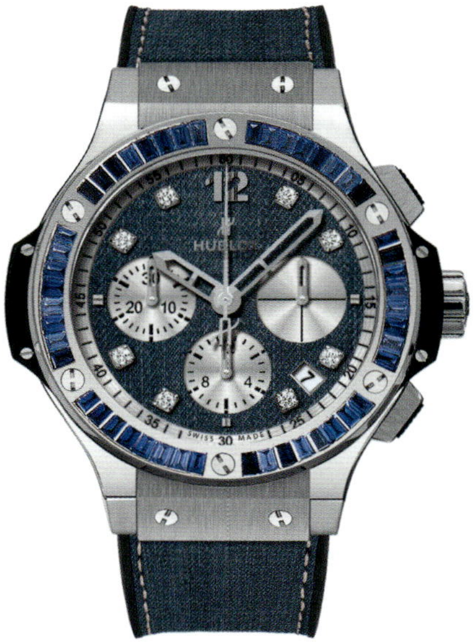

2013

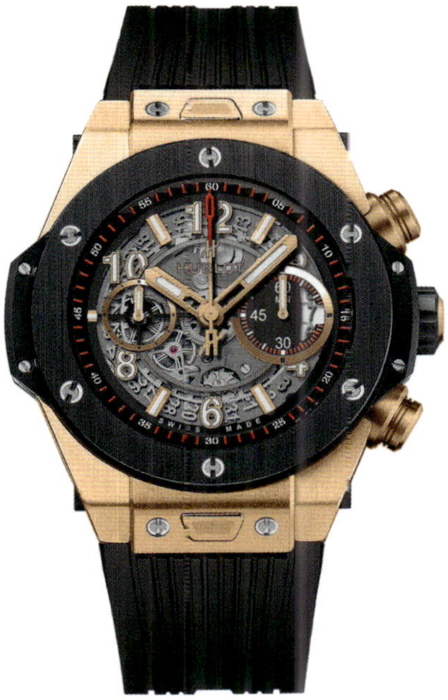

2013

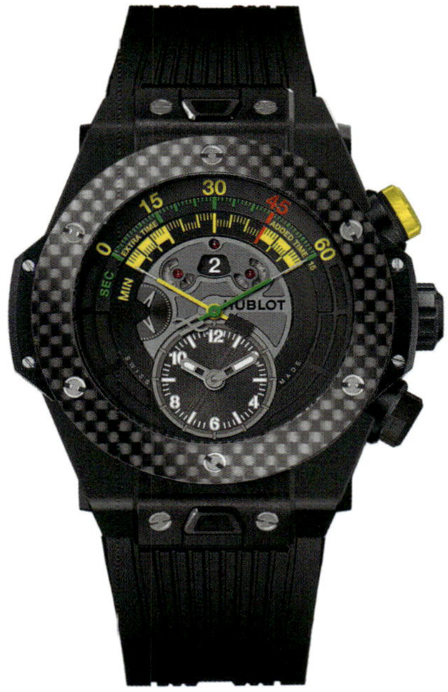

2014

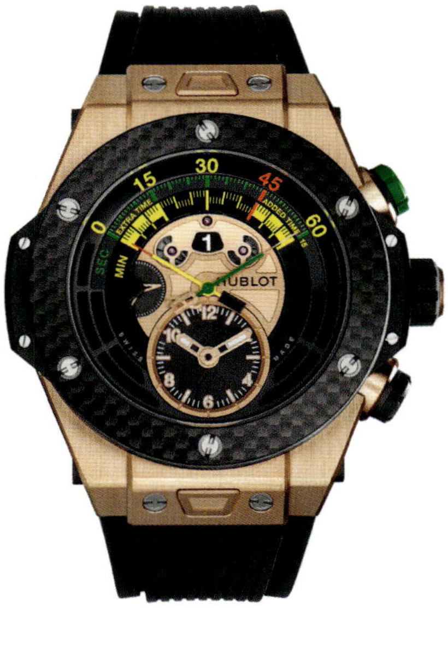

2014

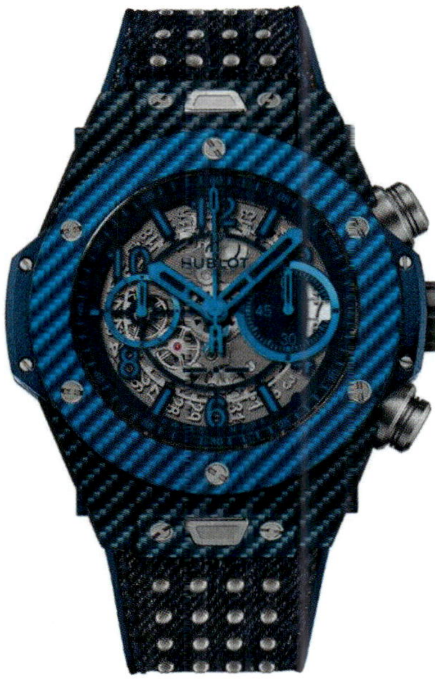

2015

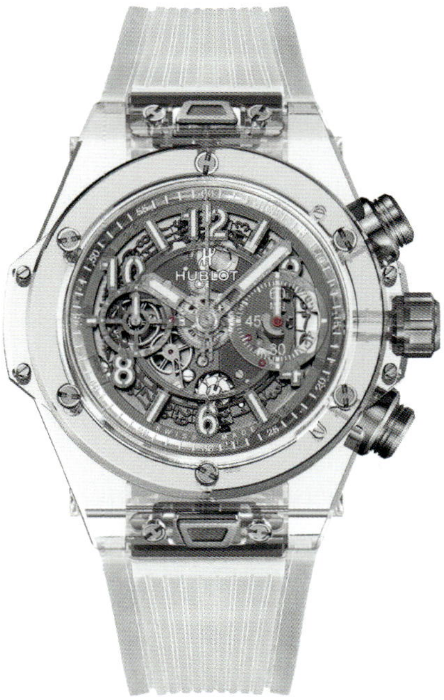

2016

IWC SCHAFFHAUSEN 1939
PORTUGIESER

SEVEN DECADES OF OPULENCE

The lavishly dimensioned IWC Portugieser is swathed in quite a few myths. To this day, the biographical background details remain something of a mystery. Very often, the names Rodriguez and Teixeira are bandied about in connection with this watch. Mr Rodriquez lived and worked in Lisbon, Mr Teixeira in Porto. Both imported watches from Schaffhausen to Portugal. In the late 1930s, they both developed a sudden passion for large, yet nonetheless elegant wristwatches. Back then, lofty principles were applied to the design and aesthetics of watches. They related to the shape, dimensions and proportions of the cases, dials and hands. For instance, on a table a watch had to rest on its lug ends. The caseback was supposed to float in the air. This ensured that the case would fit snugly around the wrist. Removed from the centre, the small second hand was to sweep as generous an arc as possible. In opulent wristwatches such as the Portugieser, the latter could only be achieved with the help of a pocket watch movement. Therefore, from 1939 onwards models produced during this period were powered by the proven hand-wound calibre 74. Starting 1944, IWC opted for the calibre 98 and since the mid-1970s the 982 with a diameter of 37.8 millimetres. By no stretch of the imagination can the story of the original Reference 325 be described as a success. Compared to the customary case dimensions of 33 to 35 millimetres at the time, it was too large as a dress watch even for men. Up until the 1980s, IWC only succeeded in selling a few hundred models. Its 1993 renaissance is owed to Günter Blümlein, who in 1981 initially worked in an advisory capacity and then as acting director for the floundering IWC. “Because in hardly any other industry do you encounter so many ugly products, we revived the minimalist wristwatch with a large hand-wound movement and named it Portugieser as a nod to its roots.” The brilliant comeback of the understated retro Reference 5441 was followed by one coup after the other. Here are just a few examples: in 1995, IWC launched a minute repeater with an affordable modular movement, and in 2000, the manufacture calibre 5000 with a week’s power reserve was the talk of the watch world. This was followed in 2003 by the Perpetual Calendar with double moon, in 2010 by the Grande Complication and in 2011 by the Sideral Scafusia, a magnificent tourbillon with constant force escapement and information displayed on both sides of the case. >

OPULENT SEIT MEHR ALS 70 JAHREN

Um die großzügig dimensionierte „Portugieser“ der IWC ranken sich einige Mythen. Letztendlich sind die exakten Hintergründe der Biographie bis heute nicht wirklich geklärt. Sehr oft sind im Zusammenhang mit dieser Armbanduhr die Namen Rodriguez und Teixeira zu hören. Herr Rodriquez lebte und arbeitete in Lissabon, Herr Teixeira in Porto. Beide importierten Uhren aus Schaffhausen nach Portugal. In den späten 1930er-Jahren begehrten sie auf einmal große, aber dennoch elegante Exemplare fürs Handgelenk. In Sachen Uhrendesign und -ästhetik galten damals sehr hehre Grundsätze. Sie bezogen sich auf die Form, Dimensionen und Proportionen der Gehäuse, Zifferblätter und Zeiger. Beispielsweise musste eine Uhr am Tisch auf den Enden ihrer Bandanstöße ruhen. Der Boden hatte also zu schweben. Auf diese Weise schmiegte sich die Schale gut ans Handgelenk. Entfernt vom Zifferblattzentrum sollte der kleine Sekundenzeiger einen möglichst großen Kreisbogen beschreiben. Letzteres ließ sich bei opulenten Armbanduhren wie der „Portugieser“ nur mit Hilfe eines Uhrwerks für Taschenuhren realisieren. In diesem Sinn beseelte die ab 1939 gelieferten Modelle das bewährte Handaufzugskaliber 74. Ab 1944 verbaute IWC das Kaliber 98 und ab Mitte der 1970er-Jahre das Kaliber 982 mit 37,8 Millimetern Durchmesser.
Als Erfolg lässt sich die Geschichte der Ursprungsreferenz 325 beim besten Willen nicht bezeichnen. Verglichen mit den damals gängigen Gehäusedimensionen von 33 bis 35 Millimetern war sie selbst Männern als Dresswatch zu groß. So konnte die IWC bis in die 1980er-Jahre nur ein paar Hundert Exemplare verkaufen. Die Renaissance im Jahr 1993 ist Günter Blümlein zu verdanken, der 1981 zunächst in beratender Funktion, dann als geschäftsführender Direktor für die heftig kränkelnde IWC tätig war: „Weil es in kaum einer anderen Branche derart viele Produkt-Hässlichkeiten gibt, haben wir die schlichte Armbanduhr mit großem Handaufzugswerk wiederbelebt und sie in Anlehnung an ihre Entstehungsgeschichte ‚Portugieser‘ getauft.“ Nach dem fulminanten Comeback der schlichten Retroreferenz 5441 ging es quasi Schlag auf Schlag. Hier nur einige Beispiele: 1995 lancierte IWC eine Minutenrepetition mit relativ kostengünstigem Modulwerk, 2000 redete die Uhrenwelt beim Manufakturkaliber 5000 von einer Woche Gangautonomie. Es folgten 2003 der ewige Kalender mit Doppelmond, 2010 die „Grande Complication“ und 2011 mit der „Sideral Scafusia“ ein spektakuläres Tourbillon mit Konstantkraft-Hemmung sowie Indikationen auf Vorder- und Rückseite. >

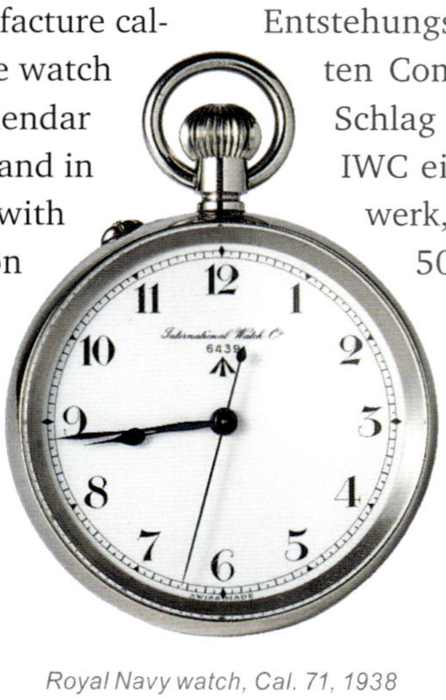

Royal Navy watch, Cal. 71, 1938

1

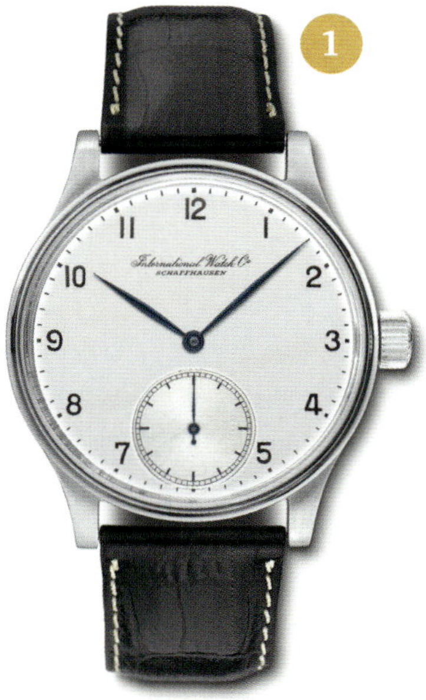

2

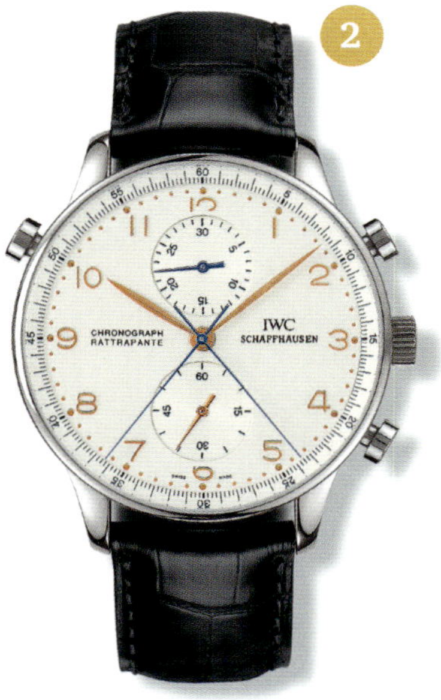

3

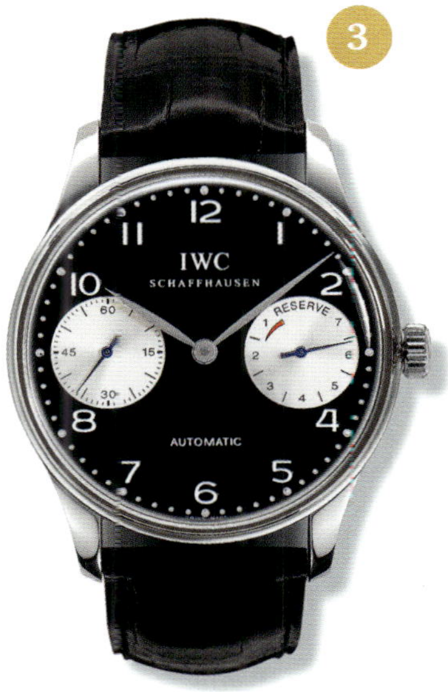

4

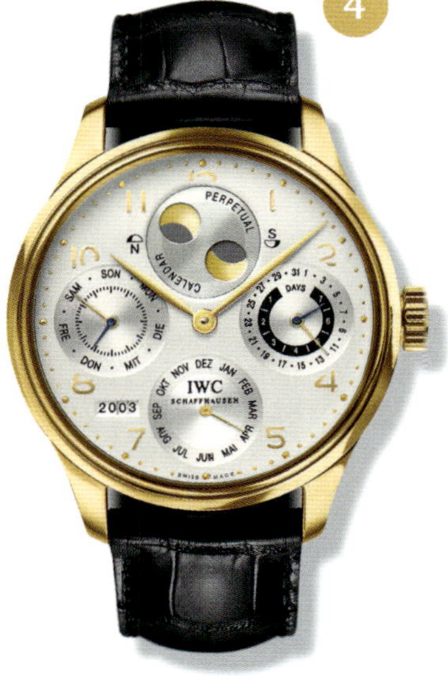

5

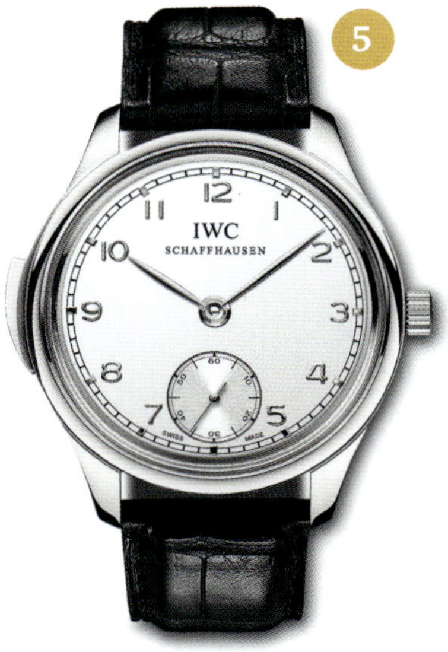

6

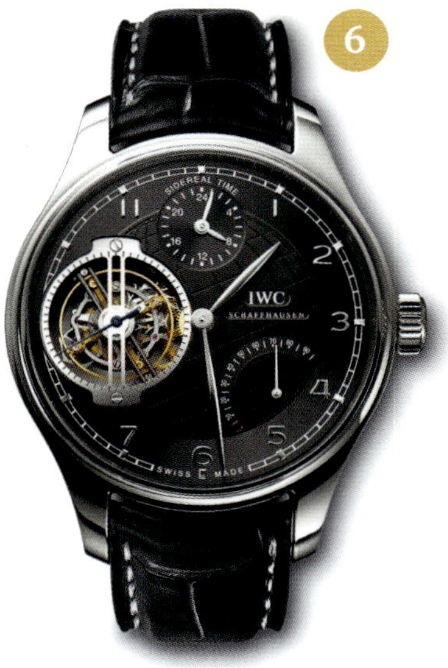

7

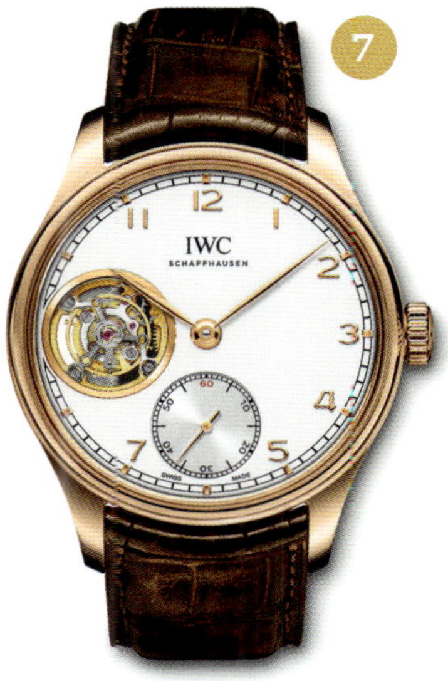

1. *Ref. 325, 1939*
2. *Rattrapante, 1995*
3. *Automatic, 2000*
4. *Perpetual Calendar, Ref. 5021, 2003*
5. *Minute Repeater, Ref. 5449, 2009*
6. *Siderale Scafusia, 2011*
7. *Tourbillon hand-wound, 2015*
8. *Perpetual Calendar 44, Cal. 52616, 2024*
9. *Hand-wound Tourbillon Day & Night, Cal. 81925, 2024*

8

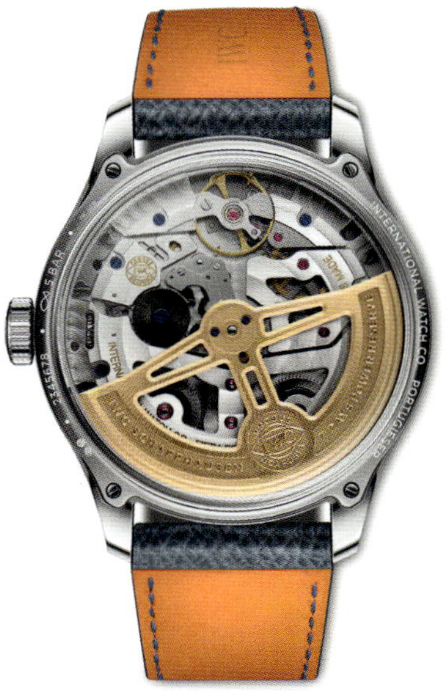

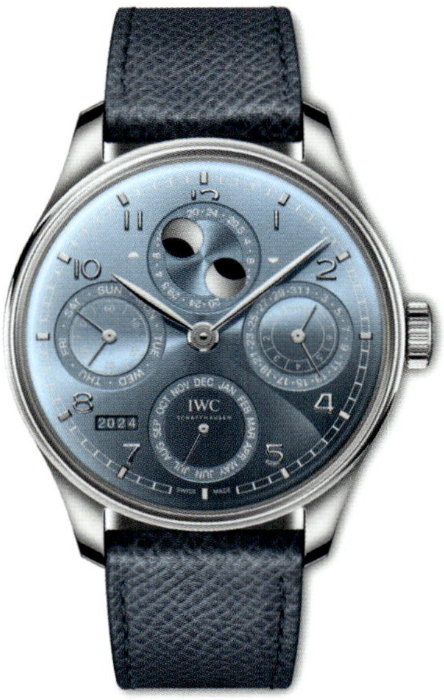

9

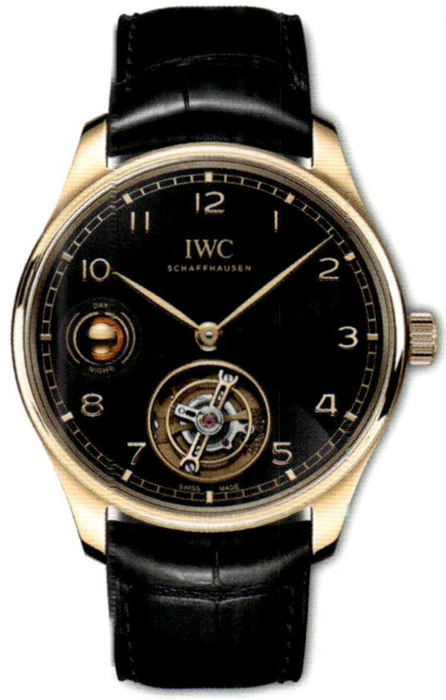

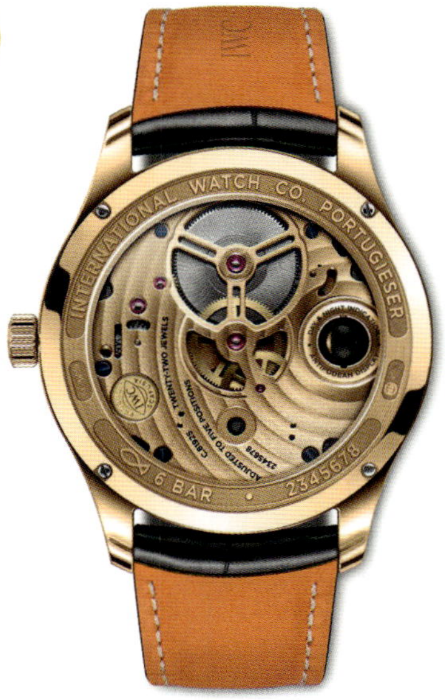

Not to mention the Portugieser presented in 2013 with proprietary chronograph calibre, and the Annual Calendar, automatic calibre 52850 launched in 2015. The jewel in the crown of the current collection is the Portugieser Eternal Calendar. It adheres to Pope Gregory's edict with a wheel that only rotates once every 400 years. In other words, the accuracy of the calendar display is guaranteed until the year 2399. The moon phase display offers even greater precision, namely one wrong day per 45 million years. The list is by no means exhaustive and shows that icons endure come what may.

Nicht zu vergessen die 2013 vorgestellte „Portugieser" mit hauseigenem Chronographenkaliber und der 2015 lancierte Jahreskalender, Automatikkaliber 52850. Die absolute Spitze der aktuellen Kollektion markiert die „Portugieser Eternal Calendar". Durch ein Rad, welches sich nur einmal in 400 Jahren dreht, befolgt diese Uhr das Edikt von Papst Gregor. Sprich, die Anzeigegenauigkeit des Kalendariums währt bis zum Jahr 2399. Noch weitaus präziser ist die Mondphasenanzeige: In 45 Millionen Jahren geht sie einen Tag falsch. Die Aufzählung ließe sich beliebig fortsetzen. So oder so zeigt sie, dass sich Ikonen auch am Rheinfall schlichtweg nicht überleben.

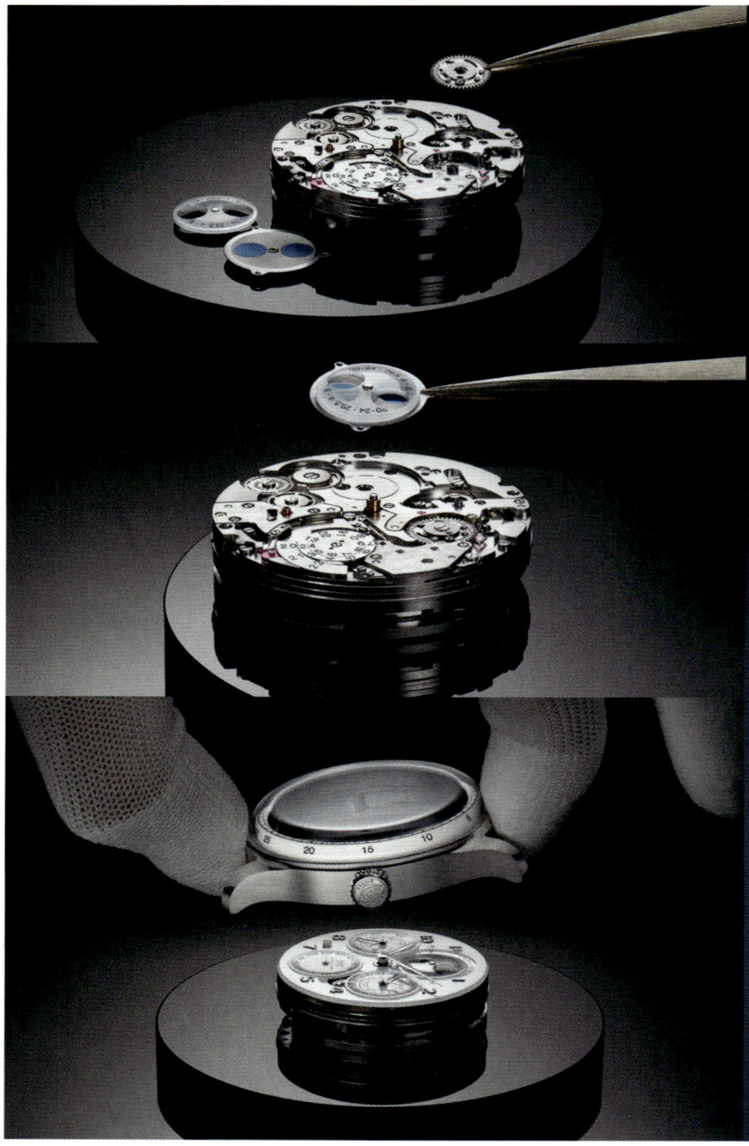

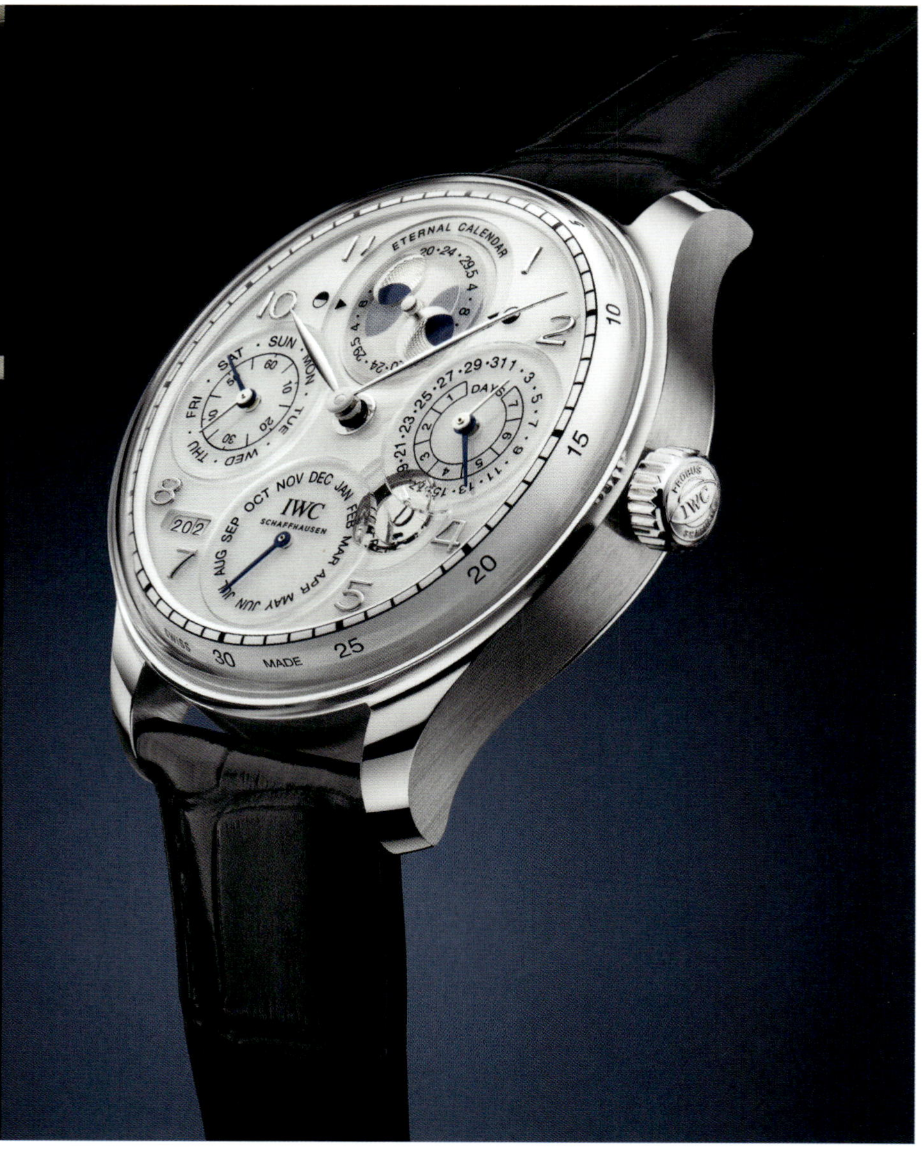

Left from top: *Assembly of the 400-years-wheel* | *Assembly of the moon phase display* | *Fitting of the case and sapphire crystal glass*
Links, von oben: *Montage des 400-Jahr-Rads* | *Montage der Mondanzeige* | *Montage des Gehäuses und Saphirglases*
Right | ***Rechts:*** *Eternal Calendar, 2024*

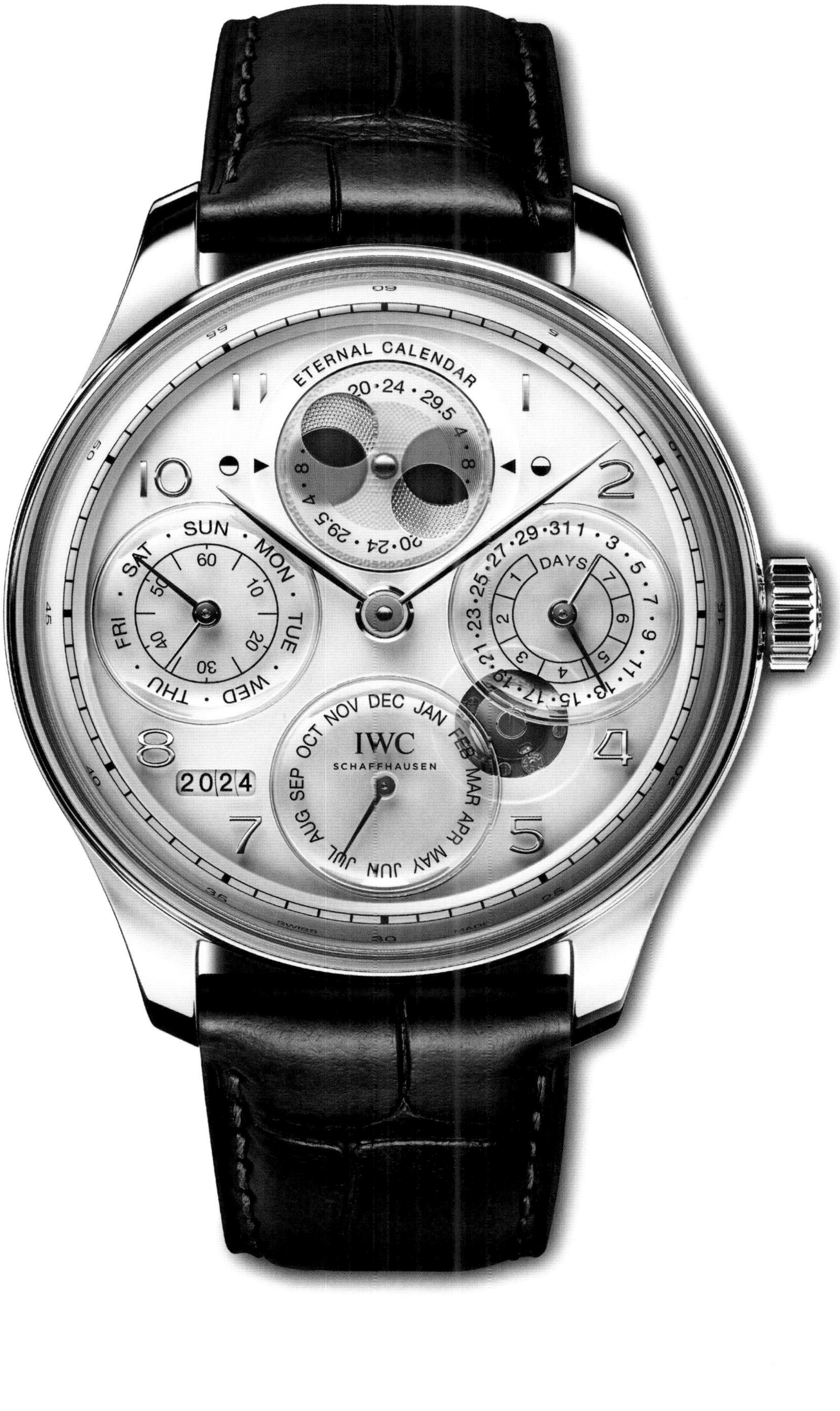
ETERNAL CALENDAR
SUN MON TUE WED THU FRI SAT
DAYS
IWC
SCHAFFHAUSEN
JAN FEB MAR APR MAY JUN JUL AUG SEP OCT NOV DEC
2024

IWC SCHAFFHAUSEN 1948

PILOT'S WATCH

UP, UP AND AWAY

In aircraft, accurate timepieces are some of the most important cockpit instruments. For decades now, they have benefitted from electronics to measure and keep time. In the past, it was quite a different story. A soft landing at the charted destination hinged on the precise coordination of distance, speed and fuel consumption. A task once painstakingly completed by conscientious cockpit crews is now discreetly performed in the background by computers. With the help of a map, compass, slide rule and, of course, a watch, those intrepid aviators had to exactly calculate the flight data in advance. There were several reasons why both military and civilian institutions opted for products from the International Watch Company. Precision and reliability have always been one of IWC's core principles. Because pocket watches were extremely impractical in the air, the watchmaker, founded in 1868, presented an easy-to-read pilot wristwatch with a black dial, large luminous numerals, prominent luminous hands and a rotating bezel with marker for setting the take-off time as early as 1936. From 1940, the Schaffhausen watchmakers worked for the German Luftwaffe. On their behalf, IWC developed a professional pilot's watch with a diameter of 55 millimetres. This opulent model, powered by the large hand-wound calibre 52 TSC, for the first time offered effective protection against magnetic fields. The military watch Mark X from the early 1940s and the legendary Mark 11, launched in 1948, displayed a more civilian look. The latter rightly deserves the qualification of being a cult watch. The models destined for the Royal Air Force sported the Broad Arrow on the dial. Before leaving the factory, every model was subjected to a 44-day test programme for navigator wristwatches. A large part of the current IWC collection is based on this tradition. Pilot's watches are to IWC what wings are to an aeroplane.

UP, UP AND AWAY

In Flugzeugen gehören genau gehende Uhren zu den wichtigsten Bordinstrumenten. Seit vielen Jahrzehnten messen und bewahren sie die Zeit auf elektronischem Wege. Früher war das noch ganz anders. Eine weiche Landung am angepeilten Zielort verlangte nach exakter Koordination von Entfernung, Geschwindigkeit und Treibstoffverbrauch. Was heute Computer unauffällig im Hintergrund erledigen, bedeutete für verantwortungsbewusste Cockpit-Crews harte Arbeit. Mit Hilfe von Karte, Kompass, Rechenschieber und natürlich einer Uhr mussten sie die Flugdaten exakt vorausberechnen. Dass militärische und zivile Institutionen vielfach auf Produkte der International Watch Co. setzten, hatte gleich mehrere Gründe: Präzision und Zuverlässigkeit gehörten seit jeher zu den Maximen der IWC. Weil Taschenuhren hoch in den Lüften ausgesprochen unpraktisch waren, präsentierte die 1868 gegründete Manufaktur schon 1936 eine bestens ablesbare Fliegerarmbanduhr mit schwarzem Zifferblatt, großen Leuchtziffern, markanten Leuchtzeigern sowie einer Merkpfeil-Drehlünette zum Einstellen der Abflugzeit.
Ab 1940 wurden die Schaffhauser Uhrmacher für die deutsche Luftwaffe tätig. In deren Auftrag entwickelte IWC eine professionelle Fliegeruhr mit 55 Millimeter Durchmesser. Dieses opulente Modell, in dem das große Handaufzugskaliber 52 TSC tickte, verfügte erstmals über einen wirksamen Schutz gegen Magnetfelder. Zivilere Ausmaße besaßen die militärische Armbanduhr „Mark X" aus den frühen 1940er-Jahren sowie die legendäre, 1948 lancierte „Mark 11". Mit Fug und Recht konnte Letztere das Attribut „Kult-Uhr" für sich in Anspruch nehmen. Die für die Royal Air Force bestimmten Exemplare trugen am Zifferblatt den sogenannten „Broad Arrow". Vor der Lieferung hatte jedes Exemplar ein 44-tägiges Testprogramm für „Navigator Wrist Watches" zu durchlaufen. Auf dieser Geschichte basiert ein wesentlicher Bestandteil der aktuellen IWC-Kollektion. Pilotenuhren gehören nämlich zu IWC wie Tragflächen zu einem Flugzeug.

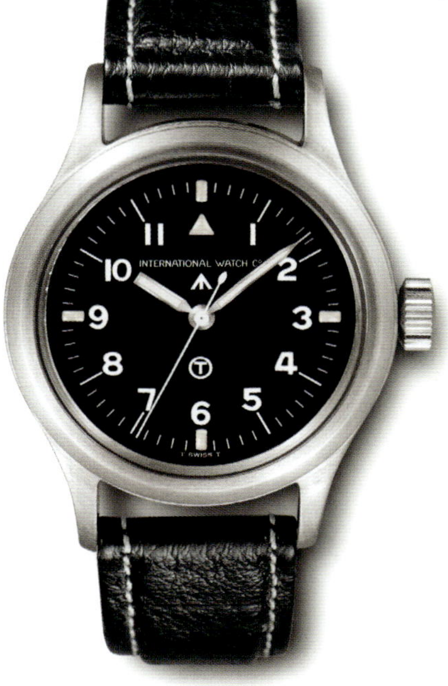

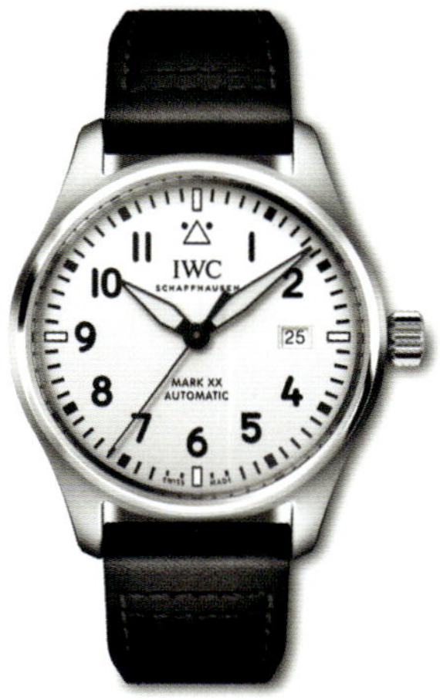

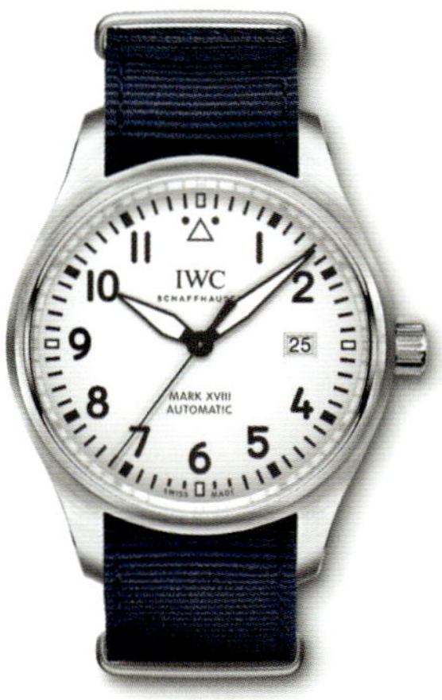

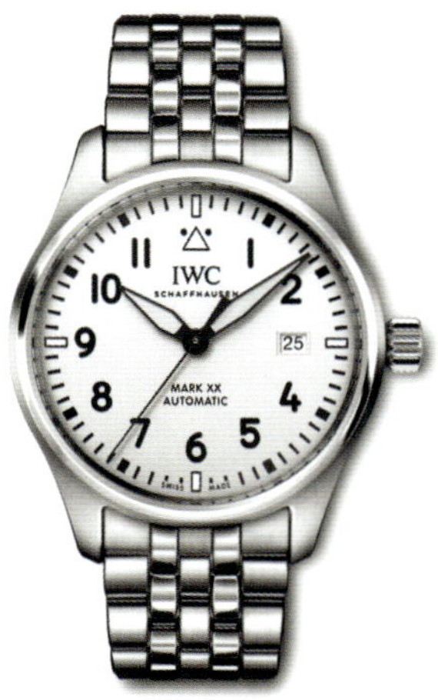

MARK XVII

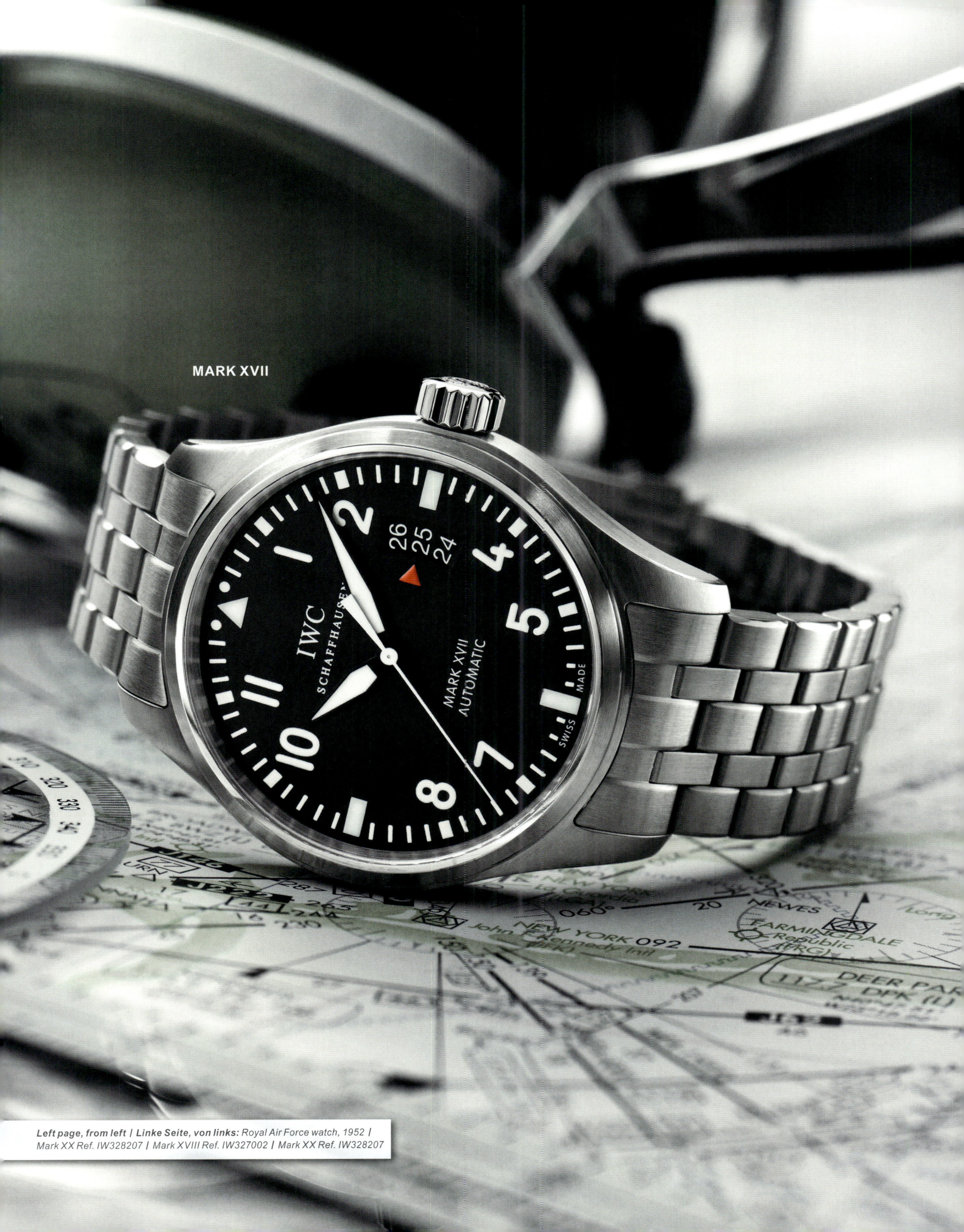

Left page, from left | Linke Seite, von links: *Royal Air Force watch, 1952 | Mark XX Ref. IW328207 | Mark XVIII Ref. IW327002 | Mark XX Ref. IW328207*

IWC SCHAFFHAUSEN 1976

INGENIEUR SL

HARD ON THE OUTSIDE AND SOFT ON THE INSIDE

IWC Schaffhausen used the Mark 11 pilot's watch, launched in 1948, as a blueprint of sorts for its Ingenieur model. Made of soft iron, the case ring, inner case and dial protect the calibre 89 from strong magnetic fields. This design principle was also applied to the Ingenieur References 666 A and 666 AD with a 36.6 mm case diameter. The former is driven by the proprietary automatic calibre 852, the latter by the 8521, with the addition of a window date. The magnetic field protection reaches up to 80,000 amperes per metre (A/m), which is the maximum possible in this case. From 1958, IWC used the optimised automatic calibres 853 and 8531 to power its Ingenieur models. The reference numbers did not change. Introduced in 1967, the Reference 866 owed its sportier look to its height of 13 millimetres and an extra half millimetre in diameter. The automatic calibre 854 came without and the 8541 with a window date, also larger thanks to its new position on the dial. Despite all this, they could not match the success enjoyed by the Reference 666.

In 1969, IWC therefore began development of an Ingenieur with a 37-millimetre case. Its lack of success on the market presumably resulted in Gérald Genta joining the fray in autumn 1972. The design of the sporty elegant Ingenieur SL, which made its debut at the 1976 Basel Watch Fair, carries his signature. The large version with a 40 mm diameter case, also known as the Jumbo Ingenieur, bore the reference number 1832. It also offered magnetic field protection of up to 1,000 Gauss field strength. The letter E in the proprietary calibre 8541ES stands for the use of amagnetic materials for the anchor, escape wheel and roller. S denotes the stop seconds. Once the bezel has been screwed from the front into the middle of the case using a special tool, the five slots are positioned differently on almost every watch. With hindsight, the Reference 1832 can be deemed the right watch at the wrong time. IWC only managed to sell fewer than a thousand of these models, which is why prices are exploding on the pre-owned market. This wristwatch inspired the 2023 Ingenieur Automatic 40. A modern take on the 50-plus Genta design was overdue. Shortcomings on the ergonomics front needed to be addressed. To increase wearing comfort, IWC pared down the case diameter to 40 millimetres and dehorned it in the process. As the newly designed middle-link attachment to the case is now flexible, the 10.8-millimetre-high wristwatch offers a snugger fit on the wrist and more curvature than the original. The tribute exacted for this sleeker look is the reduction of the magnetic field resistance to 40,000 A/m or 500 Gauss. The glass bezel is now screwed to the middle of the case so that the five distinctive slots are always in the same position. The flank guards protecting the crown owe nothing to Gérald Genta, but are part of the Ingenieur's evolution at IWC. The exclusive automatic calibre 32111 withstands water pressure of up to ten bar.

AUSSEN HART UND INNEN WEICH

Bei IWC Schaffhausen diente die 1948 lancierte Fliegeruhr „Mark 11" sozusagen als Blaupause für das Modell „Ingenieur". Werkhaltering, Innenboden und Zifferblatt aus Weicheisen schützten das Kaliber 89 vor starken Magnetfeldern. Dieses Konstruktionsprinzip kam auch bei den 1955 vorgestellten „Ingenieur"-Referenzen 666 A und 666 AD mit 36,6 Millimeter Gehäusedurchmesser zur Anwendung. Erstere beseelte das hauseigene Automatikkaliber 852, letztere das um ein Fensterdatum erweiterte 8521. Der Magnetfeldschutz reichte bis 80.000 Ampere pro Meter (A/m). Mehr war auf diese Weise schlichtweg nicht möglich. Ab 1958 stattete IWC die „Ingenieur" mit den optimierten Automatikkalibern 853 bzw. 8531 aus. An den Referenznummern änderte sich nichts. Die 1967 eingeführte, insgesamt rund 13 Millimeter dicke und um einen halben Millimeter vergrößerte Referenz 866 brachte neben dem etwas sportlicheren Look die Automatikkaliber 854 ohne und 8541 mit nun weiter nach außen verlagertem und deswegen größerem Fensterdatum. Gleichwohl konnte sie nicht an die beachtlichen Erfolge der Referenz 666 anknüpfen.

Daher startete bei IWC 1969 die Entwicklung einer „Ingenieur" mit 37 Millimeter großem Gehäuse. Weil ein vermarktungsfähiger Erfolg ausblieb, gelangte Gérald Genta vermutlich ab Herbst 1972 ins Boot. Von ihm stammte das Design der sportlich-eleganten „Ingenieur SL", welche ihren Einstand während der Basler Uhrenmesse 1976 gab. Die große Version mit 40 Millimeter zählendem Gehäusedurchmesser, auch „Jumbo Ingenieur" tituliert, trug die Referenznummer 1832. Einmal mehr reichte der Magnetfeldschutz bis 1.000 Gauss Feldstärke. Beim Manufakturkaliber 8541ES signalisierte der Buchstabe E die Verwendung amagnetischen Materials für Anker, Ankerrad und Hebelscheibe. S wies auf den Sekundenstopp hin. Weil die Lünette mit Hilfe eines Spezialwerkzeugs von vorne ins Gehäusemittelteil geschraubt wurde, nahmen die fünf Bohrungen bei nahezu jeder Uhr eine andere Position ein. Doch der Erfolg blieb aus. IWC verkaufte weniger als 1.000 Stück, weshalb die Preise am heutigen Gebrauchtuhrenmarkt geradezu explodieren.

An diese Armbanduhr knüpft die „Ingenieur Automatik 40" des Jahres 2023 an. Natürlich verlangte das rund 50 Jahre alte Genta-Design nach zeitgemäßer Adaption. Für einen höheren Tragekomfort hat IWC bei der 40-Millimeter-Schale die Gehäuselänge über die Hörner gekürzt. Und das neu gestaltete Gliederband ist nun beweglich am Gehäusekorpus befestigt. Dadurch schmiegt sich die 10,8 Millimeter hohe Armbanduhr besser ans Handgelenk als das Original. Ein Tribut an den schlankeren Auftritt ist die Reduktion der Magnetfeldresistenz auf 40.000 A/m oder 500 Gauss. Der Glasrand ist nun so mit dem Gehäusemittelteil verschraubt, dass sich die fünf charakteristischen Bohrungen immer an der gleichen Stelle befinden. Unabhängig von Gérald Genta ergab sich der nun vorhandene Flankenschutz für die Krone bei IWC im Laufe der „Ingenieur"-Evolution. Bis zu zehn bar Wasserdruck reicht der Schutz für das exklusive Automatikkaliber 32111.

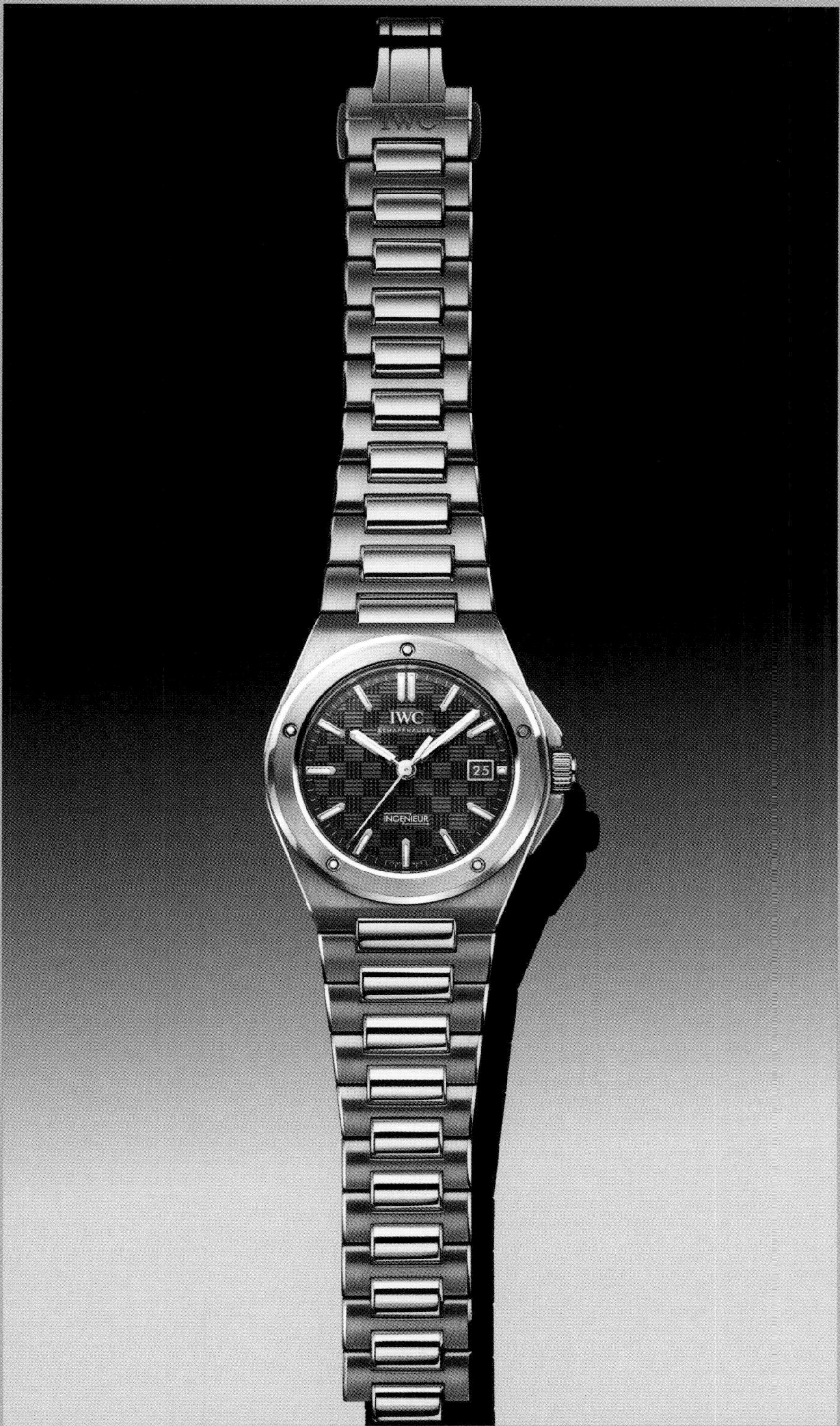

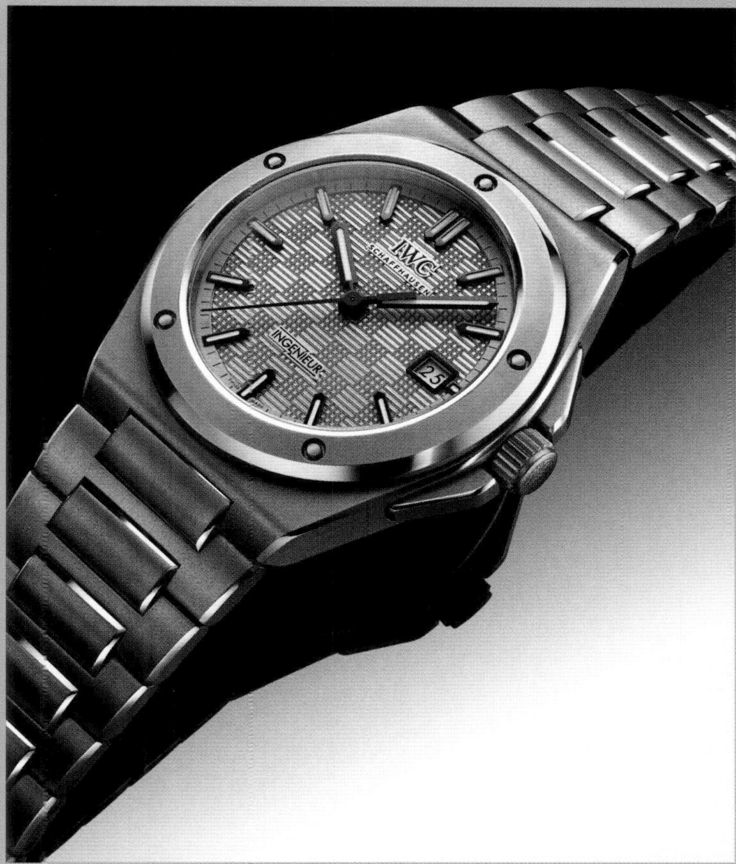

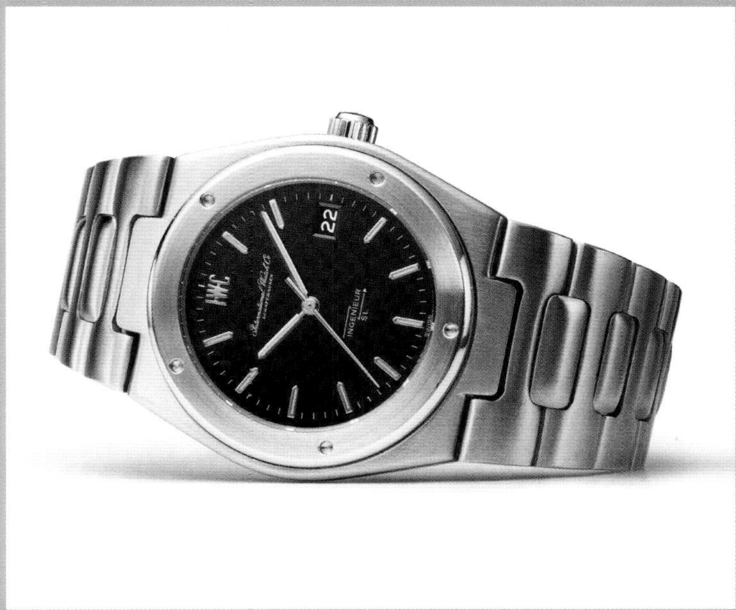

Left | Links: *Ingenieur Automatic 40 Ref. IW328903, 2023* | **Right, from top | Rechts von oben:** *Ingenieur Automatic 40 Ref. IW328904 titanium, 2023* | *Ingenieur SL, 1976*

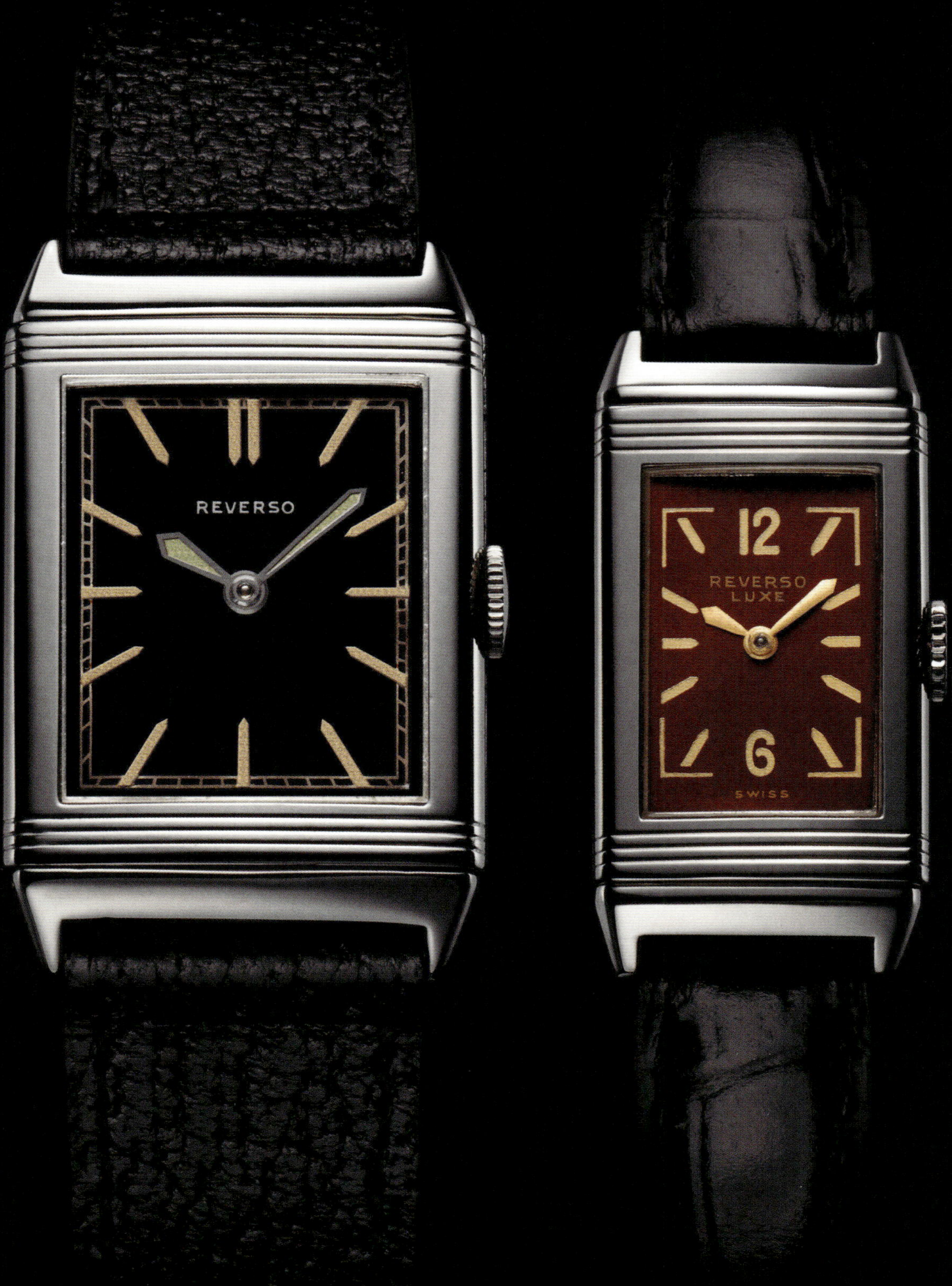

FIRST REVERSO, 1931

1931 JAEGER-LECOULTRE

REVERSO

TURNING TIME AROUND

"If you have time for someone, then don't look at your watch." These words of wisdom are ascribed to a Belgian priest. Advice that is easy to follow if you happen to be wearing a Jaeger-LeCoultre Reverso. Turn over the rectangular timepiece and the hours, minutes and seconds disappear from view. When René-Alfred Chauvot developed this masterpiece at the beginning of the 1930s, his ingenious reversal was motivated by other considerations. The Parisian engineer had been commissioned to protect the sensitive crystal glass from breakage during sporting activities such as polo. He fulfilled this remit with flying colours for watch importer César de Trey and movement maker Jacques-David LeCoultre. The making of the reversible case was entrusted to the Genevan case manufacturer Wenger. Initially, Chauvot designed a square case, into which the round movements were inserted. The rectangular Reverso followed in 1931. Due to the lack of a suitable proprietary calibre, to begin with, LeCoultre had to resort to a Tavannes moulded movement. No manufacturer is mentioned on the dial, only the name Reverso. The innovative timepiece was an immediate hit with polo players, golfers and tennis aces. In addition to its protective function, a high recognition value distinguished owners as connoisseurs. The Reverso was soon to be spotted where the reversible case was not needed at all; at concerts, soirées and high-society venues. In the 1940s, that fickle goddess fashion decreed a fad for waterproof cases and technical advances in watch glass. This led to a temporary reversal of the timepiece's fortunes. In 1972, Giorgio Corvo ordered the 200 remaining cases in stock on a visit to Le Sentier. The Italian Jaeger-LeCoultre importer sold them all at the turn of a hand. It was therefore not long before this alleged oldie made its comeback in 1979. The success of its comeback remains unabated to the present day. In the wake of this renaissance, the flip-over rectangular case was discovered to unlock almost limitless horological and creative possibilities. A glance at the collection says it all.

UHR MIT DREH

Von einem belgischen Ordenspriester stammt die weise Erkenntnis: „Wenn du Zeit hast für jemanden, dann schau nicht auf die Uhr." Dieser Maxime lässt sich mit einer Jaeger-LeCoultre „Reverso" am Handgelenk ohne Anstrengung Rechnung tragen. Wendet man nämlich die Rückseite des rechteckigen Werkscontainers nach oben, verschwinden die Stunden, Minuten und Sekunden aus dem Blickfeld. Als René-Alfred Chauvot dieses Œuvre zu Beginn der 1930er-Jahre entwickelte, ging es beim genialen Dreh um ganz andere Dinge. Der Pariser Ingenieur hatte den Auftrag erhalten, das empfindliche Kristallglas bei sportlichen Aktivitäten wie beispielsweise Pferdepolo vor Bruch zu schützen. Und diesen erledigte er für den Uhrenimporteur César de Trey sowie den Werkefabrikanten Jacques-David LeCoultre mit Bravour. Die Herstellung der intelligenten Wendeschale übernahm der Genfer Gehäusefabrikant Wenger. Zunächst hatte Chauvot übrigens ein quadratisches Gehäuse entwickelt, in das runde Uhrwerke passten. Erst im zweiten Anlauf folgte 1931 die rechteckige „Reverso", in der anfangs selbst LeCoultre mangels passendem eigenen Kaliber ein Formwerk von Tavannes verbauen musste. Am Zifferblatt stand kein Hersteller, sondern nur der Name „Reverso" zu lesen. Spontan begeisterte der neuartige Zeitmesser Polospieler, Golfer und Tennis-Asse. Zur schützenden Funktionalität gesellte sich ein hoher Wiedererkennungswert, welche die Besitzerinnen und Besitzer als Connaisseurs auszeichnete. Bald schon tauchte die „Reverso" auch dort auf, wo man das Wendegehäuse gar nicht brauchte: bei Konzerten, Soireen oder Veranstaltungen der besseren Gesellschaft. In den 1940er-Jahren führten die wankelmütige Göttin Mode, der Wunsch nach wasserdichten Schalen und technische Fortschritte beim Uhrenglas zum vorübergehenden „Reverso"-Aus. 1972 kaufte Giorgio Corvo bei einem Besuch in Le Sentier die 200 letzten Exemplare. Der italienische Jaeger-LeCoultre-Importeur verkaufte sie sozusagen im Handumdrehen. So konnte es nicht ausbleiben, dass der vermeintliche Oldtimer 1979 auf den Uhrenmarkt zurückkehrte und dort eine zweite Erfolgsgeschichte auslöste, welche bis in die Gegenwart anhält. Im Zuge der Renaissance zeigte sich, dass die Rechteckschale „mit dem Dreh" beinahe grenzenlose uhrmacherische und gestalterische Möglichkeiten eröffnet. Den Beweis liefert ein Blick in die Kollektion.

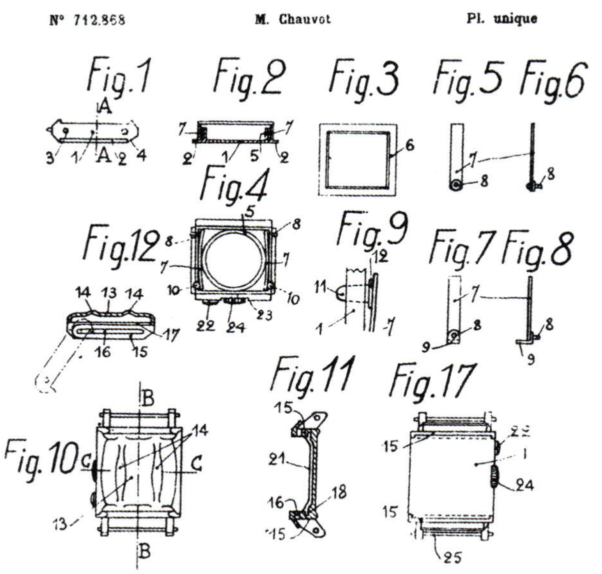

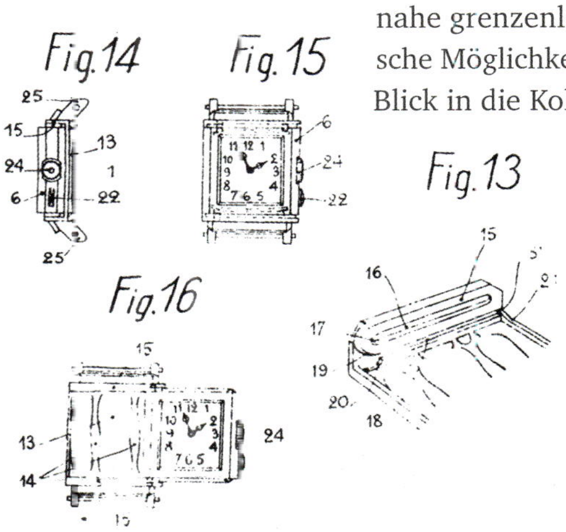

Patent N°712.868, 1931

Gr. 12

René Alfred Chauvot

Brevet N° 159982
2 feuilles. N° 1

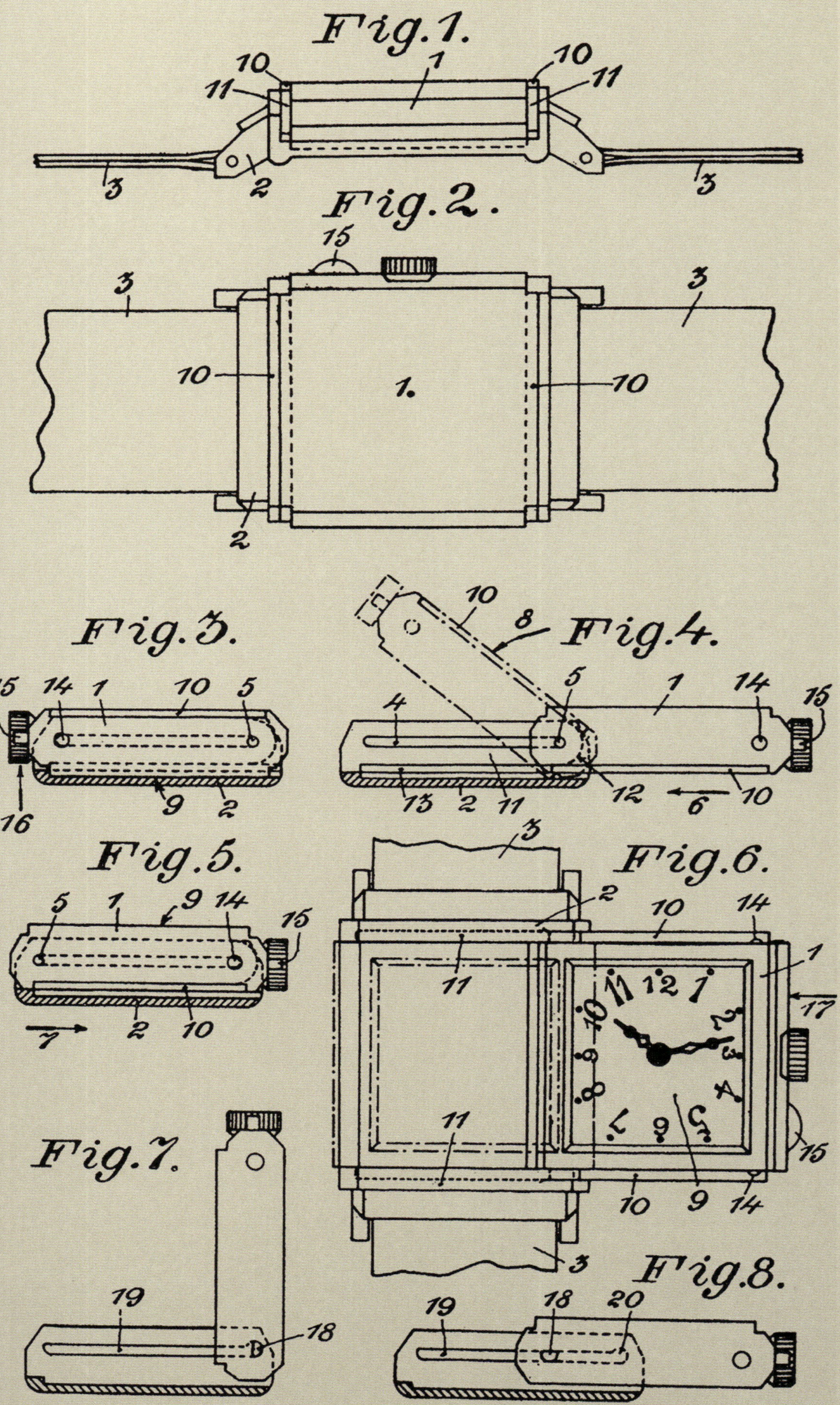

83 a | 37/00

28.217

G04b

RÉPUBLI

MINISTÈRE DU COMM

DIRECTION DE LA P

BREVET I

Gr. 1

Montre susceptible de coulisse
retourner complètement sur

M. René-Alfred CHAUVOT résidant en Fra

Demandé le 4 mars
Délivré le 3 août 1931.

La présente invention a pour objet ur
montre susceptible de coulisser dans sc
support et pouvant se retourner complèt
ment sur elle-même.

Cette montre comportant un boîtier, u
support dudit boîtier et éventuellement d
moyens permettant l'adaptation de ce su
port à tout objet convenable, peut être con
binée utilement et grâce à ces moyen
particulièrement avec un bracelet, une cei
ture, un sac ou tout autre article.

L'invention est caractérisée par les poin
principaux suivants, pouvant exister sépa
rément ou en combinaison :

a. Le support formant glissière dar
laquelle peut coulisser le boîtier lorsque
cadran est en position de visibilité, ce su
port et ce boîtier sont en prise par deu
ergots-pivots disposés sur les parois latérale
de l'une de ces deux pièces, boîtier ou su
port et engagés coulissant et pivotant sou
certaines conditions dans deux rainure
aménagées dans les parois latérales en regar
de l'autre pièce, de telle sorte que le boîtie
rabattu normalement entre les parois lat
rales, formant glissière du support,
cadran étant ainsi dissimulé en positio
d'escamotage, puisse être retourné par piv
tement d'un demi-tour sur les ergots-pivot
de manière à rendre visible le cadran, pui
engagé à nouveau dans cette position, entr

Prix du fas

a. Le support formant glissière dans

RANÇAISE.

E ET DE L'INDUSTRIE.

IÉTÉ INDUSTRIELLE.

NVENTION.

Cl. 1. N° 712.868

ıs son support et pouvant se ême.

ıe).

à 13ʰ 15ᵐ, à Paris.
blié le 14 octobre 1931.

s parois latérales du support, par coulissment des ergots-pivots dans les rainures réciproquement.

b. Un ou plusieurs dispositifs d'encliquege, bille pressée par un ressort par exeme, est ou sont disposés de manière à lidariser de façon amovible le boîtier et support dans leurs deux positions permis : escamotage ou visibilité du cadran.

Grâce à l'invention, la montre peut être rtée facultativement dans l'une ou l'autre ces deux positions, et le passage de l'une l'autre de ces positions est extrêmement nple et facile.

L'invention comporte en outre d'autres ractéristiques qui ressortent tant de la scription suivante, que du dessin annexé titre d'exemple et dans lequel :

La fig. 1 est une vue en élévation d'un îtier de montre selon la présente invenn;

La fig. 2 en est une coupe suivant A-A la fig. 1;

La fig. 3 est une vue en plan de la nette;

La fig. 4 est une vue en plan du boîtier ontrant les organes de coulissement et encliquetage;

Les fig. 5 et 6 montrent en élévation et de côté l'un des ressorts portant l'ergotvot;

: 5 francs.

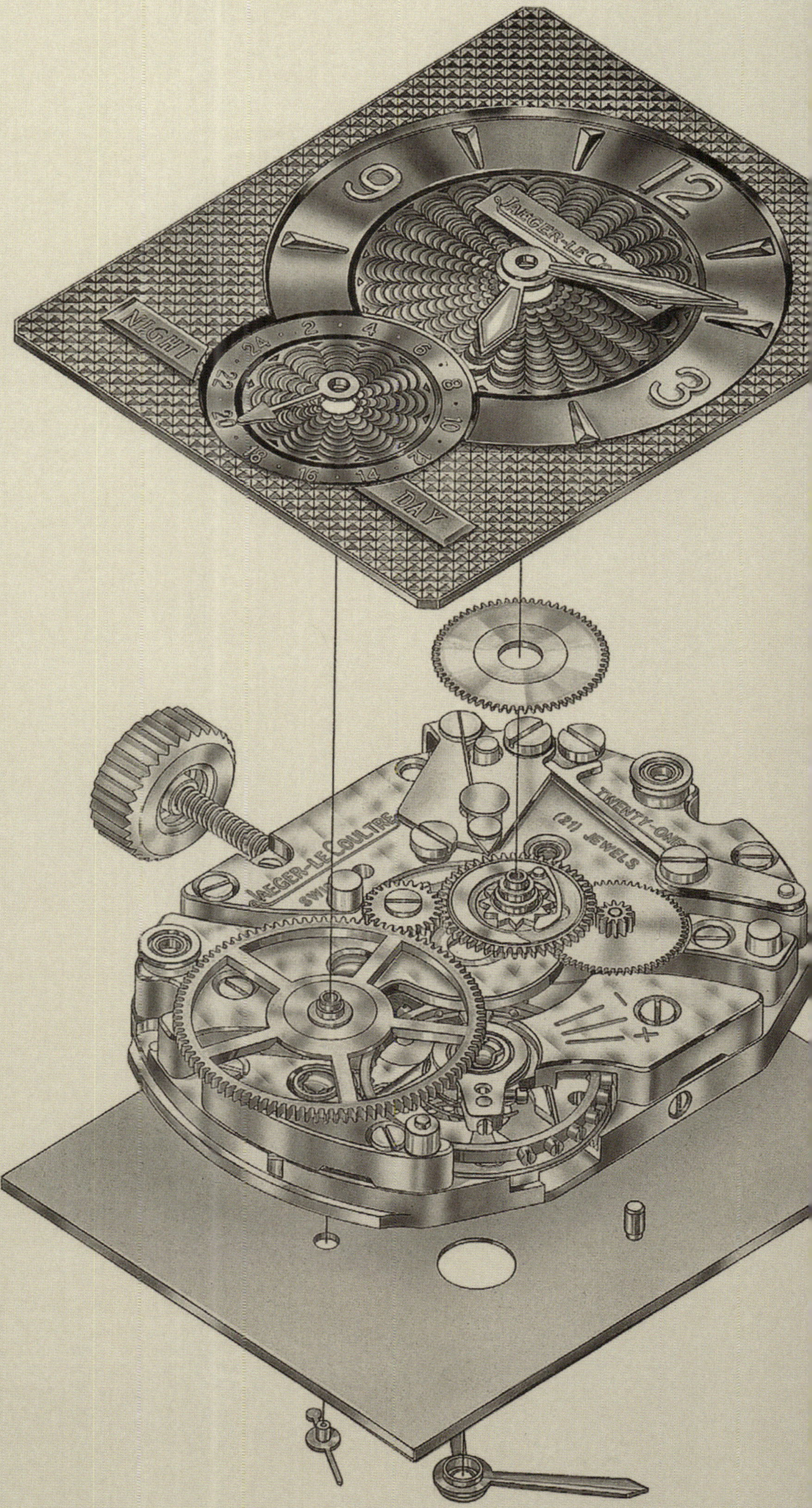

L'invention comporte en outre d'autres

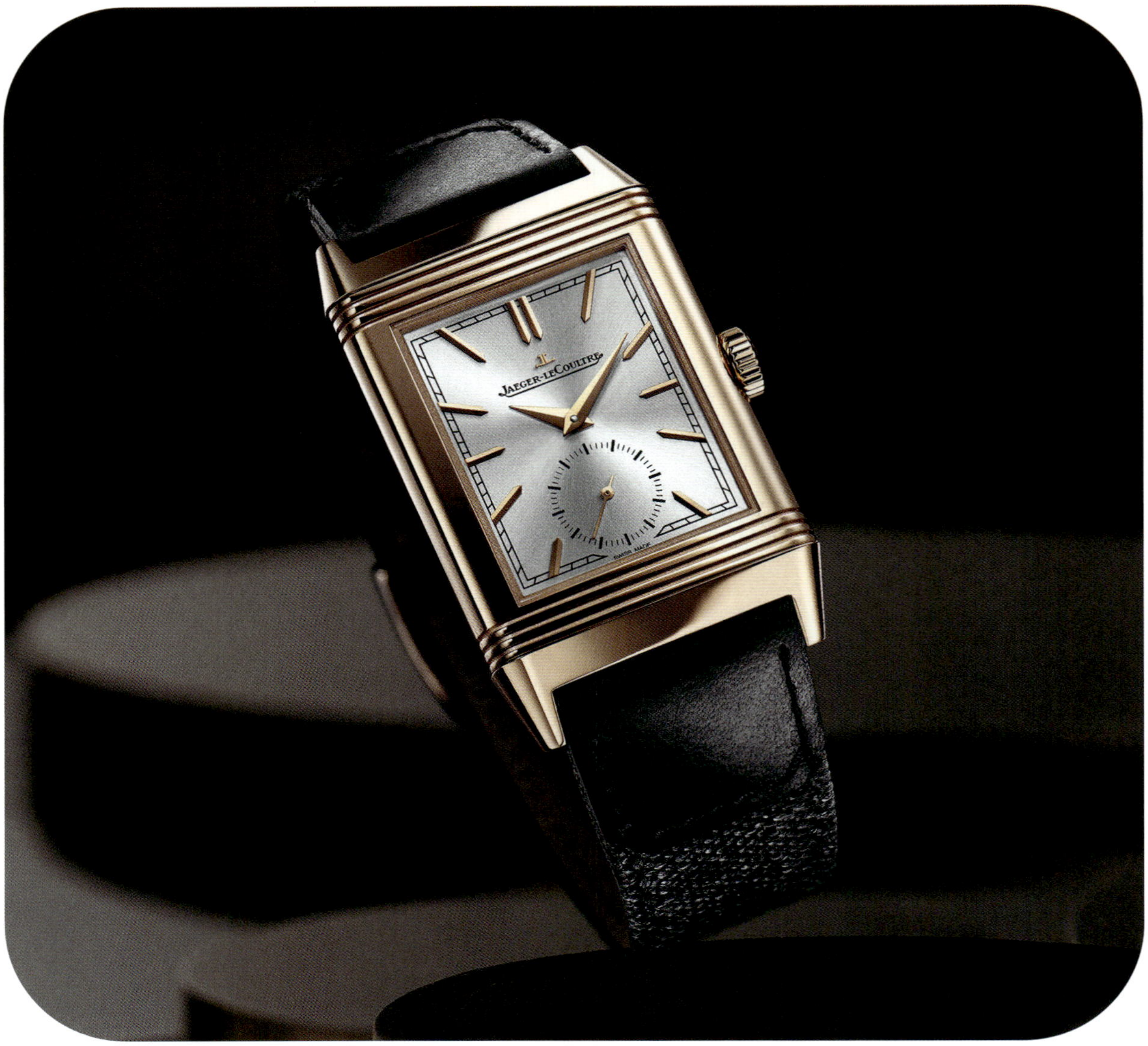

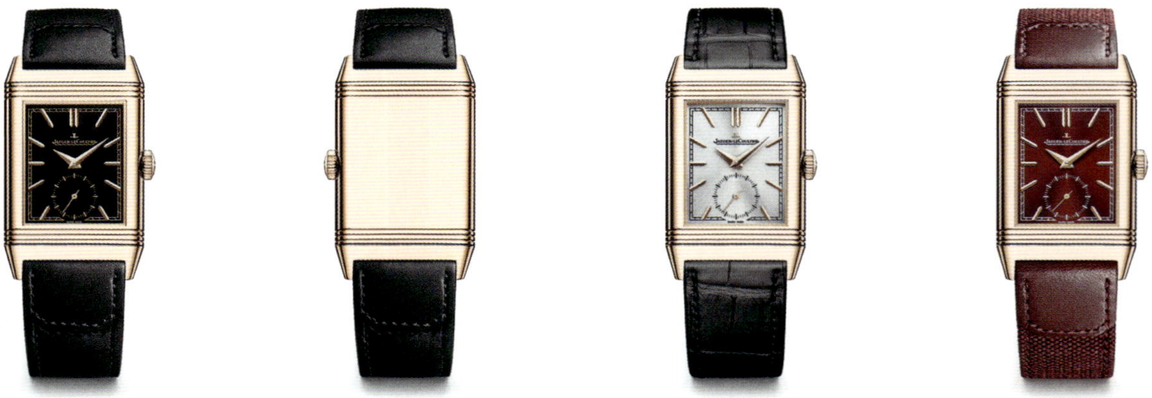

TRIBUTE MONOFACE SMALL SECONDS

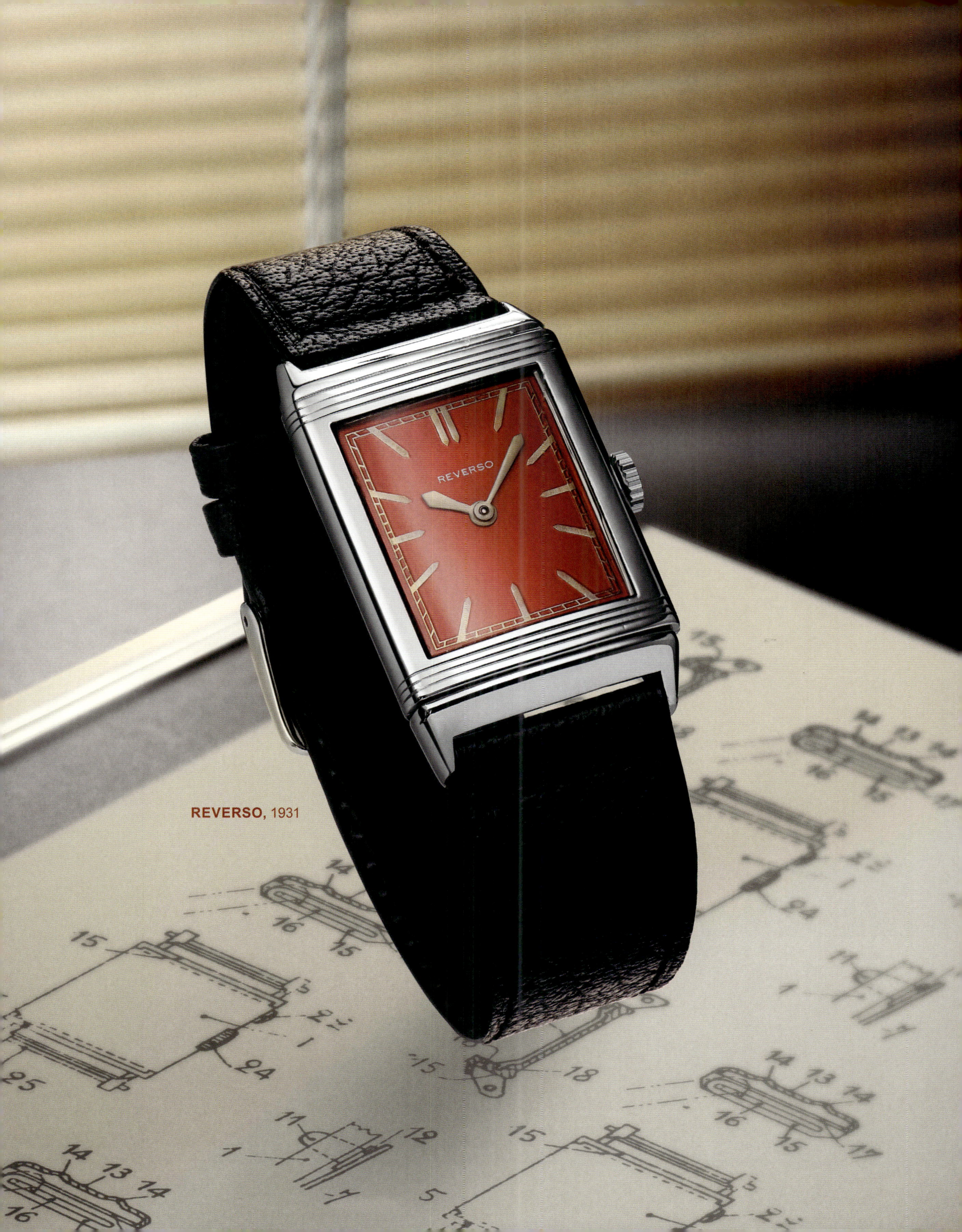

REVERSO, 1931

JAEGER-LECOULTRE 1929

CALIBRE 101

MINIMALISM WITH A FEMININE TWIST

In 1877, the sons of Antoine LeCoultre began to design and manufacture complex gear trains for mechanical watches. They also focused attention on ultra-flat and small movements. With decreasing dimensions, the effort involved increased exponentially. An eloquent example was the 1907 hand-wound Calibre 145 with its sleek height of just 1.38 millimetres. Internally, it was nicknamed the knife. In the 1920s, the engineers addressed the question of how to increase precision by using larger components, while keeping the movement dimensions small. In 1925, in the remote Joux valley, the philosopher's stone was found: it consisted of arranging the components on two levels and moving the winding and adjusting crown to the back. The Duoplan Calibre 104 or 7BF was engineered according to this model. Its diminutive dimensions of 16 x 5.8 x 3.4 millimetres merit the qualification superlative. Customers included Cartier. After some intermediate steps with slightly larger movements of this type, the legendary Calibre 101 caused a sensation in 1929.

To the present day, this baguette movement embodies a world-first of peerless achievement. It is of course listed in the *Guinness Book of Records*. Arranged on two levels, the hand-wound movement impressed with dainty dimensions of 14 x 4.8 x 3.4 millimetres and a total volume of 228.48 mm³. Despite its 74 components, the tiny timepiece, including the dial and hands, weighed less than a gram.

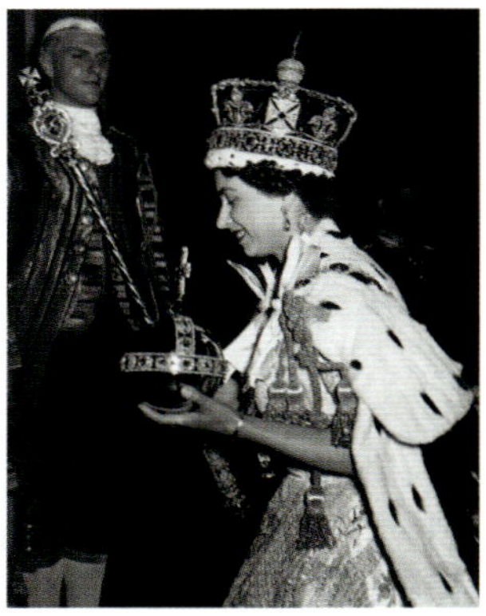

Upon merging with the French watchmaker Edmond Jaeger, the two companies became Jaeger-LeCoultre in 1937. The 2nd of June 1953 marked an historical date for the company and the crowning of the Calibre 101's career. At her coronation in Westminster Abbey, the bejewelled watch worn by Queen Elizabeth II had one inside.

A further achievement was to follow. In 1991, the watchmakers succeeded in skeletonising this minimalist microcosm.

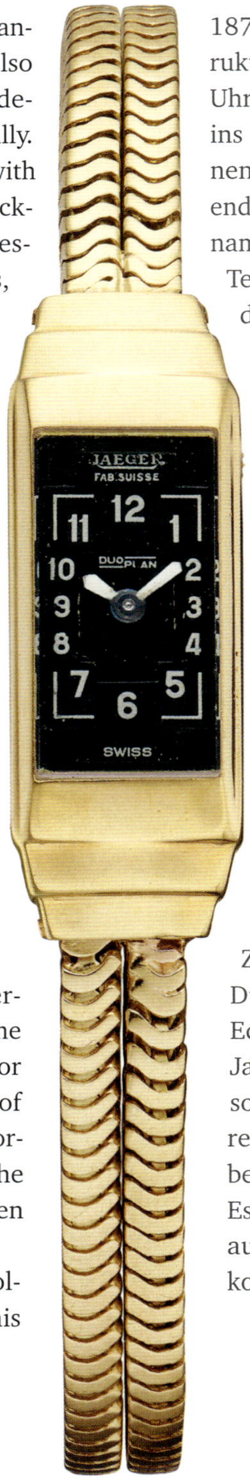

MINIMALISMUS FÜR WEIBLICHE HANDGELENKE

1877 begannen die Söhne von Antoine LeCoultre mit der Konstruktion und Herstellung komplexer Schaltwerke für mechanische Uhren. Außerdem rückten besonders flache oder kleine Uhrwerke ins Blickfeld. Mit abnehmenden Maßen stieg der Aufwand exponentiell. Beredtes Beispiel war das nur 1,38 Millimeter hoch bauende Handaufzugskaliber 145 von 1907. Intern trug es den Spitznamen „Rasierklinge“. In den 1920er-Jahren beschäftigten sich die Techniker mit der Frage, wie man die Präzision durch Verwendung größerer Bauteile steigern, gleichwohl aber die Werksdimensionen geringhalten konnte. 1925 hatte man im abgeschiedenen Vallée de Joux den Stein der Weisen gefunden: Anordnung der Komponenten auf zwei Ebenen sowie die Verlegung der Aufzugs- und Zeigerstellkrone auf die Rückseite. Unter diesen Vorzeichen wurde das „Duoplan“-Kaliber 104 oder 7BF geboren. Seine zierlichen Maße von 16 x 5,8 x 3,4 Millimetern rechtfertigten den Begriff Superlativ. Zu den Kunden gehörte auch Cartier. Nach einigen Zwischenschritten mit etwas größeren Uhrwerken dieser Bauart sorgte 1929 das legendäre „Kaliber 101“ für Furore.

Bis in die Gegenwart verkörpert dieses Baguettewerk eine nicht mehr übertroffene Weltbestleistung. Natürlich hielt sie Einzug ins Guinness-Buch der Rekorde. Das wiederum zweigeschossige Handaufzugswerk bestach durch Mini-Dimensionen von 14 x 4,8 x 3,4 Millimetern. Daraus berechneten sich 228,48 mm³ Gesamtvolumen. Trotz seiner 74 Komponenten brachte der tickende Winzling inklusive Zifferblatt und Zeigern weniger als ein Gramm auf die Waage.

Durch die Fusion mit dem französischen Uhrenfabrikanten Edmond Jaeger verschmolzen beide Unternehmen 1937 zu Jaeger-LeCoultre. Für diese Manufaktur brachte der 2. Juni 1953 sozusagen die Krönung der Karriere ihres „Kalibers 101“. Während der Zeremonie in Westminster Abbey trug Queen Elizabeth II. von England eine damit ausgestattete Schmuckuhr.

Es sollte noch besser kommen: 1991 gelang den Uhrmachern auch noch die Skelettierung dieses minimalistischen Mikrokosmos.

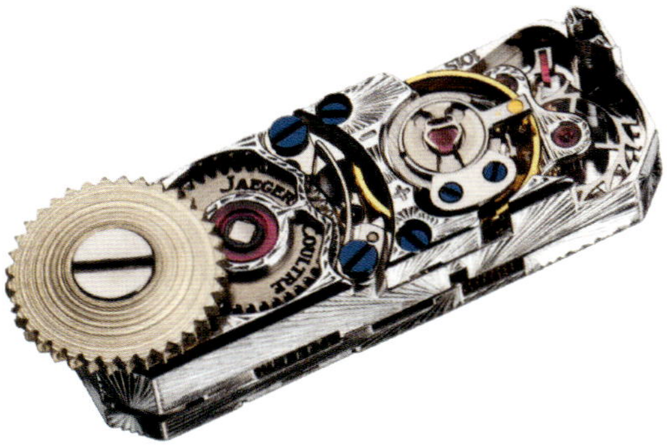

From left / Von links: *Queen Elizabeth II at her coronation, 1953 / Königin Elizabeth II. bei ihrer Krönung, 1953 / Duoplan / Calibre 101*

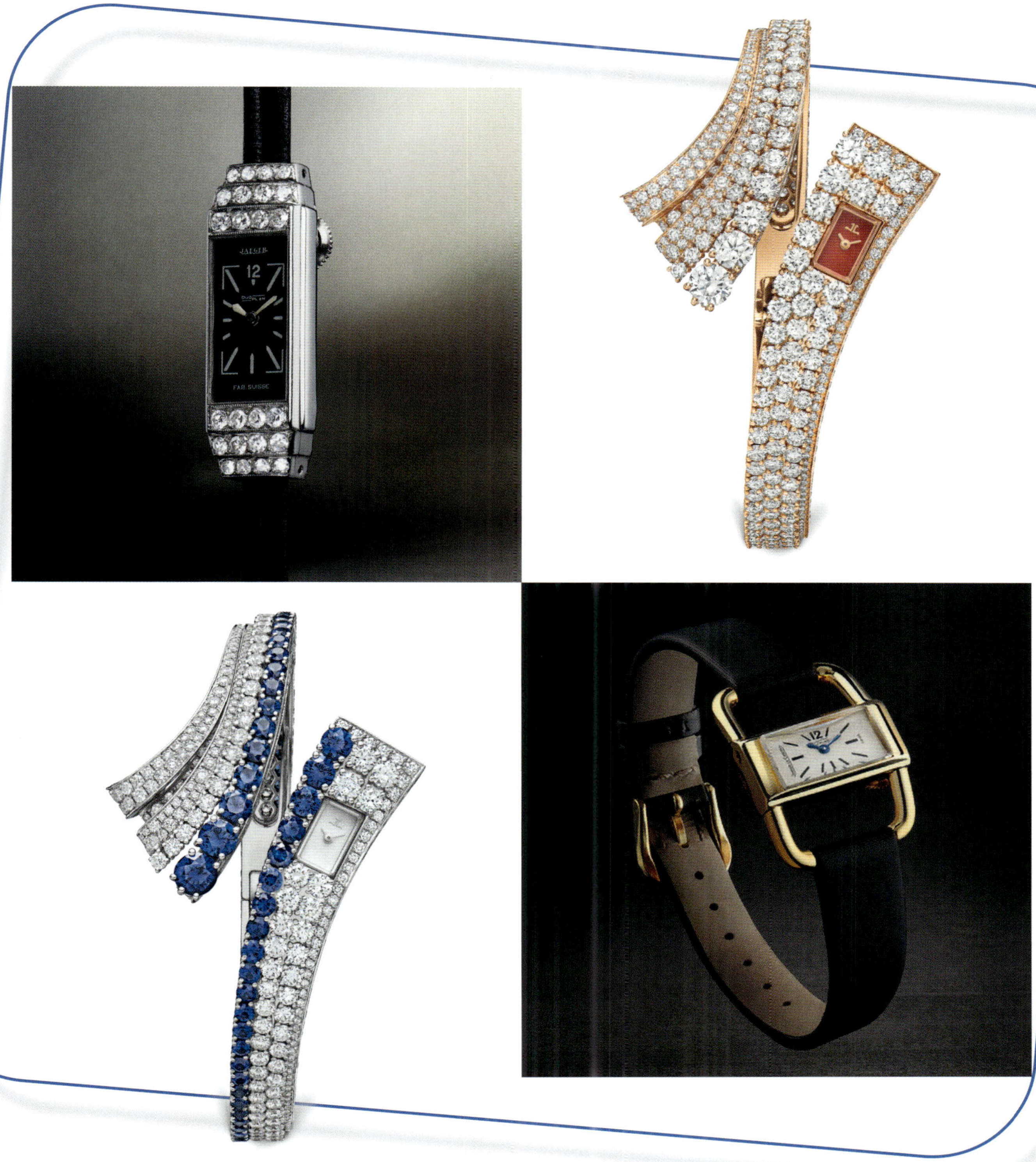

From top left, clockwise | Von oben links im Uhrzeigersinn:
Duoplan Secret, 2023 | 101 Bangle, 2023 | Duoplan | 101 Bangle, 2023

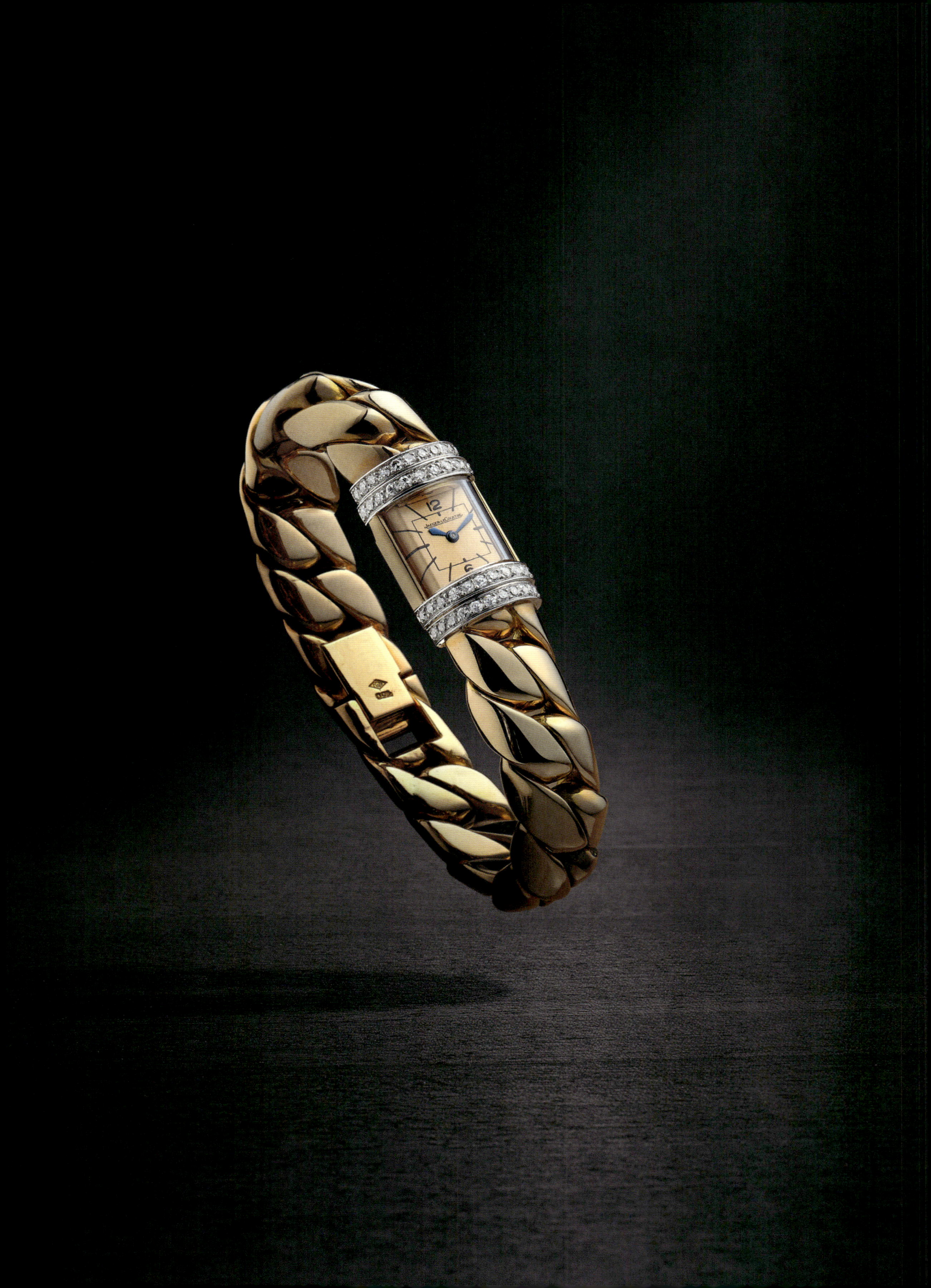

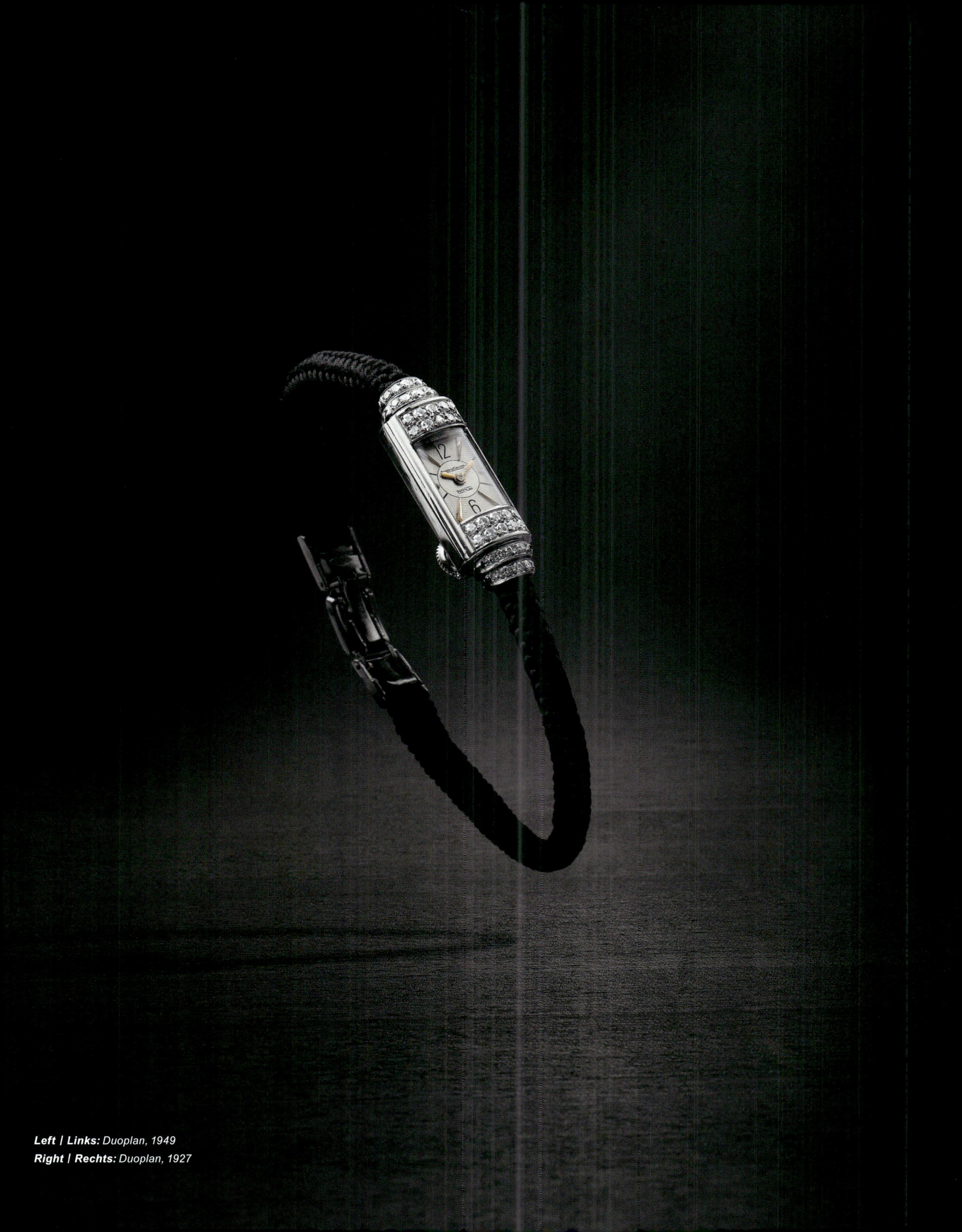

Left | Links: *Duoplan, 1949*
Right | Rechts: *Duoplan, 1927*

JUNGHANS 1961

MAX BILL

DIALS À LA MAX BILL

At the end of the twentieth century, the traditional German watch brand Junghans had also come to the conclusion that there was a lot to be said for the good old mechanical watch. A perusal of the brimming archives of the Schramberg company lighted on the outstanding legacy of the company's cooperation with Max Bill. The Swiss architect, sculptor, painter and product designer (1908–1994) is universally regarded one of the most important representatives and theorists of concrete art. In 1927 and 1928, he studied at the legendary Dessau Bauhaus. Its style was to significantly influence his versatile work, and Max Bill championed the school's ideals all his life. From 1956, Junghans embraced Bill's philosophy of creating useful objects that were "modest in a beautiful way". The first cooperation for a kitchen clock resulted in 1961 in wristwatches. Max Bill had impressed his distinctive signature on their dials. Particularly impressive is the typography of the "4" on the versions with Arabic numerals. The timepieces, which were both minimalist and functional, were available in versions for men and women, with proprietary hand-wound calibres J 73 and 84/S3.
A comeback was launched in 1997. The retro models with stainless-steel cases were henceforth powered by a purchased hand-wound calibre Eta 2801-2. A feature retained from the original 1960s models, much sought after by collectors, is the diameter. It is a modest 34 millimetres. Success was immediate and continues unabated to this day. Reason enough to revisit these icons and create a collection dedicated to them, the JUNGHANS max bill. Along with the hand-wound watches and various automatic models, it also includes the self-winding Chronoscope with a practical stopwatch function.

ZIFFERBLÄTTER VON MAX BILL

Zum Ende des 20. Jahrhunderts merkten auch die Verantwortlichen der traditionsreichen deutschen Uhrenmarke Junghans, dass die gute alte Mechanik nicht umzubringen ist. Beim Blick in die prall gefüllten Archive des Schramberger Unternehmens stach die Hinterlassenschaft der einstigen Kooperation mit Max Bill heraus. Bekanntlich zählt der Schweizer Architekt, Bildhauer, Maler und Produktgestalter (1908–1994) zu den bedeutendsten Vertretern und Theoretikern der konkreten Kunst. 1927 und 1928 hatte er am legendären Dessauer Bauhaus studiert, dessen Stil sein vielseitiges Werk später entscheidend prägte und für dessen Ideale sich Max Bill zeitlebens einsetzte. Seine Philosophie, das Nützliche, das „auf schöne Art Bescheidene" zu schaffen, nutzte Junghans ab 1956. Die erste Kooperation in Sachen Küchenuhr mündete 1961 in Armbanduhren. Deren Zifferblättern hatte Max Bill seine sehr spezielle Handschrift verliehen. Besonders eindrucksvoll bei den Versionen mit arabischen Ziffern war die Typographie der „4". Erhältlich waren die gleichermaßen reduziert wie sachlich gestalteten Zeitmesser in Versionen für Damen und Herren, ausgestattet mit den damals noch selbst entwickelten und gefertigten Handaufzugskalibern J 73 und 84/S3.
Das Comeback erfolgte im Jahr 1997. Bei den Retromodellen umfängt die Edelstahlschale ein nunmehr zugekauftes Handaufzugskaliber Eta 2801-2. Nicht geändert gegenüber den von Sammlern sehr gesuchten Originalen aus den 1960er-Jahren hat sich der Durchmesser. Er liegt bei moderaten 34 Millimetern. Der Erfolg stellte sich spontan ein und hält bis heute an. Nicht zuletzt deshalb führte die Rückbesinnung auf Ikonen zur eigenständigen Kollektion „JUNGHANS max bill". Neben besagten Handaufzugsuhren und ganz unterschiedlichen Automatikmodellen umfasst sie auch das „Chronoscope" mit Selbstaufzug und praktischer Stoppfunktion.

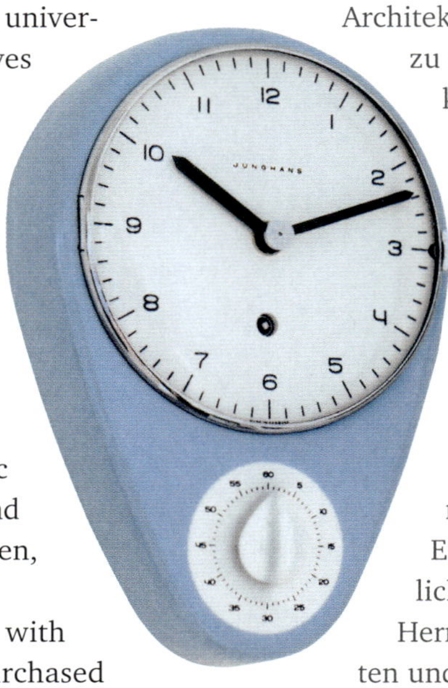

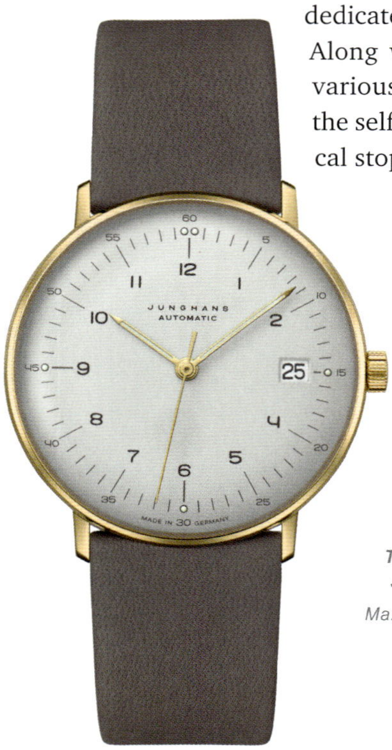

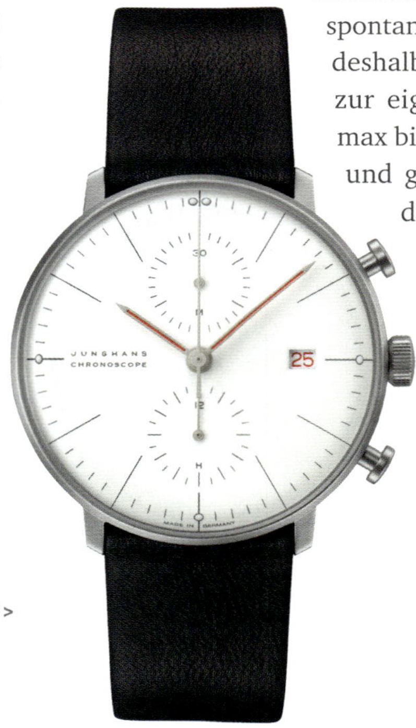

Top: *Kitchen clock, 1956* **| Oben:** *Küchenuhr, 1956*
< Max Bill Kleine Automatic Ref. 27/7108.02, 2020
Max Bill Chronoscope Bauhaus Ref. 27/4303.02, 2023 >

ORIGINAL MAX BILL, 1961

MAX BILL AUTOMATIC
REF. 27/4700.02

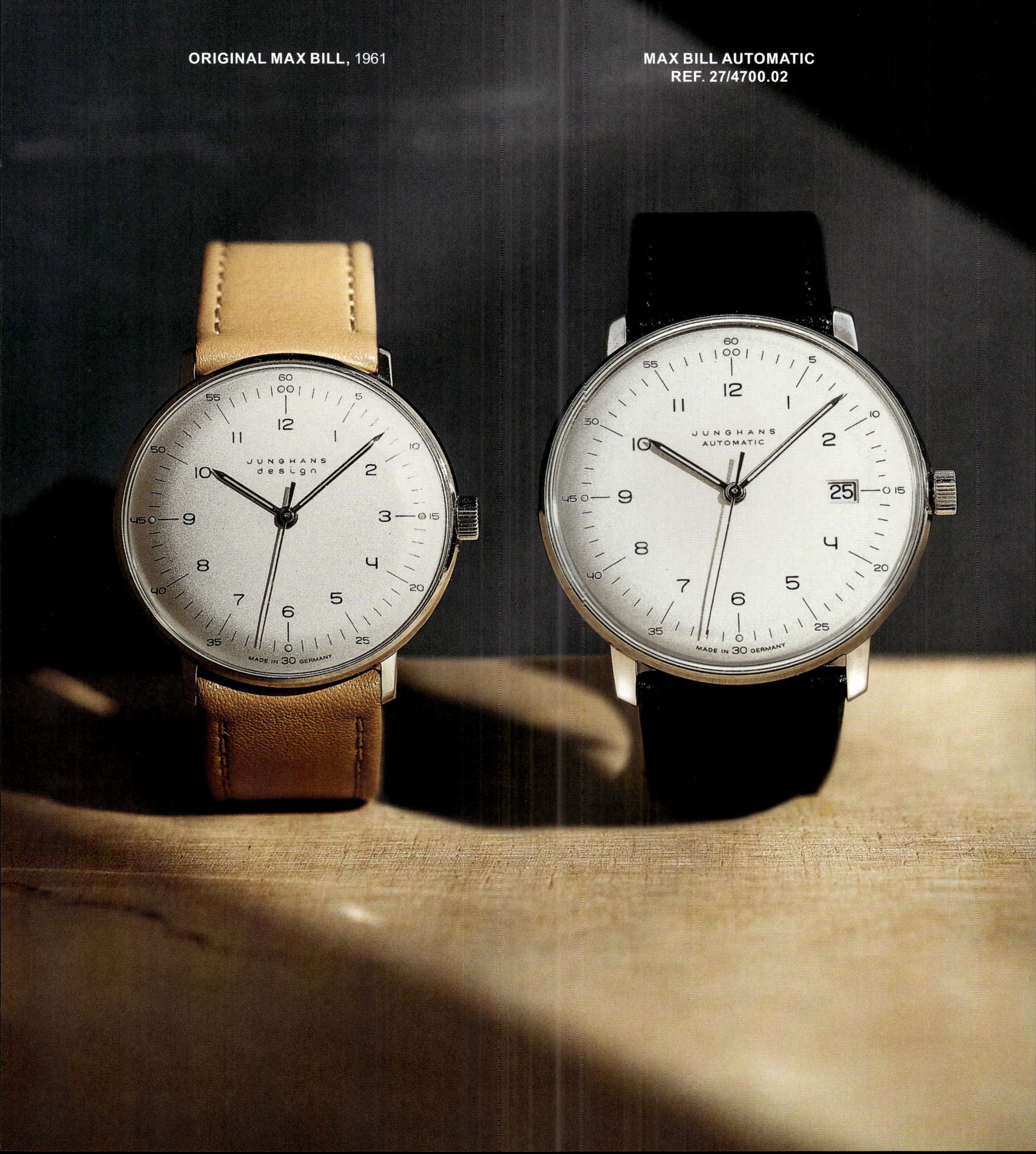

A. LANGE & SÖHNE 1994

LANGE 1

PURE GEOMETRY AND LARGE DATE COMPLICATION

For lovers of German watch culture, 24 October 1994 is a red-letter day. The date marks the curtain-up in Dresden of the first post-war watch collection by A. Lange & Söhne. Of the four timepieces presented, the Lange 1 was undoubtedly the star of the show. Since then, it has been continuously on the market, enjoys unabated popularity, and with multiple iterations, it epitomises Saxon horology. From the very first minute, the watch's extraordinary appearance with the geometrical purity of its dial design has held a magical attraction for connoisseurs. One reason is that the ratio between the diameters of the entire dial and the hour and minute disc follow the golden ratio. The same applies to the proportions of the eye-catching large-date window. In addition, the indications form an isosceles triangle. The hand-wound calibre L.901.0 with two barrels, nine screwed gold chatons, three-quarter plate made of German silver decorated with Glashütte ribbing, hand-engraved balance cock and whiplash spring to regulate beat adjustments send lovers of mechanical watches into raptures. A reassuring 72-hour power reserve enables this classic timepiece to spend the weekend in the safe while its owner enjoys the rough and tumble of sporting activities. The wound state of the two mainsprings is prominently displayed by the power reserve indicator. From 2015, the Saxons equipped their classic with the new contemporary manufacture calibre L121.1. Developed over three years, the hand-wound movement, which still has two barrels and a three-day power reserve, plays a key role in the future success of the flagship product. As one of the significant changes, the large date now changes instantaneously due to pre-tensioned springs. The balance wheel oscillates at three hertz and has a variable moment of inertia. Small cams ensure rate accuracy. The proprietary balance spring can thus oscillate freely. The index finger with whiplash spring to regulate beat adjustments is used to set the pallet beat for an even tick-tock. The potential promised by the Lange 1 is amply illustrated by the 17 movements of various designs with and without complications created for this watch line over the past three decades. And its history is still very much in the making.

ASYMMETRIE UND GROSSDATUM

Für Liebhaberinnen und Liebhaber deutscher Uhr-Kultur ist der 24. Oktober 1994 ein wichtiges Datum. Da nämlich öffnete sich in Dresden der Vorhang für die erste Nachkriegs-Uhrenkollektion von A. Lange & Söhne. Von den vier vorgestellten Zeitmessern kam der „Lange 1" zweifellos die größte Bedeutung zu. Sie ist seitdem ununterbrochen am Markt, erfreut sich anhaltender Beliebtheit und repräsentiert durch zahlreiche Modellvarianten sächsische Uhrmacherkunst par excellence. Von der ersten Minute an lenkte das außergewöhnliche Erscheinungsbild mit asymmetrisch gestaltetem und trotzdem aufgeräumt wirkendem Zifferblatt Kennerblicke magisch auf sich. Ein Grund ist darin zu sehen, dass das Verhältnis zwischen den Durchmessern des ganzen Zifferblatts sowie des Stunden- und Minuten-Kreises dem Goldenen Schnitt entspricht. Gleiches gilt auch für die Proportionen des augenfälligen Fensters für das Großdatum. Überdies bilden die Indikationen ein gleichschenkliges Dreieck. Uhrmacherisch lässt das Handaufzugskaliber L.901.0 mit zwei Federhäusern, neun verschraubten Gold-Chatons, neusilberner Dreiviertelplatine mit Glashütter Bandschliff, handgraviertem Unruh-Kloben sowie Schwanenhals-Feinregulierung die Herzen der Mechanikfans höherschlagen. Eine beruhigende Gangautonomie von 72 Stunden gestattet es, den Klassiker während eines zeitlosen oder sportlich aktiven Wochenendes im Tresor zu verwahren. Den Spannungszustand des Zugfeder-Duos stellt eine unübersehbare Gangreserveanzeige dar. Ab 2015 statteten die Sachsen ihren Klassiker mit dem neuen zeitgemäßen Manufakturkaliber L121.1 aus. Das während dreier Jahre entwickelte Handaufzugswerk mit weiterhin zwei Federhäusern und drei Tagen Gangautonomie leistet einen wichtigen Beitrag zur erfolgreichen Zukunft des Leaders. Als eine der signifikanten Änderungen springt das Großdatum dank vorgespannter Federn pünktlich um Mitternacht. Die mit drei Hertz oszillierende Unruh besitzt ein variables Trägheitsmoment. Kleine Exzenter gestatten die Regulierung des Gangs. Somit kann die selbst gefertigte Unruhspirale völlig frei schwingen. Der Rückerzeiger mit Schwanenhals-Feinregulierung dient zur Einstellung des Anker-Abfalls, sprich eines gleichmäßigen Tick-Tack. Welches Potenzial in der „Lange 1" steckt, beweist die Tatsache, dass für diese Uhrenlinie im Laufe von knapp 30 Jahren 17 Uhrwerke unterschiedlicher Ausprägung ohne und mit Komplikationen entstanden sind. Und die Geschichte ist definitiv noch nicht zu Ende.

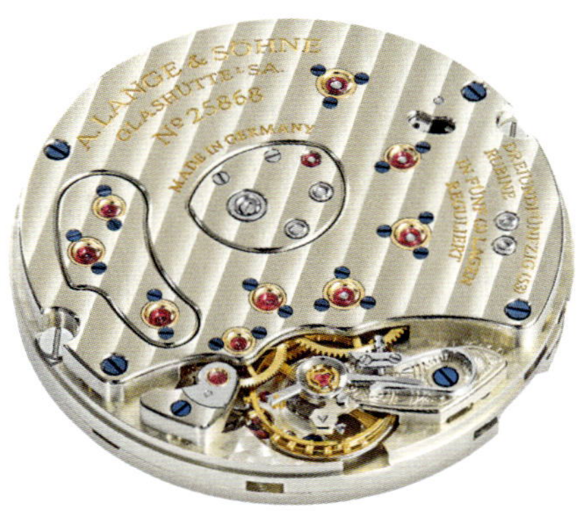

Cal. L901.0, 1994

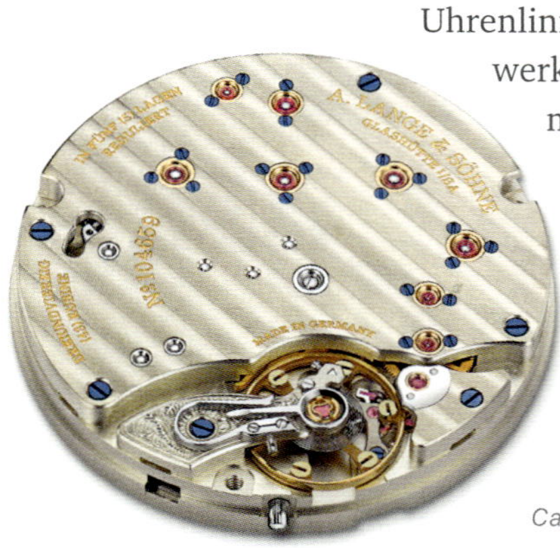

Cal. L121.1, 2015

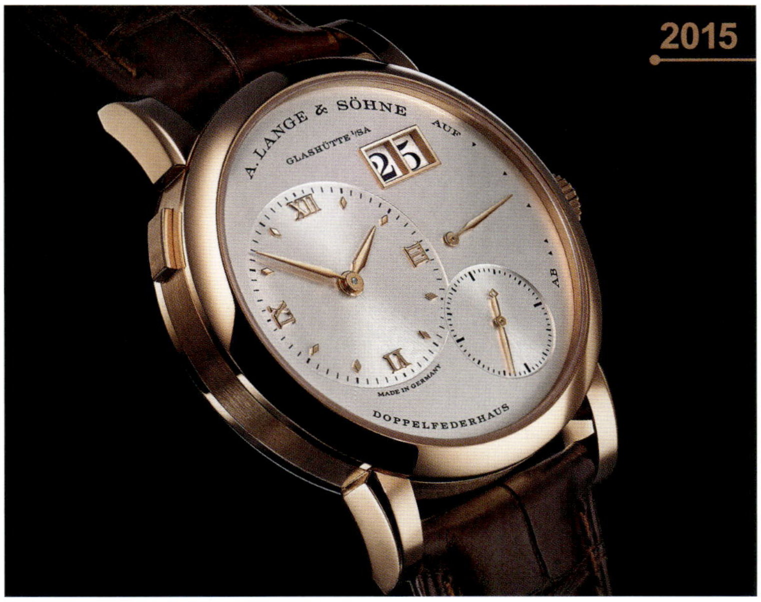

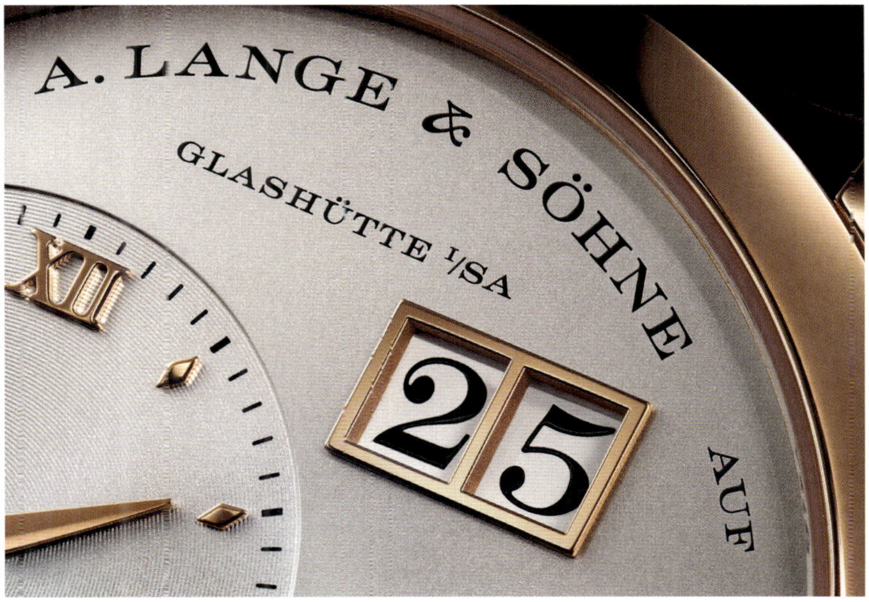

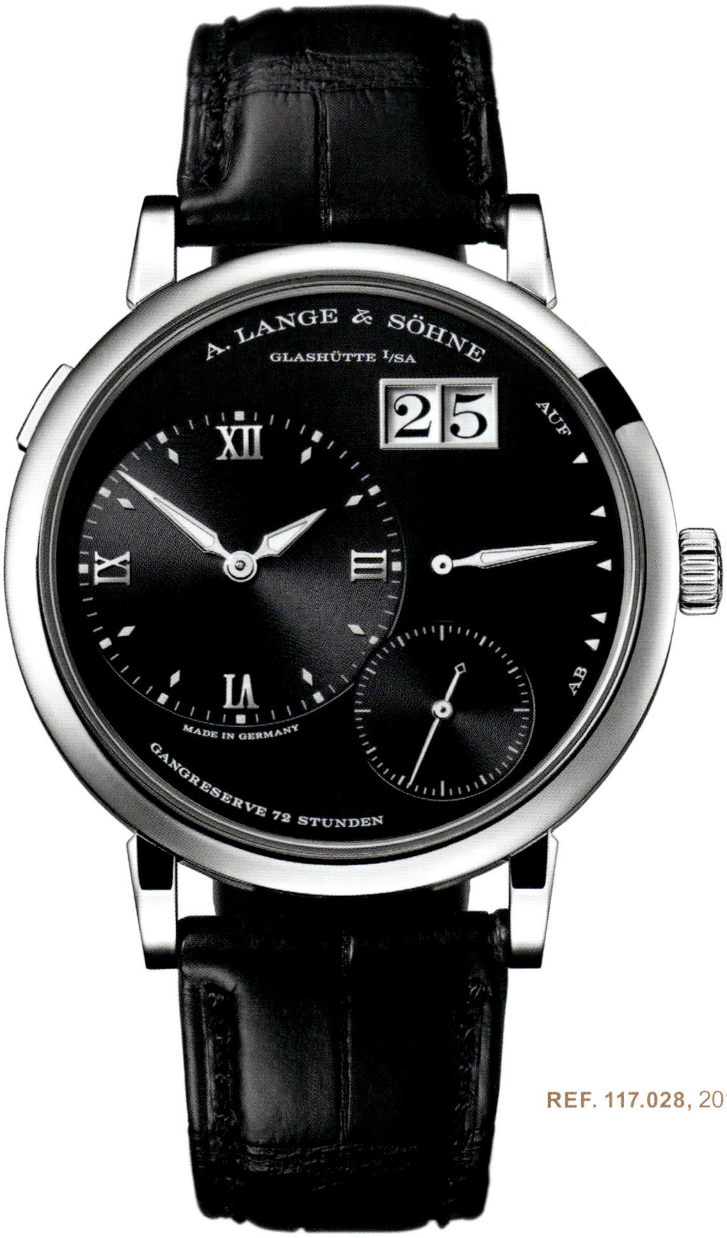

REF. 117.028, 2013

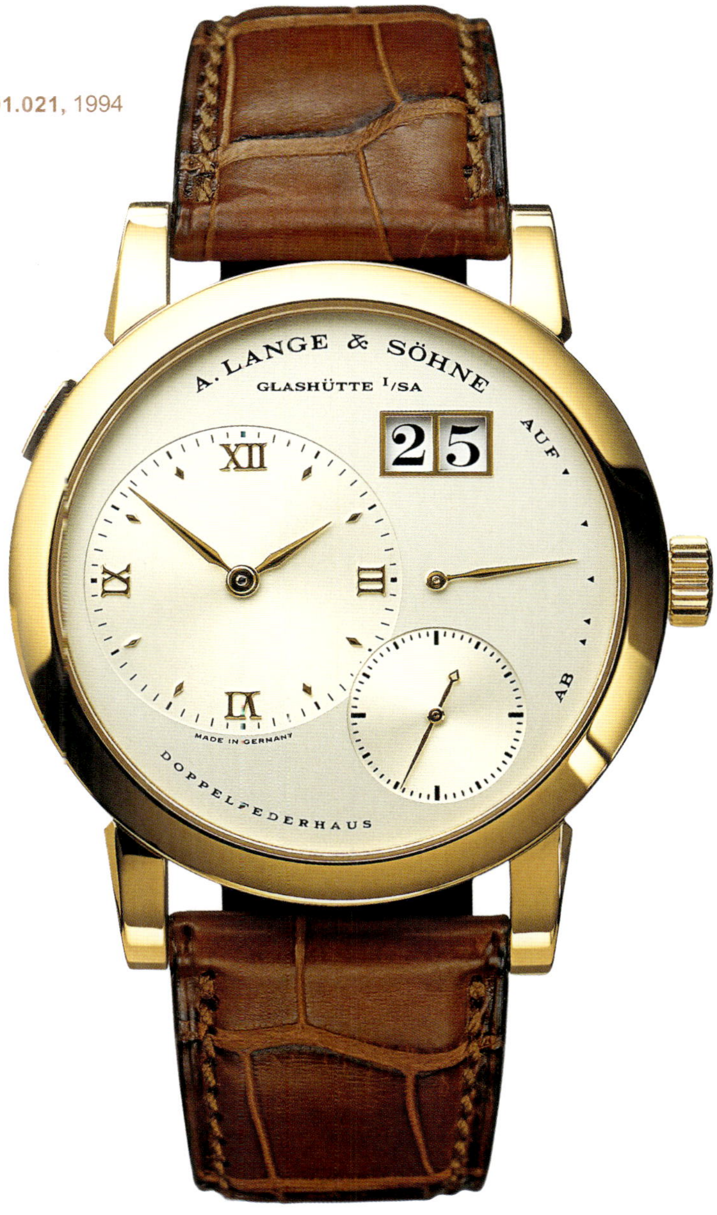

REF. 101.021, 1994

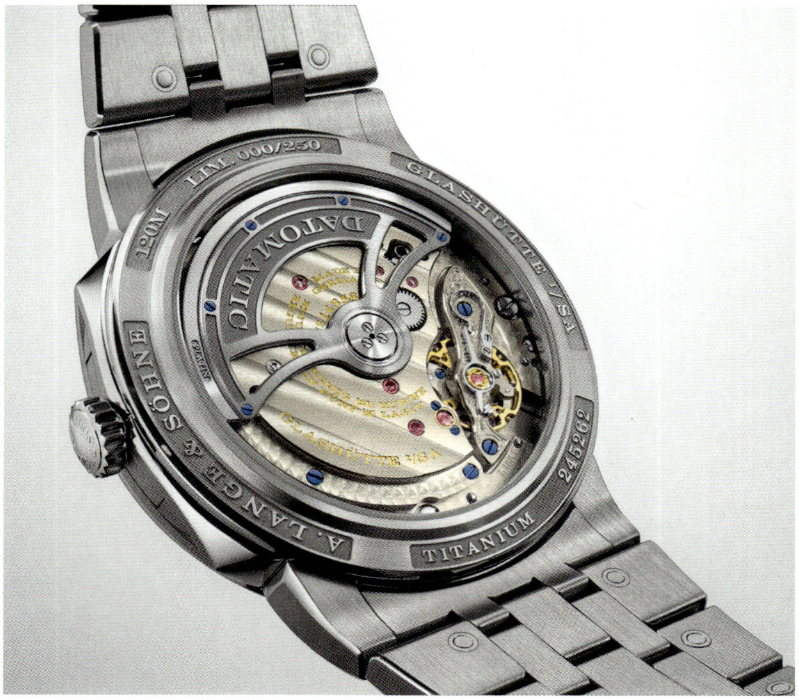
LIM. 000/250
GLASHÜTTE I/SA
A. LANGE & SÖHNE
TITANIUM
245262
DATOMATIC

2020

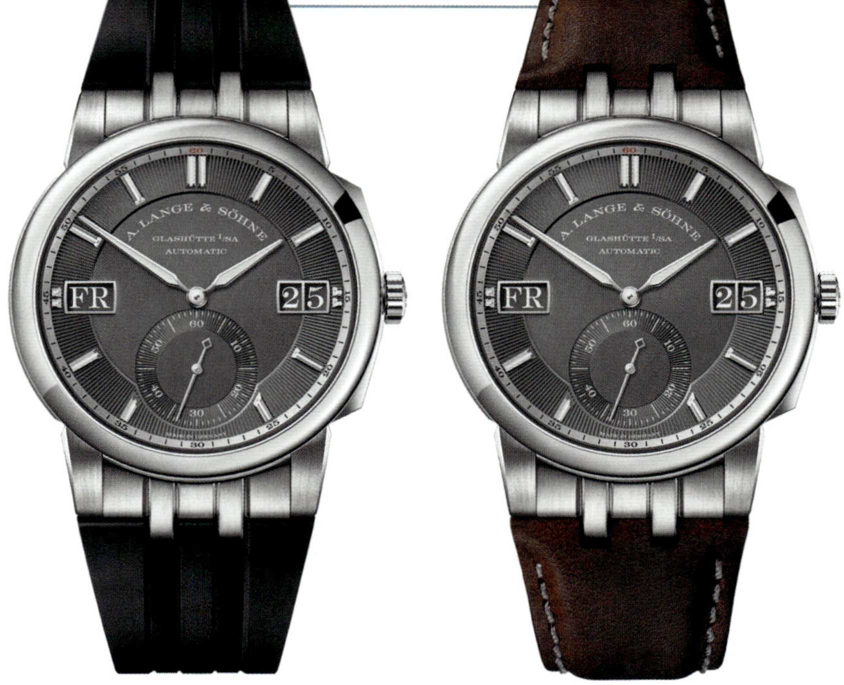
A. LANGE & SÖHNE
GLASHÜTTE I/SA
AUTOMATIC
FR
25
A. LANGE & SÖHNE
GLASHÜTTE I/SA
AUTOMATIC
FR
25

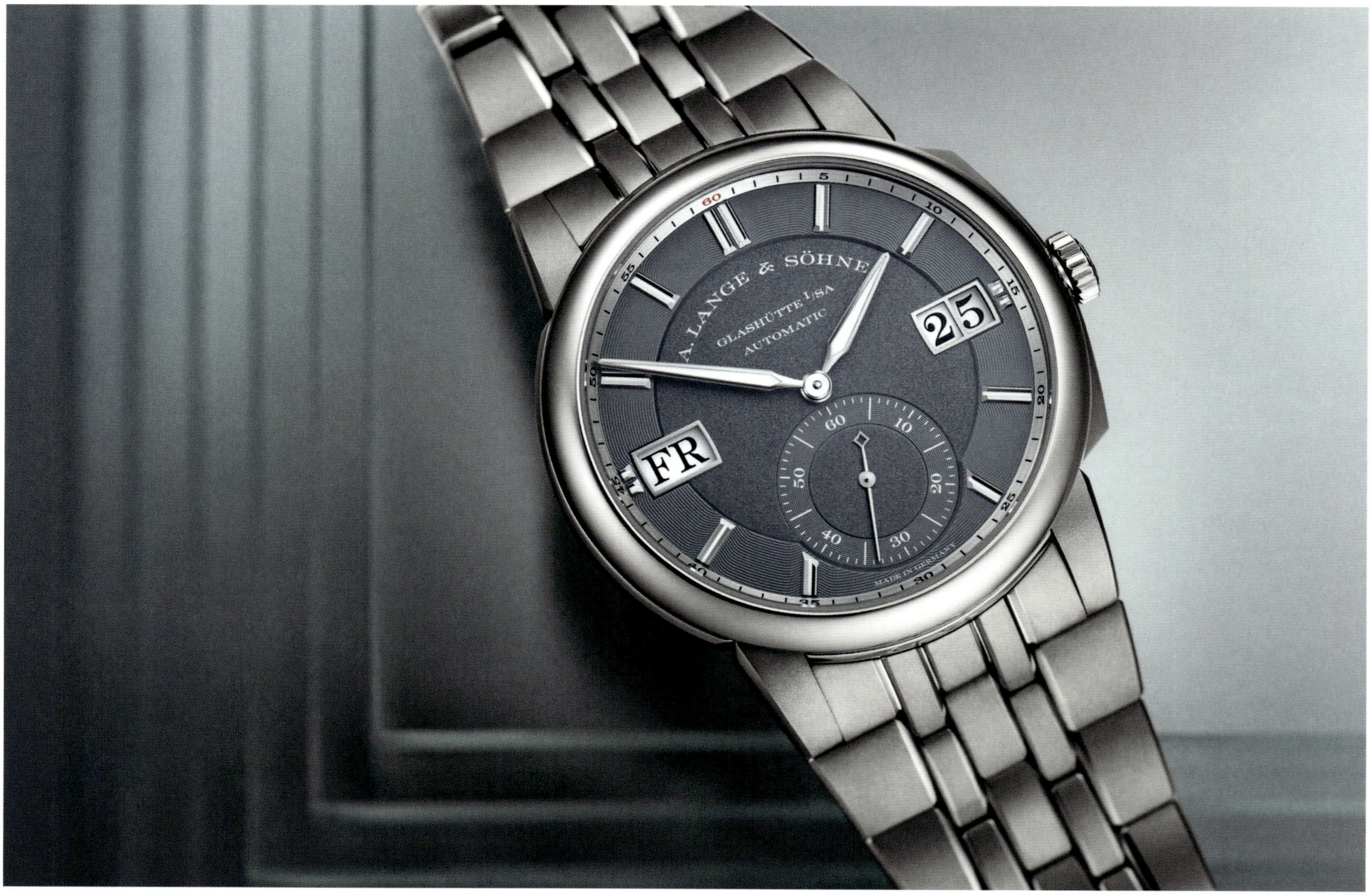
A. LANGE & SÖHNE
GLASHÜTTE I/SA
AUTOMATIC
25
FR
MADE IN GERMANY

A. LANGE & SÖHNE ODYSSEUS

ODYSSEUS CONQUERS TIME

When A. Lange & Söhne rebirthed in October 1994, no one considered base metal as a suitable case material. Exceptions, of course, prove the rule. In 2001, the late Günter Blümlein and then CEO had a very small number of Lange 1 models produced with a steel case. The change of heart did not set in until 2015. It prompted a four-year development of a completely new watch line. Despite the envisioned otherness, the result of the creative and engineering efforts was still to be immediately recognisable as bearing the Lange signature. Shape was not negotiable. Round was a must for the Odysseus, which was waterproof up to 12 bar. Neither did the future-focused angular-shaped push pieces inserted around the crown shake this resolve. On the firstling named after the Greek hero Odysseus, they were used to set and adjust the indications for the day and date located on the left and right of the dial. On the dial, luminescent white-gold hour, minute and second hands ensured legibility in the dark. It goes without saying that the 40.5 mm wide and 11.1 mm thick case had a transparent bottom. Visible behind the sapphire crystal was the proprietary automatic calibre L155.1 Datomatic, consisting of 312 components assembled by experienced watchmakers. However, precious metal could not be dispensed with entirely. A platinum outer ring on the rotor increased winding performance. Initially, the link bracelet produced in Switzerland was detachable. Now it is permanently fixed to the case. From the first, demand for the steel Odysseus, available only in the company's own boutiques, exceeded supply. In 2020, it was joined by a white-gold version with a rubber strap. And in 2022 Lange launched the Odysseus in titanium.
The current pinnacle of horological art is embodied by a chronograph presented in 2023. It comes with a stainless-steel case and bracelet. The new calibre L156.1 with centred rotating minute counter is the first from Lange with a stopwatch and self-winding system. By pulling the crown, the two pushers can be switched from the chronograph function to setting the date and day of the week. The elaborate production, limited to 100 models, is long sold out.

ODYSSEUS FINDET ANS HANDGELENK

Als A. Lange & Söhne im Oktober 1994 neu durchstartete, war unedles Metall als Gehäusematerial grundsätzlich kein Thema. Indessen gab es eine Ausnahme von der Regel. 2001 ließ der inzwischen verstorbene CEO Günter Blümlein ganz wenige Exemplare der „Lange 1" mit stählerner Schale fertigen. Erst 2015 setzte ein Sinneswandel ein. Danach entstand während vier Jahren eine gänzlich neue Uhrenlinie. Trotz der intendierten Andersartigkeit musste das Resultat gestalterischer und uhrmacherischer Bemühungen unverzüglich als Lange identifizierbar sein. Bei der Form gab es keine Kompromisse. Hier kam für die bis zwölf bar wasserdichte „Odysseus" nur ein klassisches Rund in Betracht.
Daran änderten auch die beiden mit Blick auf die Zukunft gewählten Ausbuchtungen ober- und unterhalb der Krone nichts. Beim Erstlingswerk mit dem Namen des griechischen Helden Odysseus dienten sie zum Einstellen und Korrigieren der links und rechts im Zifferblatt positionierten Indikationen für Wochentag und Datum. Vor dem Zifferblatt drehten nachtleuchtende Weißgoldzeiger für Stunden und Minuten sowie ein weiterer zur Anzeige der Sekunden. Selbstredend verfügte die 40,5 Millimeter große und 11,1 Millimeter hoch bauende Schale über einen transparenten Boden. Hinter Saphirglas agierte das eigens konstruierte Automatikkaliber L155.1 „Datomatic", welches einschlägig erfahrene Uhrmacher aus 312 Teilen zusammenfügten. Ganz ohne Edelmetall ging es jedoch auch in diesem Fall nicht. Ein äußeres Platinsegment des Rotors steigerte die Aufzugsleistung. Zu Beginn ließ sich das in der Schweiz produzierte Gliederband abnehmen. Inzwischen ist es fest mit dem Gehäuse verbunden. Von Anbeginn überstieg die Nachfrage nach der nur in eigenen Boutiquen erhältlichen Stahl-„Odysseus" das verfügbare Angebot. 2020 gesellte sich eine Weißgoldversion mit Kautschukband hinzu. Und 2022 lancierte Lange die „Odysseus" in Titan. Den vorläufigen Höhepunkt verkörpert ein 2023 vorgestellter Chronograph. Er findet mit Edelstahlgehäuse und -band ans Handgelenk. Das neue Kaliber L156.1 mit zentral drehendem Minutenzähler ist das erste der Sachsen mit Stoppfunktion und Selbstaufzug. Durch Ziehen der Krone lassen sich die beiden Drücker umschalten von Chronographenfunktion auf Einstellen von Datum und Wochentag. Nach 100 Exemplaren endet die aufwändige und im Grunde genommen längst ausverkaufte Produktion.

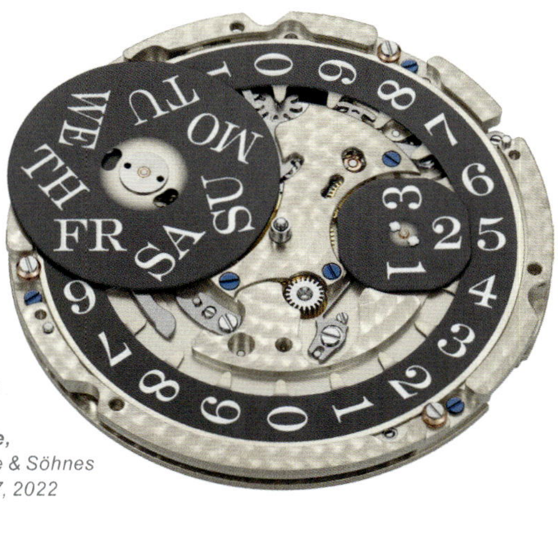

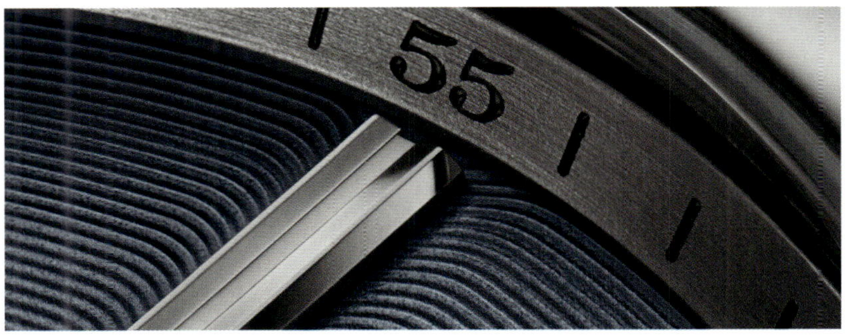

Left page, top left and bottom: *The brand's first titanium watch, Ref. 363.117, 2022* **/** ***Linke Seite, oben links und unten:*** *A. Lange & Söhnes erste Uhr aus Titan, Ref. 363.117, 2022*
Right / Rechts: *Cal. L155.1*

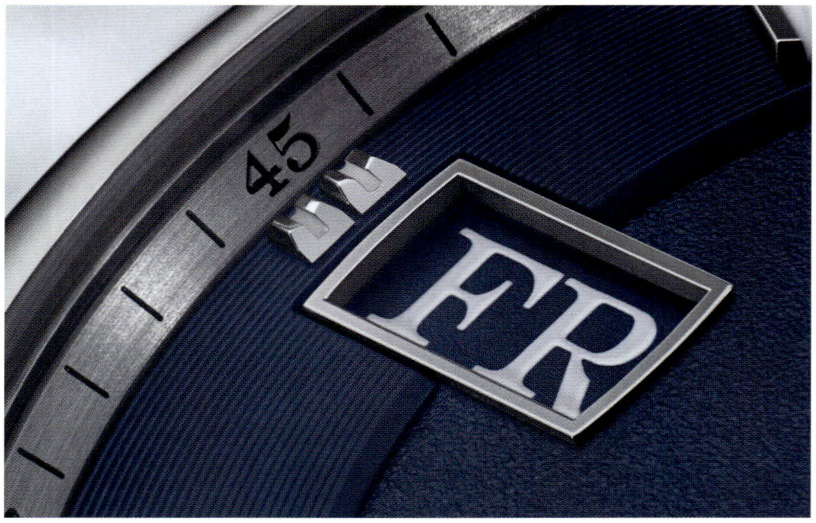

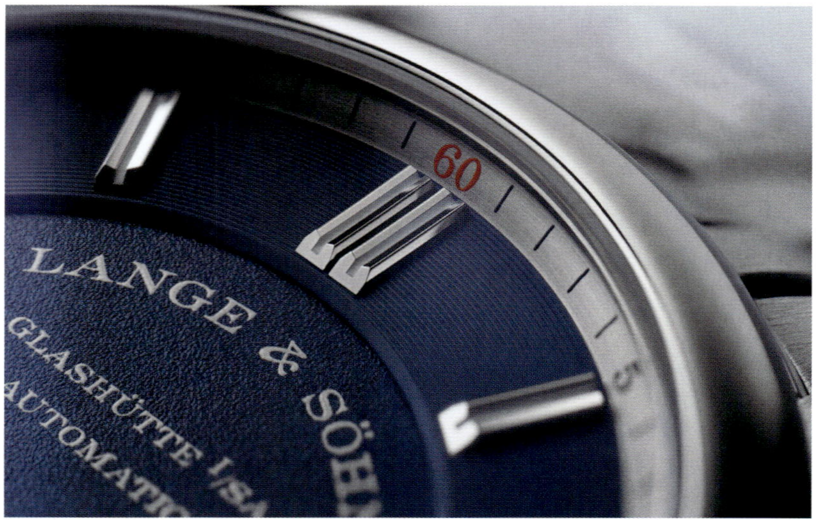

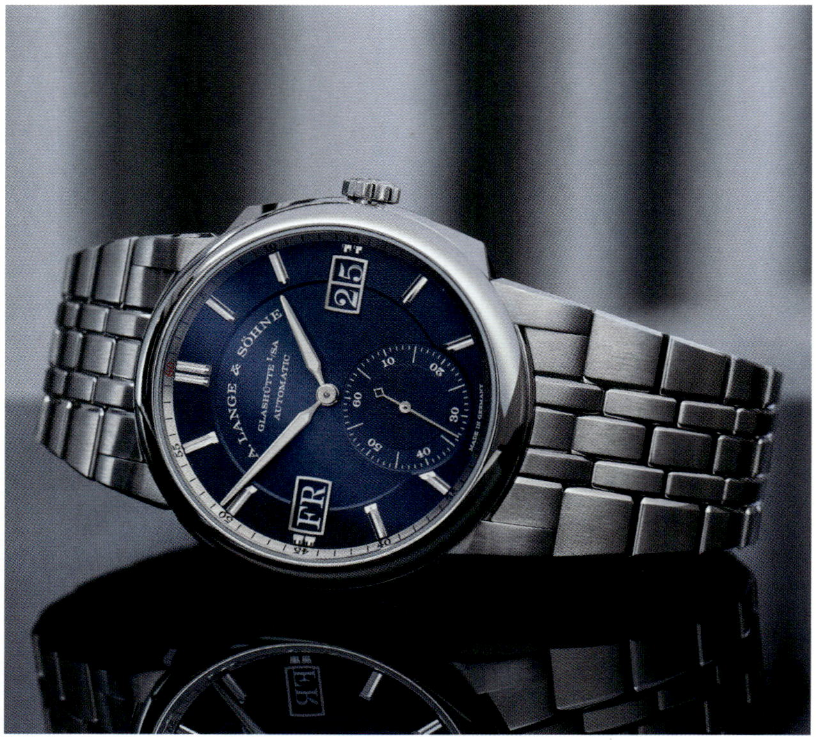

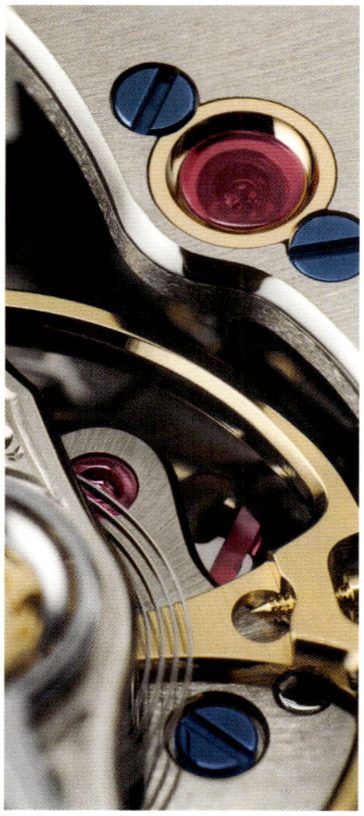

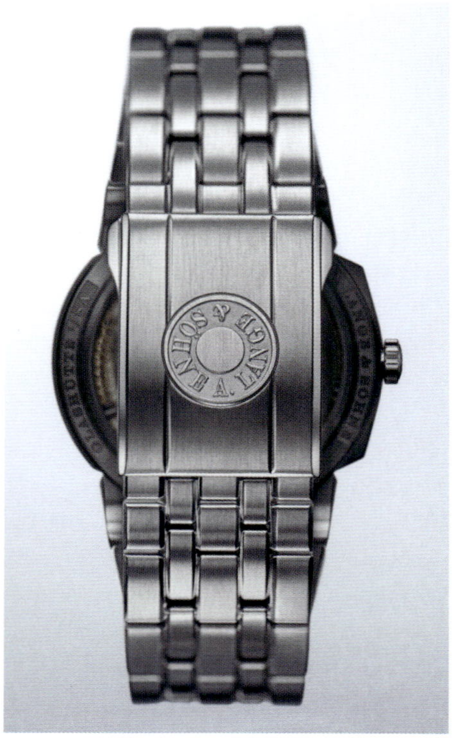

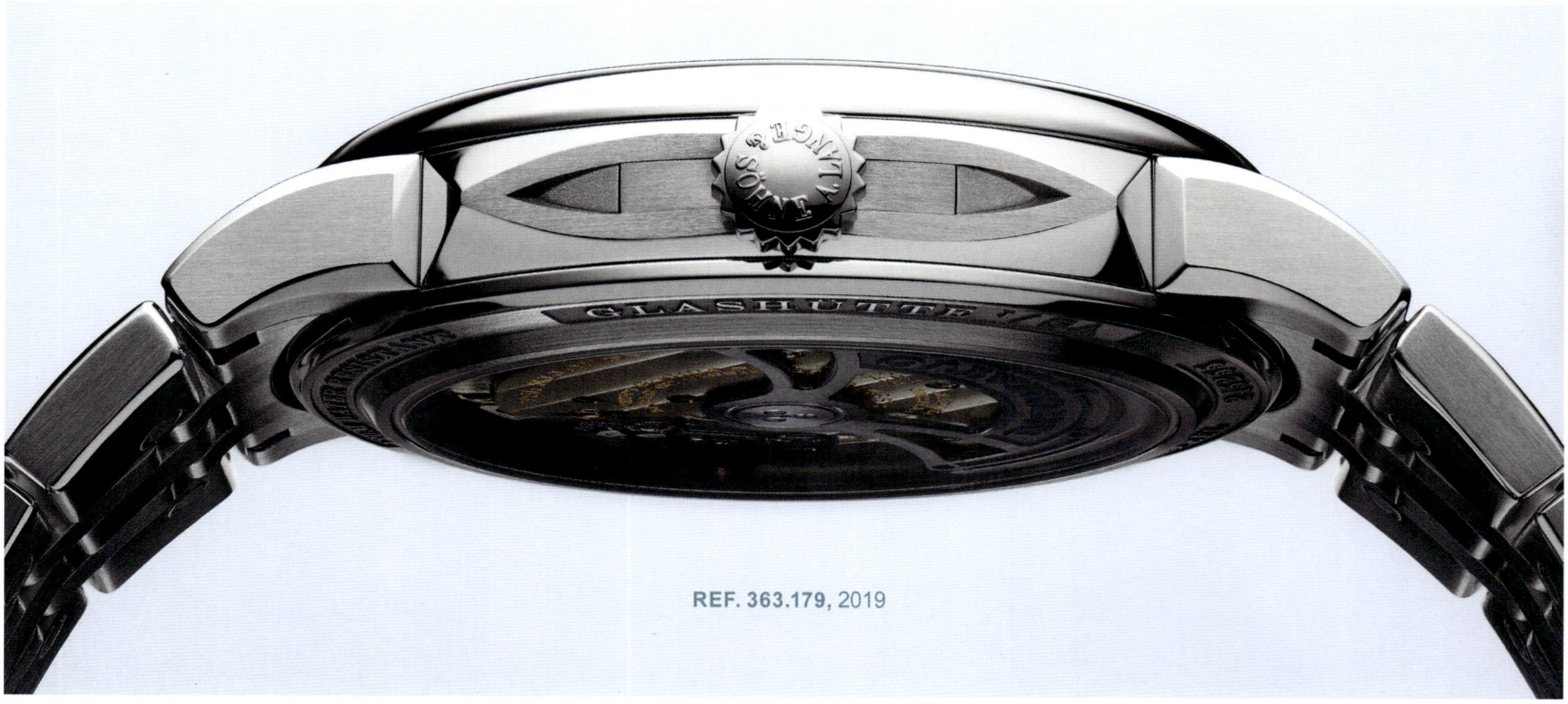

REF. 363.179, 2019

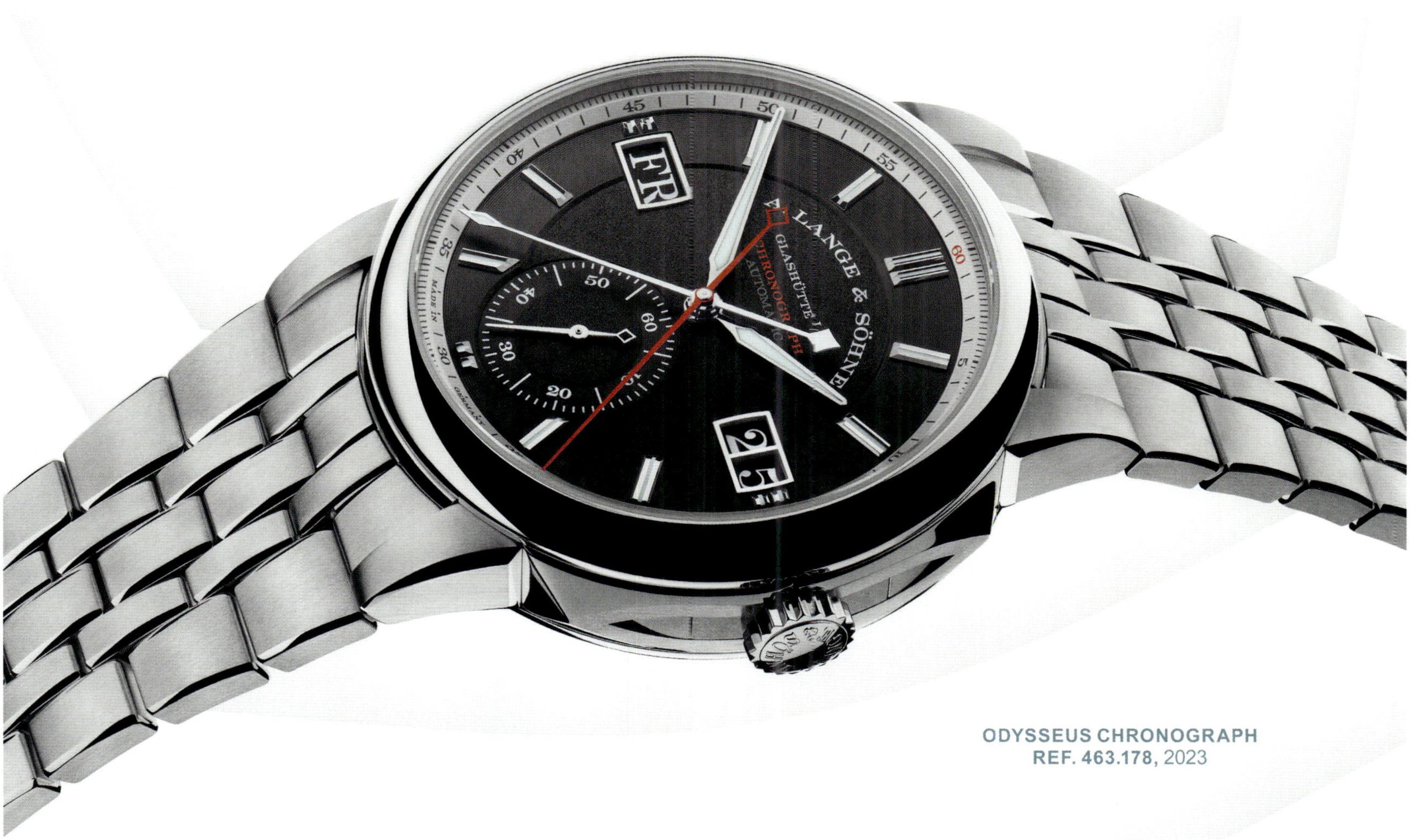

ODYSSEUS CHRONOGRAPH
REF. 463.178, 2023

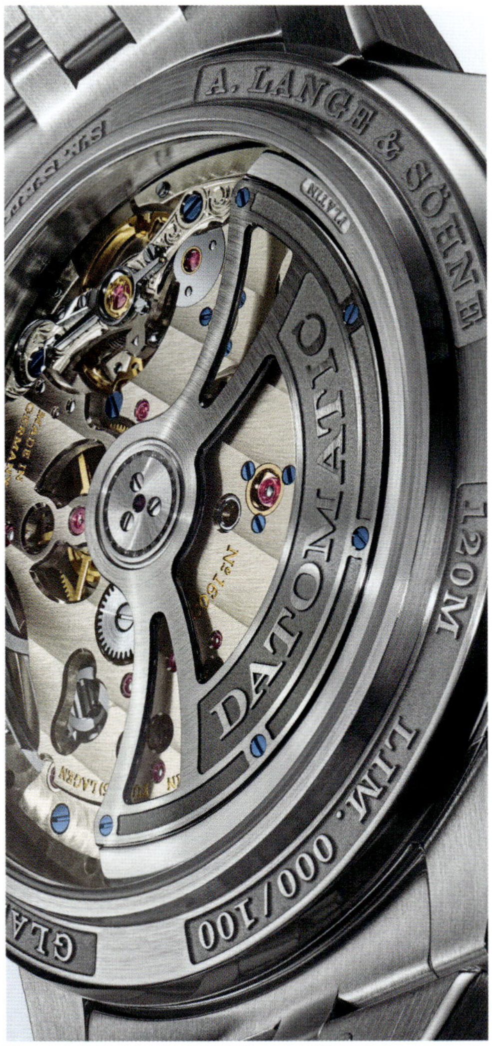

Cal. L156.1

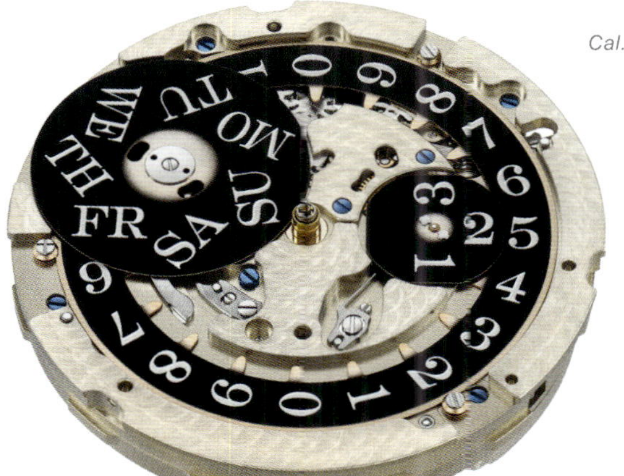

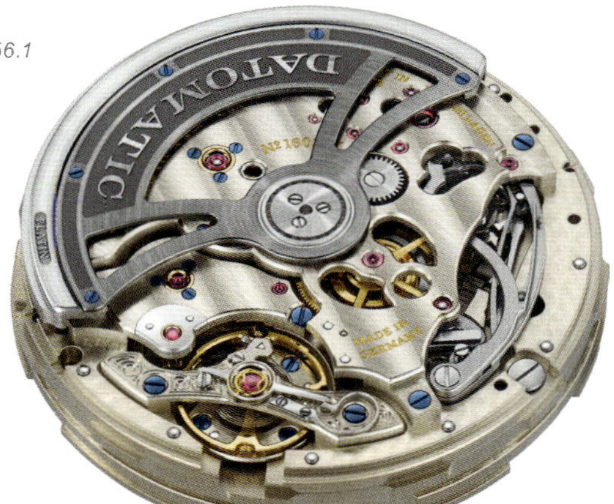

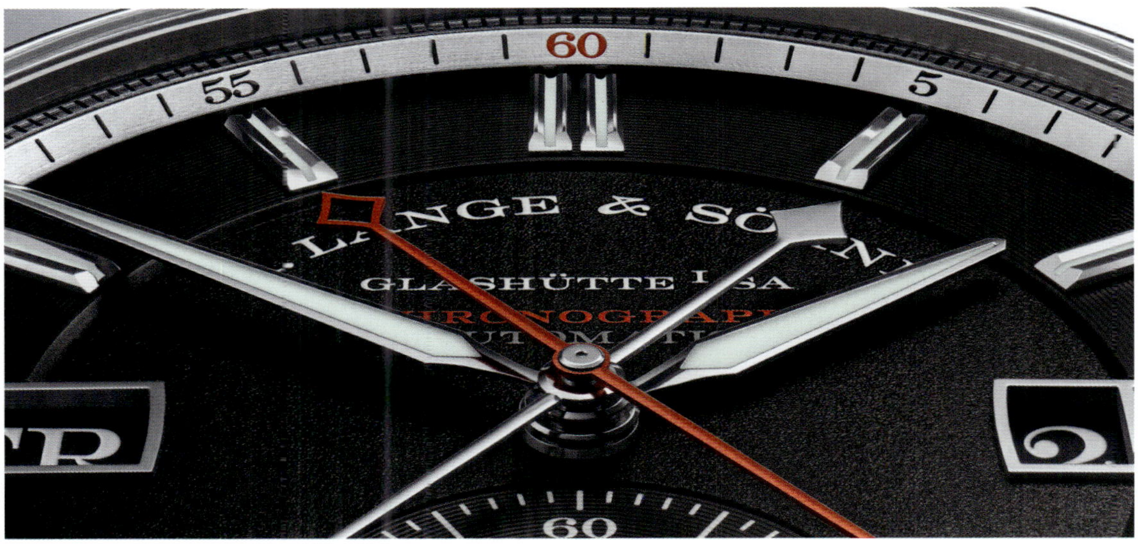

A. LANGE & SÖHNE 2009

ZEITWERK

LEAP TO FAME

At first glance of this watch, the name A. Lange & Söhne does not immediately spring to mind. Since its inception, the traditional Saxon watchmaker has focused on an analogue time display. This timepiece must have drawn gasps of surprise from many a fan of the brand. In 2009, the opulent Zeitwerk made an unexpected leap, ringing in a new era on several fronts all at once.

A significant expression of the new departure is embodied by the wide time bridge in German silver spanning the dial. Its left pillar provides a worthy frame for the digits for the large hour display. Its counterpart on the right showcases the minutes, which are also displayed in digits. Positioned at 6 o'clock is the indexed disc, on which the second hand rotates. At the top, a power reserve indicator signals when the proprietary – what else – calibre L043.1 would like to be manually replenished with fresh energy. This is the case after 36 hours at the latest. The 33.6-millimetre mechanical masterpiece consists of 388 components, including 66 functional jewels. Its regulator with in-house balance spring oscillates at a subdued rate of 2.5 hertz. As you would expect, the most remarkable innovation is constituted by the large-format digital displays for hours and minutes. For a mechanical movement, which requires only one billionth of an HP for its rotations, moving these indications requires a huge amount of effort. Impossible to achieve directly with conventional cycles, the engineers opted for a large twin barrel and the proven technique of a constant-force escapement. The additional mechanism patented for A. Lange & Söhne manages the power bursts required for the quick acceleration of the digital discs, which are lavishly sized to ensure optimum legibility. In the space of one minute, a small spiral spring accumulates so much force that the jumps are made at lightning speed. An air blade ensures a gentle shift. High noon is displayed in widescreen format one minute after 11:59. On this stroke, the Zeitwerk executes three small jumps. For A. Lange & Söhne of course they represent immense leaps. The version with a minute repeater displays even greater complexity. An impressive performance is rendered by the 2010 Zeitwerk Luminous. Its unique digital displays glow in the dark.

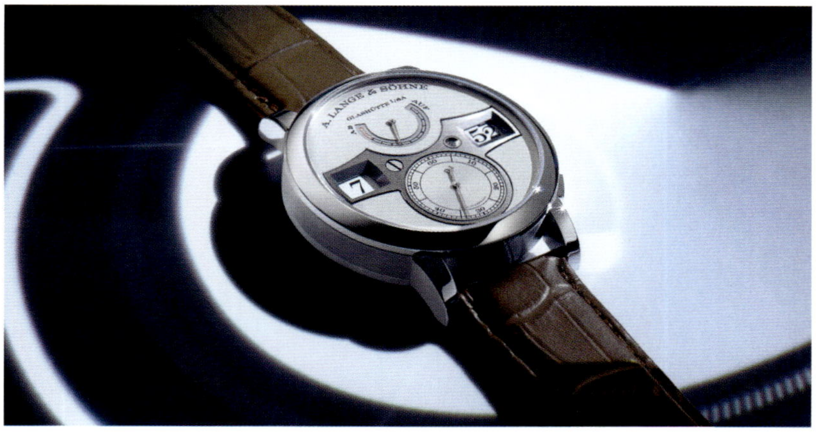

SPRUNGHAFTES WESEN

Auf den ersten Blick bringt man diese Armbanduhr gar nicht mit A. Lange & Söhne in Verbindung. Bei der sächsischen Traditionsmanufaktur gibt die analoge Zeitanzeige seit jeher den Ton an. Viele Markenfans hätten auch niemals einen derartigen Zeitmesser erwartet. 2009 belehrte sie das opulente „Zeitwerk" eines Besseren. Und es läutete gleich in mehrfacher Hinsicht eine neue Epoche ein.

Ein signifikantes Signal des Aufbruchs verkörpert die breite, aus Neusilber gefertigte Zeitbrücke über die gesamte Breite des Zifferblatts. Ihr linker Pfeiler bietet den Ziffern für die große digitale Stundenanzeige einen würdigen Rahmen. Sein rechtes Pendant umrahmt die ebenfalls digital dargestellten Minuten. Und bei „6" umfängt sie die indexierte Scheibe, vor der ein Sekundenzeiger seine Runden dreht.

Geographisch „im Norden" signalisiert eine Gangreserveanzeige, wann das selbstverständlich hauseigene Kaliber L043.1 manuell mit frischer Energie versorgt werden möchte. Spätestens nach 36 Stunden ist das der Fall. Das 33,6 Millimeter große Mechanik-Œuvre besteht aus 388 Komponenten, darunter 66 funktionale Steine. Sein Gangregler mit selbst produzierter Unruhspirale oszilliert mit entschleunigten 2,5 Hertz. Die markanteste Neuerung ist logischerweise in den großformatigen digitalen Anzeigen für Stunden und Minuten zu sehen. Für ein mechanisches Uhrwerk, welches für seine Rotationen nur ein Milliardstel PS benötigt, bedeutet das Fortbewegen dieser Indikationen einen beträchtlichen Kraftakt. Weil regelmäßige Schaltvorgänge auf direktem Wege unmöglich wären, setzen die Techniker auf ein großes Doppelfederhaus und die bewährte Technik eines sogenannten Nachspannwerks. Die für A. Lange & Söhne patentierte Zusatzmechanik bewerkstelligt die Kraftflüsse zum kurzfristigen Beschleunigen der für optimale Ablesbarkeit üppig dimensionierten Ziffernscheiben. Während einer Minute akkumuliert eine kleine Spiralfeder so viel Kraft, dass die Sprünge blitzschnell erfolgen. Ein Windflügel sorgt für sanftes Schalten. „High Noon" im Breitbildformat gibt es eine Minute nach 11:59 Uhr. Dann vollzieht das „Zeitwerk" gleich drei kleine Sprünge, die für A. Lange & Söhne allerdings immens große Schritte bedeuten. Noch weitaus höhere Komplexität bietet die Version mit Minutenrepetition. Und beeindruckende Performance bei Dunkelheit gewährleistet das 2010 vorgestellte „Zeitwerk Luminous" mit seinen einzigartigen nachtleuchtenden Digitalanzeigen.

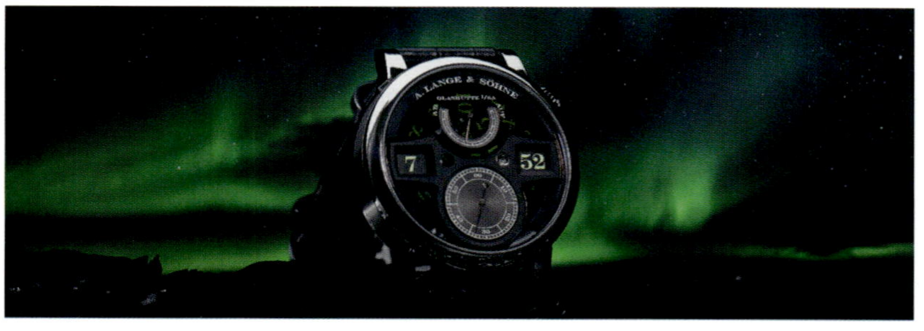

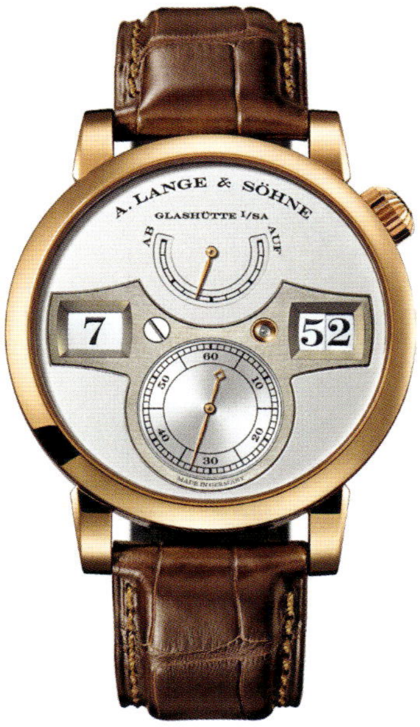

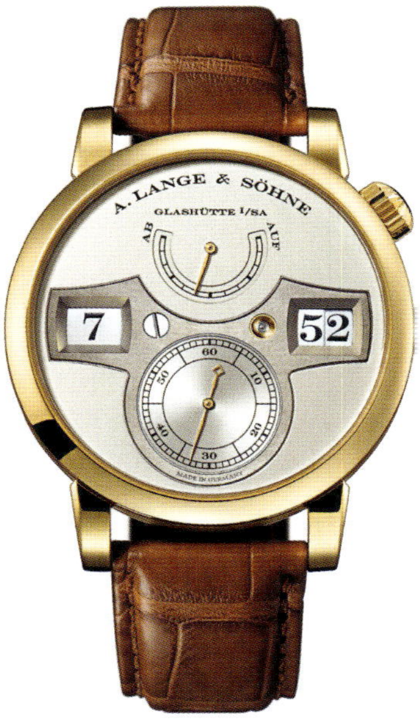

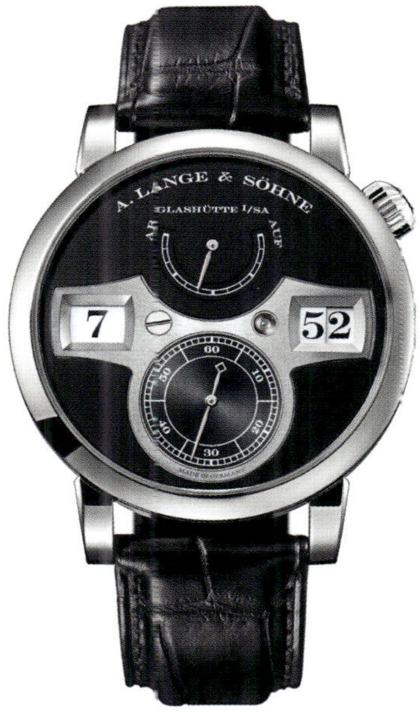

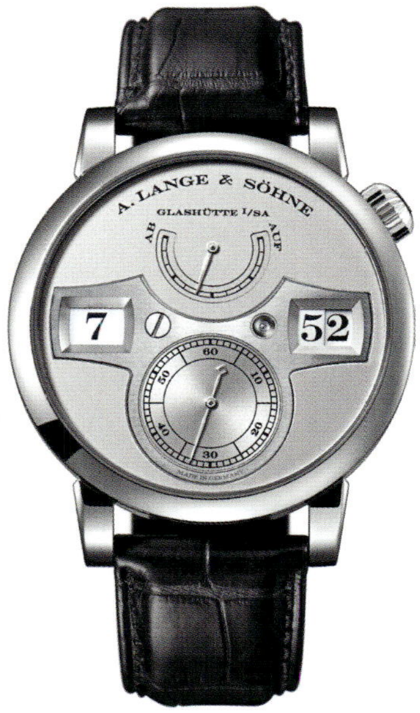

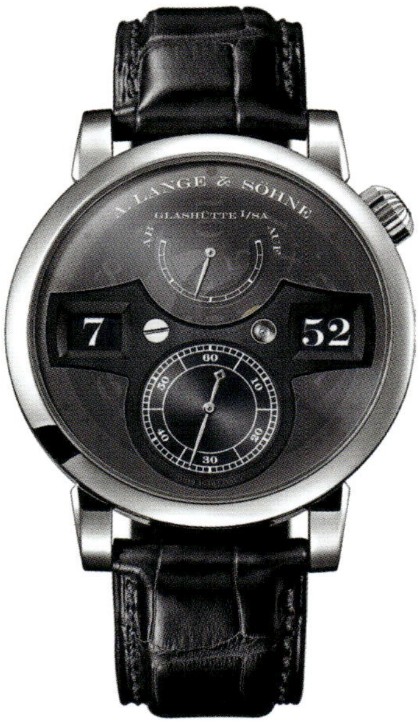

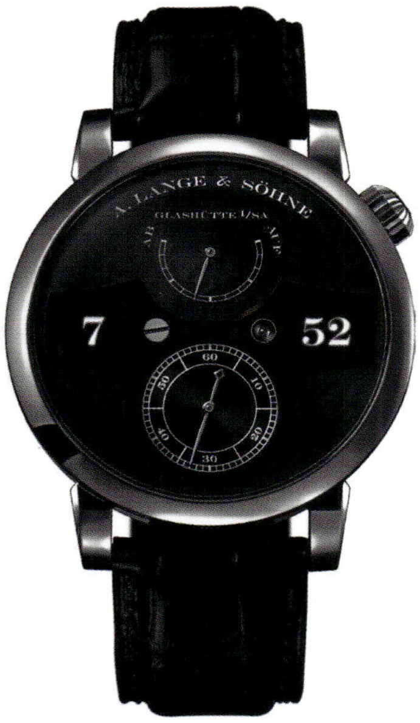

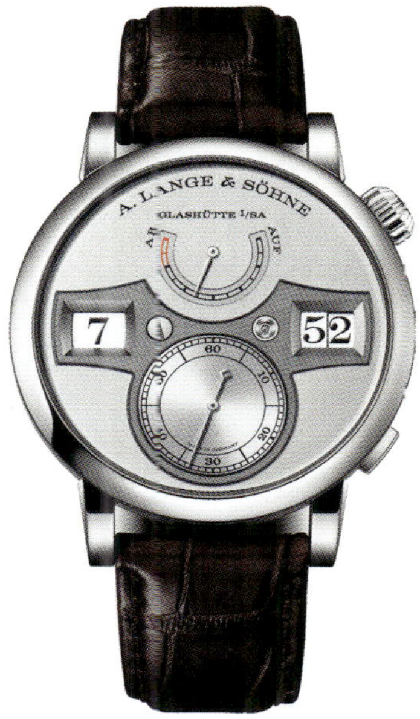

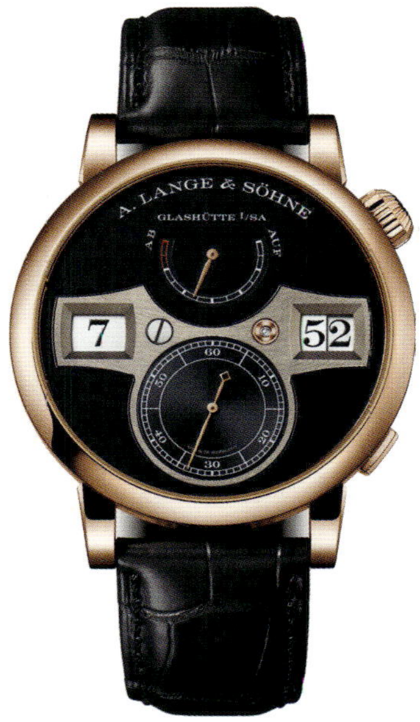

Cal. L043.1, 2009

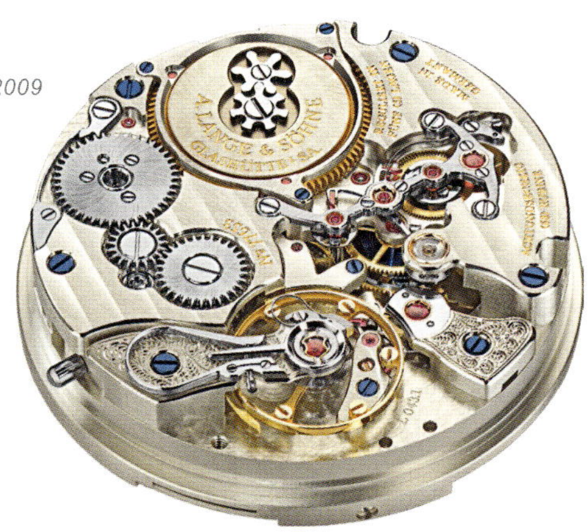

Cal. L043.6, 2022

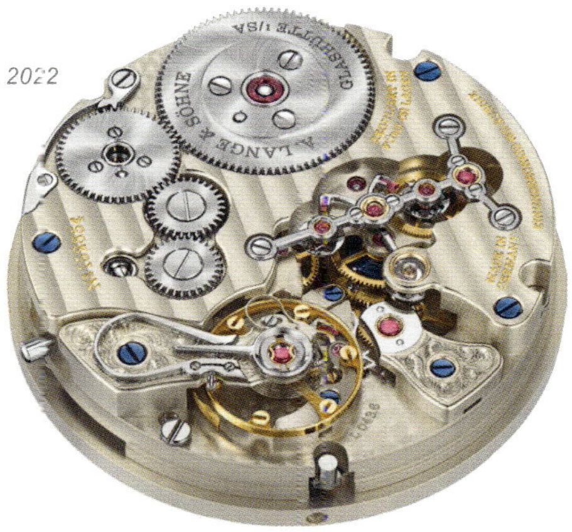

Top, from left | Oben von links: *Ref. 140.032 | Ref. 140.021 | Ref. 140.029 | Ref. 140.025*

Middle, from left | Mitte von links: *Ref. 140.035 Zeitwerk "Luminous", 2010 | Ref. 140.035 Zeitwerk "Luminous", 2010, nightview | Nachtansicht | Ref. 142.025, 2022 | Ref. 142.031, 2022*

LANG & HEYNE 2017
GEORG

GEORG FROM DRESDEN

The outstanding is rare. Connoisseurs know this only too well. When it comes to watches, this realisation applies to Lang & Heyne in a very special way. In Dresden, the august capital of Saxony, fewer than 200 watches are produced every year, and they are usually all sold in advance. Because even before placing their order, the select international clientele knows that the manufactory in Radeberg near Dresden only crafts proprietary timepieces that satisfy the highest horological standards. The mechanical movements are hand-crafted in small workshops characterised by a homely atmosphere, true to the motto that the whole is far more than the sum of its many exquisitely finished parts. Most of the models are traditionally round in shape, and are named after scions of the Saxon dynasties. The exception, strange to say, is the brand icon, which is resolutely rectangular. The Georg, featuring the Calibre VIII enjoys increasing popularity among lovers of mechanical watches. This hand-wound rectangular movement combines tradition and avant-garde design to form a perfect synthesis. All components interact on a rigorously rectangular mainplate. Metaphorically speaking, they are showcased in action on a serving plate. Measuring 34 by 26.5 millimetres, this masterly timepiece can be deservedly dubbed a feast for the eyes with its beautiful contours, outstanding finishing of the balance bridge and the four parallel polished steel cocks that hold the gear train. The energy stored in the large barrel enables the watch to run for a reassuring 55 hours without having to be rewound. This figure is purely theoretical, of course. The butter-soft rotation of the crown is so irresistible that owners will be tempted to wind their watch far more often than is actually necessary. Characterised by refined beauty and exquisite engineering, the Georg is available in platinum, rose gold and stainless steel. Signature features include the oversized small seconds hand and triple lug for attaching the leather strap.

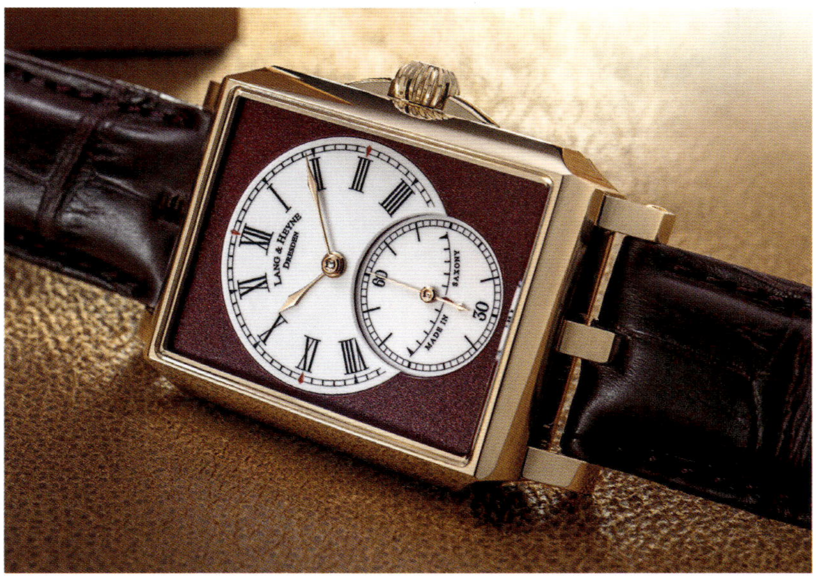

GEORG AUS DRESDEN

Alles Vorzügliche ist rar. Kenner wissen das nur zu genau. Wenn sich die Dinge um Armbanduhren drehen, trifft diese Erkenntnis auf Lang & Heyne in ganz besonderer Weise zu. Per annum entstehen in der illustren sächsischen Landeshauptstadt weniger als 200 Exemplare. Und die sind in aller Regel bereits vorab verkauft. Schon vor ihrer Bestellung ist sich der erlesene internationale Kundenkreis der Tatsache bewusst, dass nur eigens entwickelte Zeitmesser auf höchstem handwerklichem Niveau die Manufaktur in Radeberg bei Dresden verlassen. Ausnahmslos alle mechanischen Uhrwerke entstehen in kleinen Ateliers und familiärer Atmosphäre weitestgehend in aufwändiger Handarbeit. Getreu dem Motto, dass das Ganze weitaus mehr ist als die Summe seiner vielen, sorgfältig finissierten Teile. Die allermeisten Modelle sind rund wie zu alter Väter Zeiten. Alle tragen Namen von Mitgliedern des sächsischen Herrscherhauses der Wettiner. Aber ausgerechnet die markante Ikone namens „Georg“ tritt streng rechteckig in Erscheinung. Immer mehr Mechanikvoyeure haben ihre helle Freude am darin verbauten Caliber VIII. Bei diesem Handaufzugsformwerk finden Tradition und avantgardistische Gestaltungselemente zu perfekter Synthese. Alle Bauteile wirken auf einer streng rechteckigen Werkplatte zusammen. Im übertragenen Sinne agieren sie auf einer Art Präsentierteller. Wegen der formalen Ästhetik und dem unübertrefflichen Grad an Feinbearbeitung bei der Unruhbrücke und den vier parallel angeordneten Räderwerkskloben aus poliertem Stahl kann das 34 x 26,5 Millimeter messende Œuvre zweifellos als Augenweide gelten. Die im großen Federhaus gespeicherte Kraft lässt den Mikrokosmos beruhigende 55 Stunden lang ohne weiteren Energienachschub ticken. Freilich ist dieser Wert nur ein rein theoretischer. Der butterweich funktionierende Handaufzug ist derart verführerisch, dass die Besitzerinnen und Besitzer viel öfter als nötig für einige Drehungen zur Krone greifen. Die von feiner Ästhetik und hoher Sachlichkeit geprägte „Georg“ ist in Edelstahl, Platin und Roségold erhältlich. Charakteristisch sind der sehr große kleine Sekundenzeiger und jeweils drei Hörner zur Befestigung des Lederbands.

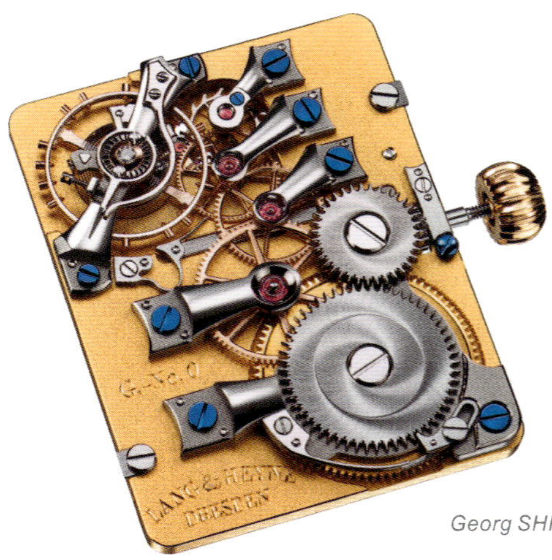

Left / Links:
Georg SHH Edition II "One of 12", 2023
Right / Rechts: *Cal. VIII*

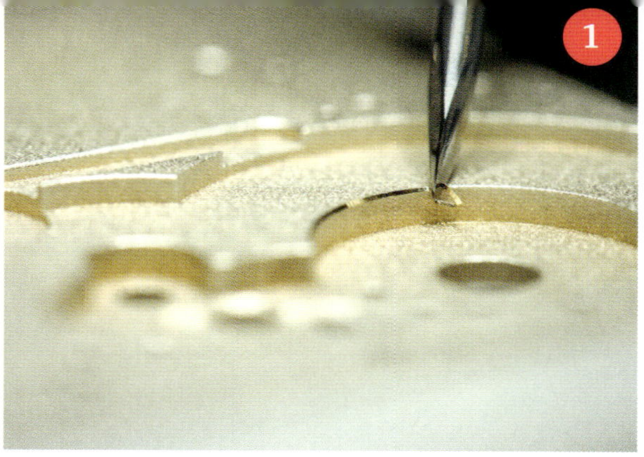

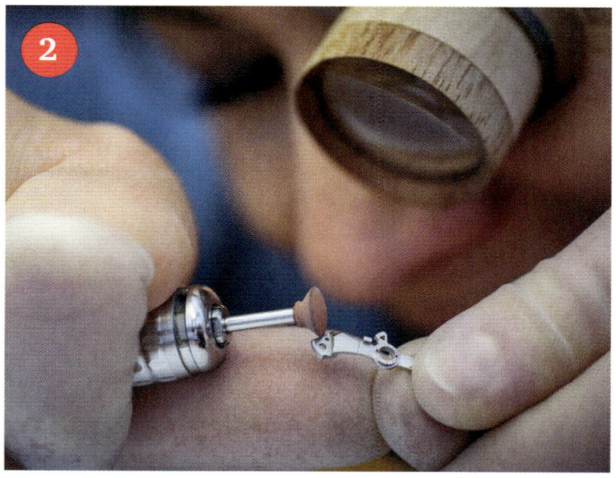

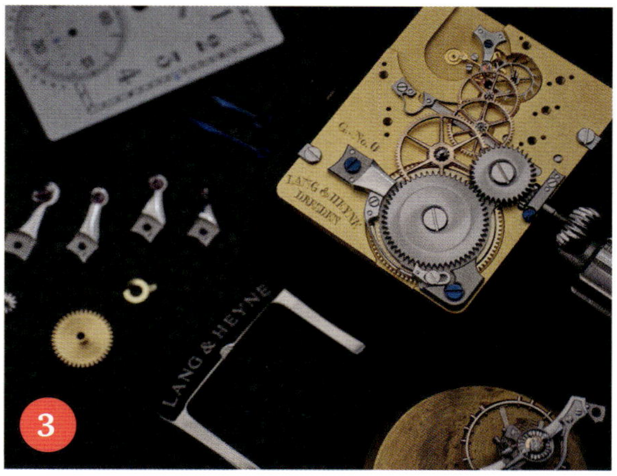

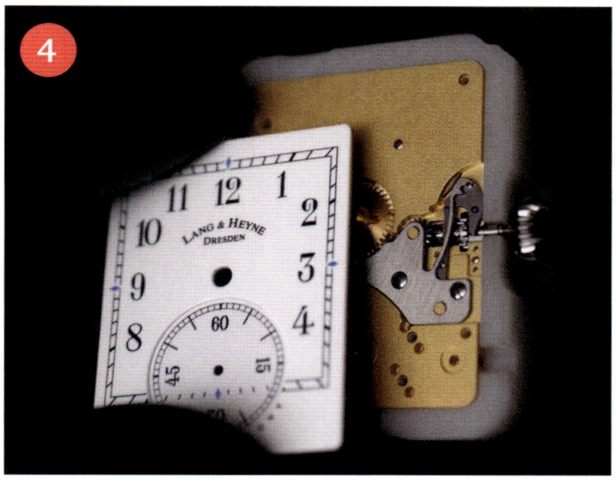

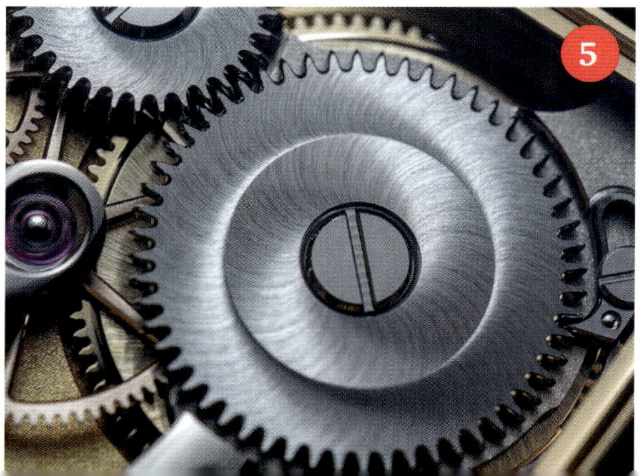

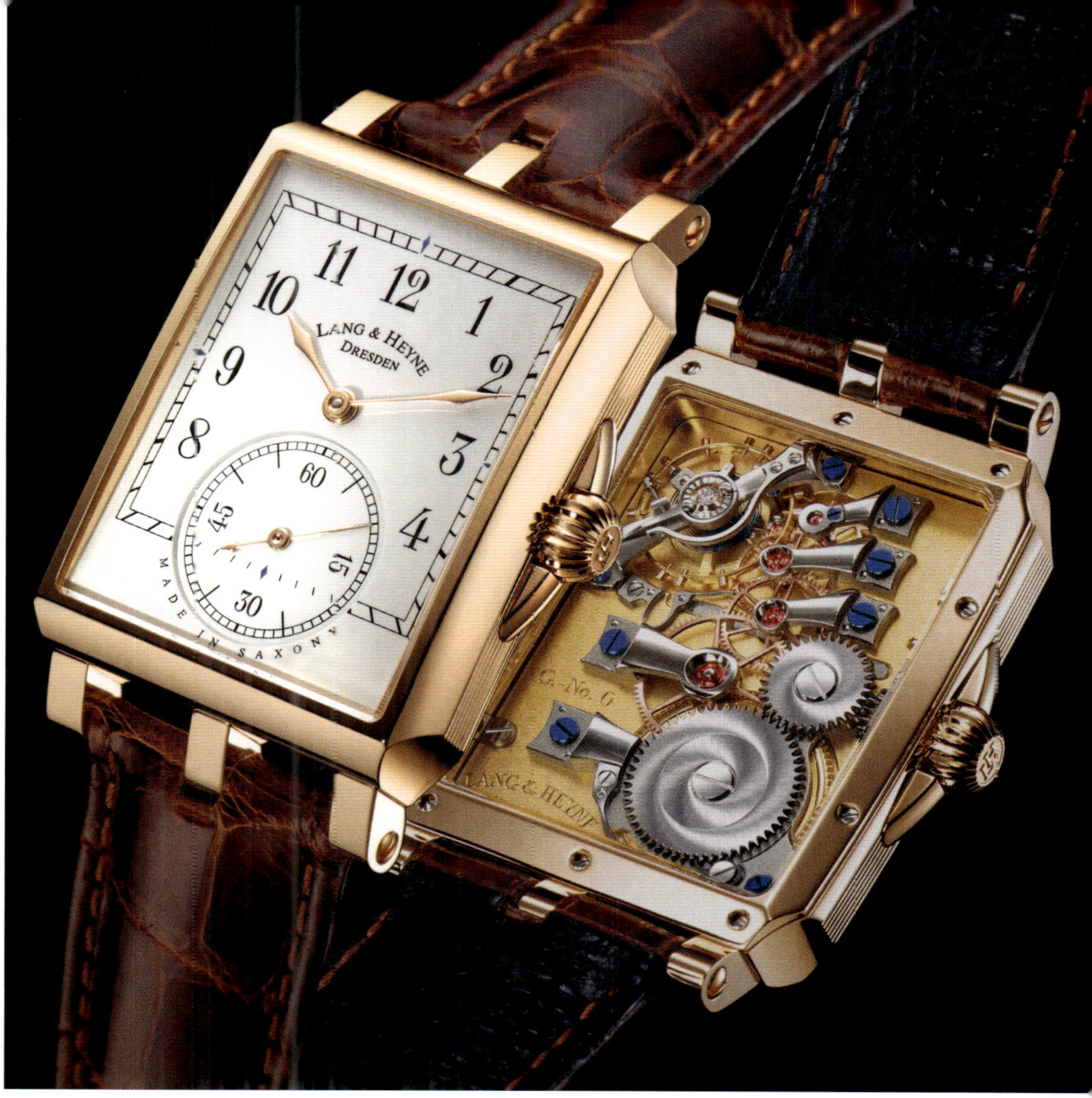

1 *Chamfering of plate | Anglieren der Platine*
2 *Finishing of the balance bridge | Klobenfinish*
3 *Components | Komponenten*
4 *Assembly | Montage*
5 *Detail of caseback | Detailaufnahme der Rückseite*

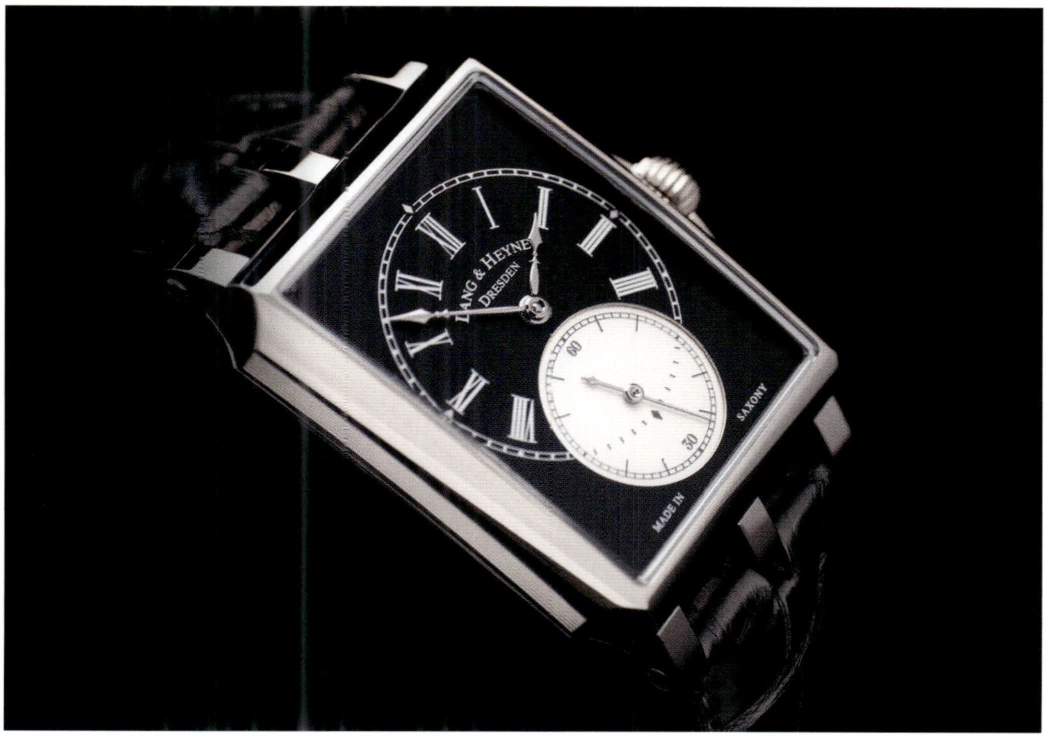

The Georg "Panda" is just one of many iterations that customers can choose from at Lang & Heyne. | Die Georg „Panda" ist nur eine von vielen Varianten, aus denen Kunden bei Lang & Heyne wählen können.

LONGINES 1932

LINDBERGH

MORE THAN A NOD TO CHARLES LINDBERGH

On 20 May 1927, Charles Lindbergh took off from an airfield near New York with his single-engine monoplane called *Spirit of St. Louis*. When he landed at Le Bourget near Paris 33 hours and 30 minutes later, he was not only the richer by a substantial $25,000, but also a deal of experience in long-distance aviation. After all, the American aviation pioneer did not have the benefit of electronics for his daring adventure on high. Lindbergh captured his insights on paper in the form of sketches. He entrusted Longines, the experienced watch brand in Saint-Imier, with the task of putting his ideas into practice. When the Lindbergh wristwatch with hand-wound Calibre 37.9 was launched in 1932, its looks and functions caused quite a stir. The large model with a diameter of 47.5 millimetres could be worn over the aviator's attire and reliably operated even wearing gloves. A sophisticated system of adjustable dials and rotating bezel henceforth provided long-distance pilots with valuable assistance in determining longitude. This eye-catching Hour Angle watch was equipped with a rotating seconds dial. This enabled pilots to constantly track the degree of rate deviations and perform the corresponding adjustments during the flight. Prior to take-off, the daily changing difference between the mean and apparent solar time could be set to the minute with the help of the rotating bezel.

Nowadays, of course, none of this is necessary. Given its romantic history, the Lindbergh continues to evoke the spirit of adventure. Anyone lucky enough to own an original from the 1930s can congratulate themselves. Those who do not will find consolation in the current collection. Fans of bygone days and aviation will discover the Heritage Avigation The Lindbergh Hour Angle Watch with a 47.5-millimetre steel case and the lavish automatic Calibre L699 based on the Eta A07.L01

CHARLES LINDBERGH LÄSST GRÜSSEN

Am 20. Mai 1927 startete Charles Lindbergh mit seiner einmotorigen „Spirit of St. Louis“ von einem Flugfeld nahe New York. Als er 33 Stunden und 30 Minuten später in Le Bourget bei Paris landete, war er nicht nur um stolze 25.000 Dollar reicher, sondern auch um sehr viel Erfahrung in der Langstreckenaviatik. Auf Elektronik konnte der amerikanische Flugpionier bei seinem gewagten Trip durch die Lüfte nämlich nicht bauen. Seine Erkenntnisse brachte Lindbergh in Gestalt von Skizzen zu Papier. Von Longines, der erfahrenen Uhrenmarke in Saint-Imier, erwartete er sich die kompetente Umsetzung seiner Ideen. Als die „Lindbergh“-Armbanduhr mit dem Handaufzugskaliber 37.9 im Jahr 1932 ans Licht der Öffentlichkeit trat, sorgte sie wegen ihrer Optik und Funktionalität für einiges Aufsehen. Das große Modell mit einem Durchmesser von 47,5 Millimetern ließ sich über der Montur tragen und selbst mit Handschuhen zuverlässig bedienen. Ein ausgeklügeltes System von einstellbaren Zifferblättern und Drehlünette leistete Langstreckenpiloten fortan wertvolle Hilfe bei der Längengradbestimmung. Beispielsweise besaß diese augenfällige Stundenwinkel-Armbanduhr ein drehbares Sekundenzifferblatt. Selbiges bot den Vorteil, dass Piloten den Grad der Gangabweichungen stets nachvollziehen und in der Luft Nachregulierungen vornehmen konnten. Die tägliche wechselnde Differenz zwischen der mittleren und der wahren Sonnenzeit ließ sich vor Abflug minutengenau mit Hilfe der Drehlünette einstellen.

All das braucht es heutzutage natürlich nicht mehr. Aber infolge ihrer Geschichte ist die „Lindbergh“ weiterhin ein echtes Statement am Handgelenk. Glücklich schätzen kann sich, wer ein Original aus den 1930er-Jahren besitzt. Fündig wird man aber auch in der aktuellen Kollektion. Nostalgiker und Flugbegeisterte entdecken dort beispielsweise die „Heritage Avigation The Lindbergh Hour Angle Watch“ mit 47,5 Millimeter großem Stahlgehäuse und dem opulenten Automatikkaliber L699 auf Basis des Eta A07.L01.

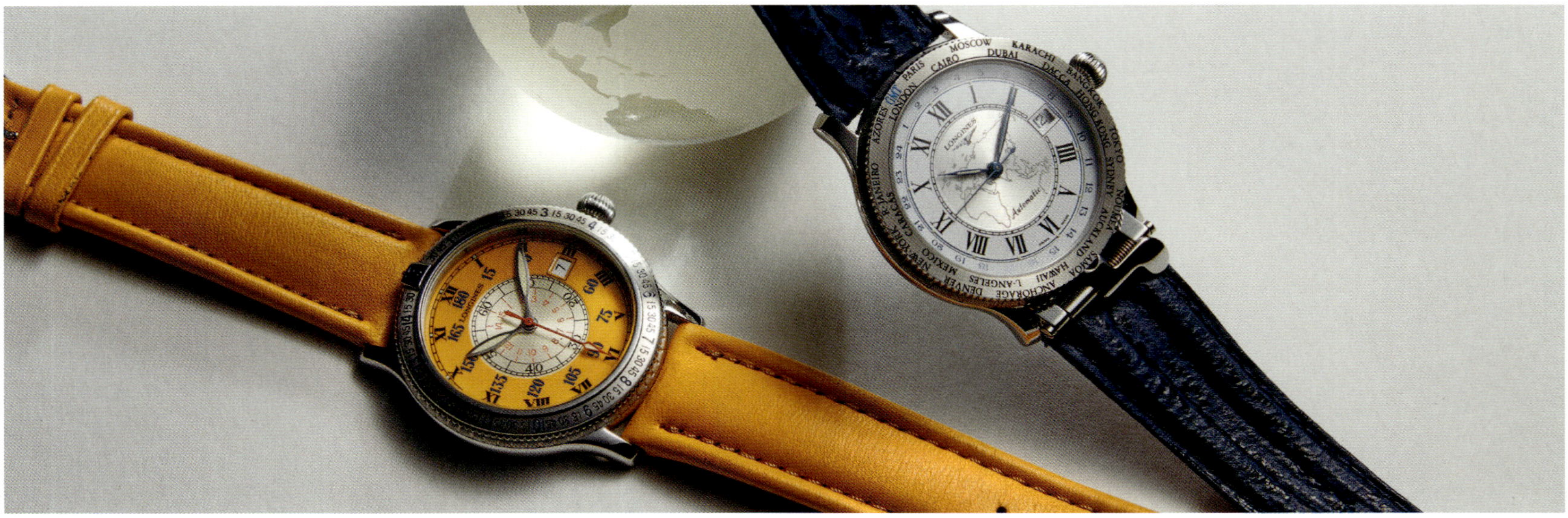

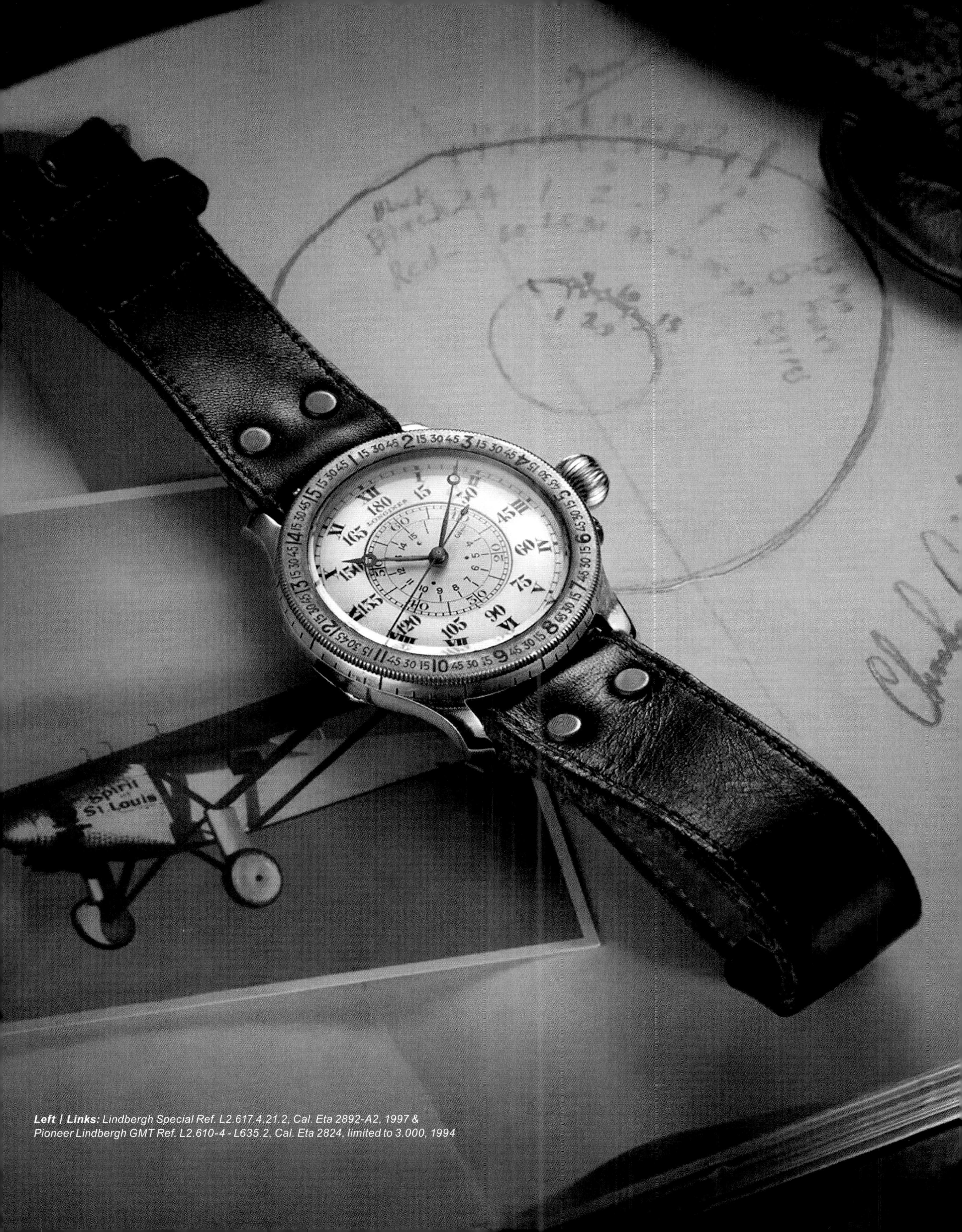

***Left | Links:** Lindbergh Special Ref. L2.617.4.21.2, Cal. Eta 2892-A2, 1997 &*
Pioneer Lindbergh GMT Ref. L2.610-4 - L635.2, Cal. Eta 2824, limited to 3.000, 1994

AIKON

FROM CALYPSO TO AIKON

On hearing the name AIKON people tend to think of the word "icon". And that is exactly what Maurice Lacroix intended when the company chose it for a new watch line. It was presented at the 2016 Basel Watch Fair. The newcomer was based on its highly successful predecessor, the 1990s Calypso. The watch line established the reputation of the business, which became a member of the Zurich DKSH Group in 2011. Since the somewhat baroque style of the Calypso had become a fashion relic, Maurice Lacroix's management entrusted it to expert hands for a makeover, the upshot of which was the AIKON. The step into the future, marked in 2016 by the launch of quartz movements, was so promising that in 2018 the next stage in the line's evolution focused on mechanical inner workings. In the early years, the product designers skirted around the line's eponymous features such as stainless steel, a six-claw bezel and five-link bracelet. These elements promoted the creation of bicolour models. They have long delighted the clientele of the Swiss company founded in 1975. Visually striking are the angular contours above and below the dial. They can also be found in the iterations with a rubber strap and in the bright affordable models with upcycled plastic case. Top-of-the-range models include those with a chronograph or skeletonised manufacture inner workings. The openwork automatic Calibre ML 234 with a classic 2.5 hertz balance frequency has a power reserve of 52 hours and a small second hand at 6 o'clock.

VON CALYPSO ZU AIKON

Wer den Namen „AIKON" hört, denkt beinahe zwangsläufig an „Icon" oder auf gut Deutsch „Ikone". Genau das war die Intention von Maurice Lacroix bei der Taufe einer neuen Uhrenlinie. Deren Vorstellung erfolgte 2016 während der Basler Uhrenmesse. Markenintern knüpfte der Newcomer an die in den 1990er-Jahren höchst erfolgreiche „Calypso" an. Diese Modellfamilie begründete den Ruf des Unternehmens, welches sich seit 2011 unter dem Dach der Zürcher DKSH-Gruppe befindet. Nachdem die inzwischen etwas barock anmutende „Calypso" sozusagen aus der Zeit gefallen war, ließen sie die Verantwortlichen im Hause Maurice Lacroix von kundiger Hand zu besagter „AIKON" weiterentwickeln. Der im genannten Jahr 2016 mit Quarzwerken gestartete Schritt in die Zukunft verlief so vielversprechend, dass schon 2018 beinahe zwangsläufig eine Evolutionsstufe mit mechanischem Innenleben folgte. In den Anfangsjahren rüttelten die Produktgestalter keinen Deut an bewährten Identifikationsmerkmalen wie Edelstahl, sechs Lünettenreitern und fünfreihigem Gliederband. Letztgenannte Stilelemente unterstützten die Kreation von Bicolor-Modellen.
Solche erfreuen sich bei der Klientel des 1975 gegründeten Schweizer Unternehmens traditionsgemäß hoher Beliebtheit. Optisch sticht beim näheren Betrachten dieser Armbanduhr die charakteristische Kante ober- und unterhalb des Zifferblatts förmlich ins Auge. Diese findet sich logischerweise auch bei den Varianten mit Kautschukband und bei den farbenfrohen Einstiegsmodellen mit upcyceltem Kunststoffgehäuse. Die Spitze der Kollektion repräsentieren Versionen beispielsweise mit Chronograph oder skelettiertem Manufaktur-Innenleben. Das durchbrochen konstruierte Automatikkaliber ML 234 mit klassischen 2,5 Hertz Unruhfrequenz besitzt eine Gangautonomie von 52 Stunden und bei der „6" einen kleinen Sekundenzeiger.

CALYPSO, 1990s

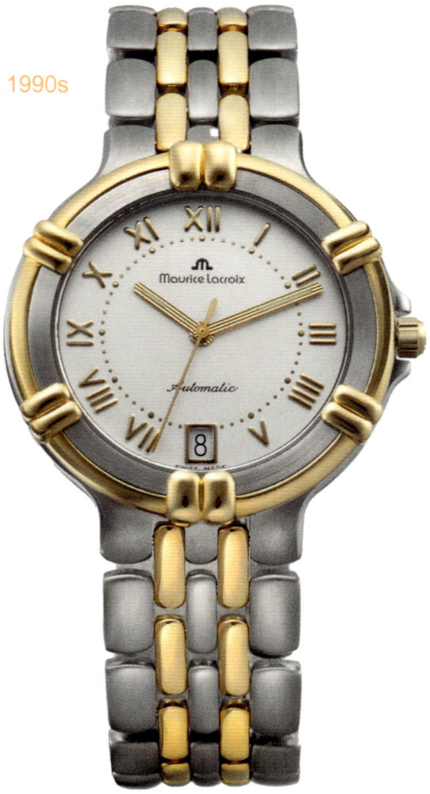

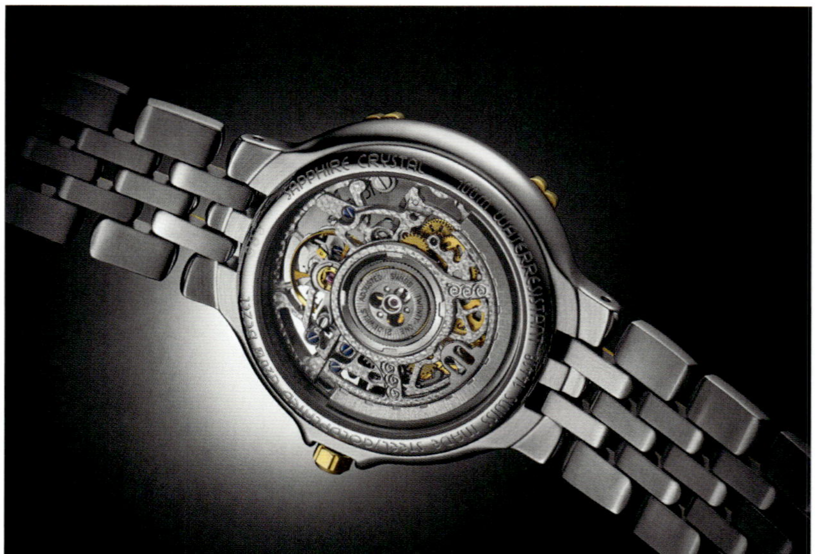

Calypso Skeleton, 1993

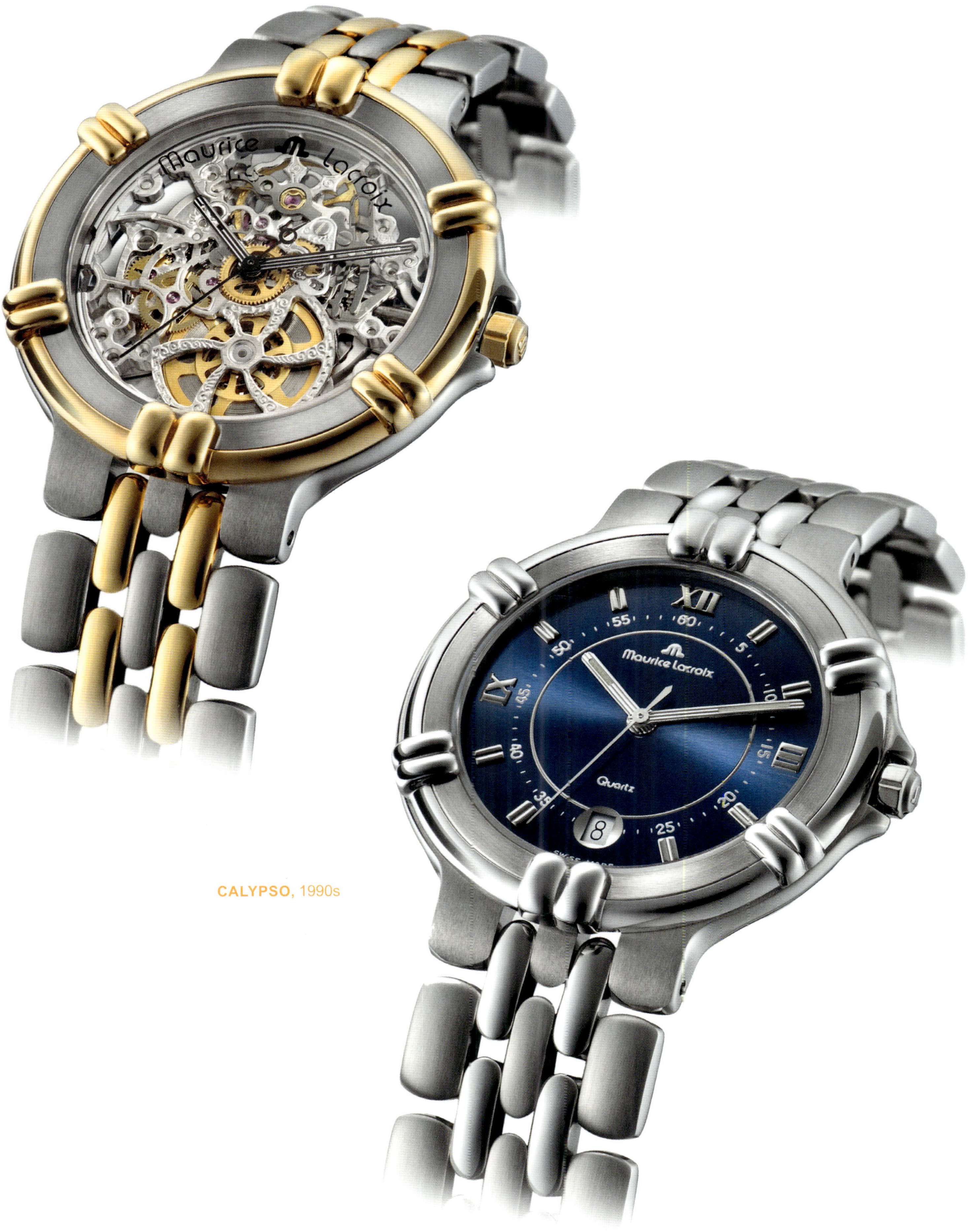

CALYPSO, 1990s

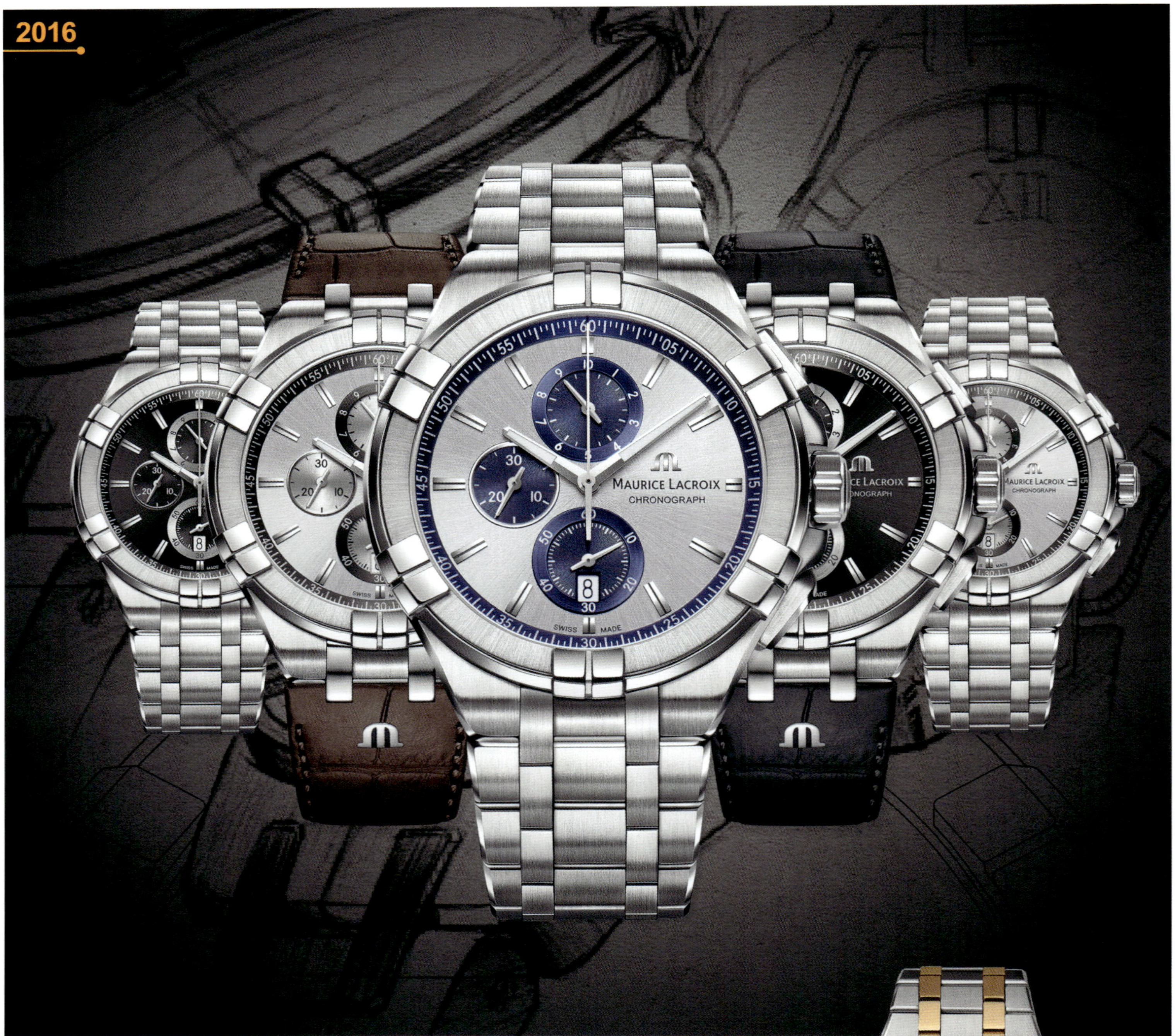

2016

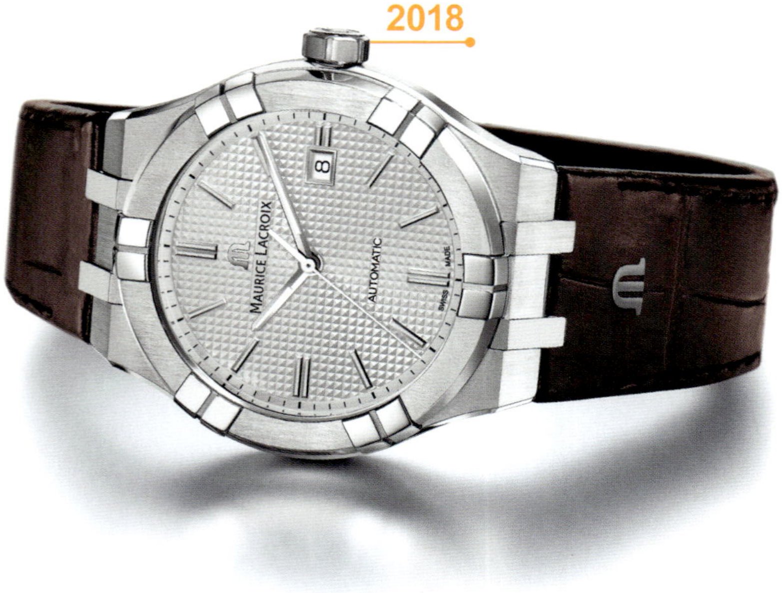

2018

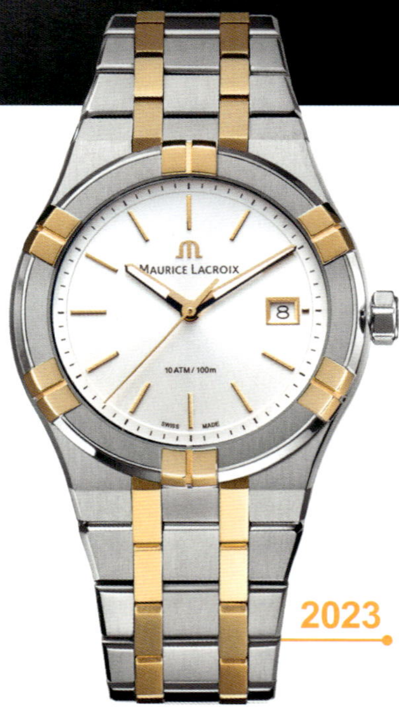

2023

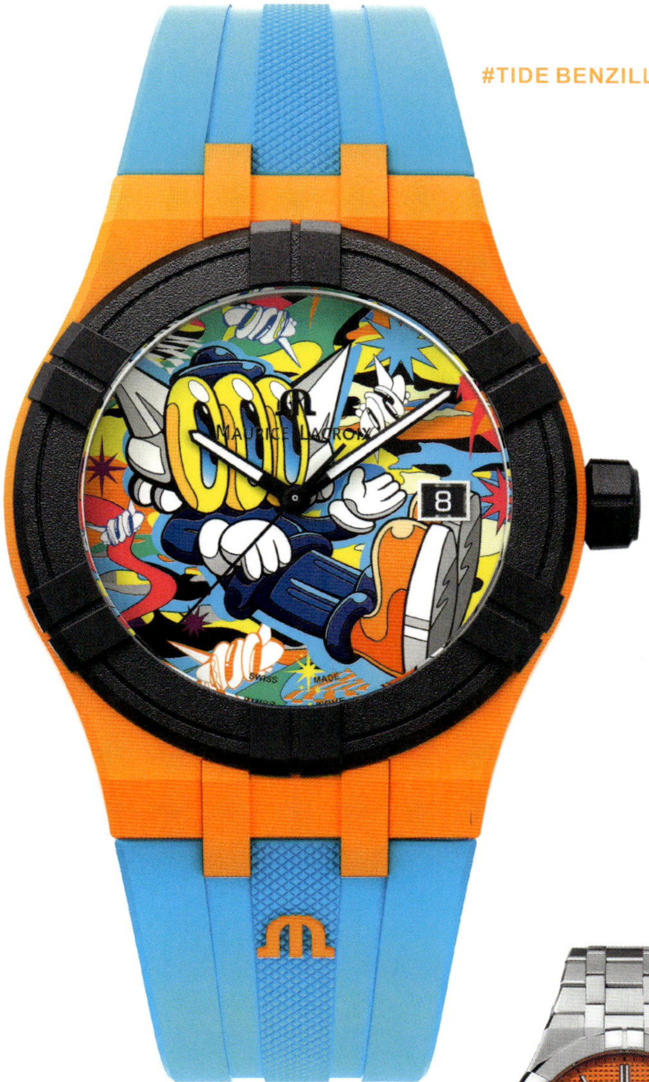

#TIDE BENZILLA, 2022

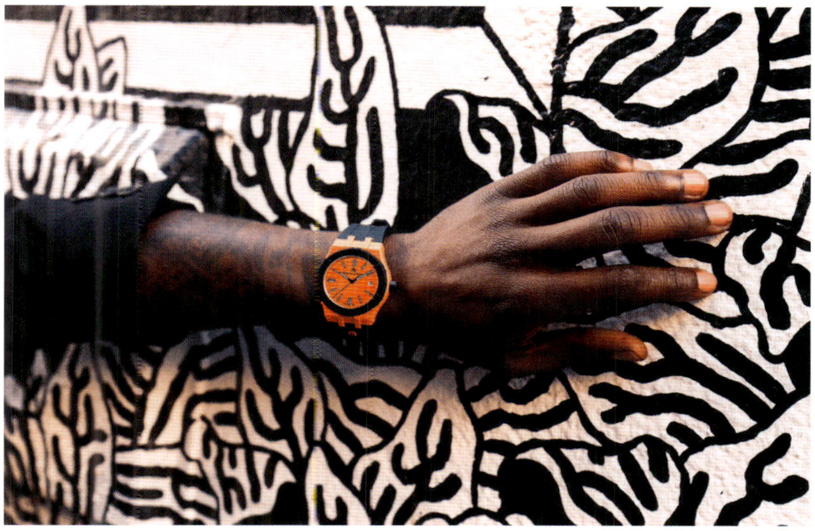

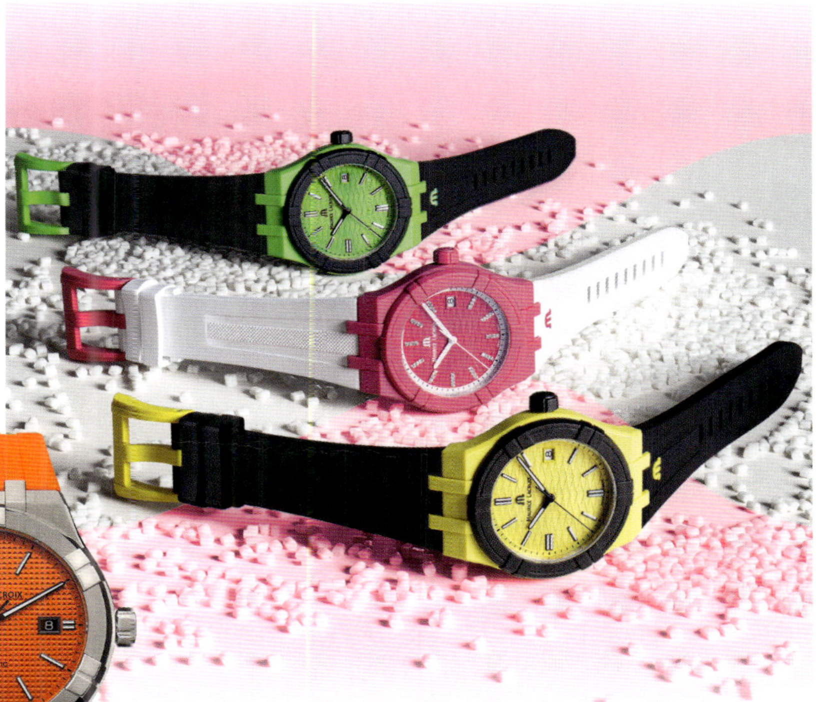

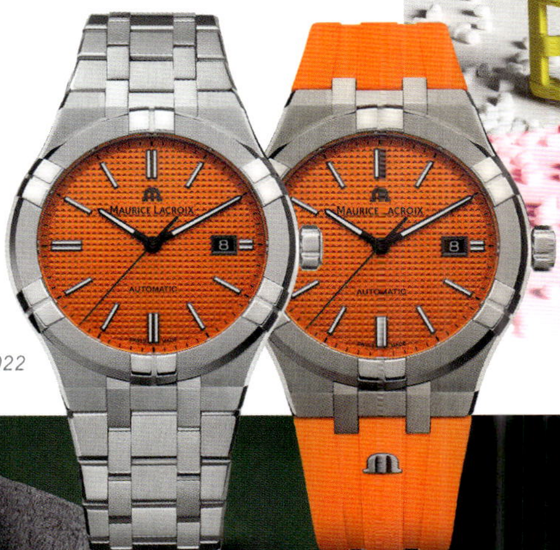

Right, from top | Rechts von oben:
AIKON #tide orange, 2022 | #tide collection, 2022 |
Summer Vibes, AIKON Automatic Limited
Summer Edition 42mm Edition, 2023

Bottom, from left | Unten von links:
AIKON Venturer, 2022 | AIKON Master Grand Date Black, 2022

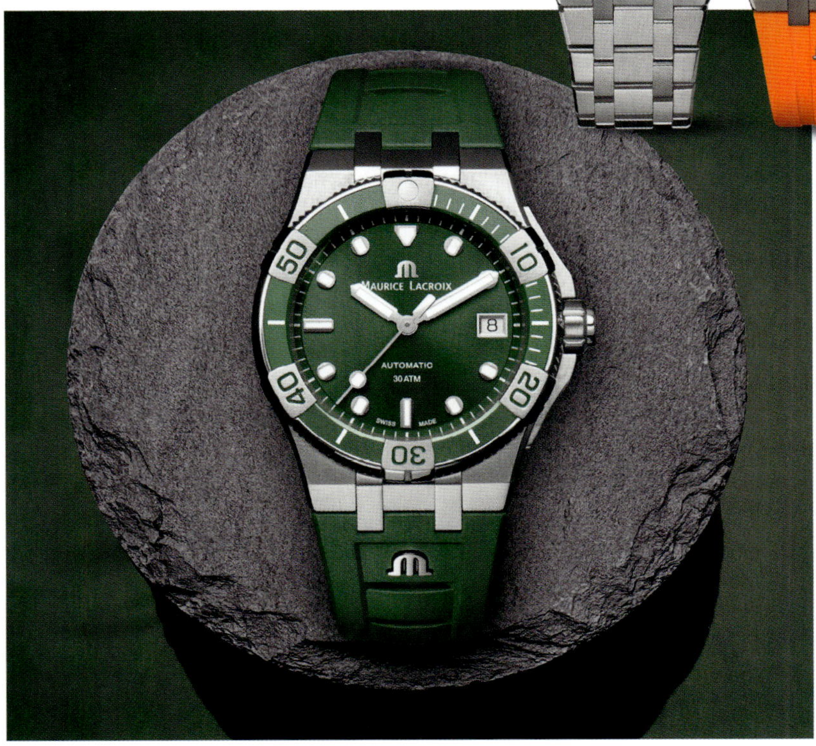

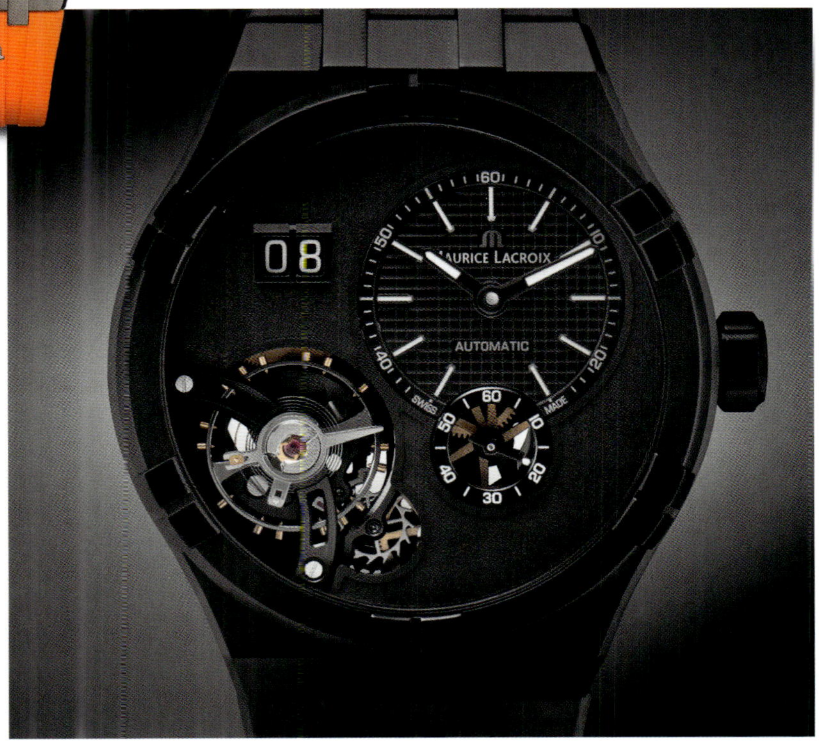

MB&F

2011

LEGACY MACHINE

WITH A LITTLE HELP FROM MY WATCH FRIENDS

Max Büsser makes his watches ably assisted by a varied group of friends. The company name Maximilian Büsser & Friends, or MB&F for short, draws attention to the fact. Brand icons include the 2011 Legacy Machine 1 with a large flying balance wheel spanning the two dials. Tradition and modernity are fused together to create a chronometric work of art. Something that equally applies to the relaxed balance pace of 2.5 hertz or 18,000 vibrations per hour. To achieve the high precision of watches of this ilk requires a great deal of fine-tuning and regulating expertise. And time is not of the essence when it comes to this process. Unsurprisingly, the annual production of this horological work of art with a power reserve indication and time zone complication is very modest. The friends nevertheless succeeded in raising the bar. The Legacy Machine 2 possesses two regulating systems, referencing the horological legacy of famous French watchmakers such as Ferdinand Berthoud, Abraham-Louis Breguet and Antide Janvier. In this watch, two flying balance wheels oscillate totally independently to compensate for any gravitational errors. The innovative inner workings of the impressive Legacy Machine 2 are protected by a high-domed sapphire crystal. A sophisticated differential wheel transfers power to each of the two regulators while averaging their individual timing rates. The hands present exactly this result to the marvelling observer.

FREUNDESKREIS FÜR UHREN

Zusammen mit einem vielfältigen Freundeskreis fertigt Max Büsser seine Uhren. Auf diesen Sachverhalt deutet der Firmenname Maximilian Büsser & Friends, kurz MB&F, zweifellos hin. Zu den Ikonen dieser Marke gehört die 2011 vorgestellte „Legacy Machine 1" mit großer, über den beiden Zifferblättern oszillierender Unruh. Hier vereinigen sich Tradition und Gegenwart zu einem chronometrischen Kunstwerk. Das gilt auch für die entschleunigte Unruhfrequenz von 2,5 Hertz. Pro Stunde sind das 18.000 Halbschwingungen. Wer solchen Armbanduhren langfristige Genauigkeit beibringen möchte, muss sich trefflich aufs Feinstellen und Regulieren verstehen. Zeit darf bei diesem Prozess ohnehin keine Rolle spielen. Kein Wunder also, dass die Jahresproduktion dieses tickenden Kunstwerks mit Gangreserveanzeige und Zeitzonen-Dispositiv sehr überschaubar ist. Aber es geht auch noch aufwändiger und exklusiver: Die „Legacy Machine 2" verfügt über gleich zwei Gangregler und knüpft damit an die uhrmacherische Hinterlassenschaft berühmter französischer Uhrmacher wie Ferdinand Berthoud, Abraham-Louis Breguet und Antide Janvier an. Bei dieser Armbanduhr dienen zwei gegenläufig schwingende Unruhn zur Kompensation eventuell auftretender Schwerpunktfehler. Bei der beeindruckenden „Legacy Machine 2" schützt ein hoch gewölbter Saphirglasdom das zeitmessende Geschehen. Ein ausgeklügeltes Planetendifferenzialgetriebe versorgt beide Schwingsysteme mit Energie. Gleichzeitig mittelt es die Gangresultate beider Gangregler. Und genau diesen Wert führen die Zeiger dem staunenden Betrachter vor Augen.

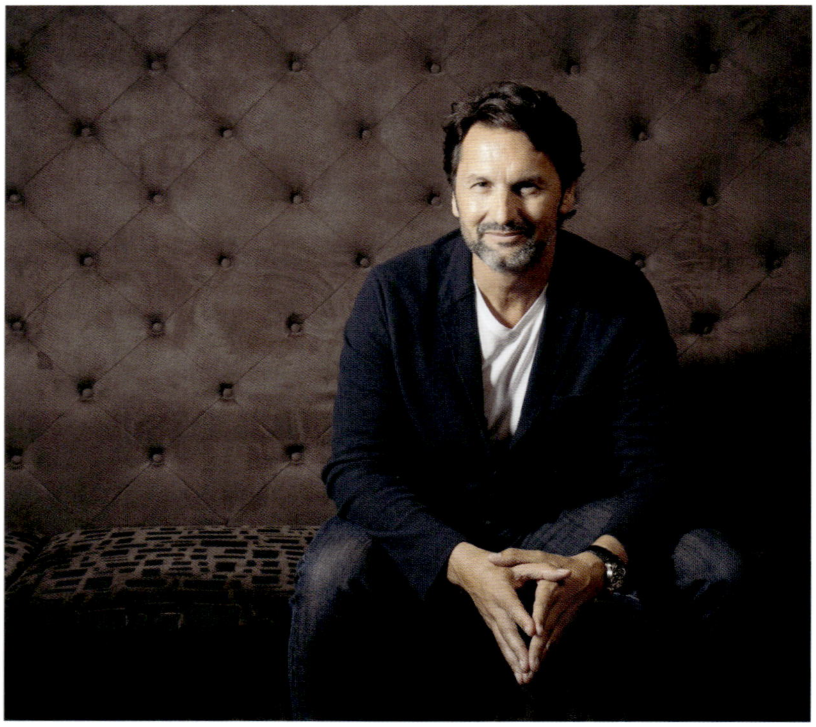

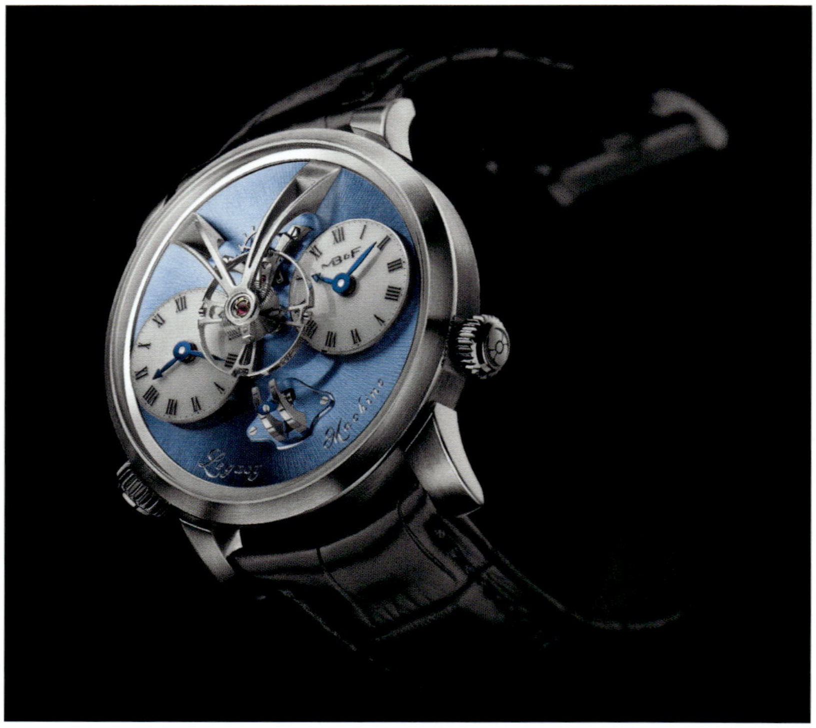

***Left / Links:** Maximilian Büsser, 2016, **Right / Rechts:** LM1, 33-piece limited edition, 2014*

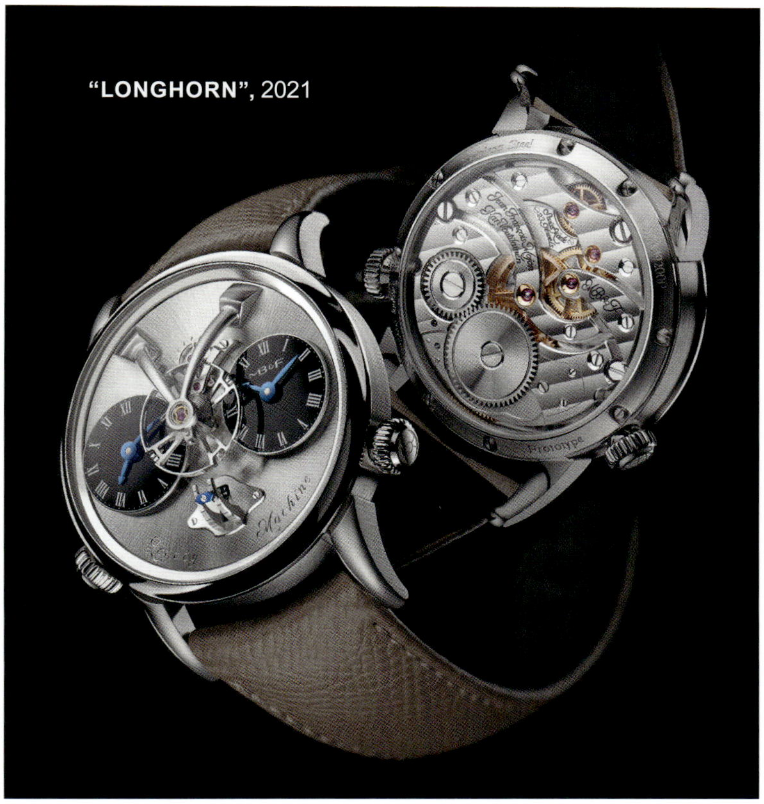

"LONGHORN", 2021

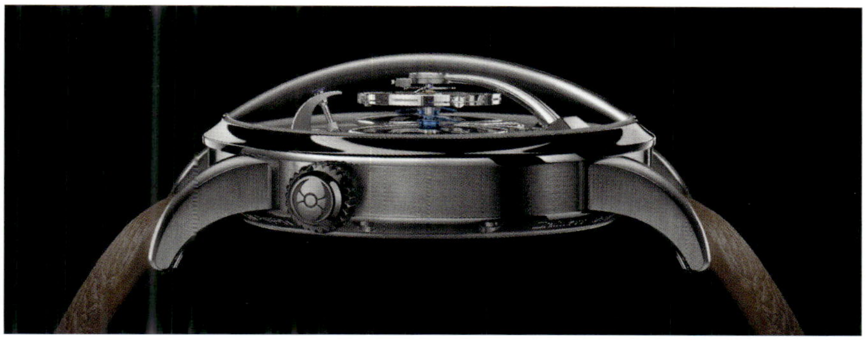

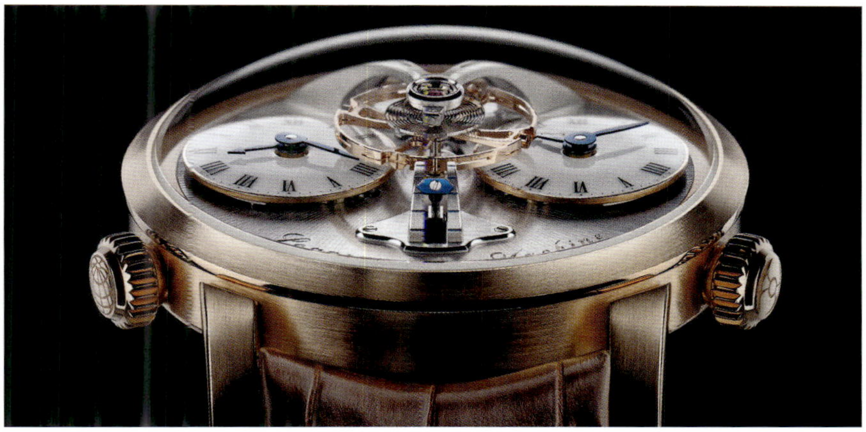

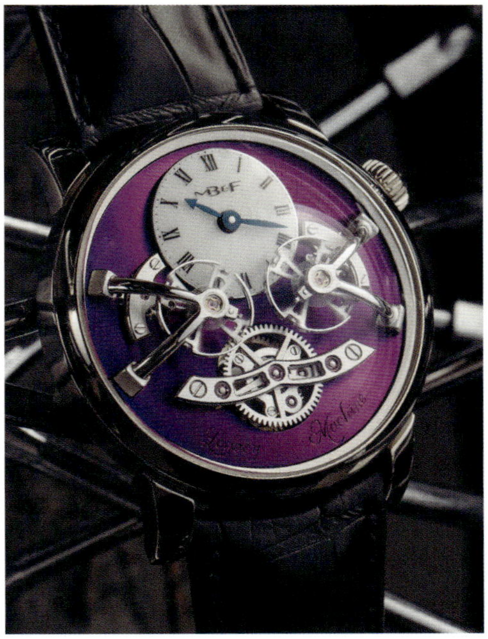

1

LEGACY MACHINE

2

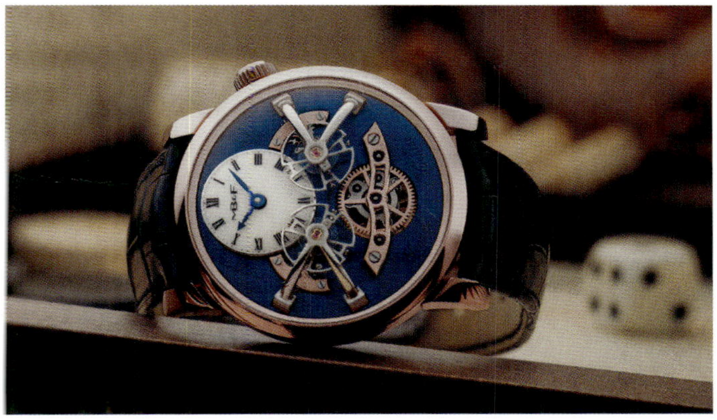

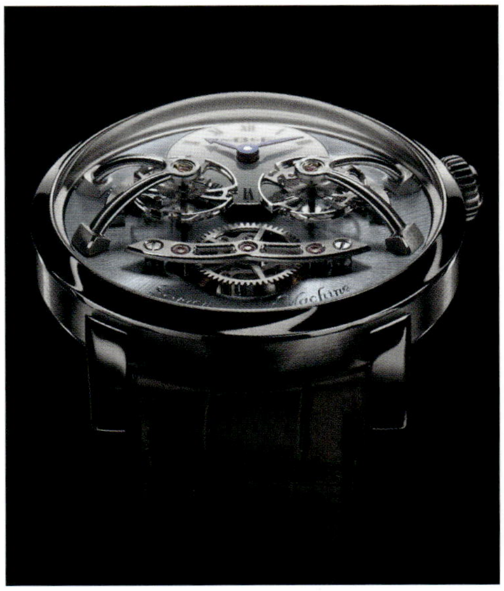

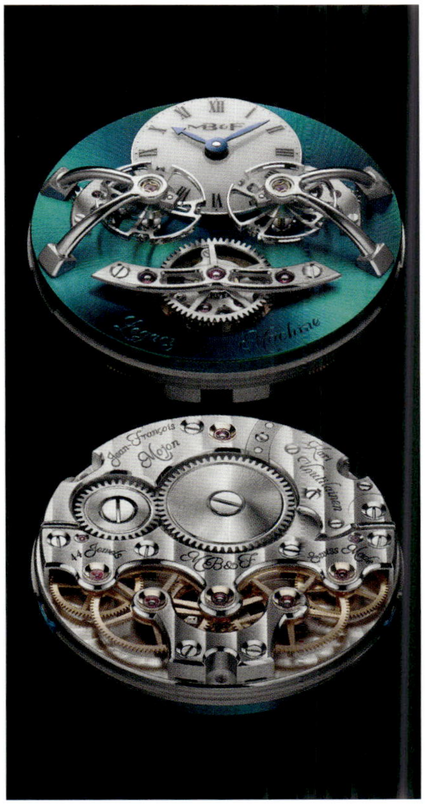

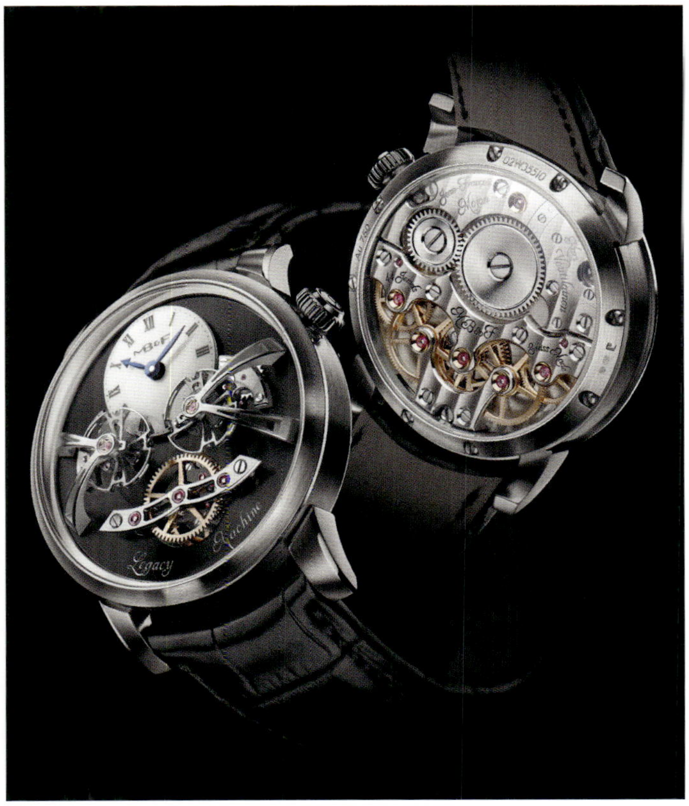

SINGLE-HAND WATCHES

SINGLE-HANDED SENTINEL

Abraham-Louis Breguet would never even have dreamed that one day in the distant future there would be a MeisterSinger watchmaker in Münster making watches inspired by him. Along with all manner of inventions aimed at improving mechanical watches, his main concern in 1798 was to keep his Paris workshops busy. With this in mind, the master watchmaker developed the Montre Souscription. As the name suggests, *souscription* being the French for downpayment, the comparatively affordable price was secured by the payment of a substantial deposit. At some point Breguet decided there was to be just one hand for the money. And that was that. This solution harked back to the thirteenth-century forerunners of modern clocks. At the turn of the millennium, Manfred Brassler from Münster was impressed by the idea of indicating the passage of time with just one hand. In the design of his watches, he therefore referenced timepieces "relaxed" in this regard, including the Nuremberg Egg by Peter Henlein. Understandably, cast in this philosophical mould, the dial of a timepiece with only an hour hand needs suitable indexing. "Whole, half and quarter hours, as well as minutes must be clearly indicated", opines Manfred Brassler. Something he achieved with aplomb on his 2001 MeisterSinger wristwatches. Distinctive features of the single-hand icons designed in Germany and made in Switzerland are the hour digits prefixed with a zero and a wedge-shaped hand. Recognition of his creativity has earned Manfred Brassler multiple design awards. Which model from the extensive range finds favour is purely a matter of personal taste. The bottom line is always one hand. There are no exceptions, not even the top models with exclusive Swiss inner workings. Then again, anyone missing their plane or train due to the lack of a minute hand only have themselves to blame.

EIN ZEIGER REICHT

Daran, dass es eines fernen Tages in Münster die Uhrenmarke MeisterSinger geben würde, dachte Abraham-Louis Breguet nicht einmal im Traum. 1798 beschäftigte ihn neben allerlei Erfindungen zur Verbesserung mechanischer Uhren auch die kontinuierliche Auslastung seiner Pariser Werkstätten. In diesem Zusammenhang entwickelte der Meisteruhrmacher die „Montre Souscription". Der Tribut für den vergleichsweise günstigen Preis bestand in einer stattlichen Anzahlung. Irgendwann lieferte Breguet fürs Geld eine Taschenuhr mit nur einem Stundenzeiger. Das war's. Derart reduziert hatten sich schon die frühen Räderuhren ab dem späten 13. Jahrhundert präsentiert. Die Darstellung der flüchtigen Zeit mit nur einem einzigen Zeiger beeindruckte um die Jahrtausendwende Manfred Brassler im westfälischen Münster. Bei der Gestaltung seiner eigenen Armbanduhren orientierte er sich deshalb an solcherart „entschleunigten" Objekten, zu denen auch das sogenannte „Nürnberger Ei" des Peter Henlein gehörte. Verständlicherweise benötigte das Zifferblatt eines philosophisch angehauchten Zeitmessers mit nur einem Stundenzeiger eine passende Indexierung.
„Ganze, halbe und viertel Stunden sowie die Minuten müssen sich klar voneinander abheben", lautet das Credo von Manfred Brassler. Und das ist ihm bei seinen MeisterSinger-Armbanduhren ab 2001 tatsächlich gelungen. Signifikante Kennzeichen der in Deutschland gestalteten und in der Schweiz gefertigten Einzeiger-Ikonen sind Stundenziffern mit vorangestellter „0" sowie ein keilförmiger Nadelzeiger. Bestätigung für sein schöpferisches Tun erfuhr Manfred Brassler durch die Zuerkennung zahlreicher Designpreise. Welches Modell aus dem breiten Produktspektrum gefällt, ist eine Frage des persönlichen Geschmacks. Mehr als einen Stundenzeiger gibt es nie. Auch nicht bei den Topmodellen mit sehr exklusivem Schweizer Innenleben. Aber Menschen, die wegen des fehlenden Minutenzeigers ihren Zug oder Flieger verpasst haben, ist ohnehin nicht zu helfen.

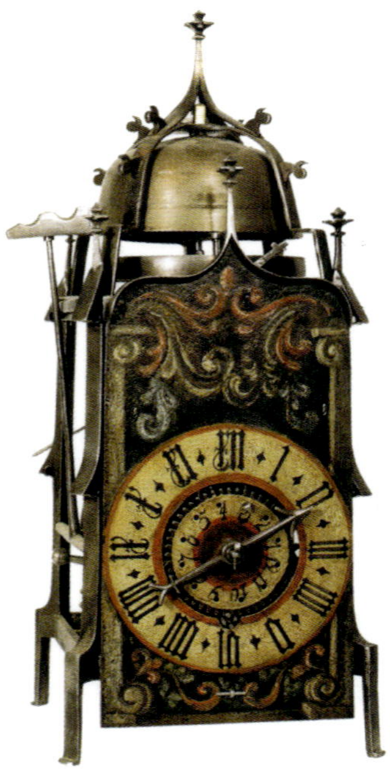

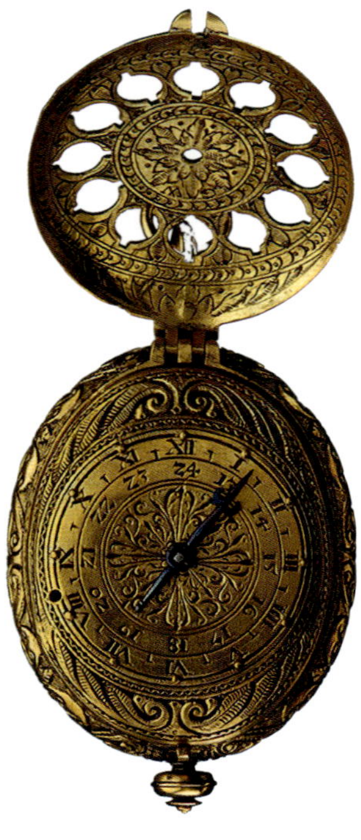

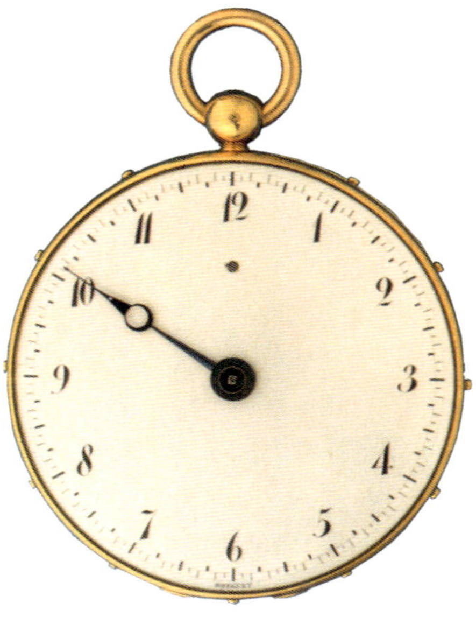

From left:
Single-hand iron clock, 1540 | Nuremberg Egg | Montre Souscription | N° 01

Von links:
Einzeiger-Eisenuhr, 1540 | Nürnberger Ei | Montre Souscription | Nr. 01

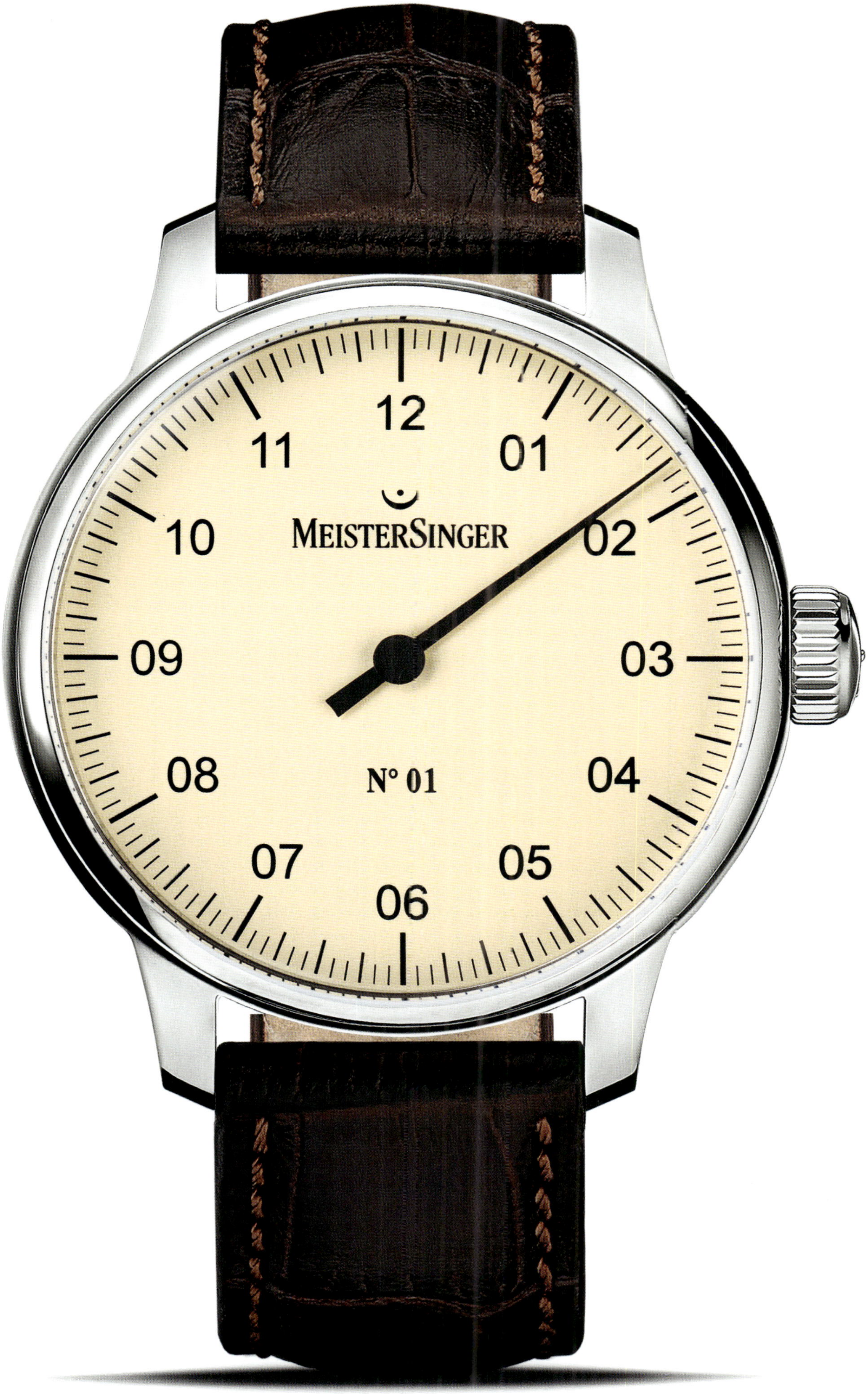
12
11
01
MeisterSinger
10
02
09
03
08
N° 01
04
07
05
06

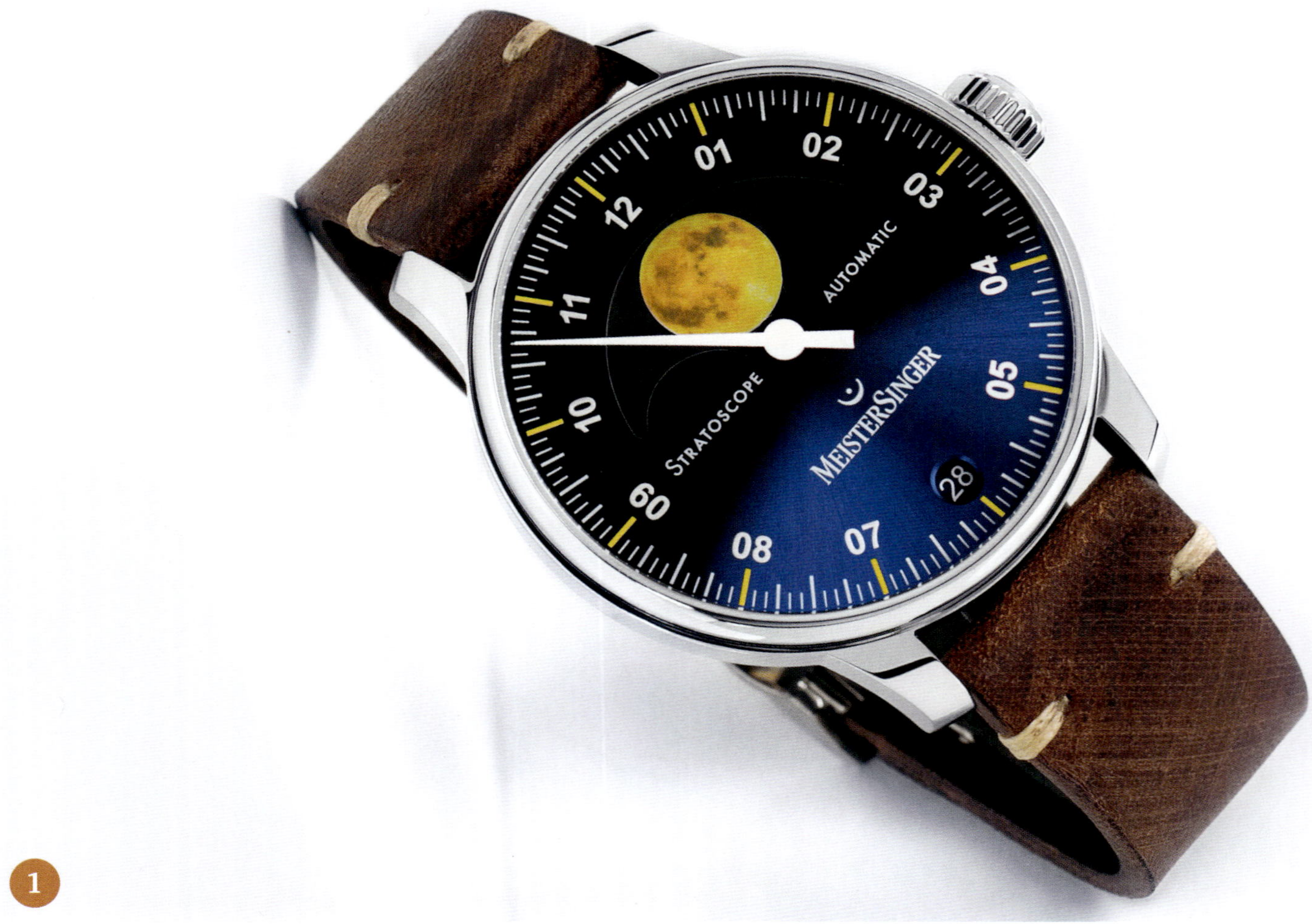
AUTOMATIC
STRATOSCOPE
MEISTERSINGER

1

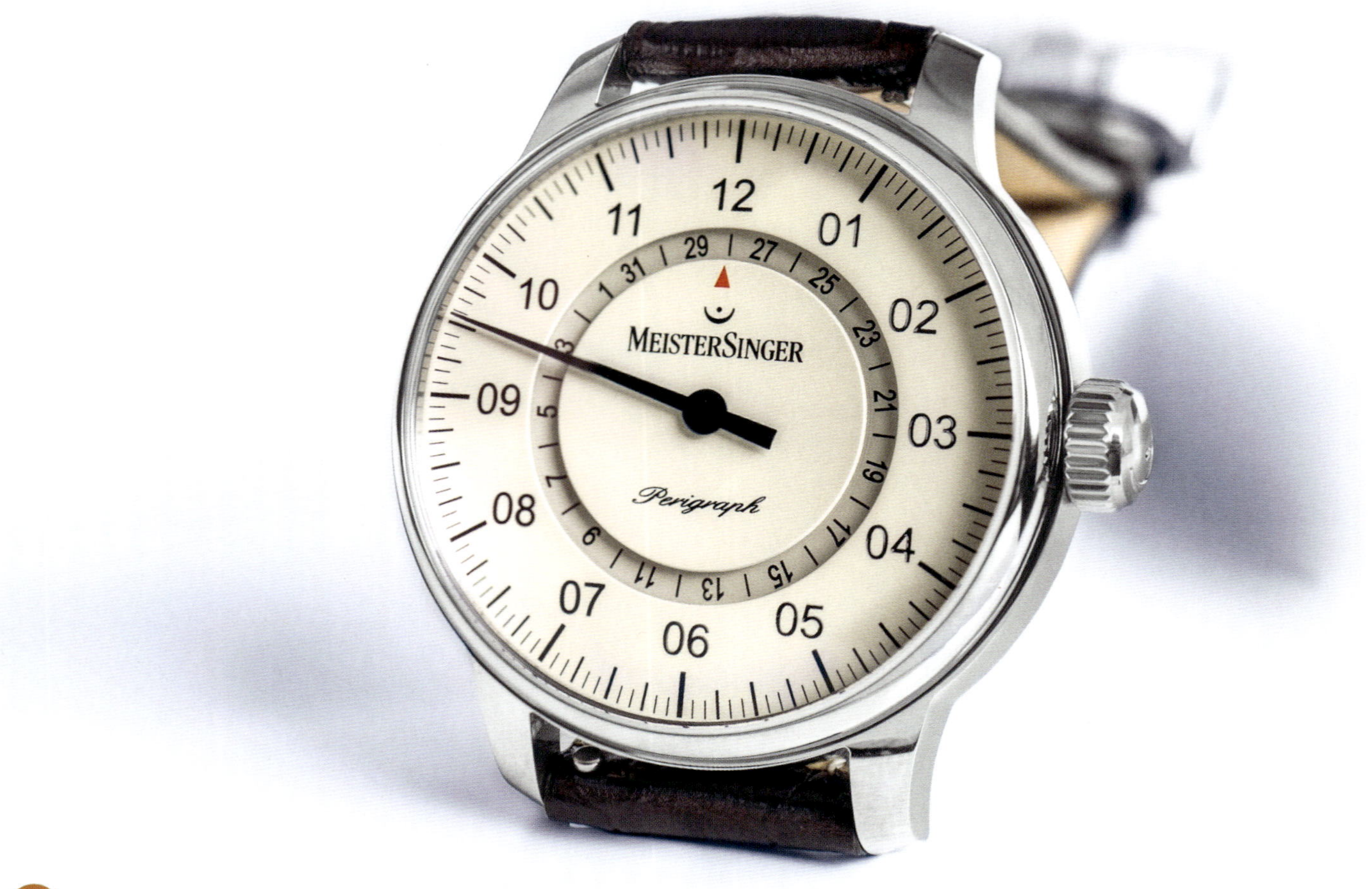
MEISTERSINGER
Perigraph

2

3

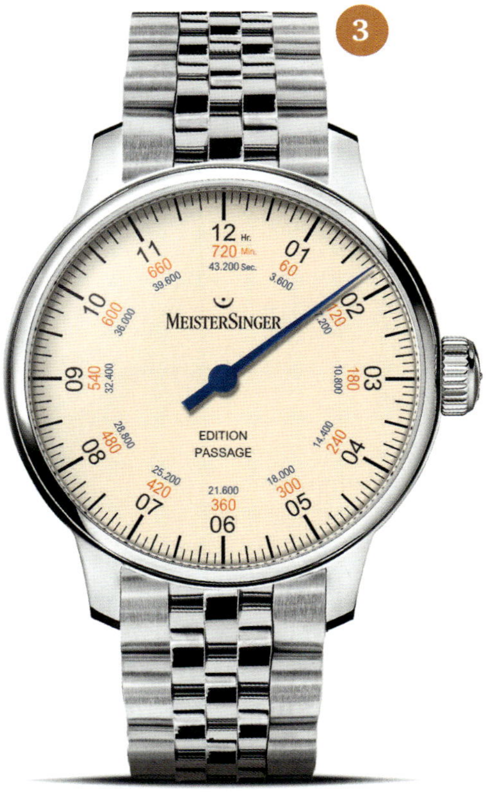

4

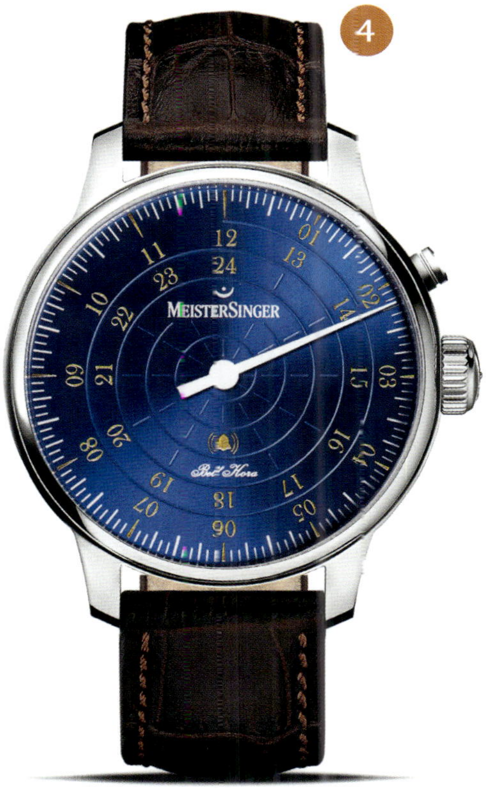

5

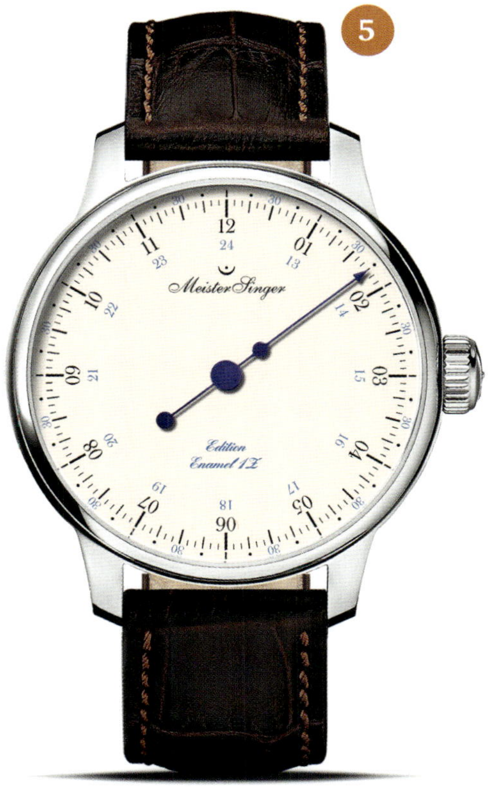

6

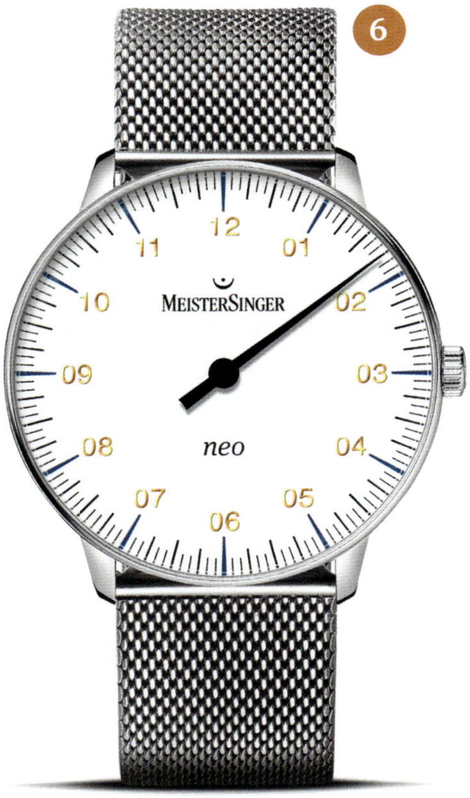

7

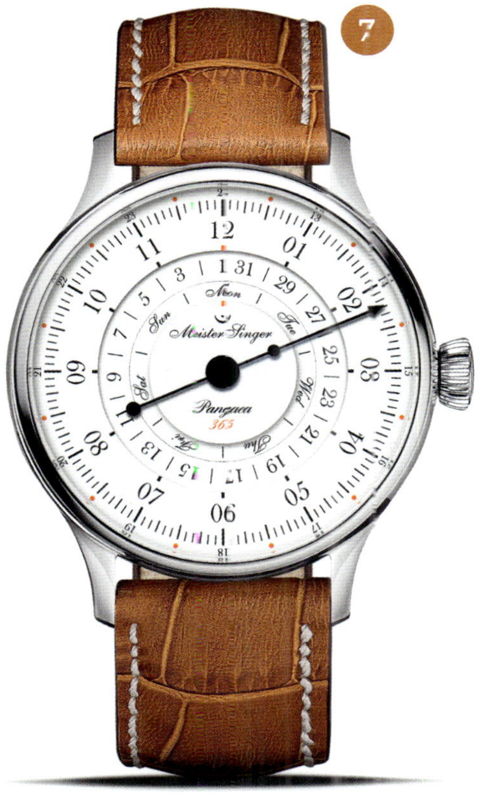

8

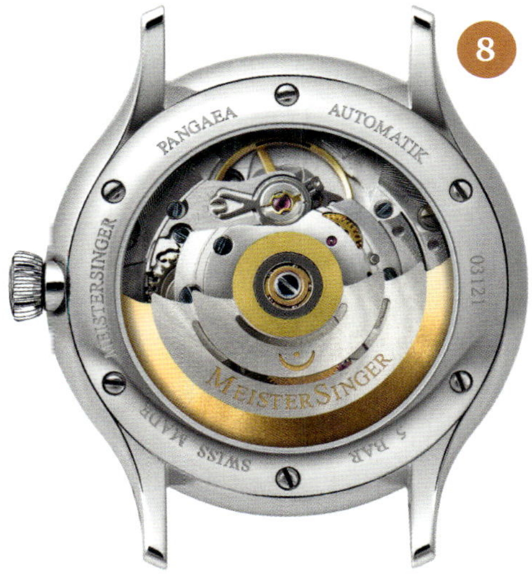

1 Stratoscope

2 Perigraph

3 Edition Passage

4 Bell Hora

5 Edition Enamel 1Z

6 neo

7 8 Pangaea Day Date

RICHARD MILLE 2022

RM UP-01 FERRARI

JUST 1.75 MILLIMETRES

Even at a third glance, the RM UP-01 Ferrari bears very little in common with what watch lovers associate with Richard Mille. However, the tagline "A racing machine on the wrist" is also an apt description of this 2022 watch. Car racing is a passion of the 73-year-old Frenchman, who founded an independent watch brand in 2001 and currently cooperates with Ferrari. From the first, the timepieces assembled in the company's workshops have flaunted design, engineering and, above all, pricing conventions. Ordinary mortals never figured in the designs of the successful watch entrepreneur. His collection with an emphasis on showcasing exquisite engineering was always destined for the upper crust. Many of his products grace the collections of affluent patrons of the brand, who snap up everything that leaves the workshops in the French-speaking Swiss Jura. The product biography begins with an edition of 17 lavish Tourbillon RM 001 timepieces, discreetly presented in Basel in 2000. At a height of more than one centimetre it offers a strong contrast to the RM UP-01 Ferrari. The latter retained the title of the world's flattest mechanical wristwatch until 2023. It adds only 1.75 millimetres to the thickness of the wearer's wrist. In April 2024, Bulgari's timepiece snatched the world record with a height of only 1.70 millimetres. The letters UP designating the Richard Mille model stand for Ultra Plat or ultra flat. Visually, as noted, this outstanding limited edition of 150 bears little resemblance to Richard Mille's signature timepieces. His hand-wound Calibre RMUP-01 with titanium baseplate and a power reserve of around 45 hours is 1.18 millimetres high. It weighs 2.82 grams. The ultra-flat movement is concealed in a titanium case measuring 51 by 39 millimetres. On the dial side, the titanium balance wheel oscillates at four hertz on the right, the hands rotate in the centre, the controls are on the left. The top control is for selecting winding or setting. The bottom control is for actual winding and to set the time. At a price of around €1.7 million (excluding tax), fastening the rubber strap of this wafer-thin watch to your wrist does not come cheap.

1,75 MILLIMETER FLACH

Selbst auf den dritten Blick hat die „RM UP-01 Ferrari" herzlich wenig gemein mit dem, was sich Uhrenliebhaberinnen und -liebhaber unter einer Richard Mille vorstellen. Obwohl: Der Slogan „Rennmaschine fürs Handgelenk" trifft auch auf diese Armbanduhr aus dem Jahr 2022 zu. Autorennen sind nämlich eine Passion des 1951 geborenen Franzosen, der sich 2001 mit einer Uhrenmarke selbständig machte und aktuell mit Ferrari kooperiert. Von Anbeginn fiel das in eigenen Werkstätten Assemblierte gestalterisch, technisch und vor allem auch preislich aus dem Rahmen. Normalsterbliche hatte der erfolgreiche Uhr-Unternehmer nie im Blick. Seine Kollektion mit deutlich zur Schau getragener Mechanik wandte sich vom Start weg an die oberen Zehntausend. Viele seiner Erzeugnisse finden in die Sammlung finanzstarker Stammkunden, welche alles erwerben, was die Ateliers im Westschweizer Jura verlässt. Die Produktbiographie beginnt mit 17 Exemplaren des opulenten, 2000 in Basel unter der Hand vorgestellten Tourbillon RM 001. Seine Bauhöhe von mehr als einem Zentimeter steht im krassen Gegensatz zur jener der „RM UP-01 Ferrari". Letztere durfte sich rühmen, bis 2023 die weltweit flachste mechanische Armbanduhr zu sein. Am Handgelenk trägt der Zeitmesser gerade einmal 1,75 Millimeter auf. Seit April 2024 gebührt Bulgari der Weltrekord mit lediglich 1,7 Millimetern Bauhöhe. Bei Richard Mille stehen die Buchstaben UP für Ultra Plat. Optisch ist der auf 150 Exemplare limitierte Ex-Superlativ wie gesagt weit entfernt von dem, was die tickenden Produkte von Richard Mille charakterisiert. Sein Handaufzugskaliber RMUP-01 mit Titanplatine und ca. 45 Stunden Gangautonomie baut 1,18 Millimeter hoch. Auf die Waage bringt es 2,82 Gramm. Der ultraflache Zeitmotor versteckt sich in einem Titangehäuse mit den Dimensionen 51 x 39 Millimeter. Vorderseitig zeigen sich ganz rechts die mit vier Hertz oszillierende Titanunruh, mittig die Zeiger und links die Bedienelemente. Mit der oberen Krone lässt sich wählen zwischen Aufziehen und Zeigerstellen. Mit der unteren werden die Uhrzeit eingestellt oder das Federhaus aufgezogen. Bei einem Preis von rund 1,7 Millionen Euro ist es verständlicherweise nicht unbedingt erschwinglich, sich diesen Hauch von Uhr per Kautschukband ans Handgelenk zu schnallen.

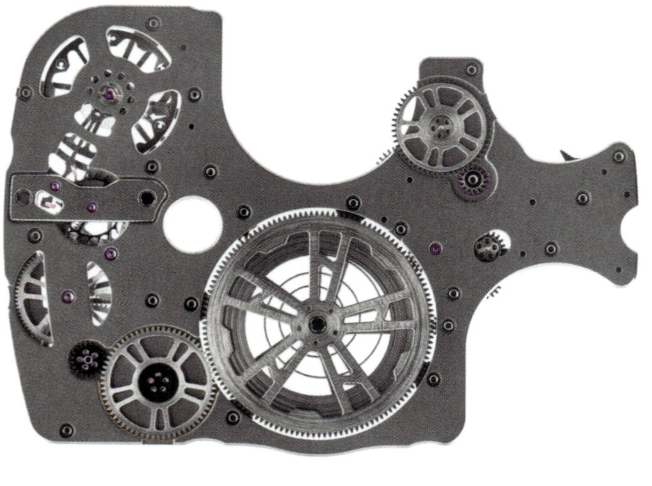

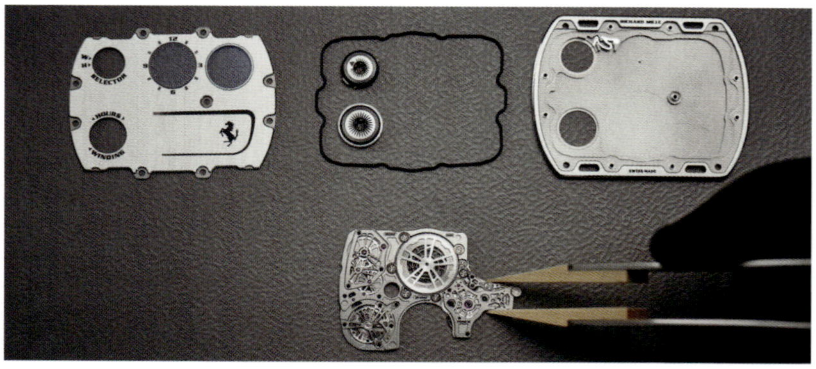

From left: Movement RMUP-01, front / Titanium case and movement /
Von links: Uhrwerk RMUP-01, Vorderseite / Titangehäuse und Uhrwerk

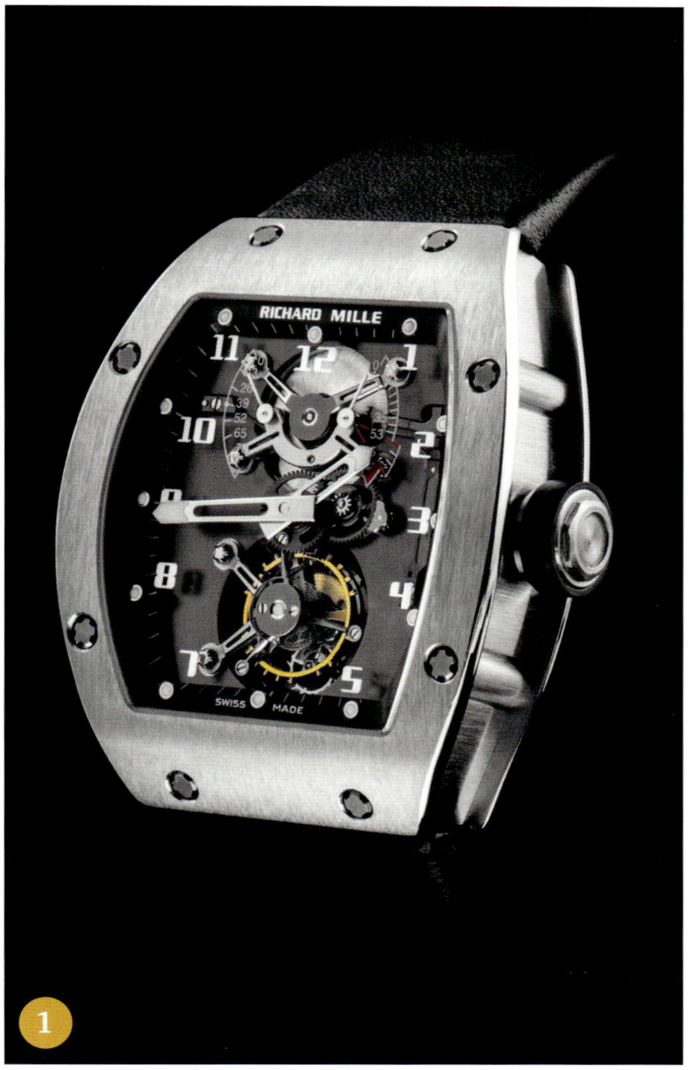

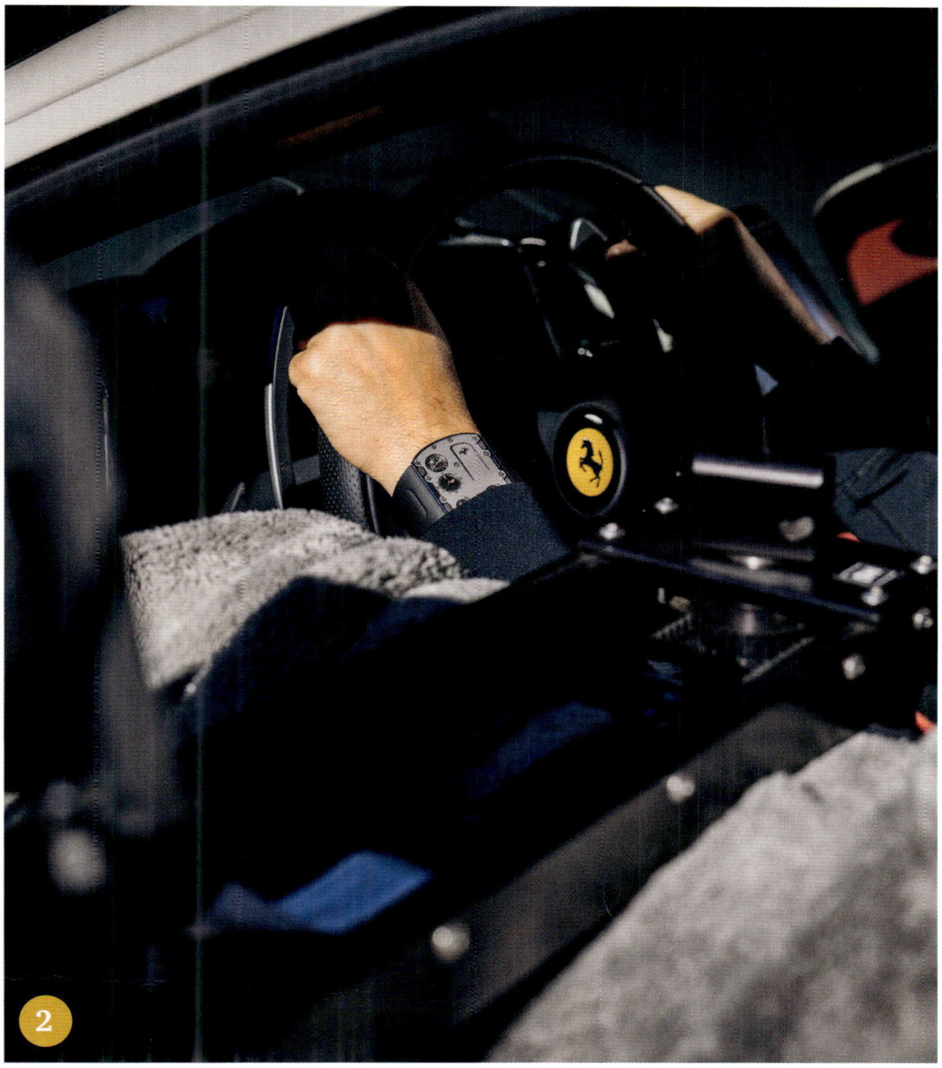

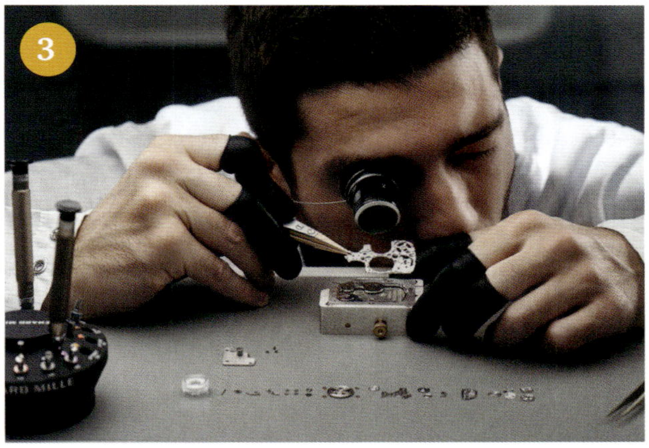

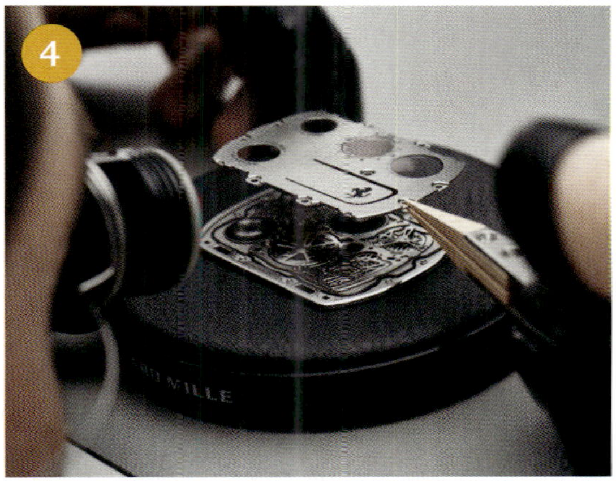

1 *Tourbillon RM 001, 2001*

2 *Richard Mille and Ferrari collaborate since 2021 | Richard Mille und Ferrari arbeiten seit 2021 zusammen*

3 4 5 6 7 *Craftsmanship: Assembly of the RM UP-01 Ferrari | Handwerkskunst: Zusammenbau der RM UP-01 Ferrari*

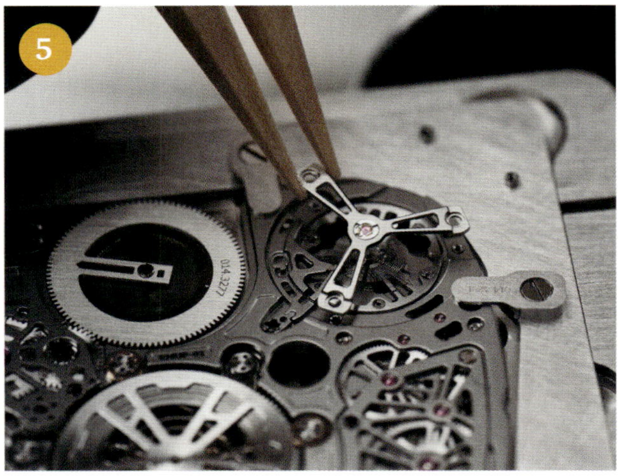

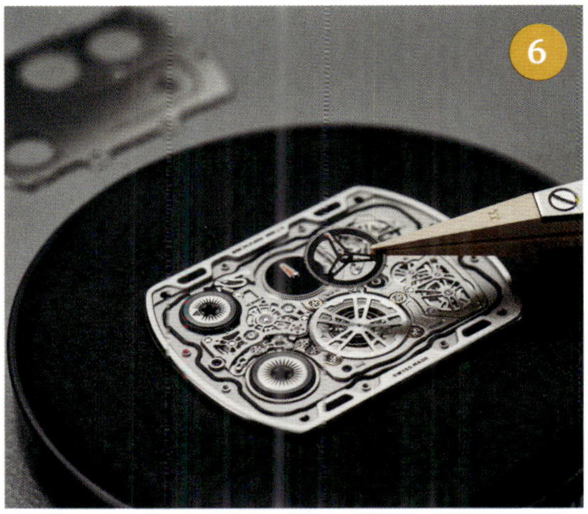

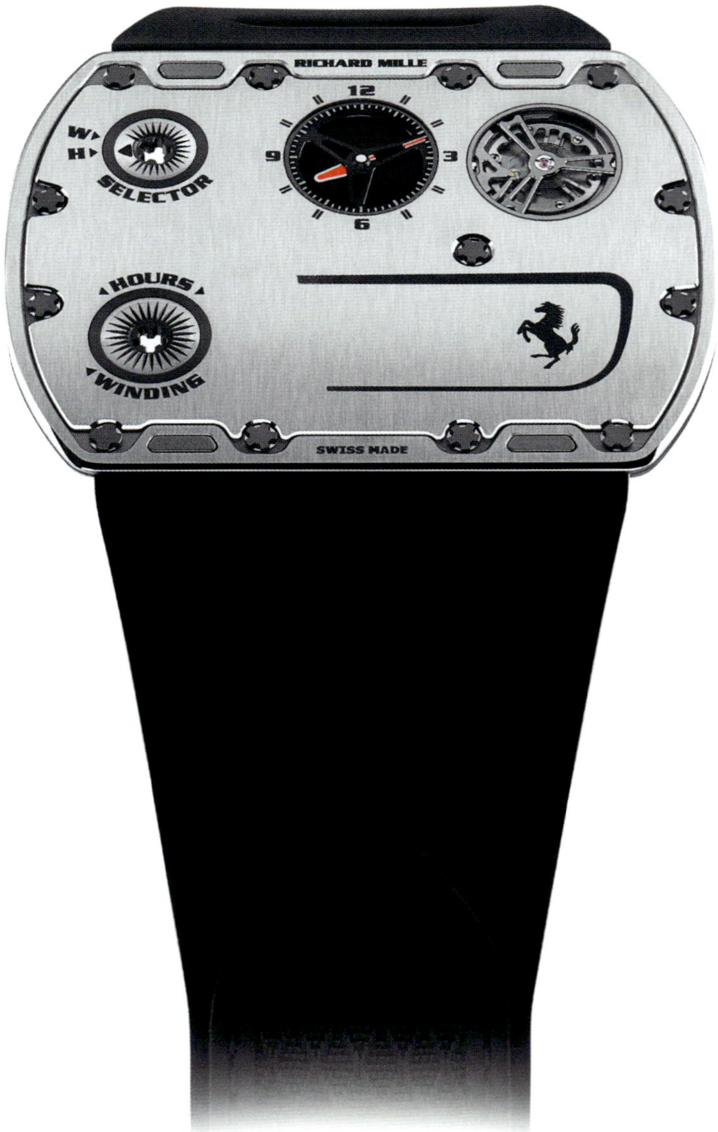

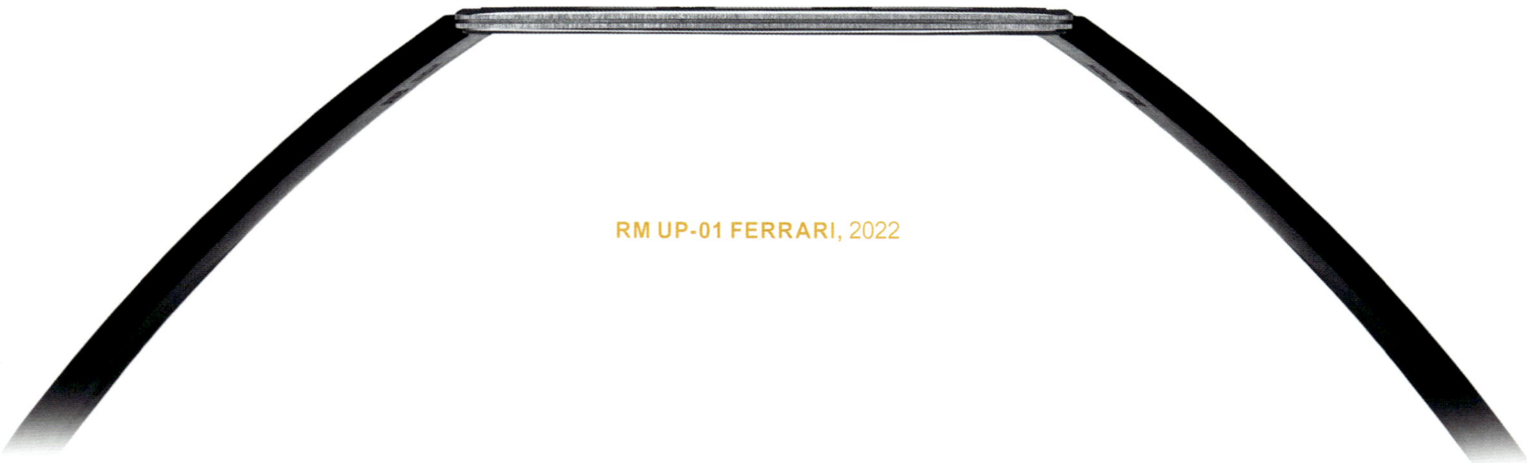

RM UP-01 FERRARI, 2022

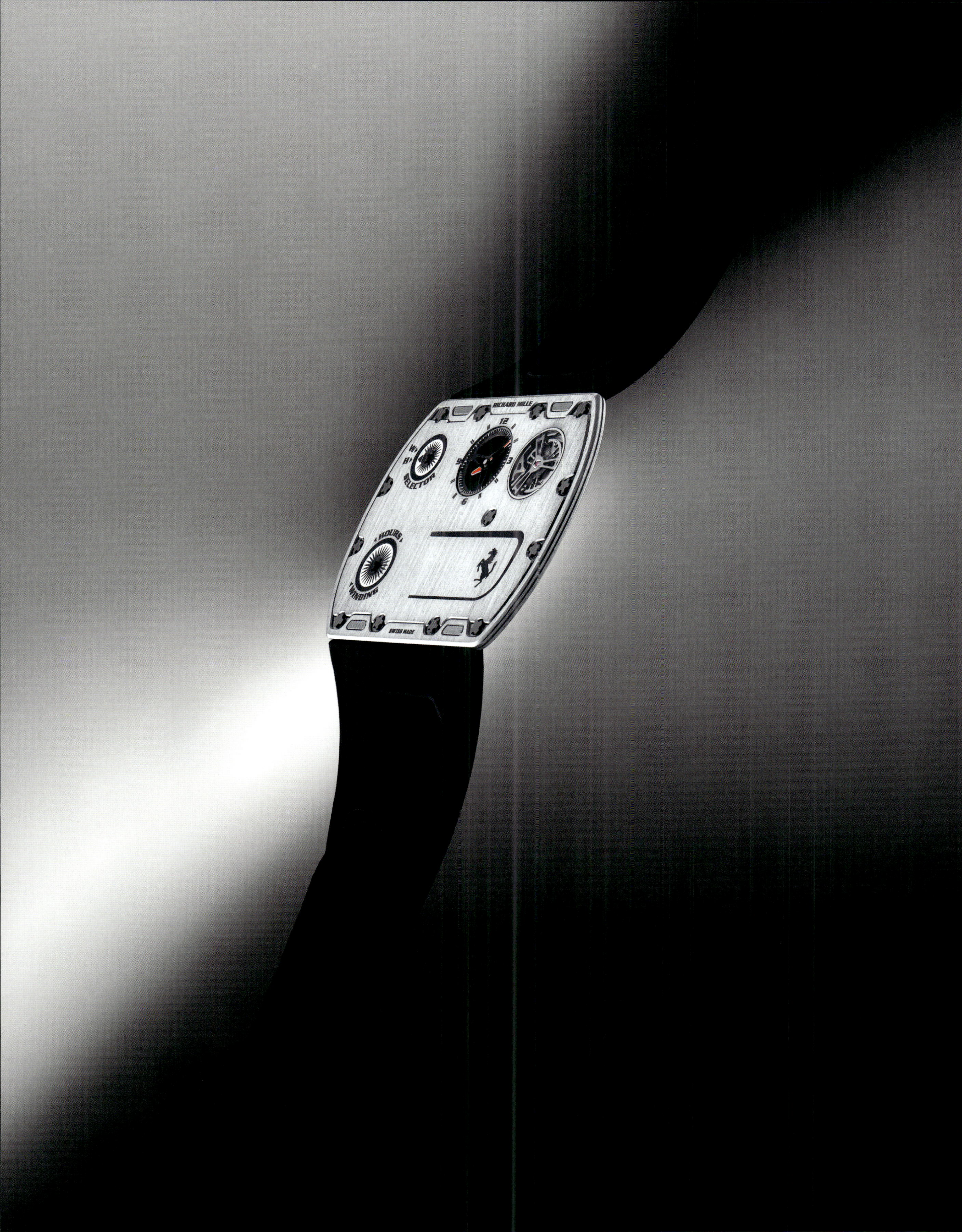
RICHARD MILLE
SELECTOR
HOURS
WINDING
SWISS MADE

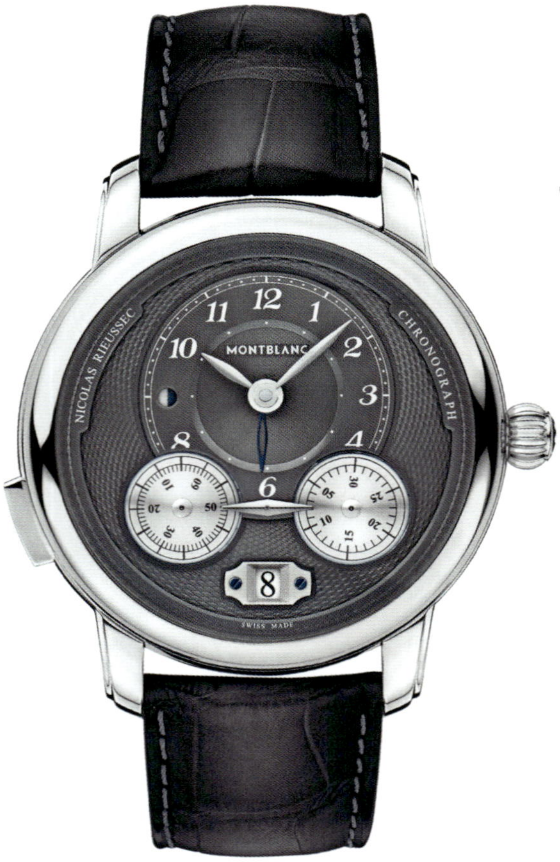

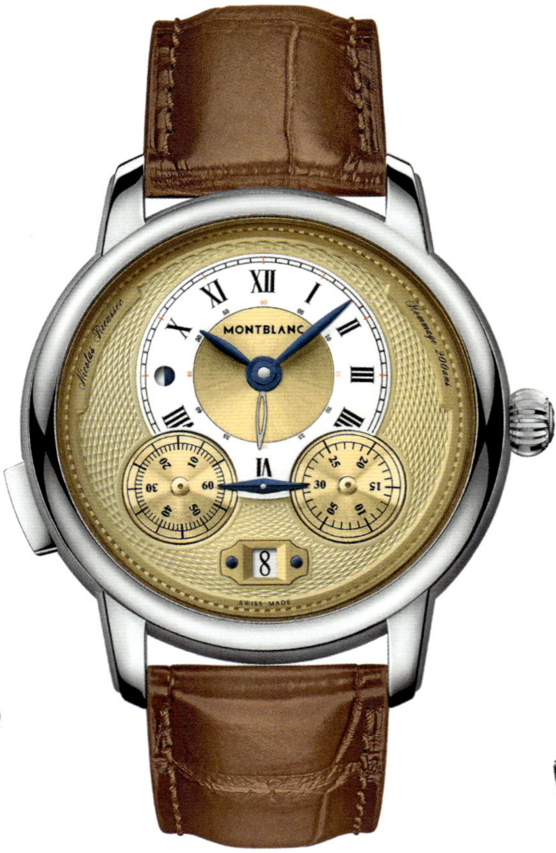

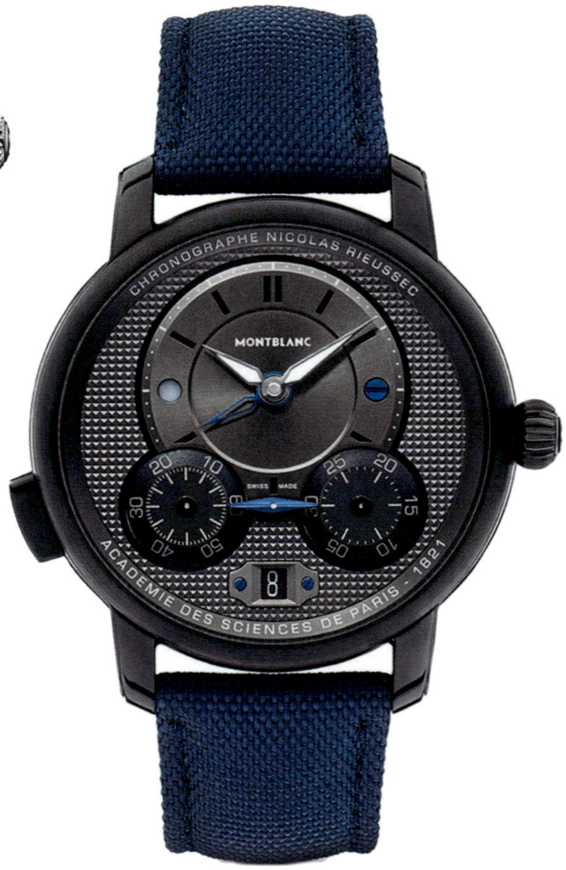

STAR LEGACY
NICOLAS RIEUSSEC
CHRONOGRAPH

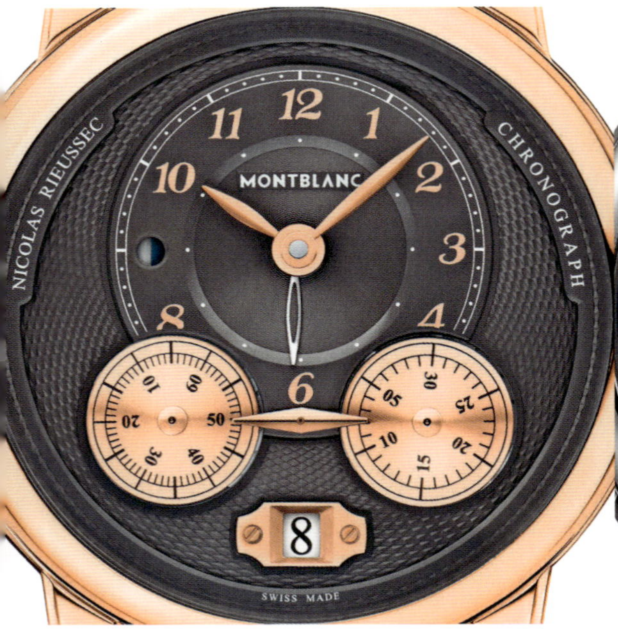

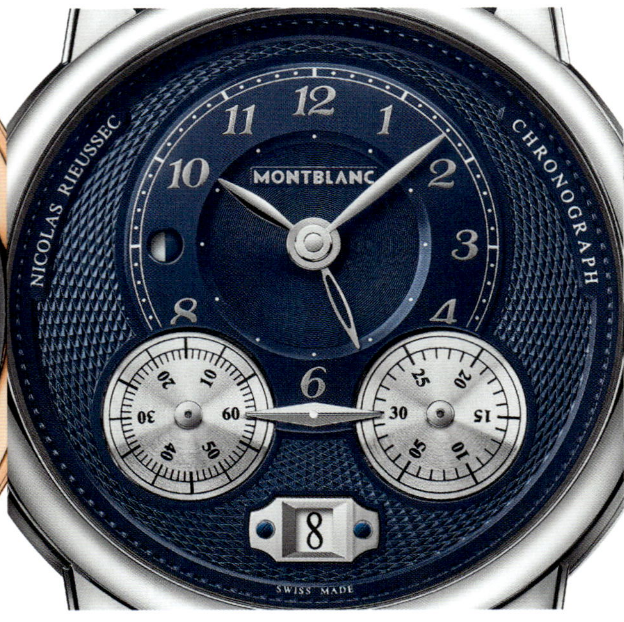

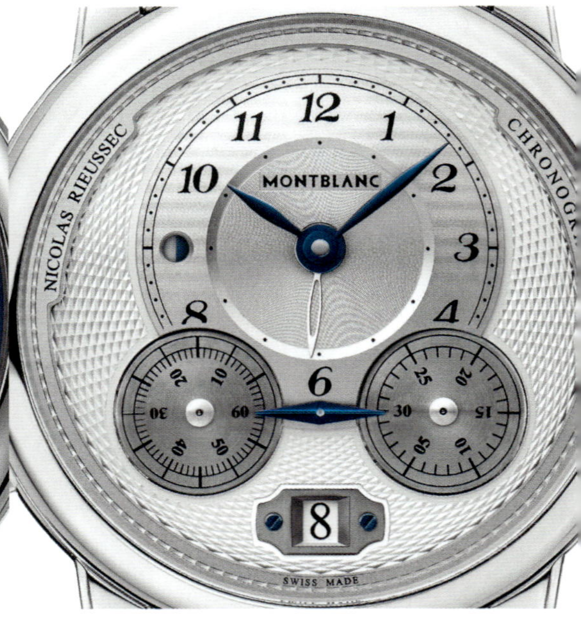

MONTBLANC RIEUSSEC CHRONOGRAPH

WRITING TIME

The word "chronograph" comes from the ancient Greek *kronos* meaning time and *grapho* meaning write. The history of this instrument unfolds in 1821 with a horse race at the Marsfeld in Paris. The forty-year-old Nicolas-Mathieu Rieussec had taken along his latest creation to the event. At the push of a button he recorded the lap times of the competitors on what he referred to as a chronograph. Everyone was impressed by the astonishingly accurate results. The upshot was a recommendation for future events of this nature and after a final report by the jury, recognition from the French Academy of Sciences. Subject of the accolade was an invention which "indicates the first and last moments of several successive time intervals without requiring the attention of the observer's eyes or ears to look at a dial or concentrate on a time signal or the oscillation of a balance wheel. … A chronograph with such a property unquestionably offers precious resources to physicians, engineers and, in general, anyone who measures phenomena." In the wake of such praise it was a foregone conclusion that the inventor received the patent for "a timepiece or distance counter, called a second chronograph" on 9 March 1822. Rieussec's chronograph had a rotating and, of course, appropriately indexed dial. The tip of a small pen marked the enamelled ring with coloured dots. The timepiece therefore had to be cleaned thoroughly for each new set of measurements.

Modern chronographs have long abandoned writing with ink. Because they now only indicate, the correct term would be chronoscope. However, the name never managed to establish itself. The German writing implement brand Montblanc, which presented its first watch collection in 1997, used the word chronograph right from the start. In the course of its horological activities, the Richemont Group member had the name Rieussec protected. There was reason to this rhyme. In 2008, Montblanc introduced new timepieces with an integrated chronograph function. Their start, stop and zeroing discs are a nod to the chronographic legacy of Nicolas-Mathieu Rieussec. The exclusive movements with column wheel and vertical clutch were co-developed with sister company ValFleurier, an acknowledged specialist. The MB R100 calibre family is manually wound. The two mainsprings in the MB R200 calibres are rotor-wound. These wristwatches stand out prominently from the wide-ranging chronograph crowd. In the course of more than 15 years, Montblanc has illustrated the development and design potential still held by the original idea of rotating discs. The many Rieussec iterations bear eloquent witness to the truth of this statement.

ZEITSCHREIBER

Aus dem Griechischen übersetzt bedeutet das Wort Chronograph nichts anderes als „Zeitschreiber". Die Geschichte dieses Instruments beginnt 1821 mit einem Pferderennen auf dem Pariser Marsfeld. Zu diesem Ereignis hatte der 1781 geborene Nicolas-Mathieu Rieussec seine neueste Kreation mitgebracht. Auf Knopfdruck erfasste das, was der Uhrmacher selber als Chronograph getauft hatte, die Rundenzeiten der Teilnehmer. Erstaunlich präzise Resultate beeindruckten „tout le monde". Sie führten nicht nur zu einer Empfehlung für künftige Begebenheiten dieser Art, sondern auch zu einer Anerkennung durch die Königliche Akademie der Wissenschaften, nachdem deren Jury darüber beratschlagt hatte. Die Würdigung bezog sich auf eine Erfindung, welche „die Dauer mehrerer aufeinander folgender Ereignisse anzeigt, ohne dass der Beobachter sich von seiner Beobachtung abwenden muss, um den Blick auf ein Zifferblatt zu werfen oder sich auf den Ton eines Zeitsignals oder der Schwingung einer Unruh zu konzentrieren. […] Ein Chronograph mit solchen Eigenschaften ist ohne Zweifel eine große Hilfe für Physiker, Ingenieure und alle anderen, die sich mit der Messung zeitlich ablaufender Ereignisse beschäftigen." Nach diesen lobenden Worten konnte es nicht ausbleiben, dass der Erfinder am 9. März 1822 das Patent erhielt für „einen Zeitmesser oder Wegzähler, genannt Sekunden-Chronograph". Rieussecs Stopper verfügte über ein rotierendes und natürlich auch entsprechend indexiertes Zifferblatt. Per Fingerdruck markierte eine Schreibspitze den emaillierten Ring mit Farbpunkten. Folglich verlangte jede neue Messreihe eine gründliche Reinigung des Zeitschreibers.

Schreiben mit Tinte gehört bei modernen Zeigerchronographen der Vergangenheit an. Weil sie nur noch anzeigen, müssten sie korrekterweise „Chronoskope" heißen. Aber dieser Terminus hat sich niemals richtig durchgesetzt. Auch die deutsche Schreibgerätemarke Montblanc, welche 1997 ihre erste Uhrenkollektion präsentierte, verwendete von Anbeginn das Wort Chronograph. Im Zuge seiner Zeit-Aktivitäten ließ sich das Mitglied des Richemont-Konzerns auch den Namen „Rieussec" schützen. Und das aus gutem Grund. 2008 stellte Montblanc neue Zeitmesser mit integrierter Stoppfunktion vor.

Deren Scheiben für Start, Stopp und Nullstellung knüpfen ganz bewusst an das chronographische Erbe von Nicolas-Mathieu Rieussec an. Die Entwicklung der exklusiven Uhrwerke mit Schaltradsteuerung und vertikaler Reibungskupplung erfolgte gemeinsam mit der einschlägig kompetenten Schwester ValFleurier. Über einen manuellen Aufzug verfügt die Kaliberfamilie MB R100, bei den Kalibern MB R200 spannt ein Rotor die beiden Zugfedern. Vom Üblichen auf dem breit gefächerten Gebiet der Chronographen heben sich diese Armbanduhren deutlich ab. Im Laufe von mehr als 15 Jahren hat Montblanc bewiesen, welche Entwicklungs- und Gestaltungsmöglichkeiten in der ursprünglichen Idee rotierender Scheiben stecken. Die Vielzahl vorgestellter „Rieussec"-Modelle lässt an Deutlichkeit nichts zu wünschen übrig.

REF. MB 130986, 2023

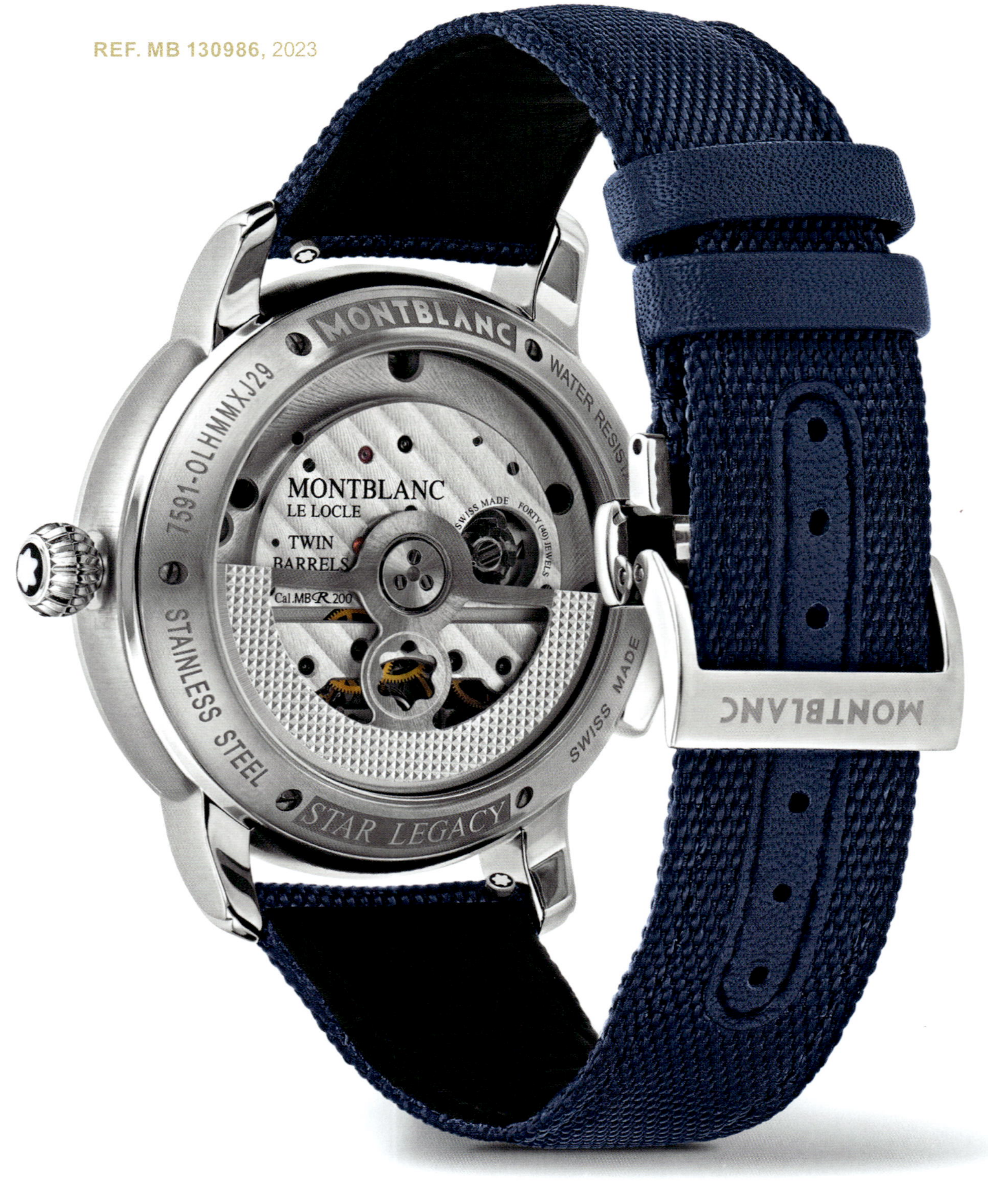

REF. MB 133232, 2024

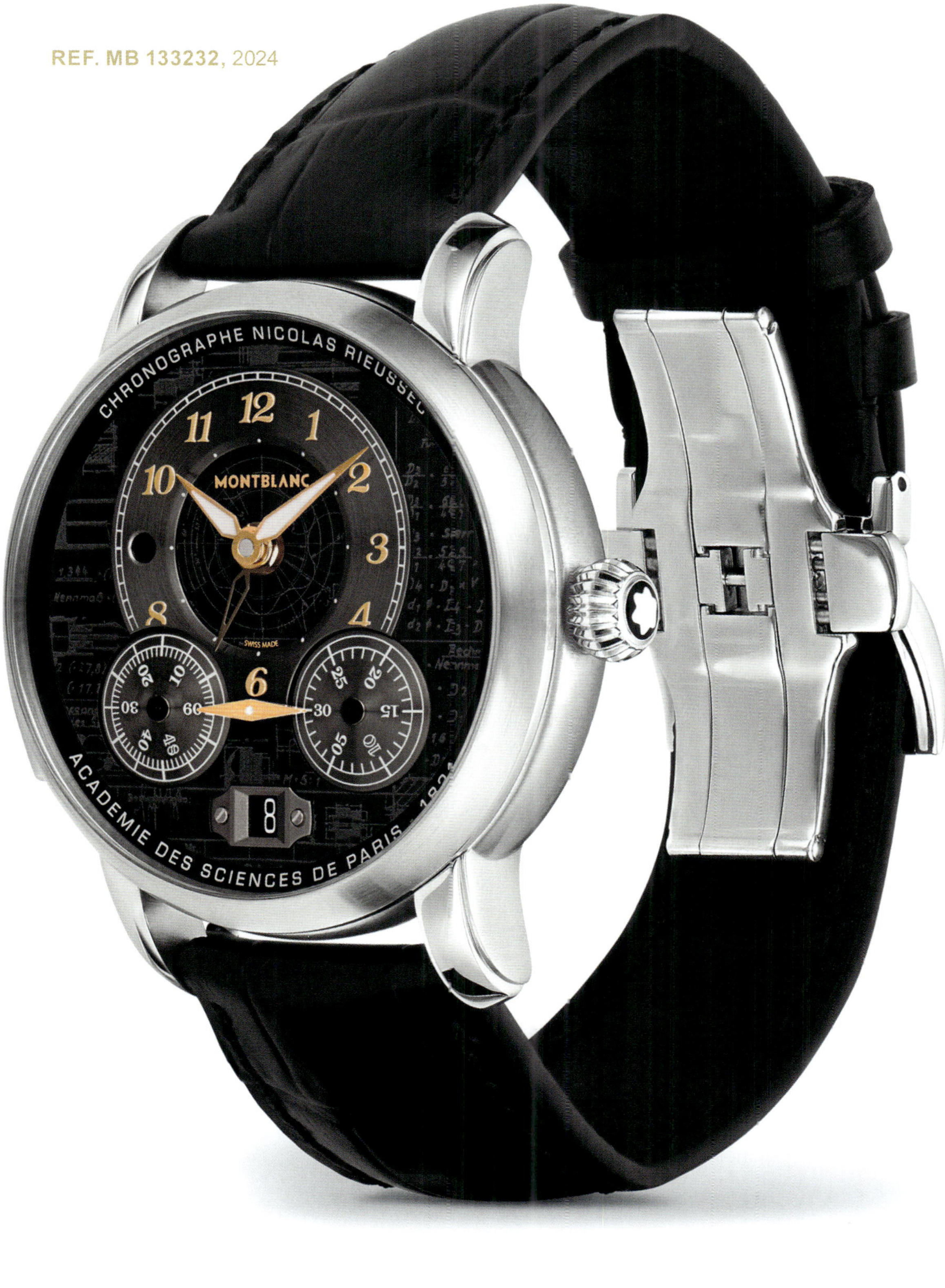

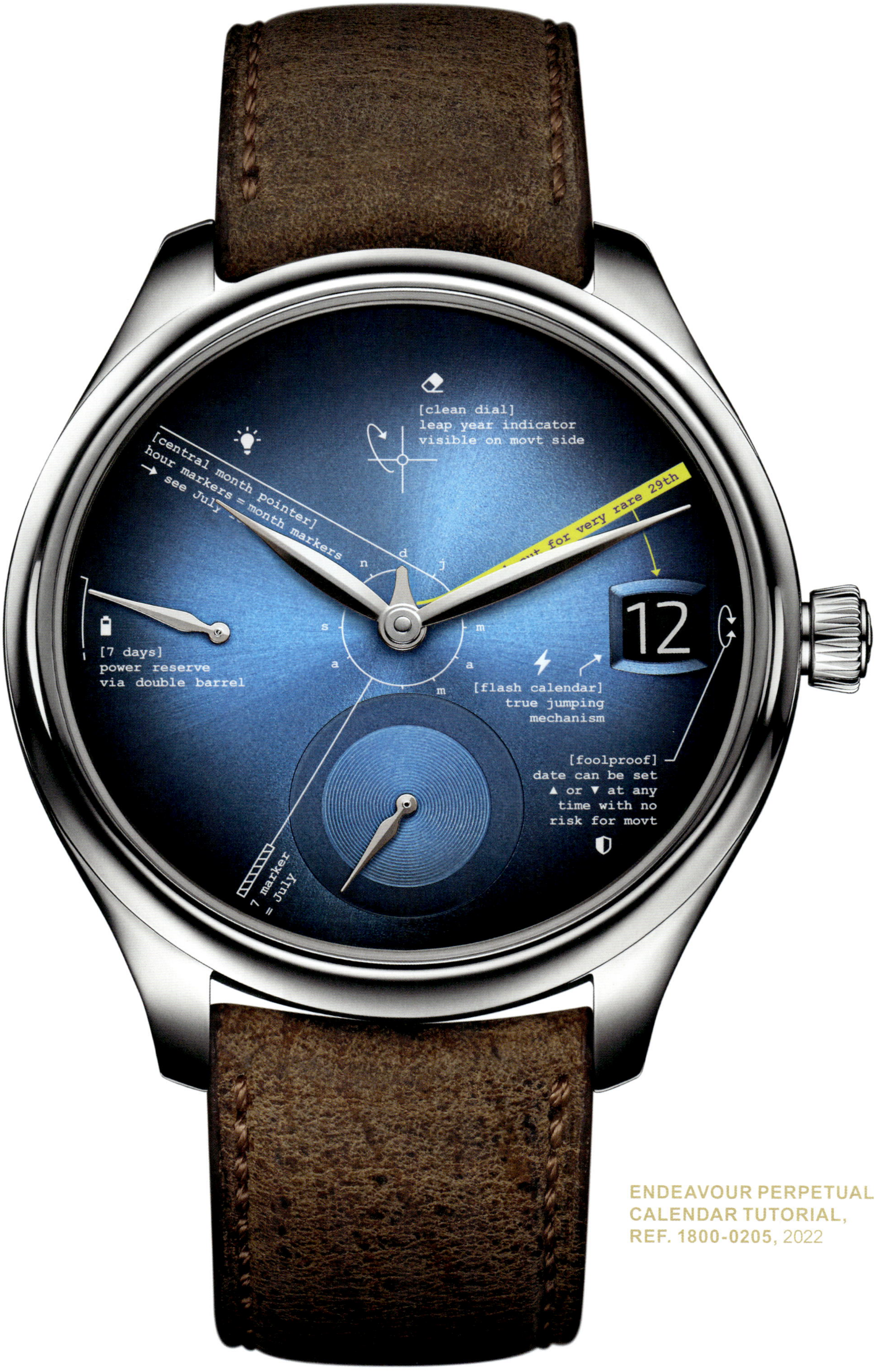

ENDEAVOUR PERPETUAL CALENDAR TUTORIAL, REF. 1800-0205, 2022

2005 H. MOSER & CIE.

ENDEAVOUR PERPETUAL CALENDAR

VIVE LA DISCRÉTION

The design of H. Moser & Cie watches is eclectic. One reason is undoubtedly that the independent manufactory enjoys flaunting conventions in the watch world. Its Endeavour line is intended to appeal to purists. The synthesis of form, colour and contours perfectly embodies the watchmaking DNA of the Swiss company. Based in Neuhausen am Rheinfall, it was founded in 1828 and made its comeback in 2005 after a protracted hiatus. A special role in its revival was played by a perpetual calendar, the complexity of which is still not apparent even at a second glance. This is down to the proprietary Calibre HMC 341 with a classic 2.5-hertz balance frequency. A self-winding system can be dispensed with in this case without a qualm. Firstly, after winding, the power reserve is good for at least a week and secondly, the lavishly sized power reserve indication at nine o'clock reminds the wearer to rewind their watch in good time. Otherwise, the customary features of watches with a perpetual calendar are absent from the dial. At three o'clock, the big date window catches the eye. By contrast, the very short hand in the centre of the dial is far more modest in appearance. Discretion being the key, it uses the 12-hour indexes to indicate the respective month of the current calendar year. Because this is not enough to set this practical complication correctly, H. Moser & Cie. also provides a leap-year indication. It is tucked away discreetly on the back of the hand-wound movement. Other credentials include a double-barrel, interchangeable Moser escapement, a proprietary Original Straumann Hairspring with a Breguet curve and a hack mechanism. The patented calendar movement can be adjusted backwards and forwards at any time of day or night via the crown. No damage can be caused in the process. This convenient detail therefore also ensures peace of mind.

VIVE LA DISCRÉTION!

Sie haben viele Gesichter, die Armbanduhren von H. Moser & Cie. Diese Vielfalt ist sicher auch darin begründet, dass die unabhängige Manufaktur in der Uhrenszene gerne ein wenig provokant auftritt. An Menschen mit Faible fürs Puristische wendet sich die Linie „Endeavour". Idealtypisch verkörpert diese Synthese aus Form, Farbe und Linienführung den uhrmacherischen Geist des 1828 gegründeten und 2005 nach längerer Pause in Neuhausen am Rheinfall wiederbelebten Unternehmens. Eine besondere Rolle spielt in diesem Zusammenhang ein ewiger Kalender, dem man die mechanische Komplexität selbst auf den zweiten Blick nicht ansieht. Der Grund ist im selbst entwickelten Kaliber HMC 341 mit sehr klassischen 2,5 Hertz Unruhfrequenz zu suchen. Auf eine Automatik kann man in diesem Fall guten Gewissens verzichten. Zum einen beträgt die Gangautonomie nach Vollaufzug mindestens eine Woche und zum anderen erinnert eine großzügig dimensionierte Gangreserveanzeige bei „9" an den rechtzeitigen Energienachschub. Ansonsten fehlen am Zifferblatt die üblichen Merkmale von Uhren mit immerwährendem Kalendarium. Bei „3" sticht das große Datumsfenster ins Auge. Weit weniger eindrucksvoll präsentiert sich ein sehr kurzer, in der Mitte des Zifferblatts drehender Zeiger. Er korrespondiert mit der Stundenindexierung, weist damit also auf den jeweiligen Monat des laufenden Kalenderjahrs. Weil das zum korrekten Einstellen dieser hilfreichen Zusatzfunktion nicht reicht, bietet H. Moser & Cie. auch noch eine Schaltjahresindikation. Die befindet sich diskret auf der Rückseite des Handaufzugswerks. Zu dessen Spezifika gehören Doppelfederhaus, austauschbare Moser-Hemmung, eine selbst produzierte Straumann-Unruhspirale mit hochgebogener Endkurve sowie ein Sekundenstopp. Per Krone lässt sich das patentierte Kalenderwerk zu jeder Tages- und Nachtzeit beliebig vor- und rückwärts korrigieren. Einen Schaden kann man dabei nicht anrichten. Das ist nicht nur komfortabel, sondern auch eine sehr beruhigende Angelegenheit.

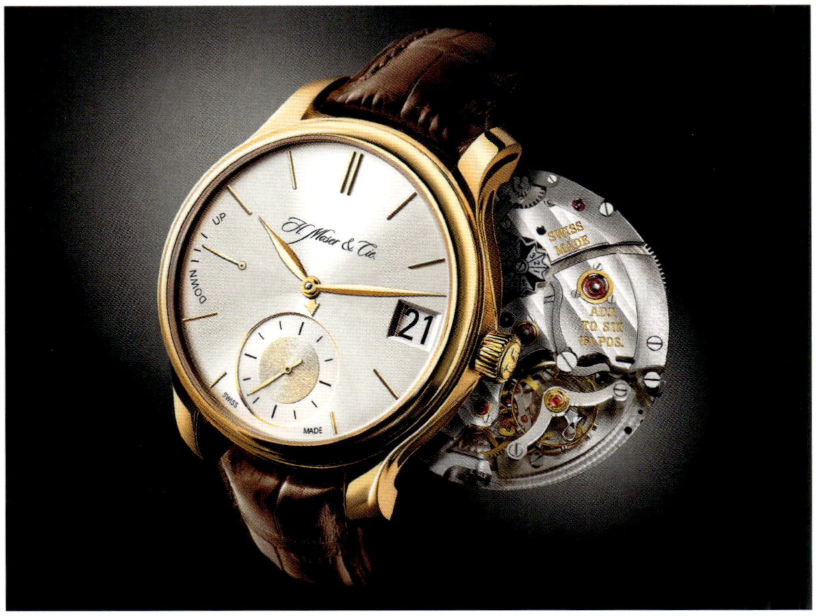

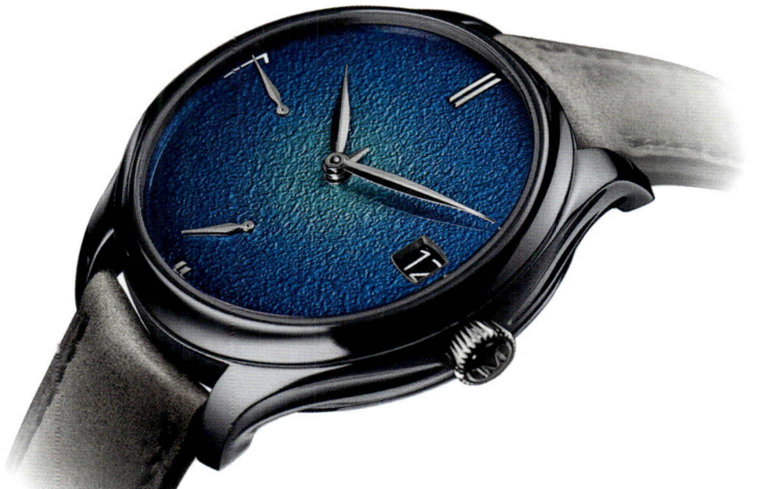

From left | Von links: *Endeavour Perpetual Calendar, Ref. 1341-0103, 2005 | Endeavour Perpetual Calendar Tantalum Blue Enamel, Ref. 1800-2000, 2023*

MÜHLE-GLASHÜTTE 2002

S.A.R. RESCUE-TIMER

FROM PRACTICE FOR PRACTICE

In the wake of the historic turnaround in Germany after the fall of the Wall in 1989, the Glashütte Mühle family company became the property of the Treuhand agency. A born entrepreneur, Hans-Jürgen Mühle founded the Mühle-Glashütte GmbH nautische Instrumente und Feinmechanik company in 1994. It seems more than logical that the small start-up opted to focus on its past expertise in nautical instruments during the first two years of its existence. Professional marine chronometers, ship clocks and other instruments such as barometers and hygrometers provided a livelihood for the small team.

In 1996, Mühle-Glashütte launched its first wristwatch on the market in the shape of the marine pilot's watch. This business sector was to become one of the mainstays of the company. The undisputed leader of the product range remains the S.A.R. Rescue-Timer presented in 2002. The creation of this wristwatch with a rubber strap, of which more than 10,000 have been produced so far, draws its inspiration from the consistent collection of insights and wishes. For nine months, the product developers in Müglitztal, Saxony, engaged in an intensive exchange with crew members from some of the German Maritime Search and Rescue Service (DGzRS) rescue cruisers. Since the service's foundation in 1865, sea rescue crews in the North and Baltic Seas have rescued more than 86,000 people in distress or danger and have done so as part of a charitable mission that is voluntary, independent and relies solely on donations. These highly resilient timepieces are on permanent duty and thus subject to continuous testing, braving the elements around the clock on the DGzRS's current fleet of 60 sea rescue cruisers and boats. Where expedient, the practical experience of the coxswains and crews is taken into account. The water resistance of the marine timepieces with their characteristic screw crown at four o'clock reaches up to 100 bar pressure or a depth of some 1,000 metres. As expected, an extremely stable and seawater-resistant stainless-steel case with four-millimetre-thick sapphire crystal is needed for the purpose. Inside, the automatic Calibre Sellita SW200 with shock-resistant woodpecker-neck fine adjustment, new regulator, counter-pressure spring and fine adjustment screw keeps the time. Special luminous indices and hands guarantee up to six hours of reliable reading even in poor light conditions.

Since 2022, Mühle-Glashütte has also offered the S.A.R. Rescue-Timer with a lightweight, anti-allergic titanium case. However, it is only waterproof up to 50 bar.

AUS DER PRAXIS FÜR DIE PRAXIS

Im Zuge der legendären „Wende", die Deutschland nach dem Fall der Mauer im Jahr 1989 erfasste, gelangte auch das Erbe der Glashütter Familie Mühle ins Eigentum der Treuhand. Weil eigenes Unternehmertum zu Hans-Jürgen Mühle gehört wie das Salz zur Suppe, rief er 1994 die „Mühle-Glashütte GmbH nautische Instrumente und Feinmechanik" ins Leben. Dass das kleine Startup in seinen ersten beiden Jahren an die einschlägige Vergangenheit anknüpfte und dabei das Thema nautische Instrumente großschrieb, klingt mehr als logisch. Professionelle Marinechronometer, Schiffsuhren sowie andere Instrumente wie Baro- und Hygrometer ernährten das kleine Team. Mit der „Marinefliegeruhr-I" brachte Mühle-Glashütte 1996 seinen ersten Zeitmesser fürs Handgelenk auf den Markt. Anschließend entwickelte sich dieser Geschäftszweig zur tragenden Säule. Unangefochtener Leader in der Produktpalette ist nach wie vor der 2002 präsentierte „S.A.R. Rescue-Timer". Die Kreation dieser Armbanduhr mit Kautschukband, von der bislang über 10.000 Exemplare entstanden sind, verknüpft sich mit dem konsequenten Sammeln von Erkenntnissen und Wünschen. Neun Monate lang tauschten sich die Produktentwickler im sächsischen Müglitztal mit Besatzungsmitgliedern einiger Seenotrettungskreuzer der Deutschen Gesellschaft zur Rettung Schiffbrüchiger (DGzRS) aus. Seit ihrer Gründung im Jahr 1865 haben die Seenotretter auf Nord- und Ostsee mehr als 86.000 Schiffbrüchige aus Seenot gerettet oder Gefahr befreit und noch viel mehr Menschen ihre gemeinnützige Hilfe angedeihen lassen – freiwillig, unabhängig und rein spendenfinanziert. Auf den gegenwärtig rund 60 Seenotrettungskreuzern und -booten der DGzRS zwischen Borkum im Westen und Ueckermünde im Osten befinden sich diese hoch belastbaren Zeitmesser nicht nur im fortwährenden Einsatz, sondern auch im Dauertest – bei jedem Wetter und rund um die Uhr. Sofern nötig, fließen praktische Erfahrungen der Vorleute und Besatzungen in die Evaluation ein. Die Wasserdichte der Rettungsboliden mit charakteristischer Schraubkrone bei „4" reicht bis zu 100 bar Druck, was etwa 1.000 Metern Tiefe entspricht. Hierzu braucht es logischerweise ein extrem stabiles und gegen Seewasser resistentes Edelstahlgehäuse mit vier Millimeter dickem Saphirglas. Die Zeit hütet das zuverlässige Automatikkaliber Sellita SW200 mit der stoßsicheren Spechthals-Feinregulierung mit neuem Rücker, Gegendruckfeder und Feinregulierschraube. Spezielle Leuchtindexe und -zeiger gewährleisten bis zu sechs Stunden lang sicheres Ablesen bei schlechten Lichtverhältnissen. Seit 2022 bietet Mühle-Glashütte den „S.A.R. Rescue-Timer" auch mit leichter, antiallergischer Titanschale an. Dem Druck des nassen Elements widersteht dieser allerdings nur bis zu 50 bar.

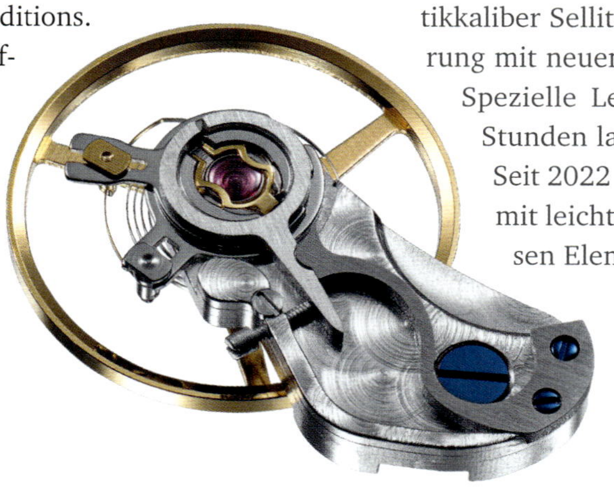

***Left:** Shock-resistant woodpecker-neck fine adjustment |*
***Links:** Stoßsichere Spechthals-Feinregulierung*

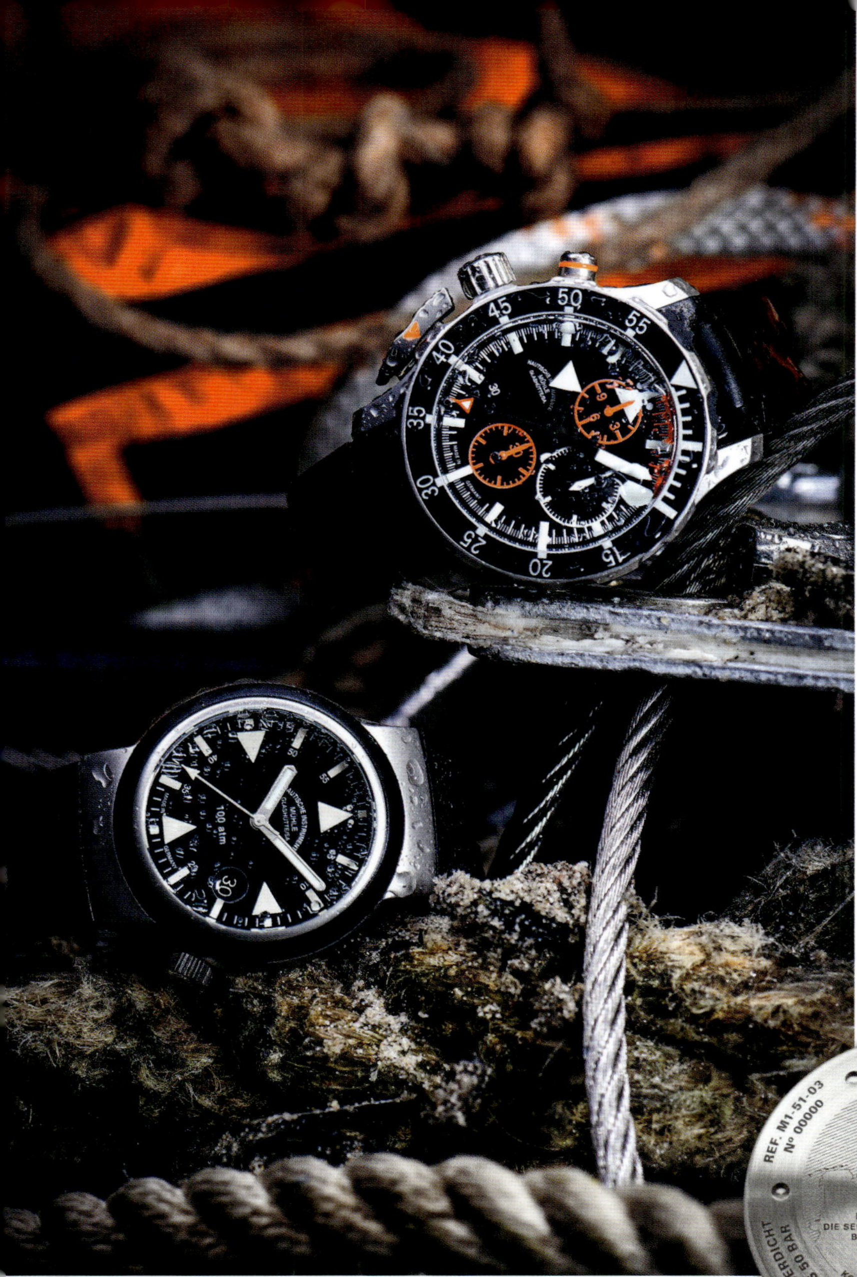

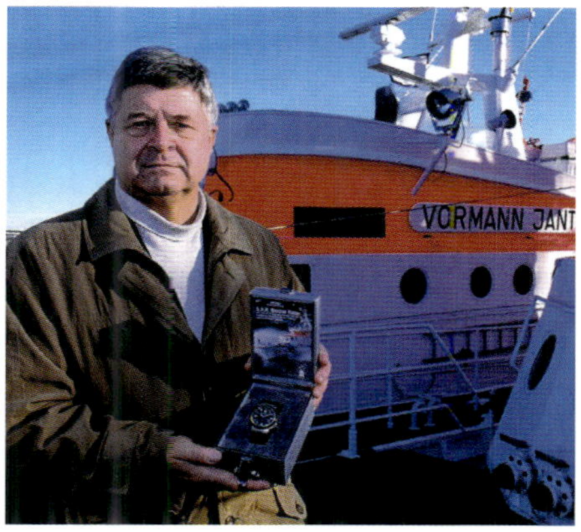

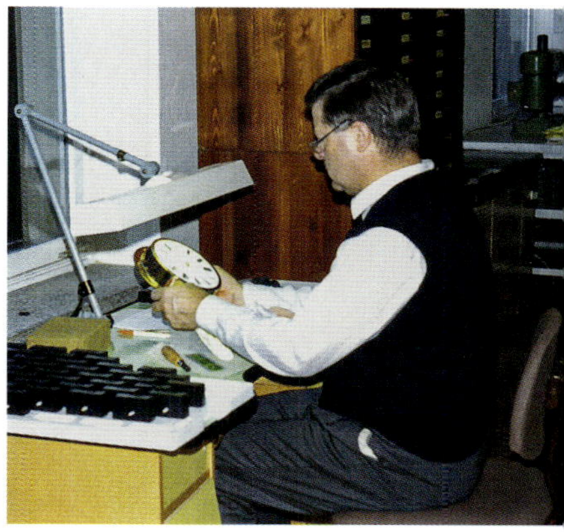

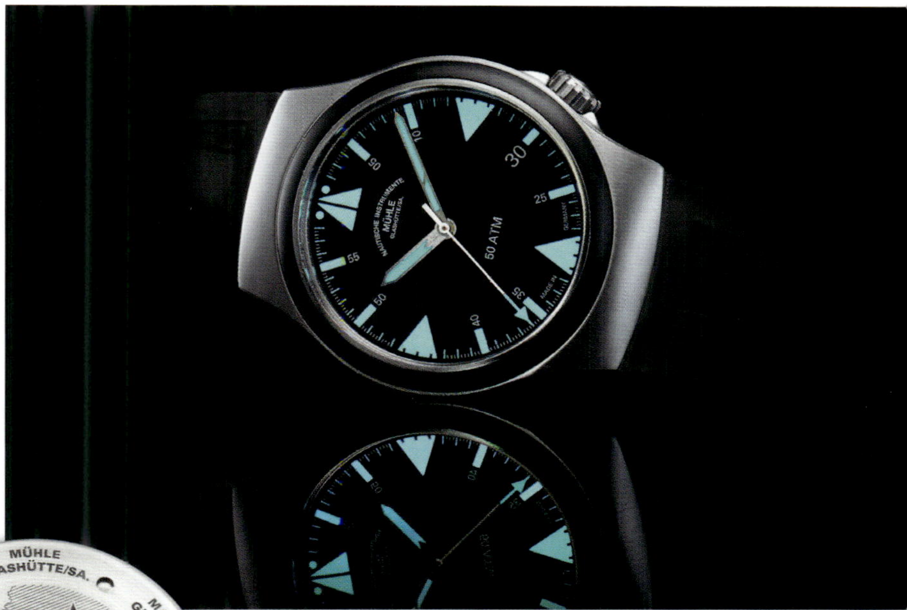

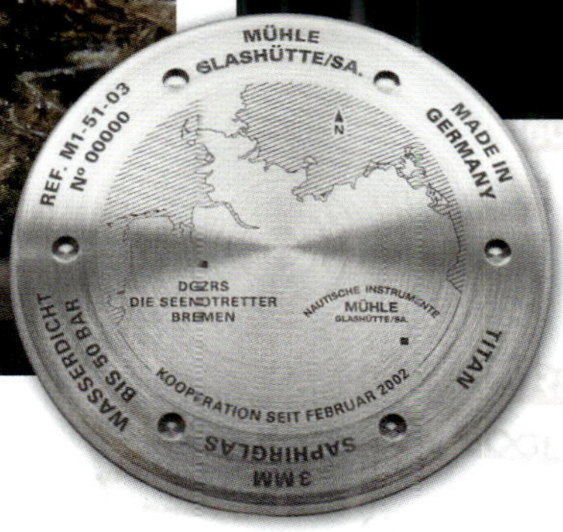

Top left | Oben links: *S.A.R. Rescue-Timer & Flieger-Chronograph*

Top middle: *Hans-Jürgen Mühle hands over the first S.A.R. Rescue-Timer to the Maritime Search and Rescue Service, 2002* **|**
Oben Mitte: *Hans-Jürgen Mühle übergibt den ersten S.A.R. Rescue-Timer an die Seenotrettung, 2002*

Top right: *Hans-Jürgen Mühle builds a ship's clock* **|**
Oben rechts: *Hans-Jürgen Mühle baut eine Schiffsuhr*

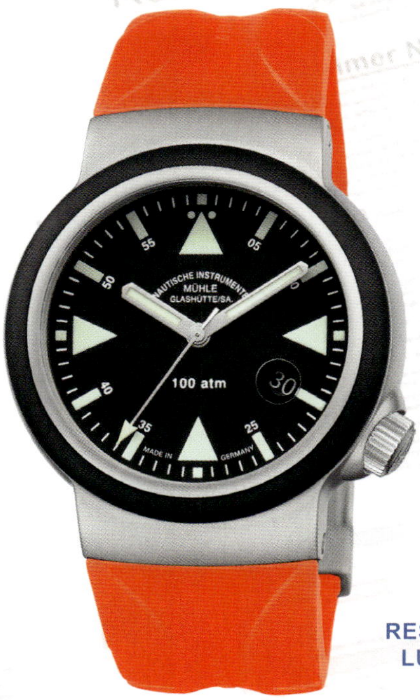

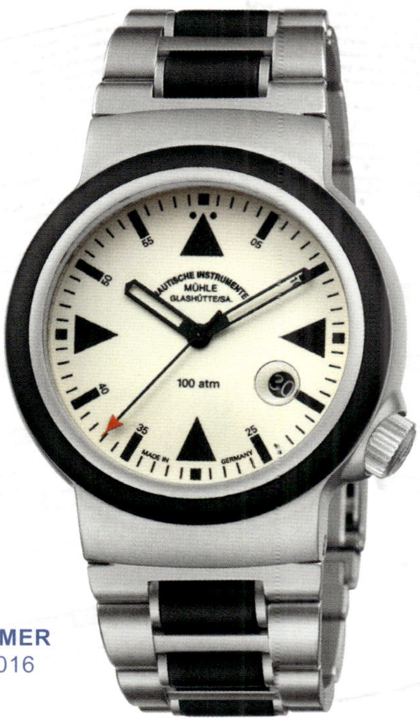

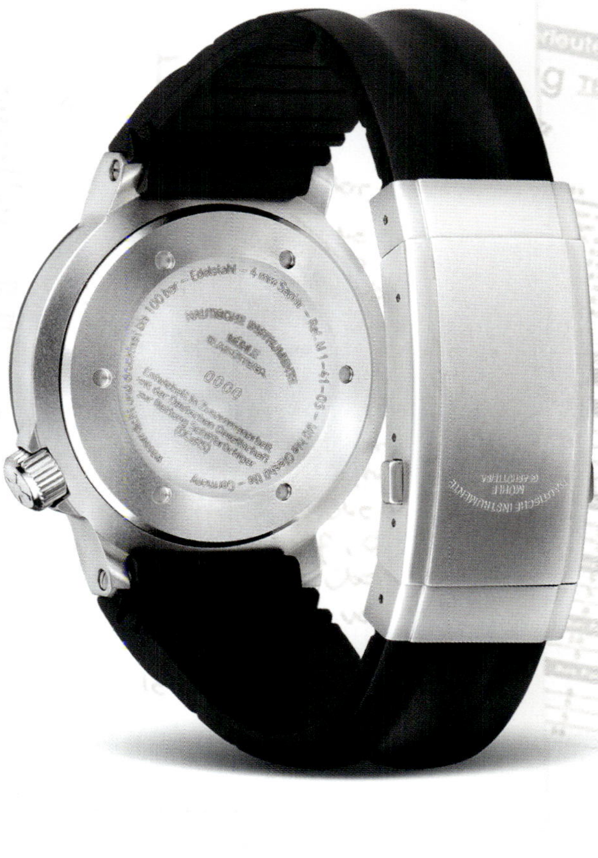

**S.A.R.
RESCUE-TIMER
LUMEN,** 2016

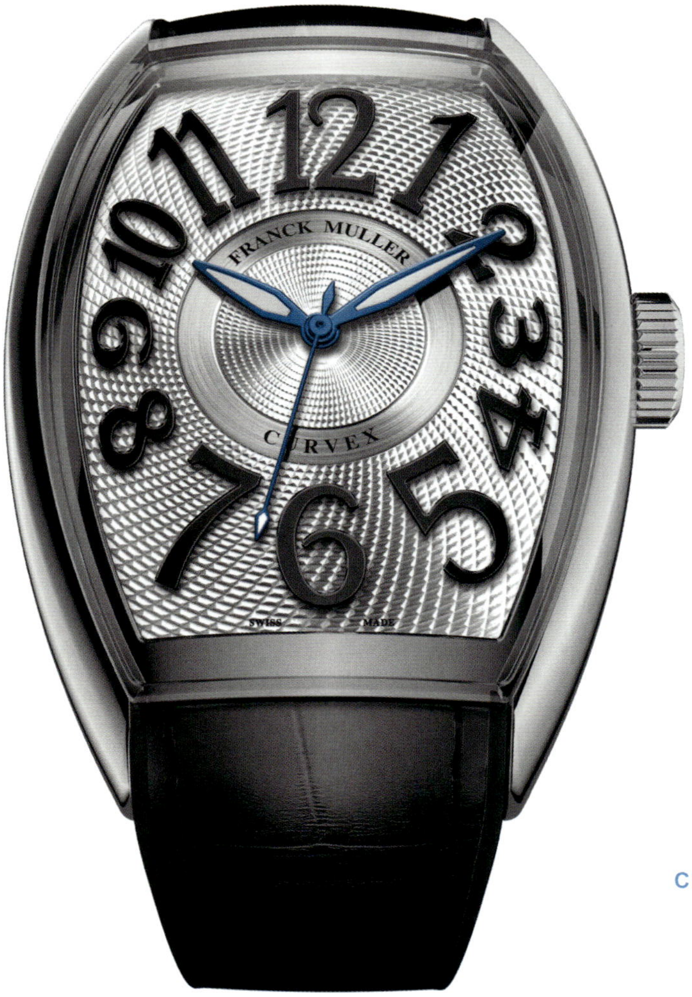

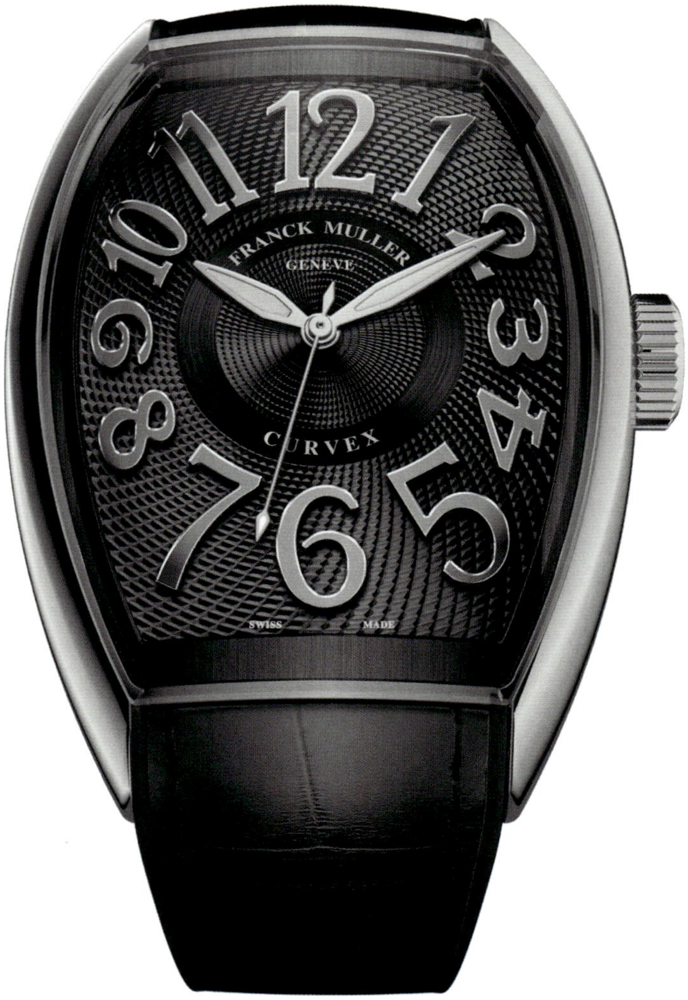

CURVEX CX

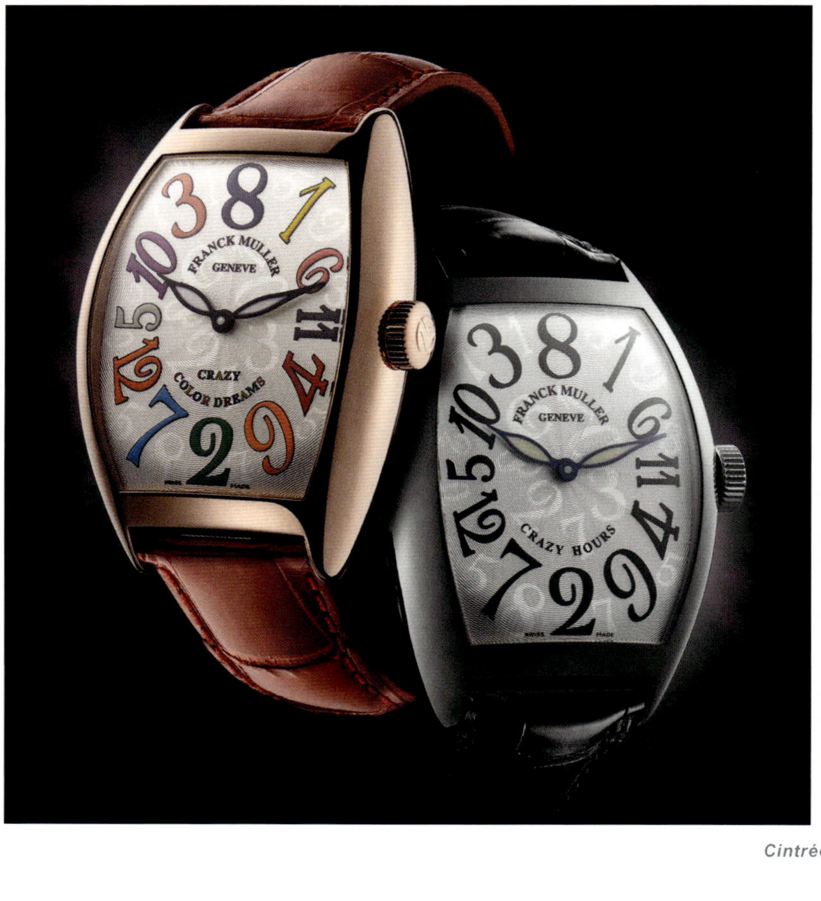

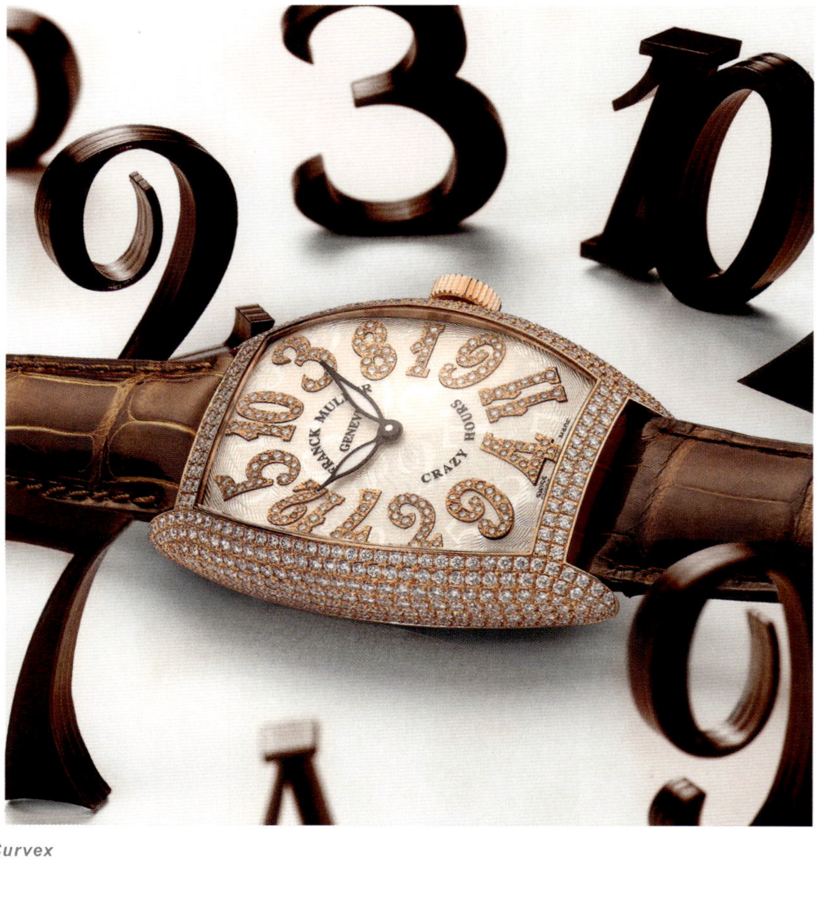

Cintrée Curvex

1992

FRANCK MULLER
CINTRÉE CURVEX

RETRO LOOK FROM GENEVA

What do Paul Anka, Franz Beckenbauer, Naomi Campbell, Bill Clinton, Celine Dion, Michael Jordan, Jennifer Lopez, Silvester Stallone, Barbara Streisand, Valentino, Bruce Willis and Robin Williams have in common? Simple: they own or have owned at least one watch by Franck Muller. And they all bought them themselves. The Geneva watch brand began production in 1991. The brand's namesake is however no longer on board.

One strand of its success story stems from a consistent cultivation of the retro look. It pays homage to the tonneau shape popularly sported on wrists during the Roaring Twenties. Franck Muller found the name "Cintrée Curvex" a more fitting appellation than "tonneau". Different sizes appeal to both men and women. The large hour numerals, some of which are vibrantly coloured, also recall bygone times. Brilliant hues take centre stage in the current Vanguard Color Dreams model. Lovers of simplicity will also find something to their taste. As a special feature, the brand has endowed the automatic movement with a platinum rotor.

RETROLOOK AUS GENF

Was haben Paul Anka, Franz Beckenbauer, Naomi Campbell, Bill Clinton, Celine Dion, Michael Jordan, Jennifer Lopez, Silvester Stallone, Barbara Streisand, Valentino, Bruce Willis und Robin Williams gemein? Ganz einfach: Sie besitzen oder besaßen mindestens eine Armbanduhr mit der Signatur Franck Muller. Und die haben sie allesamt selber erworben. 1991 nahm die Genfer Uhrenmarke ihre Produktion auf. Der Namensgeber ist allerdings nicht mehr an Bord.

Die Erfolgsgeschichte gründet sich unter anderem auf einer konsequenten Pflege des Retrolooks. Selbiger huldigt der Fassform, die in den „Roaring Twenties" an Handgelenken reüssierte. Franck Muller fand für das „Tonneau" den in seinen Augen treffenderen Namen „Cintrée Curvex". Verschiedene Größen sprechen Menschen beiderlei Geschlechts an. An vergangene Epochen erinnern auch die großen, teilweise sehr farbig ausgeführten Stundenziffern. Diese brillieren beispielsweise in der aktuellen „Vanguard Color Dreams". Wer es schlichter liebt, kommt natürlich auch zu seinem Recht. Als Besonderheit spendiert die Marke den verbauten Automatikwerken einen Platinrotor.

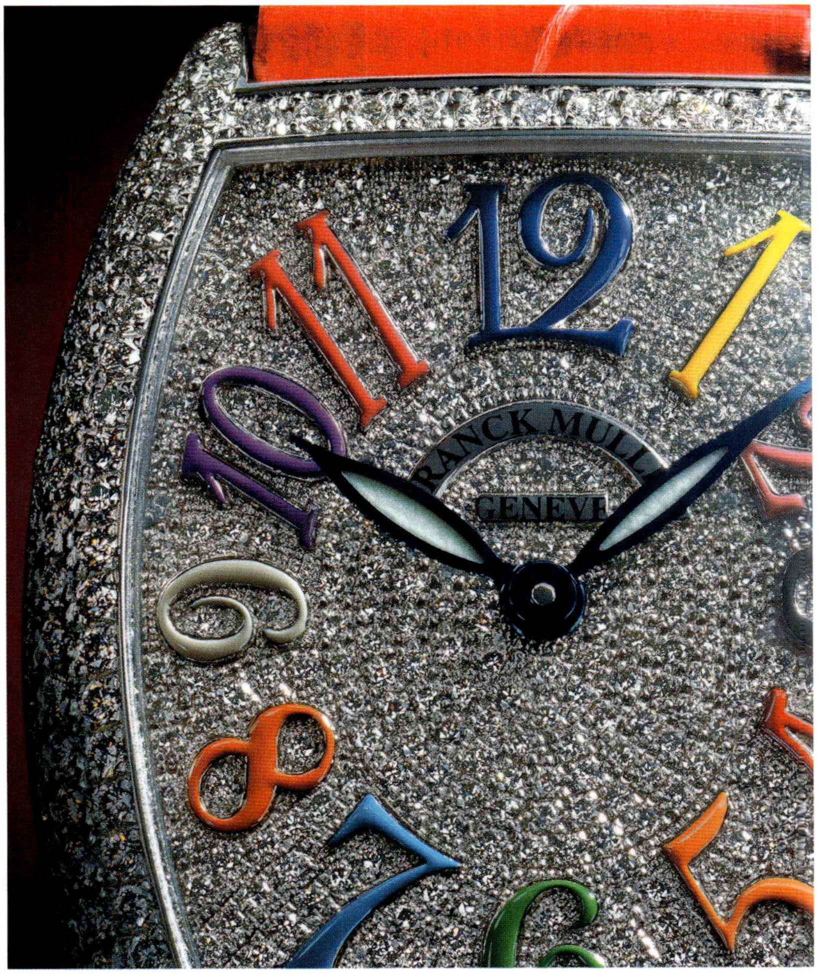

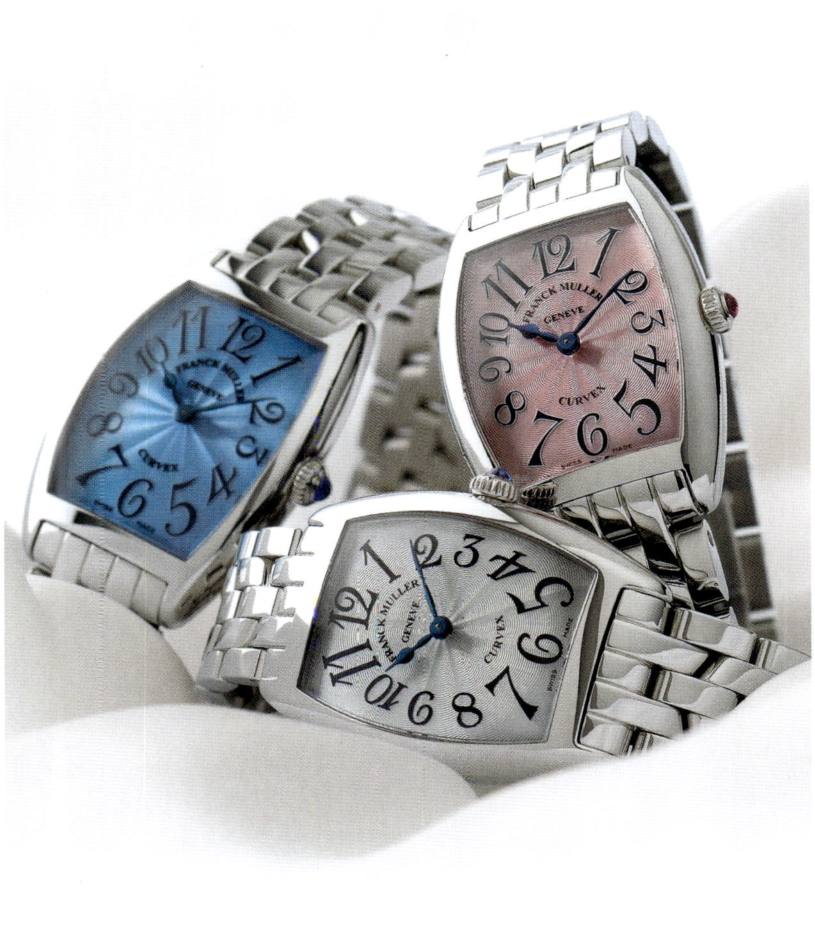

From left / Von links: *Color Dreams / Cintrée Curvex Lady*

NOMOS GLASHÜTTE 1992
TANGENTE

A SYMBIOSIS OF ART AND ENGINEERING

Glashütte is not renowned for its cosmopolitan flair. Nor has becoming Germany's watchmaking hub changed this. In the German Democratic Republic, the Müglitz valley in which the small Saxon town nestles was nicknamed the "Valley of the Clueless". This was because the area was not able to receive television broadcasts from West Germany. A state of affairs that soon changed after the fall of the Wall in 1989. In January 1991, Roland Schwertner rebirthed the name of the former "pirate" brand Nomos. The name goes back to 1906, the year in which Clemens Guido Müller and his brother-in-law Karl Nierbauer began to import and sell Swiss watches boasting the prestigious Glashütte appellation. The local watchmaking establishment filed legal proceedings. The first chapter of the Nomos biography ended in 1910 with a settlement. To begin with, IT specialist Schwertner had to contend with similar issues. Competitors protested strongly against Swiss calibres in Nomos watches, although from the outset the Glashütte content was higher than 50 per cent. From 1992 onwards, the classical styling of the Tangente with its minimalist design became synonymous with Nomos. The icon pays tribute to the motto of Bauhaus founder Walter Gropius from 1922: "Art and technology – a new unity." It is based on a sketch by the designer Susanne Günther, who in turn was inspired by a similar wristwatch from the 1930s. Multiple awards and sales of more than 50,000 models are ample proof that all involved had struck the right chord with the design revival. An important member of the Tangente collection is the classic version with the exclusive hand-wound Alpha calibre. Those who wish spare themselves the bother of manual winding can opt for the equally purist Tangente neomatik with the proprietary automatic Calibre DUW 3001. The three initials stand for Deutsche Uhren Werke (German Watch Movements), signalling to connoisseurs the use of the Nomos swing system, part of which is a proprietary balance spring. Either or any other way, the Nomos Glashütte Tangente is both an icon and cult object.

KUNST UND TECHNIK, EINE EINHEIT

Von kosmopolitischem Flair ist Glashütte weit entfernt. Daran hat auch die Entwicklung zum deutschen Uhrenmekka nichts geändert. In der Deutschen Demokratischen Republik (DDR) bezeichnete man das Müglitztal, in dem die sächsische Kleinstadt liegt, sogar als „Tal der Ahnungslosen". Der Grund: Westliches Fernsehen war nicht zu empfangen. Nach dem Fall der Mauer im Jahre 1989 änderte sich das schnell. Bereits im Januar 1991 bescherte Roland Schwertner dem Namen der ehemaligen Piratenmarke Nomos eine Renaissance. Der Name reicht zurück bis ins Jahr 1906. Da nämlich starteten Clemens Guido Müller und sein Schwager Karl Nierbauer den Import und Vertrieb von Schweizer Uhren mit der imageträchtigen Zusatzsignatur „Glashütte". Dagegen wiederum klagte das dortige Uhr-Establishment. Mit einem Vergleich endete 1910 das erste Kapitel der Nomos-Biographie. Anfangs musste EDV-Spezialist Schwertner mit ähnlichen Problemen kämpfen. Wettbewerber protestierten heftig gegen Schweizer Kaliber in Nomos-Uhren, obwohl die Glashütter Wertschöpfung von Beginn an höher lag als 50 Prozent. Ab 1992 entwickelte sich die klassisch gestaltete, optisch auf das unabdingbar Notwendige reduzierte „Tangente" zu einem Synonym für Nomos. Die Ikone huldigt dem 1922 formulierten Postulat des „Bauhaus"-Gründers Walter Gropius: „Kunst und Technik, eine neue Einheit". Sie basiert auf einem Entwurf der Designerin Susanne Günther, welche sich ihrerseits an einer ähnlichen Armbanduhr aus den 1930er-Jahren orientiert hatte. Zahlreiche Auszeichnungen und schätzungsweise mehr als 50.000 verkaufte Exemplare belegen, dass alle Beteiligten mit der gestalterischen Rückbesinnung richtig lagen. Ein wichtiges Mitglied der „Tangente"-Kollektion ist die klassische Ausführung mit dem exklusiven Handaufzugskaliber „Alpha". Mag man die Zugfeder nicht selbst in regelmäßigen Abständen spannen, bietet sich die nicht minder reduziert auftretende „Tangente neomatik" mit dem hauseigenen Automatikkaliber DUW 3001 an. Die drei Buchstaben stehen für „Deutsche Uhren Werke" und signalisieren Kennern die Verwendung des so genannten Nomos-Swing-Systems, zu dem auch die selbst entwickelte Unruhspirale gehört. So, so, oder auch noch anders: Die „Tangente" von Nomos Glashütte ist Ikone und Kultobjekt zugleich.

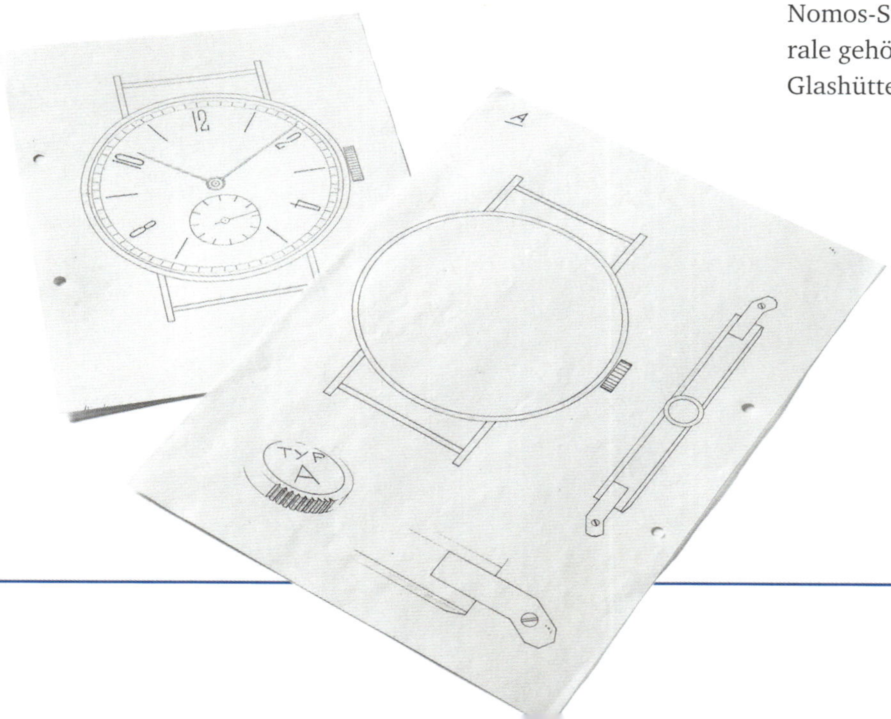

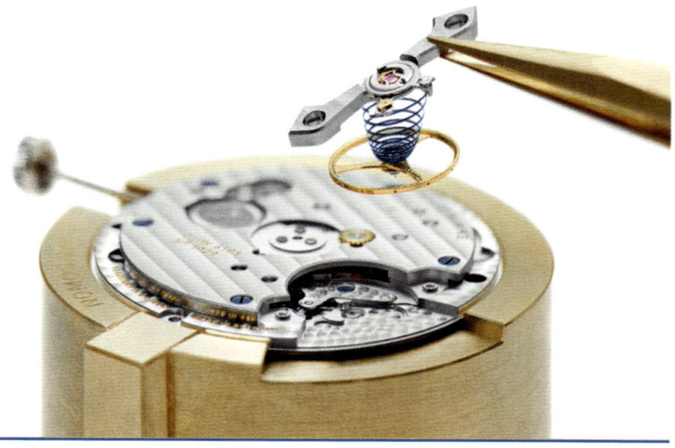

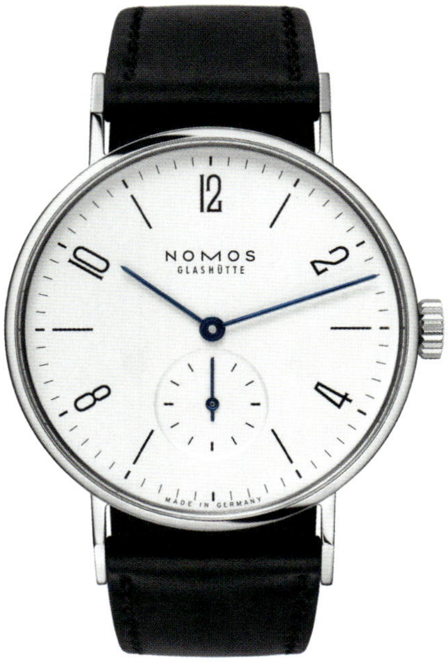

REF. 101/139

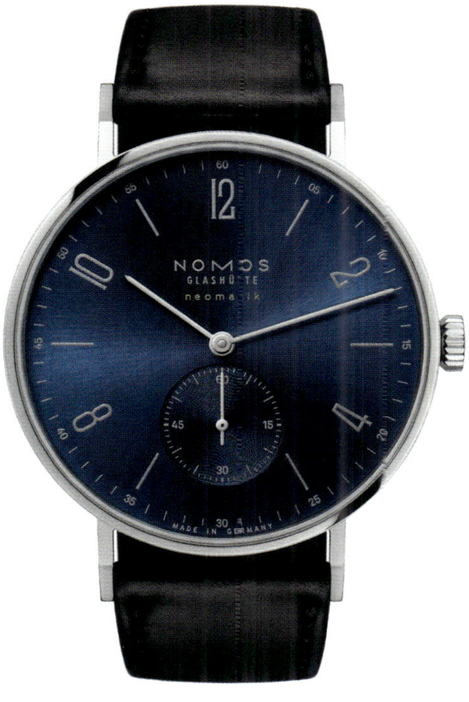

REF. 145/146

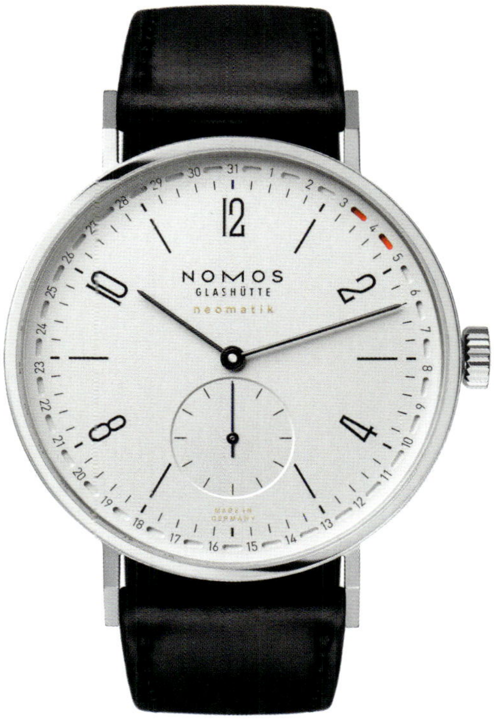

REF. 180

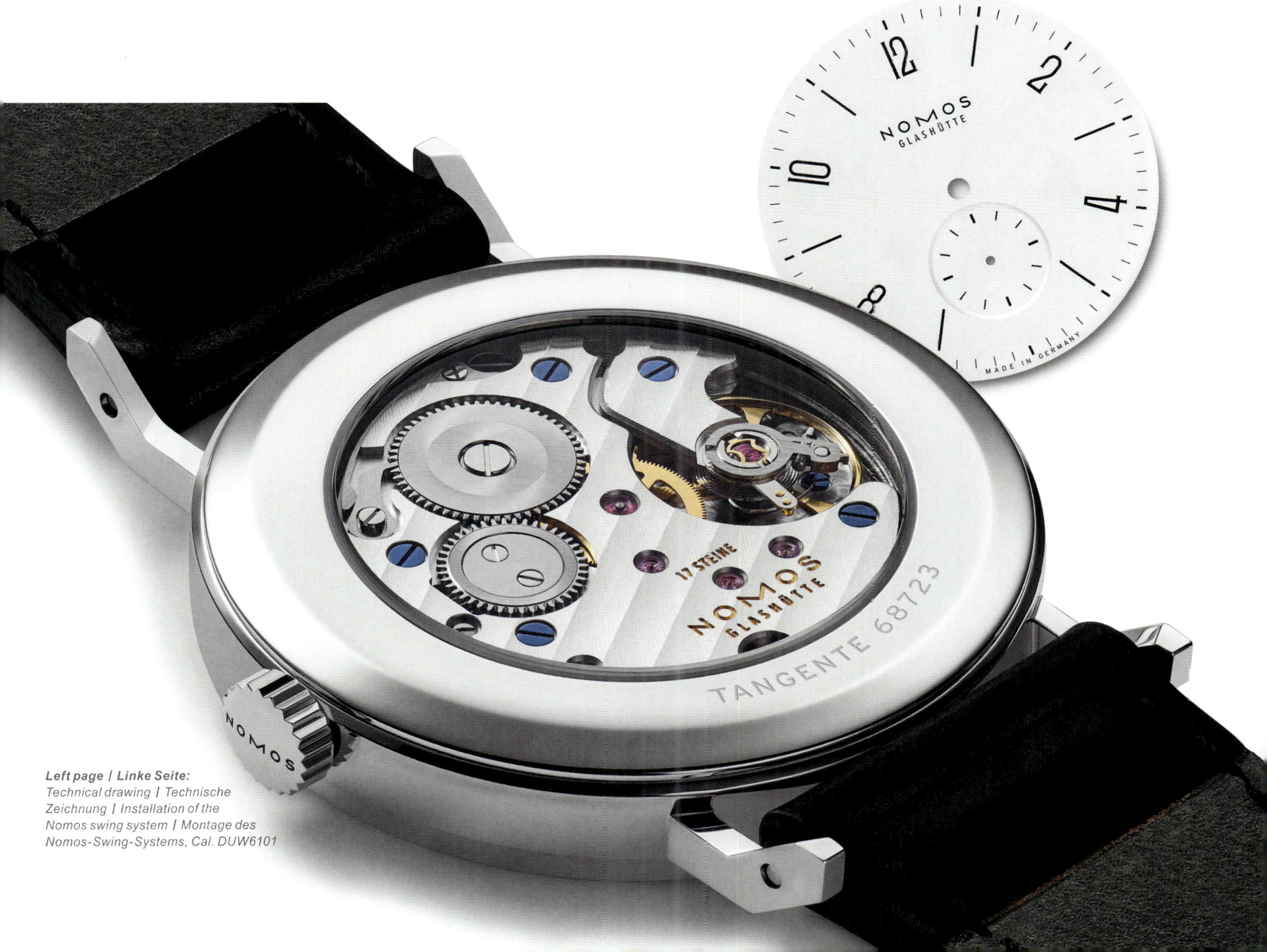

Left page | Linke Seite:
Technical drawing | Technische Zeichnung | Installation of the Nomos swing system | Montage des Nomos-Swing-Systems, Cal. DUW6101

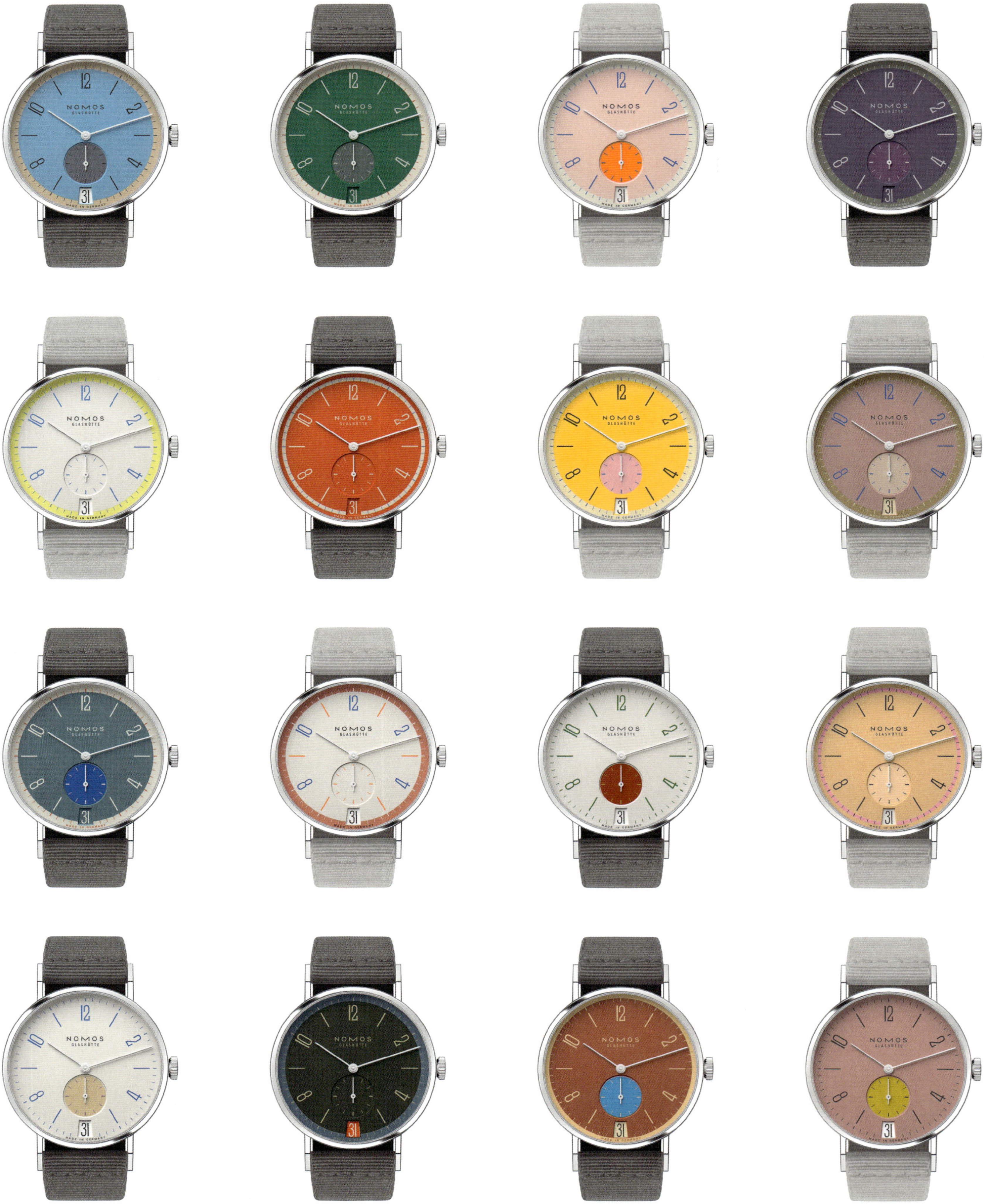
NOMOS
GLASHÜTTE

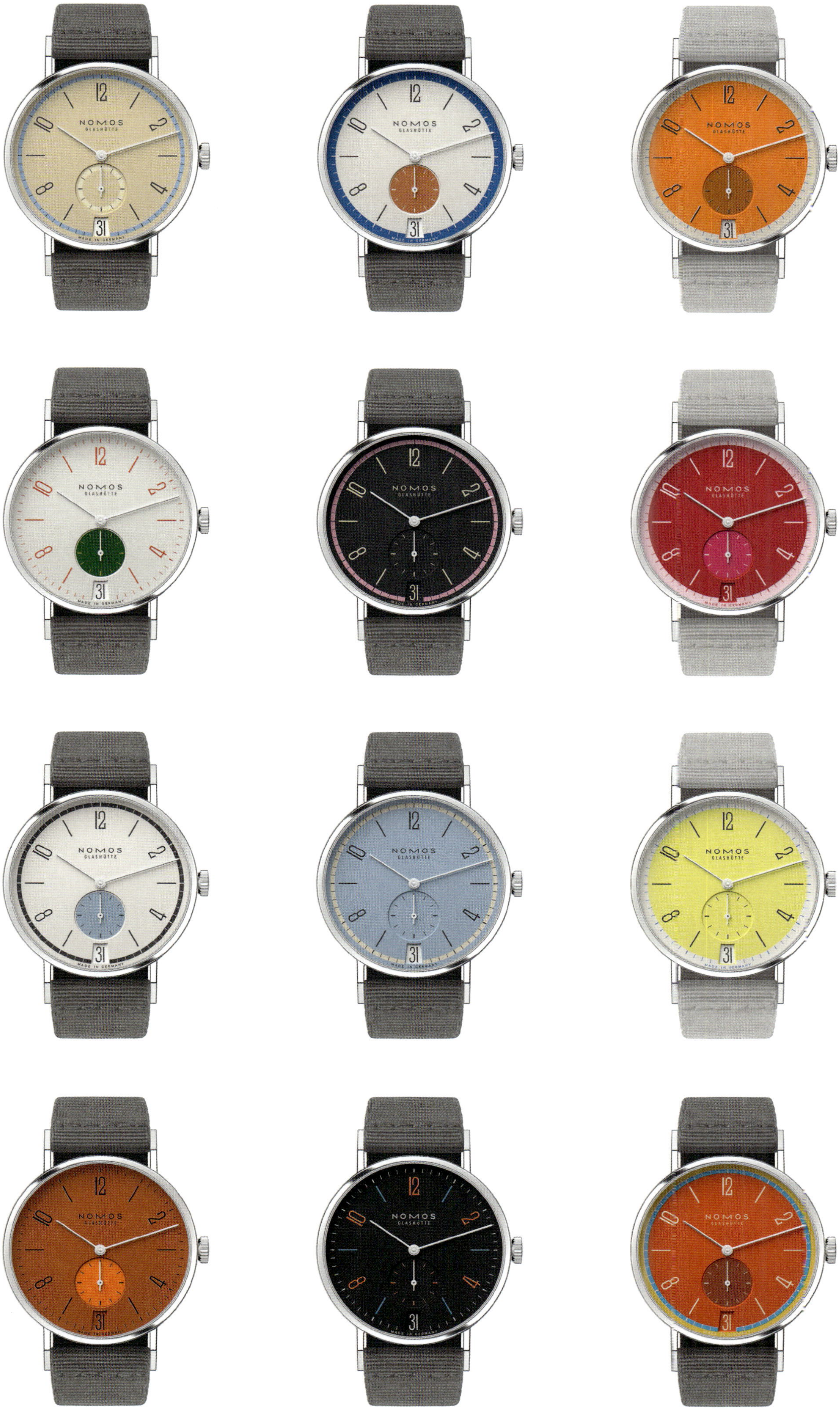

NOMOS GLASHÜTTE 2013
LAMBDA

TOWERING POWER RESERVE

Nomos Glashütte's premium calibre is the Lambda. Since 2013, it has been used in watches of the same name. The DUW 1001 calibre was birthed by the engineers and watchmakers in Müglitztal valley. The timepiece's architecture was the upshot of a cooperation between the experts of the Saxon family manufacture and renowned designers including Simon Husslein and the studio of the late Hannes Wettstein. The striking dial bears the signature of Nomos product designer Michael Paul and Professor Axel Kufus. Behind it, the 32-millimetre movement embodies outstanding Glashütte watchmaking culture. It is partially concealed by the three-quarter plate with fine sunbeam ribbing and no less than six screw-mounted gold chatons on which the gear train jewels are bedded. The screwed balance wheel completes 21,600 semi-oscillations every hour. An elegant swan-neck fine adjustment mechanism tunes the regulator in tiny steps. Although only 3.6 millimetres high, the movement accommodates generously sized components. Two barrels guarantee a power reserve of 84 hours or three and a half days. The remaining supply of energy is shown on an opulent power reserve indicator at the top of the dial. A hack mechanism ensures exact synchronisation of the small seconds dial at six o'clock. Lovers of beauty will appreciate the exquisite craftsmanship. This is revealed in the polished tempered blue steel screws, the hand-engraved balance cock, ratchet wheel with Glashütte sunburst as well as in bevelled, polished edges. Naturally, such a gem was initially only available in a solid gold case with a diameter of 42 millimetres. Marking 175 years of Glashütte watchmaking, three limited editions with steel cases were brought out in 2020. The 525 models were snapped up immediately. Today, the watches are traded at sums far higher than the original asking price.

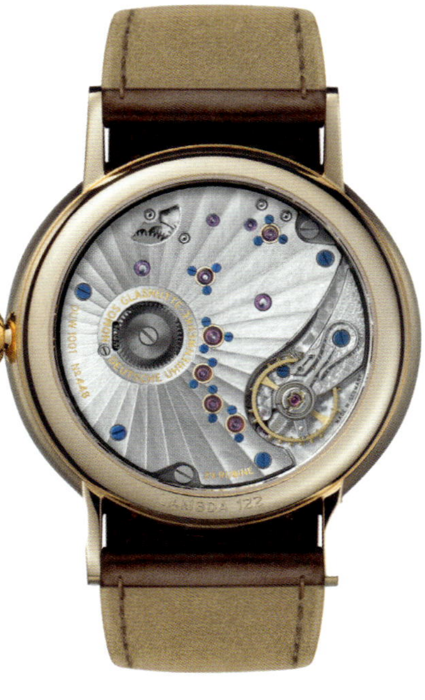

GANGRESERVE GROSSGESCHRIEBEN

„Lambda" nennt es sich, das Top-Kaliber von Nomos Glashütte. In Armbanduhren mit gleichem Namen findet es seit 2013 Verwendung. Väter des Kalibers DUW 1001 sind Techniker und Uhrmacher aus dem Müglitztal. Bei der Gestaltung des damit ausgestatteten Zeitmessers kooperierten die Experten der sächsischen Familienmanufaktur mit namhaften Designern. Dazu zählen Simon Husslein und das Büro des mittlerweile verstorbenen Hannes Wettstein. Für das augenfällige Zifferblatt zeichnen Nomos-Designer Michael Paul und Professor Axel Kufus verantwortlich. Zurück zum Handaufzugswerk mit 32 Millimeter Durchmesser. Diese Mechanik verkörpert Glashütter Uhr-Kultur par excellence. Die Platine mit vornehmem Sonnenstrahlenschliff überdeckt drei Viertel des Mechanismus. Sie trägt nicht weniger als sechs verschraubte Goldchatons zur Lagerung der Zapfen des Räderwerks. Die Unruh mit echten Masseschrauben vollzieht stündlich 21.600 Halbschwingungen. Eine elegante Schwanenhals-Feinregulierung vermag den Rückerzeiger in minimalen Schritten zu bewegen. Trotz einer Bauhöhe von nur 3,6 Millimetern bietet dieses Uhrwerk hinreichend Platz für großzügig dimensionierte Bauteile. Zwei Federhäuser gewährleisten 84 Stunden oder dreieinhalb Tage Gangautonomie. Den aktuell verbleibenden Energievorrat bildet eine nachgerade opulente Gangreserveanzeige im oberen Teil des Zifferblatts ab. Exaktes Synchronisieren der kleinen, bei „6" positionierten Sekundenanzeige mit einem Zeitsignal ermöglicht ein Unruhstopp. Augenmenschen erfreuen sich an der handwerklichen Vollendung. Die zeigt sich zum Beispiel in polierten und temperaturgebläuten Stahlschrauben, im handgravierten Unruhkloben, im Sperrrad mit Glashütter Sonnenschliff sowie in anglierten und polierten Kanten. Verständlich, dass Nomos Glashütte ein derartiges Kleinod zunächst nur in massive Goldgehäuse mit 42 Millimeter Durchmesser verpackte. Zu Ehren von 175 Jahren Glashütter Uhrmacherei gab es 2020 drei entsprechend limitierte Serien mit stählerner Schale. Die insgesamt 525 Exemplare waren im Handumdrehen ausverkauft. Mittlerweile werden sie schon deutlich über dem einstigen Publikumspreis gehandelt.

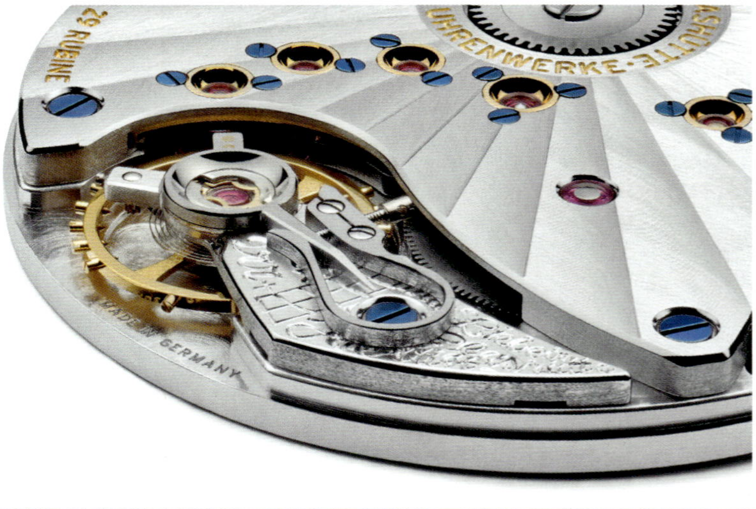

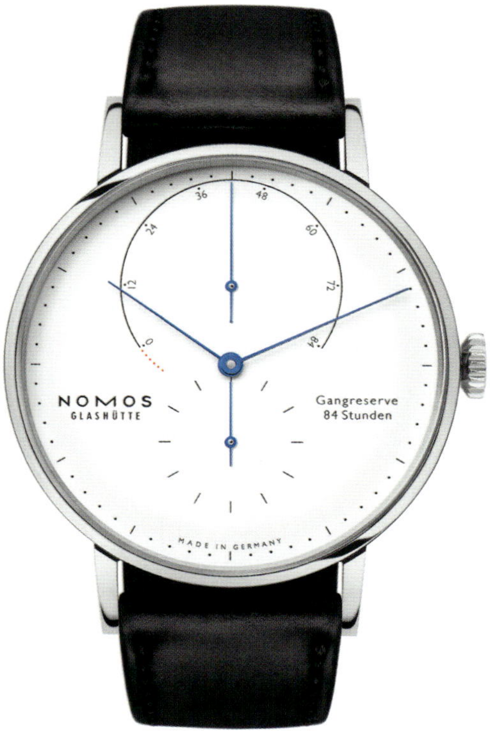

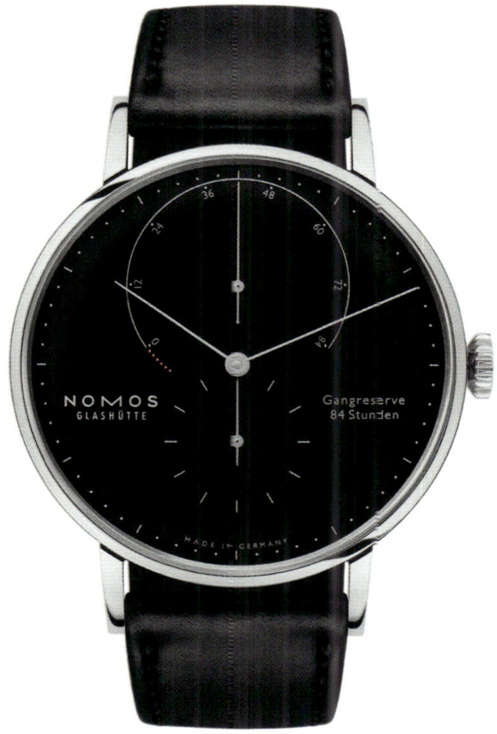

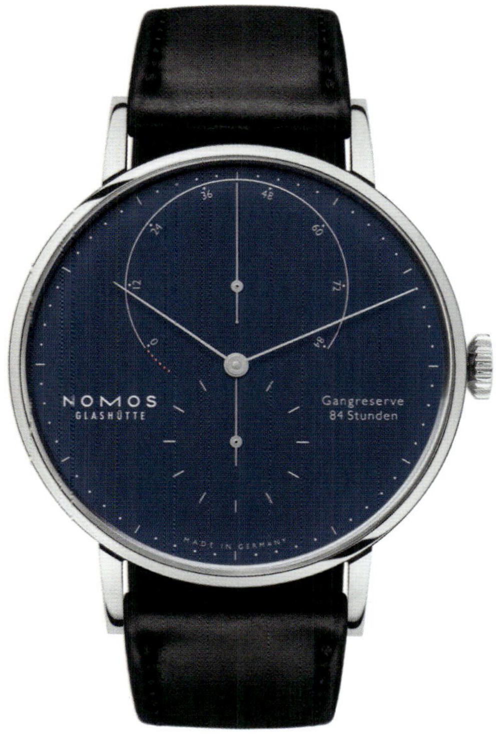

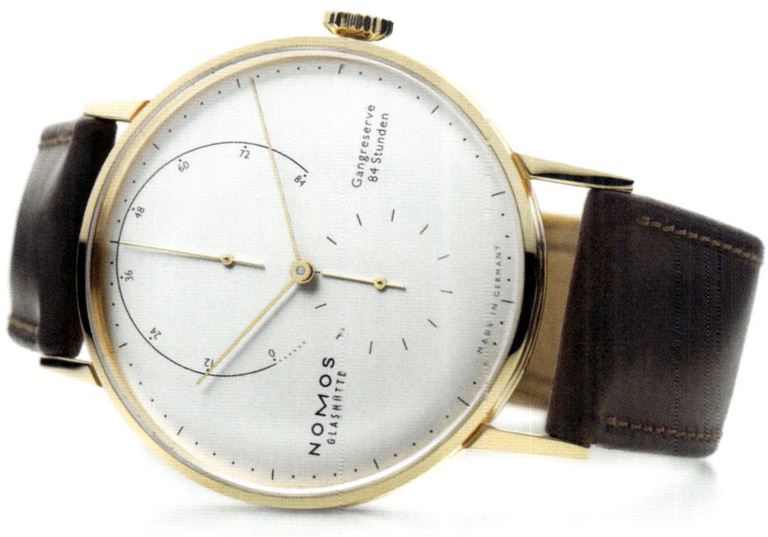

Left page* | *Linke Seite: *Ref. 930, back* | *Cal. DUW 1001*

This page, top, from left* | *Diese Seite oben von links:
Limited Edition Ref. 960.S1 | *Ref. 960.S2* | *Ref. 960.S3, 2020*

Bottom* | *Unten: *Ref. 930*

OMEGA 1957
SPEEDMASTER PROFESSIONAL

MOONSTRUCK

The fact that shocks sometimes have a stimulating effect was evidenced on 4 October 1957. The launch of the first Soviet satellite caused a furore in the USA. The *Sputnik* shock unleashed all available forces. The first sense of relief came with the launch of the *Explorer I* in February 1958. After that, the American dream of conquering space gradually became a reality. In 1961, the first humans entered orbit. In space, precise and, above all, highly resilient mechanical watches were essential to safety. In 1964, chronographs of various makes were purchased from a specialist dealer. Incognito, of course. However, because irregular conduct did not sit well with the style and rules of an official American institution, a precisely regulated procedure was initiated. Based on clearly defined specifications, the NASA Manned Spacecraft Center, Gemini & Flight Support Procurement Office, Houston, Texas 77058 wrote to several watch brands. Only four answered … and delivered. The receipt of the chronographs was followed by a strict selection procedure which took place from 21 October 1964 to 1 March 1965. The testers optimised the test procedure spanning several weeks for the final selection round. It included exposure to temperature fluctuations, heavy impact and vibration, and waterproofness, decompression, corrosion and wear resistance tests. On 1 March 1965, the Omega Speedmaster chronograph with the Lémania column wheel Calibre 321 was nominated the official mission timepiece of NASA. From 23 March 1965, the chronograph accompanied American astronauts on their space explorations. Omega celebrated the nomination by qualifying the name with "Professional". When the Eagle landed safely on the surface of the moon at 20:17 UTC on 20 July 1969, there was still a slight problem. Buzz Aldrin noted the failure of the electronic on-board timepiece. Neil Armstrong, whose statement, "That's one small step for man, one giant leap for humanity" went down in history, left his watch in the shuttle for safety's sake.
Armstrong left the shuttle watchless, but when Aldrin followed him 20 minutes later, he was the first person with a watch on the moon. The present whereabouts of the Speedmaster, the legendary Moonwatch, priceless because of its history, are unknown. In the 1970s, it disappeared without trace along with other items on its way to Washington. Due to the success of this watch in the civilian sector, Omega switched to a less expensive movement. Until 2021, the Lémania hand-wound Calibre 1861 with a slotted lever shift powered the timepiece. Since then, inspired by the fourth generation of the Speedmaster Moonwatch, the 42-millimetre chronograph with Hesalite crystal has featured the proprietary column-wheel Calibre 3861 with coaxial escapement and 30-minute and 12-hour counters. Anyone with a mind to acquiring a much more affordable Moonwatch than the steely original will find that the cooperation model between Omega and its sister company Swatch fits the bill. Inside the bioceramic case is a quartz movement.

AUF DEN MOND GEKOMMEN

Dass Schocks mitunter stimulierend wirken, zeigte sich am 4. Oktober 1957. Da wirbelte der Start des ersten sowjetischen Satelliten die USA kräftig durcheinander. Der „Sputnik-Schock" setzte alle verfügbaren Kräfte frei. Erste Erleichterung brachte der Start des „Explorer I" im Februar 1958. Danach wurde der amerikanische Traum von der Beherrschung des Weltraums sukzessive Realität. 1961 begaben sich erste Erdenbürger in den Orbit. Dort brauchte es zur Sicherheit präzise und vor allem auch hoch belastbare mechanische Uhren. 1964 erfolgte der Kauf von Armbandchronographen unterschiedlicher Provenienz bei einem Fachhändler. Inkognito versteht sich. Weil ungeordnetes Handeln jedoch nicht zum Stil und Reglement einer offiziellen amerikanischen Institution passte, startete ein exakt geregeltes Verfahren. Auf der Grundlage klar definierter Spezifikationen schrieb das „NASA Manned Spacecraft Center, Gemini & Flight Support Procurement Office, Houston, Texas 77058" mehrere Uhrenmarken an. Lediglich vier antworteten … und lieferten. Auf das Eintreffen der Chronographen folgte zwischen dem 21. Oktober 1964 und dem 1. März 1965 ein strenges Ausleseverfahren. Für die Endausscheidung optimierten die Tester das mehrwöchige Prüf-Prozedere, welches beispielsweise Temperaturwechsel, heftige Stöße, Vibrations-, Wasserdichtigkeits- und Dekompressionsprüfungen oder Korrosions- und Abnützungstests umfasste. Am 1. März 1965 erfolgte die Ernennung des Omega „Speedmaster"-Chronographen mit dem Lémania-Schaltradkaliber 321 zum offiziellen Missionszeitmesser der NASA. Ab dem 3. März 1965 begleitete der Stopper US-amerikanische Astronauten bei Ausflügen ins All. Omega trug der Kür durch den Namenszusatz „Professional" Rechnung. Als der „Adler" am 20. Juli 1969 um 20:17 UTC glücklich auf der Oberfläche des bleichen Erdtrabanten gelandet war, gab es noch ein kleines Problem: Buzz Aldrin konstatierte ein Versagen des elektronischen Bordzeitmessers. Zur Sicherheit ließ Neil Armstrong, dessen Satz „Es ist nur ein kleiner Schritt für einen einzelnen Menschen, aber ein großer Schritt für die Menschheit" Geschichte geschrieben hat, seinen Zeitmesser in der Mondfähre zurück. Armstrong stieg ohne Uhr aus, doch als Aldrin ihm 20 Minuten später folgte, war er der erste Mensch mit einer Uhr auf dem Mond. Wo die wegen ihrer Historie unendlich wertvolle „Speedmaster", die legendäre „Moonwatch" heute ist, weiß niemand. In den 1970er-Jahren verschwand sie spurlos auf dem Weg nach Washington, zusammen mit anderen Requisiten. Wegen des Erfolgs dieser Uhr auch im zivilen Bereich wechselte Omega zu einem kostengünstigeren Werk. Bis 2021 tickte im Stahlgehäuse das Lémania-Handaufzugskaliber 1861 mit Kulissenschaltung. Seitdem findet sich in dem, von der vierten Generation der „Speedmaster Moonwatch" inspirierten 42-Millimeter-Chrono mit Hesalitglas das hauseigene Schaltradkaliber 3861 mit koaxialer Ankerhemmung, 30-Minuten- und 12-Stunden-Zähler. >

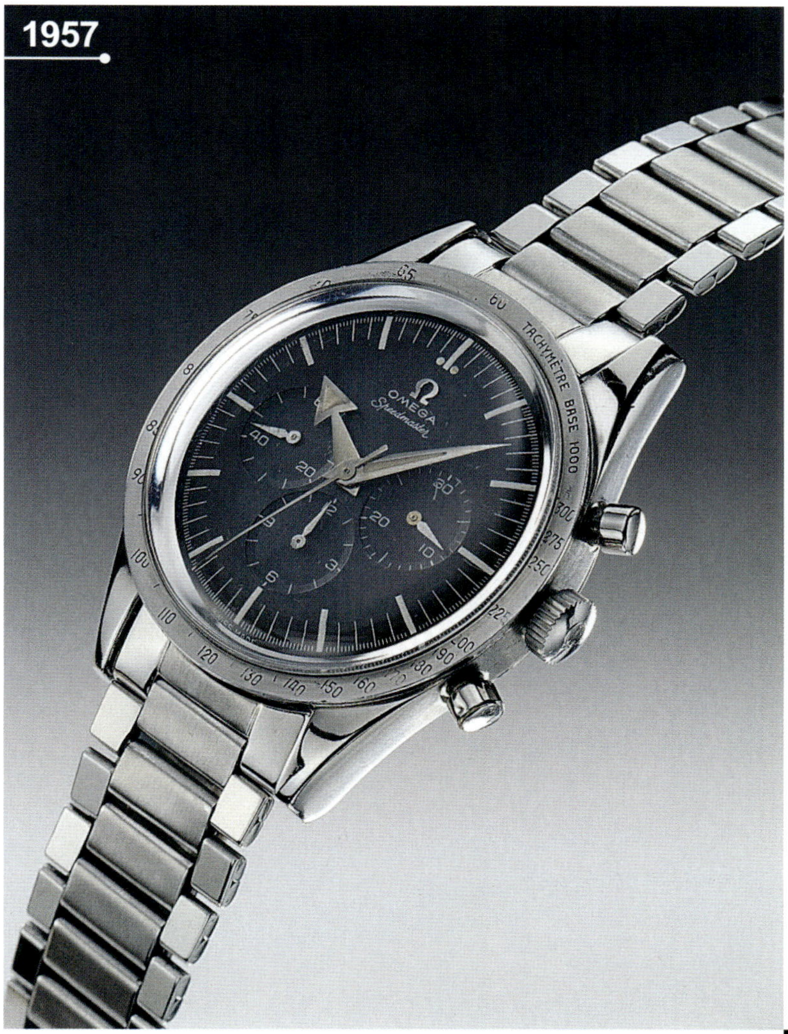

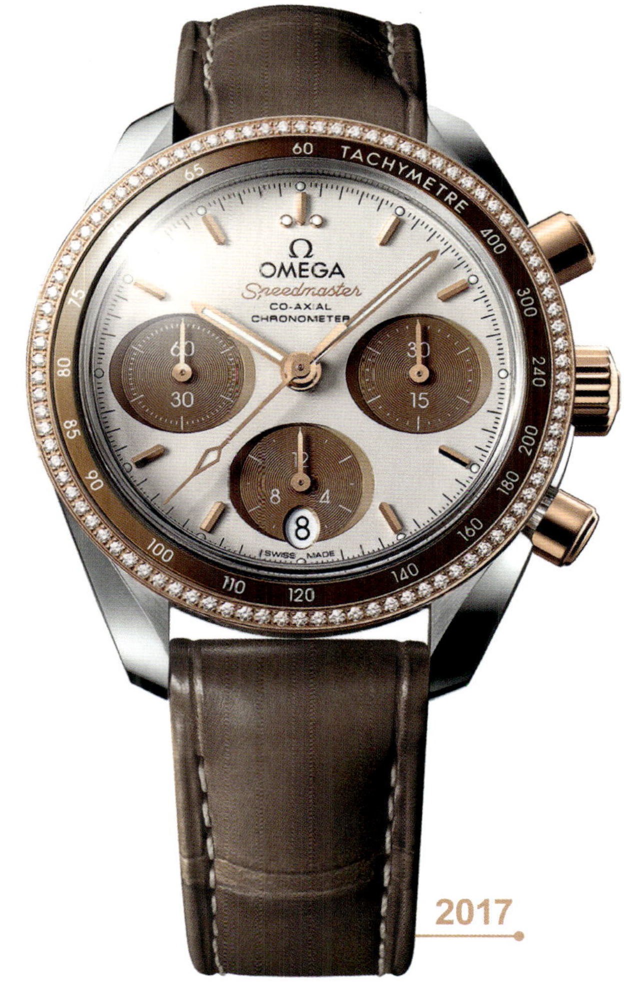

Moonswatch, 2024

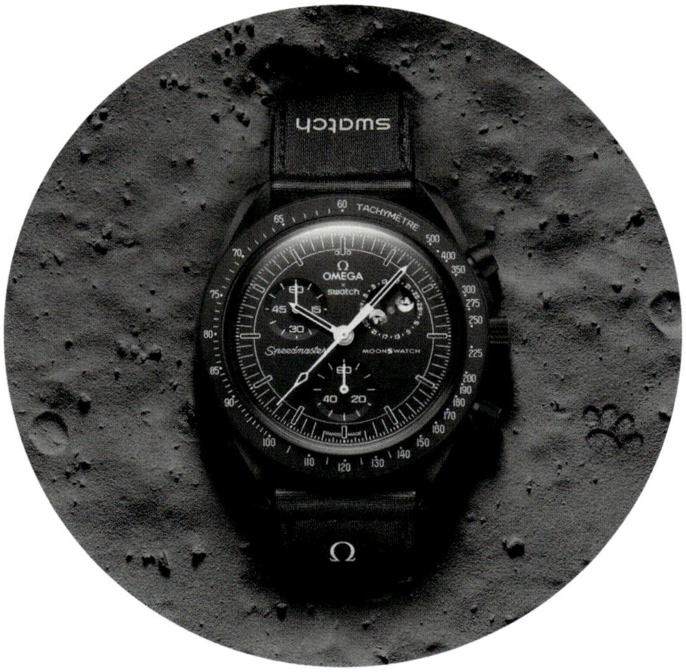

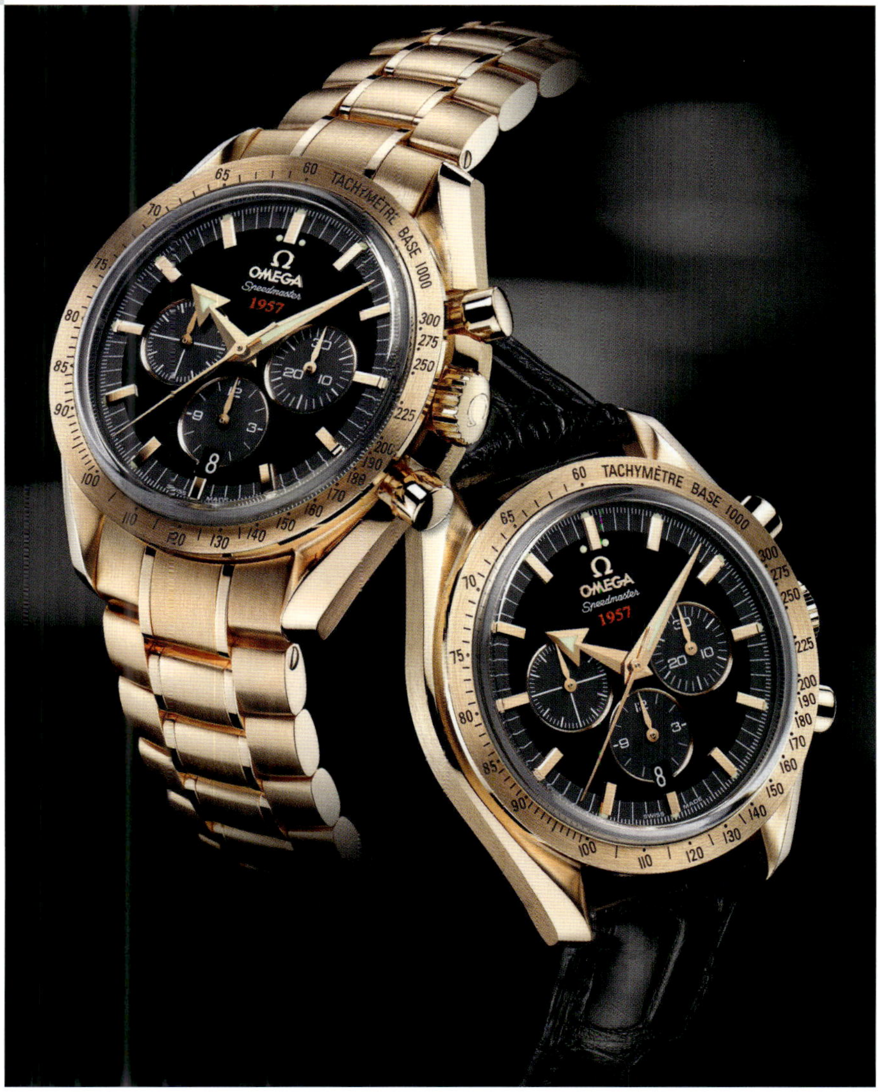

Wer die „Moonwatch“ sehr viel günstiger kaufen mag als das stählerne Original, kann das „Moonswatch“-Kooperationsmodell zwischen Omega und der Schwester Swatch wählen. Im Bioceramic-Gehäuse findet sich ein Quarzwerk.

1957
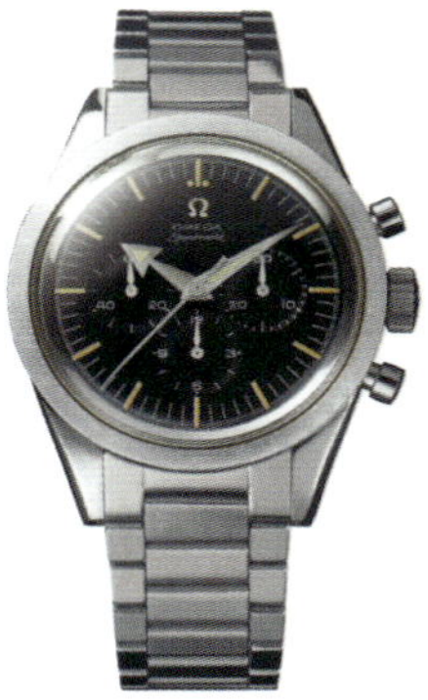
1959
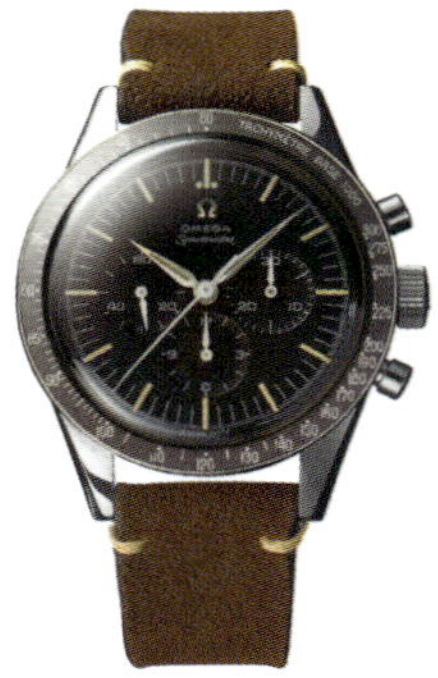
1962
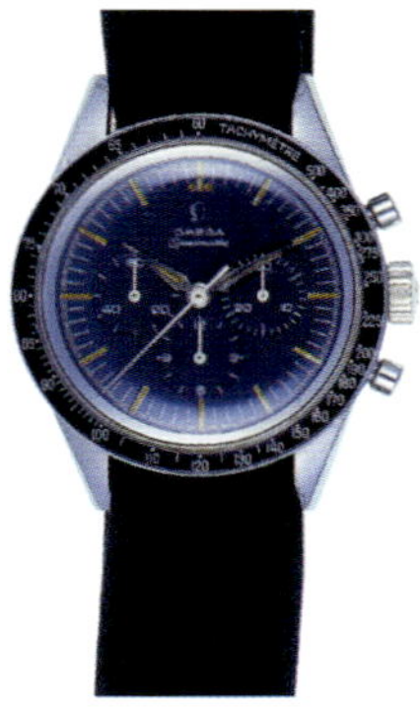
1963
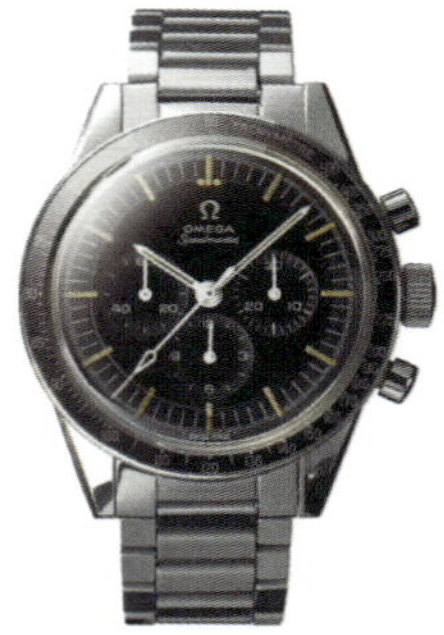
1964
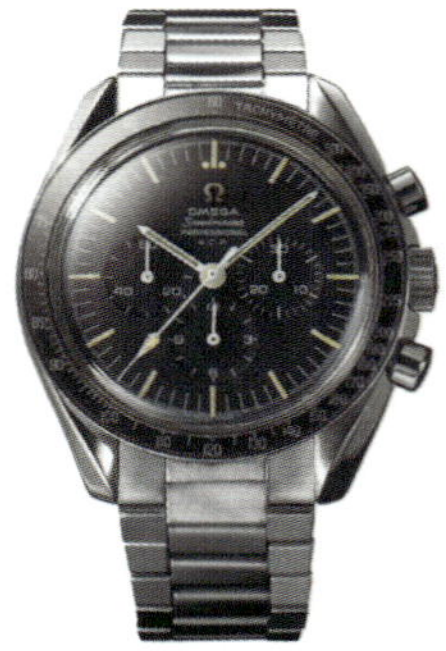
1975
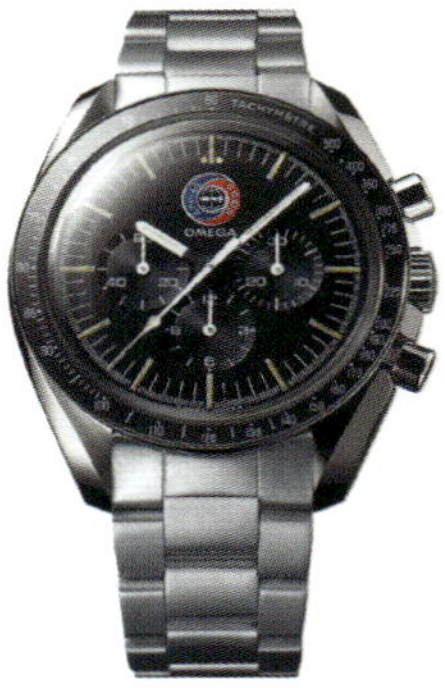
1978
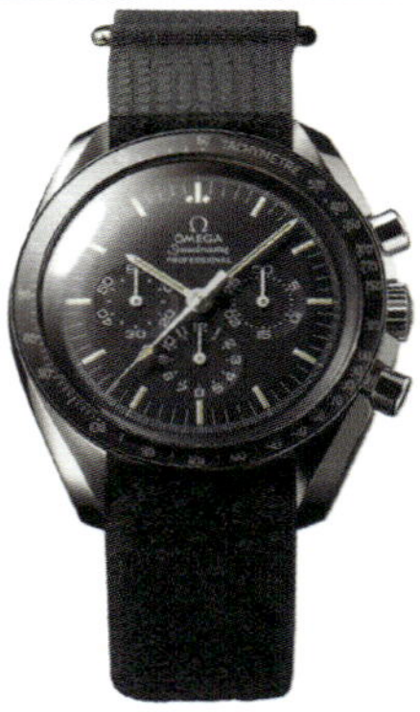
1979
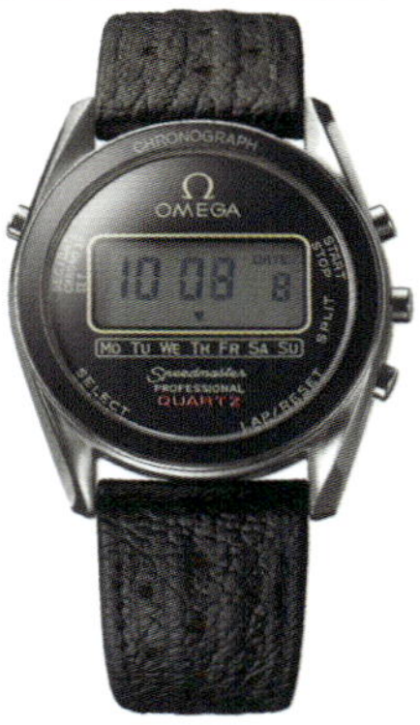
1980
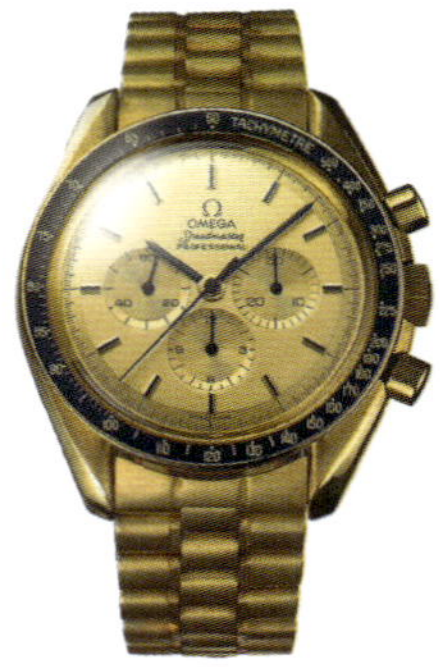
1983
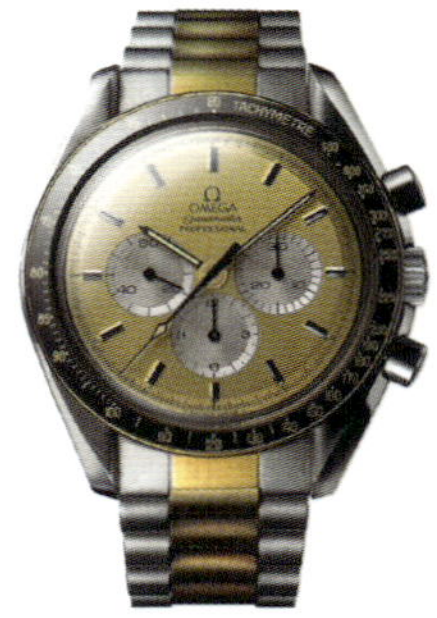
1995
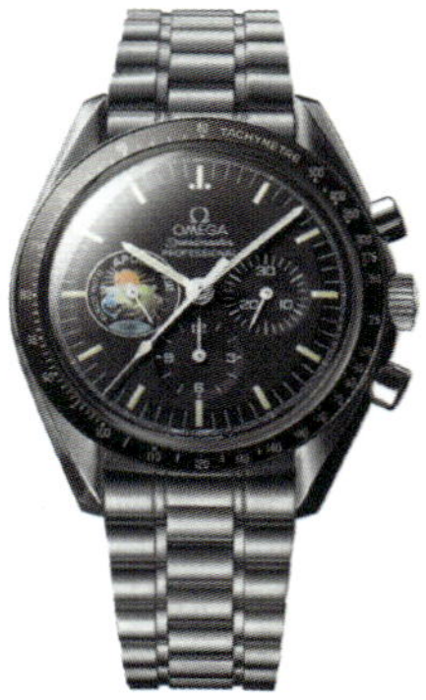
1996
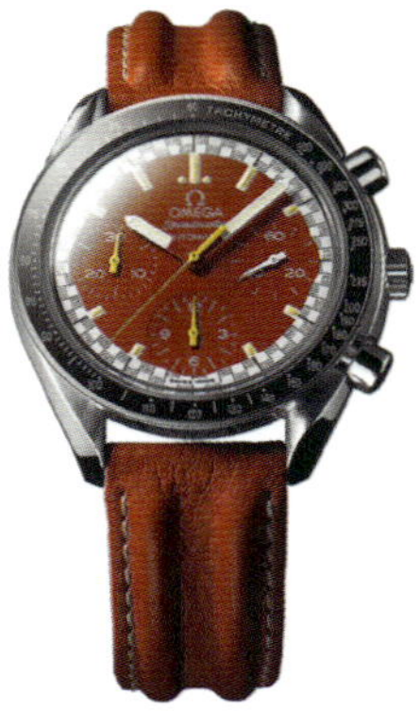
1997
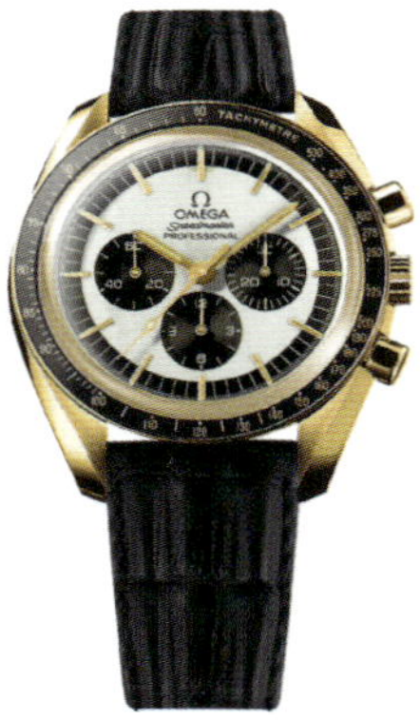
1998
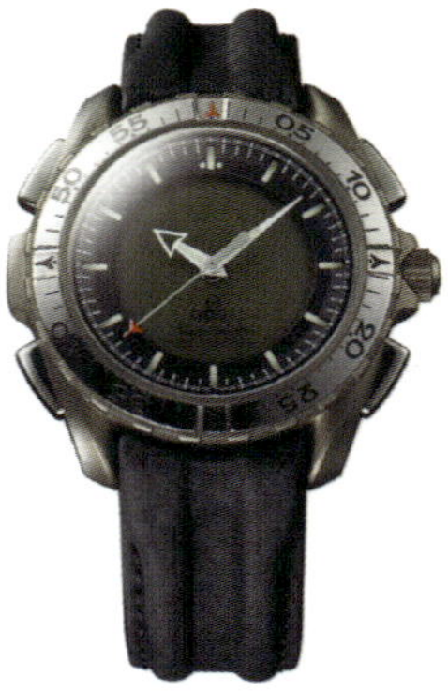
1999
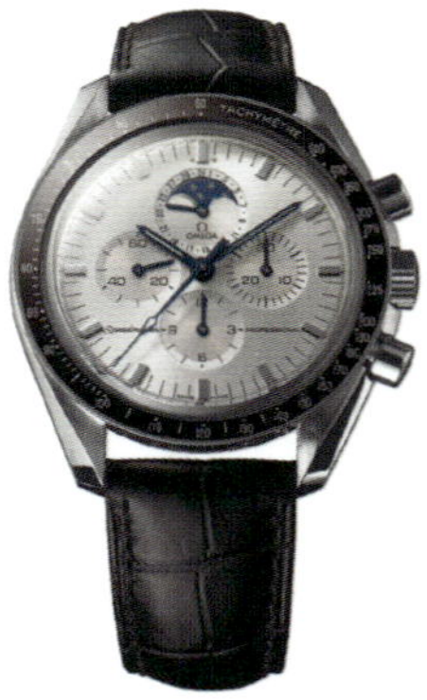
2007
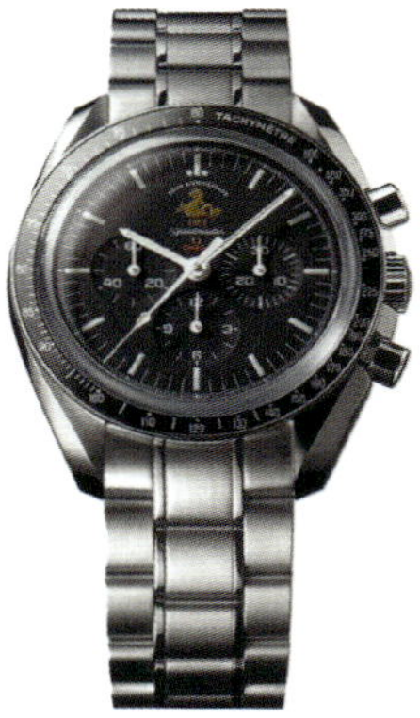
2008
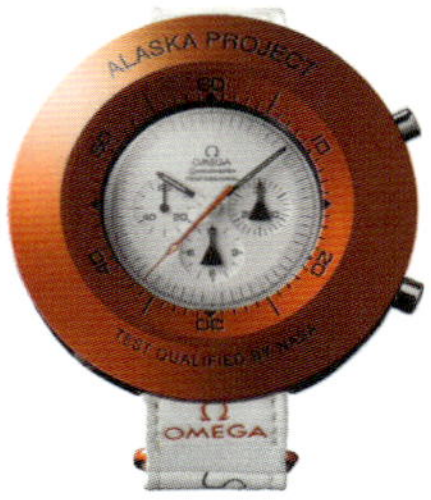
2009
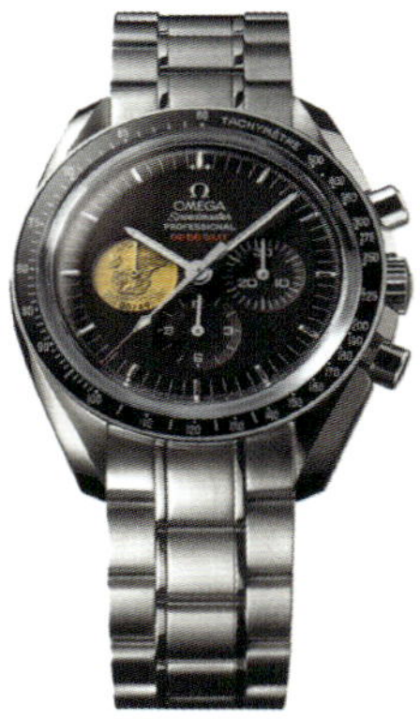
2010
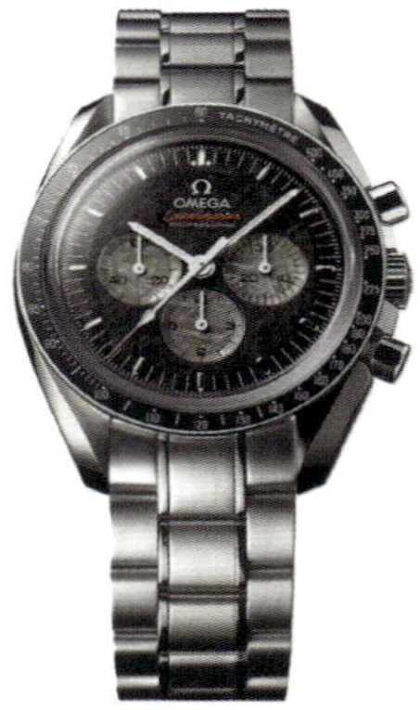
2011
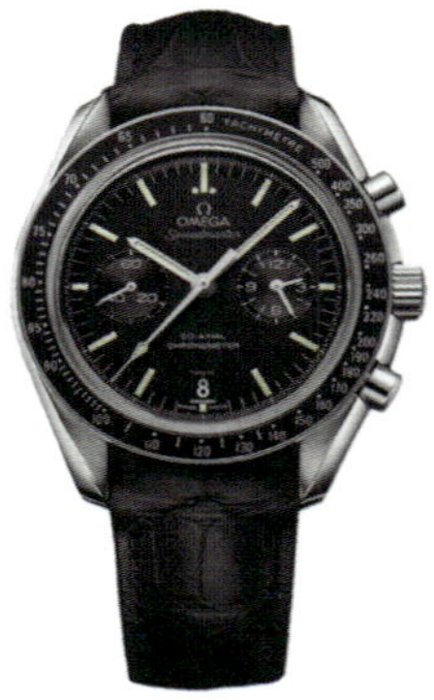
2017
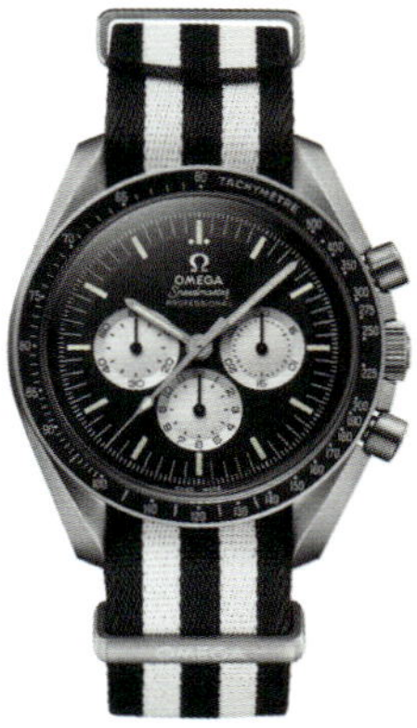
2019
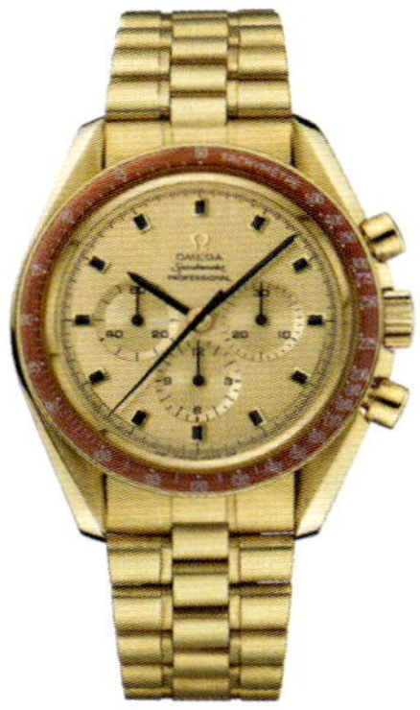
2023
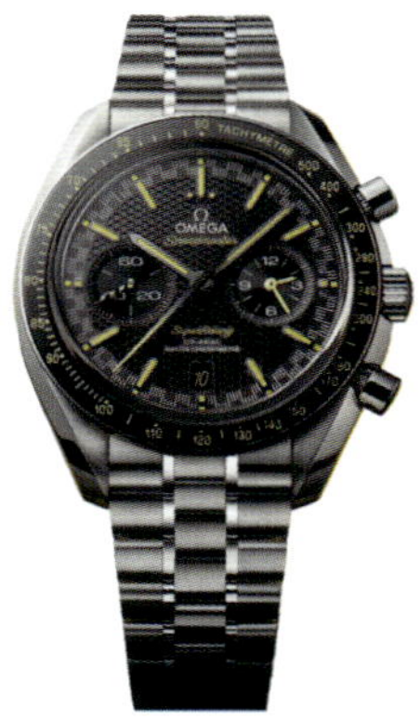

1957 *Broad Arrow*
1959 *First Omega in Space, worn by Walter Schirra / Erste Omega im Weltall, getragen von Walter Schirra*
1963 *Third Generation*
1964 *A.C.P.*
1965 *The Moonwatch*
1968 *Racing Dial*
1969 *Alaska I (Prototype)*
1971 *Mark III*

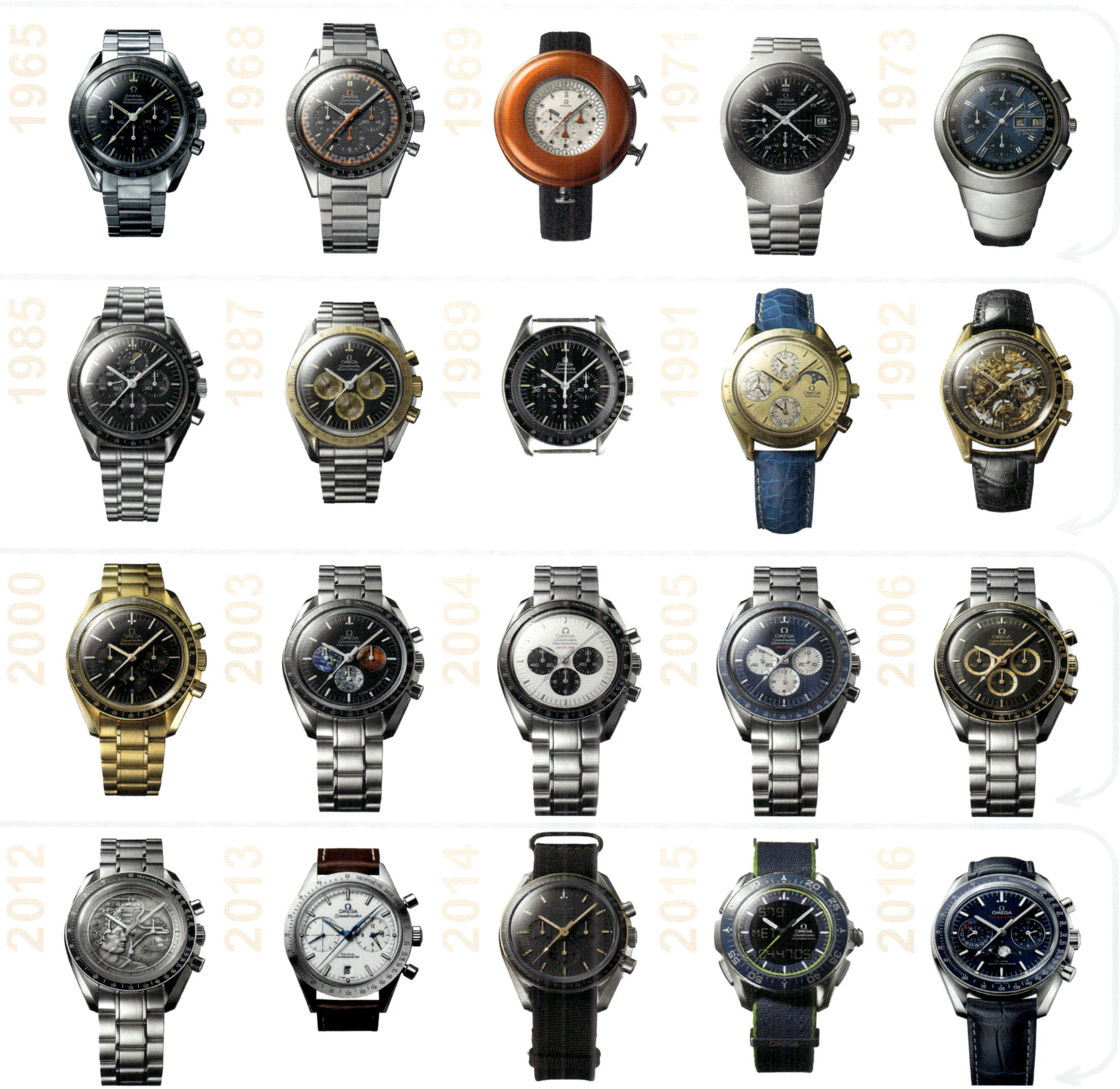

1973 *Speedsonic*
1975 *Apollo-Soyuz*
1978 *Alaska III*
1979 *Alaska IV*
1980 *Apollo 11*
1983 *Bicolor*
1985 *Speedymoon*
1987 *Italy*
1991 *Perpetual Calendar*
1992 *Skeleton*
1995 *Apollo 13*

1996 *Racing Schumacher*
1997 *The Golden Panda*
1998 *X-33*
1999 *Apollo 11*
2000 *Apollo-Soyuz*
2003 *From the Moon to Mars*
2004 *Apollo 11*
2005 *Gemini IV*
2006 *Apollo 15*
2007 *50th Anniversary Patch*
2008 *Alaska Project*

2009 *Apollo 11*
2010 *Apollo-Soyuz*
2011 *Co-Axial*
2012 *Apollo 17*
2013 *Speedmaster '57*
2014 *Apollo 11*
2015 *X33 Si2*
2016 *Moonphase*
2017 *#Speedytuesday*
2019 *Apollo 11*
2023 *Super Racing*

SEAMASTER

1948

RULER OF THE SEAS

René Bannwart can be rightly regarded as the father of the first Seamaster line. As early as 1940, the later Corum founder had set up his own design department at Omega. This was not something to be taken for granted at the time, because case makers generally supplied the designs for new models. However, at Omega the project turned out to be a decisive step in a new direction, and a path later to be trodden by many competitors. To mark the company's centenary, the sales director of the time expressed his wish for a watch that was in equal measure sporty, sturdy and waterproof. Pressed by Adolphe Vallat, Bannwart showed him one of his case designs, commenting laconically as he did so that the timepiece was "far too clumsy". Contrary to the product designer's unflattering comment, his boss was highly enamoured. Thus, in 1948, the first Omega watch with the "Seamaster" signature on the dial made its début. The name was well-founded. This version of the proven pilot's watch was able to withstand water pressure up to a depth of 50 metres. The model with a small second was powered by the new automatic Calibre 28.10 RA RG-343. Its counterpart with a central seconds hand was called 28.10 RA SC-350. Both movements were also available with official chronometer certificates. The first entry in the curriculum vitae of the Seamaster was thus penned. In the decades that followed, it ran like a golden thread through Omega's history.
Today, the portfolio of Swatch Group's largest member in terms of turnover would be unthinkable without it. Its biography is studded with countless modifications and iterations. In 1995, on payment of a commensurate fee for product placement, the Seamaster entered the world of film on the wrist of Pierce Brosnan as James Bond 007. In 2006, Daniel Craig did the honours. In the current Omega collection, the Seamaster is represented by the Aqua Terra, Diver, Planet Ocean and Heritage iterations. The pedigree of all models is instantly recognisable, not least on account of their distinctive lugs.

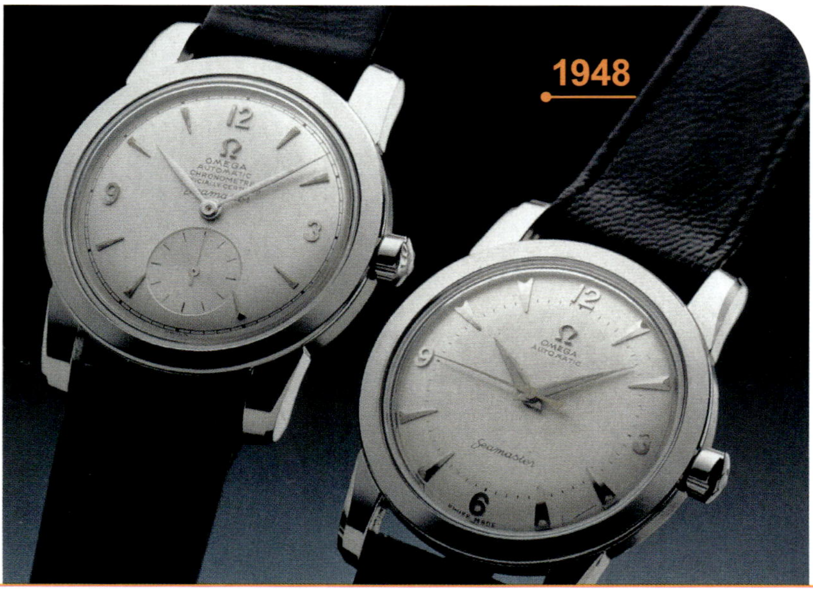

1948

MEISTER DER SEE

Als Vater der ersten „Seamaster“-Linie kann mit Fug und Recht René Bannwart gelten. Bereits 1940 hatte der spätere „Corum“-Gründer im Hause Omega eine eigene Designabteilung aufgebaut. Eine solche war damals keineswegs selbstverständlich, weil die Entwürfe zum Outfit neuer Modelle im Allgemeinen von Gehäusefabrikanten stammten. Bei Omega erwies sich das Unterfangen jedoch als entscheidender Schritt in eine neue, später von zahlreichen Wettbewerbern ebenfalls eingeschlagene Richtung. Zum 100-jährigen Firmenjubiläum verlangte der damalige kaufmännische Direktor nach einer gleichermaßen sportlichen, robusten wie auch wasserdichten Armbanduhr. Weil Adolphe Vallat drängte, präsentierte Bannwart einen seiner Gehäuseentwürfe mit der lakonischen Bemerkung, dieser Zeitmesser sei „viel zu plump“. Im Gegensatz zur Auffassung des Produktgestalters zeigte sich sein Chef hellauf begeistert. So entstand bis 1948 die erste Omega-Armbanduhr mit der zusätzlichen Signatur „Seamaster“ am Zifferblatt. Der Name ergab natürlich Sinn. Bis zu fünf Atmosphären Druck widerstand diese Version tausendfach bewährter Fliegerarmbanduhren dem Druck des Wassers. Die Ausführung mit kleiner Sekunde verfügte über das neue Automatikkaliber 28.10 RA RG-343. Das Pendant mit mittigem Sekundenzeiger nannte sich 28.10 RA SC-350. Beide Werke gab es auch mit offiziellem Chronometerzeugnis. Damit begann das Curriculum Vitae besagter „Seamaster“. In den anschließenden Dekaden zog es sich durch die Omega-Geschichte wie ein roter Faden.
Aus dem Produktportfolio des nach Umsatz größten Mitglieds der Swatch Group ist diese Armbanduhr nicht mehr wegzudenken. Unzählige Modifikationen und Deklinationen begleiten ihre Biographie. Gegen angemessene Bezahlung für das Product-Placement fand die „Seamaster“ 1995 erstmals am Handgelenk von Pierce Brosnan als James Bond 007 ihren Weg auf die Leinwand. Daniel Craig tat es ihm ab 2006 gleich. In der aktuellen Omega-Kollektion ist die „Seamaster“ in den Varianten „Aqua Terra“, „Diver“, „Planet Ocean“ oder „Heritage“ vertreten. Alle Ausführungen einigt der unter anderem durch die markanten Bandanstöße hervorgerufene Wiedererkennungswert.

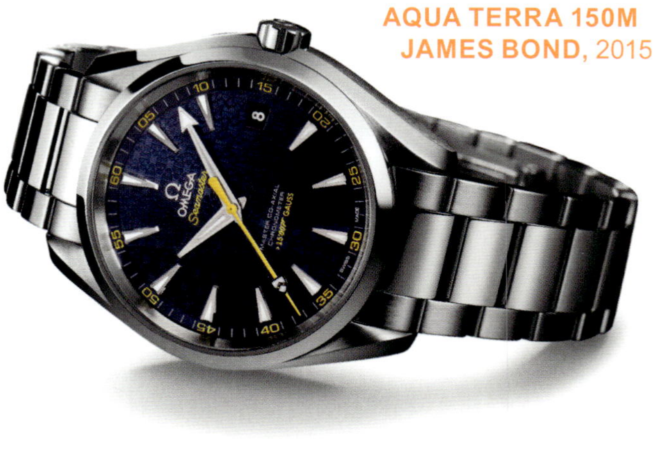

AQUA TERRA 150M
JAMES BOND, 2015

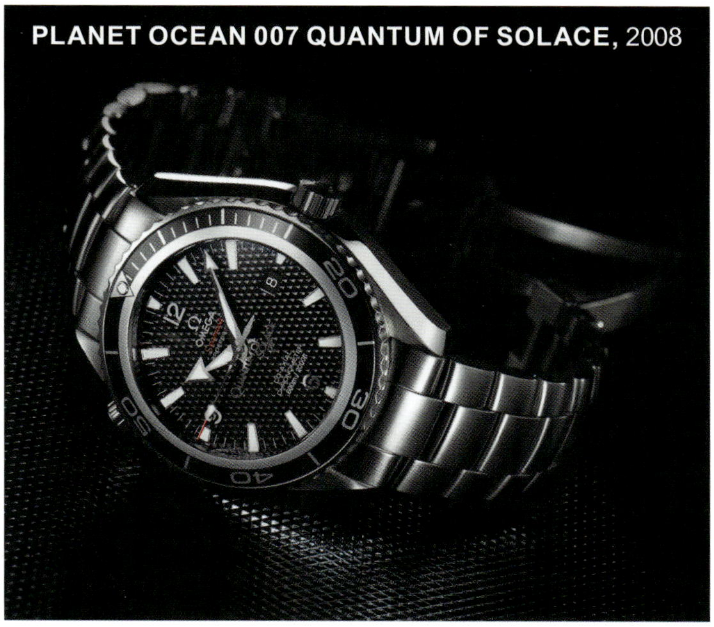
PLANET OCEAN 007 QUANTUM OF SOLACE, 2008

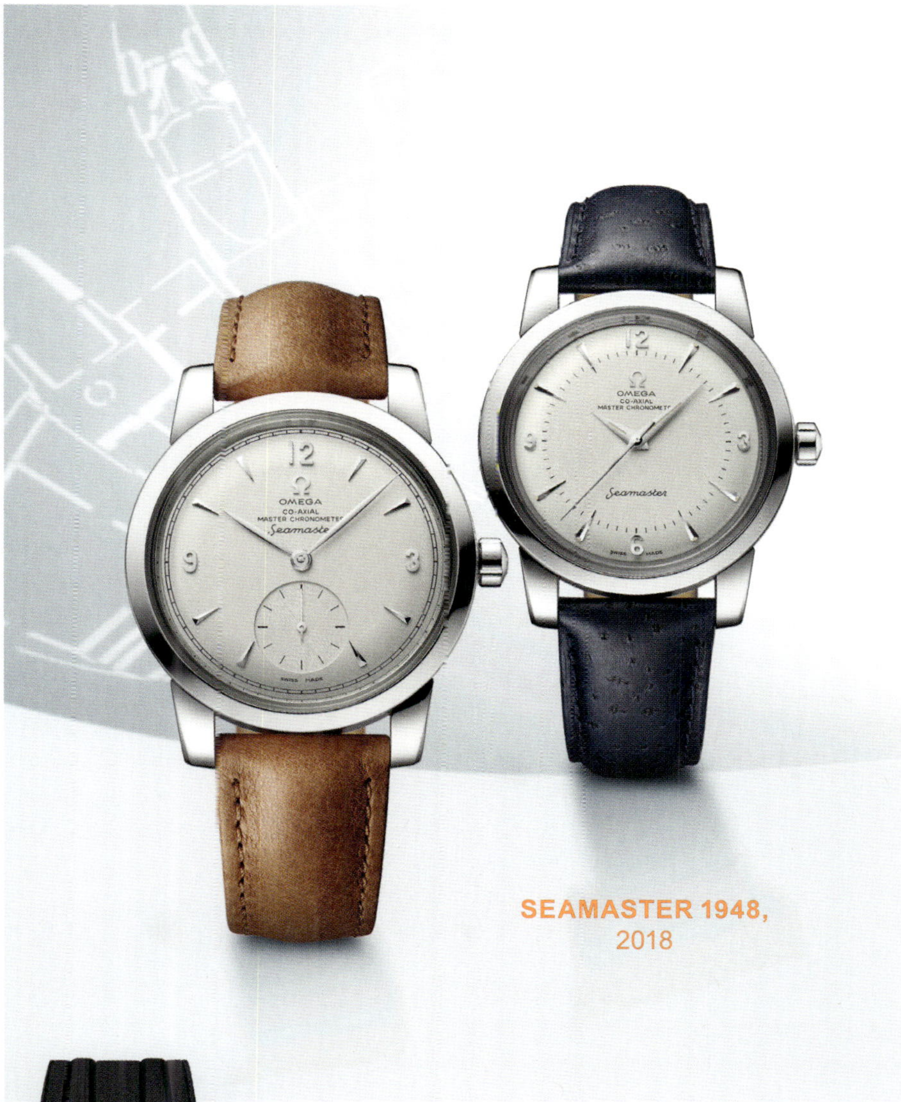

SEAMASTER 1948,
2018

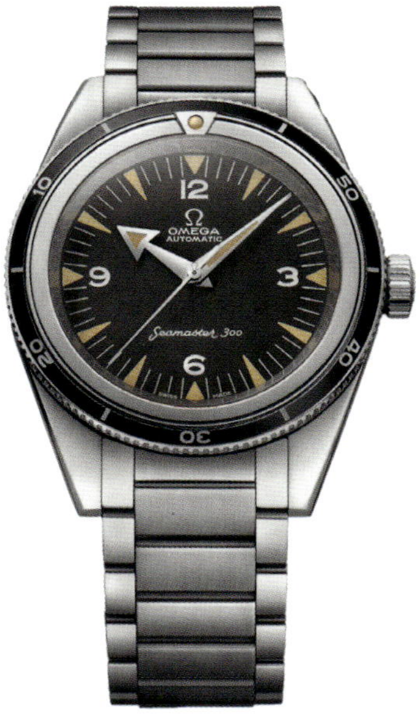

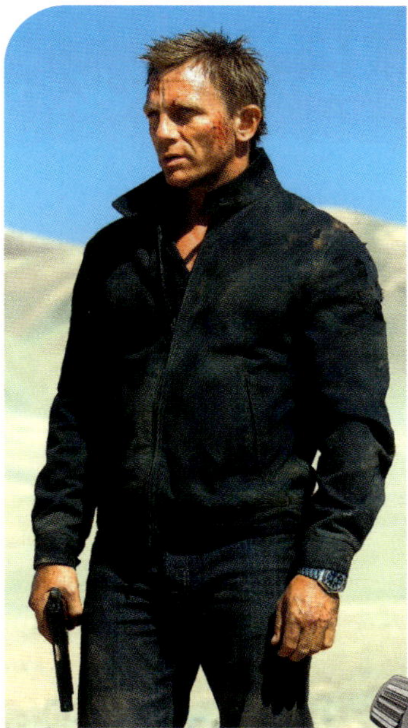

Above | Oben: *Seamaster 300, 2017 |*
Daniel Craig as James Bond 007 in Quantum of Solace *|*
Daniel Craig als James Bond in Ein Quantum Trost, *2008*

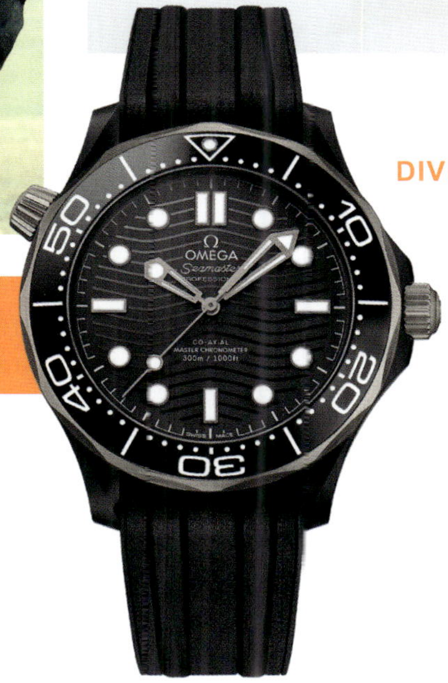

DIVER 300M

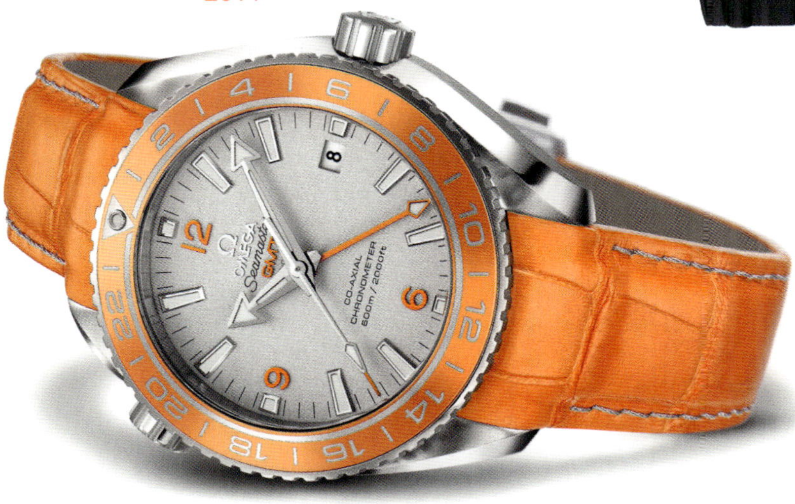

PLANET OCEAN,
2014

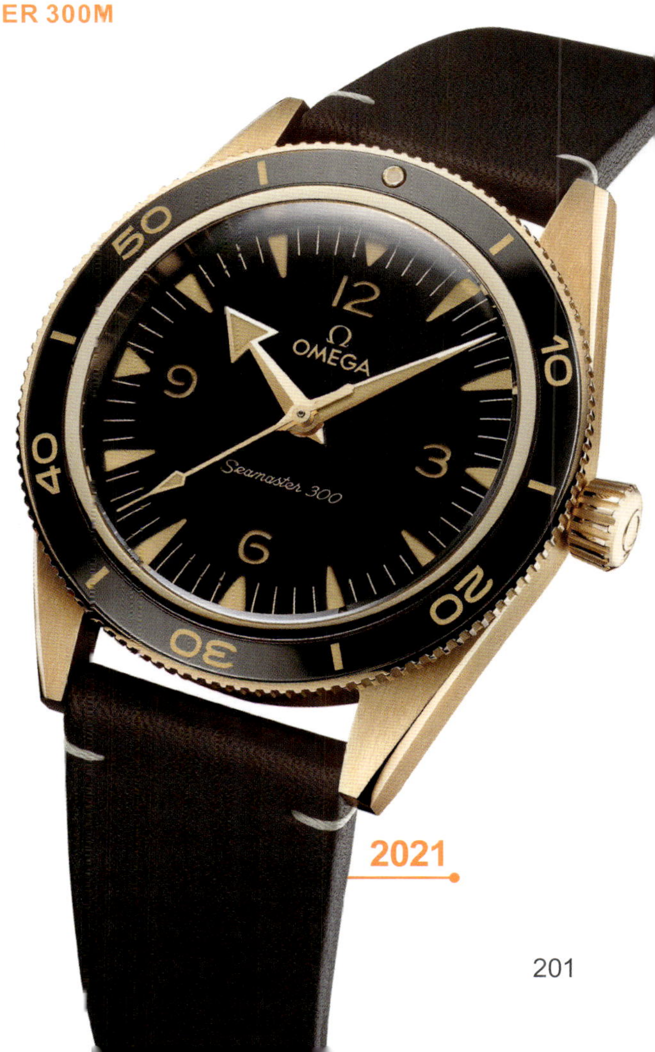

2021

POINTER DATE

POINTER DATE

Traditionally, mechanical watches can display the date in different ways, by window and date disc or by a hand. In 1938, Oris opted for the latter and at the time very popular option with its Pointer model. The launch coincided with the readiness for serial production of a proprietary movement. It featured the pin-pallet escapement, an Oris must at the time. The date hand rotates in the centre. The Arabic numeral track around the periphery of the dial is generously sized. This wristwatch was not only highly popular, but also accredited itself well in a variety of situations. Its resilience is evidenced by an Oris Pointer Date that was lost in Hölstein in 1954. Found on a road covered in deep snow, once wound up, it ticked away undaunted by its adventure. Astonishingly, no one reported the loss of this precious item to the police. The old adage thus proved true in this case, finders keepers, losers weepers.

Therefore, it comes as no surprise that at least one Pointer model has been included in every Oris collection up to the present day. This also applied to the period after the management buyout and the relaunch of the Oris brand in 1985.

Today, a modern automatic movement with a Swiss lever escapement is preferred. It is either based on the proven Sellita SW200, which Oris calls Number 754, or on the proprietary Calibre 400 with a five-day power reserve. This version with a small second hand and pointer date bears the number 403. Buyers enjoy the peace of mind afforded by a ten-year guarantee.

ZEIGERDATUM

Traditionsgemäß können mechanische Uhren das Datum auf unterschiedliche Weise anzeigen: per Fenster und Datumsscheibe oder per Zeiger. Bei der 1938 vorgestellten „Pointer" hat sich Oris für die letztgenannte und damals sehr populäre Möglichkeit entschieden. In jenem Jahr erlangte ein entsprechendes Handaufzugswerk seine Serienreife. Ausgestattet mit der seinerzeit für Oris obligatorischen Stiftankerhemmung. Der Datumszeiger drehte im Zentrum. Entsprechend groß konnten die am Rand des Zifferblatts angeordneten Ziffern und Zahlen für das Datum ausfallen. Diese Armbanduhr kam beim Publikum nicht nur bestens an, sondern sie bewährte sich auch in ganz unterschiedlichen Lebenslagen. Beleg für die robuste Ausführung ist eine 1954 in Hölstein verloren gegangene Oris „Pointer Date". Entdeckt auf einer Straße in tiefem Schnee, tickte sie nach dem Aufziehen wieder unverdrossen.

Erstaunlicherweise reklamierte bei der Polizeistation niemand sein Eigentum an diesem guten Stück. Deshalb durfte es der glückliche Finder am Ende behalten.

Kein Wunder also, dass bis in die Gegenwart jede Oris-Kollektion mindestens ein „Pointer"-Modell beinhaltete. Das galt auch für die Zeit nach dem Management-Buyout und dem 1985 eingeleiteten Relaunch der Marke Oris.

Heutzutage findet sich in den Gehäusen ein modernes Automatikwerk mit Schweizer Ankerhemmung. Entweder basiert dieses auf dem bewährten Sellita SW200, das Oris mit der Nummer 754 bezeichnet, oder auf dem hauseigenen Calibre 400 mit fünf Tagen Gangautonomie. Hier nennt sich die Ausführung mit kleinem Sekundenzeiger und Zeigerdatum 403. In diesem Fall bekommen die Käuferinnen und Käufer zur Sicherheit sogar zehn Jahre Garantie mit auf den Weg.

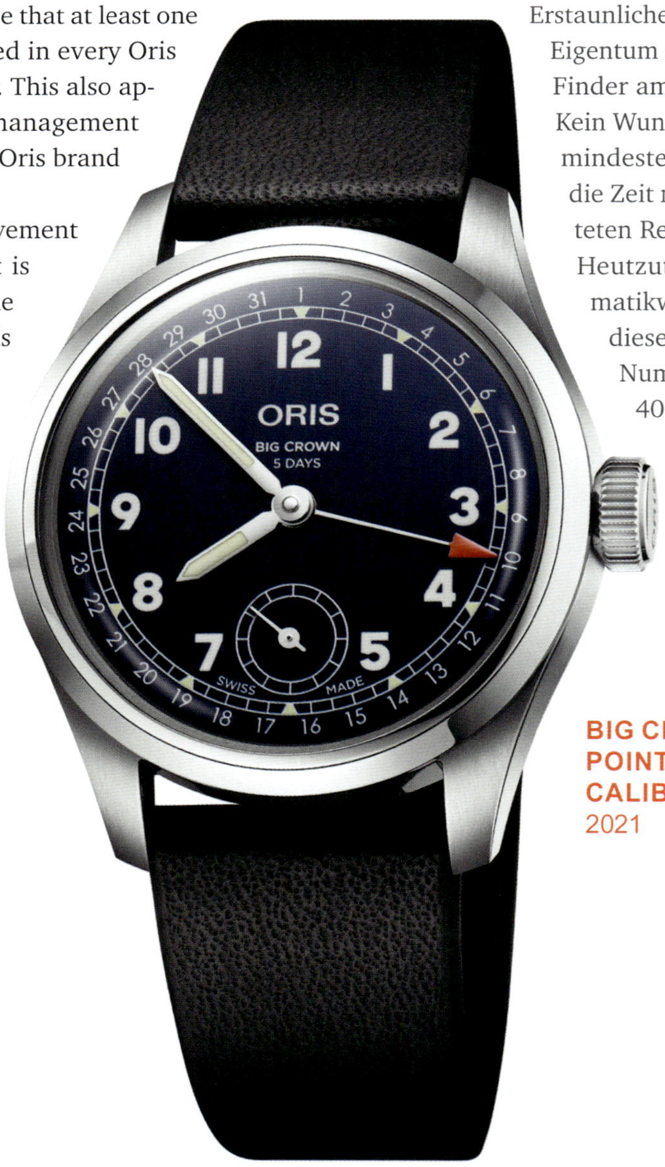

BIG CROWN POINTER DATE CALIBRE 403,
2021

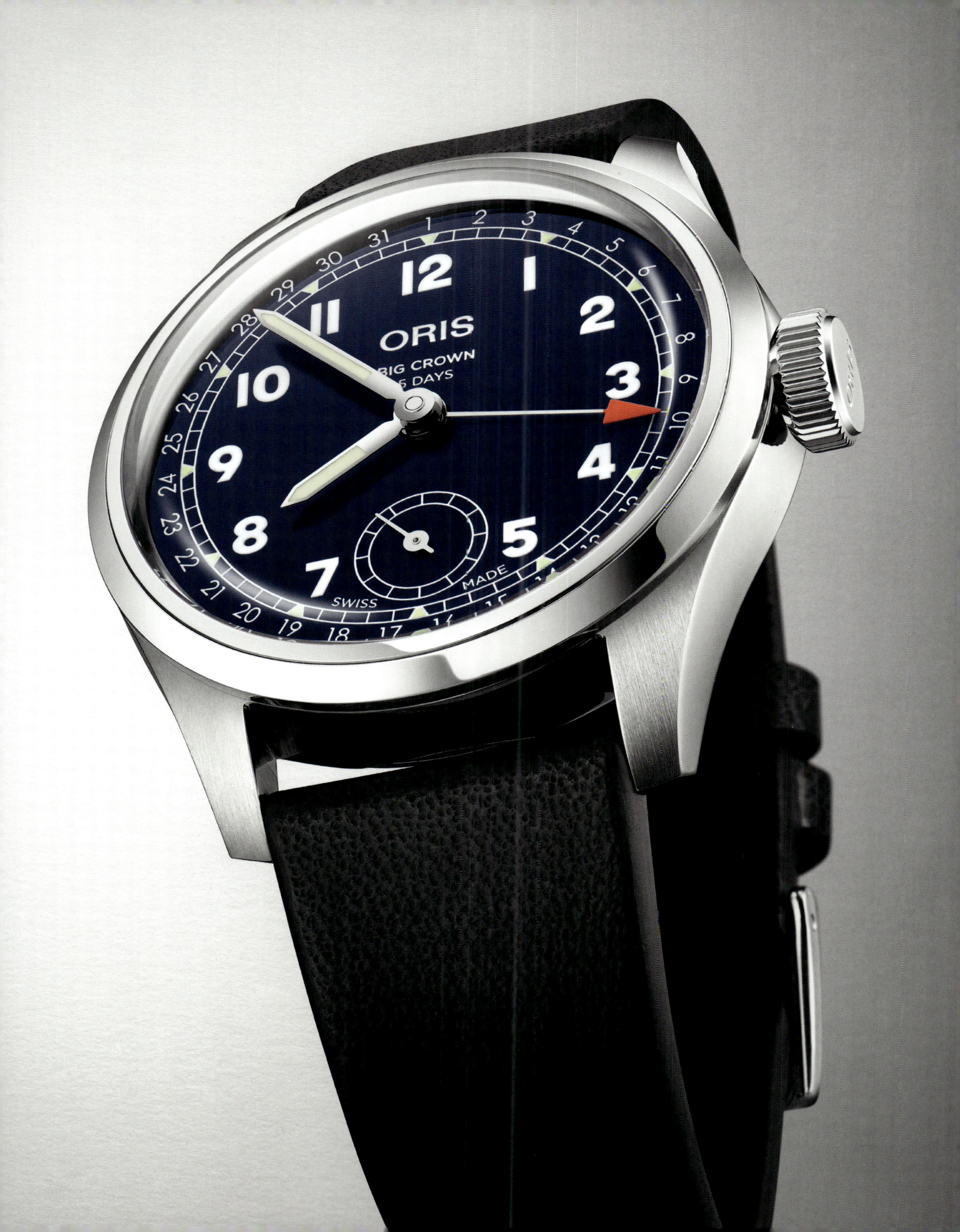
ORIS
BIG CROWN
SWISS
MADE

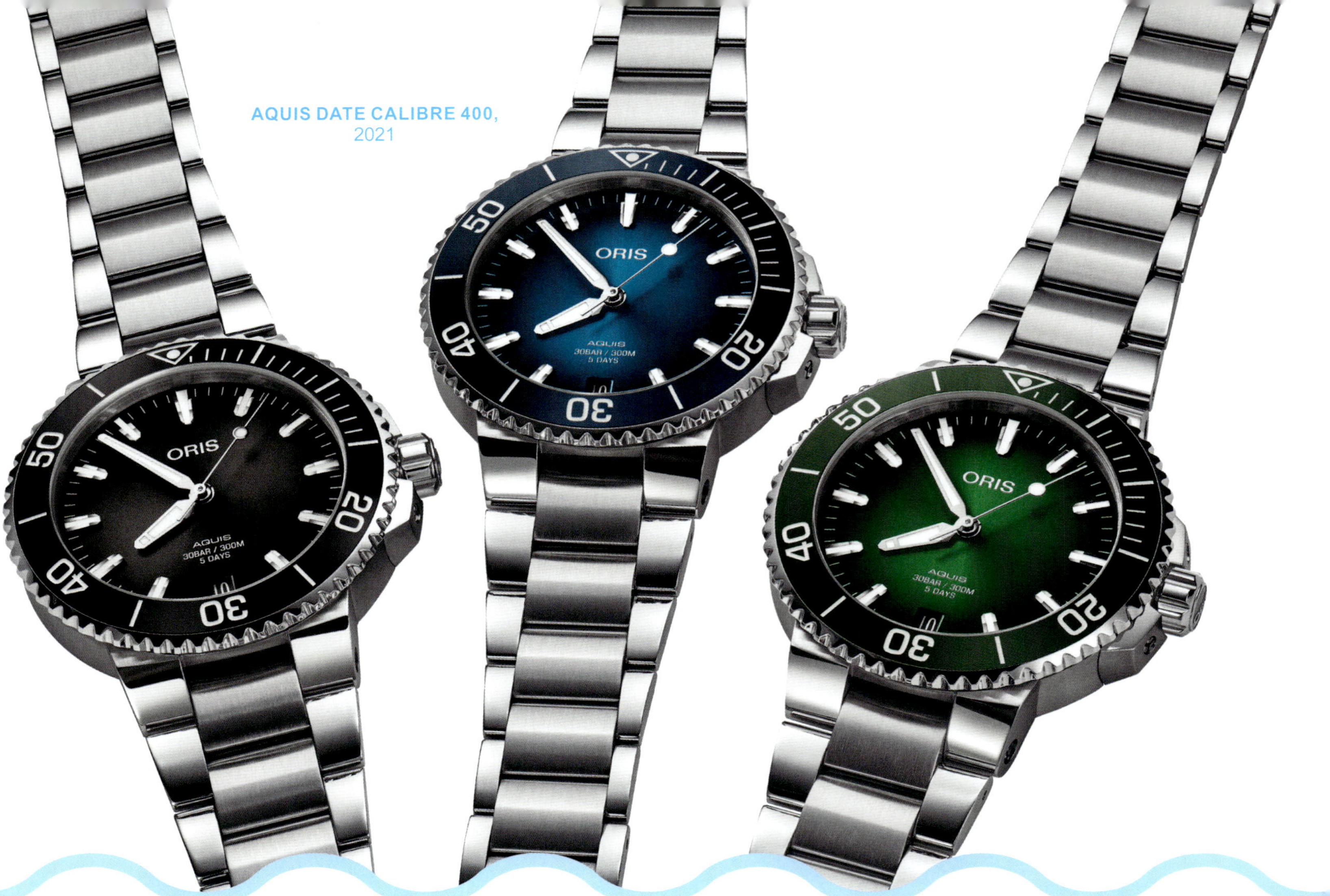

AQUIS DATE CALIBRE 400, 2021

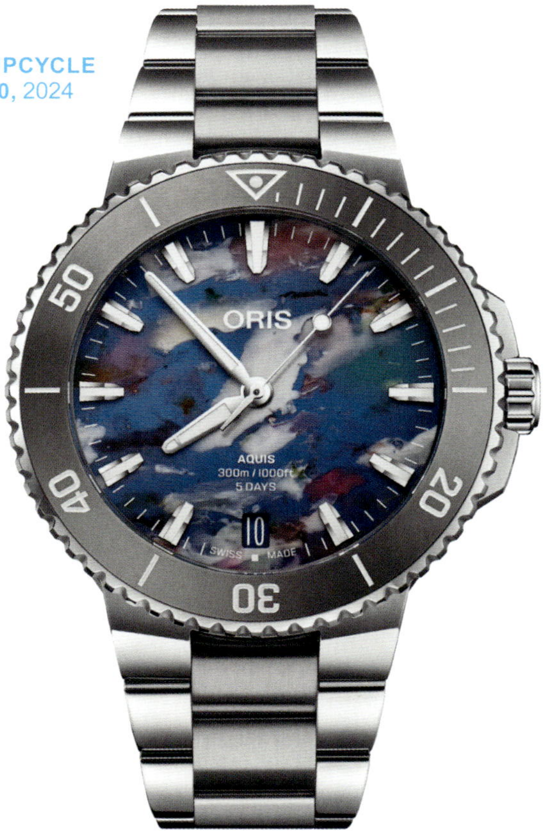

AQUIS DATE UPCYCLE CALIBRE 400, 2024

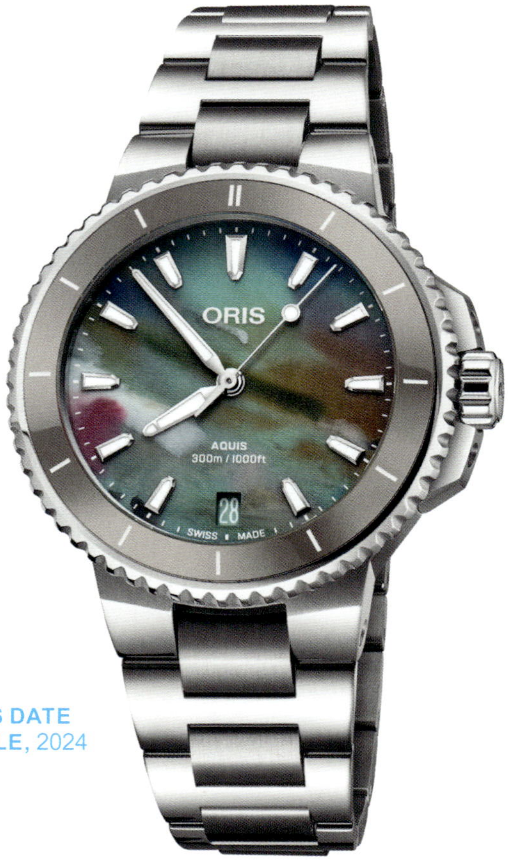

AQUIS DATE UPCYCLE, 2024

ORIS AQUIS

ABOVE AND BELOW THE WAVE WITH ORIS

Oris and Aquis are almost synonyms. As the name of this watch model suggests, it was designed for use underwater. Once submerged, the heights achieved by the functional ensemble of the case, movement, dial, hands and bracelet are soon revealed in the deep. If a diver's timepiece fails while they are underwater, the consequences can be grave. When oxygen reserves are running low and the surface is still a long way off, for instance. This and many more aspects were meticulously addressed by Oris product designers when they developed watches for the silent world. After complying with the exceedingly rigorous criteria set out in the relevant standards, the TT1 Divers débuted in 2001. It was followed in 2006 by the Divers collection, named after Carlos Coste. In 2009, the ProDiver chronograph was launched. Finally, in 2011 the diver's watch icon Aquis came onto the market. The launch phase of this visually impressive and functional watch line focused on limited models dedicated to protecting endangered marine regions.

These include the Maldives as well as the Great Barrier Reef, the Tubbataha Reef and the wreck of the Kittiwake battleship, which provides a habitat for endangered marine species. Oris donates a part of the revenue from the sale of these watches to preserve their namesakes. It goes without saying that the bezel of the Aquis can only be moved in one direction. This effectively precludes any unpleasant surprises, since if adjusted unintentionally, the remaining time underwater that is indicated will be shorter, never longer. Simply the fact that the proprietary automatic Calibre 400 was launched in an Aquis case shows how important this watch line is to the company. The début took place in 2020. In spring 2024, Oris launched a redesign of its Aquis model. Features include slimmer hands and indices on the dial. Of course, the Aquis also performs well out of water.

UNTER UND ÜBER WASSER MIT ORIS

Oris und „Aquis" sind fast schon Synonyme. Wie der Name dieses Uhrenmodells andeutet, wurde es für Einsätze unter Wasser konzipiert. Dort zeigt sich sehr schnell, was das funktionale Ensemble aus Gehäuse, Uhrwerk, Zifferblatt, Zeiger und Armband tatsächlich zu leisten vermag. Bei Tauchgängen kann das Versagen des mitgeführten Zeitmessers äußerst folgenreich sein. Beispielsweise dann, wenn die Sauerstoffvorräte zu Ende gehen und der Weg nach oben noch ein weiter ist. Dies und vieles mehr haben die Oris-Produktgestalter bei der Entwicklung von Armbanduhren für die Welt des Schweigens penibel bedacht. Nach überaus strengen, in spezifischen Normen eindeutig definierten Kriterien entstand ab 2001 zunächst der „TT1 Divers". 2006 folgte die nach Carlos Coste benannte „Divers"-Kollektion. 2009 ging der „ProDiver"-Chronograph an den Start. Die Taucheruhren-Ikone namens „Aquis" gelangte schließlich 2011 auf den Markt. Die Startphase dieser gleichermaßen markanten wie funktionalen Uhrenlinie unterstrichen limitierte Modelle zum Schutz existenzbedrohter Meeresregionen.

Dazu gehören die Malediven ebenso wie das Great Barrier Reef, das Tubbataha-Riff und das Wrack des Schlachtschiffs „Kittiwake", welches gefährdeten Meerestierarten Lebensraum bietet. Aus dem Verkaufserlös dieser Uhren stiftet Oris einen Teil zum Erhalt der Namensgeber. Von selbst mag sich verstehen, dass sich die Lünette der „Aquis" nur in einer Richtung bewegen lässt. Auf diese Weise gibt es keine bösen Überraschungen, denn bei unbeabsichtigtem Verstellen wird die verbleibende Tauchzeit allenfalls kürzer, aber niemals länger. Die Bedeutung dieser Uhrenlinie für Oris ergibt sich allein schon aus der Tatsache, dass das selbst entwickelte Automatikkaliber 400 in einem „Aquis"-Gehäuse seinen Einstand gab. So geschehen im Jahr 2020. Im Frühjahr 2024 ging bei Oris eine gestalterisch modifizierte Version der „Aquis" an den Start. Kennzeichen sind zum Beispiel schlankere Zeiger und Indexe am Zifferblatt. Natürlich funktioniert die „Aquis" auch über Wasser zuverlässig.

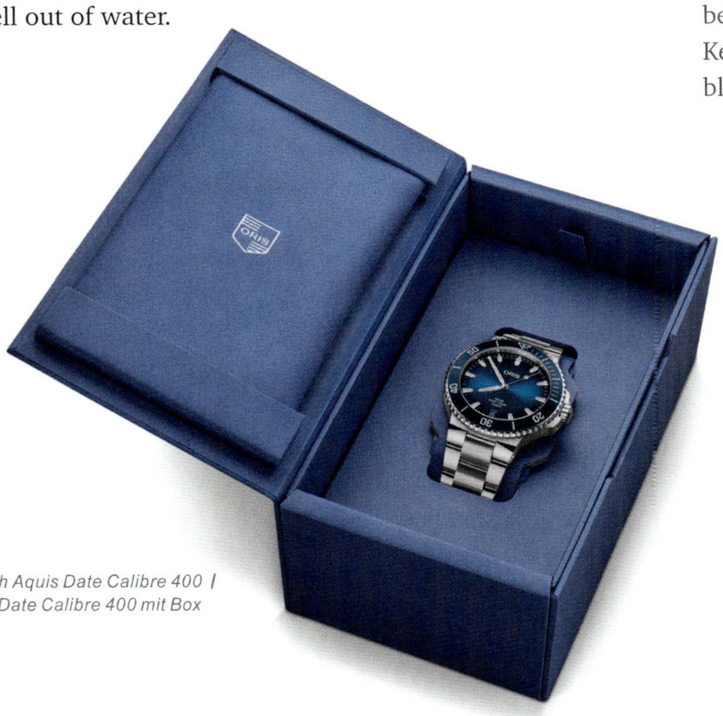

Box with Aquis Date Calibre 400 / Aquis Date Calibre 400 mit Box

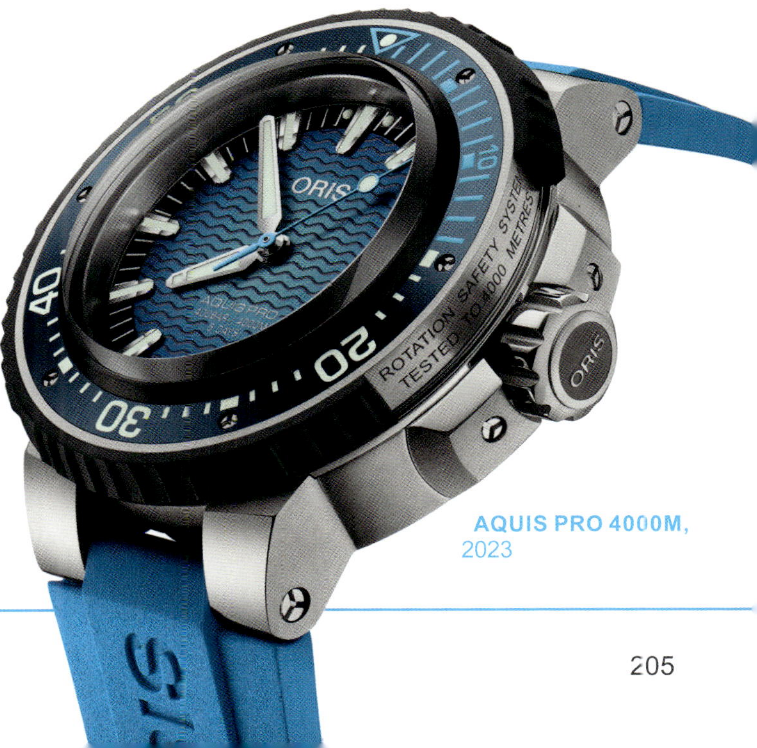

AQUIS PRO 4000M, 2023

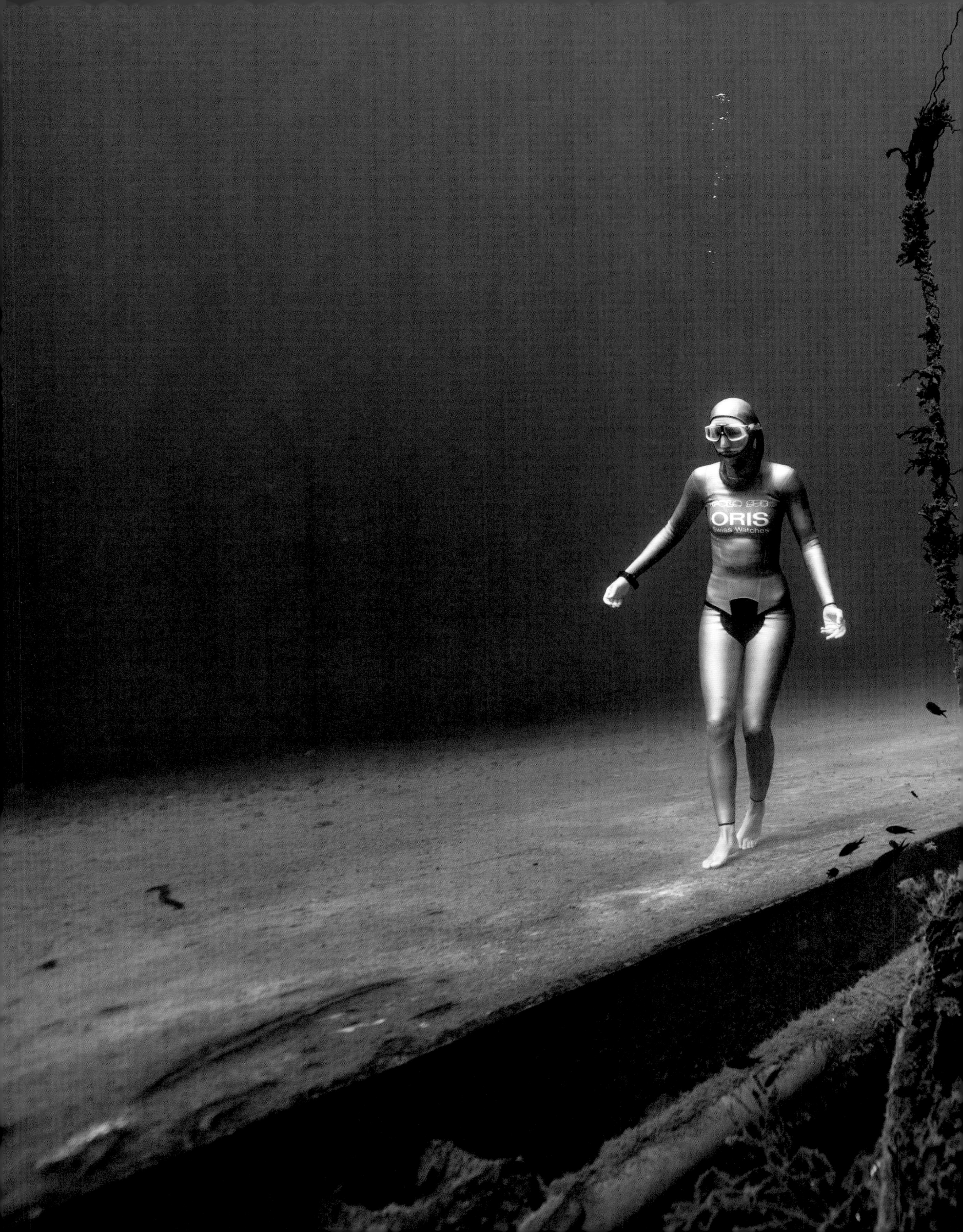
ORIS
Swiss Watches

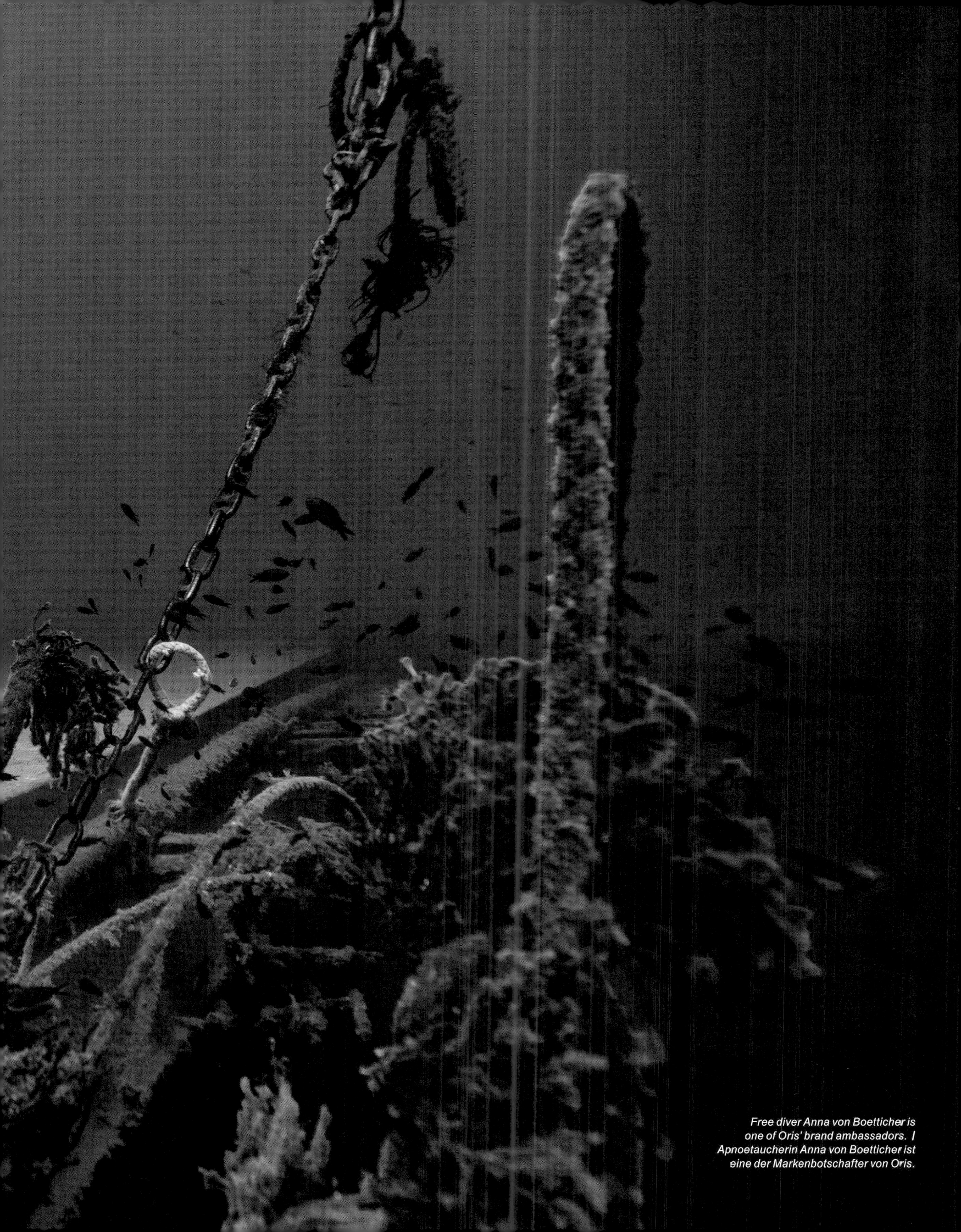

Free diver Anna von Boetticher is one of Oris' brand ambassadors. | Apnoetaucherin Anna von Boetticher ist eine der Markenbotschafter von Oris.

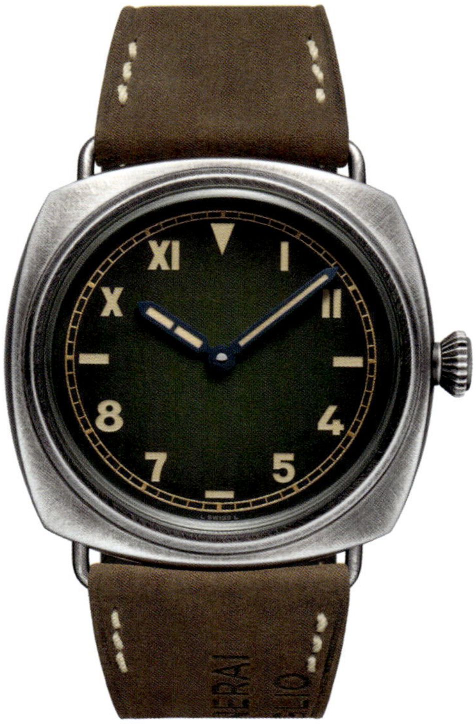

RADIOMIR CALIFORNIA DIAL
REF. PAM01349,
2023

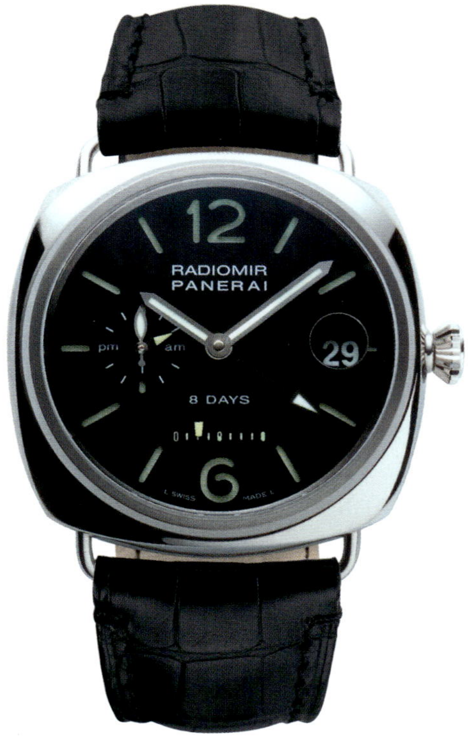

RADIOMIR 8 DAYS
REF. PAM200,
2010

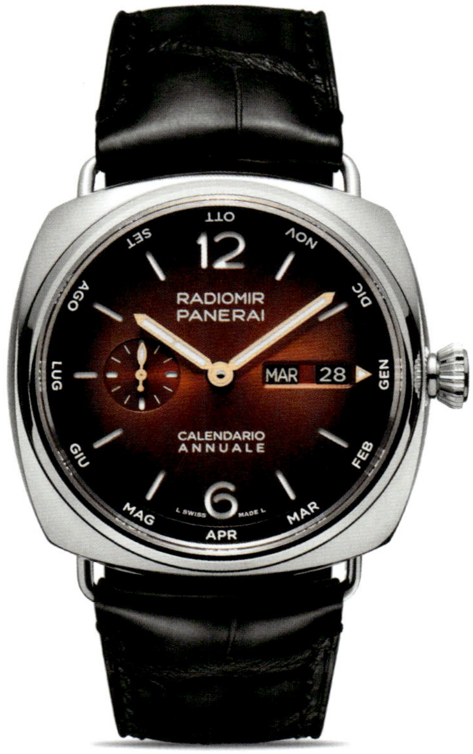

RADIOMIR ANNUAL CALENDAR
REF. PAM01432,
2023

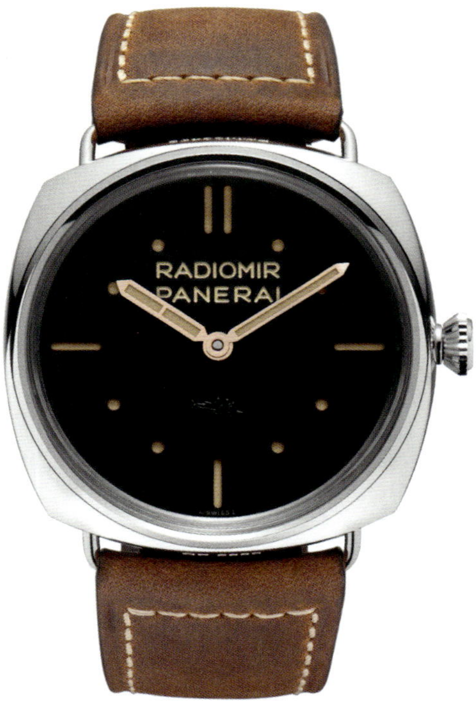

RADIOMIR 3 DAYS
REF. PAM00425

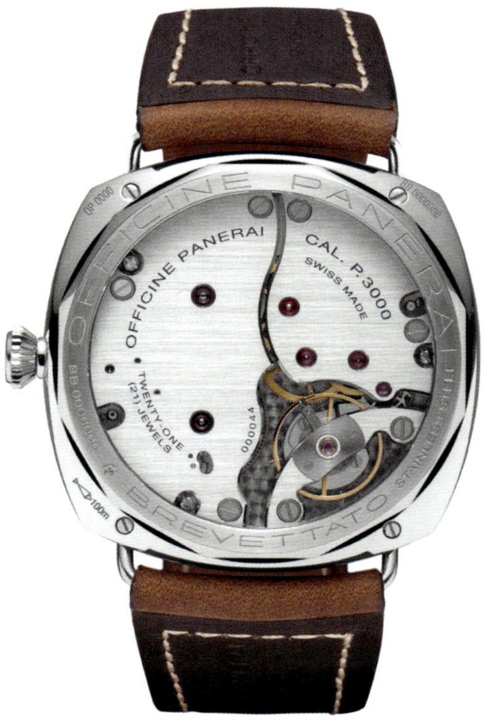

RADIOMIR 3 DAYS
REF. PAM00425
CAL. P.3000

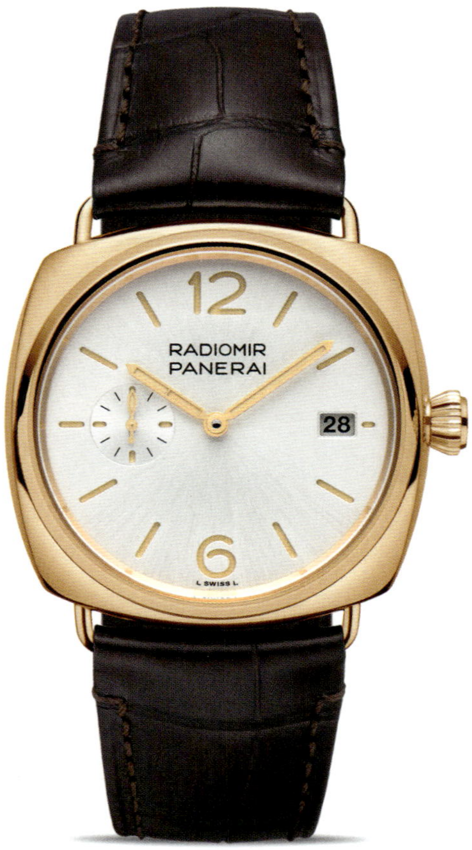

RADIOMIR QUARANTA
REF. PAM01026,
2023

1938

PANERAI RADIOMIR

APPROVED BY THE ITALIAN NAVY

At its inception, the family business founded in Florence in 1860 by Guido Panerai had nothing to do with watches. Its history unfolds in connection with high-precision navigational instruments. Panerai soon gained a reputation for the quality of its products. Enough at any rate for the company founder to win the trust of the Italian authorities. Its Navy placed orders for depth gauges, timed detonators and target-homing devices for torpedoes. All Panerai devices adhered to one guiding principle: reliability even in the most adverse situations. In the 1930s, Guido's children Guiseppe and Maria opened a specialist shop "Orologeria svizzera" close to Florence Cathedral, with high-quality timepieces by Patek Philippe and Rolex. In 1935, they presented the first prototype of an innovative diver's watch developed for the war ministry. A commission had previously rigorously tested other models and found them all to be lacking. Either the dials were not luminous enough or the cases were unable to withstand the water pressure to an adequate extent. Only the cushion-shaped Panerai with large Rolex Oyster case fitted the bill. Inside its case with screw-down crown, it was powered by a Rolex hand-wound calibre, based on a Cortébert movement 618. After extensive testing, the ministry ordered ten watches. They proved their mettle in the military field with exemplary performance during a secret special mission by the Italian Navy. From 1938, the timepieces were worn by members of the RMM "Incursori", who carried out daring attacks on enemy ships. The name Radiomir is composed of "radio" for radium and "mira" for visor. The use of a special mixture of zinc sulphate, radium bromide and mesothorium for numerals and hands yielded a particularly high level of luminosity. In 1997, Richemont Group bought the Panerai brand from the family. Since then, the Radiomir line has been one of the mainstays of the collection.

FÜR DIE ITALIENISCHE MARINE

Mit Uhren hatte das 1860 von Guido Panerai in Florenz gegründete Familienunternehmen anfangs nichts zu tun. Vielmehr begann die Geschichte mit präzisen Instrumenten für die Nautik. Offensichtlich sprach sich die Qualität der Panerai-Produkte rasch herum. Immerhin konnte der Firmengründer schon bald das Vertrauen der italienischen Behörden gewinnen. Zum Beispiel erteilte die Marine Aufträge für Tiefenmesser, zeitgesteuerte Zünder oder Zielvorrichtungen für Torpedos. Alle Geräte von Panerai gehorchten nur einer Prämisse: Zuverlässigkeit selbst in schwierigsten Situationen. In den 1930er-Jahren eröffneten Guidos Kinder Guiseppe und Maria nahe dem Florentiner Dom das Fachgeschäft „Orologeria svizzera" mit hochwertigen Zeitmessern von Patek Philippe und Rolex. 1935 präsentierten sie den ersten Prototypen einer neuartigen Taucheruhr, entwickelt für das Kriegsministerium. Zuvor hatte eine Kommission andere Modelle auf Herz und Nieren geprüft. Zufriedenstellen konnte keine. Entweder leuchtete ihr Zifferblatt nicht stark genug oder die Gehäuse vermochten dem Wasserdruck nicht hinreichend zu widerstehen. Erst die kissenförmige Panerai „Radiomir" mit großer „Oyster"-Schale von Rolex genügte den Anforderungen. Im Inneren des Schraubkronengehäuses tickte ein Rolex-Handaufzugskaliber, dessen Rohwerk 618 von Cortébert stammte. Nach eingehender Erprobung orderten die Behörden zehn Exemplare, die sich bei einem geheimen Spezialkommando der italienischen Marine in allen militärischen Lebenslagen bewährten. Ab 1938 fand man die Zeitmesser u. a. an den Handgelenken von Männern der „Incursori"-Gruppe, welche gewagte Angriffe auf feindliche Schiffe unternahmen. Der Name „Radiomir" beinhaltet „Radio" für Radium und „Mira" für Visier. Durch die Verwendung einer Spezialmischung aus Zinksulfat, Radiumbromid und Mesothonum für Ziffern und Zeiger ergab sich eine besonders hohe Leuchtkraft. 1997 erwarb der Richemont-Konzern die Marke Panerai von der Familie. Seitdem gehört die Linie „Radiomir" zu den tragenden Säulen der Kollektion.

Left | Links: *First | Erste Radiomir*
Right | Rechts: *Radiomir, 1956*

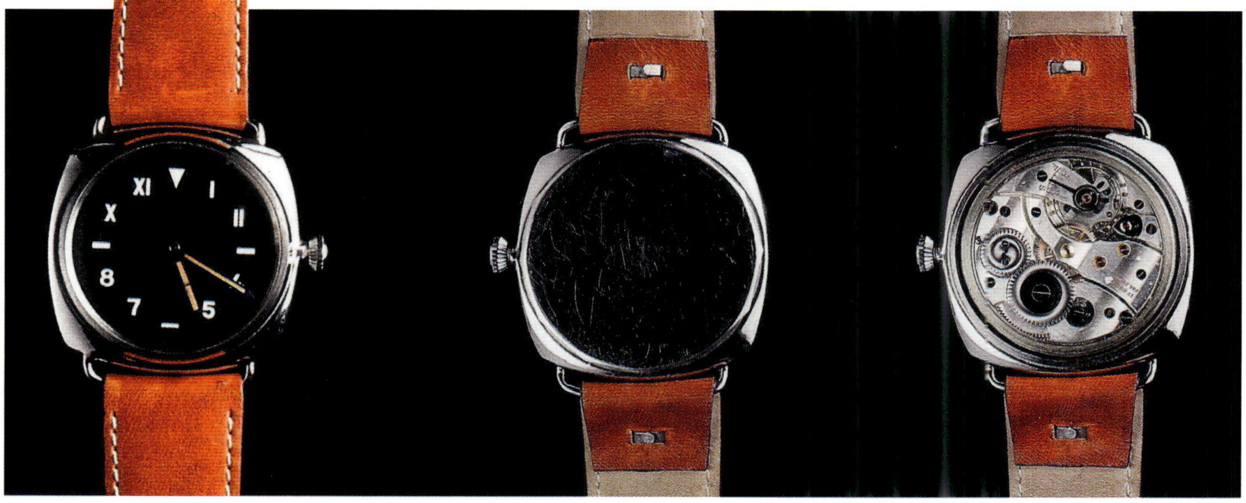

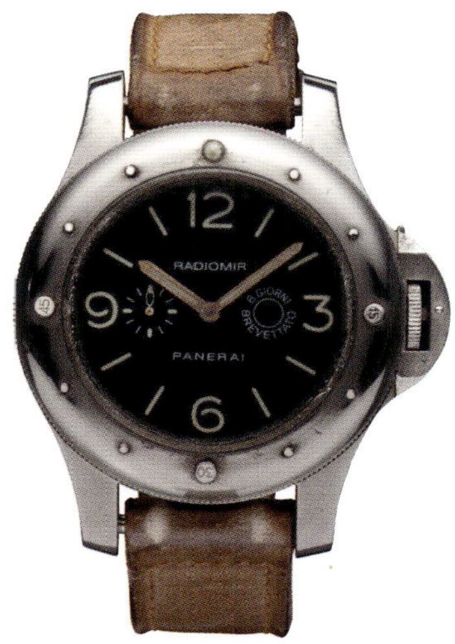

PANERAI LUMINOR

1949

STATEMENT CROWN GUARD

The sumptuous Panerai Radiomir undoubtedly represents a milestone in the history of military diver's watches. However, chinks in the armour of the cushion-shaped case were soon revealed in the harsh underwater world. Its thin, loop-shaped lugs soldered to the case could break or become detached when exposed to extreme stress. This meant the loss of a vital timekeeping instrument underwater. For this reason, the models with which the Royal Italian Navy fleet was kitted out in the early 1950s featured a new cushion-shaped case with integrated lugs. The Rolex Cortébert Calibre 618 was used throughout. Its screwed crown satisfied expectations: screwed tight, it reliably kept water away from the movement. Occasionally, however, in his haste the diver might not close the case properly after winding up the watch. The impact of salt water penetration was severe. The use of automatic movements would have provided a solution to the problem. Because Officine Panerai had no access to such movements, they decided to develop a completely new sealing system. The bridge mounted on the side above the crown held a movable lever. This pressed the crown and seal firmly against the case flank. Winding the watch and setting the hands required the wearer to flip up the lever. After winding, it was simply impossible to overlook the projecting component. In Italy, the patenting, documented by prominent engraving on the bridge, took place in July 1956 after successful tests by the military. The model name Luminor is a reference to a new tritium-based luminous paint for the dial and hands.

As part of Richemont Group, the history of the distinctive Panerai Luminor continues to unfold in a variety of iterations.

UNÜBERSEHBARER KRONENBÜGEL

In der Geschichte militärischer Taucheruhren verkörperte die opulente Panerai „Radiomir" zweifellos einen Meilenstein. In der harten Unterwasserwelt zeigten sich jedoch auch schnell die Problemstellen des kissenförmigen Gehäuses. Seine dünnen, schlingenförmig ausgeführten und ans Gehäuse gelöteten Bandbügel konnten bei extremen Belastungen brechen oder gar abreißen. Unter Wasser bedeutete das den Verlust des immens wichtigen Zeitmessers. Deswegen erhielt das Flottenkommando der königlich-italienischen Marine in den frühen 1950er-Jahren ein neues kissenförmiges Gehäuse mit integrierten Bandanstößen. Im Original tickte abermals das von Rolex gelieferte Cortébert-Kaliber 618. Seine Schraubkrone tat das, was man von ihr erwartete: Ordentlich zugedreht, hielt sie Wasser zuverlässig fern vom Uhrwerk. Gelegentlich jedoch erfolgte das Verschließen der Schale nach dem manuellen Spannen der Zugfeder im Eifer des Gefechts nicht mit der nötigen Sorgfalt. Die Folgen durch eindringendes Salzwasser waren gravierend. Eine Lösung des Problems hätte in der Verwendung von Automatikwerken bestanden. Weil die Officine Panerai darauf keinen Zugriff hatten, machten sie sich an die Entwicklung eines völlig neuen Dichtsystems. Die seitlich über der Krone montierte Brücke hielt einen beweglichen Hebel. Dieser presste das Bedienelement samt Dichtung fest gegen die Schalenflanke. Aufziehen und Zeigerstellen verlangten nach bewusstem Hochklappen. Das dann weit abstehende Bauteil ließ sich nach der Energiezufuhr beim besten Willen nicht übersehen. In Italien erfolgte die in einer unübersehbaren Brückengravur dokumentierte Patentierung im Juli 1956 nach erfolgreichen Tests durch die Militärs. Der Modellname „Luminor" resultierte aus einer neuen, Tritium-basierten Leuchtsubstanz für Zifferblatt und Zeiger.

Unter dem Dach der Richemont-Gruppe schreibt Panerai die Geschichte der markanten „Luminor" heutzutage in ganz unterschiedlichen Ausprägungen fort.

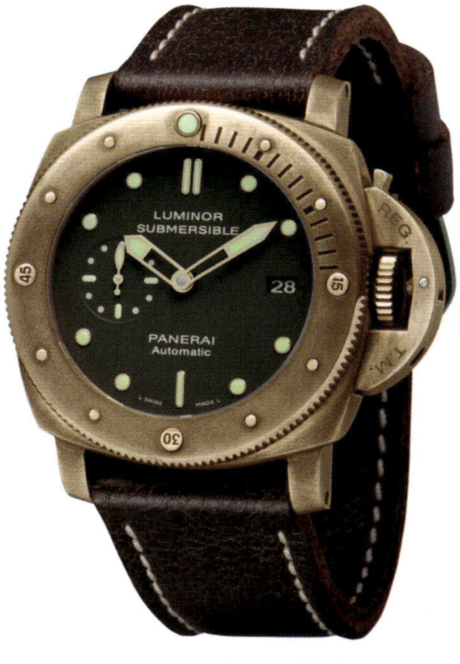

Luminor Submersible Bronzo
Ref. PAM00382, 2011

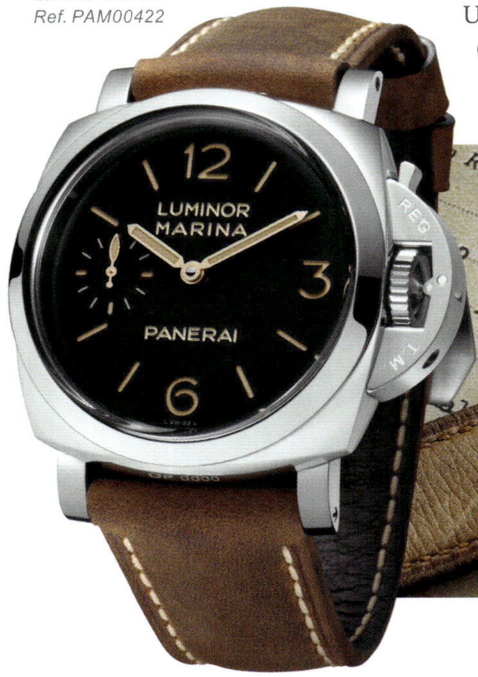

Luminor Marina
Ref. PAM00422

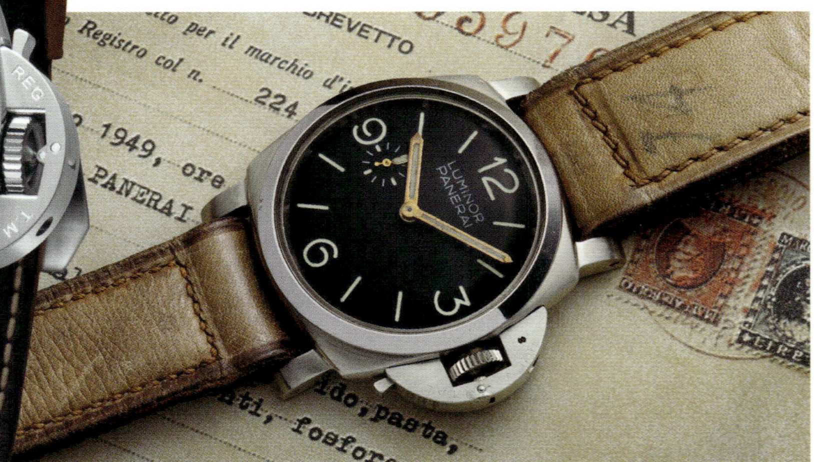

Top / Oben: Luminor Ref. 6152-1, 1960
Right page / Rechte Seite: Luminor Submersible Luna Rossa Ref. PAM01565, 2024

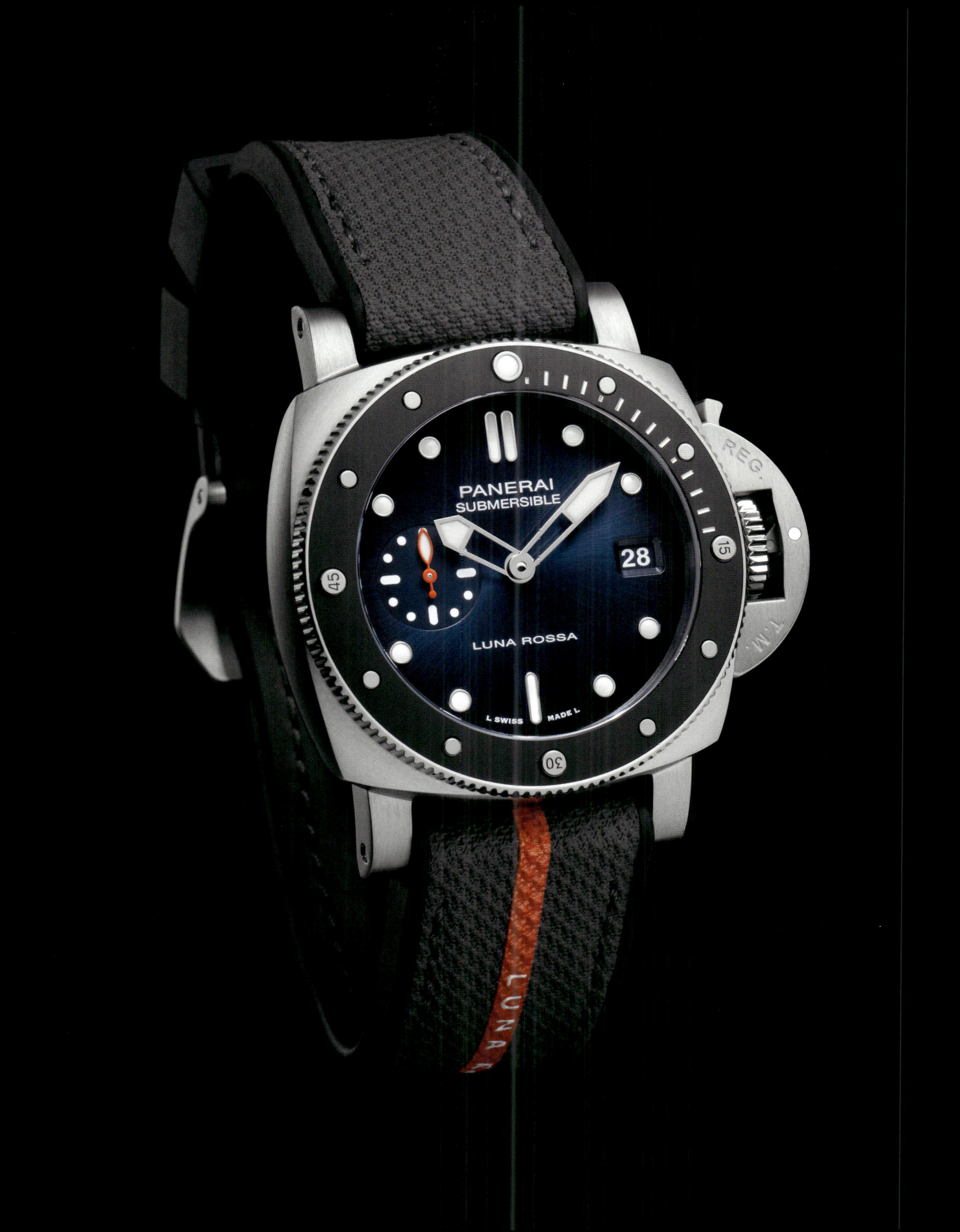
PANERAI
SUBMERSIBLE
LUNA ROSSA
SWISS MADE
28
REG. T.M.
LUNA

PARMIGIANI FLEURIER 2021
TONDA PF

ROUND, ELEGANT AND SPORTY

Tonda means round, as anyone with a smattering of Italian will know. Watches have been round for centuries because lathes are among the oldest tools known to man. In addition, the rotating hands symbolise the infinite cycle of coming and going. It therefore seemed natural that Parmigiani Fleurier, the Swiss watch manufacturer with Italian origins, would name its round watch line Tonda. Since the company was founded in 1996, the word has stood for understated elegance. In 2020, the Tonda GT exuded a sporty flair. The newcomer was designed by master watchmaker Dino Modolo. Some twelve months later, the less sporty Tonda PF made its début in the remote watchmaking town of Fleurier. True to the philosophy that Parmigiani Fleurier is traditionally more than just a sports watch.

By omitting the name and using only a logo, the company appeals to a community of connoisseurs. Although in the top price bracket, the Tonda PF ethos is not luxury, but expertise in the art of fine watchmaking. This is conveyed in the subtle texture of the dial and the very short indices. The surface of the dial is therefore pristinely uncluttered. The luxurious understatement of the steel models is accentuated by the solid platinum bezel. In 2023, the new Tonda PF Sport joined the Tonda PF at Parmigiani Fleurier. The latter's more distinctive appearance, which only becomes apparent at second glance, is referenced by its successor, the Tonda GT. This model differs from the Tonda PF in its expressive fluted glass bezel, the more obvious ribbing of the glass rim, guilloché patterning of the dial and additional rotating second hand. The two top models come in an impressive number of iterations. No matter whether purist without complications or complex, precious metal or steel, lovers of this watch line are sure to find a model to their taste.

RUND, ELEGANT UND SPORTLICH

„Tonda" bedeutet rund. Solches wissen Menschen, die der italienischen Sprache ein wenig mächtig sind. Uhren besitzen diese Form seit Jahrhunderten, denn Drehbänke gehören zu den ältesten Maschinen. Außerdem symbolisieren die rotierenden Zeiger den unendlichen Kreislauf des Kommens und Gehens. Insofern lag es auf der Hand, dass Parmigiani Fleurier, die Schweizer Uhrenmanufaktur mit italienischen Wurzeln, seine runde Uhrenlinie „Tonda" taufte. Seit der Gründung im Jahr 1996 stand das Wort für zurückhaltende Eleganz am Handgelenk. 2020 brachte die „Tonda GT" eine sportive Attitüde. Gestalterischer Vater des Newcomers war Altmeister Dino Modolo. Gut zwölf Monate später gab im abgeschiedenen Uhrmacherstädtchen Fleurier die weniger sportliche „Tonda PF" ihren Einstand. Getragen von der Philosophie, dass Parmigiani Fleurier traditionsgemäß mehr ist als nur eine Sportuhr.

Durch das Weglassen des Namenszugs und die ausschließliche Verwendung eines Logos wendet sich das Unternehmen an eine Gemeinschaft sachkundiger Zeitgenossinnen und -genossen. Obwohl im höchsten Preissegment angesiedelt, versteht sich die „Tonda PF" nicht als Symbol für Reichtum, sondern als Ausdruck für Wissen rund um feine Uhrmacherkunst. Dazu trägt die subtile Textur des Zifferblatts ebenso bei wie die sehr kurze Indexierung. Letztere öffnet die Fläche, vor der die Zeitzeiger drehen. Zum luxuriösen Understatement der Stahlmodelle trägt eine Lünette aus massivem Platin bei. 2023 gesellte sich bei Parmigiani Fleurier zur „Tonda PF" die neue „Tonda PF Sport". Durch ihren erst auf den zweiten Blick markanteren Auftritt knüpft letztere an die „Tonda GT", welche sie in der Kollektion ersetzt. Unterschiede zur „Tonda PF" bestehen u. a. in der augenfälligeren Riffelung des Glasrands, der andersartigen Guillochierung des Zifferblatts und dem zusätzlich drehenden Sekundenzeiger. Die Wahlmöglichkeiten bei beiden Leadern sind beachtlich. Egal ob puristisch ohne Zusatzfunktionen oder kompliziert, ob aus Edelmetall oder Stahl: Wer diese Uhrenlinie mag, wird definitiv fündig.

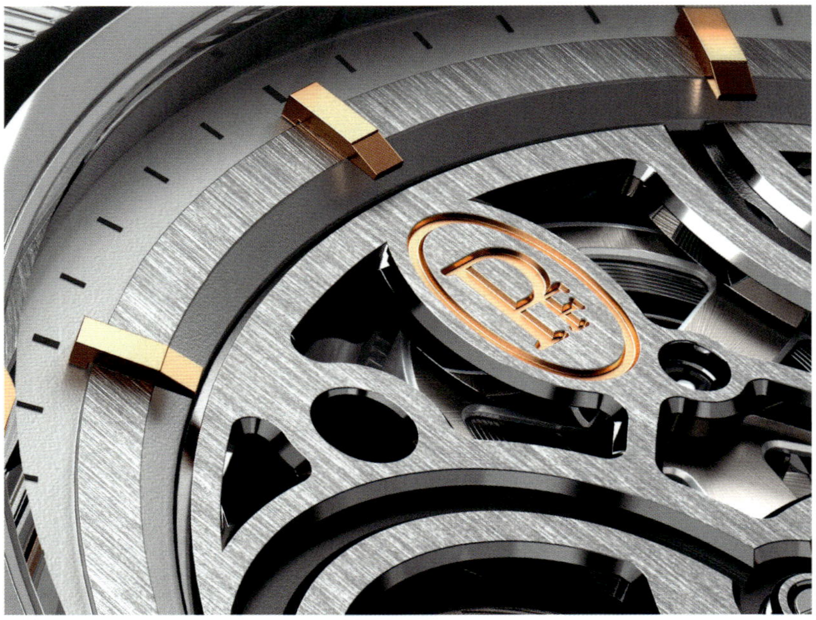

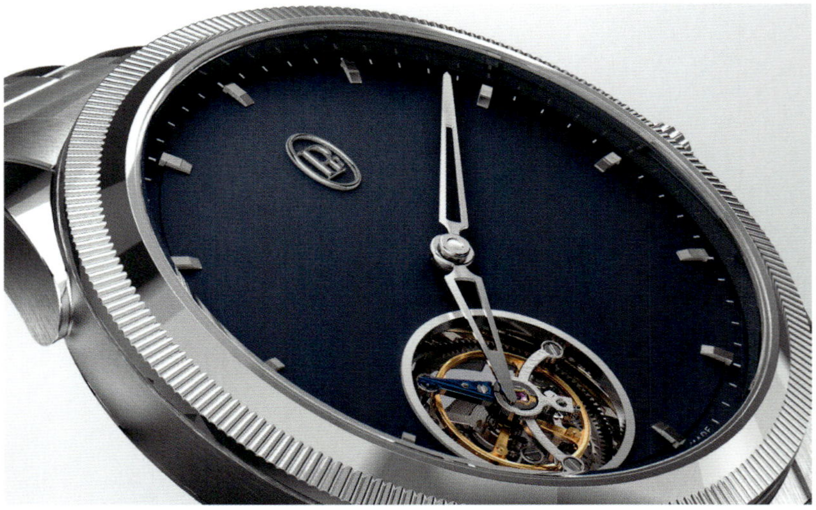

From left | Von links: *Tonda PF Skeleton, detail, 2024 | Tonda PF Flying Tourbillon, 2023*

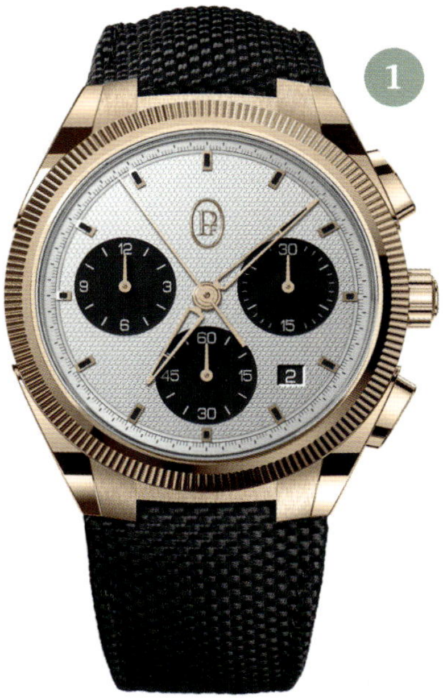

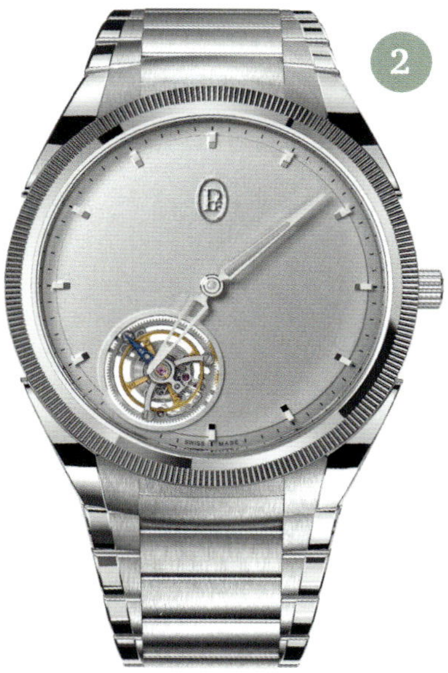

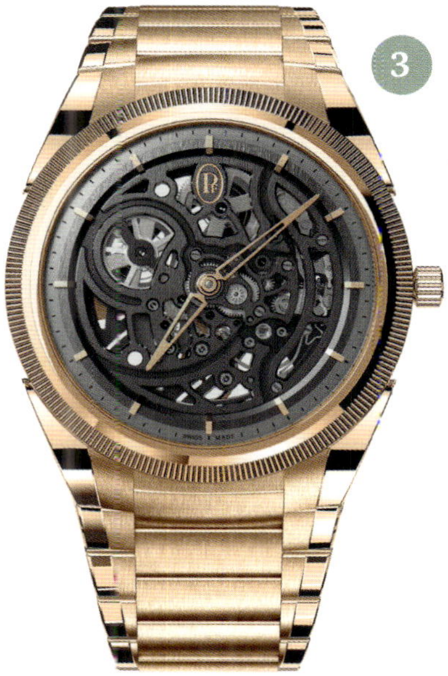

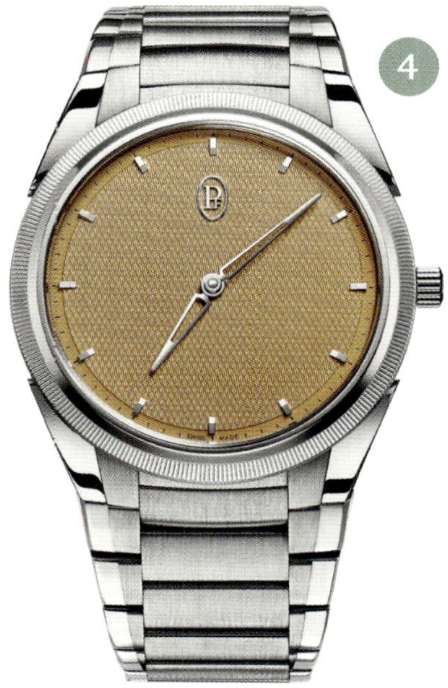

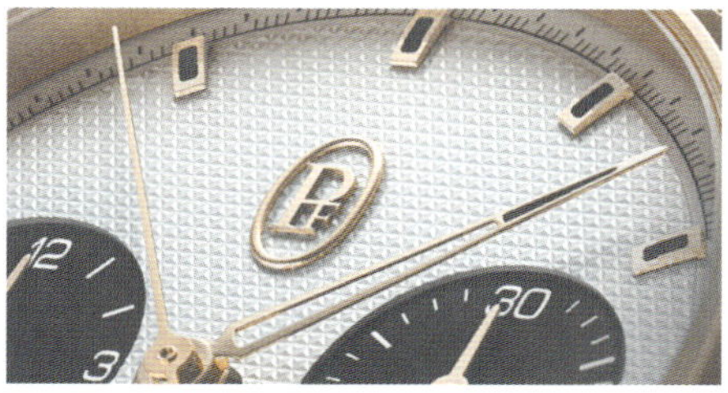

1 *Sport Chronograph Gold, 2023*
2 *Flying Tourbillon, 2023*
3 *Skeleton, 2024*
4 *Micro Rotor Siena, front & back, 2024*
5 *Annual Calendar*

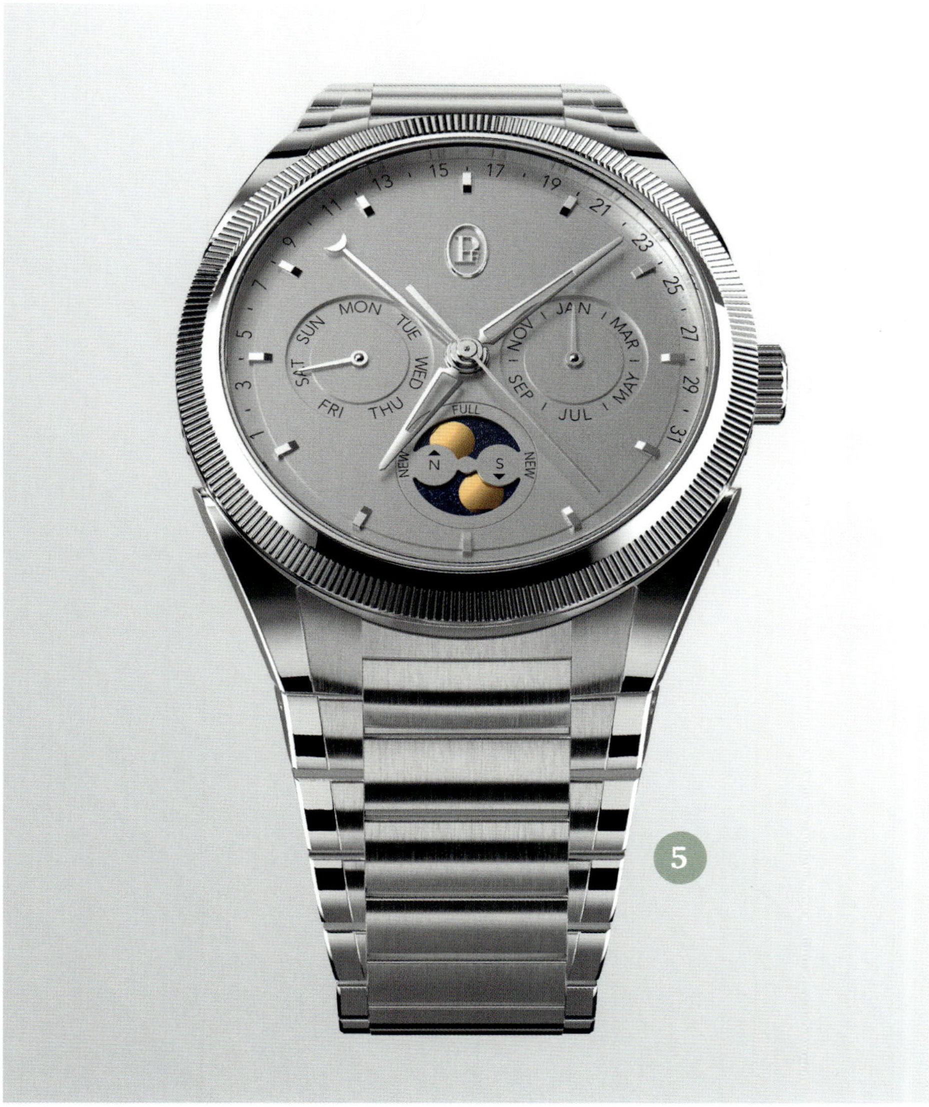

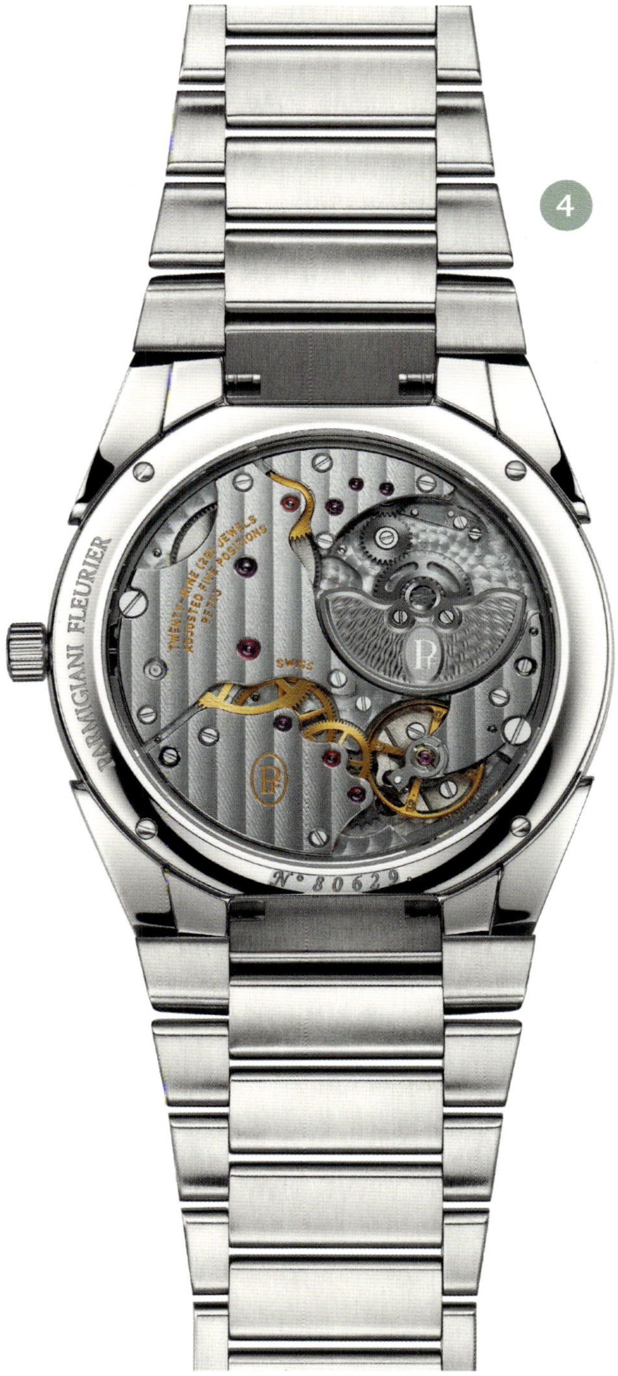

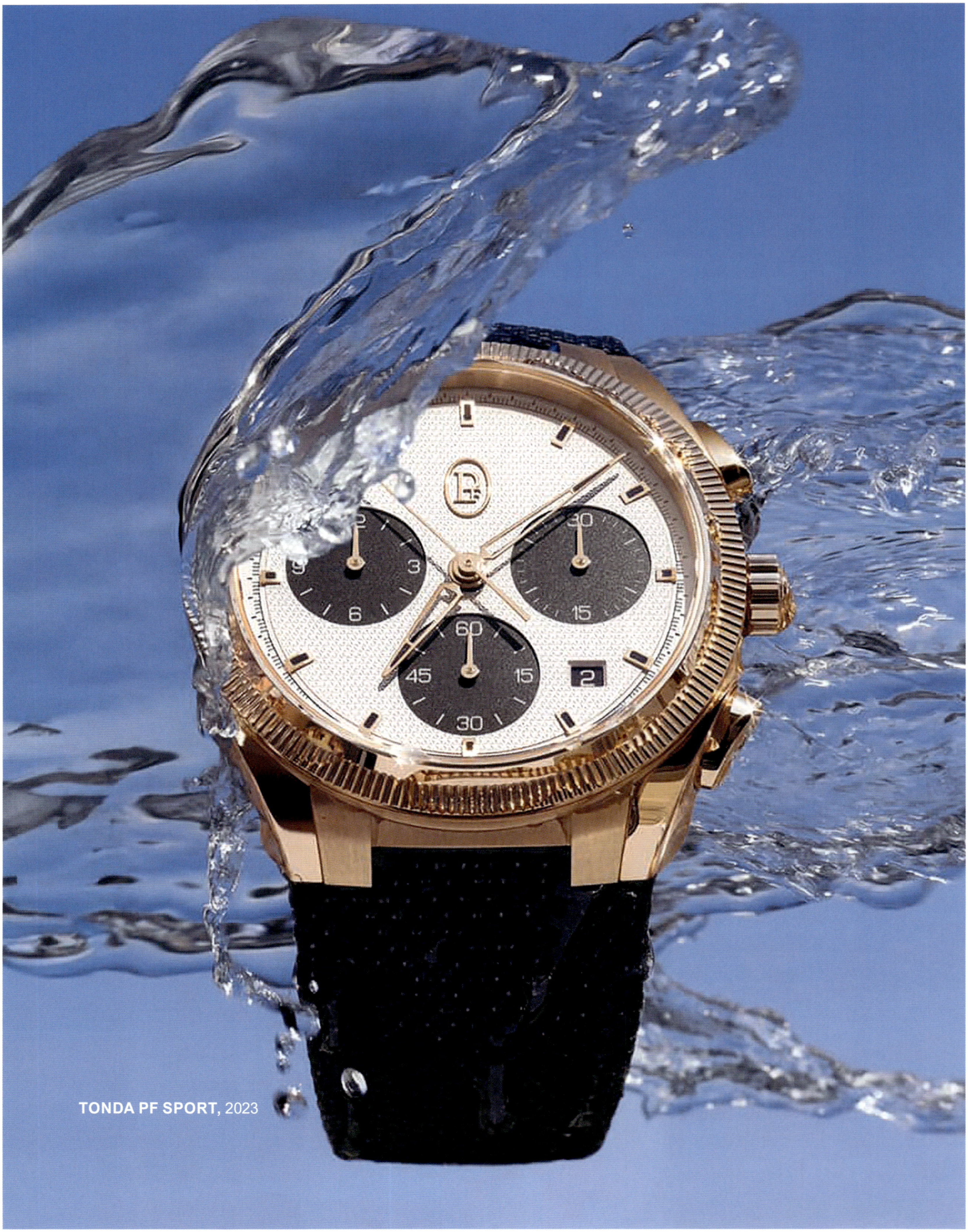

TONDA PF SPORT, 2023

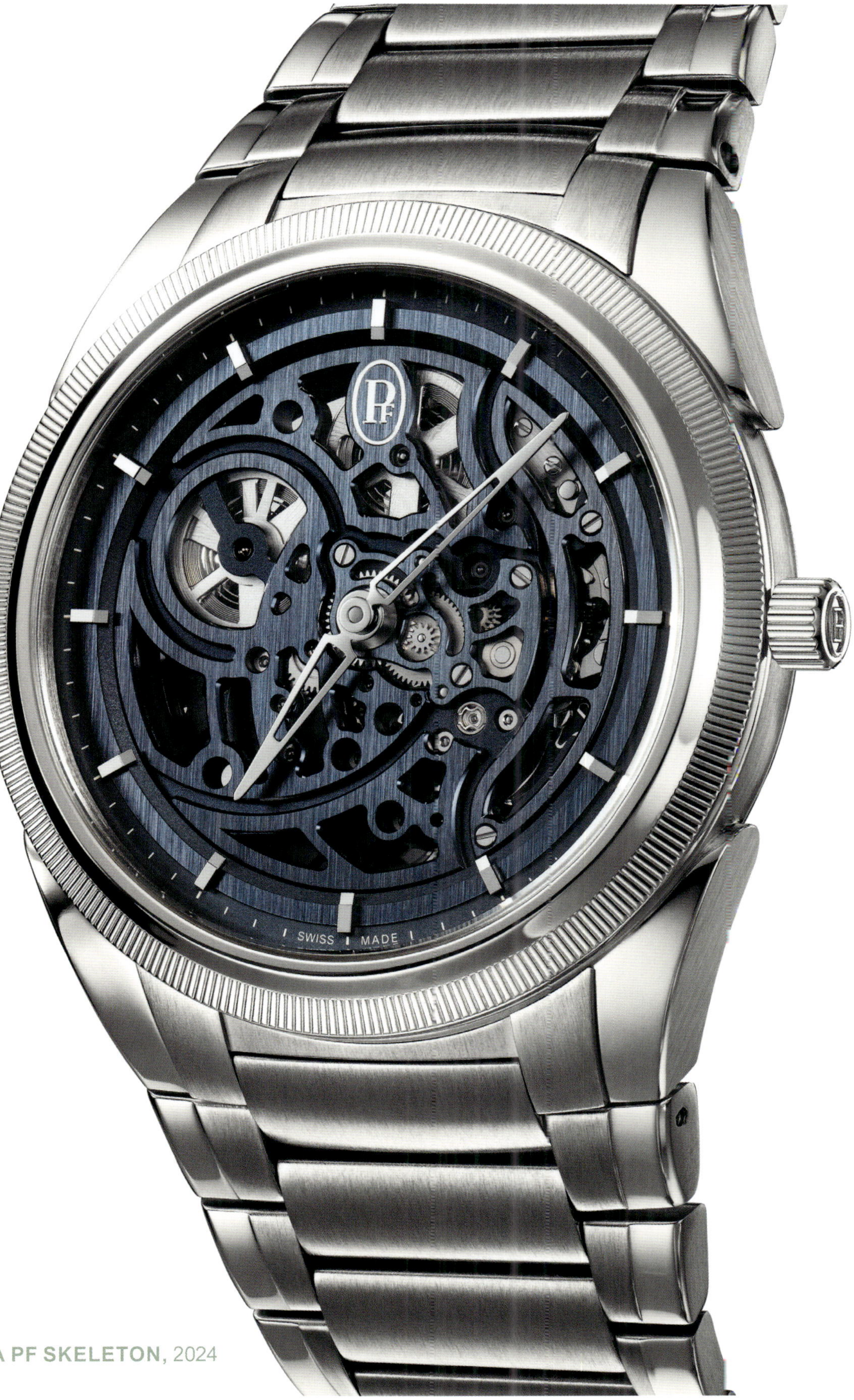

TONDA PF SKELETON, 2024

PATEK PHILIPPE AQUANAUT

1997

DESIGNED BY THIERRY STERN

The history of the Patek Philippe Aquanaut unfolds in the mid-1990s with the company's current president Thierry Stern. In 1997 his first creation débuted as Reference 5060A. The manufactory offered the stainless-steel wristwatch as an entry-level model with a price tag of around €4,250. A consummate combination of porthole-like rotundity and octagonality, the newcomer boasted its pedigree as a descendant of the famed neoclassic Nautilus.

With a screwed case and screw-down crown, the model is waterproof up to three bar. Distinctive features include a prominent glass bezel and the rich interplay of polished and satin-finished surfaces. With this avant-garde design Thierry Stern aimed to create a coherent look with high recognition value. The most eye-catching characteristic of the bold dial are Arabic numerals in white gold. The three central hands for hours, minutes and seconds are also made of gold. A comfortable grip on the wrist is ensured by an anti-allergic, tear-resistant Tropical strap that is resistant to salt water, ultraviolet rays, heat and cold. Along with a steel folding clasp, it consists of 20 components. The proprietary Calibre 330 SC is responsible for timekeeping. Watchmakers must assemble a total of 228 components for the automatic movement with a unidirectional gold rotor and 48-hour power reserve. When the steely Nautilus was phased out in 2021, the Aquanaut immediately assumed its role as flagship model. Reflecting this promotion, prices also rose accordingly on the parallel market. The successor to the original reference is the Aquanaut 5167 with a case diameter of 40.8 millimetres and the automatic Calibre 26-330 SC. It is also available with a steel link bracelet.

GESTALTET VON THIERRY STERN

Die Geschichte der Patek Philippe „Aquanaut“ beginnt Mitte der 1990er-Jahre mit dem aktuellen Präsidenten Thierry Stern. 1997 debütierte sein gestalterisches Erstlingswerk als Referenz 5060A. Mit einem Preis von umgerechnet rund 4.250 Euro offerierte die Manufaktur diese Edelstahlarmbanduhr als eine Art Einstiegsmodell. Dank einer gekonnten, Bullaugen-ähnlichen Synthese aus Achteck und Rund präsentierte sich der Newcomer als unverwechselbarer Spross des anerkannten Neoklassikers „Nautilus“.

Die Ausstattung mit Schraubboden und -krone gestattet eine Wasserdichte bis zu drei bar Druck. Ins Auge stechen unter anderem ein markanter Glasrand sowie das kontrastreiche Dekor aus satinierten und polierten Flächen. Sein avantgardistisches Design betrachtete Thierry Stern ganz bewusst als optische Einheit mit hohem Wiedererkennungswert. Hervorstechendstes Merkmal des ausdruckstarken Zifferblatts sind arabische Ziffern, allesamt ausgeführt in weißem Gold. Aus diesem Material bestehen auch die drei in der Mitte angeordneten Zeiger für Stunden, Minuten und Sekunden. Für komfortablen Halt am Handgelenk sorgt ein gleichermaßen antiallergisches wie reißfestes Tropical-Armband, welches sich Salzwasser, ultravioletten Strahlen, Hitze oder Kälte problemlos widersetzt. Samt stählerner Faltschließe besteht es aus 20 Komponenten. Die Zeitanzeige bewerkstelligt das hauseigene Kaliber 330 SC. Uhrmacher benötigen für das Automatikwerk mit einseitig wirkendem Goldrotor und maximal 48 Stunden Gangautonomie summa summarum 228 Komponenten. Nach dem Auslaufen der stählernen „Nautilus“ im Jahr 2021 hat die „Aquanaut“ unverzüglich deren herausragende Position eingenommen. Dementsprechend entwickelte sich auch das Preisniveau am Parallelmarkt. Nachfolger der ursprünglichen Referenz ist die „Aquanaut 5167“ mit 40,8 Millimeter Gehäusedurchmesser und dem Automatikkaliber 26-330 SC. Sie ist auch mit einem stählernen Gliederband verfügbar.

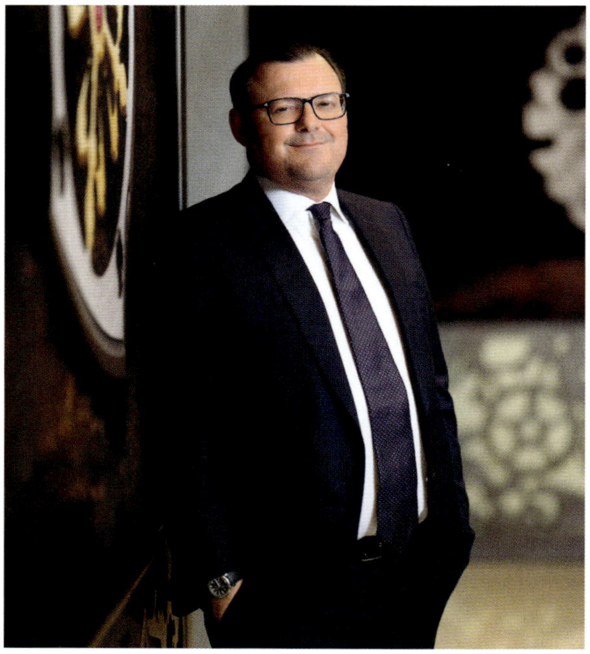

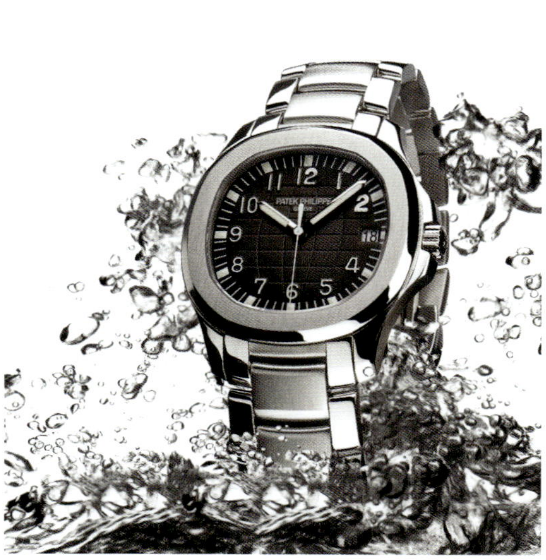

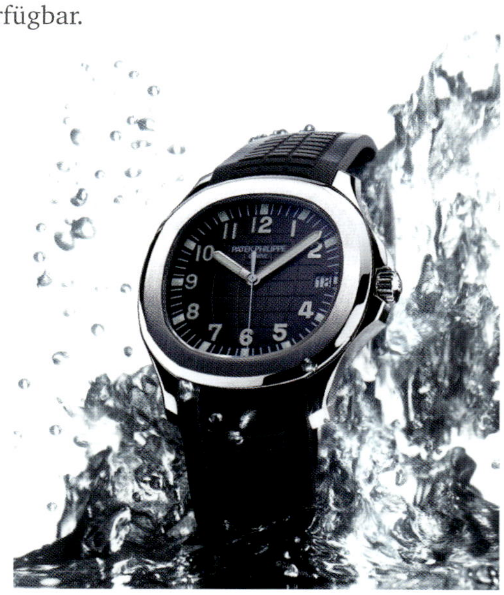

From left: *Thierry Stern | Ref. 5167/1A-001 | Ref. 5167A-001*

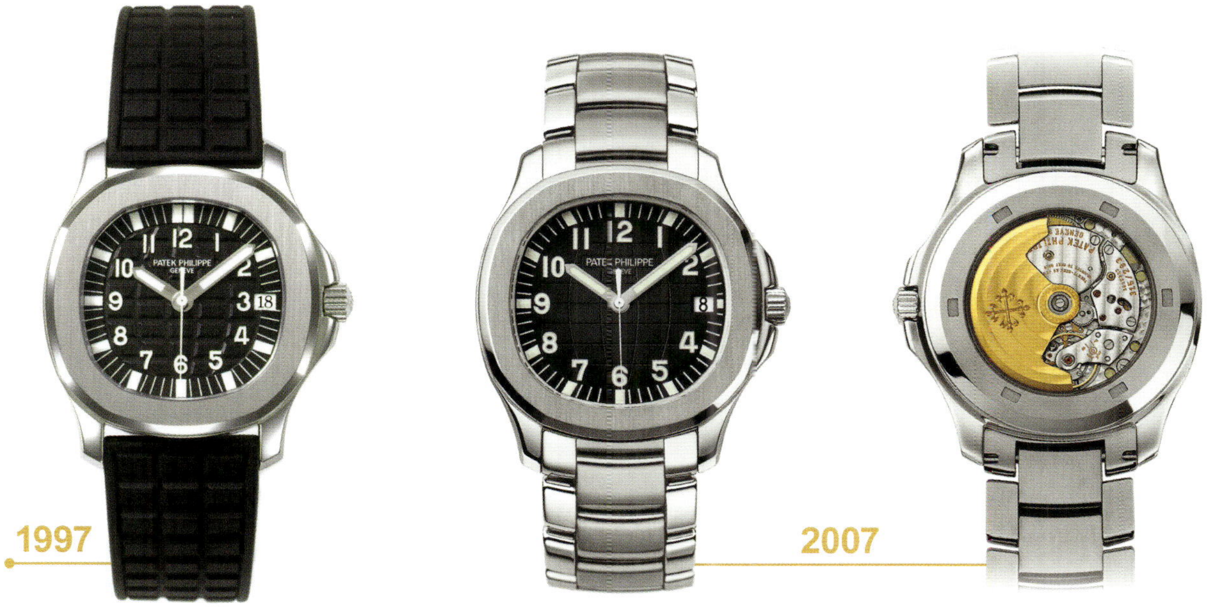

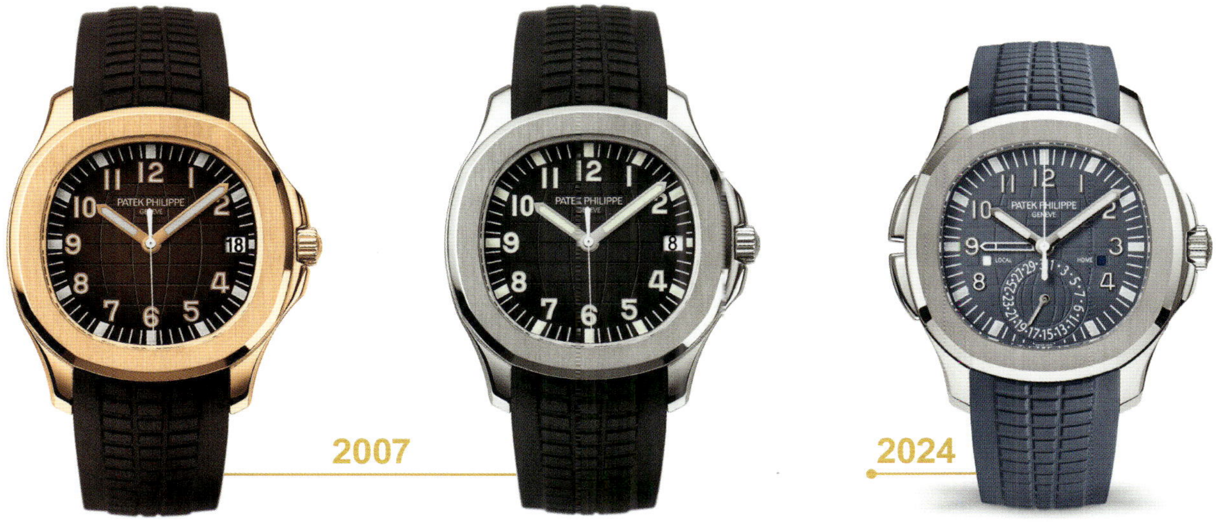

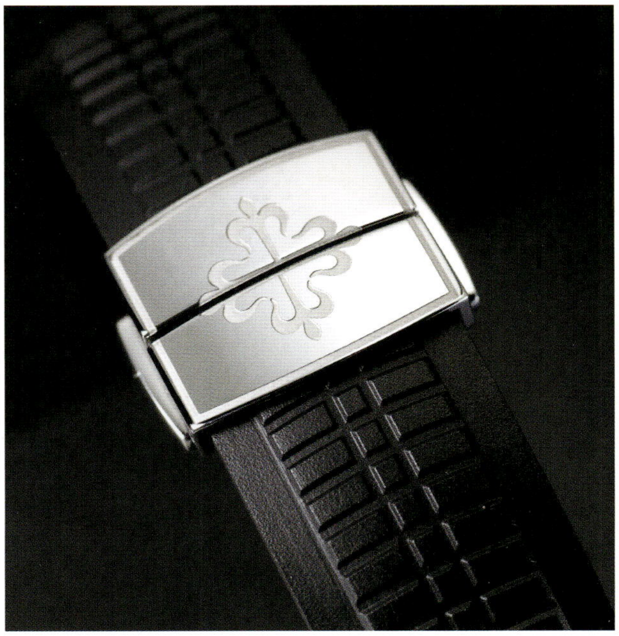

Aquanaut folding clasp with engraved brand logo | *Aquanaut Faltschließe mit graviertem Markenlogo*

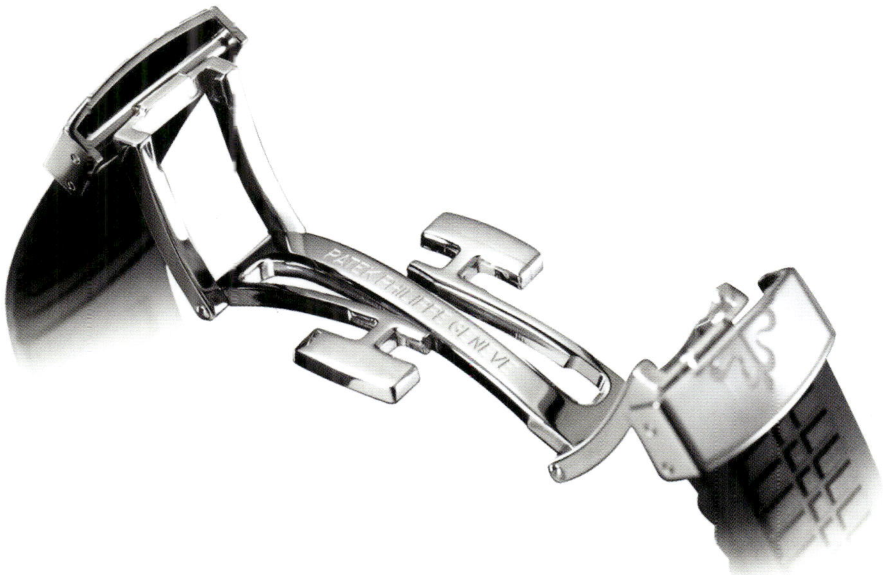

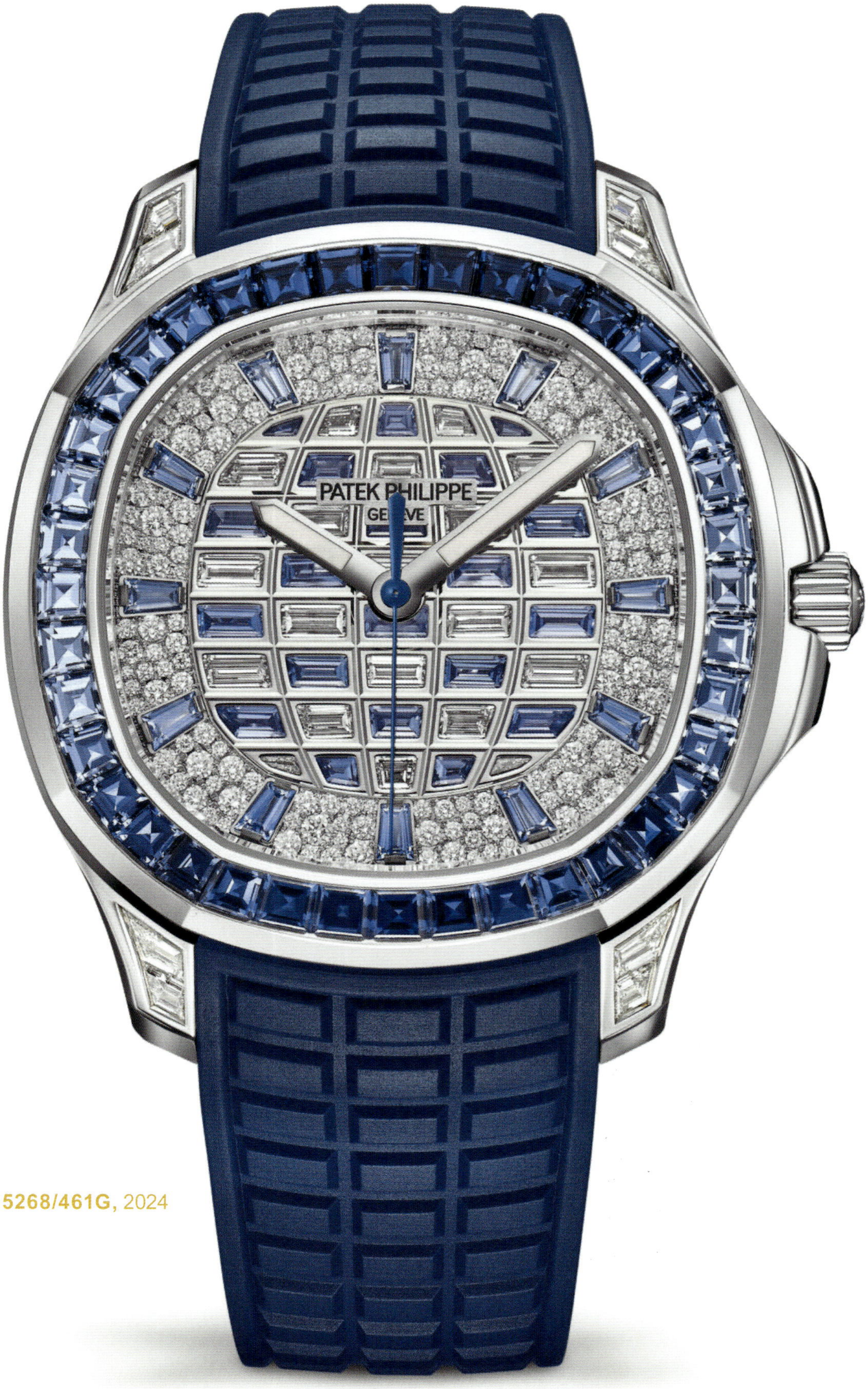

REF. 5268/461G, 2024

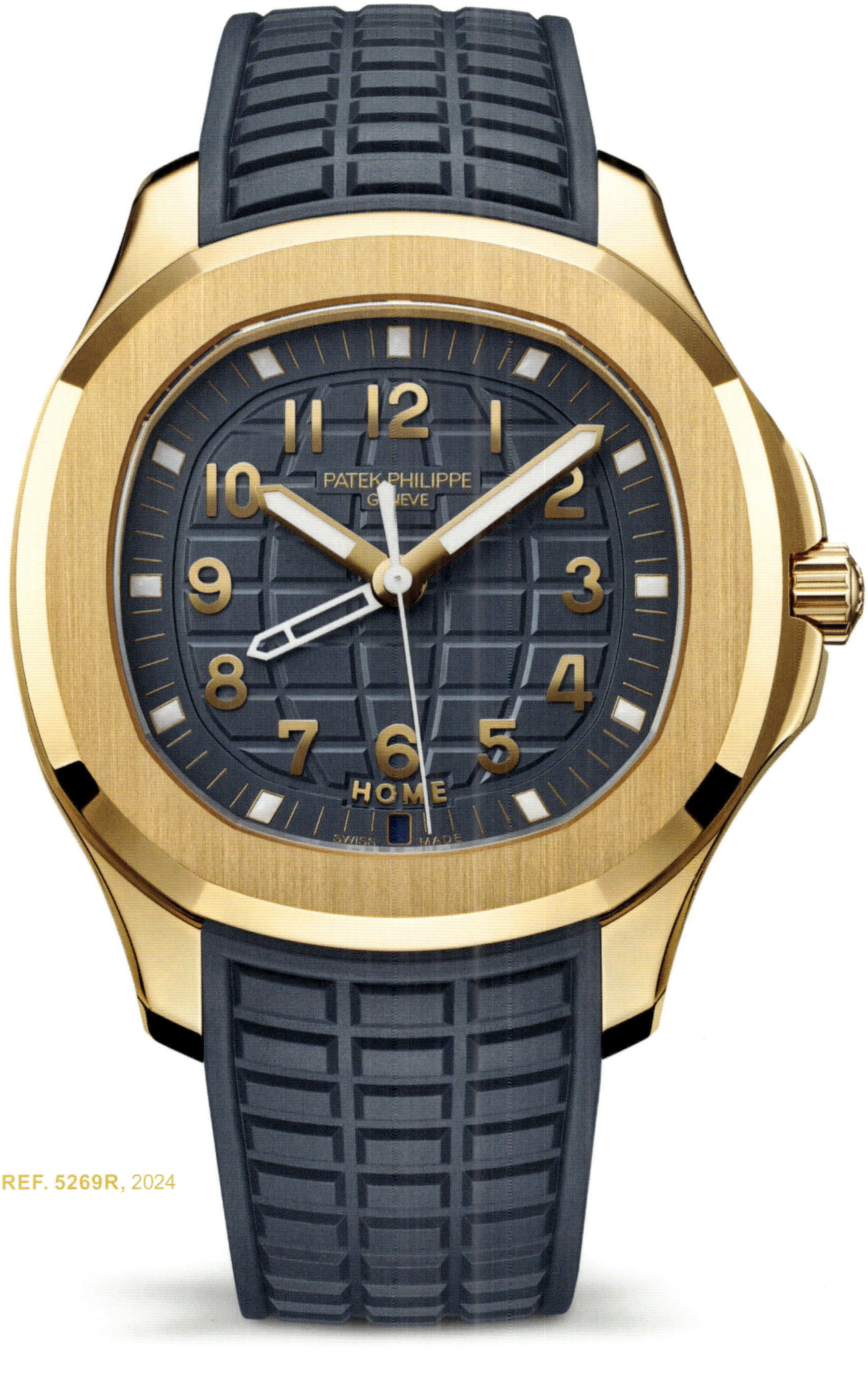

REF. 5269R, 2024

PATEK PHILIPPE 1996
ANNUAL CALENDAR

A ONCE-A-YEAR AFFAIR

The watchmakers at Patek Philippe have devoted themselves to the date display on watches for generations. All were agreed that the manual corrections needed in all months with fewer than 31 days are a perfect nuisance. One solution to the problem is the sophisticated and thus expensive perpetual calendar, which only causes a minor problem that can be quickly remedied in secular years that cannot be divided by 400, such as 2100. In the 1990s, the engineers thought that price-conscious watch lovers would be happy to comply once a year. Based on this conviction, in 1996 the world's first mechanical watch with a practical annual calendar was launched.

Protected by Swiss patent CH 685585 G of 1 March 1996, the mechanism is programmed so that it automatically takes account of the different month lengths from 1 March to the end of February the following year. Then a small intervention is required. Afterwards, the calibre is all set for the next twelve months. Along with the date, the days of the week and the months can also be read on the dial. In all, an automatic movement with an annual calendar consists of considerably more individual parts than the flatter Patek Philippe Calibre 240Q with a perpetual calendar. The latter consists of numerous shift levers, pawls, lock springs and cams. By contrast, the annual calendar, consisting of 134 components, runs almost exclusively with the help of wheels and pinions. The sophisticated gear chain is mounted on its own calendar plate. An ingenious isolator ensures that the date display progresses almost without any mechanical resistance. In total, 134 components are required for an annual calendar movement. Watchmakers attach it to the front of the base movement. A 24-hour display eases correct setting to ensure that all indications progress at midnight not noon. The moon moves exactly around its axis. It only deviates from the astronomical norm by a day every 122 years or so. During the 1996 Basel Watch Fair, Patek Philippe presented the Annual Calendar 5035, the first reference with this innovation.

In 2005, when Patek Philippe used a silicon escape wheel in the Advanced Research Annual Calendar Reference 5250, produced in a limited edition of 100 models, an annual calendar was added to the Calibre 315. The same applies to the Advanced Research Reference 5350 with Spiromax balance spring made of temperature-stable Silinvar, which was launched in 2007. Patek Philippe produced 300 of these. In 2008, the same quantity was produced of the Reference 5450 Advanced Research. Its hallmark is the Pulsomax escapement with Silinvar escape wheel and pallets.

In recent years, the Geneva family manufactory has continuously developed its annual calendar. The analogue calendar has now been joined by a digital version with apertures for day and month.

NUR EINMAL JÄHRLICH

Über die Datumsanzeige bei Armbanduhren machen sich die Uhrmacher von Patek Philippe seit Generationen ihre Gedanken. Dass es überaus lästig ist, in allen Monaten mit weniger als 31 Tagen manuelle Korrekturen vorzunehmen, stand für aller außerhalb jedes Zweifels. Eine Lösung des Problems besteht im aufwändigen und entsprechend teuren ewigen Kalender, der nur in nicht durch 400 teilbaren Säkularjahren, also beispielsweise 2100, ein kleines, rasch behebbares Problem bereitet. In den 1990er-Jahren glaubten die Techniker, preisbewussten Uhrenliebhaberinnen und -liebhabern einmal jährlich etwas zumuten zu können. Aus diesem Bewusstsein heraus entstand bis 1996 die weltweit erste mechanische Armbanduhr mit praktischem Jahreskalender.

Der durch das Schweizer Patent CH 685585 G vom 1. März 1996 geschützte Mechanismus ist so programmiert, dass er die unterschiedlichen Monatslängen vom 1. März bis Ende Februar des kommenden Jahres selbsttätig berücksichtigt. Dann wird ein kleiner Handgriff fällig. Anschließend ist die Kadratur gewappnet für weitere knapp zwölf Monate. Vom Zifferblatt lassen sich neben dem Datum auch die Wochentage und die Monate ablesen. Insgesamt erfordert ein Automatikwerk mit Jahreskalender deutlich mehr Einzelteile als das flachere Patek-Philippe-Kaliber 240Q mit ewigem Kalender. Bei letztgenanntem besteht dieser aus zahlreichen Schalthebeln, -klinken, Sperrfedern und Nocken. Demgegenüber funktioniert der aus 134 Komponenten assemblierte Jahreskalender fast ausschließlich mit Hilfe von Rädern und Trieben. Die ausgeklügelte Getriebekette ist auf einer eigenen Kalenderplatine montiert. Ein durchdachter Isolator sorgt für die Fortschaltung der Datumsanzeige beinahe ohne mechanischen Widerstand. Alles in allem sind 134 Bauteile für eine Jahreskalender-Kadratur erforderlich. Uhrmacher befestigen diese auf der Vorderseite des Basisuhrwerks. Eine 24-Stunden-Anzeige erleichtert das richtige Einstellen, damit alle Indikationen nachts und nicht mittags weiterschalten. Genauestens bewegt sich der Mond um seine Achse. Von der astronomischen Norm weicht er erst nach gut 122 Jahren um nur einen Tag ab. Im Rahmen der Basler Uhrenmesse 1996 zeigte Patek Philippe die „Jahreskalender 5035“ als erste Referenz mit dieser Innovation.

Als Patek Philippe 2005 in der auf 100 Exemplare limitierten Referenz 5250 „Advanced Research Jahreskalender“ ein Silizium-Ankerrad verwendete, erhielt das Kaliber 315 auch einen Jahreskalender. Gleiches gilt für die 2007 lancierte Referenz 5350 „Advanced Research“ mit „Spiromax“-Unruh aus temperaturstabilisiertem „Silinvar“. Hiervon fertigte Patek Philippe 300 Stück. Die gleiche Quantität gab es 2008 von der Referenz 5450 „Advanced Research“. Ihr Kennzeichen ist die „Pulsomax“-Hemmung mit „Silinvar“-Ankerrad und -Anker.

In den zurückliegenden Jahren hat die Genfer Familienmanufaktur ihren „Jahreskalender“ kontinuierlich weiterentwickelt. Neben dem analogen Kalendarium gibt es mittlerweile auch ein digitales.

1 *Ref. 5035J-001, 1996*
2 *Ref. 5035G-039, 2022*
3 *Ref. 5035P-001, 1997*
4 *Ref. 5450P, 2008*
5 *Ref. 5250G, 2005*
6 *Ref. 5350, 2006*

1

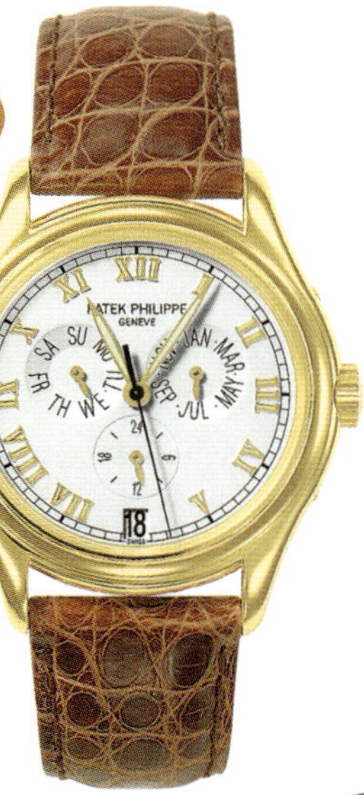

2

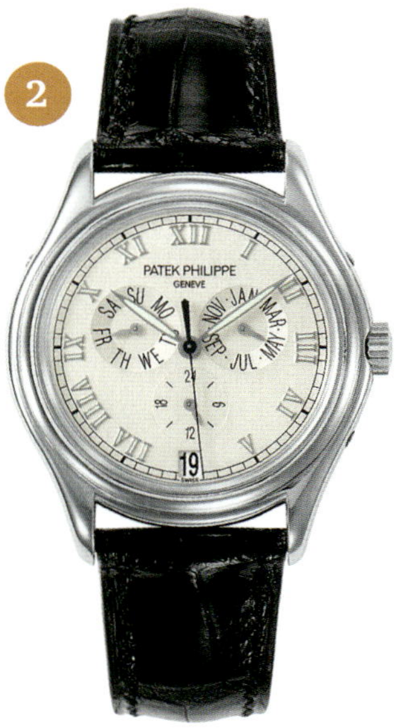

3

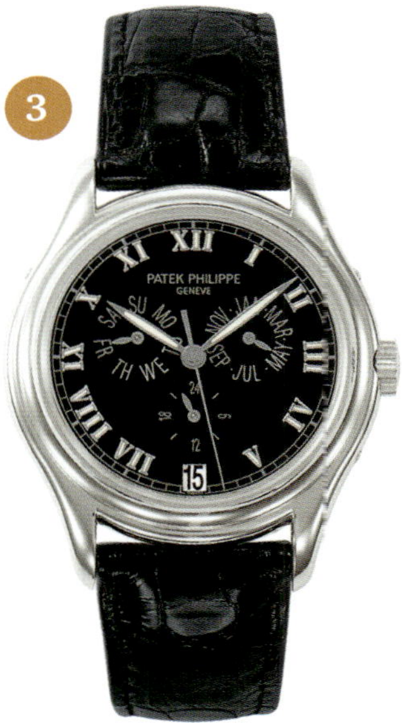

4

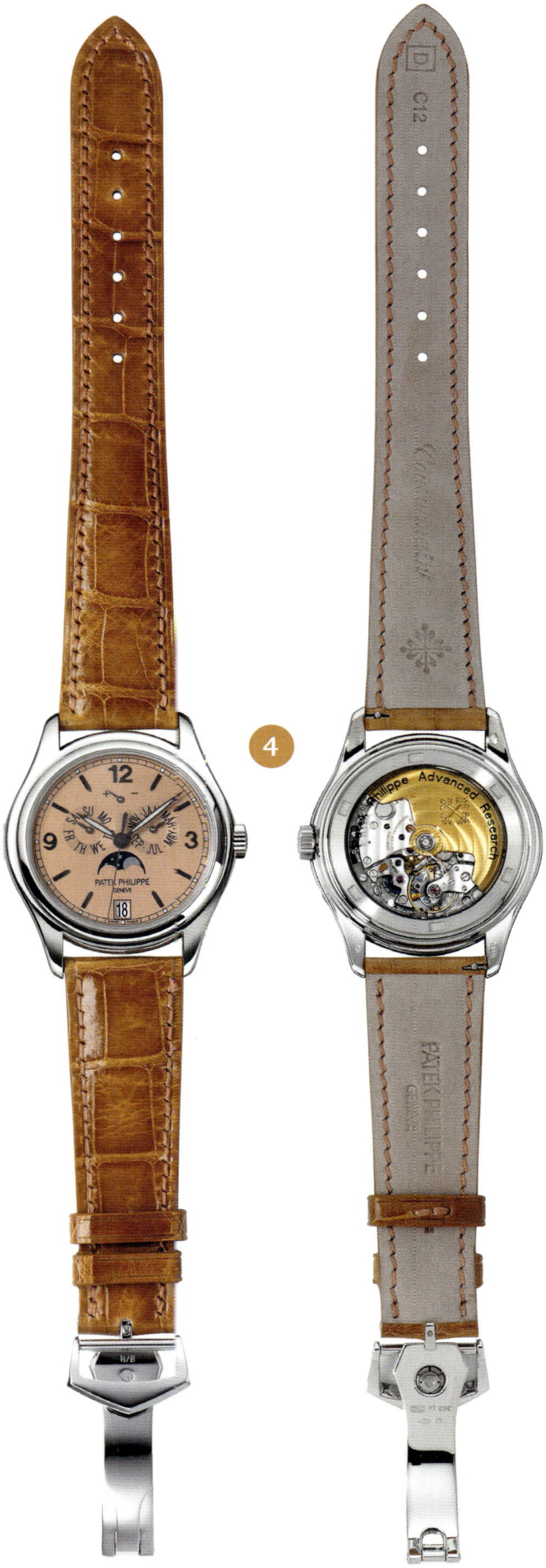

5

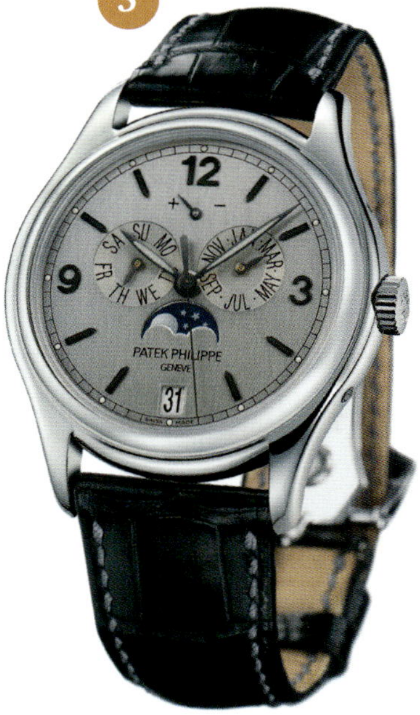

6

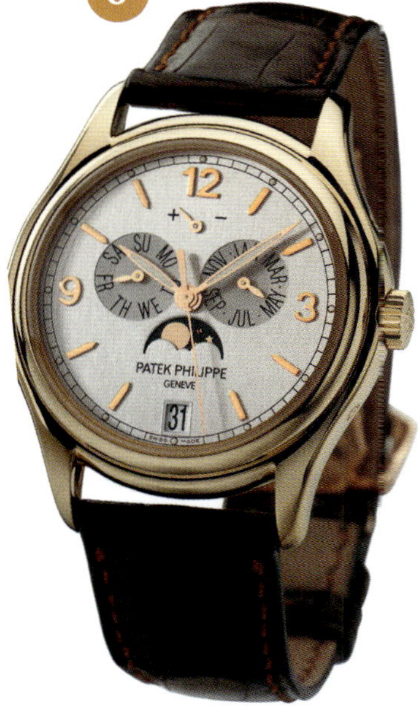

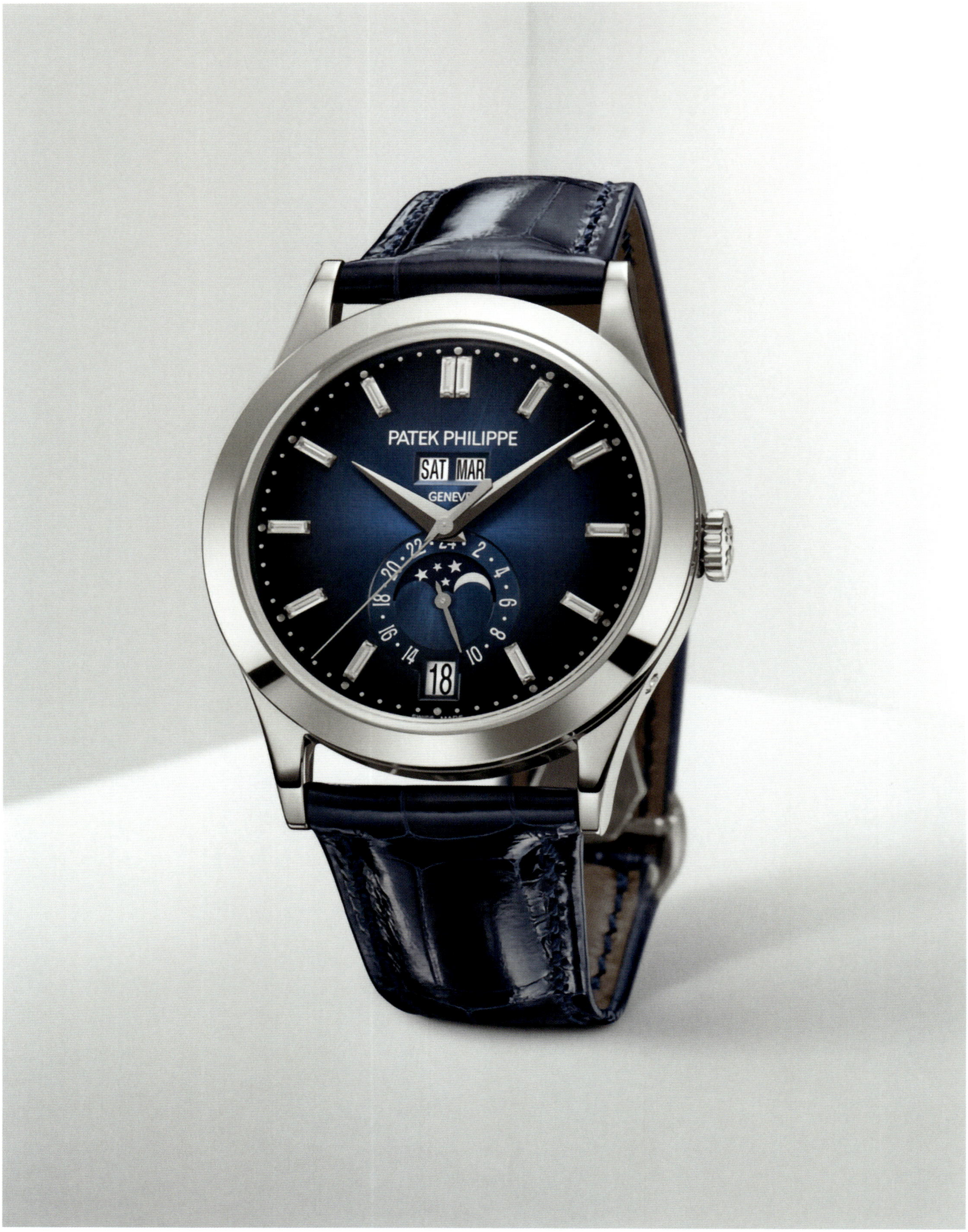

Ref. 5396G-017, 2024

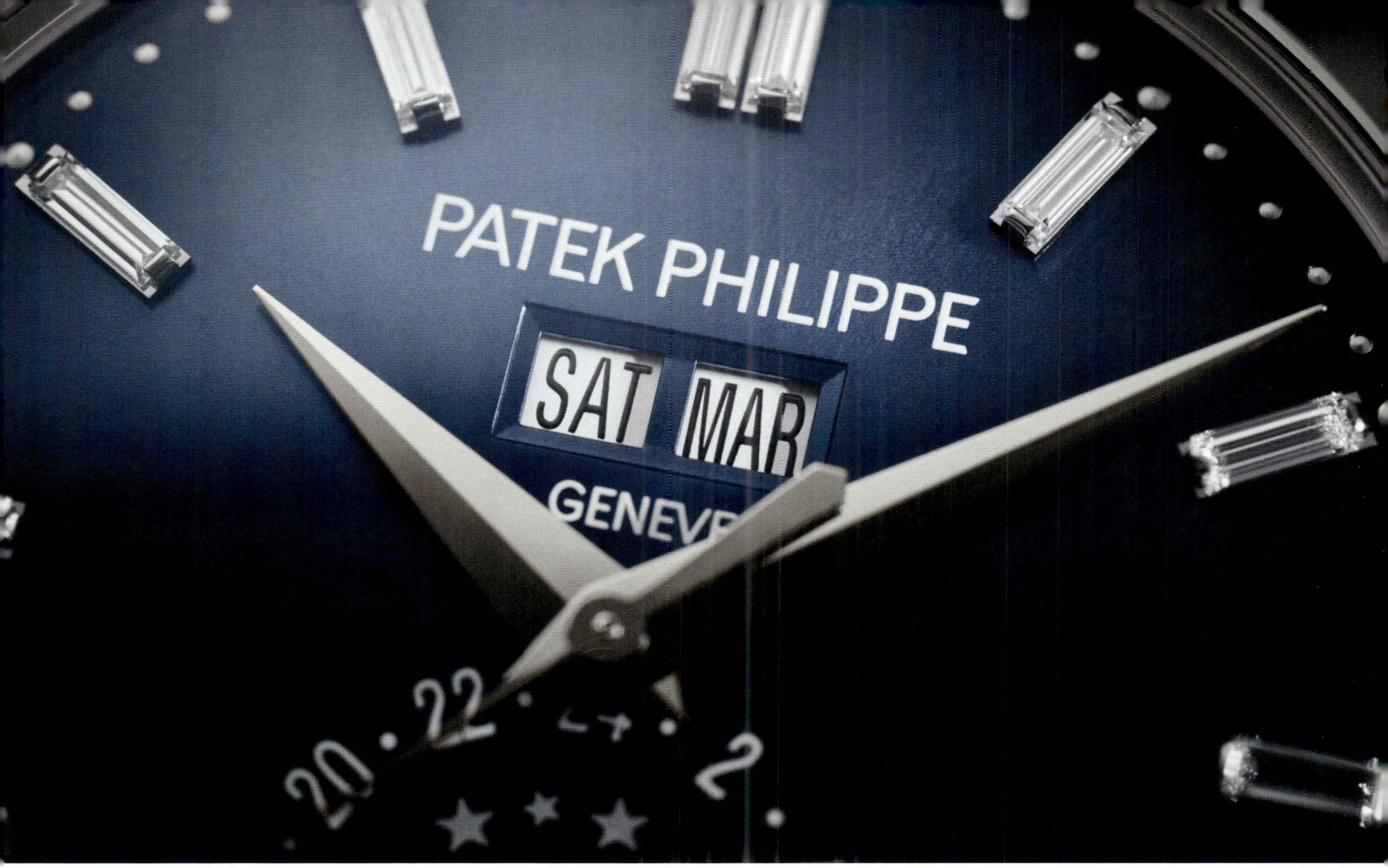
PATEK PHILIPPE
SAT
MAR

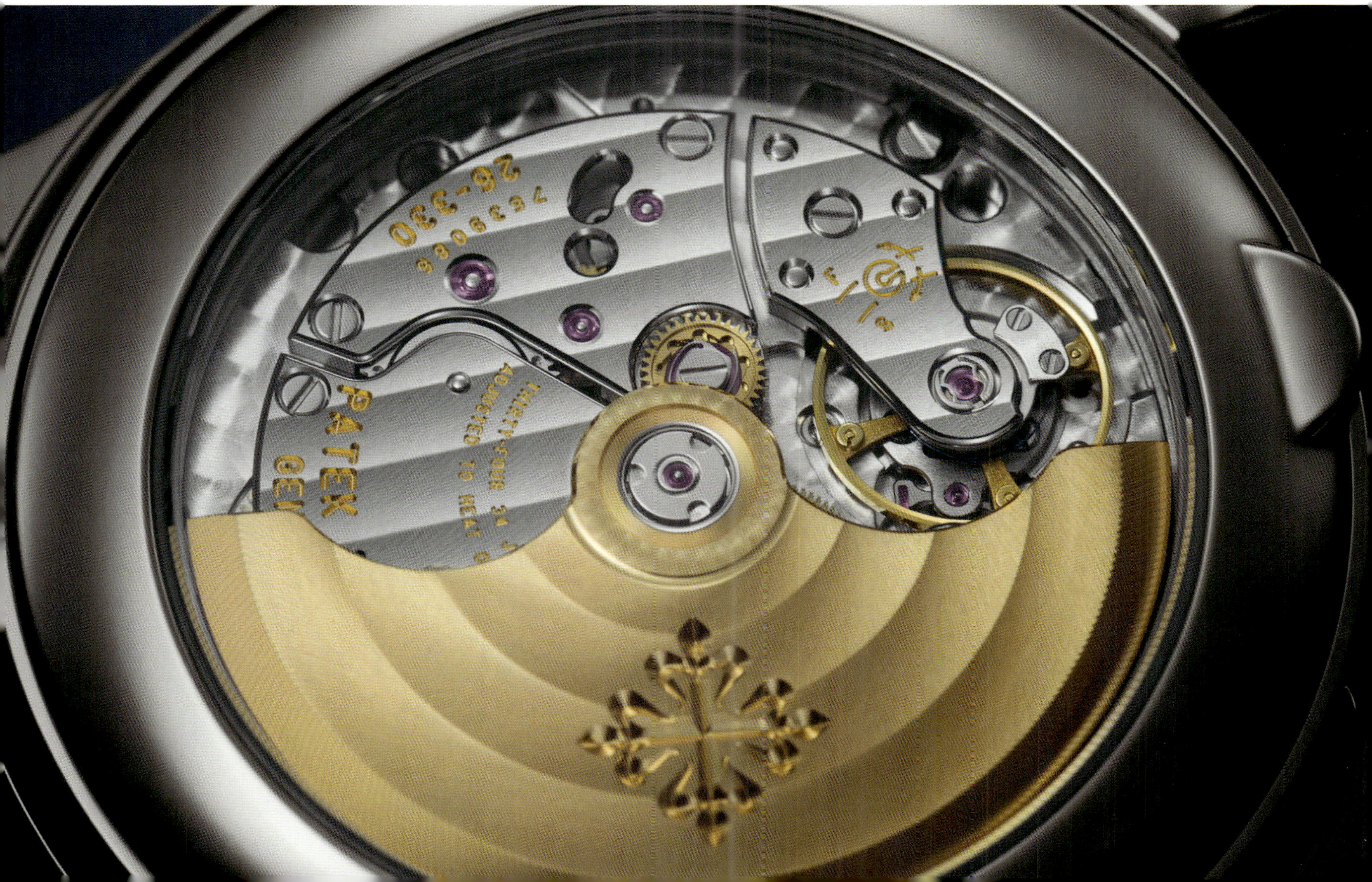

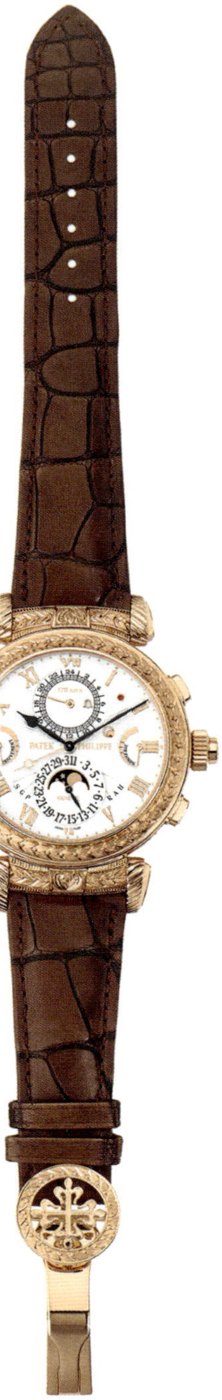

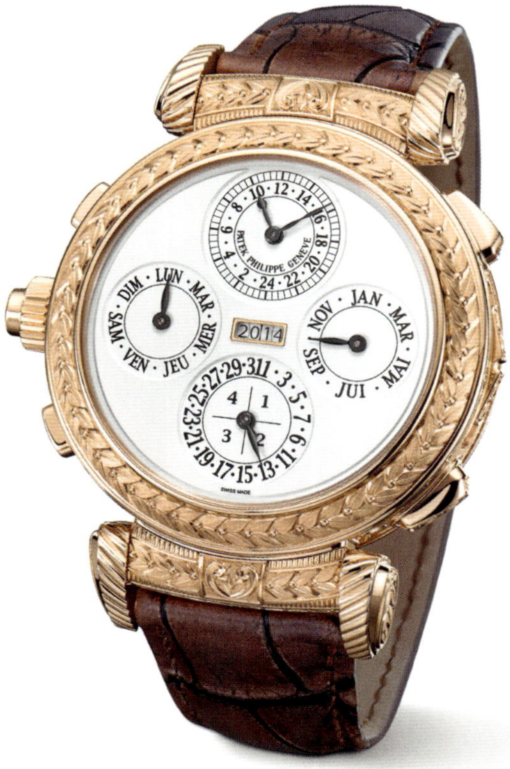

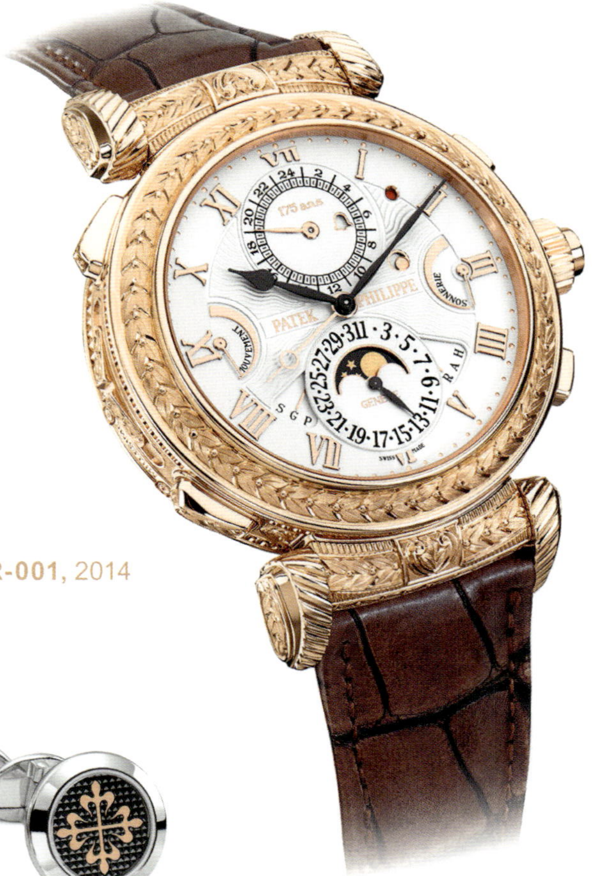

REF. 5175R-001, 2014

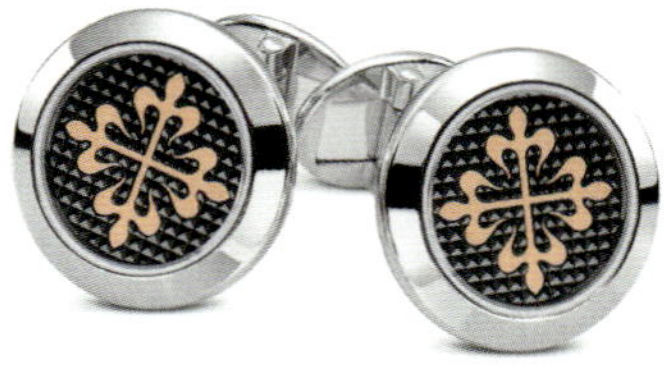

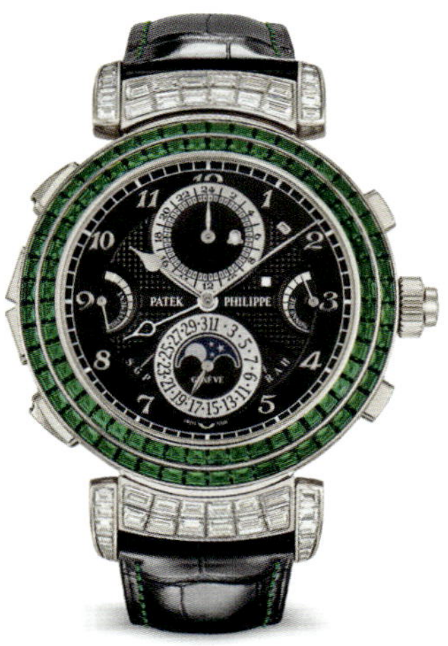

Ref. 6300/403G-001

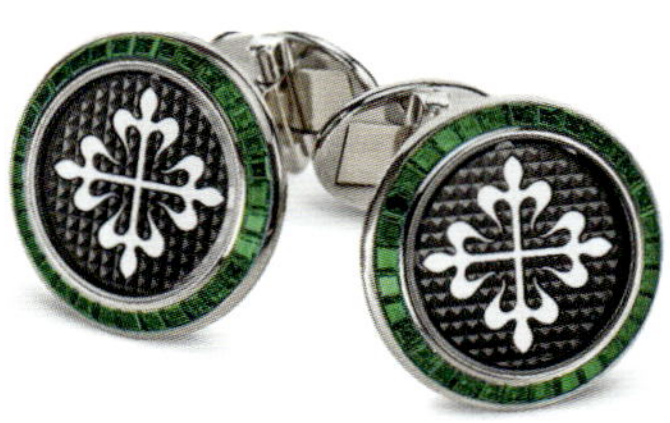

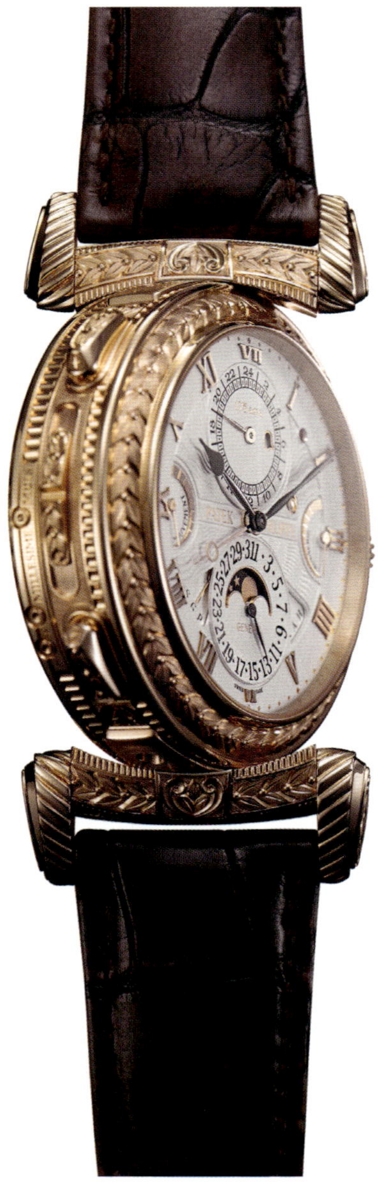

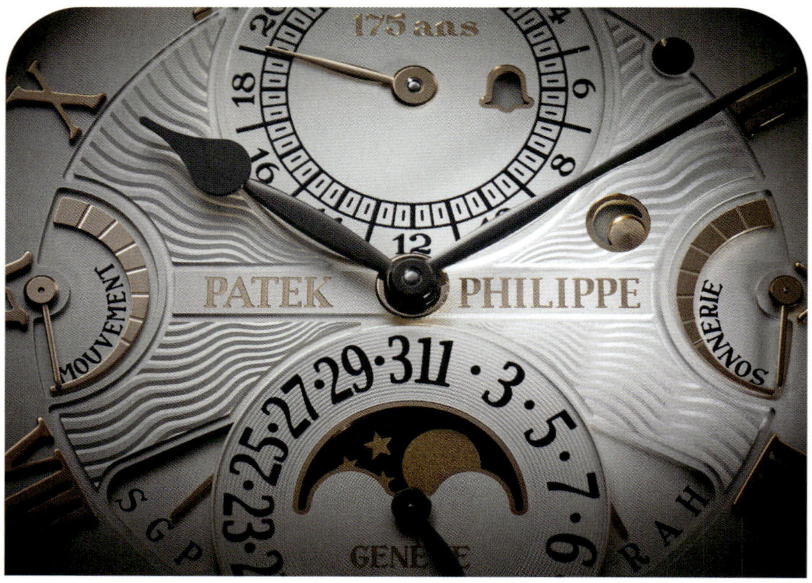

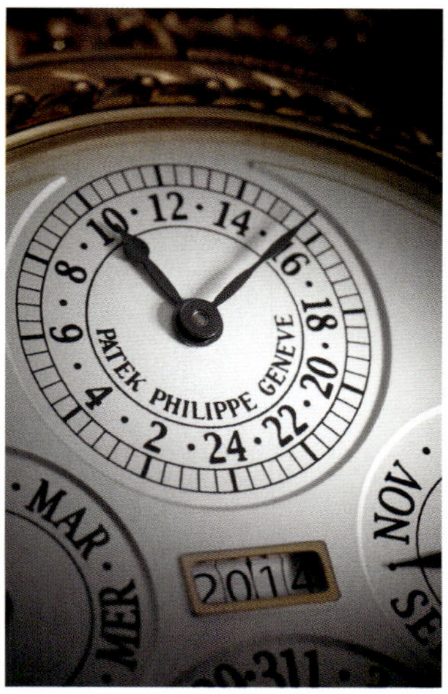

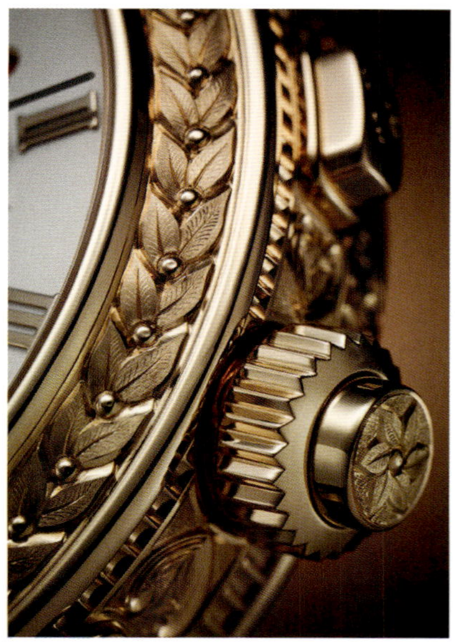

PATEK PHILIPPE
GRANDMASTER CHIME

ONE WATCH, TWO FACES

Patek Philippe's most complex watch is the Grandmaster Chime. It first stepped into the limelight as Reference 5175 in a limited edition of seven in 2014 to mark the manufactory's one-hundred-and-seventy-fifth anniversary. More than 100,000 working hours were devoted to the development, making and assembly of the double-faced timepiece with a reversible case. Of these, 60,000 hours were dedicated to the 1,366 components of the highly complex hand-wound Calibre GS AL 36-750 QIS FUS IRM. Including the 214 individual parts for the case, the watch consists of a total of 1,580 components.

The front with the crown pointing to the right was designated by the developers to indicate and chime out the time. On the reverse side, the equally exquisite dial sports the perpetual calendar with jumping indications, a window with a four-digit year display and the moon phase display.

The watch boasts not one but two patented world premières.

- The integrated alarm mechanism uses the minute repeater to strike the entire time sequence. This was the first time ever that such a function had featured on a mechanical wristwatch.
- The timepiece also possesses a date repeater. It sounds the date with a double high-low chime for ten-day intervals and a high chime for single days.

A further four patents highlight the watchmaker's innovative genius. Isolation prevents friction between the various mechanisms and thus costly damage. The case and pushers are engraved with inscriptions and symbols for ease of operation. The Grandmaster Chime is therefore more or less self-explanatory.

In addition to the above-mentioned complications, the strikework spectrum also includes a grande and petite sonnerie. A power reserve of 30 hours ensures that the suppressible self-strike does not flag as the day progresses. In the perpetual calendar, the day, date, month and leap year advance simultaneously at midnight.

Along with the local time, the main dial also indicates the hour of a second time zone, including day and night information. Also featured are power reserve indicators for the movement and strikework, the indication for the position of the crown (winding, alarm and hand-setting), and the indication for the choice of strikework mode: grande sonnerie (G), petite sonnerie (P) or silence (S). A bell-shaped window shows whether the alarm is on or off. A small round window is used for the strikework isolator on/off display.

In 2016, the Grandmaster Chime Reference 6300 joined the regular collection as an unlimited model. It sports a white gold case with hand-guilloché Clous de Paris décor. Since 2023, on request Patek Philippe will deliver matching cufflinks in white and rose gold with the Reference 6300GR-001.

EINE UHR, ZWEI GESICHTER

„Grandmaster Chime" heißt sie, die komplizierteste Armbanduhr von Patek Philippe. Ans Licht der Öffentlichkeit trat die auf nur sieben Exemplare limitierte Erstreferenz 5175 im Jahr 2014 zum 175. Manufaktur-Jubiläum. Mehr als 100.000 Arbeitsstunden wurden für Entwicklung, Fertigung und Montage des Zeitmessers mit zwei Zeitgesichtern und Wendegehäuse benötigt. Davon entfielen 60.000 Stunden auf die 1.366 Komponenten des hoch komplexen Handaufzugskalibers GS AL 36-750 QIS FUS IRM. Inklusive der 214 Einzelteile für das Gehäuse besteht die Uhr aus insgesamt 1.580 Bauteilen.

Das Zifferblatt bei rechts positionierter Krone haben die Entwickler der Uhrzeit und dem Schlagwerk zugedacht. Die gegenüberliegende, nicht minder kunstvoll gestaltete Scheibe widmet sich dem ewigen Kalender mit springenden Indikationen, voll ausgeschriebener Jahreszahl sowie der Mondphasenanzeige.

Zwei patentierte Weltpremieren muss man förmlich erlauschen:

- Das eingebaute Weckerwerk schlägt die eingestellte Alarmzeit in der vollständigen Tonfolge der Minutenrepetition. Eine derartige Funktion hat es in einer mechanischen Armbanduhr noch nie zuvor gegeben.
- Auf Wunsch repetiert dieser Zeitmesser auch das Datum. Nach der Auslösung erklingen erst die Zehntagesintervalle per Doppelschlag hoch-tief und dann die weiteren Tage mit hohem Schlag.

Innovatives Handeln unterstreichen vier weitere Patente. Isolatoren unterbinden den Kraftschluss zwischen verschiedenen Mechanismen und verhindern so kostspielige Beschädigungen. Benutzerfreundlich präsentieren sich gravierte Inschriften und Symbole am Gehäuse und auf den Funktionsdrückern. Damit ist die „Grandmaster Chime" quasi selbsterklärend.

Über das Genannte hinaus umfasst das gesamte Schlagwerkspektrum auch eine „Grande und Petite Sonnerie". 30 Stunden Gangautonomie sorgen dafür, dass der abstellbare Selbstschlag über den Tag hinweg keinesfalls schlappmacht.

Das ewige Kalendarium schaltet Datum, Wochentag, Monat und Schaltjahreszyklus mitternächtlich springend fort.

Außer der Lokalzeit bildet das Hauptzifferblatt auch noch die Stunde einer zweiten Zeitzone samt Tag- und Nachtinformation ab. Ferner finden sich dort die Gangreserveanzeigen für Geh- und Schlagwerk, die Indikation für die Position der Krone (Aufziehen, Alarm stellen, Zeiger stellen), jene für die Wahl des selbsttätigen Schlagwerks Grande Sonnerie (G), Petite Sonnerie (P) oder Silence (S). In einem glockenförmigen Fenster ist zu sehen, ob das Weckerwerk ein- oder ausgeschaltet ist. Ein kleines rundes Fenster gilt der Schlagwerkisolationsanzeige „Ein/Aus".

2016 folgte die „Grandmaster Chime Referenz 6300" als unlimitierter Bestandteil der regulären Kollektion. Sie verfügt über ein Weißgoldgehäuse mit handguillochiertem Clous de Paris-Dekor. Passende Manschettenknöpfe liefert Patek Philippe seit 2023 zur Referenz 6300GR-001 in Weiß- und Roségold auf Wunsch dazu.

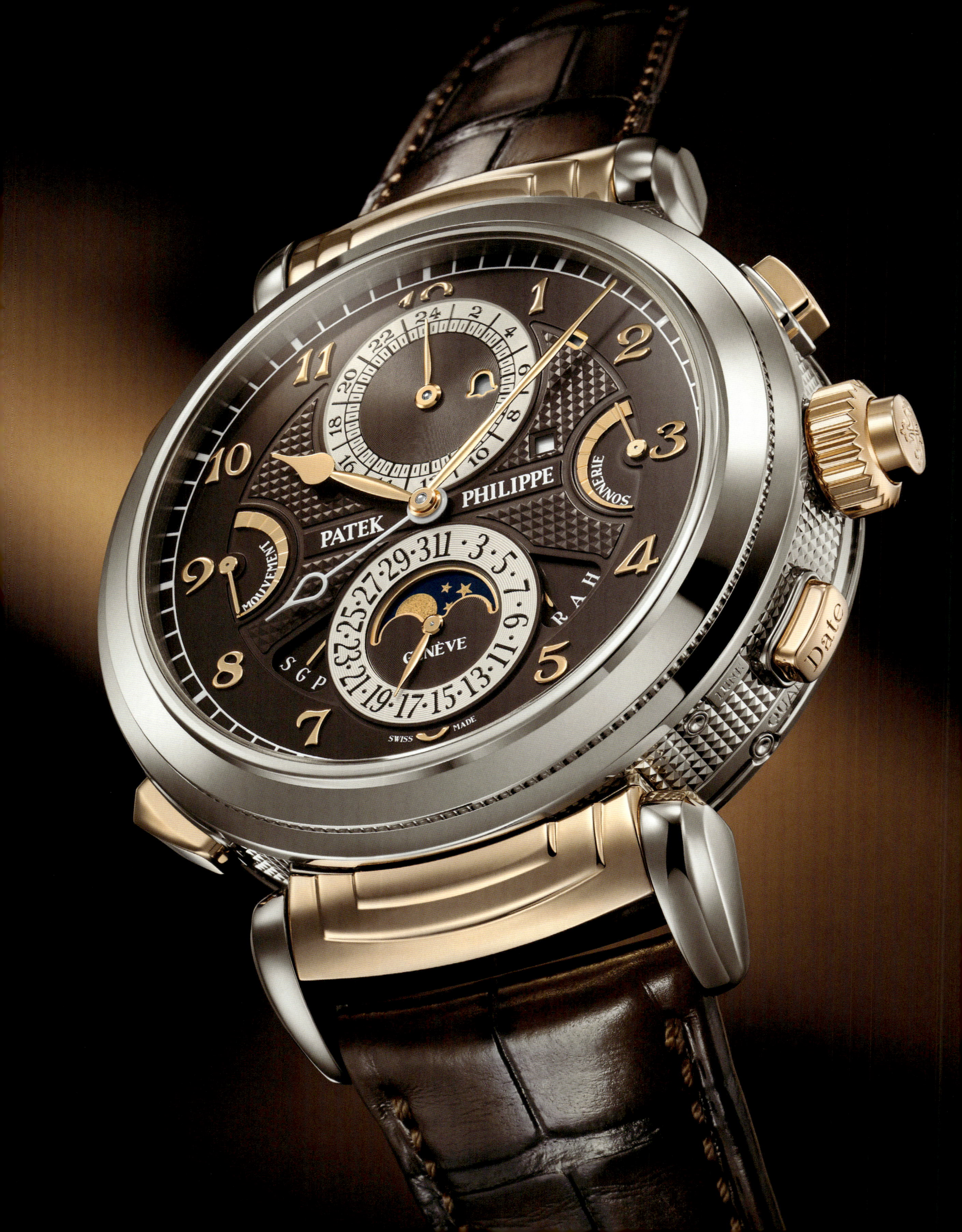
PATEK PHILIPPE
MOUVEMENT
GENÈVE
SWISS MADE
Date

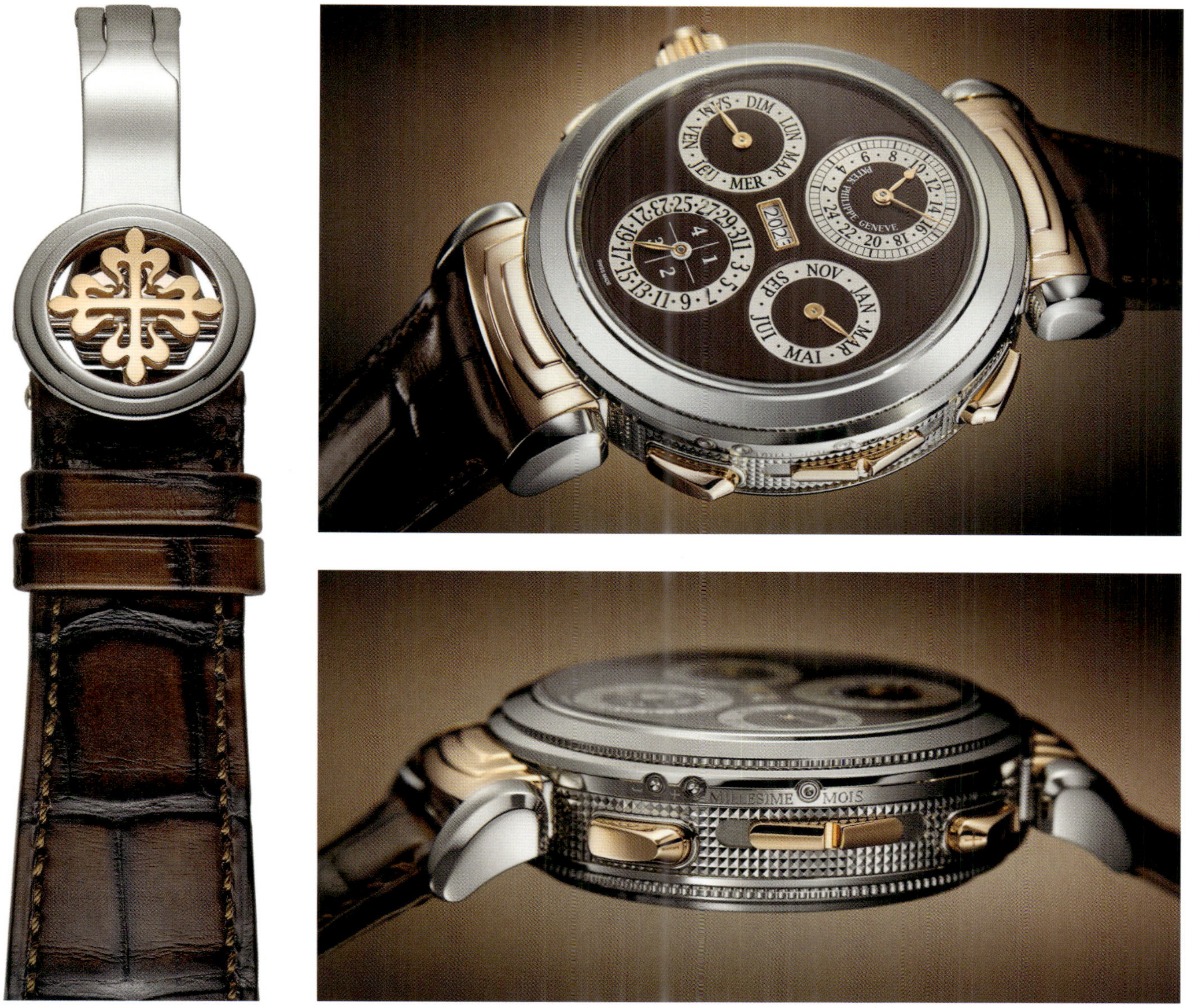

Ref. 6300GR-001, 2023

PATEK PHILIPPE NAUTILUS

1976

PATEK PHILIPPE AND THE NAUTILUS

Along with creative talent, Gérald Genta also possessed a great deal of business acumen. When he encountered key employees of Patek Philippe in the hotel restaurant during the 1974 Basel Watch Fair, he lost no time in expertly sketching his alternative to the much-discussed Royal Oak. Genta knew of Philippe Stern's passion for sailing. This is why Genta's sketch bore a resemblance to the porthole of a ship. Born in 1938, Philippe Stern had been working since 1966 for the company owned by his family. He remembered his first meeting with the maverick product designer on this subject. "Genta came to us with this idea. He believed very much in the idea and was sure that it would be a once-in-a-century success." His conviction was not far off the mark. The project however was time-consuming. It was two years before the sketches were to germinate and take on concrete shape. Certain similarities to the Royal Oak were of no consequence. "We were plagued by recurrent doubts. Sometimes we wondered whether we had strayed too far from the Patek Philippe line and if we should really make such a large sporty steel watch." Fortunately, the doubts were not acted on.

After several design modifications and prototypes, the wristwatch, dubbed the Nautilus, was showcased at the 1976 Basel Watch Fair. With no flat automatic movement of its own, Patek Philippe delivered the Reference 3700/1A with the purchased Calibre 28-255 C. This was a modified automatic movement from LeCoultre with the Gyromax balance wheel. At Audemars Piguet it was called Calibre 2121. In the beginning, Patek Philippe's authorised retailers evinced little enthusiasm. Noses were wrinkled at the nasty steel, excessive size and extortionate price. After all, at the time, for the equivalent of around €2,250 you could buy a small car. Despite scrupulous development, Patek Philippe had to correct the first series of patented cases. Due to oxidation, it was difficult to loosen the stainless-steel screws for service purposes. Solid gold screws were the remedy in this case. Customer service replaced the screws. The modification went into serial production from case number 536.400 and in 1980, the women's Nautilus in gold, steel or steel/gold reaped greater success. It was powered by the proprietary quartz Calibre E 19. In 1988, men finally got the message as well. >

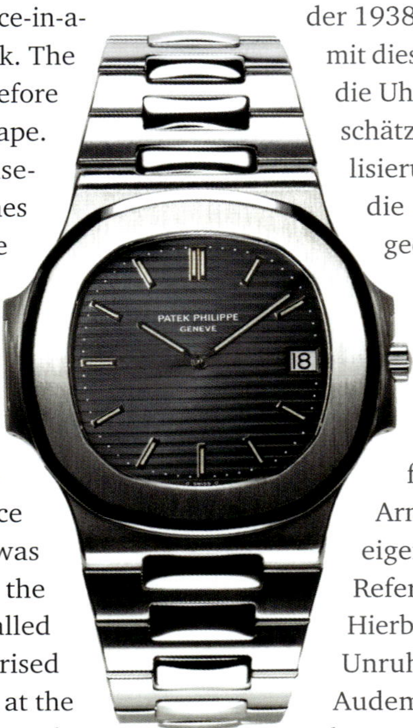

PATEK PHILIPPE UND DIE NAUTILUS

Neben seinen gestalterischen Fähigkeiten besaß Gérald Genta auch jede Menge Geschäftssinn. Als er 1974 während der Basler Uhrenmesse im Hotelrestaurant maßgebliche Angestellte von Patek Philippe entdeckte, skizzierte er mit flotter Hand eine Alternative zur viel diskutierten „Royal Oak". Genta wusste um die Segelsport-Passion von Philippe Stern, der schon seit 1966 in den Diensten des seiner Familie gehörenden Unternehmens stand. Deshalb verfügte Gentas Entwurf über gewisse Ähnlichkeiten mit der Pfortluke eines Schiffs. Seine erste diesbezügliche Begegnung mit dem durchaus eigenwilligen Produktgestalter erinnerte der 1938 geborene Philippe Stern wie folgt: „Genta kam zu uns mit dieser Idee. Er war sehr überzeugt von der Idee und glaubte, die Uhr würde ein Jahrhunderterfolg werden." Mit seiner Einschätzung lag der Designer keineswegs falsch. Aber die Realisierung des Projekts verlangte nach Zeit. Zunächst mussten die Zeichnungen im Laufe zweier Jahre zur fertigen Uhr gedeihen. Gewisse Parallelen zur „Royal Oak" störten keineswegs. „Zwischendurch zögerten wir immer wieder. Und wir fragten uns manchmal, ob das noch eine Patek Philippe sei, ob wir wirklich eine so große sportliche Stahluhr machen sollten." Zum Glück blieb es bei gedanklichen Zweifeln. Nach mehreren Designmodifikationen und Prototypen lag die „Nautilus" getaufte Armbanduhr 1976 in den Basler Messevitrinen. Mangels eigenem flachen Automatikwerk lieferte Patek Philippe die Referenz 3700/1A mit dem zugekauften Kaliber 28-255 C. Hierbei handelte es sich um ein modifiziertes, mit „Gyromax"-Unruh ausgestattetes Automatikwerk von LeCoultre, das Audemars Piguet unter der Kaliberbezeichnung 2121 verbaute. Anfänglich ließen die Patek-Philippe-Konzessionäre wenig Euphorie erkennen. Schnöder Stahl, übermäßig groß und vor allem zu teuer lauteten naserümpfende Kommentare. Immerhin gab es seinerzeit für umgerechnet rund 2.250 Euro schon einen Kleinwagen. Trotz größter Sorgfalt bei der Entwicklung musste Patek Philippe die erste Serie der patentierten Gehäuse nachbessern. Bedingt durch Oxidation ließen sich die Edelstahl-Verschlussschrauben beim Service nur mühevoll lösen. Abhilfe schafften massivgoldene Schrauben. Der Austausch erfolgte im Zuge des Kundendienstes. In Serie ging die Modifikation ab Gehäusenummer 536.400. Deutlich besser reüssierte die 1980 vorgestellte Damen-„Nautilus" in Gold, Stahl, oder Stahl/Gold. Bestückt mit dem hauseigenen Quarzkaliber E 19. 1988 platzte der Knoten endlich auch bei den Männern. >

Top / Oben: *First Nautilus Ref. 3700, 1976*

1
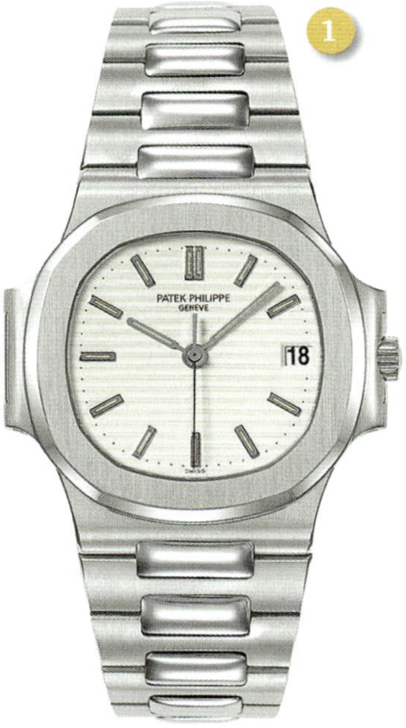

2
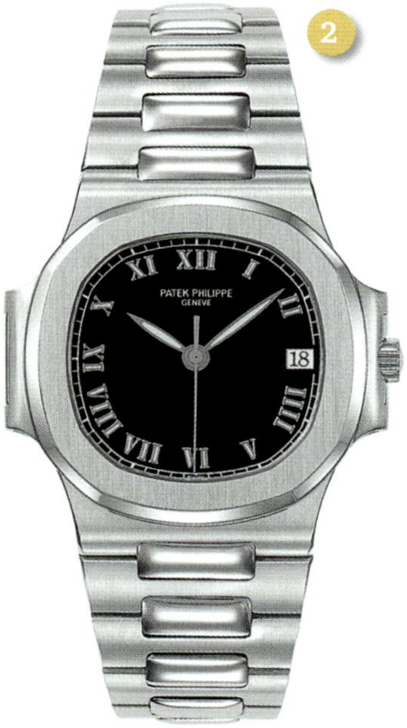

3
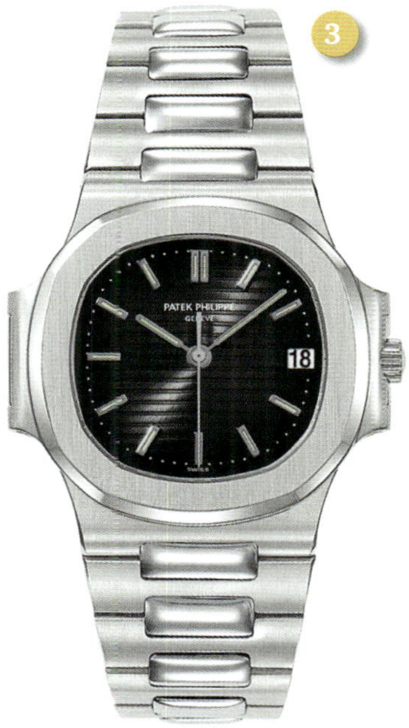

4
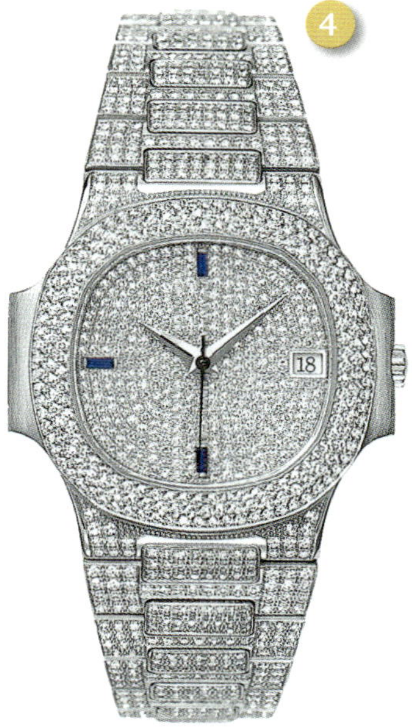

5
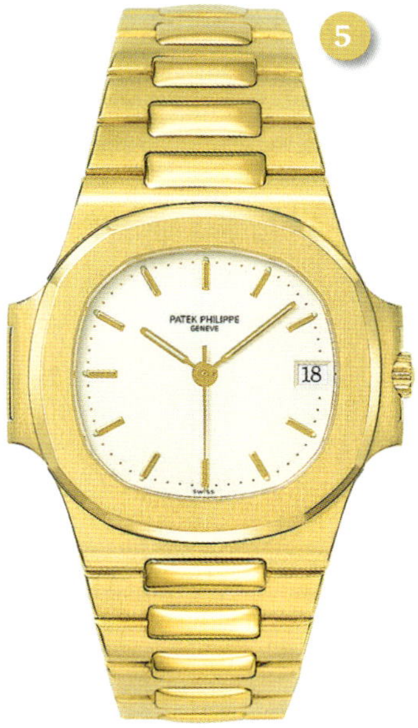

6
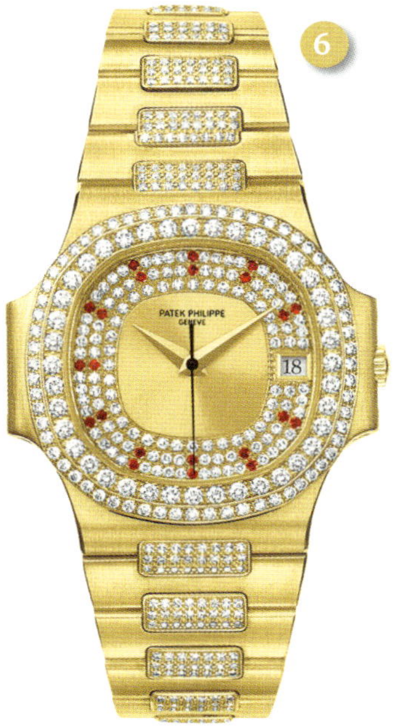

1 *Ref. 3800/1A*

2 *Ref. 3800/1A*

3 *Ref. 3800/1A*

4 *Ref. 3800/108*

5 *Ref. 3800/1J*

6 *Ref. 3800/105J*

7 *Ref. 4700/158G-001*

8 *Ref. 4700/54P-011*

9 *Ref. 4700/530J-001*

10 *Ref. 4700/172J-001*

7
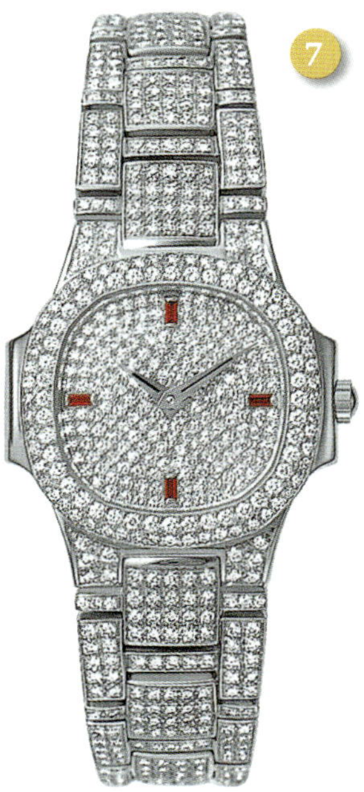

8
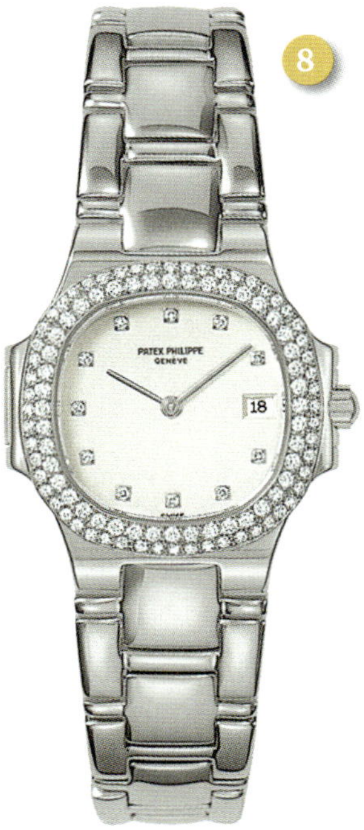

9
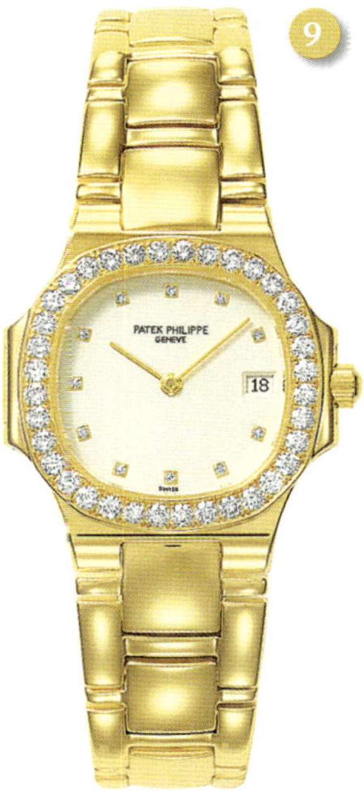

10
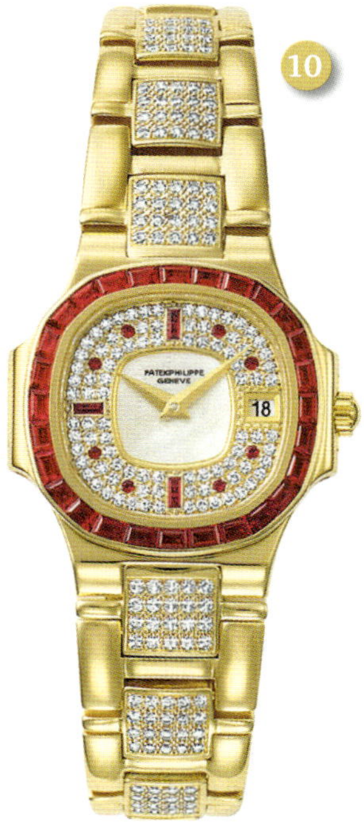

Almost overnight, Patek Philippe could barely keep up with demand for the medium-sized steel Reference 3800/1. The thirtieth Nautilus anniversary in 2006 was marked by the legendary Reference 5711/1A with a three-part, 40-millimetre transparent caseback. From 2008, this steel reference, since discontinued, with the automatic Calibre 324 SC, was so popular that impatient prospective buyers were happy to pay more on the grey market than the official recommended selling price.

In 2021, President Thierry Stern stopped production of the unlimited steel Nautilus 5711/1A. The same year, two limited series were launched. One with a green dial and a second in turquoise for the one-hundred-and-seventieth anniversary of Patek Philippe's long-standing customer Tiffany & Co.

Since 2022, the classic Nautilus without additional function is only available in white gold as Reference 5811/1G. Waiting lists are once again the order of the day.

Quasi über Nacht konnte Patek Philippe nicht genug von der mittelgroßen Stahlreferenz 3800/1 liefern. Der 30. „Nautilus"-Geburtstag im Jahr 2006 brachte die legendäre Referenz 5711/1A mit dreiteiligem, 40 Millimeter messendem Sichtbodengehäuse. Ab 2008 erfreute sich diese mittlerweile eingestellte Stahlreferenz mit dem Automatikkaliber 324 SC derart hoher Käufergunst, dass Ungeduldige im Graumarkt gerne mehr zahl(t)en als den offiziell genannten Publikumspreis.

2021 stoppte Präsident Thierry Stern die Produktion der unlimitierten stählernen „Nautilus" 5711/1A. Im gleichen Jahr folgten noch zwei limitierte Serien. Eine mit grünem Zifferblatt und eine zweite in Türkis zum 170. Geburtstag des langjährigen Patek-Philippe-Kunden Tiffany & Co.

Seit 2022 kann man die klassische „Nautilus" ohne Zusatzfunktion nur noch in Weißgold als Referenz 5811/1G erwerben. Wartelisten sind auch hier wieder an der Tagesordnung.

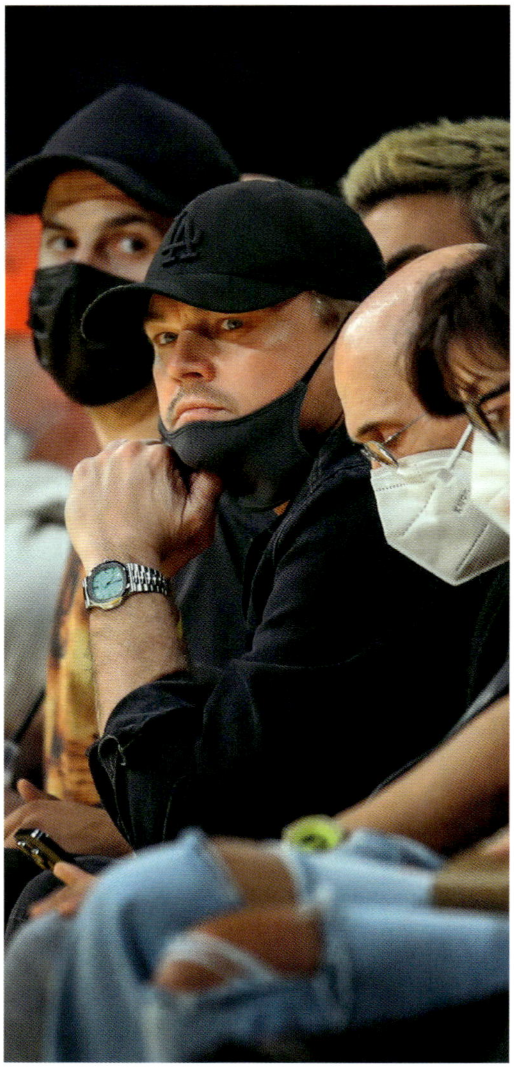

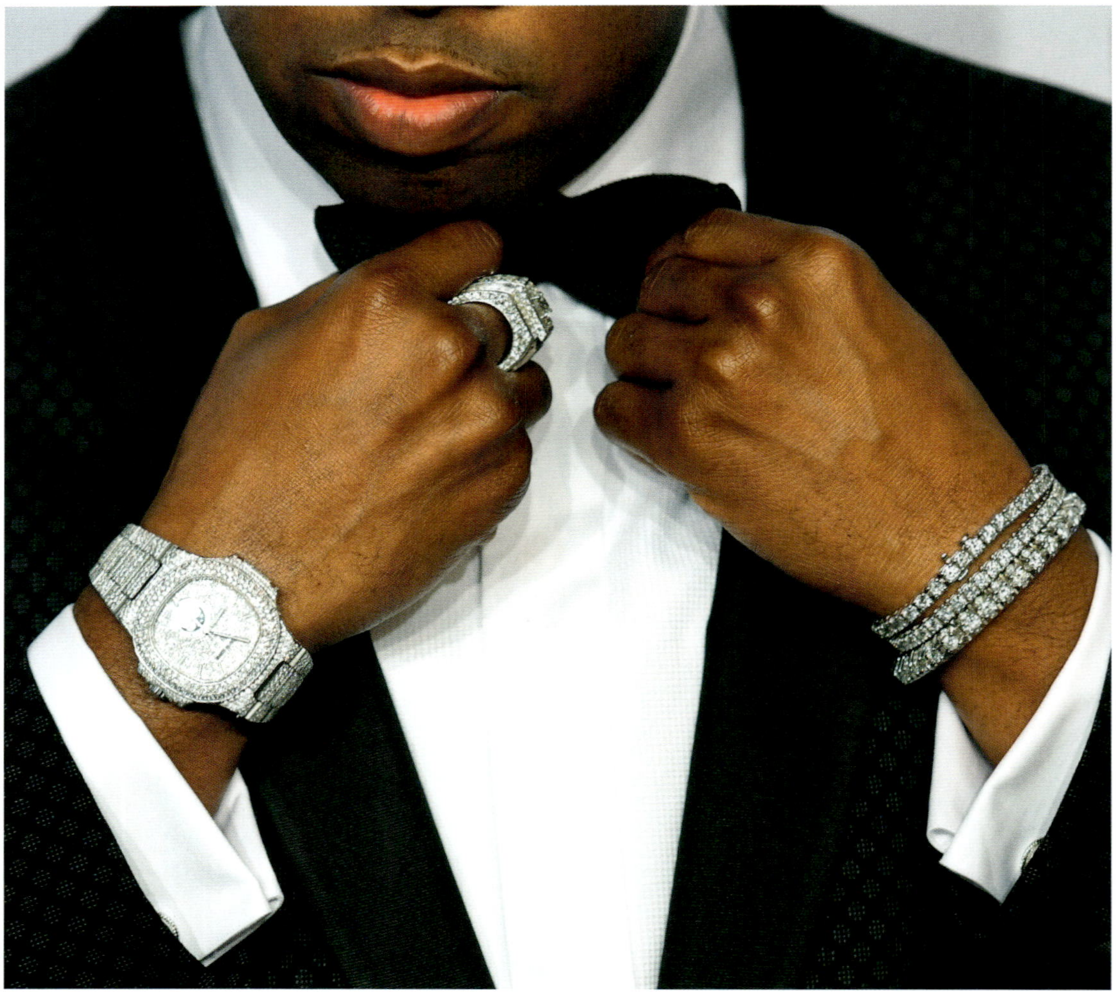

From left: *Leonardo DiCaprio with a Nautilus Tiffany & Co.* | *US rapper Yo Gotti, wearing a Nautilus at an event in New York City, 2017* | **Von links:** *Leonardo DiCaprio mit einer Nautilus Tiffany & Co., 2022* | *US-Rapper Yo Gotti mit einer Nautilus bei einem Event in New York, 2017*

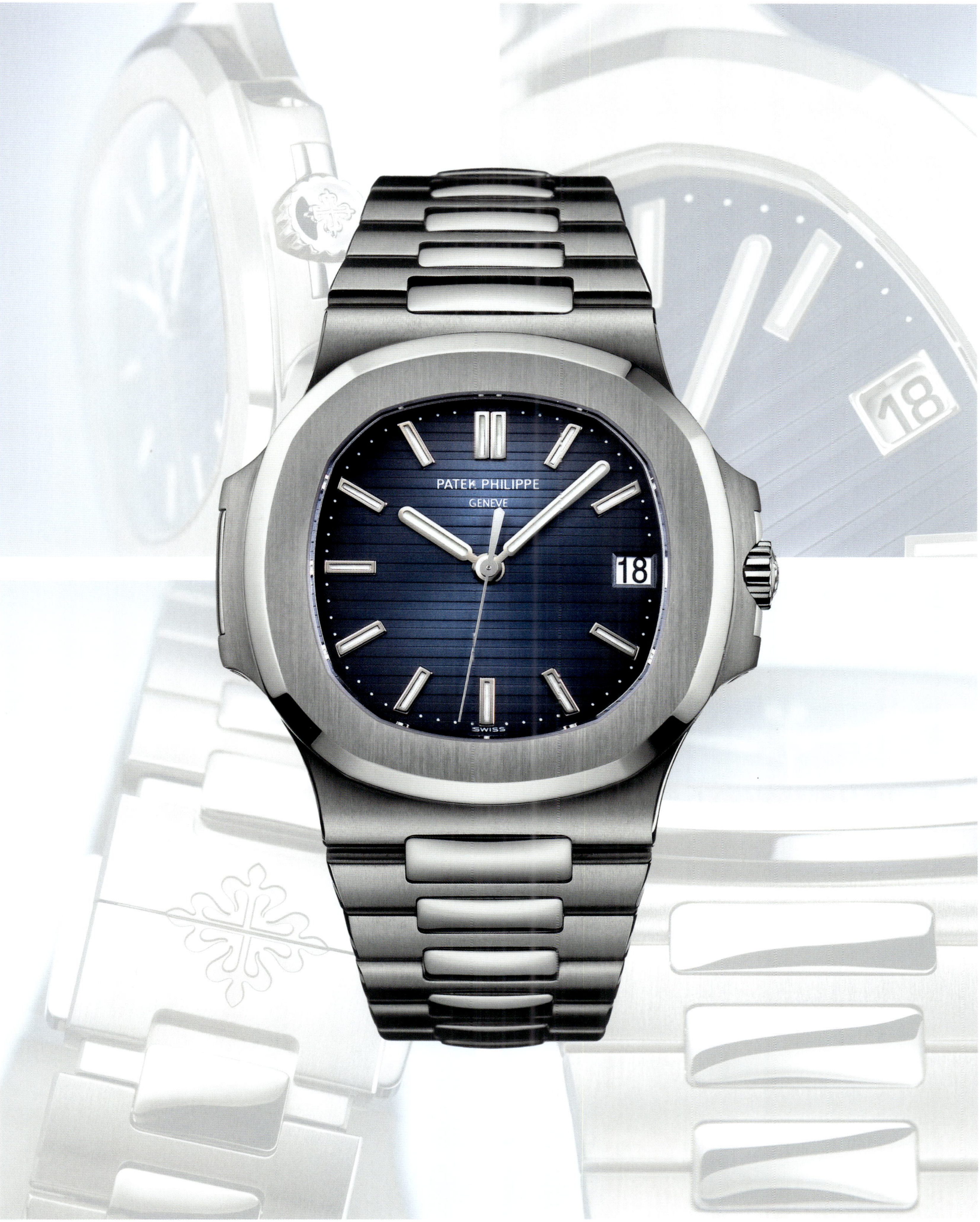

Ref. 5711/1A

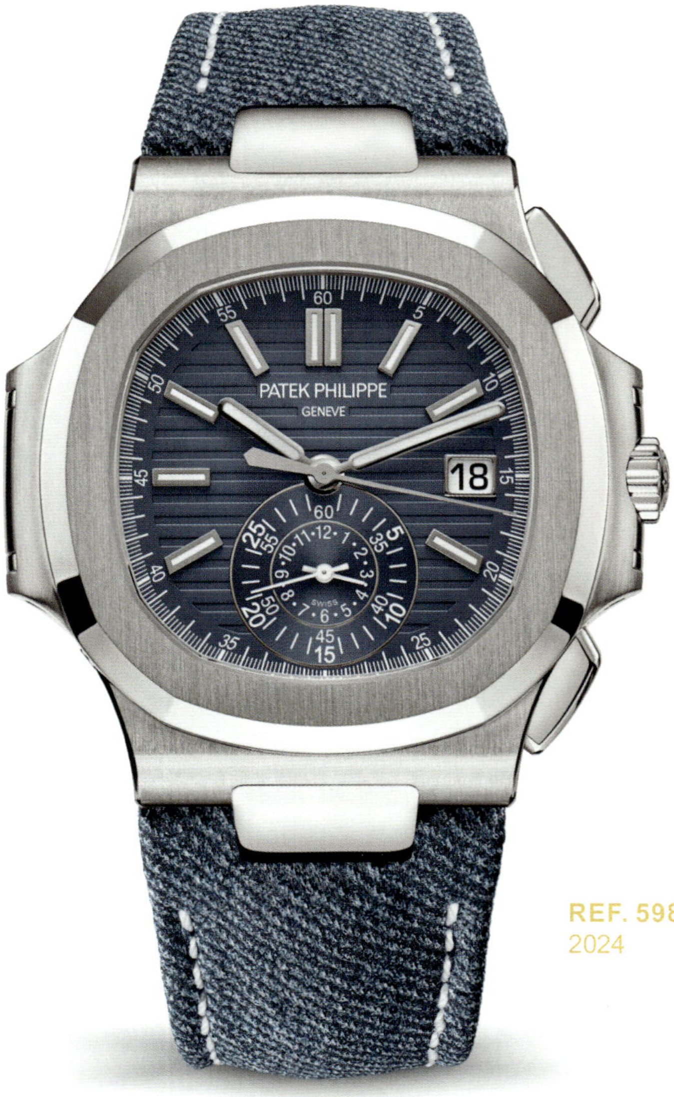

REF. 5980/60G-001,
2024

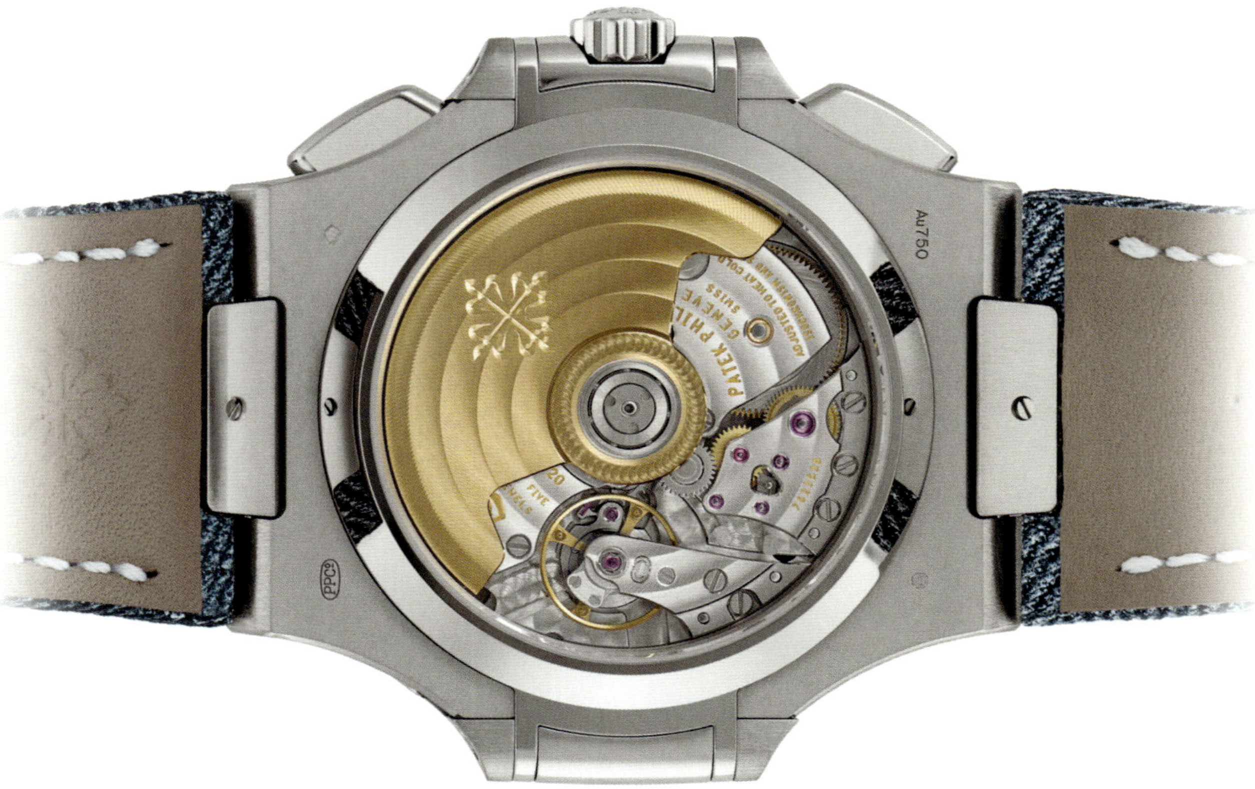

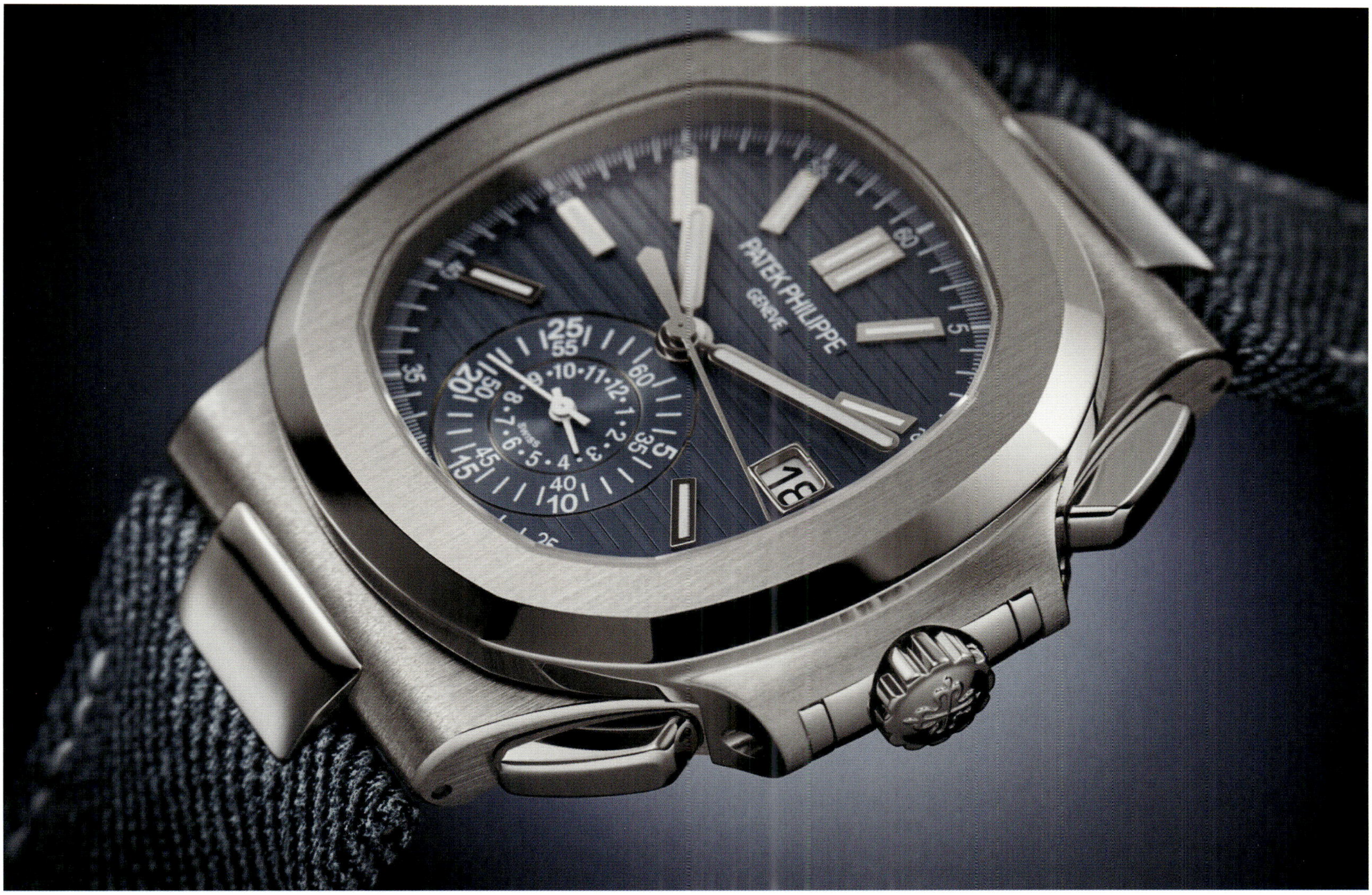

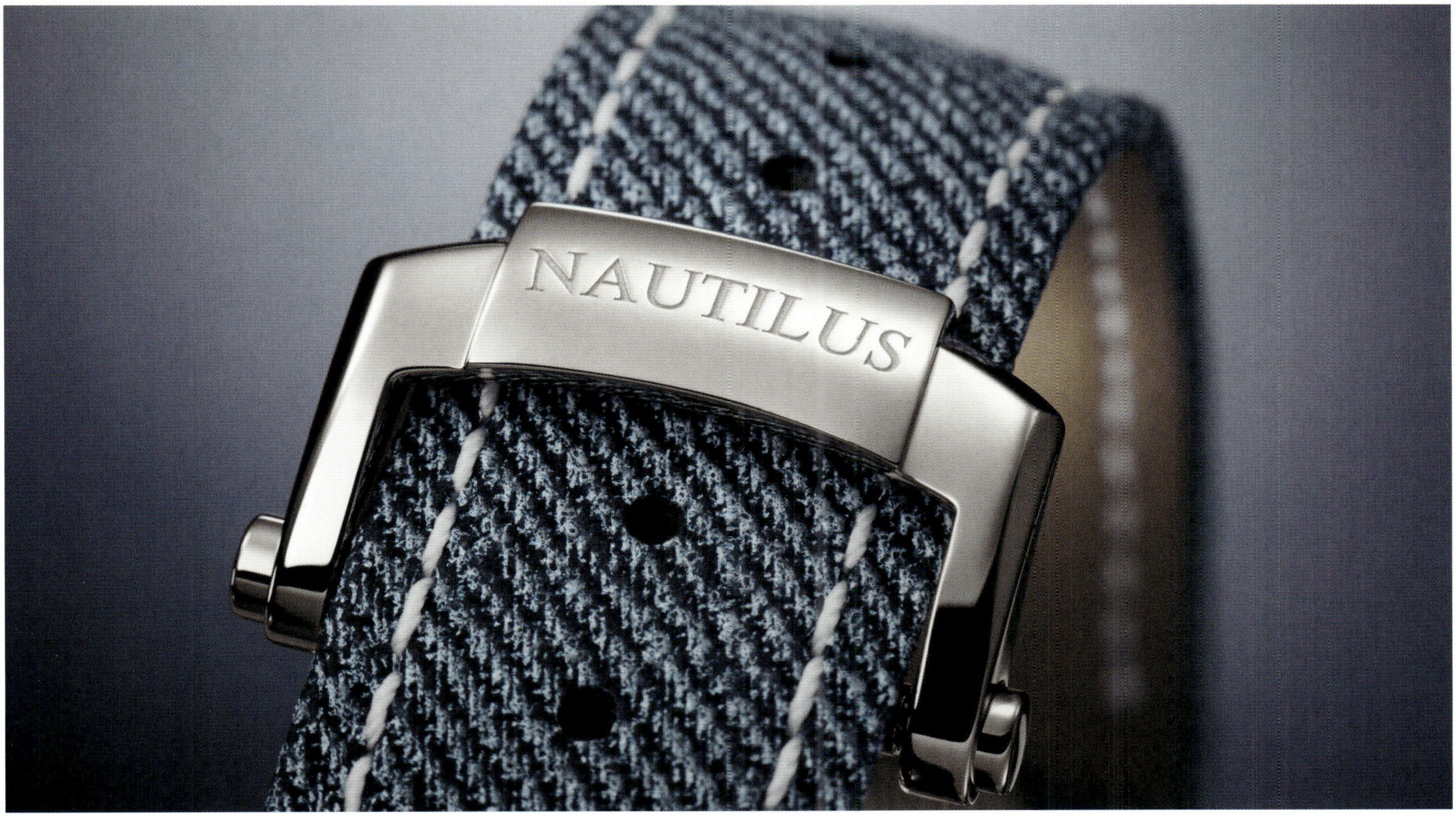

Left page | Linke Seite: *Cal. CH 28-520 C* **| This page, bottom:** *Calfskin strap with a blue-gray denim-pattern and a white gold Nautilus fold-over clasp |*
Diese Seite unten: *Kalbslederarmband mit „Jeans"-Optik und Nautilus-Faltschließe in Weißgold*

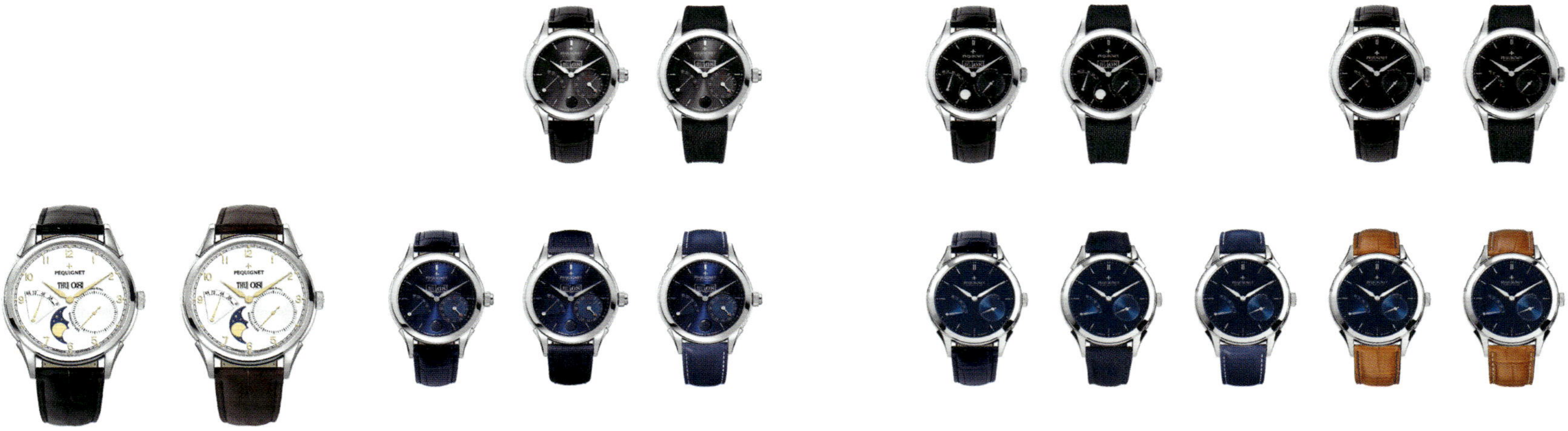

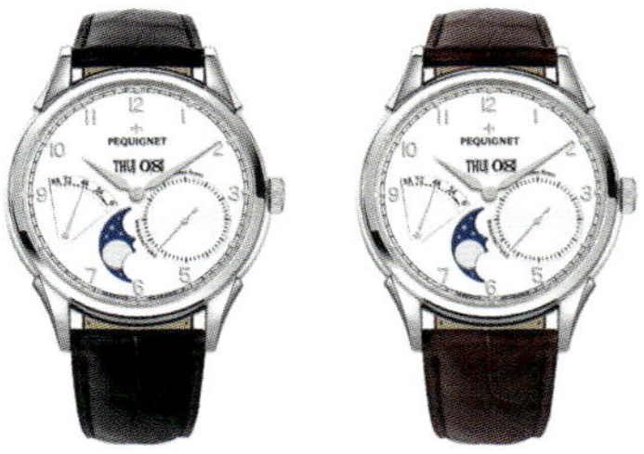

The Royale collection includes the models:

GMT, Grand Sport, Manuelle, Origine, Paris, Saphir and Titane | Die Royale-Kollektion umfasst die Modelle: GMT, Grand Sport, Manuelle, Origine, Paris, Saphir und Titane

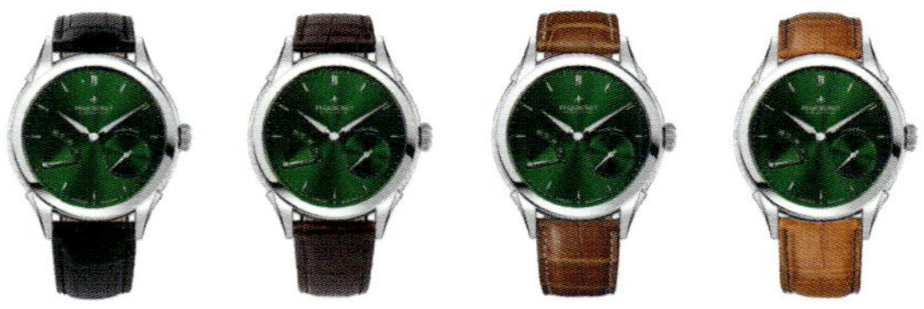

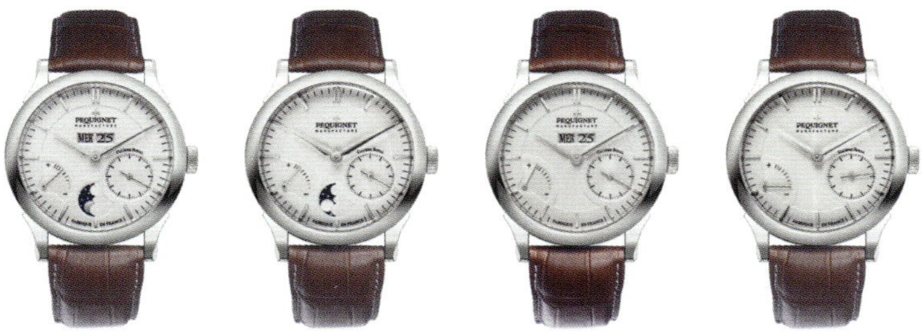

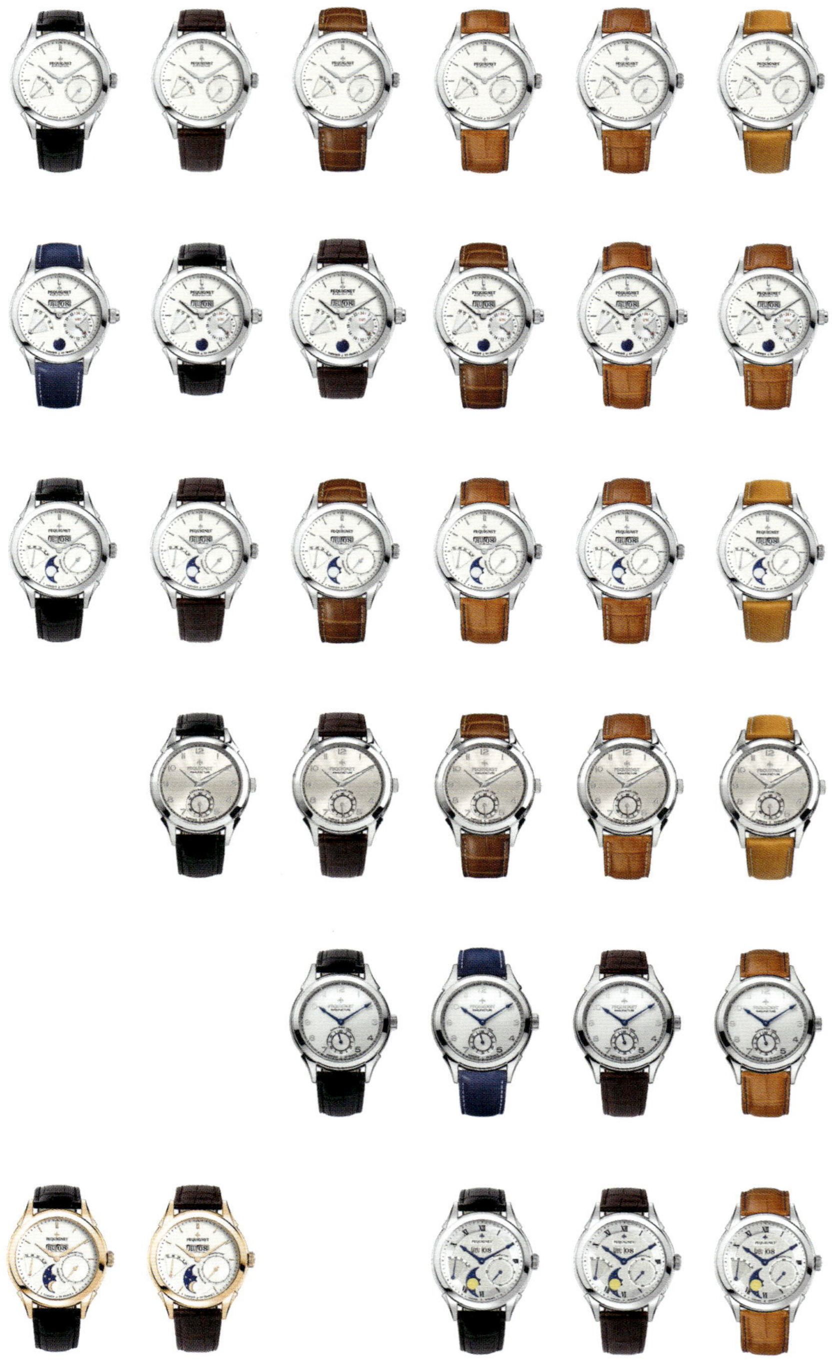

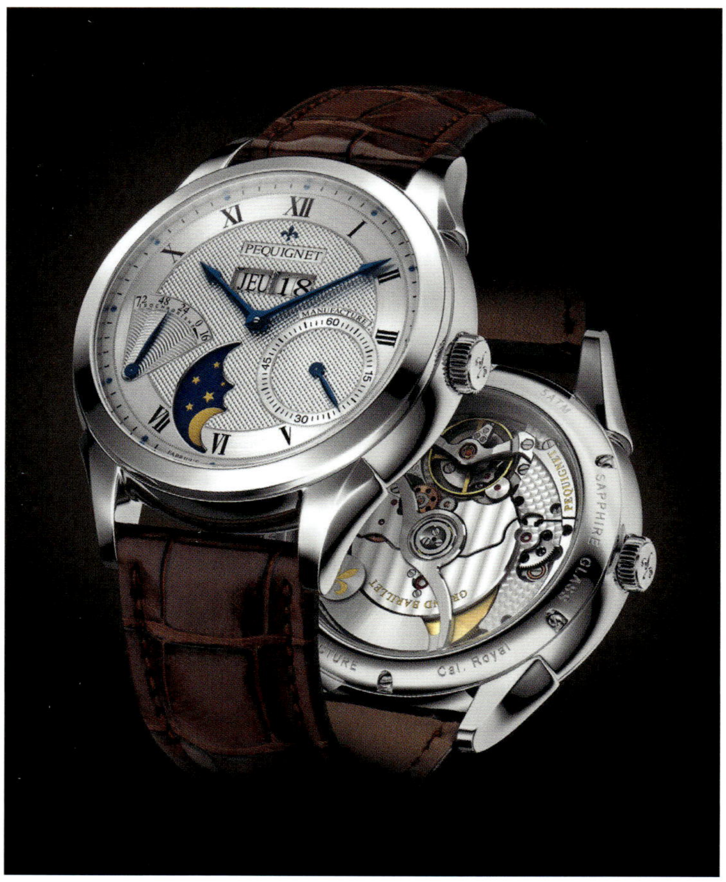

Royale Origine

PEQUIGNET CALIBRE ROYAL®

ROYAL CALIBRE

The biography of this French watch brand begins in 1973 in Morteau with a certain Monsieur Émile Péquignet. A self-taught designer, he developed fashionable collections of midrange timepieces. On 15 January 2004, a former Zenith manager took over the company. In 2011, it became a true manufactory with the glorious Calibre Royal®. However, this audacious step marked the beginning of a few difficult years. In 2021, French Entrepreneur Hugues Souparis became majority shareholder and president of the company. An in-house Pequignet creation protected by eight international patents, the Calibre Royal® EPM01 was unaffected by these business aspects. The brand had engaged two experienced engineers for this ambitious project in 2011. They fulfilled the brief entrusted to them by creating a platform for horological innovations. Even the basic model with small seconds is formidable. The integrated module requires no fewer than 318 components to display the large date, day of the week and remaining power reserve. The technical documentation consists of 279 drawings, all produced in-house. A large barrel delivers a power reserve calculated at 100 hours, of which 88 hours are guaranteed. After 72 hours, in a vertical position, the balance wheel amplitude still measures 220 degrees. Winding is ensured by a bidirectional rotor eccentrically mounted on ball bearings made of tungsten carbide. An innovative transmission uncouples manual winding with Breguet toothing when automatic winding is active. A frequency of only three hertz permits the use of a relatively large Glucydur balance wheel. A glance through the watchmaker's magnifying glass reveals small screws to change the mass inertia. There is thus no need for a regulator mechanism. Mechanical stability is ensured by a curved balance bridge. On the top Royale Origine model, the Pequignet icon with the Calibre Royal® offers an unusual vertical display of the various moon phases, indicated with high precision.

KÖNIGLICHES KALIBER

Die Biographie dieser französischen Uhrenmarke begann 1973 in Morteau mit Monsieur Émile Péquignet. Der Design-Autodidakt entwickelte modisch geprägte Kollektionen mittelpreisiger Zeitmesser fürs Handgelenk. Am 15. Januar 2004 übernahm ein ehemaliger Zenith-Manager die Firma. Mit dem königlichen „Calibre Royal®" mauserte sie sich 2011 zur waschechten Manufaktur. Aber mit diesem mutigen Schritt begannen Existenzprobleme und turbulente Jahre. 2021 wurde der französische Unternehmer Hugues Souparis Mehrheitseigner und Präsident der Firma. Unberührt von diesen wirtschaftlichen Aspekten blieb das von Pequignet selbst entwickelte und durch acht internationale Patente geschützte „Calibre Royal®" EPM01. Für dieses ehrgeizige Projekt hatte sich die Marke 2011 zwei erfahrene Ingenieure an Bord geholt. Auftragsgemäß kreierte das Duo eine Plattform mit Potenzial für künftige Aufbauten und Erweiterungen.

Schon die Basis mit kleiner Sekunde hat es in sich. Nicht weniger als 318 Komponenten braucht es für die integrierte Baugruppe zur Anzeige von Großdatum, Wochentag und verbleibender Gangreserve. Die technische Dokumentation umfasst 279 Zeichnungen, allesamt gefertigt im eigenen Büro. Ein großes Federhaus bietet 100 Stunden kalkulierte und 88 Stunden garantierte Gangautonomie. Nach 72 Stunden zeigen die Messgeräte in vertikaler Position noch 220 Bogengrade Unruh-Amplitude. Den Aufzug besorgt ein exzentrisch angeordneter Kugellagerrotor aus Wolframkarbid-Vollhartmetall in beiden Drehrichtungen. Ein neuartiges Getriebe schaltet den Handaufzug mit Breguet-Verzahnung bei aktiver Automatik ab. Nur drei Hertz Unruhfrequenz gestatten die Verwendung einer relativ großen Glucydur-Unruh. Der Blick durch die Uhrmacherlupe zeigt kleine Schrauben zur Veränderung der Masseträgheit. Dadurch ist ein Rückermechanismus obsolet. Mechanische Stabilität gewährleistet eine geschwungene Unruhbrücke. Im Topmodell „Royale Origine", der Pequignet-Ikone mit besagtem „Calibre Royal®", bildet eine ungewöhnliche, weil senkrecht stehende Indikation die verschiedenen Lichtphasen des Mondes mit hoher Präzision ab.

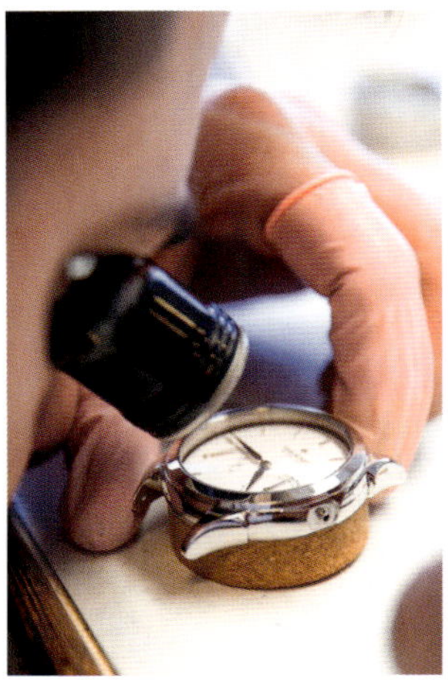

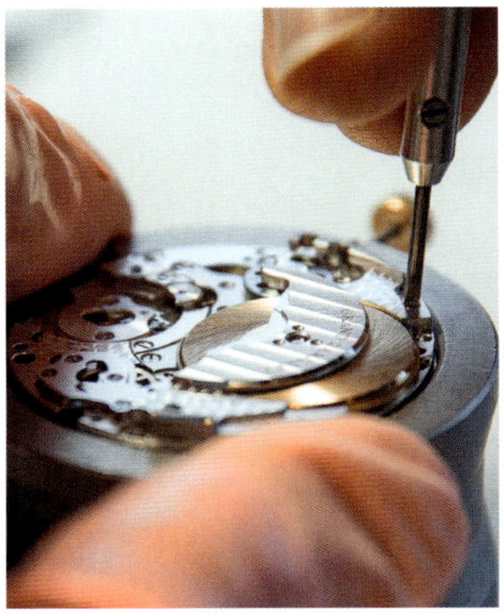

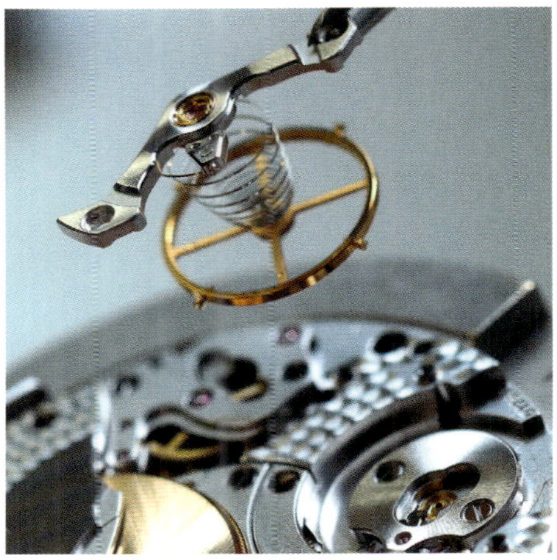

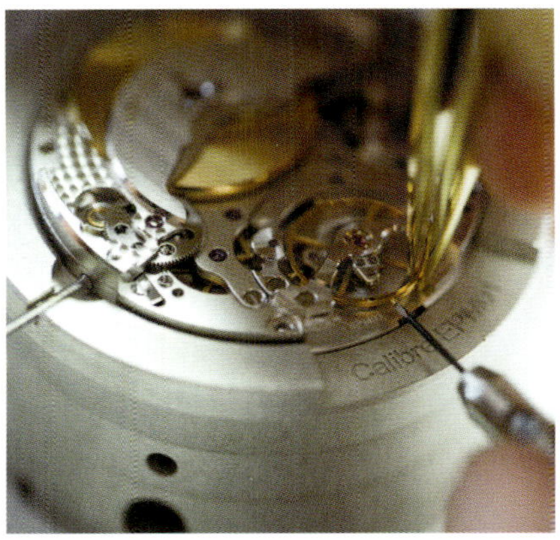

At the heart of the Calibre Royal® / Im Herzen des Calibre Royal®

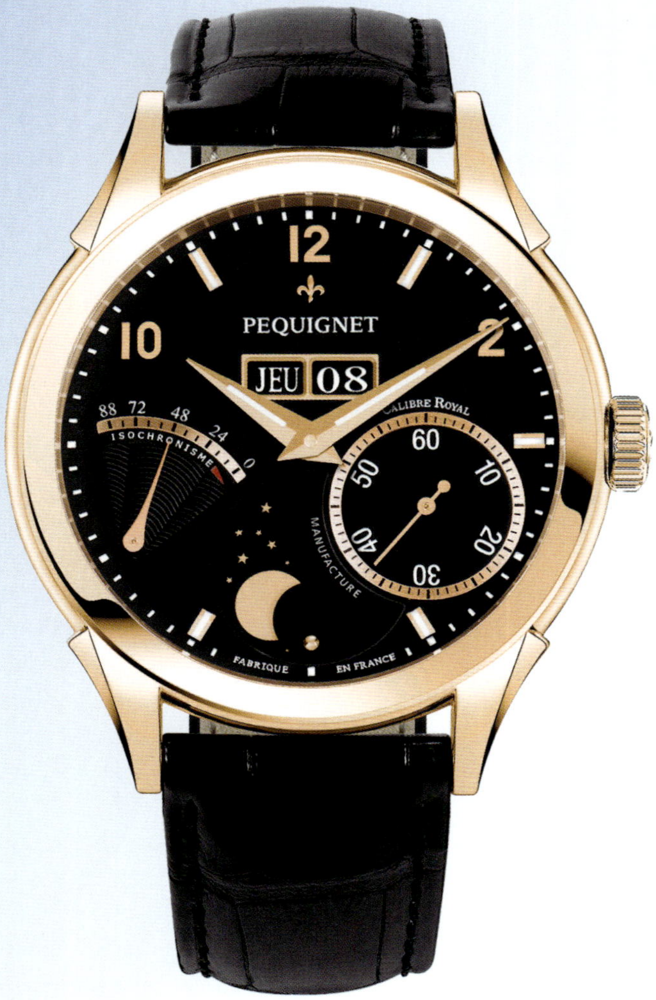

ROYALE ORIGINE
REF. 9011548CN

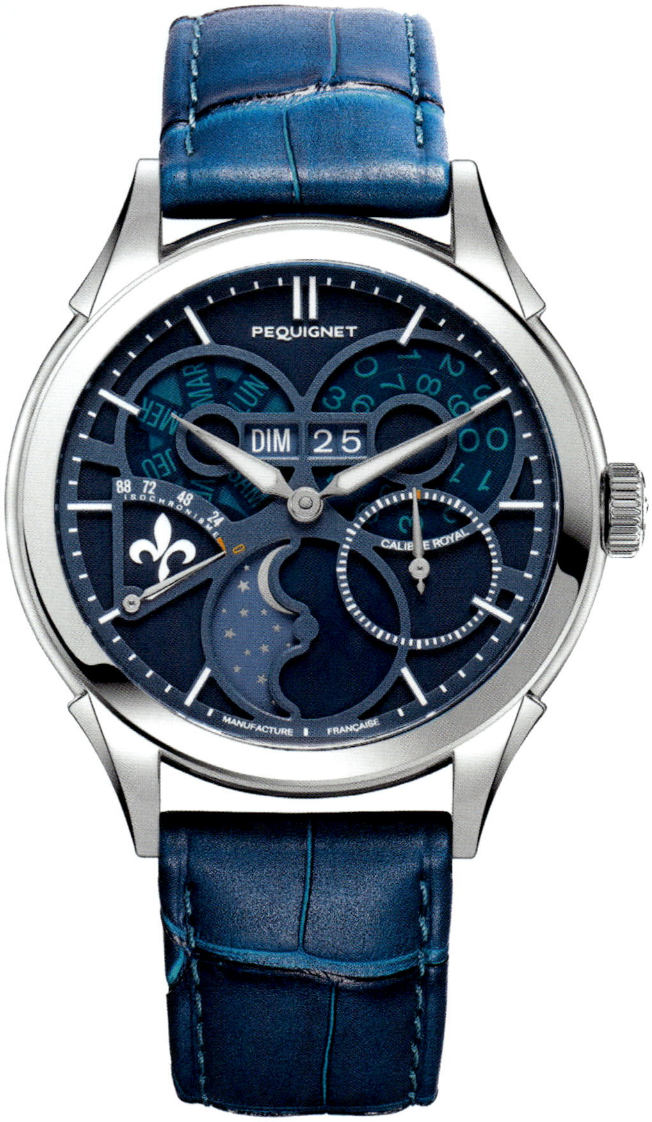

ROYALE SAPHIR
REF. 9010883CB

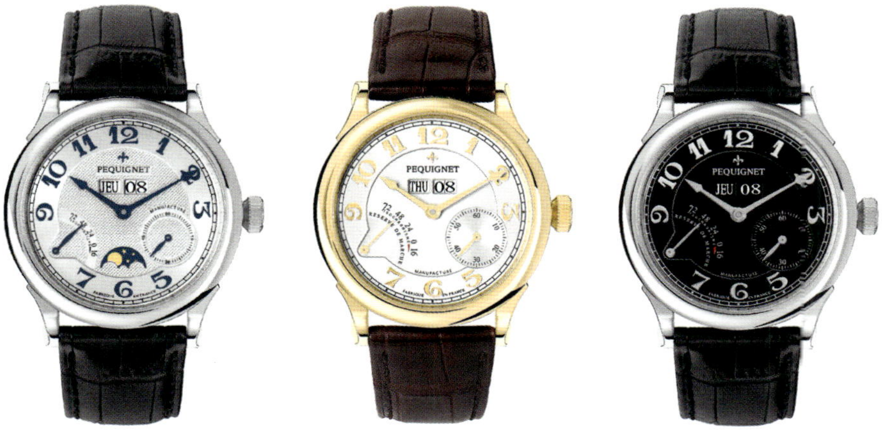

ROYALE PARIS MODELS

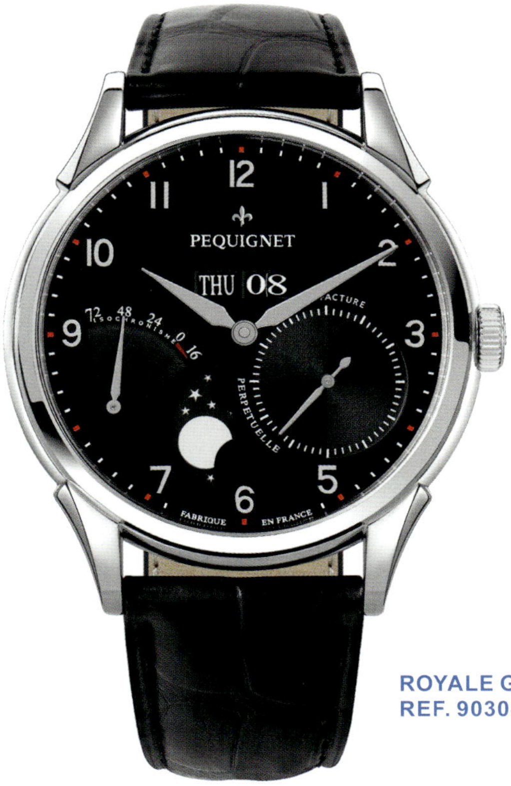

ROYALE GRAND SPORT
REF. 9030443CN

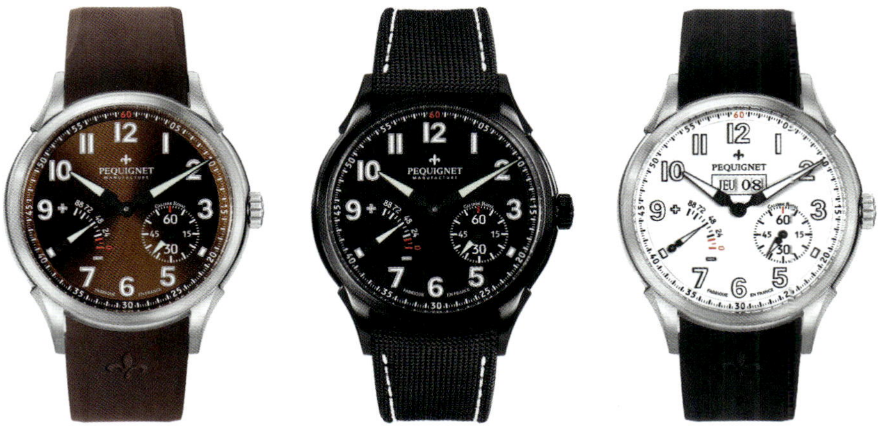

ROYALE TITANE MODELS

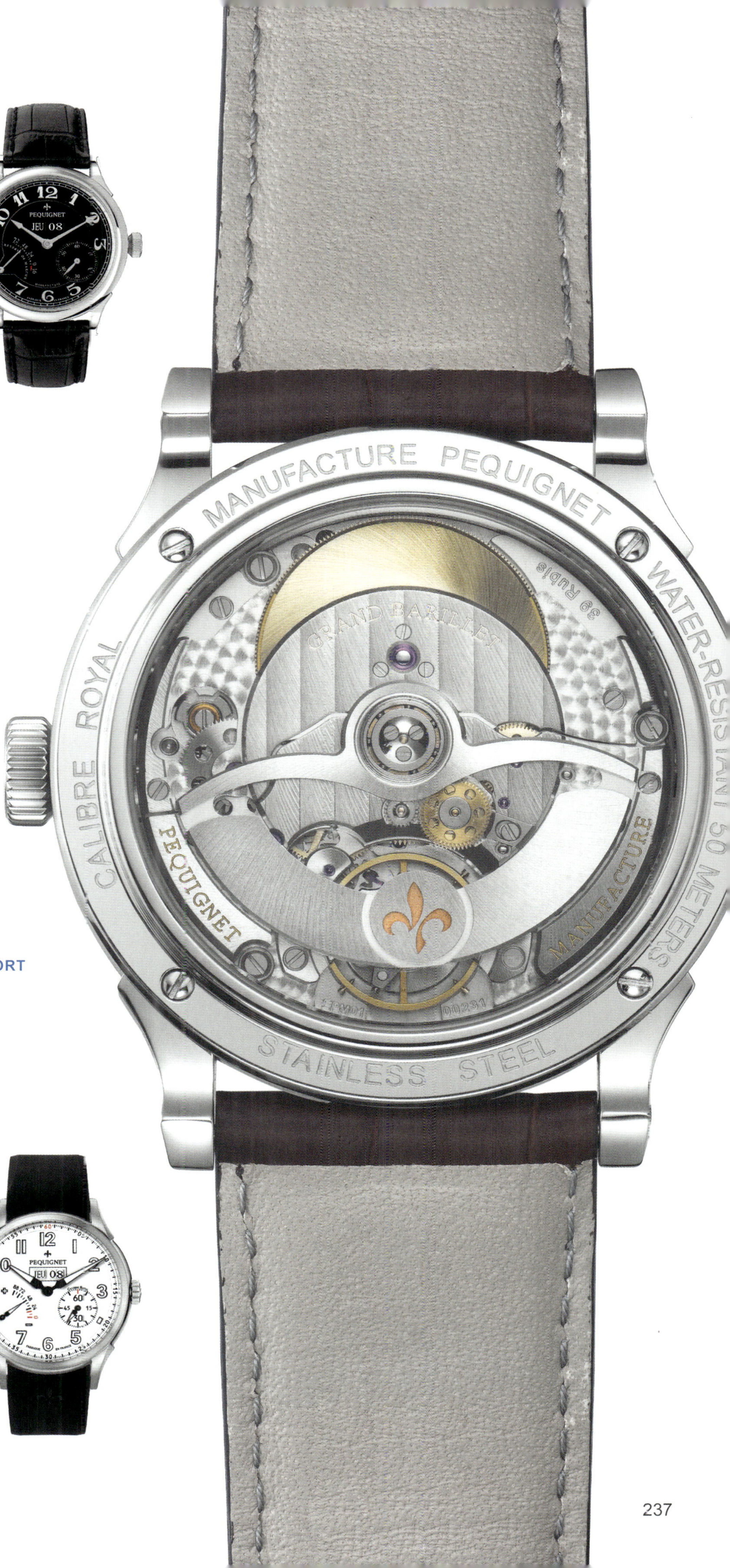

PIAGET
ALTIPLANO

1998

HIGH PLATEAU

The famous Altiplano plateau extends from Peru to Chile. At an average of 3,500 metres above sea level, it ranks as one of the highest inhabited areas of the world. Piaget chose this region as the namesake for its classic ultra-thin wristwatches. The Altiplano was launched in 1998. When a Richemont Group member talks about 1957 however, it is in reference to the launch of its proprietary manufacture calibre. The era began with the 9P, the mainspring of which was best wound daily. In 1960, it was the turn of the 12P with a solid-gold bidirectional micro-rotor. Its height of 2.3 millimetres earned it an entry in the Guinness Book of Records. It was to retain this record-beating status until 1978. Marking the fiftieth anniversary of the 12P in 2010, Piaget presented a totally new calibre, the 1208P. Its ultra-thin 2.35 millimetres could not be rivalled by any other automatic calibre. Its flatness was owed in part to some gears being just 0.12 millimetres thick. Despite this, the slim 30-millimetre newcomer did not disappoint when it came to its energy reserves. The comparatively small balance wheel only ceased to oscillate after around 40 hours without replenishment. This micro-rotor movement and its iterations are ideal for elegant wristwatches that can be worn on many occasions. These do not include during sports and other outdoor activities. In 2014, Piaget set another world record for flatness with the Altiplano 38 mm 900P. With the hand-wound Calibre 900P, case and movement are melded into a functional unit of only 3.65 millimetres in height. Initially presented in 2018 and produced in miniscule quantities since 2020, the Piaget Altiplano Ultimate Concept set only temporary records in the height stakes.

After four years of research and development, the total height including the case made of a cobalt and chromium alloy was honed down to an amazing two millimetres. If you deduct 0.2 millimetres for the sapphire crystal, it leaves around 1.8 millimetres for the caseback, which doubles as the main plate, at a thickness of only 1.2 millimetres and the parts mounted on it. This ultra-flat microcosm with a power reserve of 40 hours consists of a staggering 283 components. The price is equally breathtaking. The same money would easily buy an apartment in a good city location. The latter statement also applies to the new Piaget world record. Launched in 2024, the Altiplano Ultimate Concept Tourbillon is just two millimetres high.

FLACHE HOCHEBENE

Von Peru bis Chile erstreckt sich die als „Altiplano" bestens bekannte Hochebene. Mit durchschnittlich 3.500 Metern über dem Meeresspiegel zählt sie zu den höchsten menschlichen Siedlungsgebieten. Piaget hat diese Region zur Namenspatin für seine ultraflachen Armbanduhren klassischer Ausprägung erkoren. An den Start ging die „Altiplano" konkret im Jahr 1998. Wenn das Mitglied des Richemont-Konzerns selbst vom Jahr 1957 spricht, dann meint es jedoch die Anfänge der Ära entsprechender Manufakturkaliber. Da ging es nämlich los mit dem 9P, dessen Zugfeder am besten täglich manuell gespannt werden musste. 1960 folgte das 12P, ausgestattet mit massivgoldenem, in beiden Drehrichtungen aufziehendem Mikrorotor. 2,3 Millimeter Bauhöhe brachten den Eintrag ins Guinness-Buch der Rekorde. Bis 1978 blieb dieser Status erhalten. Zum 50. Geburtstag des 12P im Jahr 2010 stellte Piaget das völlig neu entwickelte 1208P vor. Seine Mini-Bauhöhe von 2,35 Millimetern unterschritt damals wiederum keine andere Automatik. Einer der angewandten Tricks: Manche Zahnräder waren gerade einmal 0,12 Millimeter dick. An alltagstauglicher Gangautonomie mangelte es dem knapp 30 Millimeter großen Newcomer trotzdem nicht. Erst nach rund 40 Stunden ohne Energienachschub stellte die vergleichsweise kleine Unruh ihre Oszillationen ein. Dieses Mikrorotor-Uhrwerk und seine Derivate eigneten sich vorzüglich für elegante Armbanduhren, die sich bei vielen Gelegenheiten tragen lassen. Nur bei Sport und anderen Outdoor-Aktivitäten sollten sie tunlichst vom Handgelenk. Mit der „Altiplano 38 mm 900P" erzielte Piaget 2014 einen weiteren Flachheits-Weltrekord. Beim Handaufzugskaliber 900P verschmolzen Gehäuse und Uhrwerk zu einer konstruktiven, nur 3,65 Millimeter hoch bauenden Einheit. Nur vorübergehend Ultimatives in Sachen Bauhöhe lieferte indessen die 2018 erstmals vorgestellte und ab 2020 in winzigen Stückzahlen produzierte Piaget „Altiplano Ultimate Concept".

Nach vier Jahren Forschungs- und Entwicklungsarbeit sank die Bauhöhe samt Gehäuse aus einer Kobalt-Chrom-Legierung auf sagenhafte zwei Millimeter. Zieht man von diesem Wert 0,2 Millimeter für das Saphirglas ab, bleiben etwa 1,8 Millimeter für den als Platine dienenden Gehäuseboden mit lediglich 1,2 Millimeter Materialstärke und die darauf befestigten Teile. Summa summarum braucht es für den ultraflachen Mikrokosmos mit 40 Stunden Gangautonomie exakt 283 Komponenten. Entsprechend hoch ist der Preis. Fürs Geld gäbe es auch eine Eigentumswohnung in guter Großstadtlage. Letztgenannte Aussage trifft auch auf den neuen Piaget-Weltrekord zu. Das „Altiplano Ultimate Concept Tourbillon" von 2024 baut lediglich zwei Millimeter hoch.

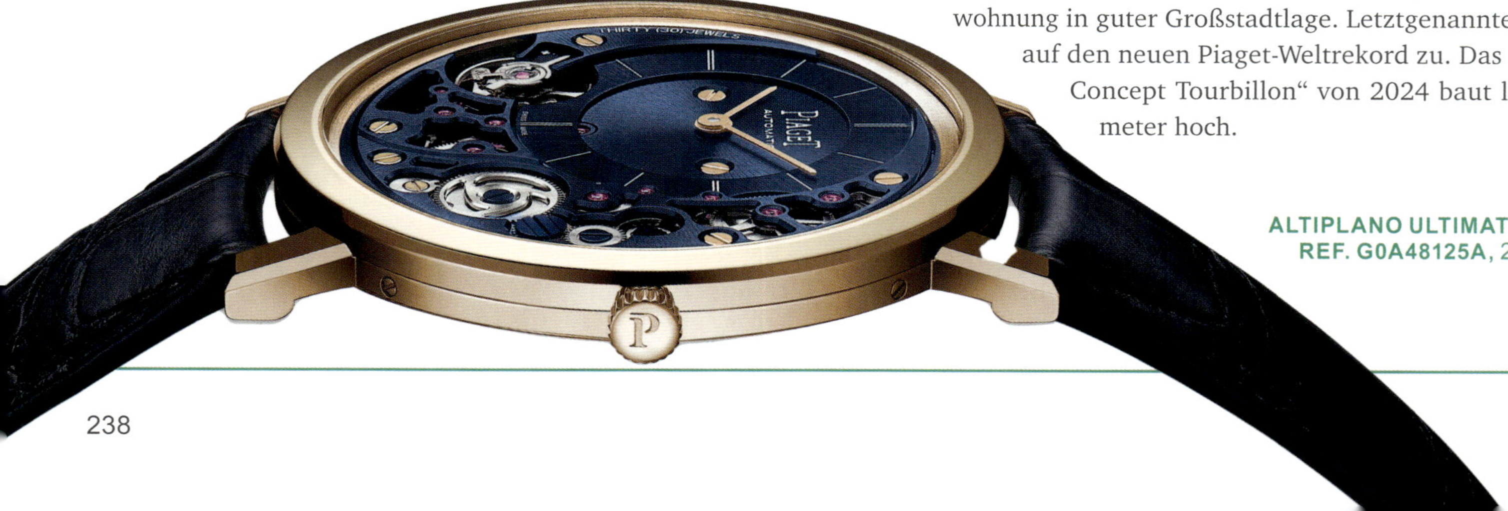

ALTIPLANO ULTIMATE AUTOMATIC
REF. G0A48125A, 2023

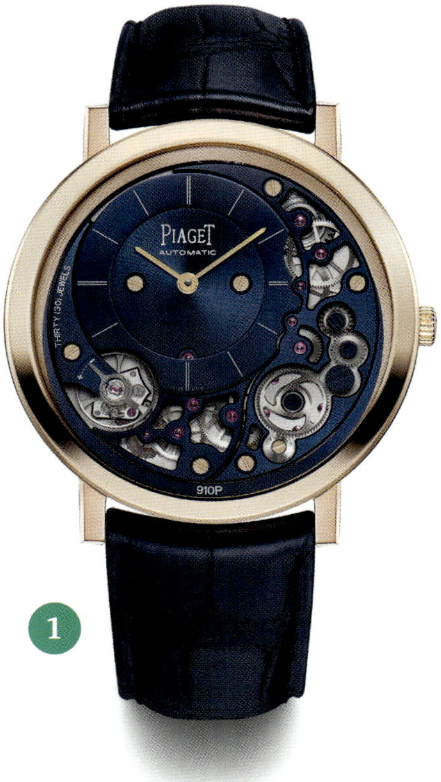

1

1 *Ref. G0A48125, 2023*

2 *Ultimate Concept Ref. G0A47506, 2020*

3 *Ultimate Concept, back*

4 *Ultimate Concept, profile*

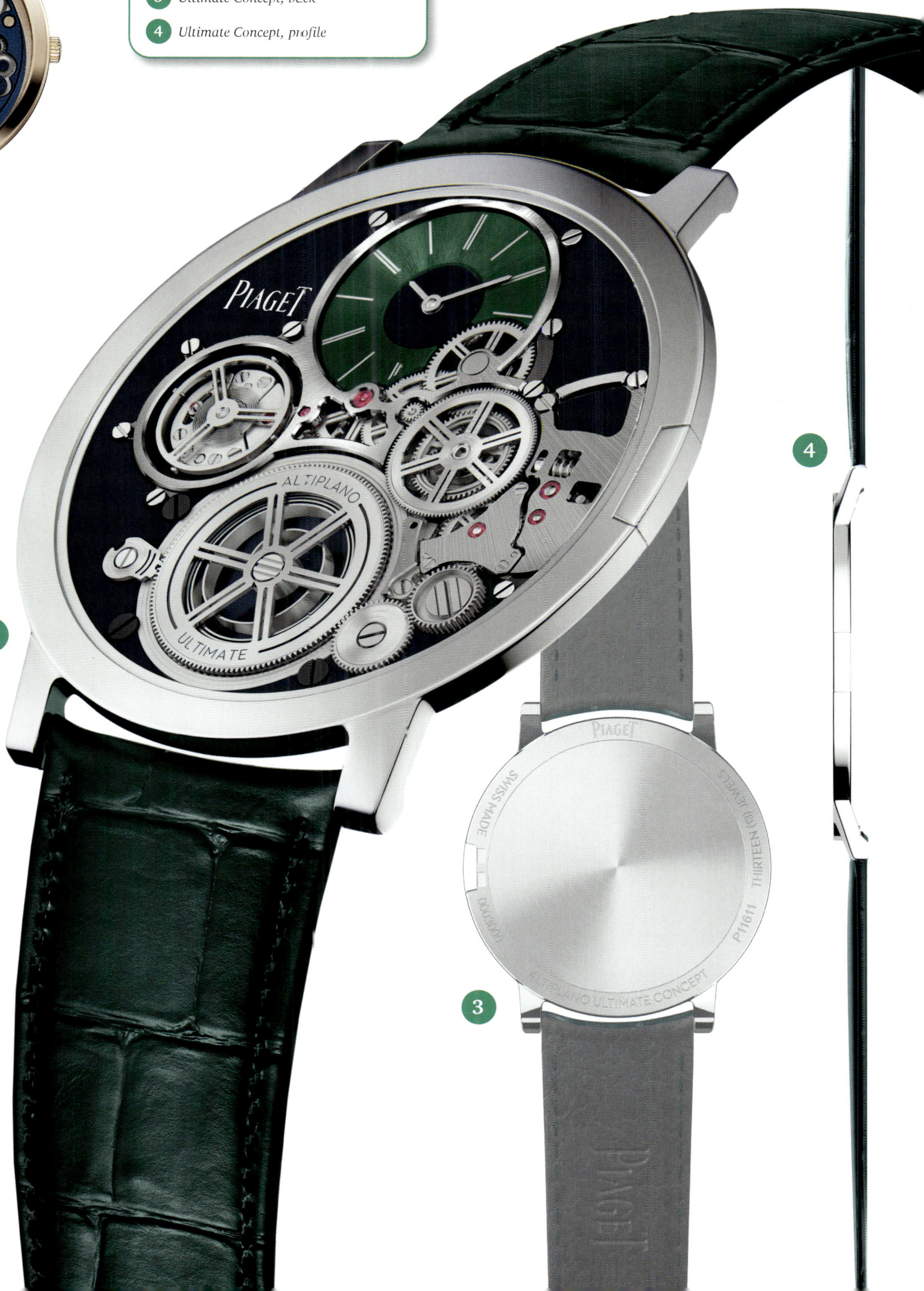

2

3

4

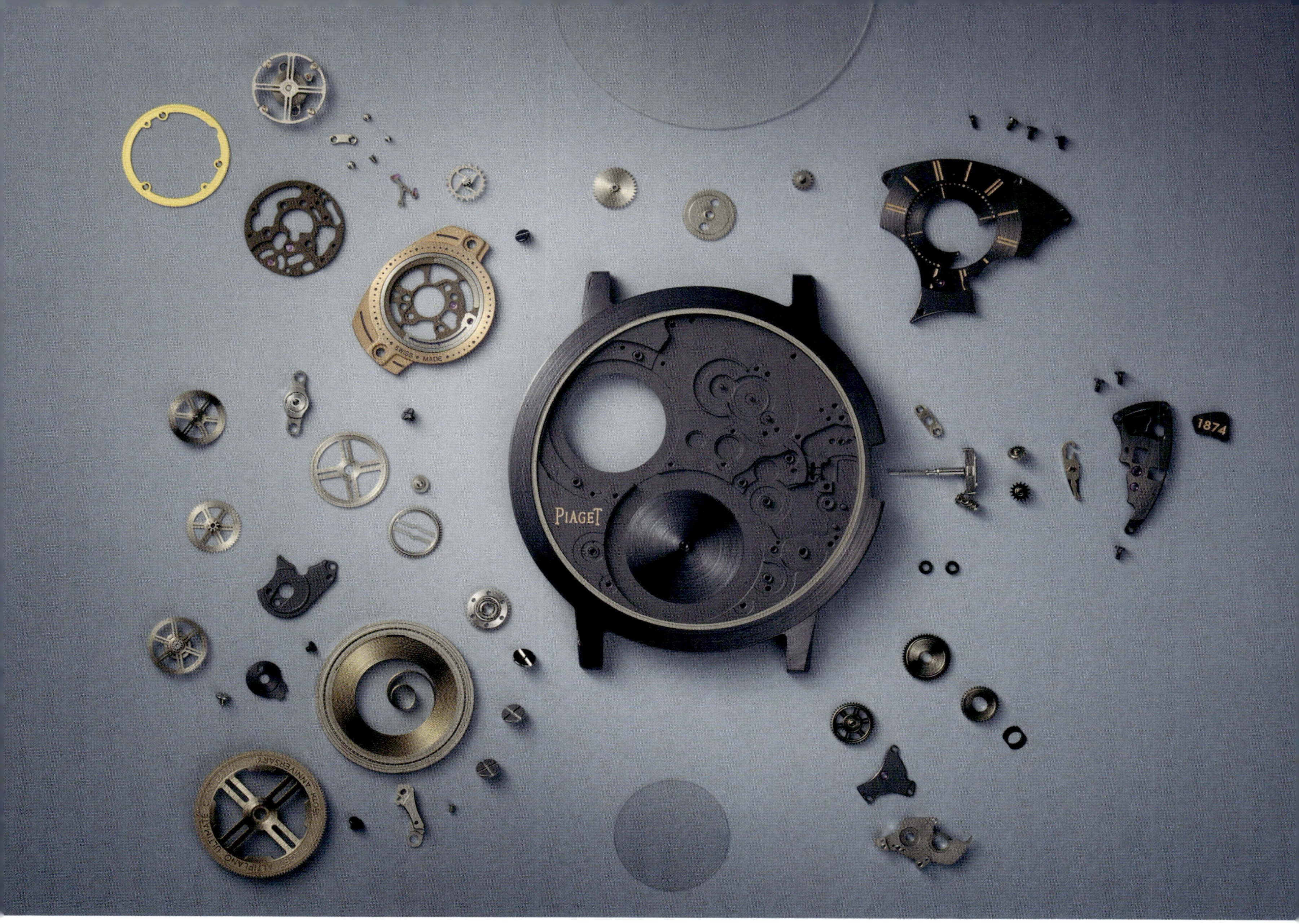
PIAGET
1874
SWISS
MADE

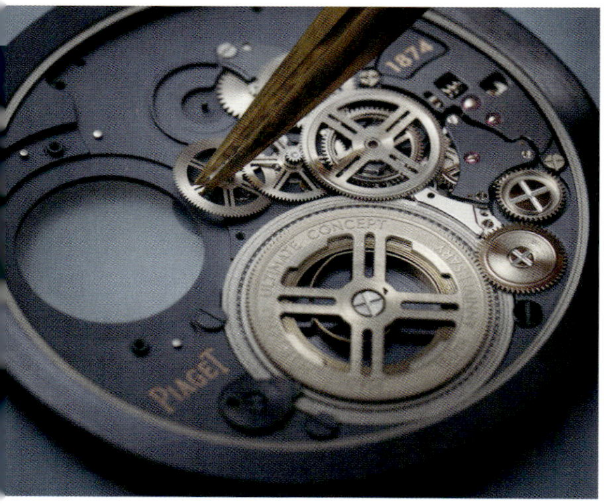
1874
PIAGET

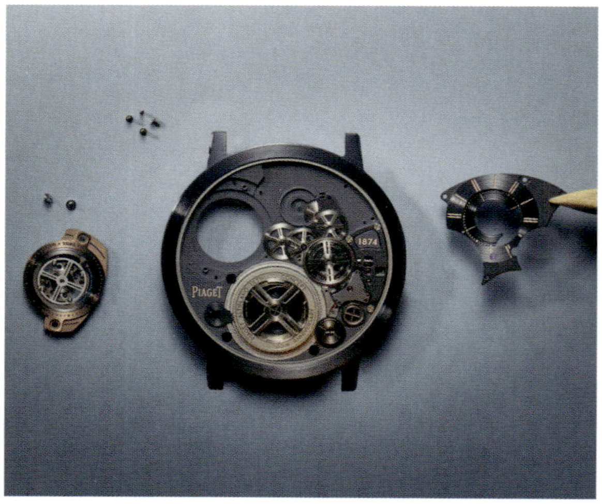
PIAGET

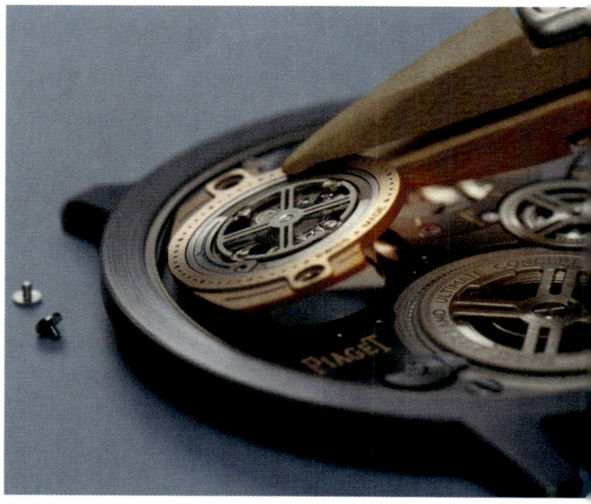
PIAGET

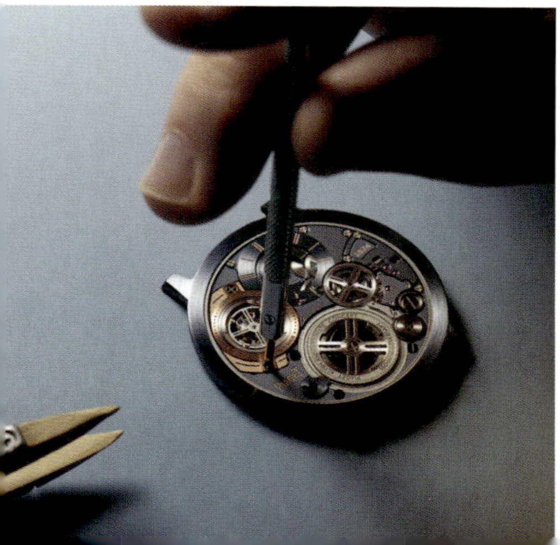

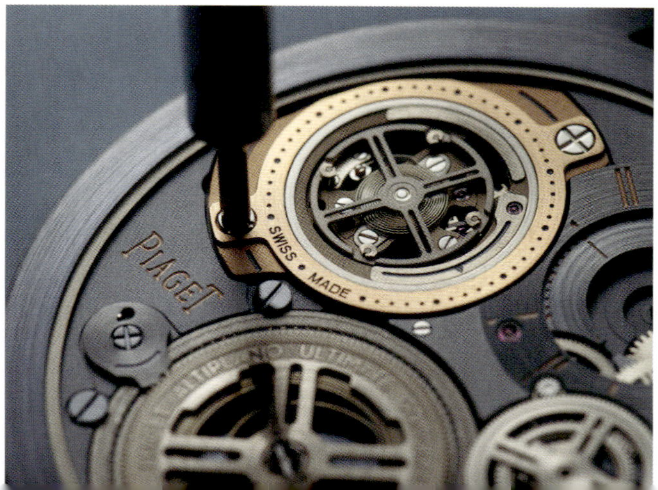
PIAGET
SWISS
MADE

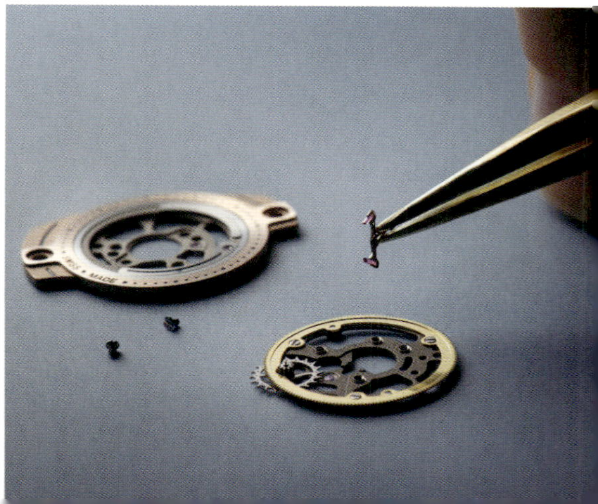

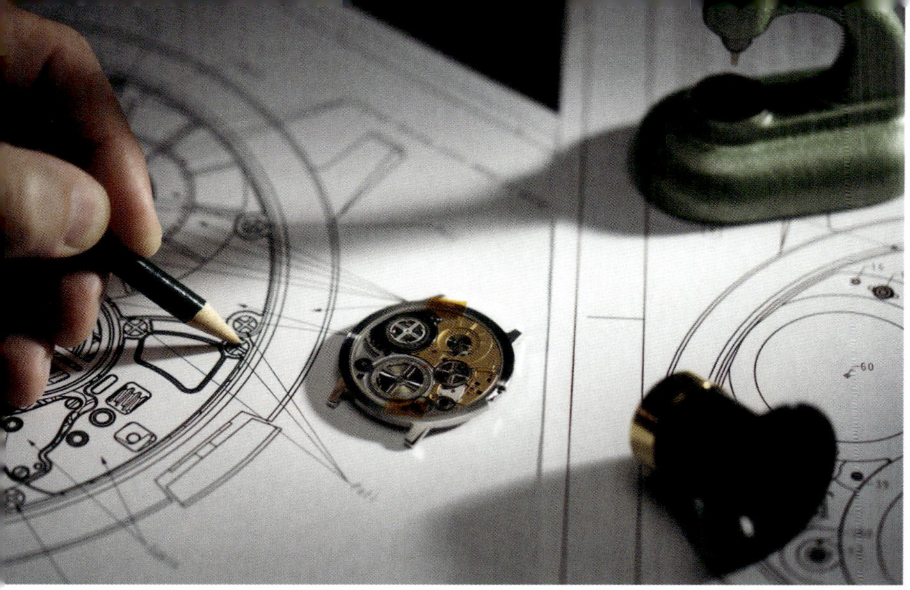

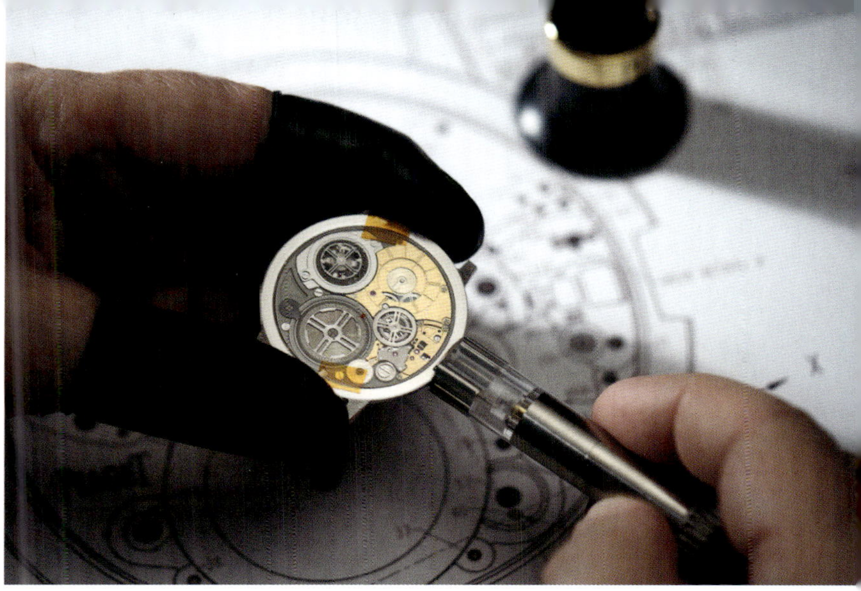

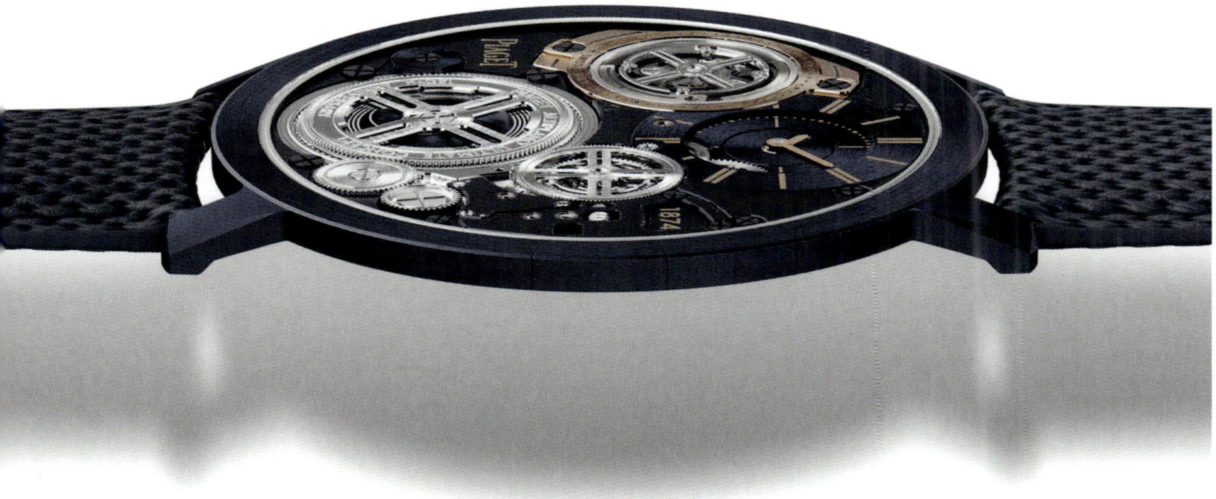

The world's thinnest tourbillon | Der flachste Tourbillon der Welt:
The Altiplano Ultimate Concept Tourbillon released in 2024, the year of Piaget's 150th anniversary, is only two millimetres thick. | Der Altiplano Ultimate Concept Tourbillon wurde pünktlich zum 150. Jubiläum Piagets 2024 präsentiert und ist nur zwei Millimeter hoch.

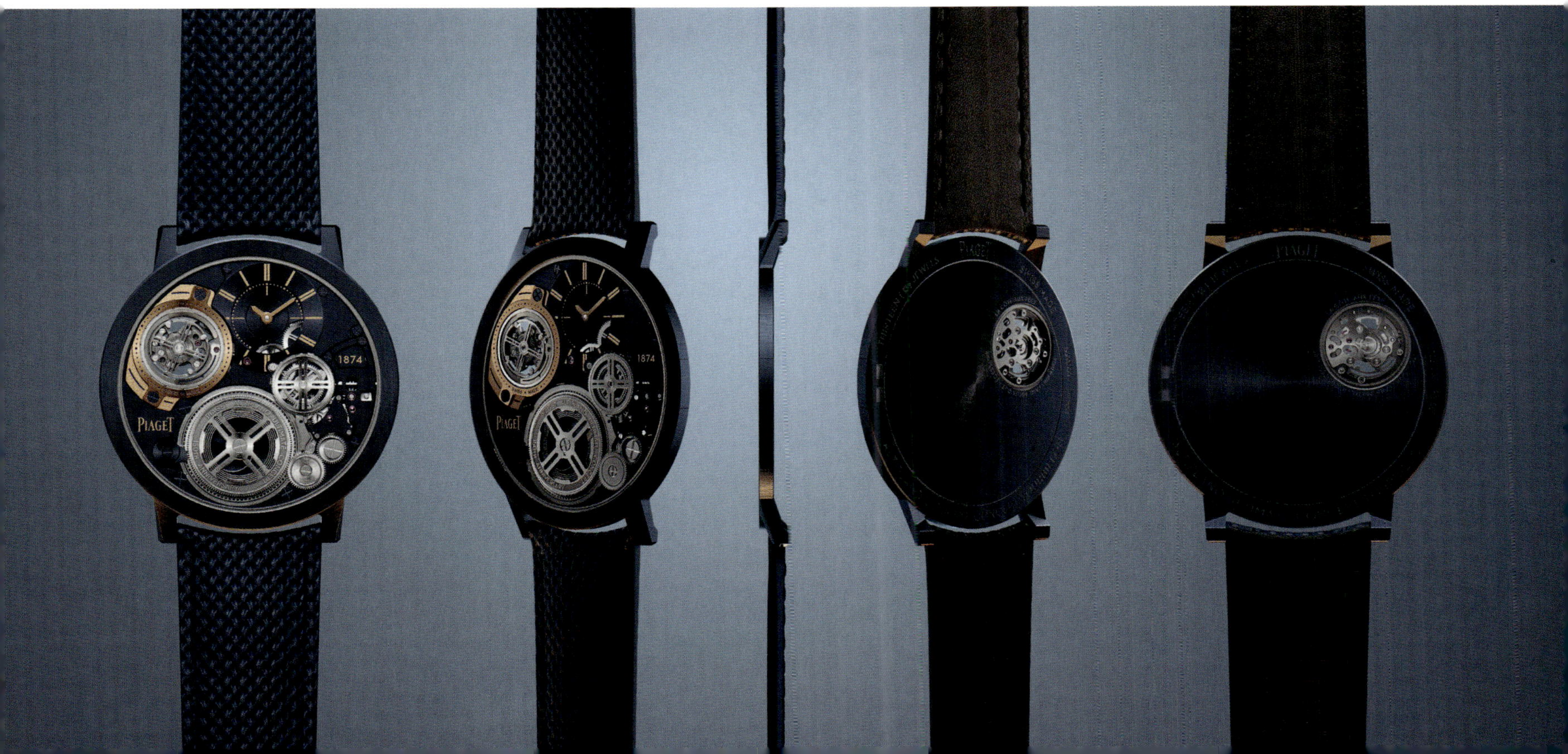

PORSCHE DESIGN CHRONOGRAPH 1

1972

PIONEER IN BLACK

To engage with this black watch icon we have to fast-track back to 1972. Regular use of leap seconds is needed to adjust the high precision achieved by atomic timepieces to imprecise solar time. The first quartz watches with a liquid crystal display delight electronics enthusiasts. At Valjoux, the new Calibre 7750 is revving up in the starting blocks. This automatic chronograph watch movement appeals to a very small clique of customers who value watchmaking tradition more than progress for its own sake, including the Swiss watch brand Orfina and Ferdinand Alexander Porsche. After attending a Waldorf school, a short intermezzo with Max Bill and Otl Aicher at the Ulm School of Design and a classical apprenticeship at Bosch, F. A. joined the Zuffenhausen family business in 1957. Until 1963, it was a hotbed for hot wheels even on the test stand. We are talking about the legendary Porsche 911.

The early 1970s was overshadowed by a crisis not only in Swiss watchmaking, but also in the automotive industry. And Porsche was no exception. Literally overnight, F. A. Porsche had to seek new horizons. His former employer's request for watches as a token of appreciation for long-serving employees could not have been more propitious. "I believe that one should not pack anything unless one knows exactly what's inside it. One must consciously concern oneself with the function of the object and grow along with it." In 1972, a first chronograph watch was created in tune with this credo. In matt black, the Chrono I replicated the distinctive look of the 911 dashboard instruments on the wearer's wrist. The red chronograph hand matched the colour of the tachometer indicator. The quest for a suitable manufacturer resulted in Orfina Uhren AG, founded in Grenchen in 1922, winning the stakes. The 50-year-old manufacturer was aware of the upcoming production launch of the Valjoux 7750 calibre. Nothing stood in the way of the joint venture that was to make history. As anticipated, the black automatic chronograph with a link bracelet divided opinion as almost no other timepiece before. Although this did not prevent more than 50,000 watch lovers around the world from acquiring this horological design manifesto. Joining their ranks were famous racing-car drivers such as Mario Andretti, Emerson Fittipaldi, Ronnie Peterson, Clay Regazzoni and Carlos Reutemann. They lent their support to the chronographic statement "Black is Beautiful". >

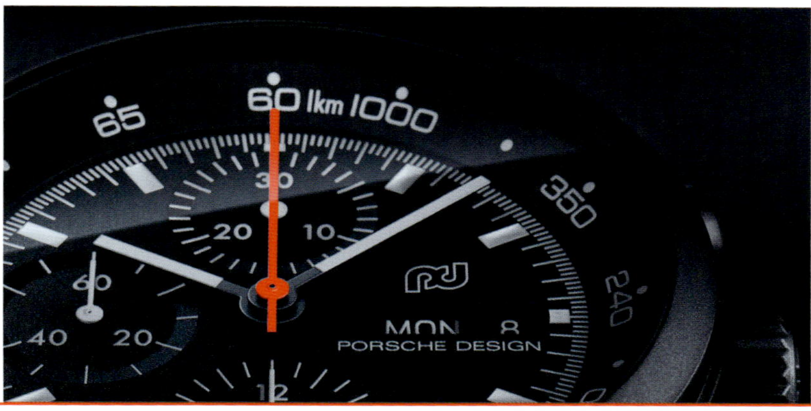

SCHWARZER PIONIER

Die Beschäftigung mit der schwarzen Uhr-Ikone bedingt eine kurze Rückblende ins Jahr 1972: Die hohe Präzision von Atomuhren verlangte in regelmäßigen Abständen nach Schaltsekunden zur Angleichung an die astronomische Zeit. Elektronikfreaks erfreuten sich an ersten Quarzarmbanduhren mit Flüssigkristallanzeige. Bei Valjoux stand das neue Kaliber 7750 endlich in den Startlöchern. Neben der Stoppfunktion besaß es auch einen automatischen Aufzug. Zum sehr überschaubaren Interessentenkreis, dem uhrmacherische Tradition mehr bedeutete als reiner Fortschrittsglaube, gehörten die Schweizer Uhrenmarke Orfina und Ferdinand Alexander Porsche.

Nach dem Besuch einer Waldorf-Schule, einem kurzen Intermezzo bei Max Bill und Otl Aicher an der Ulmer Hochschule für Gestaltung sowie einer klassischen Lehre im Hause Bosch nahm „Eff-A" 1957 einen Job beim Zuffenhausener Familienbetrieb an. Dort gedieh bis 1963 das, was fährt, selbst wenn es steht. Gemeint ist der legendäre Porsche 911.

Anfang der 1970er-Jahre kriselte es nicht nur bei Schweizer Uhrenfabrikanten, sondern auch in der Automobilindustrie. Porsche machte da keine Ausnahme. Quasi über Nacht musste sich F. A. Porsche nach neuen Aufgabenfeldern umsehen. Dabei kam ihm zupass, dass sein ehemaliger Arbeitgeber nach Armbanduhren als Geschenke für verdiente Mitarbeiterinnen und Mitarbeiter verlangte. „Ich meine, dass man nichts verpacken sollte, von dem man nicht weiß, was in ihm steckt. Man muss sich ganz bewusst mit der Funktion des jeweiligen Objekts befassen und damit wachsen."

Unter Einbeziehung dieses Credos entstand 1972 ein erster Armband-Chronograph. Ausgeführt in mattem Schwarz, brachte der „Chrono I" den unverwechselbaren Look der Bordinstrumente des Porsche 911 ans Handgelenk. Der rote Chronographenzeiger korrespondierte mit den Pendants des Drehzahlmessers oder Tachometers. Auf der Suche nach einem geeigneten Hersteller machte die 1922 in Grenchen gegründete Orfina Uhren AG das Rennen. Der damals exakt 50 Jahre alte Fabrikant wusste um den bevorstehenden Produktionsstart des Kalibers Valjoux 7750. Folglich stand dem zeitschreibenden Gemeinschaftsprojekt nichts mehr im Wege. Wie nicht anders zu erwarten, polarisierte der schwarze Automatik-Stopper mit Gliederband wie kaum eine Armbanduhr zuvor. Gleichwohl ließen sich mehr als 50.000 Uhrenliebhaberinnen und Uhrenliebhaber rund um den Globus nicht davon abbringen, auf dieses chronometrische Design-Manifest zu blicken. Auch namhafte Autorennfahrer wie Mario Andretti, Emerson Fittipaldi, Ronnie Peterson, Clay Regazzoni und Carlos Reutemann unterstrichen das chronographische Statement „Black is Beautiful". >

50th Anniversary Limited Edition: Chronograph 1 – 1972, 2022

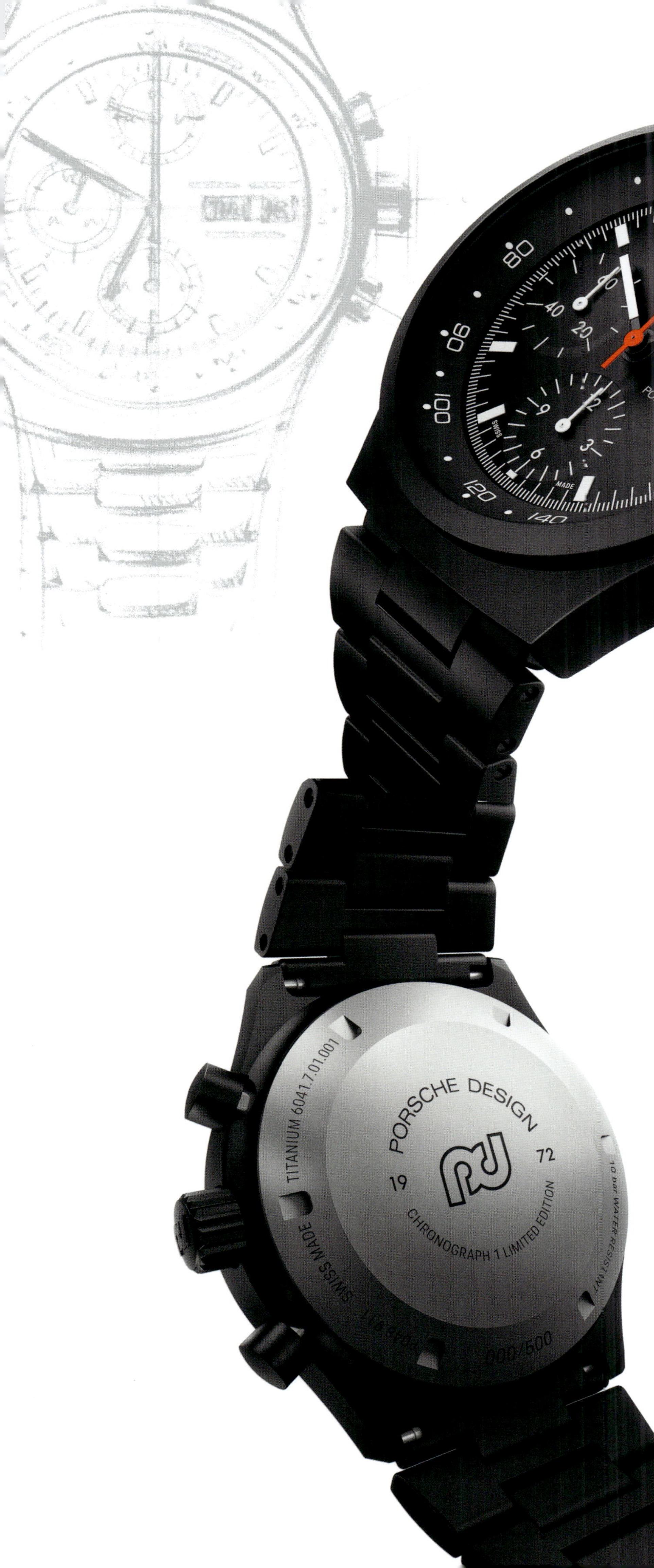

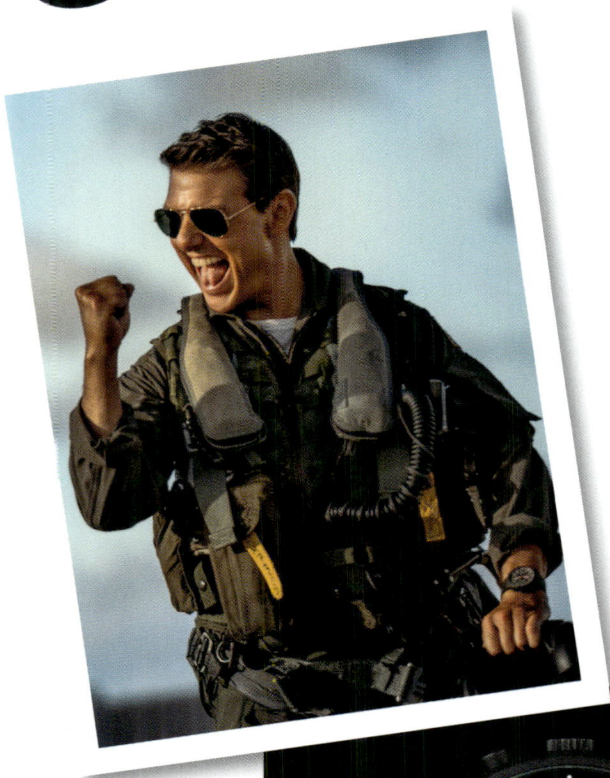

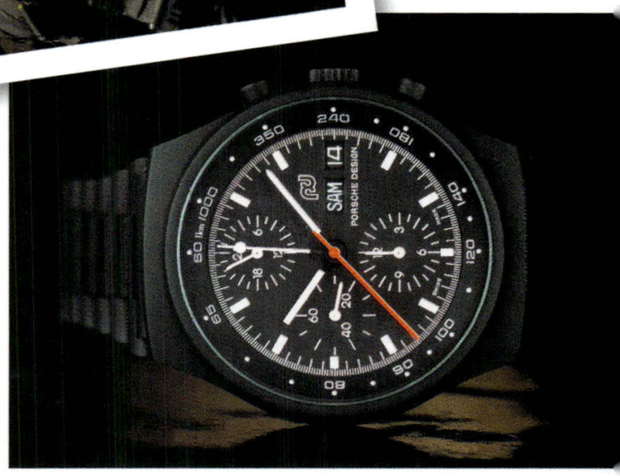

Middle: *Tom Cruise in* Top Gun: Maverick, *2022. He also wore Chronograph 1 in the first* Top Gun *movie in 1986* **|** ***Mitte:*** *Tom Cruise in* Top Gun: Maverick, *2022. Er trug auch im ersten* Top Gun*-Film von 1986 einen Chronograph 1.*

Bottom | Unten: *Chronograph 1 by Orfina, 1972*

Imitations abounded like mushrooms shooting out of the ground. F. A. Porsche might have been annoyed by this. But with his customary sangfroid he shrugged it off. He knew he was a trendsetter in the watch business with its tendency towards conservatism. It was a role F. A. Porsche retained even after his death in 2012. It lives on in his 40-year-old creation that is still going strong. In 2022, it celebrated its 50th anniversary in robust health. Now attired in a lightweight, anti-allergic titanium outfit. The watch world also has F. A. Porsche to thank for the use of this material on chronographs. By contrast, the Valjoux 7750 calibre is no more. Porsche cooperates with the specialist Concepto on the current inner workings. The movement goes by the name of WERK.

Fast schon wie Pilze schossen Plagiate aus dem Boden. F. A. Porsche hätte sich darüber aufregen können. Mit der ihm zeitlebens eigenen Gelassenheit tat er es nicht. Ihm war klar, Trendsetter im gestalterisch eher konservativen Uhrenbusiness gewesen zu sein. Diese Rolle blieb „Eff-A" auch nach seinem Tod im Jahr 2012. Da war seine Kreation 40 Jahre alt und immer noch am Markt. 2022 feierte sie den 50. Geburtstag bei bester Gesundheit.

Ausgestattet nun mit leichtem, antiallergischen Titan-Outfit. Auch die Verwendung dieses Werkstoffs für Chronographen verdankt die Uhrenwelt F. A. Porsche. Ausgedient hat hingegen das Kaliber Valjoux 7750. Für das aktuelle Innenleben kooperiert Porsche Design mit dem Spezialisten Concepto. Die tickenden und zeitschreibenden Resultate nennen sich WERK.

CHRONOGRAPH 1 UTILITY – LIMITED EDITION, 2024

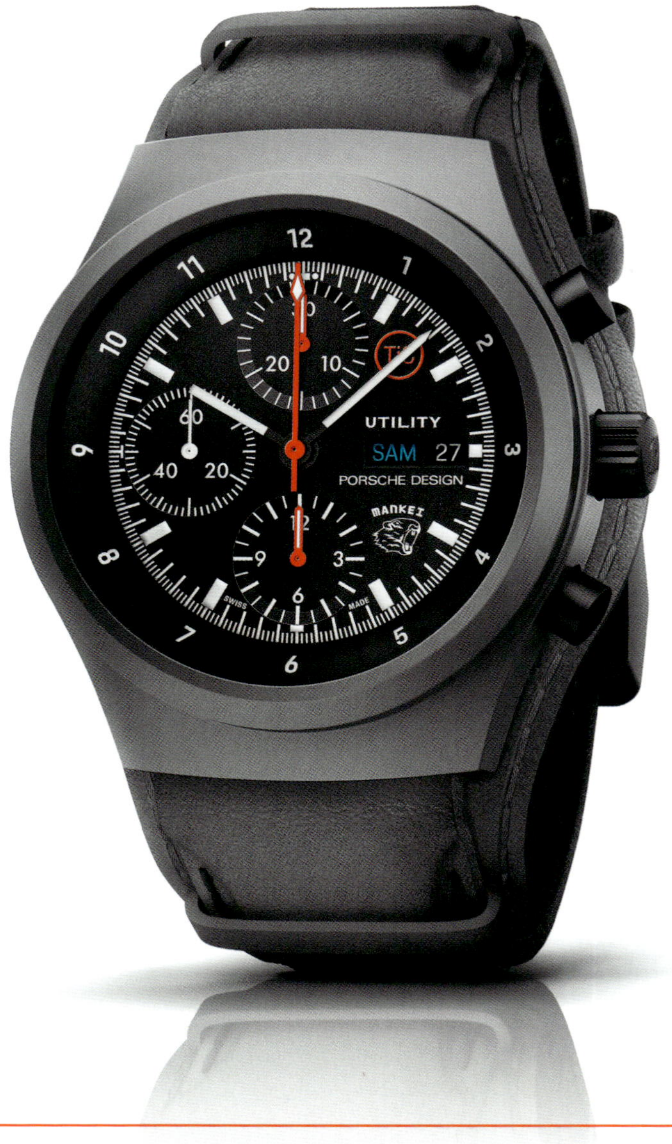

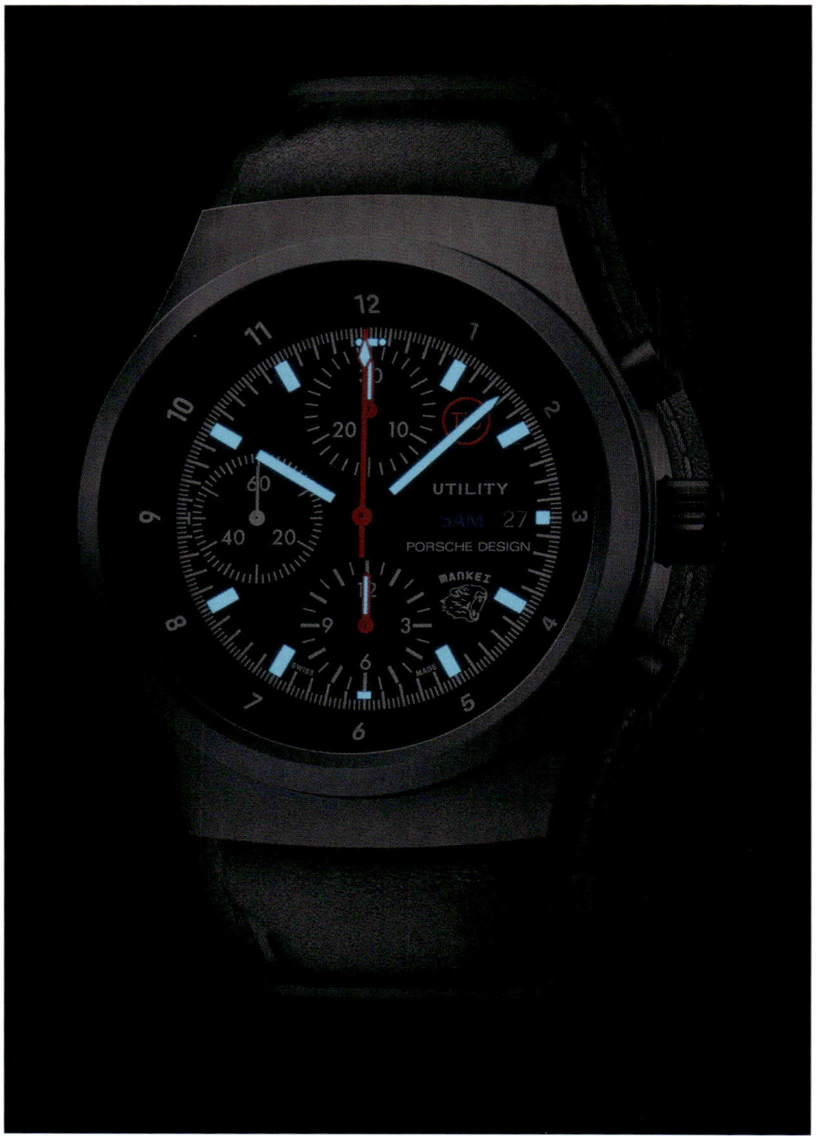

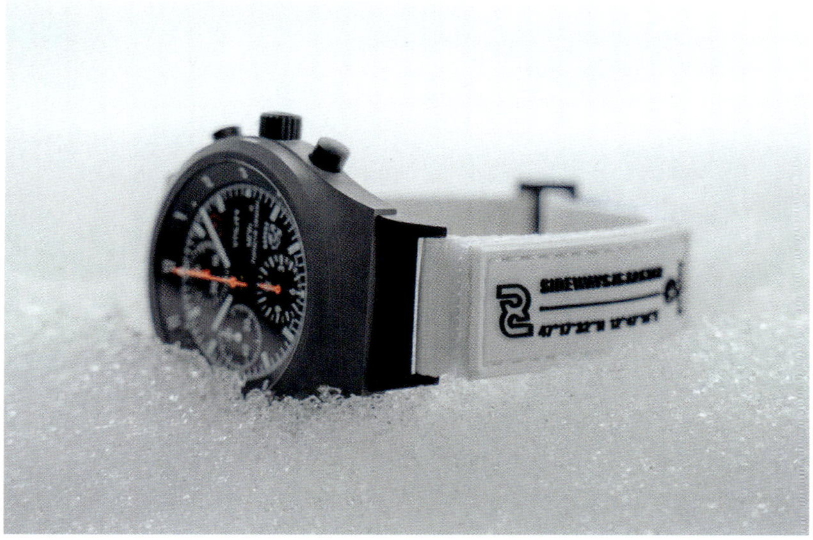

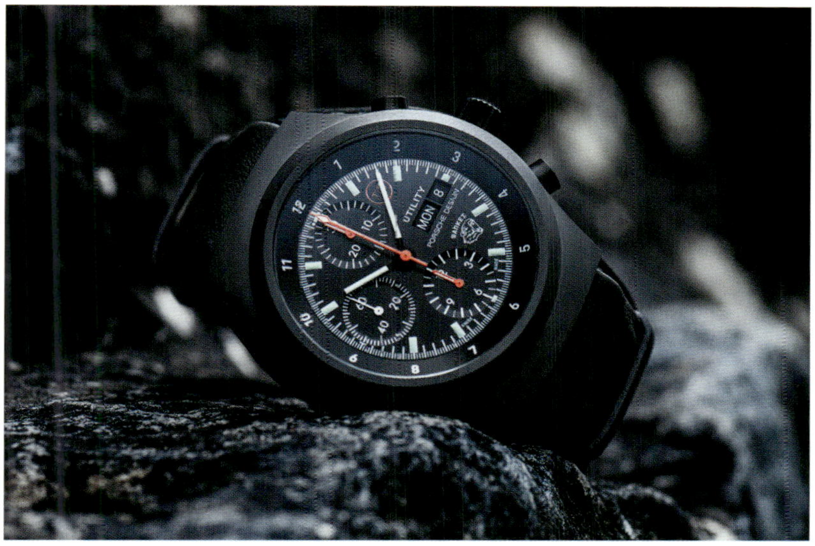

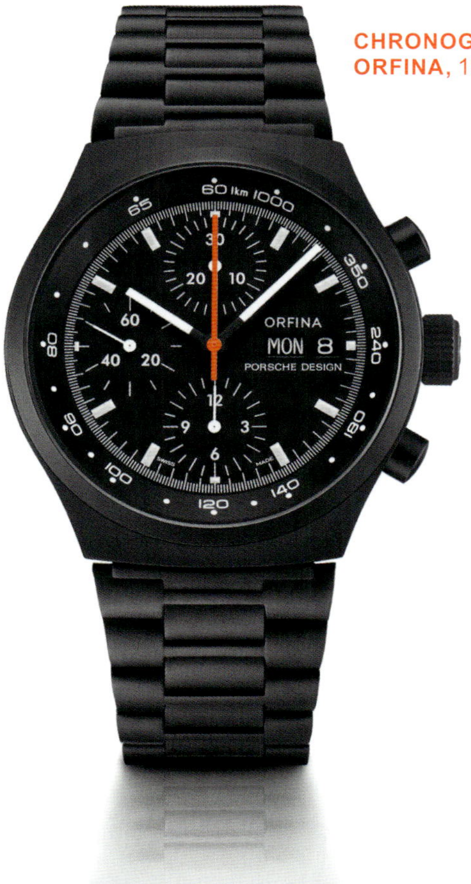

CHRONOGRAPH 1
ORFINA, 1972

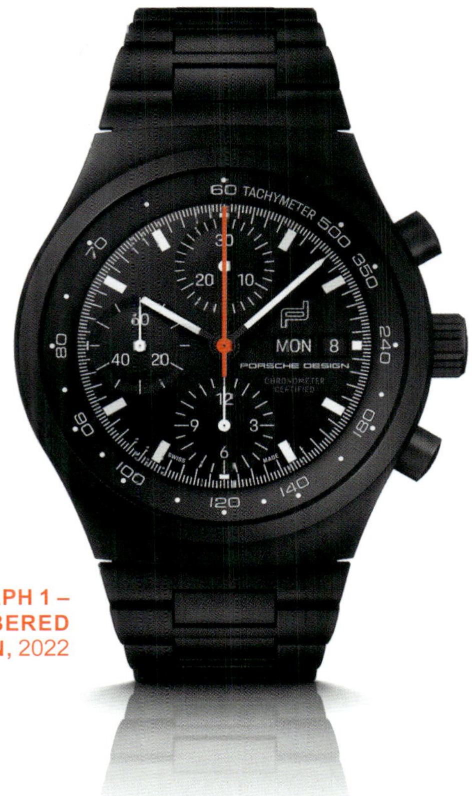

CHRONOGRAPH 1 –
ALL BLACK NUMBERED
EDITION, 2022

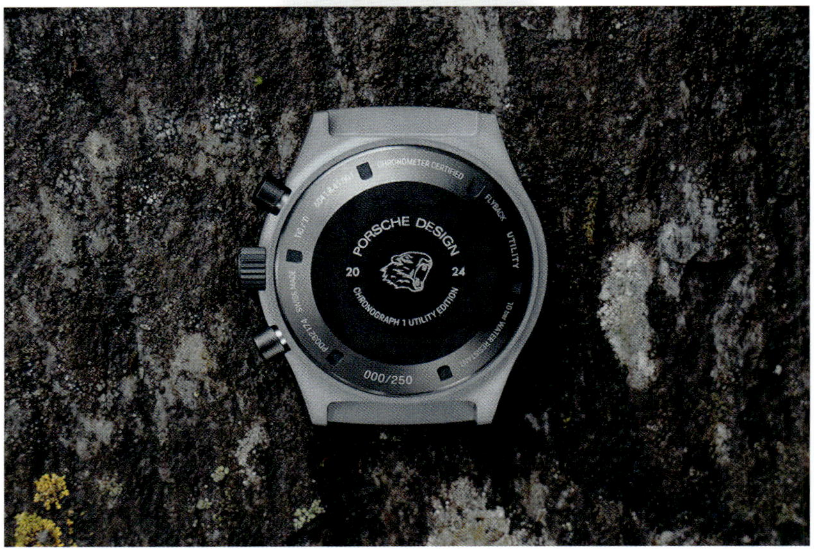

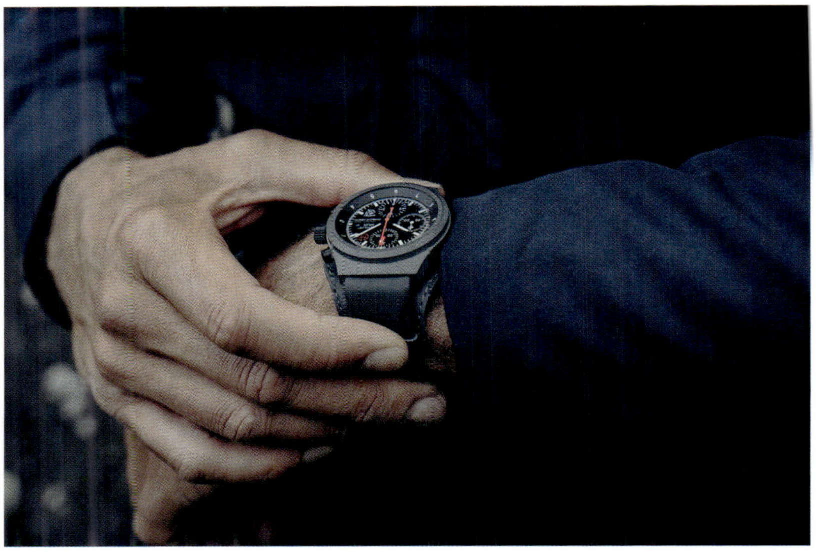

TYPE 3 B, 2013

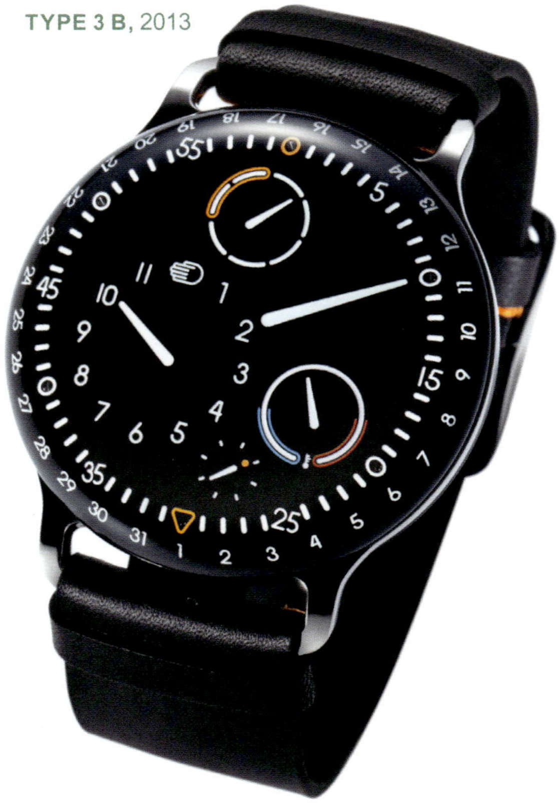

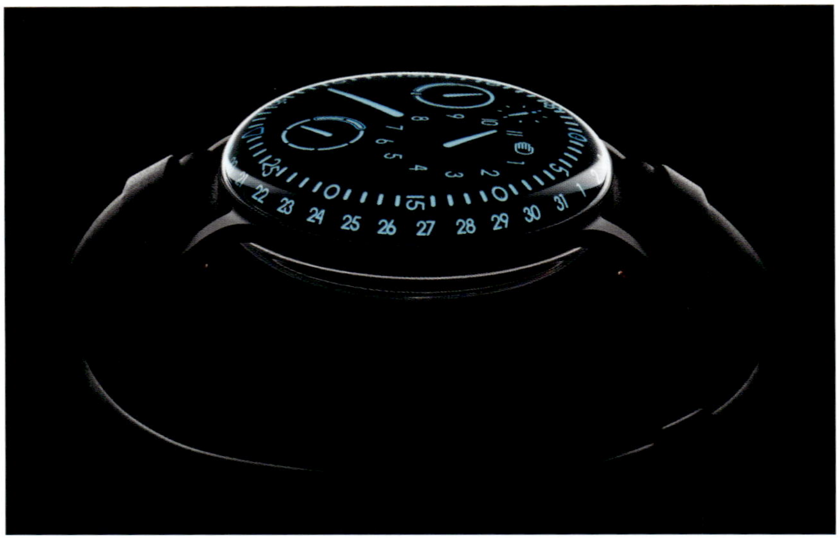

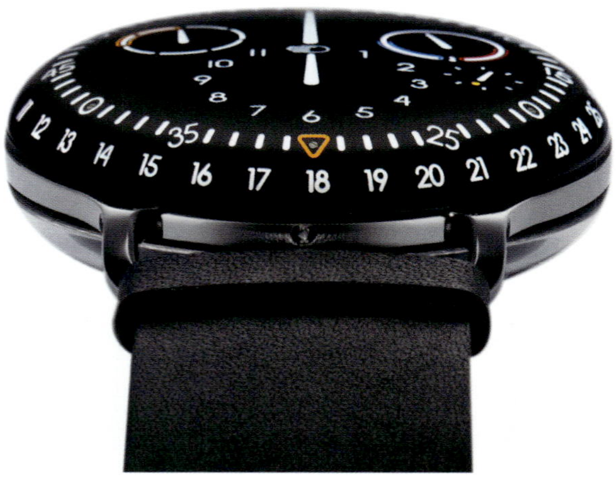

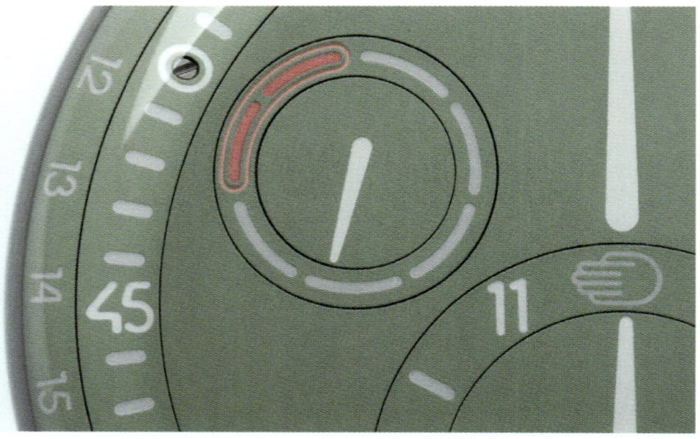

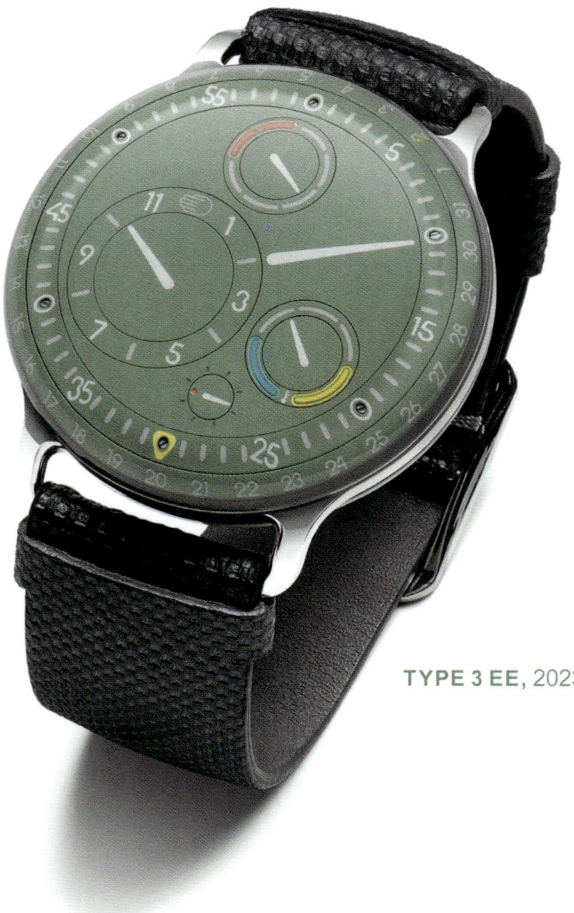

TYPE 3 EE, 2023

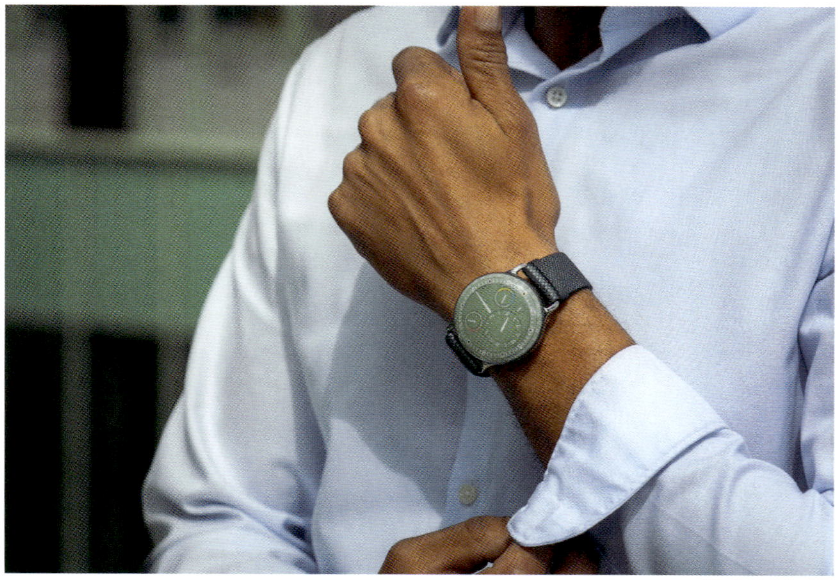

2013

RESSENCE TYPE 3

HAND FLATTERER

When it comes to the hand-flattering qualities of a watch case, Ressence is in a class of its own. The name of this watch brand abbreviates and combines renaissance and essence. The former stands for the rebirth of what constitutes the essence of a watch. Namely the time display, which on watches with an ultra-smooth surface enjoys the limelight. The bold engineering is completely concealed in the round case. A modified standard automatic movement powers the ingenious and therefore patented ROCS (Ressence Orbital Convex System). A backroom player, it lets the highly unusual display take the stage. Along with the hours, minutes and seconds, the TYPE 3 model also indicates the date on the edge. The rather cryptic indication with five strokes and two zeros displays the days of the week. All rotational movements are transmitted from the Eta calibre to the display module by means of tiny magnets. For optimum legibility, the oil-filled dome of the TYPE 3 can be easily removed. A total of 387 components are required for one of these wristwatches. The attention-catching display alone requires 28 gears and 47 jewels. The hand-winding mechanism is equally ingenious and can also be used to set the time and date. Do not look for a crown because there isn't one. Its role is also taken over by the deep-domed sapphire crystal caseback. Various indications show which of the several functions is currently active. Owners of a Ressence will have to get used to their wrist becoming a conversation starter, while shirt lovers will appreciate its cuff-friendly smoothness.

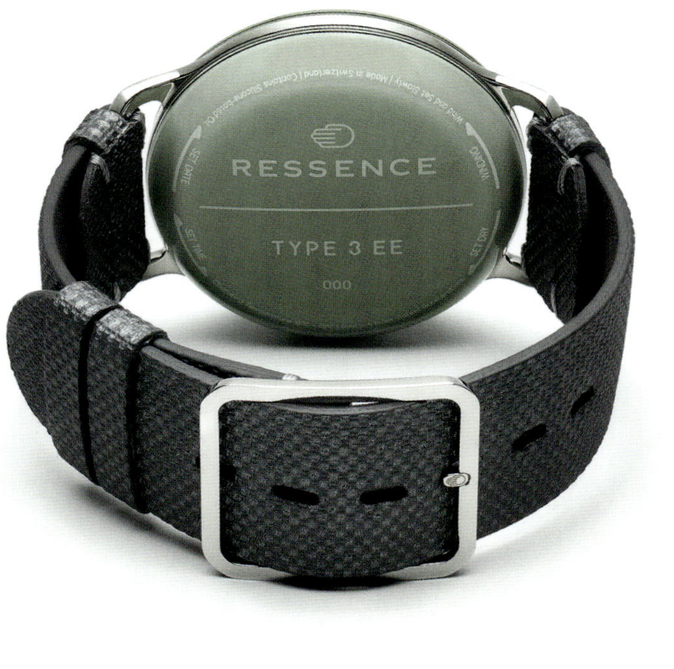

TYPE 3 EE, the engraved indications are filled with green grade A Super-LumiNova® / die Indikatoren auf dem Zifferblatt sind graviert und mit grüner Super-LumiNova® von höchster Qualität gefüllt

HANDSCHMEICHLER

Wenn es um handschmeichlerische Qualitäten eines Uhrengehäuses geht, macht Ressence so schnell niemand etwas vor. Der Name dieser Uhrenmarke versteht sich als Kürzel aus „Renaissance" und „Essence". Ersteres steht für die Wiedergeburt dessen, was zweitens die Essenz einer Uhr ausmacht. Und das ist die Zeitanzeige, welche bei den Armbanduhren mit extrem glatter Oberfläche unübersehbar in den Vordergrund rückt. Die durchaus komplexe Mechanik verschwindet komplett in der runden Schale. Den Antrieb für das ausgeklügelte und deswegen auch patentierte ROCS (Ressence Orbital Convex System) liefert ein modifiziertes Standardautomatikwerk. Es agiert im Hintergrund und überlässt die Bühne dem stark aus der Reihe des Üblichen fallenden Display. Neben den Stunden, Minuten und Sekunden stellt das Modell „TYPE 3" ganz außen auch das Datum dar. Den Wochentagen gilt die etwas kryptisch gestaltete Indikation mit fünf Strichen sowie zwei Nullen. Alle Drehbewegungen werden mit Hilfe winziger Magnete vom Eta-Kaliber auf das Anzeigemodul übertragen. Die zum Zweck optimaler Ablesbarkeit beim „TYPE 3" mit Öl befüllte Kuppel lässt sich als Ganzes mit wenigen Handgriffen entfernen. Insgesamt 387 Komponenten sind für eine dieser Armbanduhren nötig. Allein das ins Auge stechende Display erfordert 28 Triebe und 57 funktionale Steine. Nicht minder intelligent präsentiert sich die Mechanik zum – wenn nötig – manuellen Aufziehen des Uhrwerks sowie zum Einstellen von Zeit und Datum. Eine Krone sucht man vergebens. Deren Job übernimmt der ebenfalls stark gewölbte Saphirglasboden. Welche von den verschiedenen Funktionen gerade aktiv ist, geht aus diversen Markierungen hervor. Bei aller Zurückhaltung sollte es mit einer Ressence am Handgelenk nicht an Aufmerksamkeit mangeln. Und die Manschetten, sofern man Hemden mit solchen trägt, werden definitiv geschont.

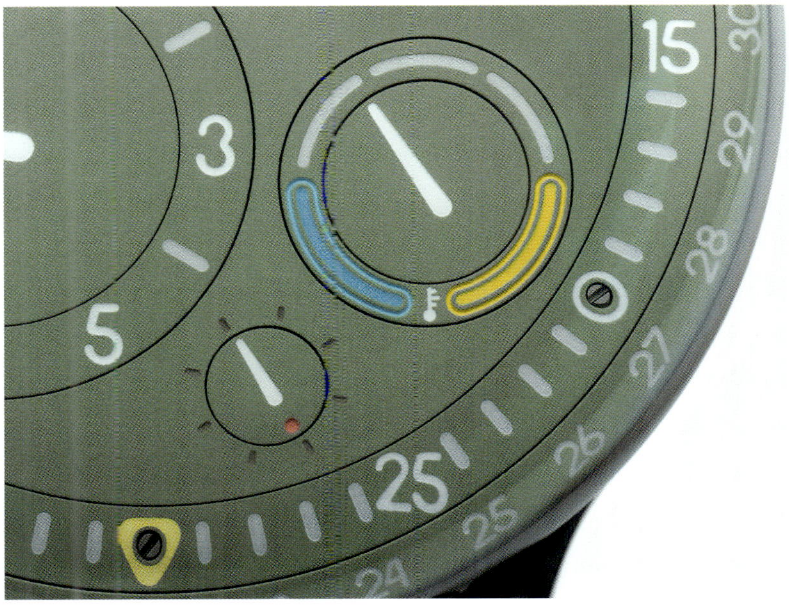

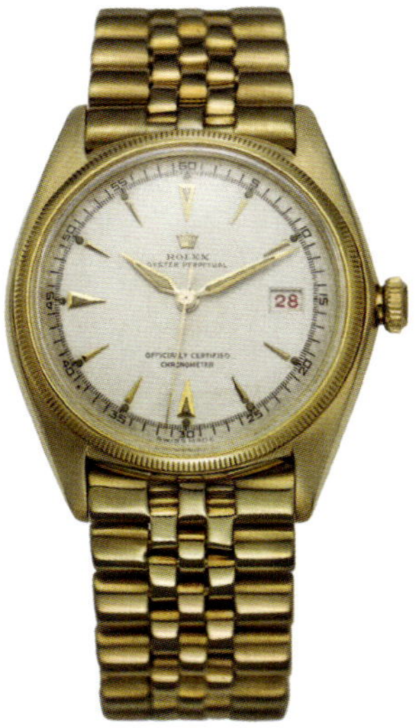

FIRST DATEJUST,
1945

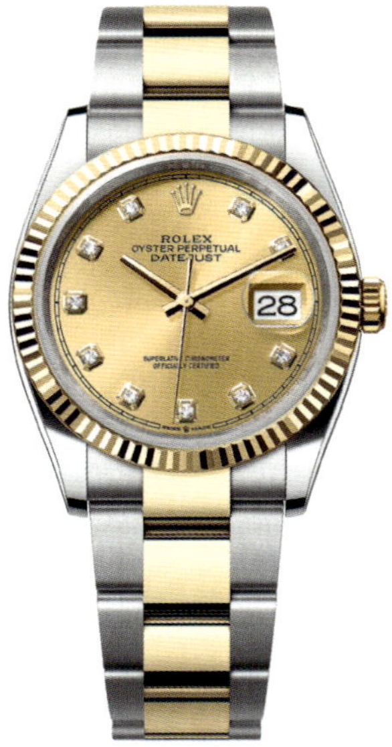

DATEJUST 36
REF. 126233

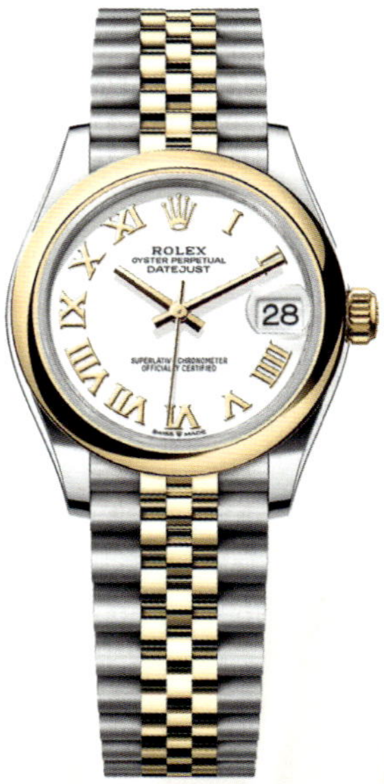

DATEJUST 31
REF. 278243

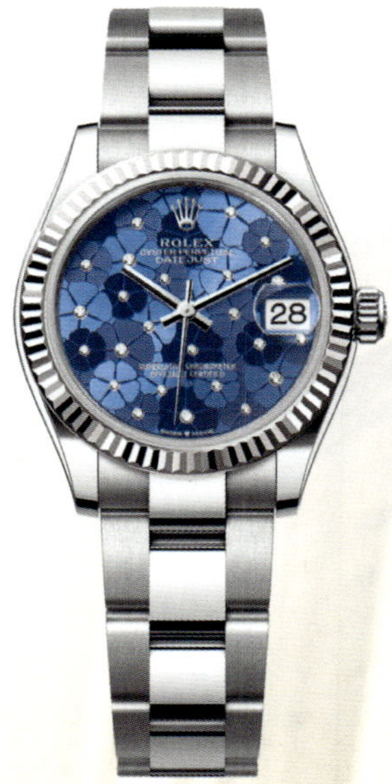

DATEJUST 31
REF. 278274

Craftsmanship: The hour markers and Rolex crown are attached by hand | Präzisionsarbeit: Die Stundenindizes und die Rolex-Krone werden per Hand am Zifferblatt angebracht

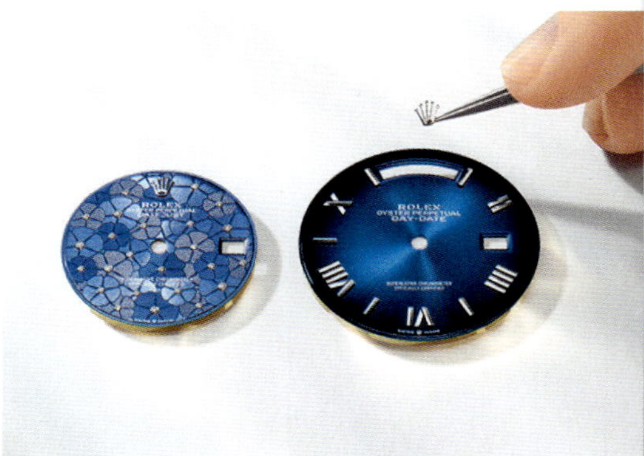

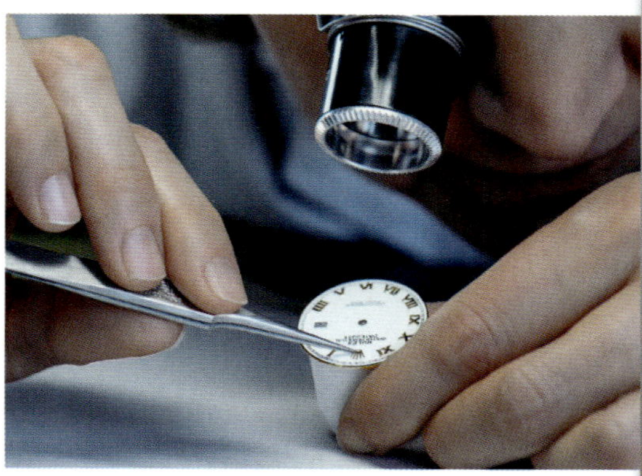

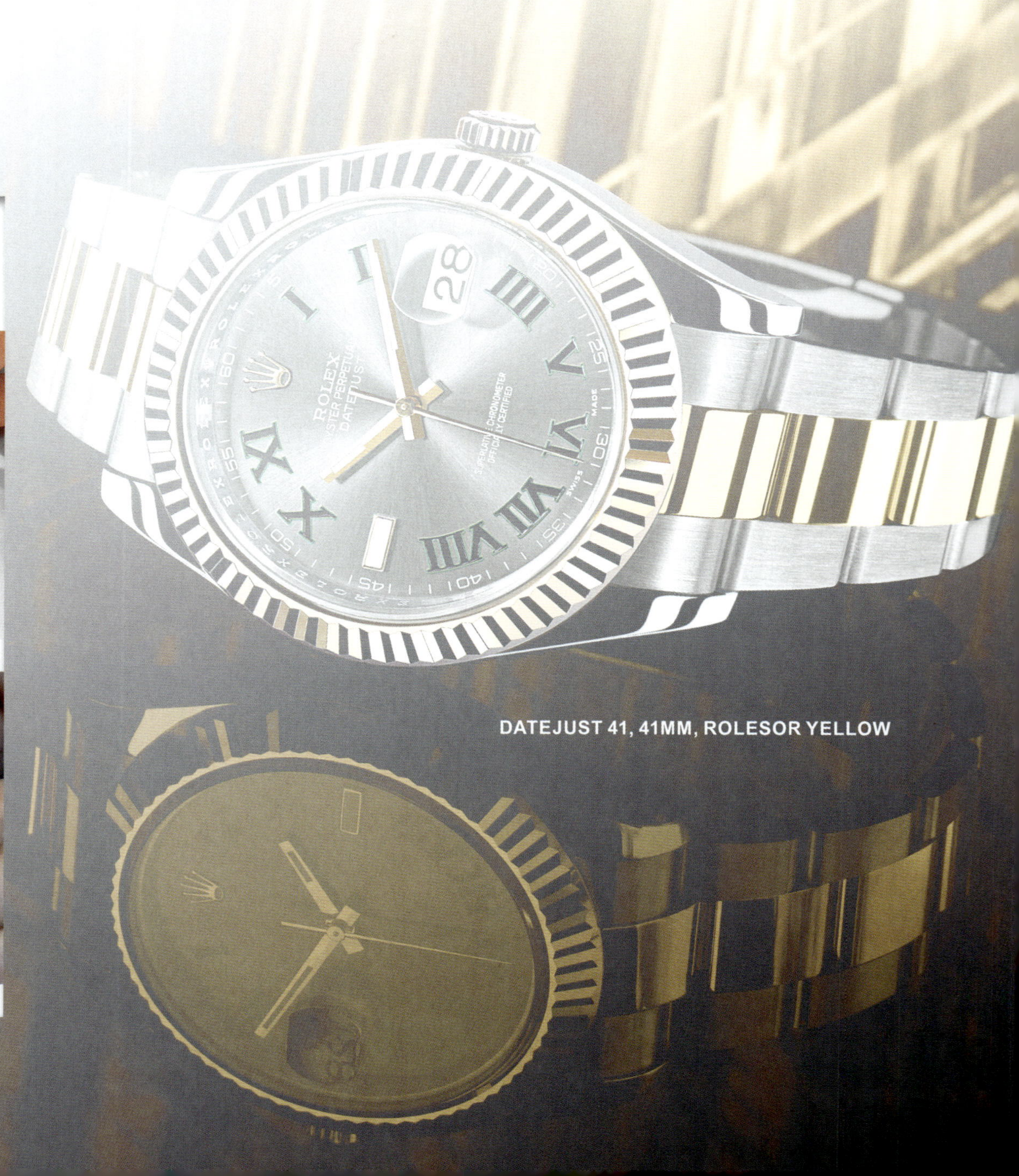

DATEJUST 41, 41MM, ROLESOR YELLOW

ROLEX DATEJUST

MIDNIGHT MOVE

Many horological inventions are ascribed to Rolex. But the date window on wristwatches was definitely not one of them. Dials with a small cut-out for reading off the date digitally have been around since 1915. At that time, Hans Wilsdorf had probably never spared a thought or dreamt of this practical complication. Reliability and precision were his main concern. On creating the Datejust, which was presented in 1945, the Bavarian had quite a different kind of world première in mind. Marking the fortieth anniversary of his company's founding in England, he aimed to combine three fundamental achievements from the eventful biography of a burgeoning brand, namely the waterproof Oyster case and Perpetual movement with self-winding rotor, and the official certification of its chronometric precision. To crown the ensemble the manufactory, whose crown logo was already famed, added a prominent easy-to-read date window at 3 o'clock. Punctually at midnight, the centre-mounted disc advanced one position to the next day. The name "Datejust" was therefore entirely justified. In the catalogue of the time, Rolex wrote, "Without exaggeration, this creation can be described as a synthesis of the entire body of advanced horological science in Switzerland." The Datejust soon assumed the status of a chronometric icon. Further evidencing the excellence of this wristwatch was the distinction of being hailed "Wristwatch of the Twentieth Century" in the USA. Naturally, Rolex has continuously developed its now octogenarian Datejust over the decades in the subtle fashion that typifies the brand. Consequently, its distinctive characteristics and high recognition value are always preserved.

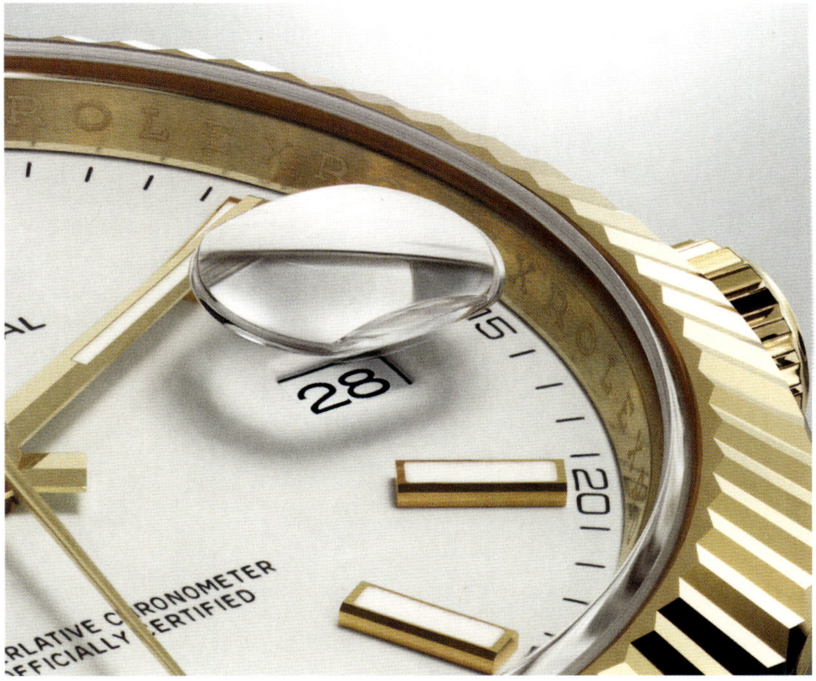

DATUM „JUST IN TIME"

Auf das Konto des Hauses Rolex gehen zahlreiche Erfindungen rund um die Uhr. Definitiv nicht dazu gehört das Fensterdatum für Armbanduhren. Zifferblätter mit einem kleinen Ausschnitt, durch den sich das Datum ablesen lässt, gibt es bereits seit ungefähr 1915. Zu diesem Zeitpunkt dachte Hans Wilsdorf vermutlich noch nicht einmal im Traum an diese hilfreiche Zusatzfunktion. Ihm ging es in erster Linie um Zuverlässigkeit und Präzision. Beim Kreieren der 1945 vorgestellten „Datejust" hatte der gebürtige Bayer eine ganz andere Art Weltpremiere im Blick. Zum 40. Jubiläum seiner englischen Firmengründung zielte er auf die Kombination dreier elementarer Errungenschaften aus der bewegten Biographie des stetig wachsenden Unternehmens: Ein wasserdichtes „Oyster"-Gehäuse, ein „Perpetual"-Uhrwerk mit Rotor-Selbstaufzug, beides zertifiziert durch ein offizielles Gangzeugnis. Sozusagen als Krönung dieses Ensembles fügte die Manufaktur mit dem damals schon weithin bekannten Kronenlogo ein gut ablesbares Fensterdatum bei „3" hinzu. Pünktlich um Mitternacht, also „just in time", bewegte sich die zugehörige, im Zentrum gelagerte Scheibe um eine Position auf den nächsten Tag weiter. Der Name „Datejust" war also keineswegs aus der Luft gegriffen. Im damaligen Katalog schrieb Rolex: „Man kann, ohne zu übertreiben, diese Schöpfung als Synthese der gesamten modernen Uhrenwissenschaft der Schweiz bezeichnen." Rasch entwickelte sich die „Datejust" zu einer chronometrischen Ikone. Als weiterer Beweis für die Exzellenz dieser Armbanduhr kann die ehrenvolle Kür zur „Armbanduhr des 20. Jahrhunderts" in den USA gelten. Natürlich hat Rolex seine bald 80 Jahre junge „Datejust" im Laufe der Jahrzehnte kontinuierlich weiterentwickelt. Und zwar in gewohnter Weise subtil. Demzufolge blieben die unverwechselbare Charakteristik und der damit verknüpfte Wiedererkennungswert stets erhalten.

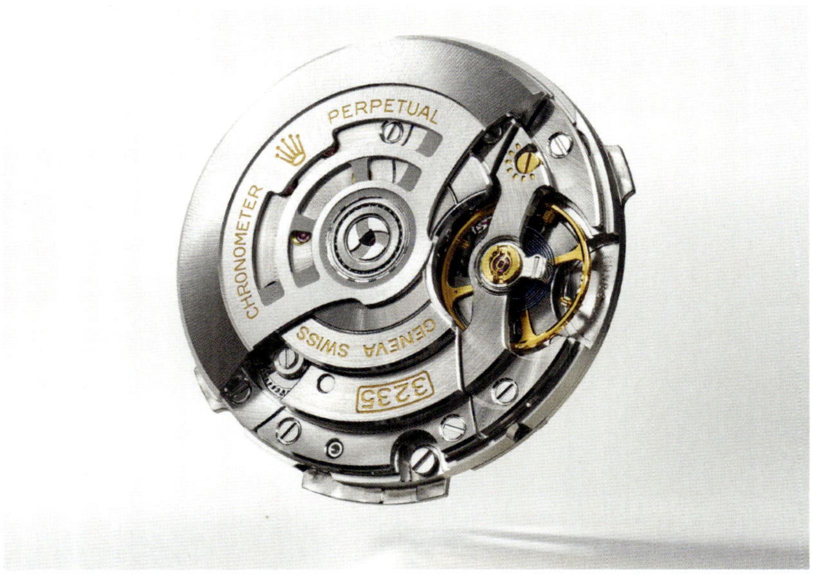

Cal. 3235

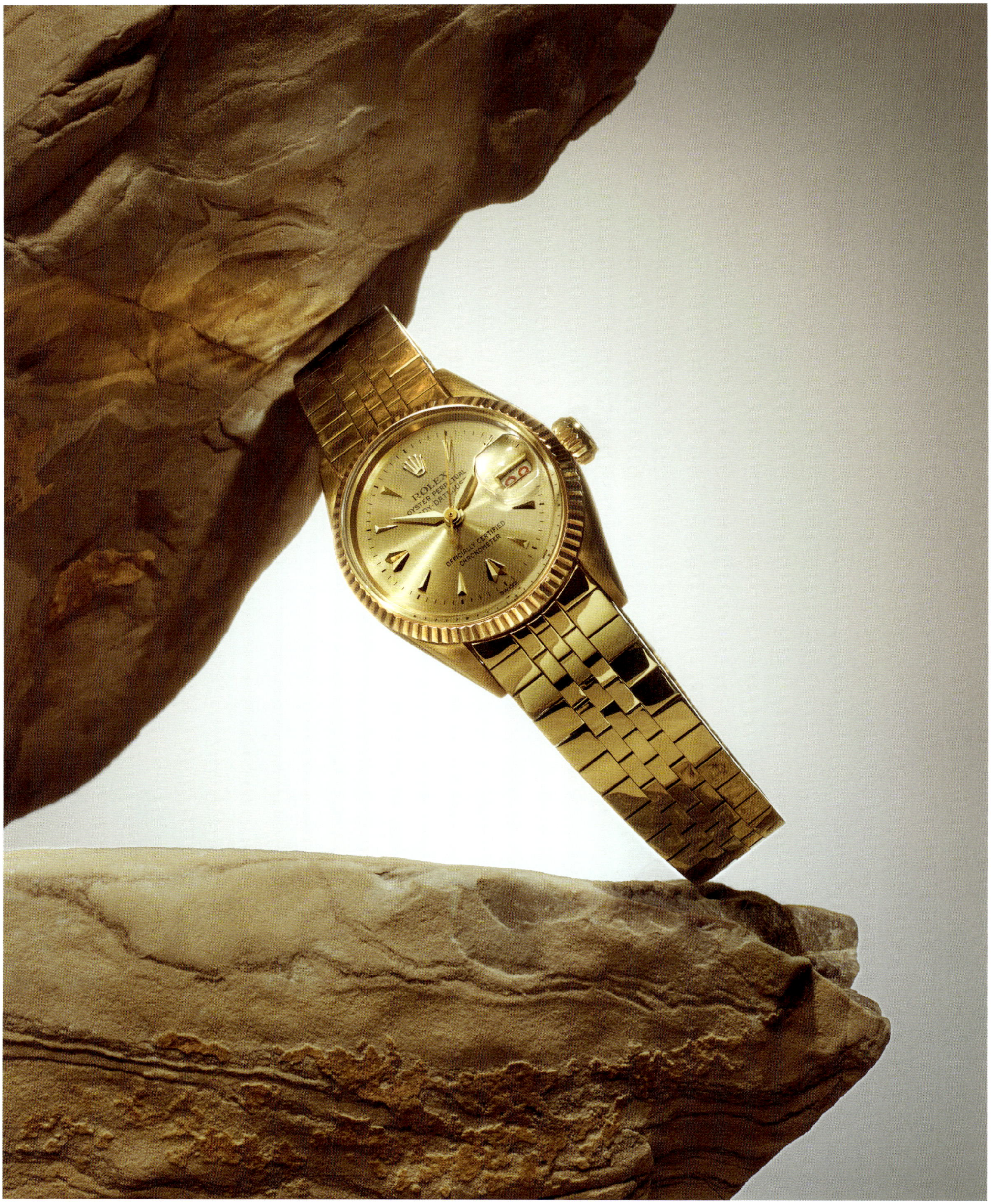
ROLEX
OYSTER PERPETUAL
OFFICIALLY CERTIFIED
CHRONOMETER

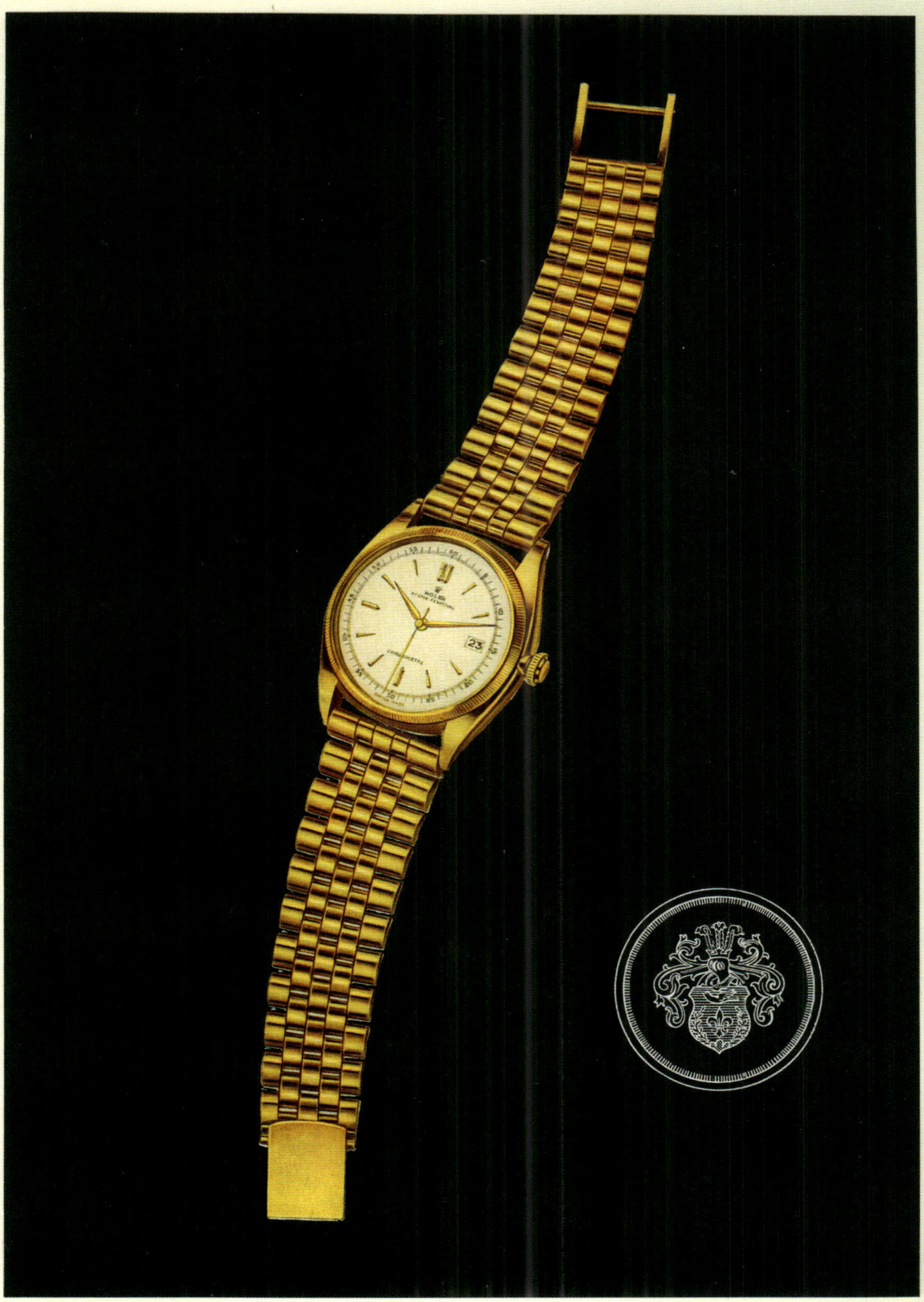

Tafel 25

Ref. 4467 (Jubiläumsmodell ges. gesch.)

Left page: *The first Lady-Datejust, launched in 1957* **|** ***Linke Seite:*** *Die erste Lady-Datejust wurde 1957 lanciert.*
This page **|** ***Diese Seite:*** *Datejust Ref. 4467, 1945*

ROLEX SUBMARINER 1953

MOTHER OF ALL PROFESSIONAL DIVER'S WATCHES

Hans Wilsdorf naturally reserved the honour of presenting the first Rolex diver's watch for himself. Among those in attendance at the 1953 product presentation was René-Paul Jeanneret. A member of the Rolex board of directors, it was he who suggested a hybrid of a waterproof sports and elegant dress watch. Obviously, with a quarter of a century's experience, the manufactory did not lack the relevant expertise. The Oyster launched in 1926 and the systematic evolution of its case for Panerai and daring combat divers from the Italian Navy had left many promising avenues to explore. Quite apart from that, Jeanneret, himself an avid sports diver, was a close friend of the legendary professional diver Jacques-Yves Cousteau. The two friends possessed any amount of know-how to create resilient timepieces that were a match for the silent world. The Submariner – the diver's friend with doubly secured Twinlock crown – was therefore a logical consequence. Prior to the timepiece's first public appearance, Rolex spent the entire year in the Mediterranean scrupulously testing the newcomer's resistance to water. The Institute for Deep Sea Research in Cannes kept a meticulous record of the results. On 26 October 1953, the Submariner notched up 132 successful dives in water depths of between twelve and sixty metres. The Submariner achieved stardom on the wrist of Sean Connery as super-agent 007. Back then, there was no question of paid product placement. Rather, novelist Ian Fleming had made his hero James Bond the proud owner of a Rolex Submariner. In the course of more than 70 years the Rolex Submariner has sustained a series of metamorphoses in respect of water proofness, appearance and safety, for instance with the unidirectional rotating bezel and consistently mechanical inner workings. No matter with or without date window and flank guard, the mother of all professional diver's watches has faithfully preserved its excellent reputation as an underwater icon.

MUTTER ALLER PROFESSIONELLEN TAUCHERUHREN

Natürlich ließ es sich Hans Wilsdorf nicht nehmen, die erste Taucherarmbanduhr des Hauses Rolex höchstpersönlich vorzustellen. Mit von der Partie bei dieser Produktpräsentation im Jahr 1953: René-Paul Jeanneret. Von diesem Mitglied des Rolex-Direktoriums stammte der Vorschlag zu einem Hybrid aus wasserdichter Sport- und dennoch eleganter Alltagsarmbanduhr. An einschlägigen, mehr als 25 Jahre zurückreichenden Erfahrungen mangelte es der Manufaktur bekanntlich nicht. Die 1926 lancierte „Oyster" und die systematische Evolution ihres Gehäuses für Panerai und wagemutige Kampftaucher der italienischen Kriegsmarine hatten viele positive Spuren hinterlassen. Ganz abgesehen davon bestand eine enge Freundschaft zwischen Jeanneret, selbst begeisterter Sporttaucher, und dem legendären Profi Jacques-Yves Cousteau. Das Duo verfügte über jede Menge Knowhow zur Kreation belastbarer Zeitmesser für die „Welt des Schweigens". „Submariner – the diver's friend" mit doppelt gesicherter „Twinlock"-Krone war also nur eine logische Konsequenz. Dass der Newcomer dem nassen Element problemlos zu widerstehen vermochte, hatte Rolex im Jahr vor dem ersten öffentlichen Auftritt im Mittelmeer ausgiebig testen lassen. Dabei führte das Institut für Tiefseeforschung in Cannes akribisch Buch. Am 26. Oktober 1953 konnte die „Submariner" auf 132 erfolgreiche Tauchversuche in Wassertiefen zwischen zwölf und 60 Metern zurückblicken. Filmische Berühmtheit erlangte die „Submariner" am Handgelenk von Sean Connery als Super-Agent 007. Von bezahltem Product-Placement war damals noch keine Rede. Vielmehr hatte Romanautor Ian Fleming seinem Helden James Bond eine Rolex „Submariner" ans Handgelenk geschrieben. Hinsichtlich der Wasserdichte, Optik, sicherheitsrelevanter Funktionalität beispielsweise durch eine unidirektional verstellbare Drehlünette und dem stets mechanischen Innenleben hat die Rolex „Submariner" während mehr als 70 Jahren logischerweise einige Metamorphosen erlebt. Egal ob ohne oder mit Fensterdatum, ohne oder mit Flankenschutz für die Schraubkrone: Ihrem exzellenten Ruf als Unterwasser-Ikone ist die „Mutter aller professionellen Taucheruhren" immer treu geblieben.

1

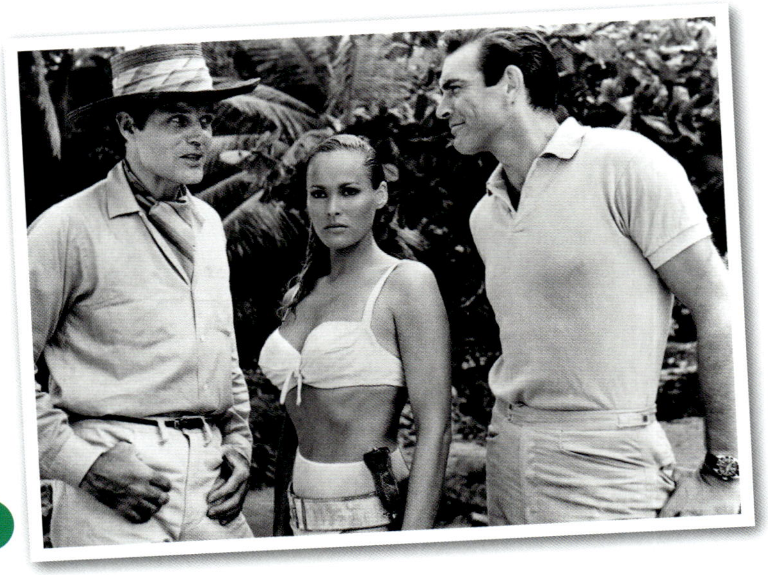

2

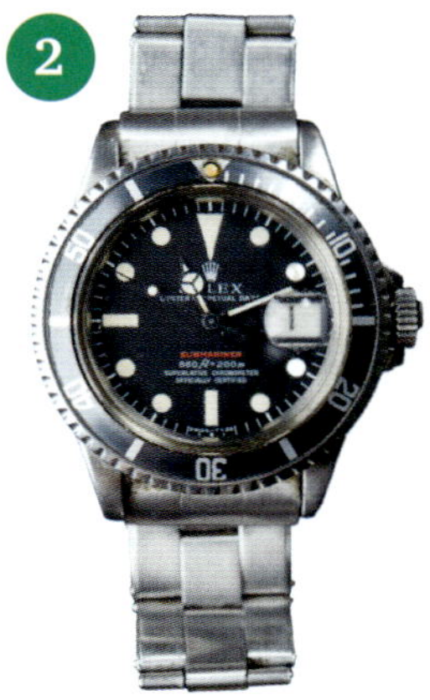

3

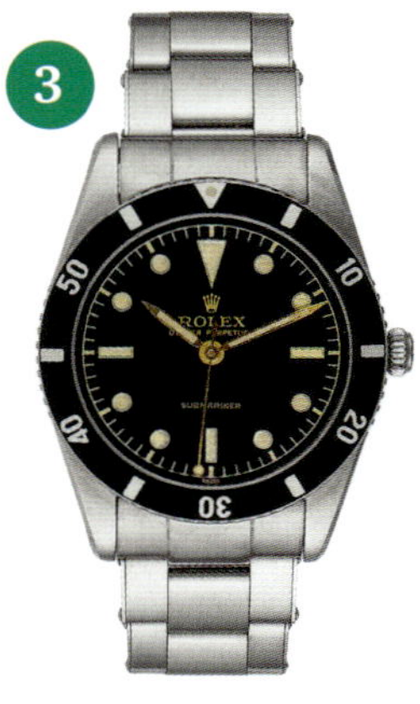

4

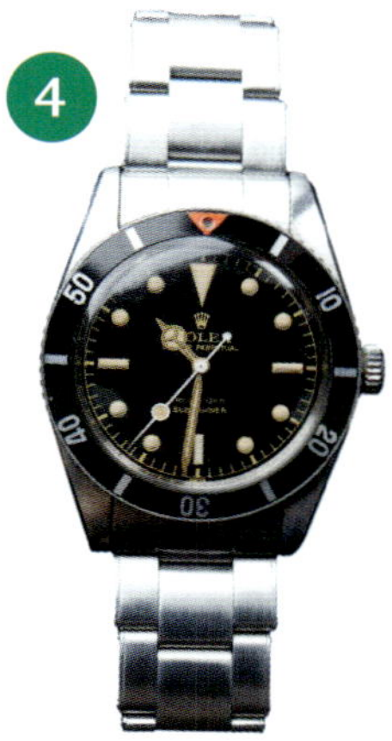

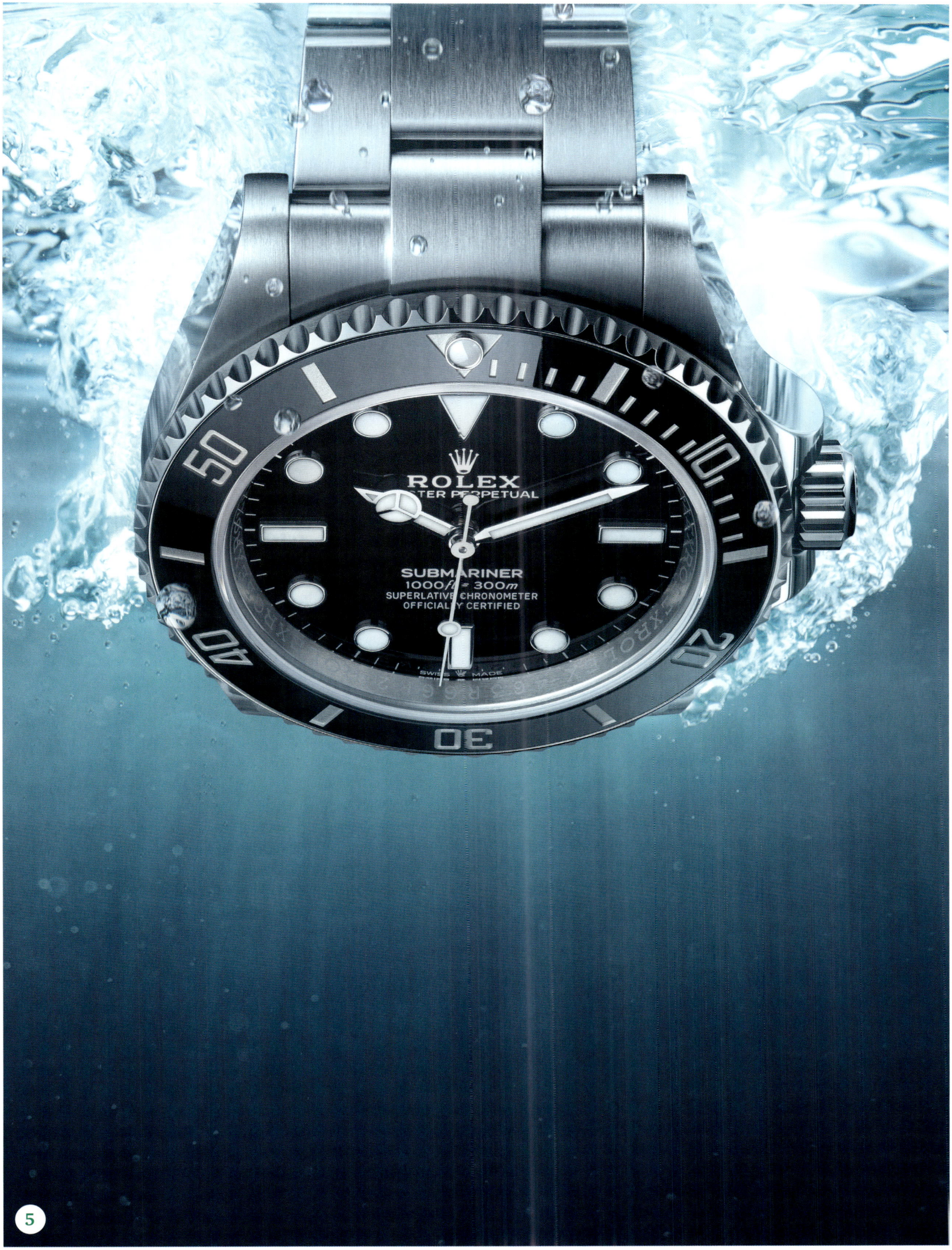
ROLEX
SUBMARINER
1000ft = 300m
SUPERLATIVE CHRONOMETER
OFFICIALLY CERTIFIED

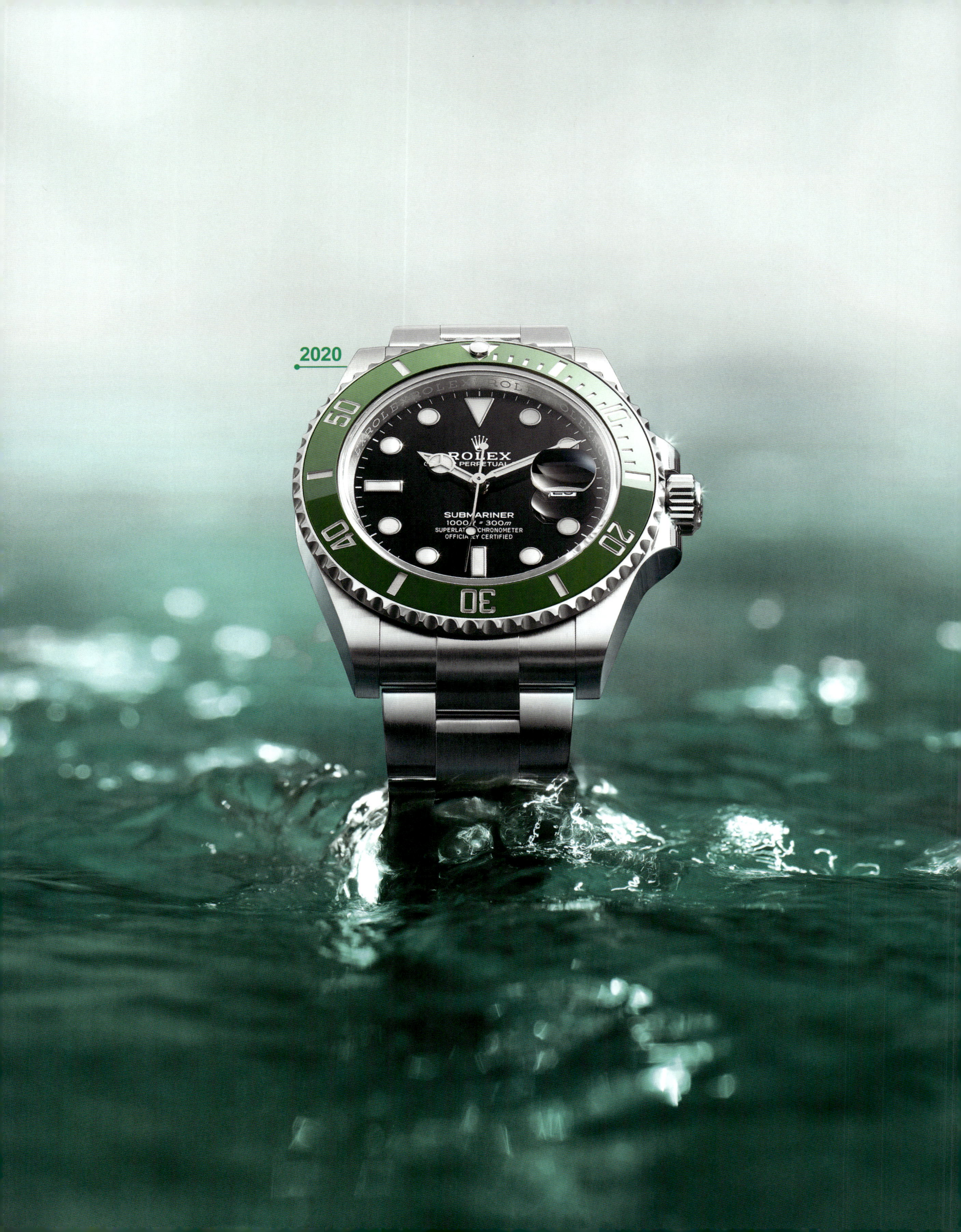
2020
ROLEX
SUBMARINER

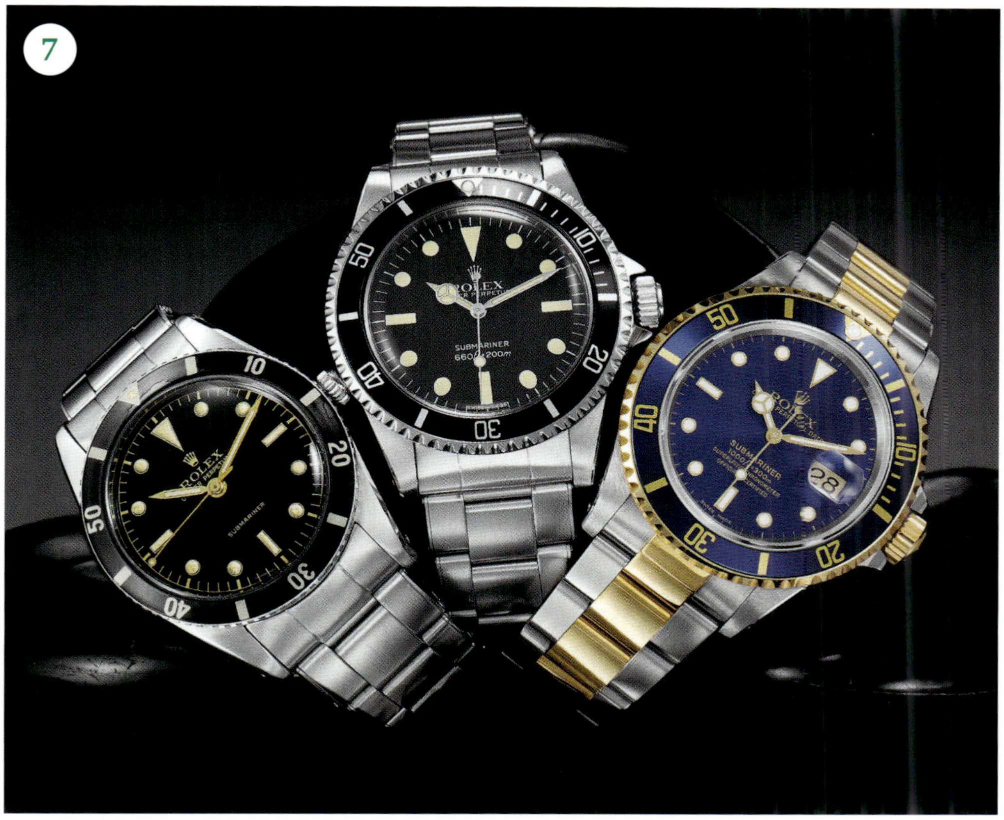

7

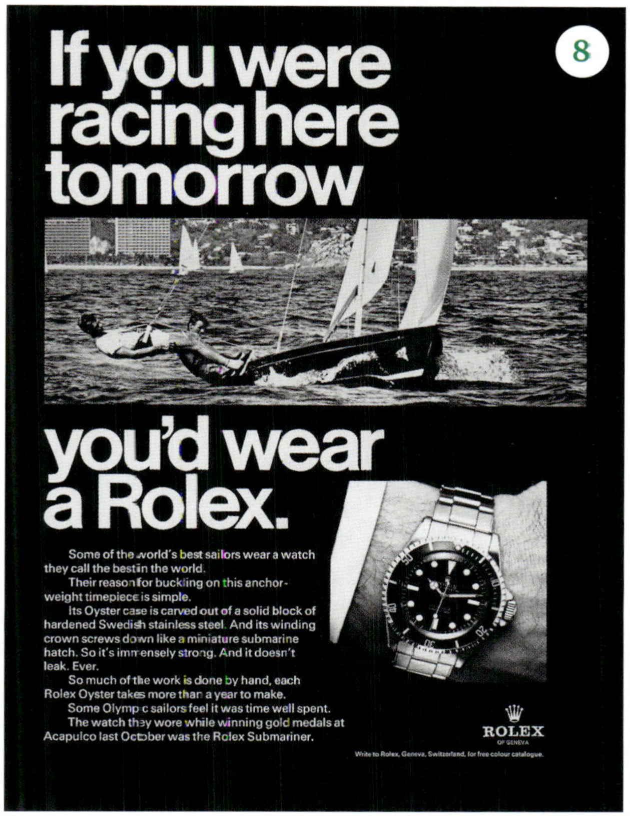

8

1. *Sean Connery as James Bond 007 in ‚Dr. No', 1962*
2. *Ref. 1680*
3. *First Submariner, 1953*
4. *Ref. 6536, 1955*
5. *Ref. 124060, 2020*
6. *Ref. 126610LV, 2020*
7. *Submariners from 1953, 1959, 1983*
8. *Advertisement, 1968*
9. *Ref. 126613LB, 2020*
10. *German brochure for the Submariner, 1953*

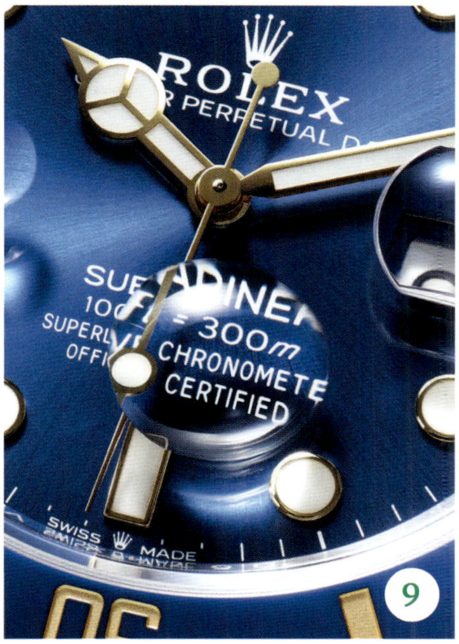

9

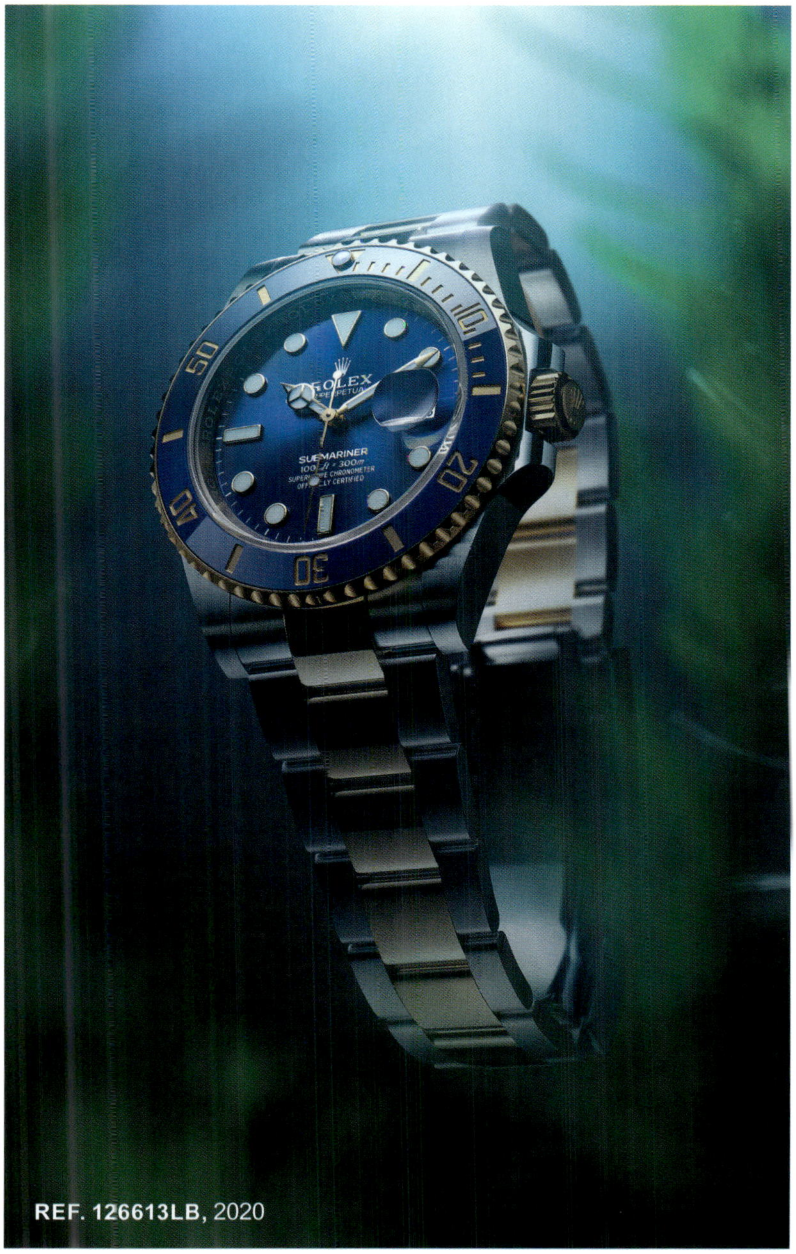

REF. 126613LB, 2020

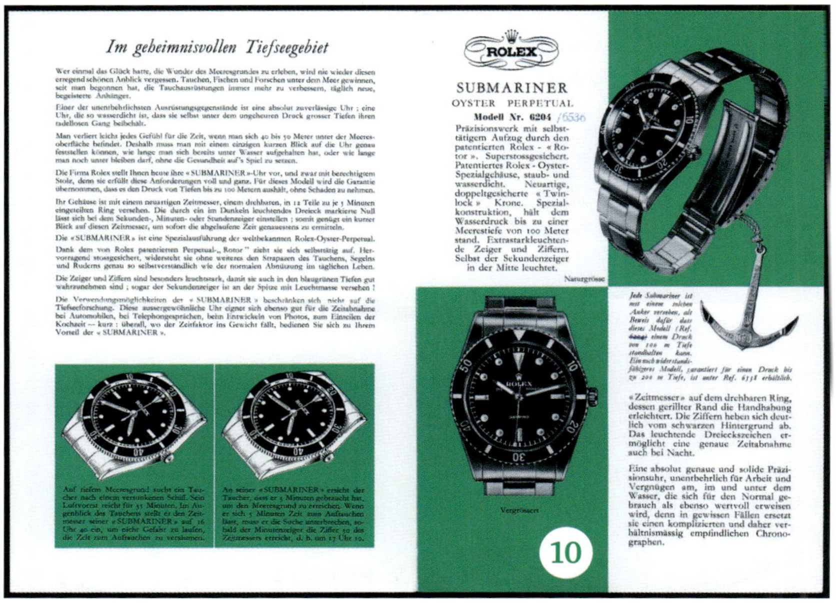

Im geheimnisvollen Tiefseegebiet

Wer einmal das Glück hatte, die Wunder des Meeresgrundes zu erleben, wird nie wieder diesen erregend schönen Anblick vergessen. Tauchen, Fischen und Forschen unter dem Meer gewinnen, seit man begonnen hat, die Tauchausrüstungen immer mehr zu verbessern, täglich neue, begeisterte Anhänger.

Einer der unentbehrlichsten Ausrüstungsgegenstände ist eine absolut zuverlässige Uhr ; eine Uhr, die so wasserdicht ist, dass sie selbst unter dem ungeheuren Druck grosser Tiefen ihren tadellosen Gang beibehält.

Man verliert leicht jedes Gefühl für die Zeit, wenn man sich 40 bis 50 Meter unter der Meeresoberfläche befindet. Deshalb muss man mit einem einzigen kurzen Blick auf die Uhr genau feststellen können, wie lange man sich bereits unter Wasser aufgehalten hat, oder wie lange man noch unter bleiben darf, ohne die Gesundheit auf's Spiel zu setzen.

Die Firma Rolex stellt Ihnen heute ihre « SUBMARINER »-Uhr vor, und zwar mit berechtigtem Stolz, denn sie erfüllt diese Anforderungen voll und ganz. Für dieses Modell wird die Garantie übernommen, dass es den Druck von Tiefen bis zu 100 Metern aushält, ohne Schaden zu nehmen.

Ihr Gehäuse ist mit einem neuartigen Zeitmesser, einem drehbaren, in 12 Teile zu je 5 Minuten eingeteilten Ring versehen. Die durch ein im Dunkeln leuchtendes Dreieck markierte Null lässt sich bei dem Sekunden-, Minuten- oder Stundenzeiger einstellen ; somit genügt ein kurzer Blick auf diesen Zeitmesser, um sofort die abgelaufene Zeit genauestens zu ermitteln.

Die « SUBMARINER » ist eine Spezialausführung der weltbekannten Rolex-Oyster-Perpetual.

Dank dem von Rolex patentierten Perpetual-„Rotor" zieht sie sich selbsttätig auf. Hervorragend stossgesichert, widersteht sie ohne weiteres den Strapazen des Tauchens, Segelns und Ruderns genau so selbstverständlich wie der normalen Abnützung im täglichen Leben.

Die Zeiger und Ziffern sind besonders leuchtstark, damit sie auch in den blaugrünen Tiefen gut wahrzunehmen sind ; sogar der Sekundenzeiger ist an der Spitze mit Leuchtmasse versehen !

Die Verwendungsmöglichkeiten der « SUBMARINER » beschränken sich nicht auf die Tiefseeforschung. Diese aussergewöhnliche Uhr eignet sich ebenso gut für die Zeitabnahme bei Automobilen, bei Telephongesprächen, beim Entwickeln von Photos, zum Einteilen der Kochzeit — kurz : überall, wo der Zeitfaktor ins Gewicht fällt, bedienen Sie sich zu Ihrem Vorteil der « SUBMARINER ».

Auf tiefem Meeresgrund sucht ein Taucher nach einem versunkenen Schiff. Sein Luftvorrat reicht für 55 Minuten. Im Augenblick des Tauchens stellt er den Zeitmesser seiner « SUBMARINER » auf 16 Uhr 40 ein, um nicht Gefahr zu laufen, die Zeit zum Auftauchen zu versäumen.

An seiner « SUBMARINER » ersieht der Taucher, dass er 3 Minuten gebraucht hat, um den Meeresgrund zu erreichen. Wenn er sich 5 Minuten Zeit zum Auftauchen lässt, muss er die Suche unterbrechen, sobald der Minutenzeiger die Ziffer 30 des Zeitmessers erreicht, d. h. um 17 Uhr 10.

ROLEX

SUBMARINER

OYSTER PERPETUAL

Modell Nr. 6204 /6536

Präzisionswerk mit selbsttätigem Aufzug durch den patentierten Rolex - « Rotor ». Superstossgesichert. Patentiertes Rolex - Oyster-Spezialgehäuse, staub- und wasserdicht. Neuartige, doppeltgesicherte « Twinlock » Krone. Spezialkonstruktion, hält dem Wasserdruck bis zu einer Meerestiefe von 100 Meter stand. Extrastarkleuchtende Zeiger und Ziffern. Selbst der Sekundenzeiger in der Mitte leuchtet.

Naturgrösse

Jede Submariner ist mit einem solchen Anker versehen, als Beweis dafür dass dieses Modell (Ref. 6204) einem Druck von 100 m Tiefe standhalten kann. Ein noch widerstandsfähigeres Modell, garantiert für einen Druck bis zu 200 m Tiefe, ist unter Ref. 6538 erhältlich.

Vergrössert

« Zeitmesser » auf dem drehbaren Ring, dessen gerillter Rand die Handhabung erleichtert. Die Ziffern heben sich deutlich vom schwarzen Hintergrund ab. Das leuchtende Dreieckszeichen ermöglicht eine genaue Zeitabnahme auch bei Nacht.

Eine absolut genaue und solide Präzisionsuhr, unentbehrlich für Arbeit und Vergnügen am, im und unter dem Wasser, die sich für den Normal gebrauch als ebenso wertvoll erweisen wird, denn in gewissen Fällen ersetzt sie einen komplizierten und daher verhältnismässig empfindlichen Chronographen.

10

ROLEX 1955/1982

GMT-MASTER & GMT-MASTER II

COSMOPOLITAN TRAVEL MATE

In 1953, Pan American Airlines approached Rolex with a request for a precise watch to indicate the time in two different time zones, namely local time and the time back at home. Hans Wilsdorf lent the request a ready ear. The managing director jotted down his initial ideas for the timepiece while staying at the luxury Ritz hotel in Paris. The letter to his Geneva staff formed the basis for the development of the GMT Master presented in 1954 and launched the following year as Reference 6542. A striking characteristic of the cosmopolitan timepiece was an additional 24-hour hand. It corresponded to a glass bezel for setting and indicating the second-zone time. Initially, the rotating bezel with a back-printed acrylic glass inlay was made of coated plastic. However, it soon transpired that this soft material was barely able to withstand high loads and temperature fluctuations. A more resilient metal as a substitute was quickly found. The impressed Pan Am managers elected the new GMT Master as the airline's official timepiece. Aircraft crews were issued with a black-dial version, ground staff with white-dial models. At the launch of the second generation in 1960, Rolex announced that 20 out of 21 aircraft navigators considered the professional qualities of the GMT Master to be particularly useful. The distinguishing feature of the Reference 1675 was the screw crown guard. From 1982, the GMT Master II, Reference 16760, cut a much more practical figure. On the Fat Lady equipped with automatic Calibre 3085, the 12-hour hand can be moved forward or backward as desired when the crown is extended by half. A date window was soon to follow. Until 1988, both the GMT Master I and GMT Master II were available from Rolex. Since then, only the GMT Master II has been on sale in different versions with steel or precious-metal cases. In 2013, Rolex introduced the scratch-resistant Cerachrom bezel.
The Rolex GMT Master has earned itself all sorts of nicknames. "Coke" refers to the black and red glass bezel, "Pepsi" for blue and red and "Batman" for a blue and black version. It comes as no surprise that in private, the American Apollo astronauts entrusted their precious time to a GMT Master. John Leonard Swigert, reserve pilot of the Apollo 13 command capsule, took his with him in his pocket. In 1970, on his happy return to earth, he thanked Rolex for his GMT Master enabling him "to always be on time".

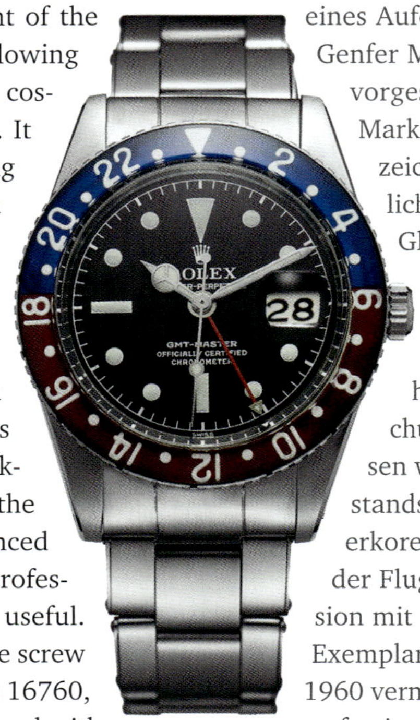

Top | Oben: *First GMT-Master, 1955*
Right page: *Advertisements from the 1950s and 1960s |*
Bottom left: *Photo of Stuart A. Roosa's GMT-Master taken during the Apollo 14 mission, accompanied by a message from the astronaut*
Rechte Seite: *Werbeanzeigen aus den 1950er- und 1960er-Jahren |*
Unten links: *Foto der GMT-Master von Stuart A. Roosa, aufgenommen während der Apollo-14-Mission, inklusive einer Nachricht des Astronauten*

KOSMOPOLIT FÜRS HANDGELENK

1953 schrieb man, als die Pan American Airlines von Rolex eine genau gehende Armbanduhr zum simultanen Ablesen zweier Zonenzeiten begehrte. Anzeigen sollte sie die jeweilige Lokal- und dazu die Heimatzeit. Bei Hans Wilsdorf fiel das Anliegen auf fruchtbaren Boden. Erste Ideen zur Realisation skizzierte der Rolex-Generaldirektor während eines Aufenthalts im Pariser Luxushotel Ritz. Sein Brief an die Genfer Mitarbeiter diente als Basis zur Entwicklung der 1954 vorgestellten und im Folgejahr als Referenz 6542 auf den Markt gebrachten „GMT-Master". Signifikantes Erkennungszeichen des kosmopolitischen Zeitmessers war ein zusätzlicher 24-Stunden-Zeiger. Er korrespondierte mit einem Glasrand zum Ein- und Darstellen der zweiten Zonenzeit. Anfangs bestand die gesamte Drehlünette mit rückwärtig bedrucktem Acrylglas-Inlay aus beschichtetem Kunststoff. Schnell kristallisierte sich jedoch heraus, dass dieses weiche Material hohen Beanspruchungen und Temperaturschwankungen kaum gewachsen war. Rasch folgte daher ein Wechsel zu deutlich widerstandsfähigerem Metall. Die begeisterten Pan-Am-Manager erkoren die neue „GMT-Master" zum offiziellen Zeitmesser der Fluggesellschaft. Flugzeugbesatzungen erhielten die Version mit schwarzem Zifferblatt, das Bodenpersonal blickte auf Exemplare in Weiß. Beim Start der zweiten Generation im Jahr 1960 vermeldete Rolex, dass 20 von 21 Flugzeugnavigatoren die professionellen Eigenschaften der „GMT-Master" als besonders hilfreich betrachteten. Als augenfälliges Erkennungsmerkmal der Referenz 1675 präsentierte sich ein Flankenschutz für die Schraubkrone. Deutlich funktionaler trat ab 1982 die „GMT-Master II" mit der Referenz 16760 in Erscheinung. Bei der mit dem Automatikkaliber 3085 ausgestatteten „Fat Lady" ließ sich der 12-Stunden-Zeiger per halb gezogener Krone beliebig vor- oder rückwärts bewegen. Das Fensterdatum folgte auf dem Fuß. Bis 1988 offerierte Rolex die „GMT-Master I" und die „GMT-Master II" nebeneinander.
Seitdem gibt es nur noch die „GMT-Master II" in unterschiedlichen Ausführungen mit Stahl- oder Edelmetall-Gehäuse. 2013 wurde die kratzfeste „Cerachrom"-Lünette eingeführt. Mit der Rolex „GMT-Master" verknüpfen sich allerlei Spitznamen. „Coke" steht für einen schwarz-roten Glasrand, „Pepsi" für das blau-rote und „Batman" für das blau-schwarze Pendant. Nicht verwundern dürfte die Tatsache, dass amerikanische „Apollo"-Astronauten ihre kostbare Zeit ganz privat einer „GMT-Master" anvertrauten. Auch John Leonard Swigert, Ersatz-Pilot der „Apollo-13"-Kommandokapsel, nahm sein Exemplar in der Hosentasche mit. Nach der glücklichen Rückkehr zur Erde dankte er Rolex im Jahr 1970 dafür, dass ihn seine GMT-Master „immer pünktlich sein" ließ.

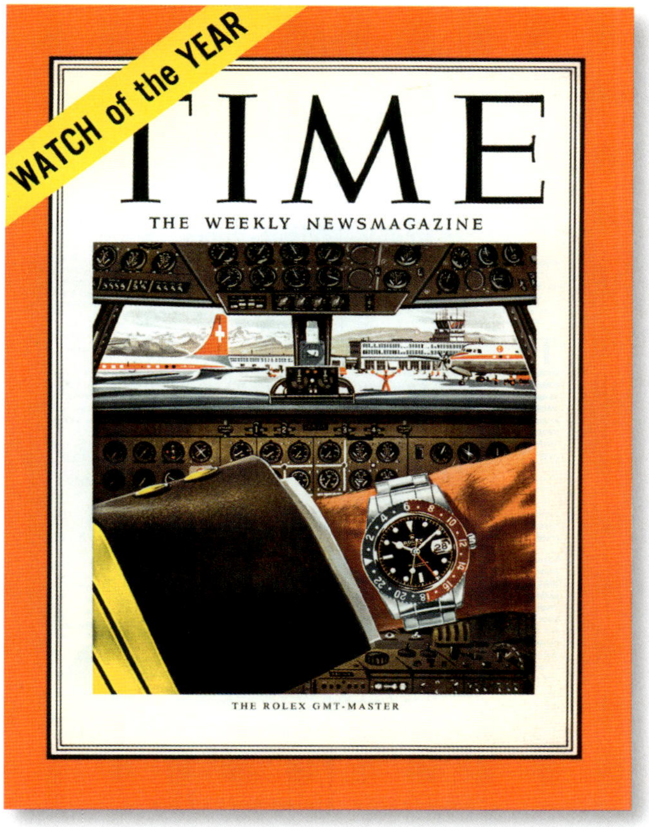
WATCH of the YEAR
TIME
THE WEEKLY NEWSMAGAZINE
THE ROLEX GMT-MASTER

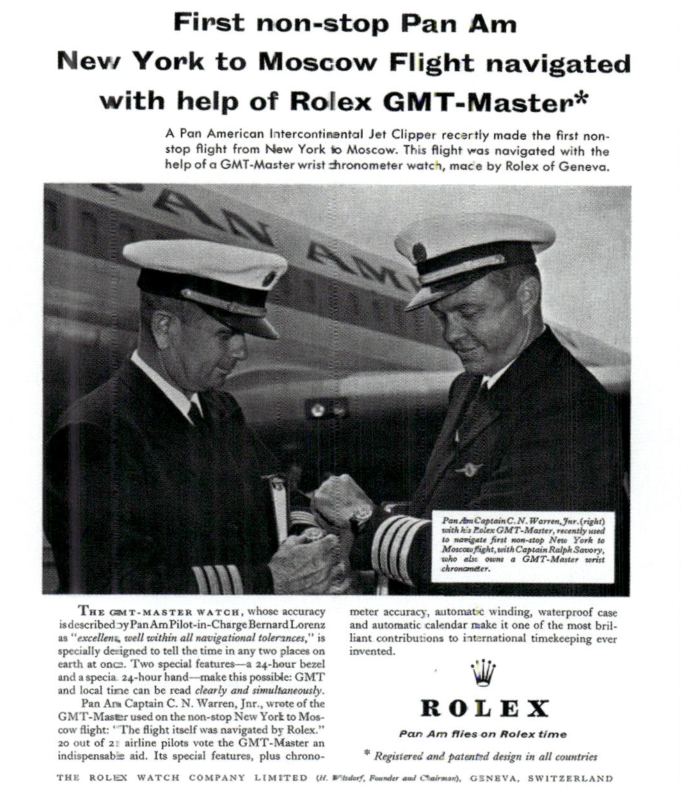
First non-stop Pan Am
New York to Moscow Flight navigated
with help of Rolex GMT-Master*
A Pan American Intercontinental Jet Clipper recently made the first non-stop flight from New York to Moscow. This flight was navigated with the help of a GMT-Master wrist chronometer watch, made by Rolex of Geneva.
Pan Am Captain C. N. Warren, Jnr. (right) with his Rolex GMT-Master, recently used to navigate first non-stop New York to Moscow flight, with Captain Ralph Savory, who also owns a GMT-Master wrist chronometer.
THE GMT-MASTER WATCH, whose accuracy is described by Pan Am Pilot-in-Charge Bernard Lorenz as "excellent, well within all navigational tolerances," is specially designed to tell the time in any two places on earth at once. Two special features—a 24-hour bezel and a special 24-hour hand—make this possible: GMT and local time can be read clearly and simultaneously.
Pan Am Captain C. N. Warren, Jnr., wrote of the GMT-Master used on the non-stop New York to Moscow flight: "The flight itself was navigated by Rolex." 20 out of 21 airline pilots vote the GMT-Master an indispensable aid. Its special features, plus chronometer accuracy, automatic winding, waterproof case and automatic calendar make it one of the most brilliant contributions to international timekeeping ever invented.
ROLEX
Pan Am flies on Rolex time
* Registered and patented design in all countries
THE ROLEX WATCH COMPANY LIMITED (H. Wilsdorf, Founder and Chairman), GENEVA, SWITZERLAND

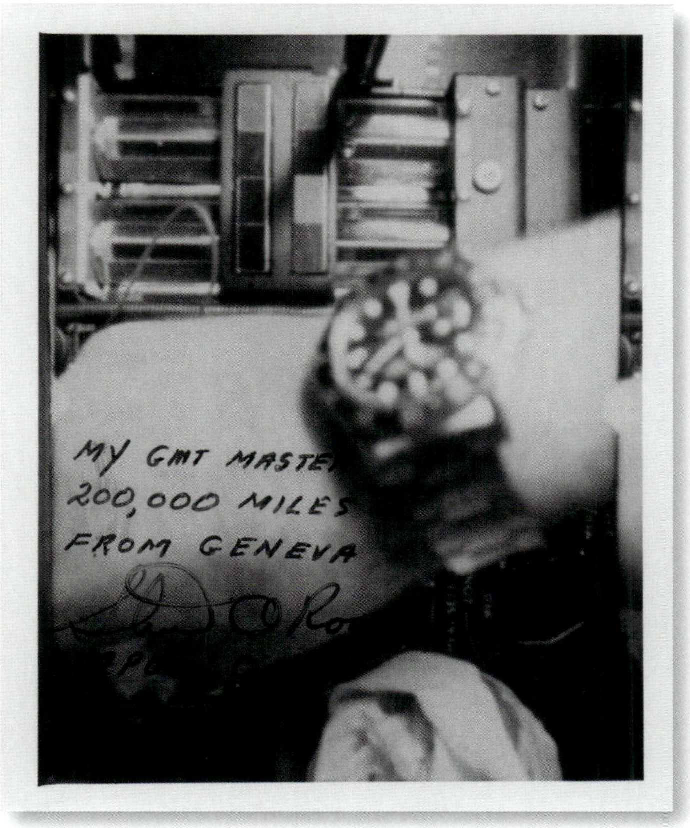
MY GMT MASTER
200,000 MILES
FROM GENEVA

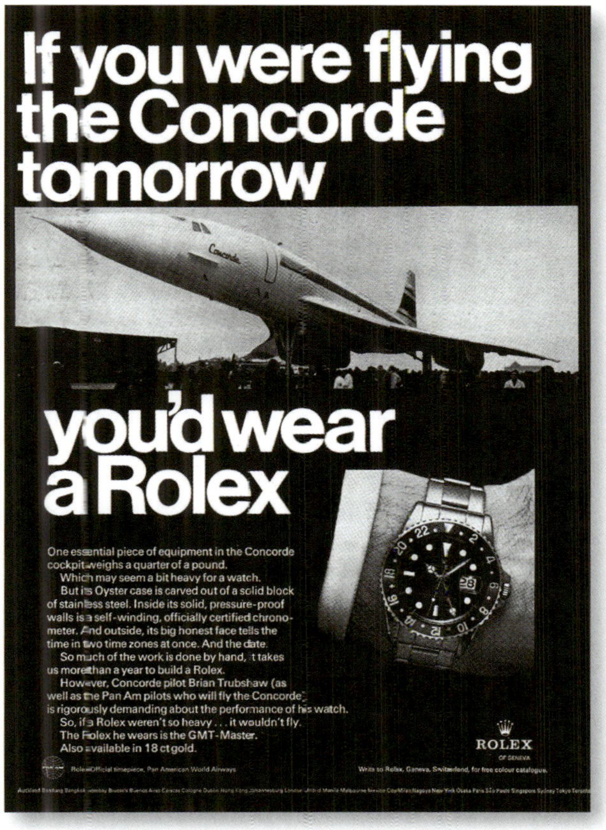
If you were flying the Concorde tomorrow
you'd wear a Rolex
One essential piece of equipment in the Concorde cockpit weighs a quarter of a pound.
Which may seem a bit heavy for a watch.
But its Oyster case is carved out of a solid block of stainless steel. Inside its solid, pressure-proof walls is a self-winding, officially certified chronometer. And outside, its big honest face tells the time in two time zones at once. And the date.
So much of the work is done by hand, it takes us more than a year to build a Rolex.
However, Concorde pilot Brian Trubshaw (as well as the Pan Am pilots who will fly the Concorde) is rigorously demanding about the performance of his watch.
So, if a Rolex weren't so heavy . . . it wouldn't fly.
The Rolex he wears is the GMT-Master.
Also available in 18 ct gold.
ROLEX
OF GENEVA
Rolex Official timepiece, Pan American World Airways
Write to Rolex, Geneva, Switzerland, for free colour catalogue.

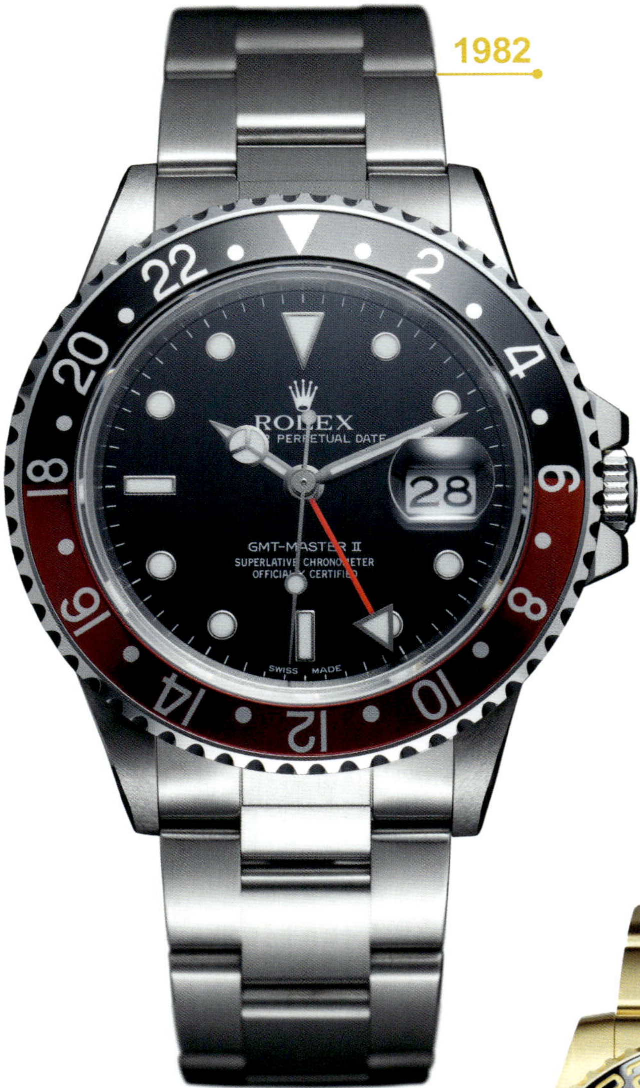

1982

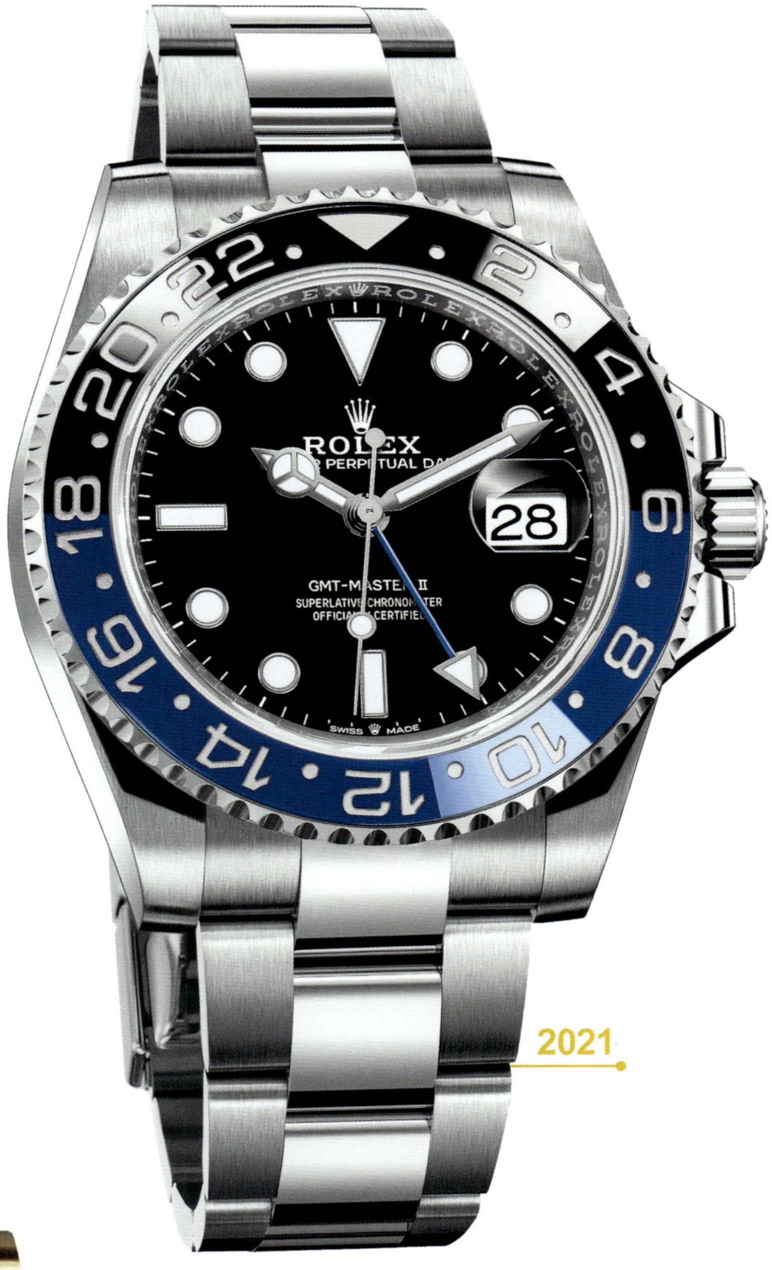

2021

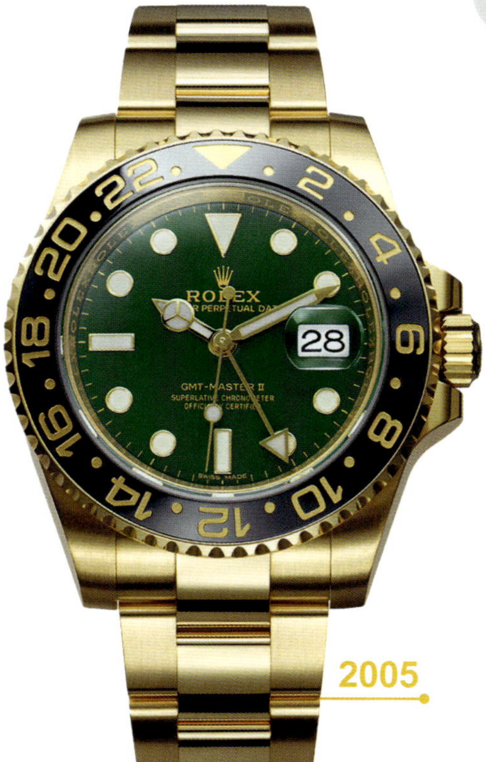

2005

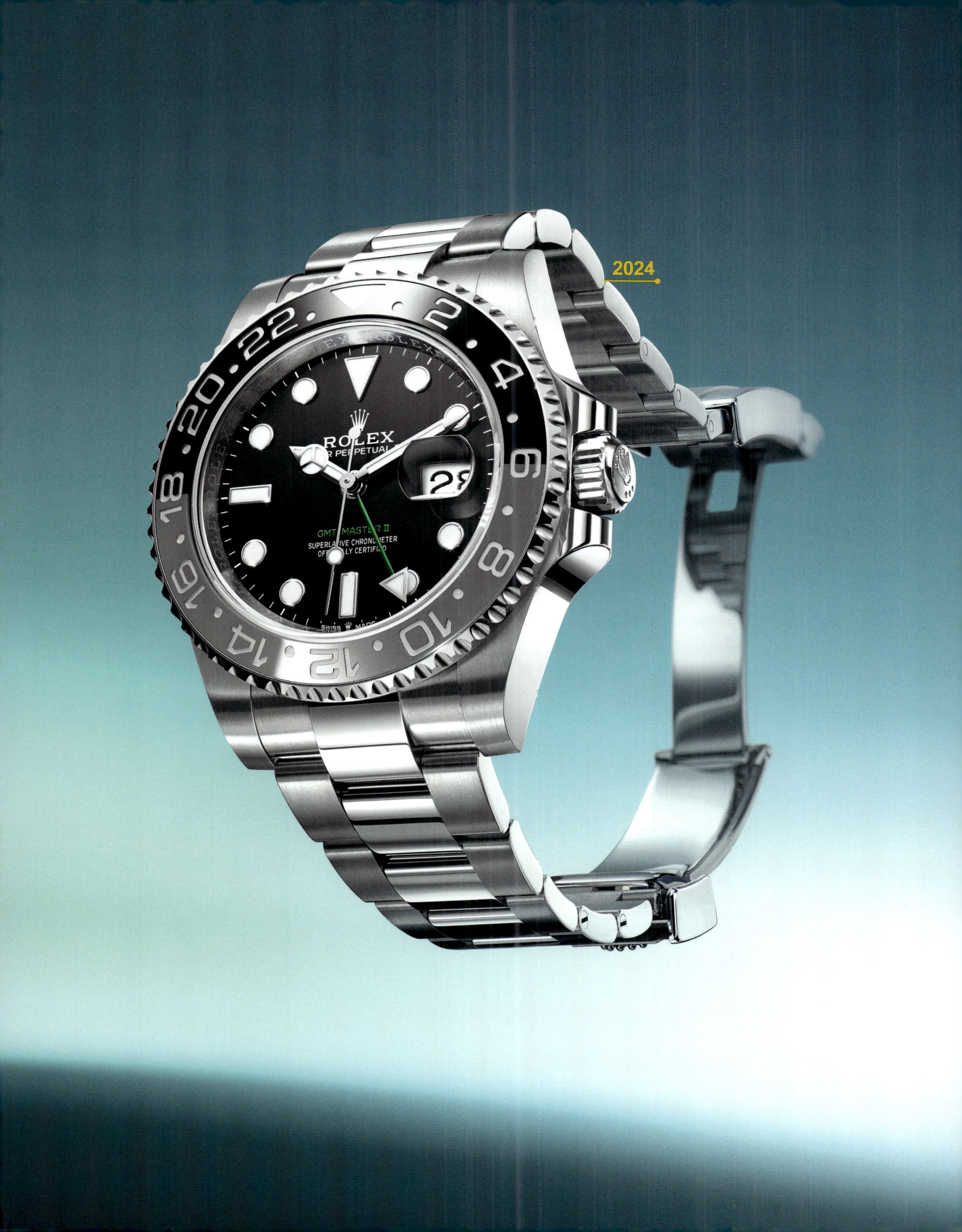
2024
ROLEX

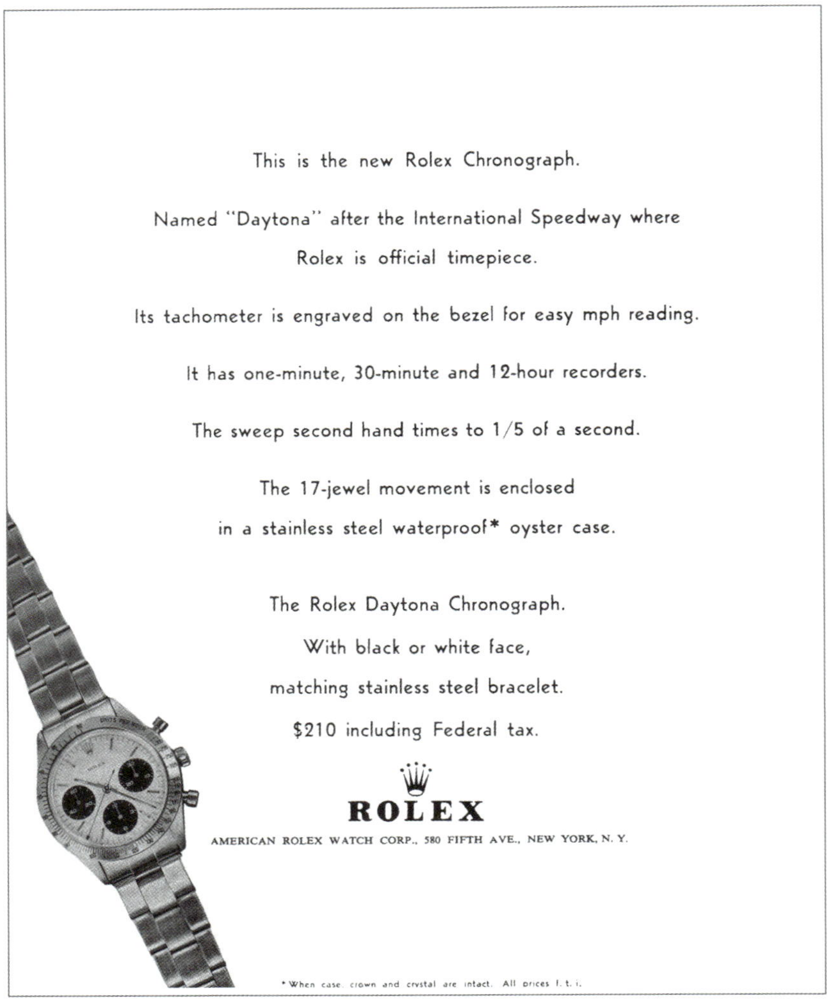

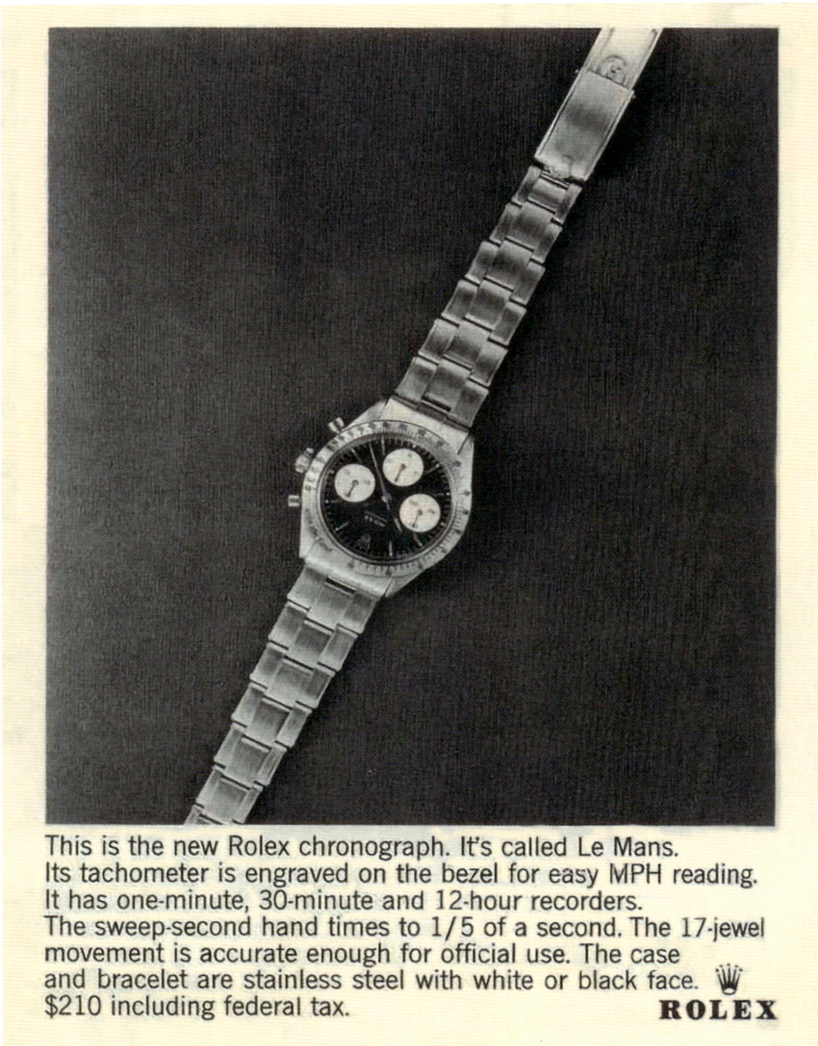

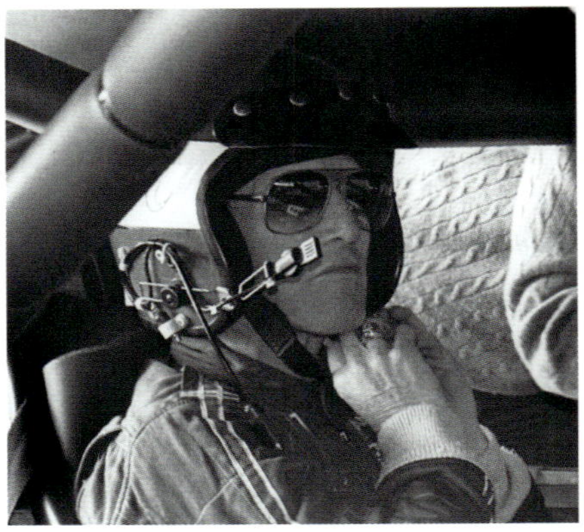

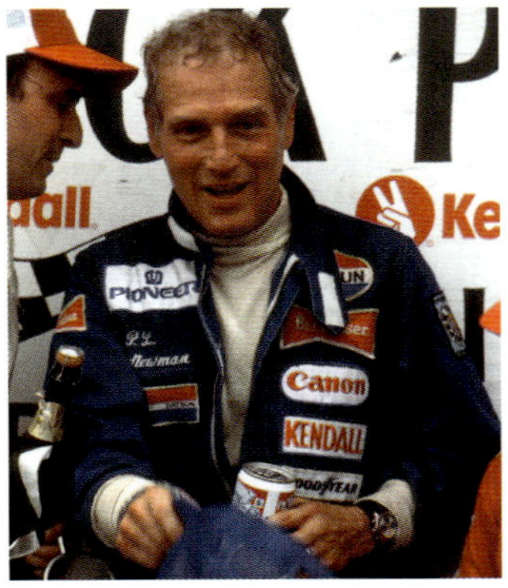

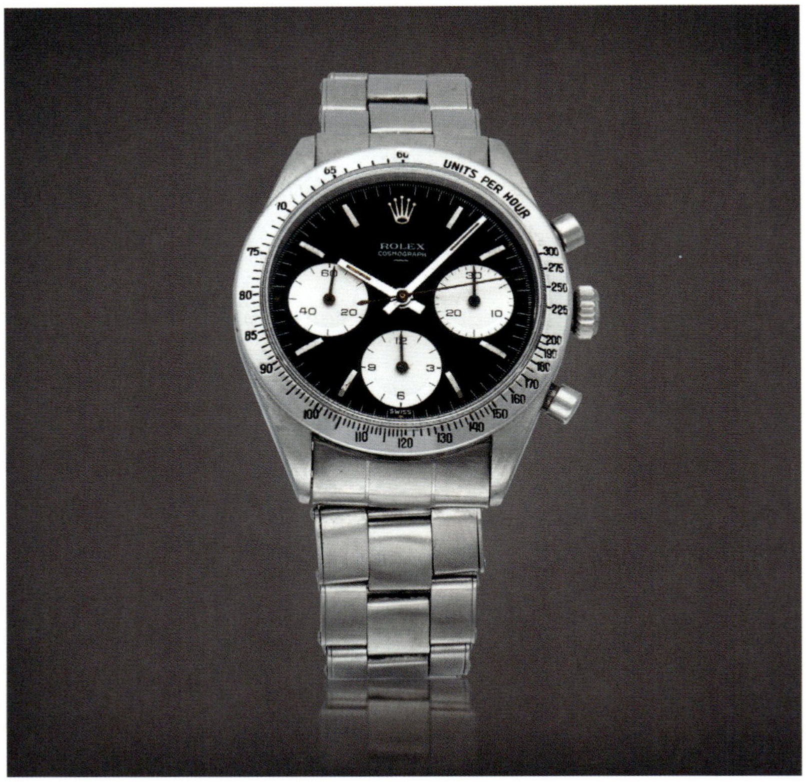

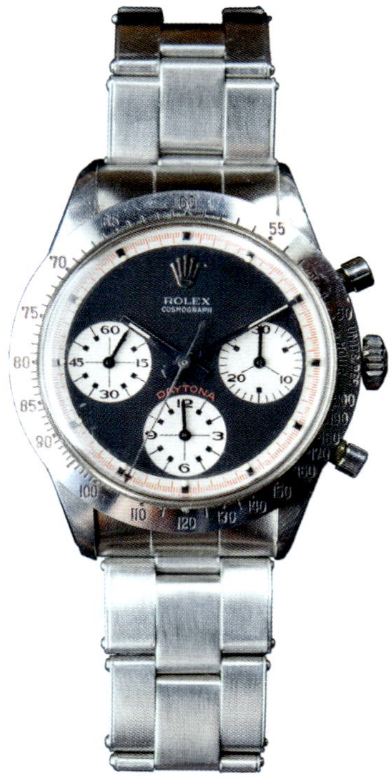

Left, from top | Links von oben: *Advertisement for the Daytona chronograph, 1964 | Advertisement for the "Le Mans" chronograph, 1963*

Right, from top | Rechts von oben: *Paul Newman | Daytona "Le Mans", 1963 | 1966 Rolex Daytona, Ref. 6239 "Paul Newman"*

COSMOGRAPH DAYTONA

TRADITION IN TIME TRACKING

Rolex published the first ads featuring the four chronograph References 2022, 2023, 2303 and 2508 in 1936. Thereafter, the brand routinely advertised wristwatches with an integrated chronograph. All the movements were engineered by the specialist Valjoux based in the Joux Valley. In the late 1940s, Rolex presented a more modern generation. The cases featured a screw-down crown and caseback and round pushers. References 6034 and 6234, which were produced until the 1960s, outstripped the others in importance. Predecessor of the iconic Dayton is deemed to be the Reference 6238 with a smooth glass bezel. In 1963, the innovative Reference 6239, nicknamed Le Mans, reached the starting blocks. This bracelet chronograph owed its striking appearance to the contrasting colours on the dial and the tachymeter scale, now relocated on the bezel. In 1964, the model was renamed Daytona in US newspaper advertisements. The name change was due to the importance of the overseas market and the hugely popular Florida circuit. In 1965, screw-on pushers were added to the new winding and hand-setting crown on the Reference 6240. Because the 37-millimetre case was widely deemed too large for many and the black-and-white look failed to convince, its career ended in 1968. A redesign was launched in September 1971. Produced until 1978, References 6263 and 6265 sported stainless-steel-engraved and black acrylic glass bezels. The Oyster case guaranteed waterproofness of up to ten bar. Particularly coveted, but also frequently manipulated, are versions with the Paul Newman dial, recognised by the hash marks with a small square on the end. With no proprietary movements to call its own, in 1986 an automatic Daytona, powered by the Zenith El Primero, was launched. Because the standard version of the Calibre 400 with a five-hertz balance frequency underwhelmed, the Rolex engineers stripped it down and re-engineered it. Completed in 1988, the Calibre 4030 comprised some 50 per cent new or modified components. The date window was dispensed with and the balance speed was reduced to just 28,800 vph. In 2000, Rolex remedied the significant gap in its calibre portfolio with its automatic Calibre 4130. The first patent applications for it are dated 31 October 1995. Column wheel, vertical clutch and totalisers shifted slightly towards 12 o'clock are just a few of the features of this flat 6.5 mm movement. From the first, there were waiting lists for the steel Reference 116520. The same goes for its counterpart model 116500LN with black Cerachrom bezel, launched in 2016. In 2020, marking its sixtieth anniversary, Rolex equipped the Daytona with the evolved automatic Calibre 4131. The rotor now has more ball bearings, 22 in all. The more efficient escapement is called Chronergy; the shock absorbers go by the name of Paraflex. For the first time in the history of the Oyster, the platinum Reference 126506 is fitted with a sapphire crystal display caseback. Celebrating the centenary of the 24 Hours of Le Mans, the white-gold Reference 126529LN features a 24-hour counter at 9 o'clock in addition to the red 100 on the tachymeter scale.

ZEIT ZUM ANHALTEN

Erste Anzeigen mit den vier Chronographen-Referenzen 2022, 2023, 2303 und 2508 veröffentlichte Rolex gegen 1936. Alle verbauten Werke stammten vom Spezialisten Valjoux. In den späten 1940er-Jahren präsentierte Rolex eine modernere Generation. Deren Gehäuse verfügten über Schraubkrone und -boden sowie runde Drücker. Deutlich wichtiger waren jedoch die bis in die 1960er-Jahre hergestellten Referenzen 6034 und 6234. Als Vorläuferin der Ikone namens „Daytona" gilt die Referenz 6238 mit glattem Glasrand. 1963 ging die innovative Referenz 6239 mit dem Beinamen „Le Mans" an den Start. Dieser Armband-Chronograph stach durch die Verwendung von Kontrastfarben am Zifferblatt und der zur Lünette verlagerten Tachymeterskala ins Auge. 1964 wechselte der Modellname in US-amerikanischen Zeitungsanzeigen zu „Daytona". Gründe waren die Bedeutung des überseeischen Markts und der immens populäre Rundkurs in Florida. 1965 gesellten sich bei der Referenz 6240 zur neuen Aufzugs- und Zeigerstellkrone auch verschraubbare Drücker. Weil die 37-Millimeter-Schale vielen zu groß erschien und die Schwarzweiß-Optik auch nicht restlos überzeugte, endete die Karriere bereits 1968. Ein weiteres Facelifting brachte das Jahr 1971: Ab September gelangten die bis 1978 produzierten Referenzen 6263 und 6265 mit Acrylglas-Lünette bzw. graviertem und schwarz ausgelegtem Glasrand zum Verkauf. Besonders begehrt, aber auch häufig manipuliert sind die Versionen mit „Paul Newman"-Zifferblatt, erkennbar an kleinen Quadraten. In Ermangelung eigener Mechanik führte der Weg zur Automatik-„Daytona" 1986 zu Zenith und dem „El Primero". Weil das dortige Kaliber 400 in der Standardausführung mit fünf Hertz Unruhfrequenz nicht überzeugte, überarbeiteten es die Rolex-Techniker bis 1988 gründlich. Heraus kam das Kaliber 4030 mit etwa 50 Prozent neuen oder modifizierten Komponenten. Das Fensterdatum fiel weg und die Unruh vollzog in der Stunde nur noch 28.800 Halbschwingungen. Die signifikante Lücke im eigenen Kaliber-Portfolio schloss Rolex im Jahr 2000 mit dem hauseigenen Automatikkaliber 4130. Erste Patentanträge hierfür datierten auf den 31. Oktober 1995. Schaltradsteuerung, vertikale Reibungskupplung und leicht in Richtung „12" verlagerte Totalisatoren waren nur einige Merkmale dieses 6,5 Millimeter flachen Uhrwerks. Von Anbeginn gab es Wartelisten für die Stahlreferenz 116520. So geschehen auch beim 2016 lancierten Pendant 116500LN mit schwarzer „Cerachrom"-Lünette. 2020, zum 60. Geburtstag, stattete Rolex die „Daytona" mit dem weiterentwickelten Automatikkaliber 4131 aus. Sein Rotor dreht nun um 22 winzige Kugeln. Die effizientere Ankerhemmung nennt sich „Chronergy" und die Stoßsicherung heißt fortan „Paraflex". Erstmals in der „Oyster"-Geschichte erhält die Platinreferenz 126506 einen Saphirglas-Sichtboden. Und die anlässlich des 100. Jubiläums der „24 Stunden von Le Mans" herausgegebene Weißgoldreferenz 126529LN besitzt neben der roten „100" in der Tachymeterskala auch einen 24-Stunden-Zähler bei „9".

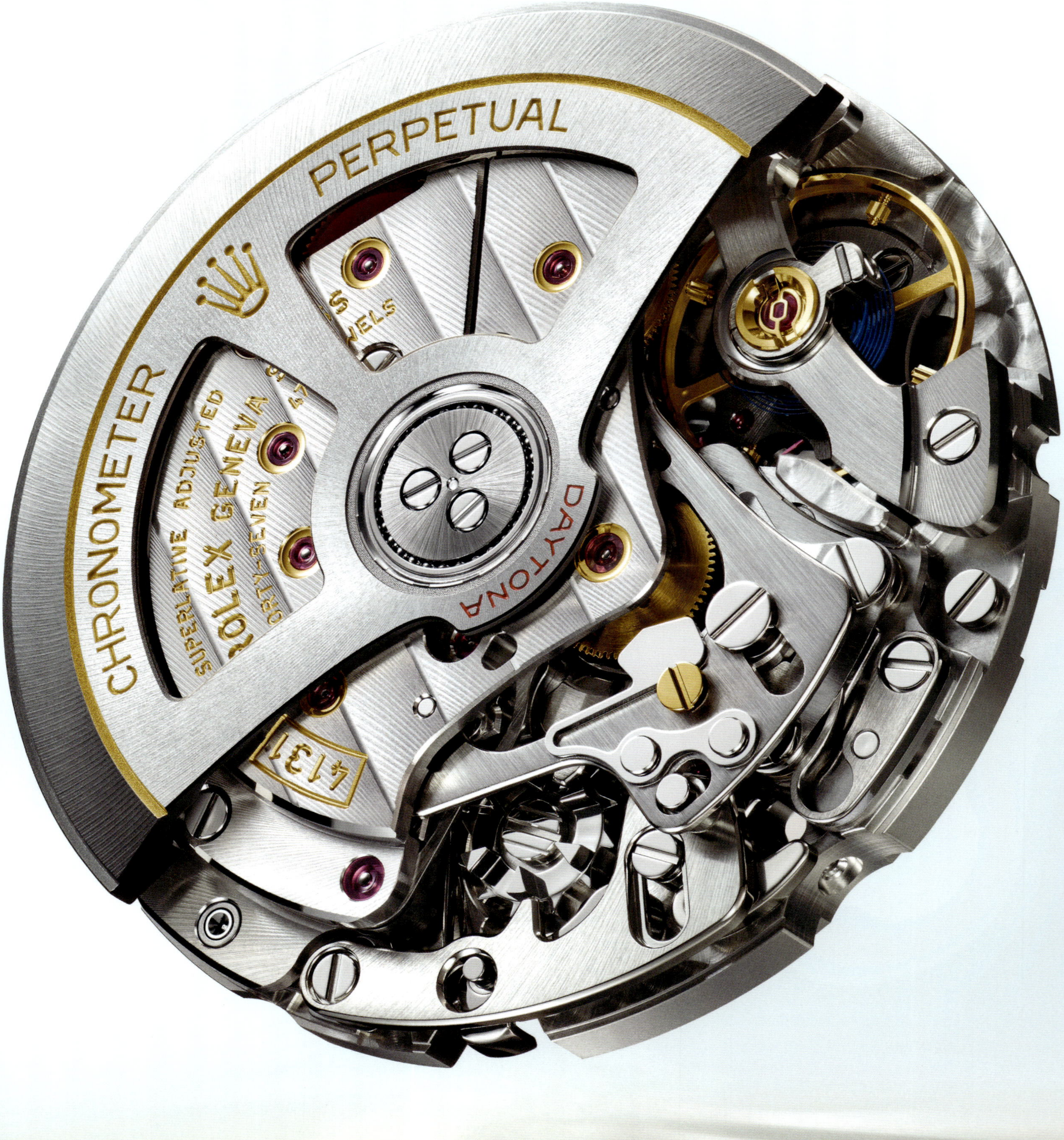
PERPETUAL
CHRONOMETER
SUPERLATIVE ADJUSTED
ROLEX GENEVA
DAYTONA
4131

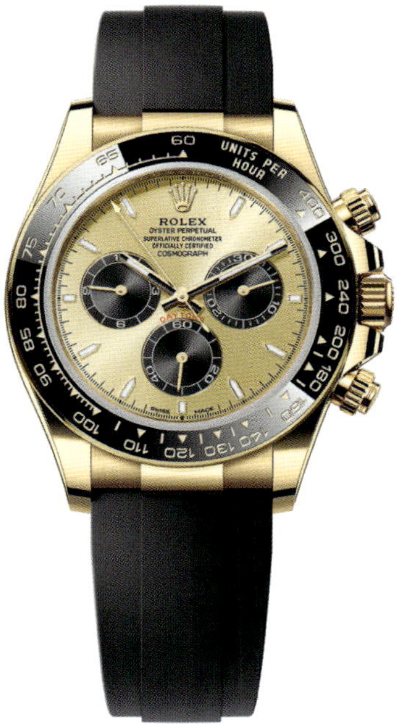

REF. 126518LN,
2023

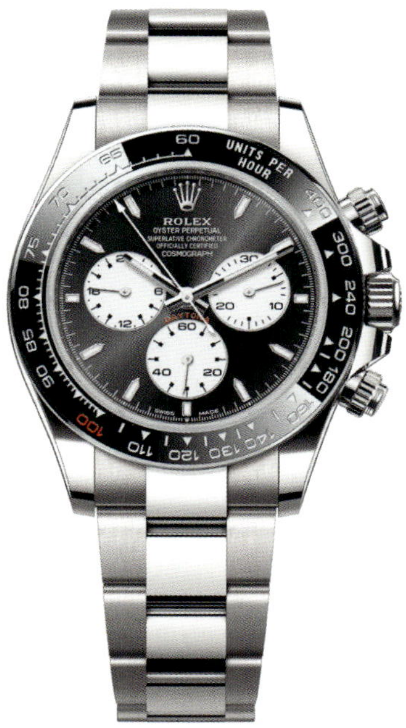

OYSTER PERPETUAL COSMOGRAPH DAYTONA "24H LE MANS"
REF. 126529, 2023

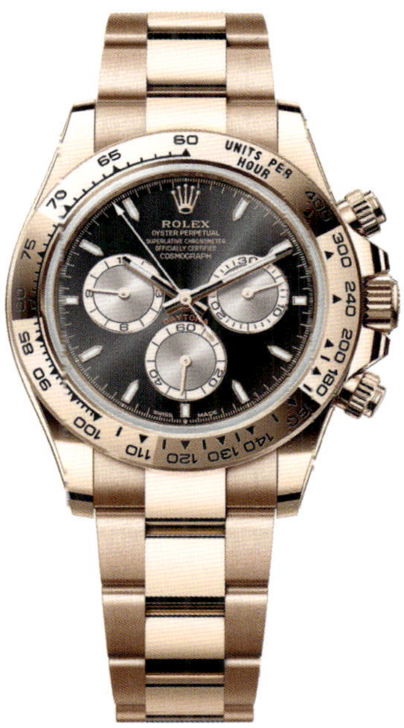

REF. 126505,
2023

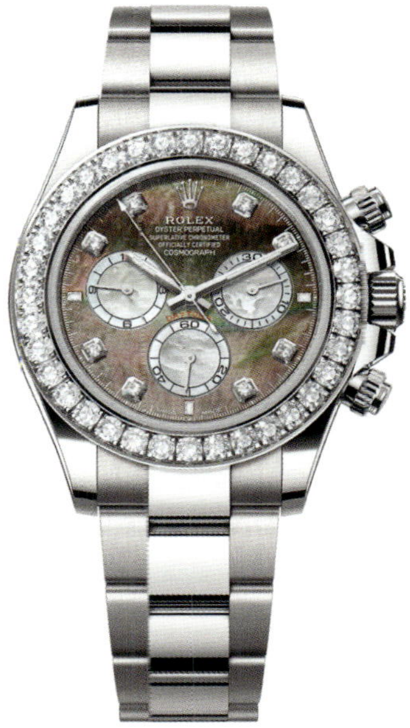

REF. 126579RBR,
2024

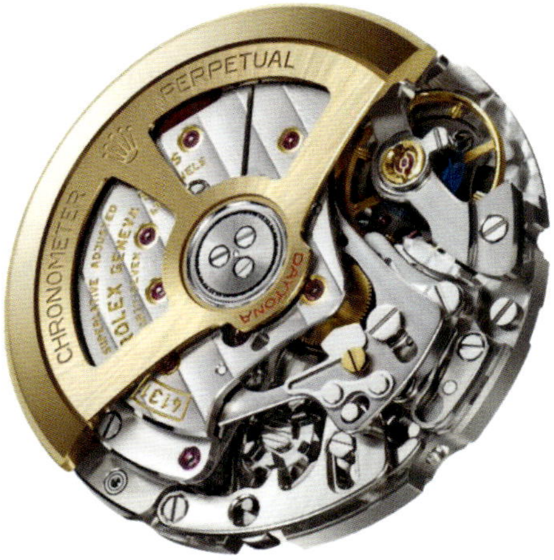

Cal. 4131, 2020

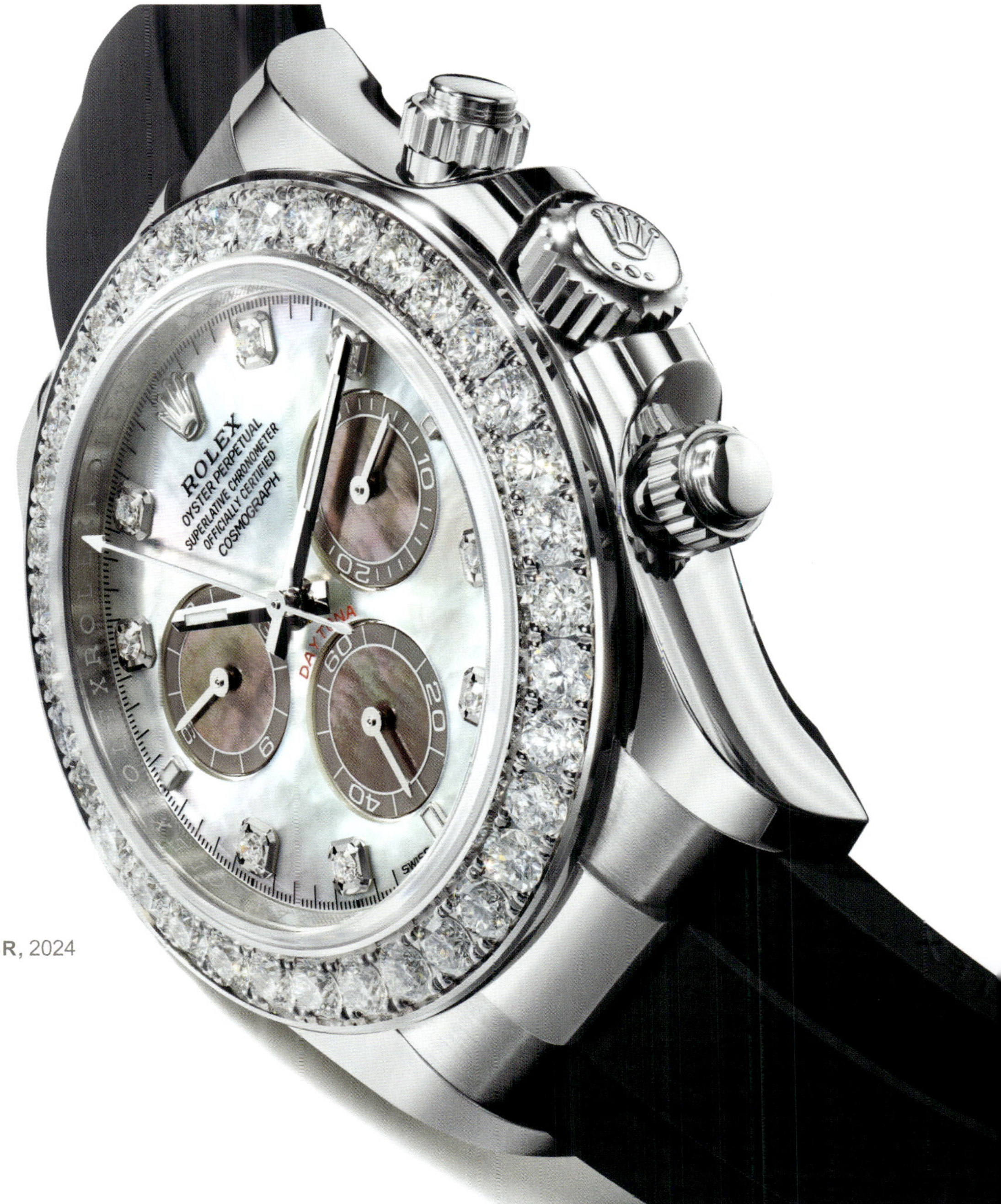

REF. 126589RBR, 2024

OYSTER PERPETUAL DEEPSEA CHALLENGE

CHALLENGING DEPTHS

It is not an easy wear, the "wristwatch" that accompanied Jacques Piccard and Don Walsh in 1960 on the outer skin of the diving boat *Trieste* in their exploration of the 10,916-metre-deep Mariana Trench. It took the two men four hours and 43 minutes to reach the bottom of the sea at 1.10 p.m. Twenty minutes later, they began their gradual return to the surface. Leaving the realm of silence claimed a further three and a half hours. Rolex was apprised of the following: "I'm delighted to inform you that your watch runs as precisely at a depth of eleven thousand metres as it does on the surface. Best regards, Jacques Piccard." No one knows for sure how many of the huge Deep Sea Specials (DSS) with ultra-high-domed crystal Rolex produced. The talk is of a total of seven prototypes made of steel. Add to that a few iterations, some of which were made of gold.

Another lavish timepiece, worn by James Cameron on his 10,908-metre dive to explore the Mariana Trench on 26 March 2012, was not intended for the general public either. Based on its rich experience, Rolex developed the experimental Deepsea Challenge specially for the director and researcher. The tests prefacing the successful dive involved applying 17 tons of pressure to the crystal.

Ten years later, Rolex presented the Oyster Perpetual Deepsea Challenge, an iconic timepiece that is available to buy and can be worn daily. With a diameter of 50 millimetres and a height of 23 millimetres, it is far from being a dainty piece. But such dimensions are required to achieve waterproofness up to 1,100 bar or a diving depth of 11,000 metres. Because of its weight, steel was not considered as an option for the case and link bracelet of the Reference 126067. Ultra-light RLX titanium, on the other hand, was. The titanium grade 5 alloy reduces the mass by around a third compared to the Deepsea Challenge. The nine-millimetre sapphire crystal alone accounts for quite a lot of the roughly 250 grams. The excess pressure arising during the decompression phases is released from the case with Triplock winding and hand-setting crown via a helium valve. Paying homage to the two historic dives in the Mariana Trench, the caseback is engraved with "Mariana Trench" and the dates "23-03-2012" and "26-01-1960". The hours, minutes and seconds are indicated by the automatic Calibre 3230 with efficient Chronergy escapement and Parachrom balance spring above and below water with a maximum daily deviation of two seconds. The pre-sale test is carried out with pressure equal to a water depth of 13,750 metres.

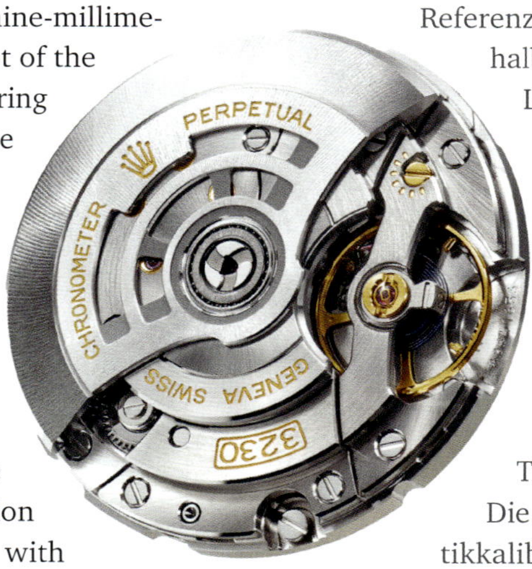

Cal. 3230

HERAUSFORDERUNG IN DER TIEFE

Nur schwerlich am Handgelenk tragen kann man jene „Armbanduhr", welche Jacques Piccard und Don Walsh 1960 an der Außenhaut des Tauchboots „Trieste" in den 10.916 Meter tiefen Marianengraben begleitete. Nach vier Stunden und 43 Minuten hatte das Duo um 13:10 Uhr den Meeresgrund erreicht. 20 Minuten später ging es allmählich wieder nach oben. Dreieinhalb Stunden nahm die Rückkehr aus der Welt des Schweigens in Anspruch. Dann erfuhr Rolex Folgendes: „Es ist mir eine Freude Ihnen mitzuteilen, dass Ihre Uhr auch in 11.000 Metern so genau geht wie an der Oberfläche. Beste Grüße, Jacques Piccard." Wie viele Exemplare der riesigen „Deep Sea Specials" (DSS) mit extrem hoch gewölbtem Glas das Haus Rolex fertigte, lässt sich nicht mit letzter Sicherheit sagen. Die Rede ist von insgesamt sieben Prototypen in Stahlausführung. Hinzu gesellten sich einige Nachbauten, für deren Outfit teilweise auch Gold Verwendung fand.

Ein weiteres opulentes Zeitmessinstrument, mit dem James Cameron am 26. März 2012 exakt 10.908 Meter wiederum in den Marianengraben tauchte, war ebenfalls nicht für die Allgemeinheit gedacht. Speziell für den Regisseur und Forscher hatte Rolex auf der Grundlage reicher Erfahrungen die experimentelle „Deepsea Challenge" entwickelt. Die Tests vor dem erfolgreichen Tauchgang erfolgten mit 17 Tonnen Druck auf dem Glas.

Zehn Jahre später, also 2022, präsentierte Rolex in Gestalt der „Oyster Perpetual Deepsea Challenge" eine käuflich erwerb- und täglich tragbare Ikone. Natürlich waren 50 Millimeter Durchmesser und 23 Millimeter Bauhöhe nicht eben zierlich. Aber diese Dimensionen brauchte es halt für eine Wasserdichte bis zu 1.100 bar Druck oder 11.000 Meter Tauchtiefe. Stahl als Werkstoff für Gehäuse und Gliederband kam für die Referenz 126067 aus Gewichtsgründen nicht in Betracht. Deshalb gelangte Titan-RLX ins Spiel. Durch die Titan-Grade5-Legierung reduzierte sich die Masse gegenüber der „Deepsea Challenge" um etwa 30 Prozent. Zu den rund 250 Gramm trug allein das neun Millimeter dicke Saphirglas einiges bei. Den während der Dekompressionsphasen entstehenden Überdruck ließ ein Heliumventil aus der Schale mit „Triplock"-Aufzugs- und Zeigerstellkrone entweichen. Als Hommage an die beiden historischen Tauchgänge in den Marianengraben trug der Gehäuseboden die Gravuren „Mariana Trench" sowie die Daten „23-01-1960" und „26-03-2012". Die Stunden, Minuten und Sekunden zeigte das Automatikkaliber 3230 mit effizienter „Chronergy"-Hemmung und „Parachrom"-Unruhspirale über und unter Wasser mit einer maximalen täglichen Gangabweichung von zwei Sekunden an. Die Prüfung vor dem Verkauf erfolgte mit einem Druck, der 13.750 Metern Wassertiefe entsprach.

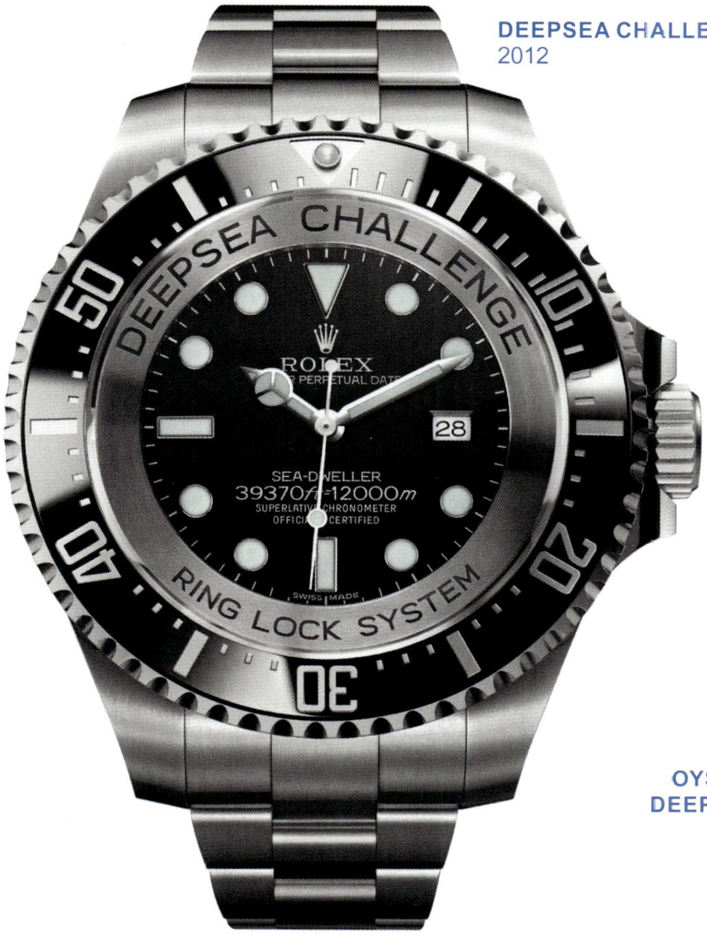

DEEPSEA CHALLENGE,
2012

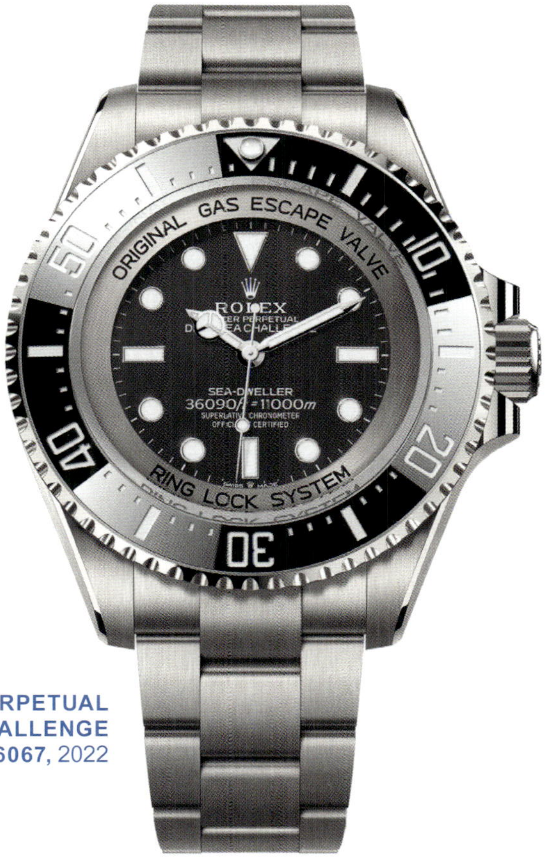

OYSTER PERPETUAL
DEEPSEA CHALLENGE
REF. 126067, 2022

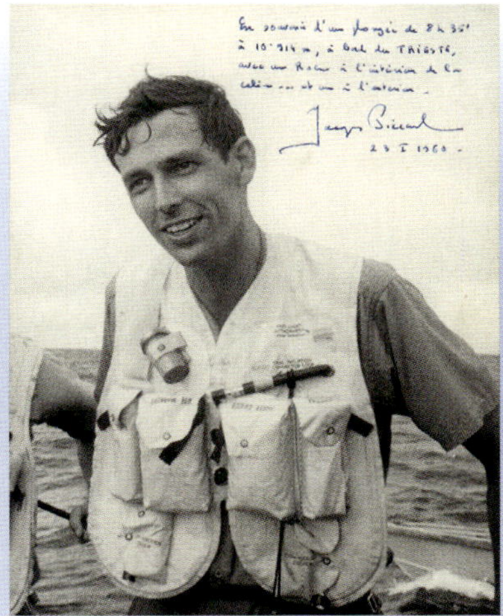

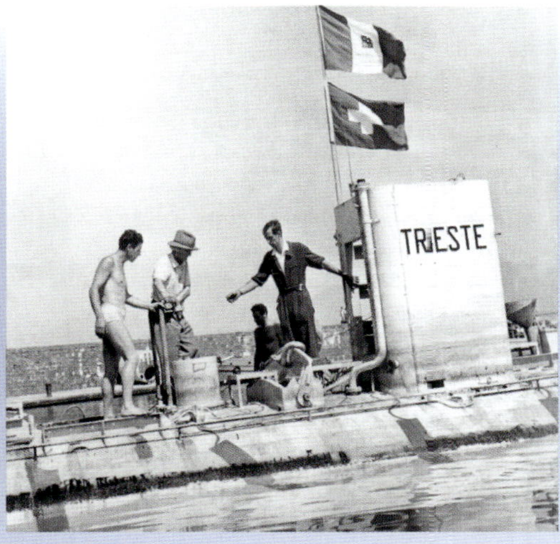

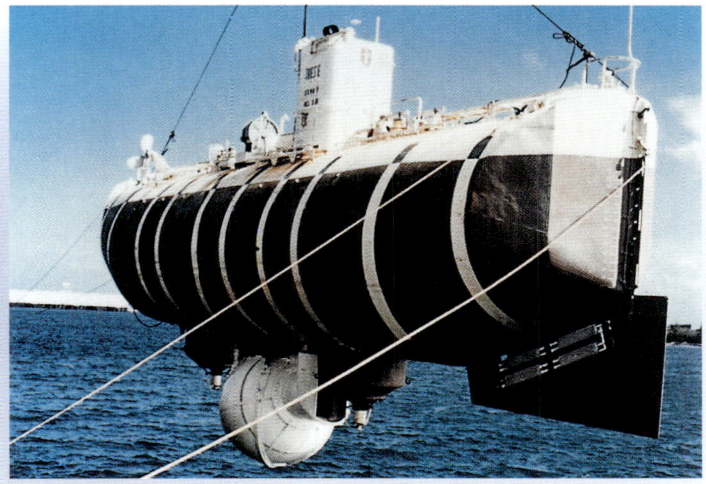

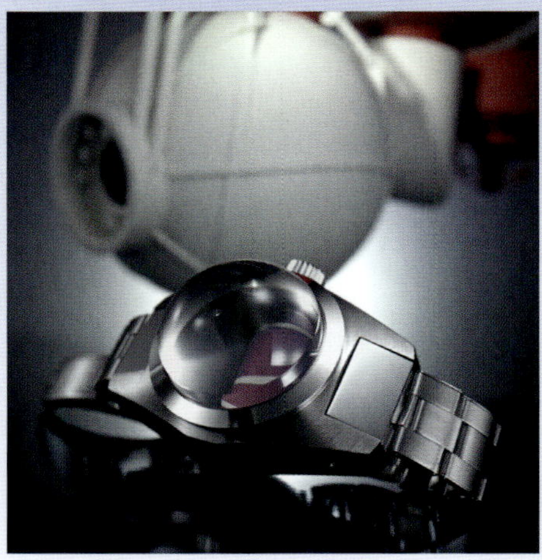

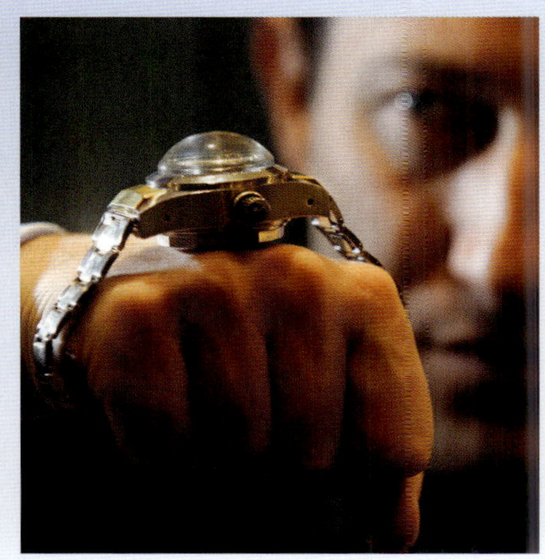

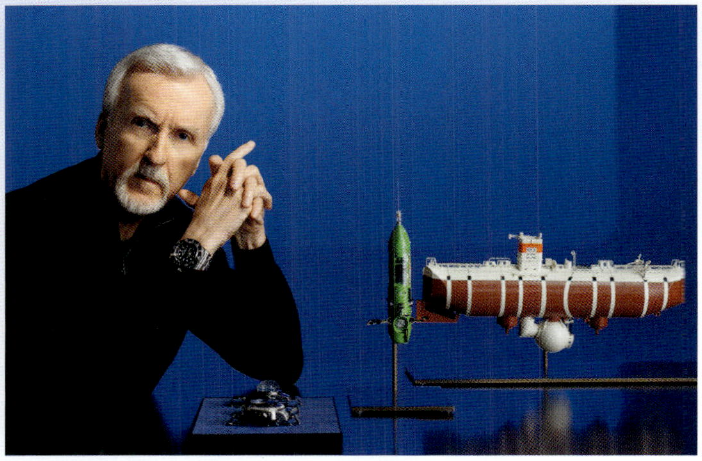

From top left, clockwise | Von oben links im Uhrzeigersinn: *Jacques Piccard | Auguste and Jacques Piccard on the Bathyscaphe "Trieste" | James Cameron | Deep Sea Special*

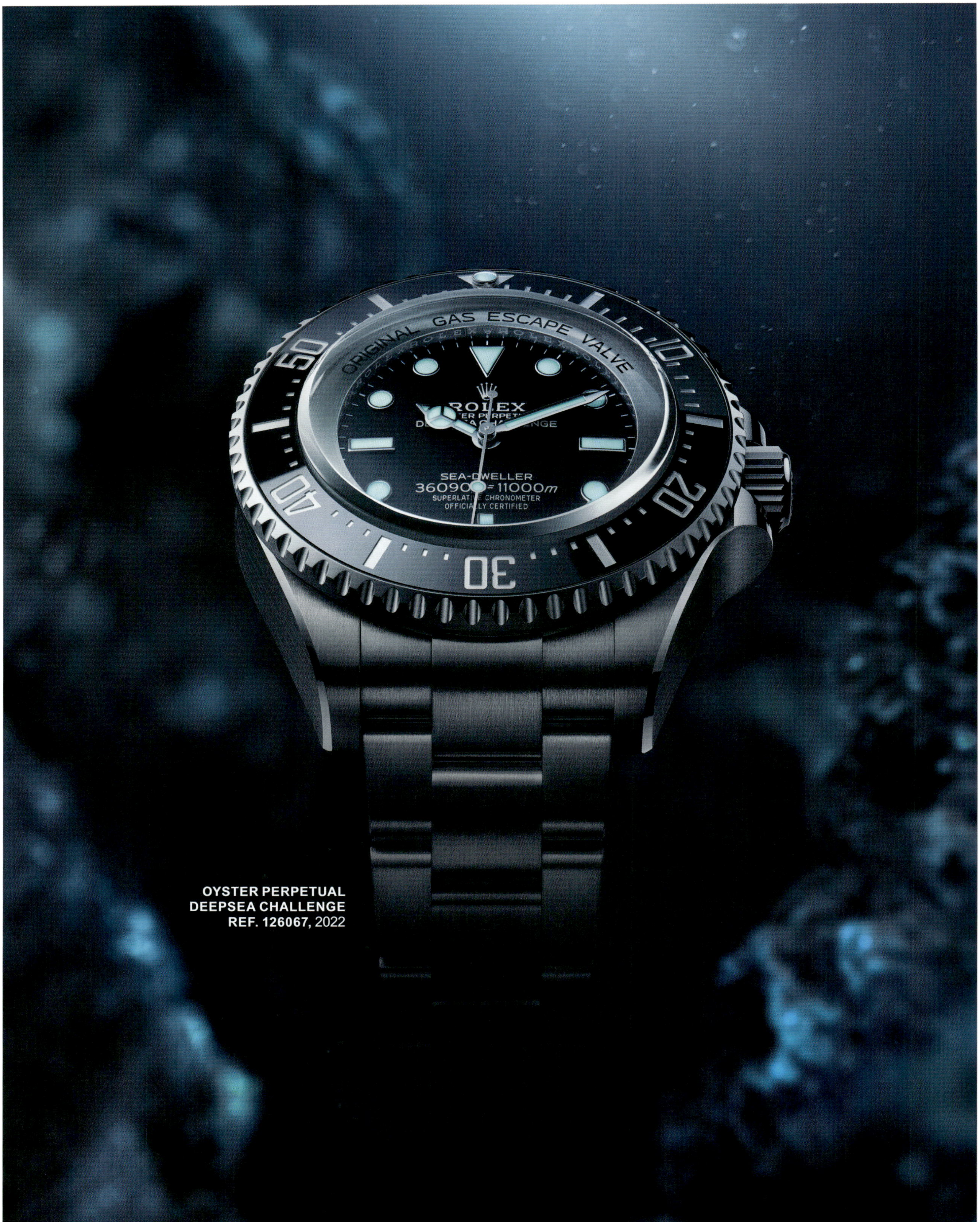

OYSTER PERPETUAL
DEEPSEA CHALLENGE
REF. 126067, 2022

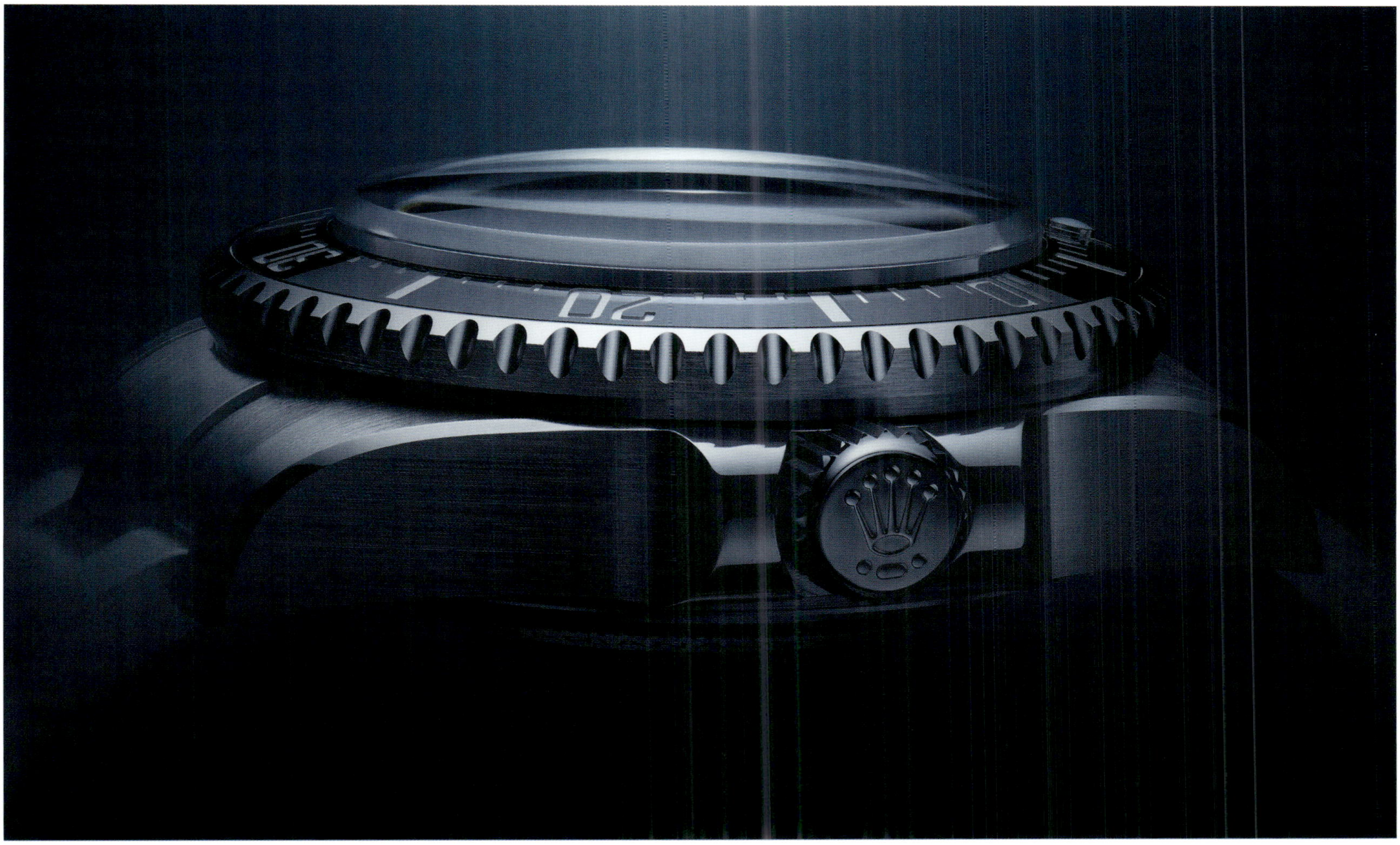

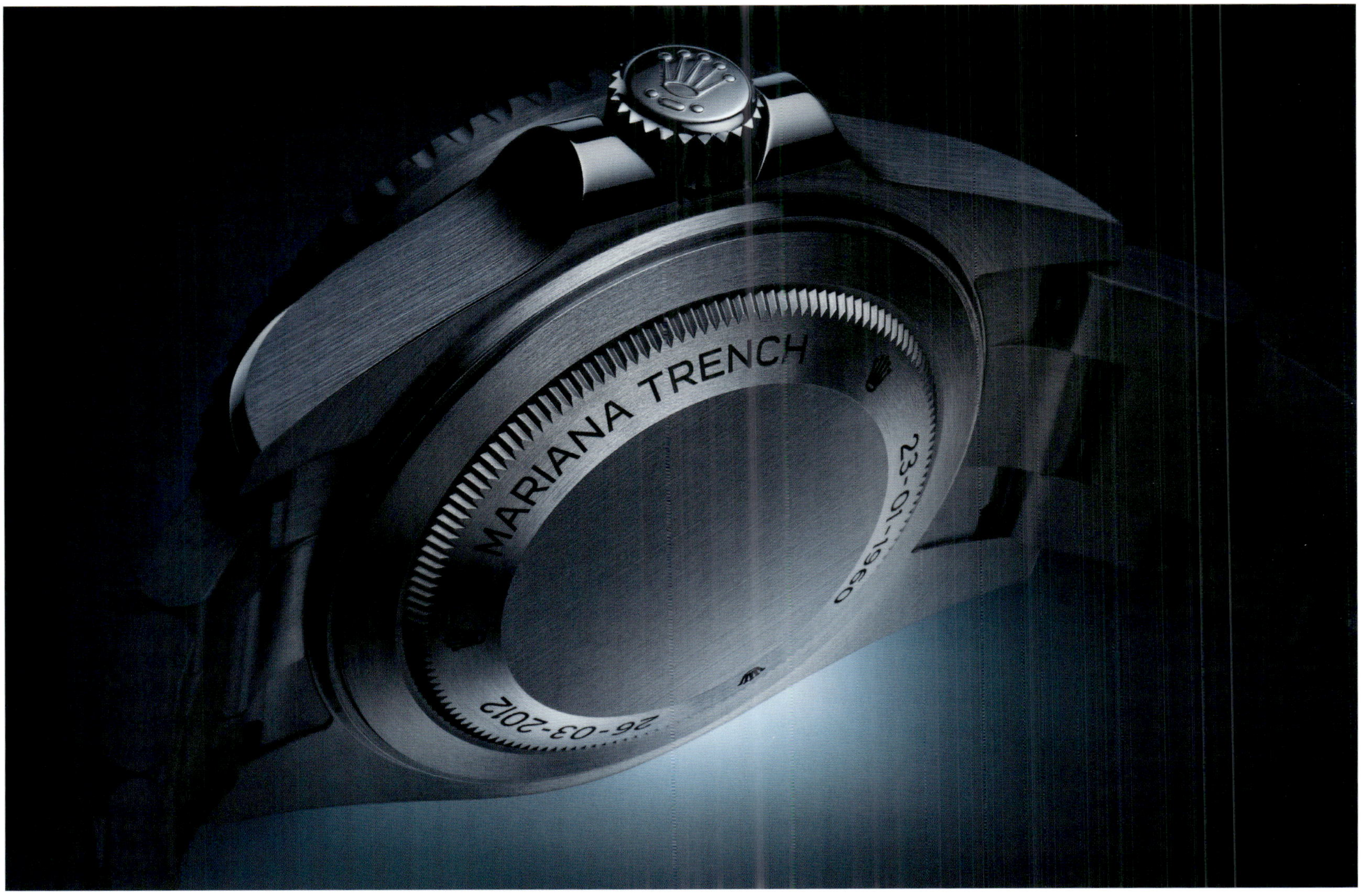
MARIANA TRENCH
23-01-1960
26-03-2012

ROLEX
SKY-DWELLER

INSPIRED BY SAROS

On the Rolex Sky Dweller launched in 2012, the engineering complexity flourishes inside the 42-millimetre Oyster case that is waterproof up to 10 bar. The gradual development of the automatic Calibre 9001 with a power reserve of 72 hours began as early as 2004. That year the Geneva manufactory received the first of 14 patents, five of which were entirely new, for this versatile timepiece. The gem in its crown is without doubt its annual calendar. It only needs to be programmed manually once a year at the end of February. This couldn't be simpler due to the ingenious Ring Command system with an easy-grip rotating bezel to preselect the relevant crown function. The first counterclockwise position enables you to set the date and local time. The design of the calendar cadrature was inspired by the Saros cycle consisting of 18.03 years. At the end of this period, the sun, earth and moon are once more in the same position in relation to one another. In addition to the traditional date calendar system, the mechanism consists of only four gear wheels and two gear ratios with a planetary gear wheel (the sun) at the centre. A satellite wheel (the moon) engages in the teeth of the planetary wheel, with its four fingers representing the four 30-day months. Logically, this circumnavigation takes one month. The date disc drives the ensemble. Connected to it is the fixed planetary gear, which rotates around its own axis. At the end of each short month, one of the four fingers receives an extra impulse, which causes the date to jump two days nearly instantly. The months are marked by 12 apertures outside the hour markers to ease setting. The current month is indicated in red. For the time-zone function the engineers emulated the Rolex GMT Master II with an eccentric 24-hour indication and separately adjustable 12-hour hands. The latter are set by the crown in the second bezel position, naturally followed by the window date in either direction. In third position, the crown is used to set the time to the exact second. The balance wheel is halted for this purpose.

INSPIRIERT VON SAROS

Die hohe mechanische Komplexität blüht bei der 2012 lancierten Rolex „Sky-Dweller“ definitiv im Inneren des 42 Millimeter großen und bis zu zehn bar wasserdichten „Oyster“-Gehäuses. Bereits 2004 startete die schrittweise Entwicklung des Automatikkalibers 9001 mit etwa 72 Stunden Gangautonomie. In besagtem Jahr konnte die Genfer Manufaktur das erste von fünf neuen und insgesamt 14 Patenten für diesen vielseitigen Zeitmesser entgegennehmen. Als Filetstück kann zweifellos dessen Jahreskalender gelten. Nur jeweils Ende Februar muss er manuell für ein weiteres Jahr programmiert werden. Dank des ausgeklügelten Ring-Command-Systems mit griffiger Drehlünette zum Vorwählen der jeweiligen Kronenfunktion geht das ganz einfach. Die erste linksseitige Position gilt dem Einstellen des Datums und der Lokalzeit. Inspirierend beim Konstruieren der Kalender-Kadratur wirkte der so genannte „Saros“-Zyklus von 18,03 Jahren. An dessen Ende stehen sich Sonne, Erde und Mond wieder in der gleichen Position gegenüber. Lediglich zwei Übersetzungsverhältnisse und vier Zahnräder, welche das altbekannte Fensterdatum ergänzen, benötigt das Schaltwerk mit einem „Sonnenrad“ im Zentrum. In dessen Zähne greift ein Planetenrad mit vier die 30-Tage-Monate repräsentierenden „Mond“-Fingern. Logischerweise dauert die Umrundung einen Monat. Als Antrieb des Ganzen dient die Datumsscheibe. Mit ihr ist das um seine eigene Achse rotierende Planetenrad fest verbunden. Jeweils am Ende aller kurzen Monate erfährt einer der vier Finger einen zusätzlichen Impuls. Dadurch vollzieht der bedruckte Datumsring in Sekundenbruchteilen gleich zwei Sprünge. Das Einstellen des Kalendariums unterstützen Monatsfenster hinter den zwölf Stundenindexen. Dunkle Färbung signalisiert den aktuellen Monat. Von der Rolex „GMT-Master II“ leiteten die Techniker die Zeitzonenfunktion mit exzentrischer 24-Stunden-Indikation sowie unabhängig verstellbarem 12-Stunden-Zeiger ab. Auf letzteren wirkt die Krone in der zweiten Lünettenposition ein. Natürlich folgt ihm das Fensterdatum in beiden Richtungen. Bei ganz nach links bewegtem Glasrand dient die Krone dem sekundengenauen Einstellen der Uhrzeit. In diesem Fall hält die Unruh nämlich an.

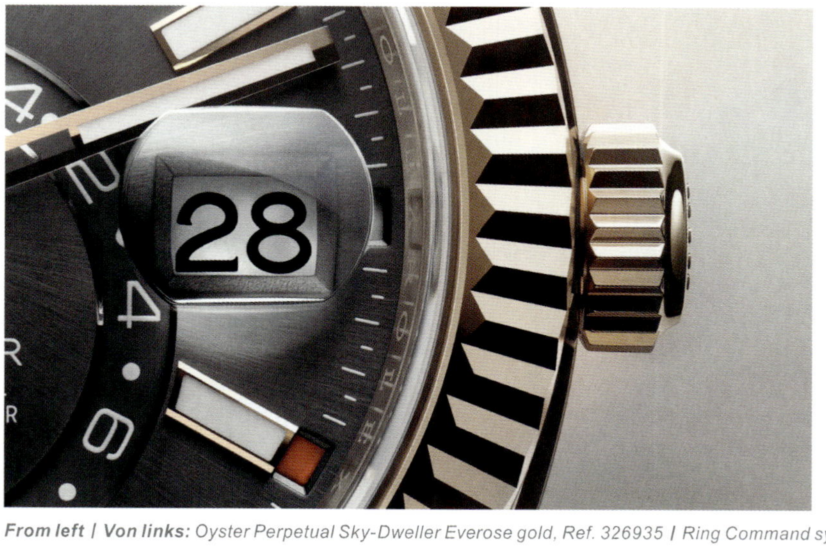

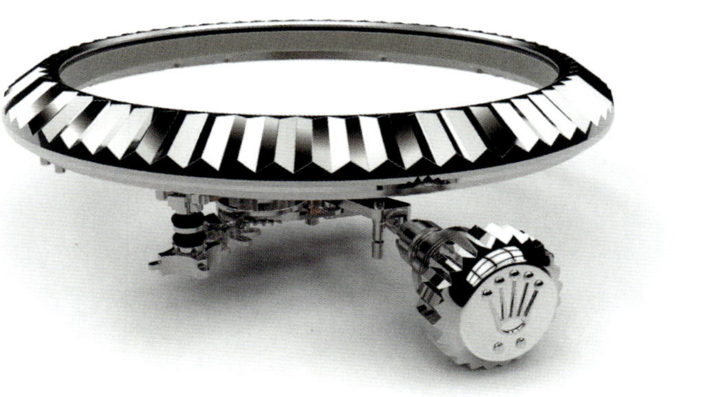

***From left | Von links:** Oyster Perpetual Sky-Dweller Everose gold, Ref. 326935 | Ring Command system, 2014*

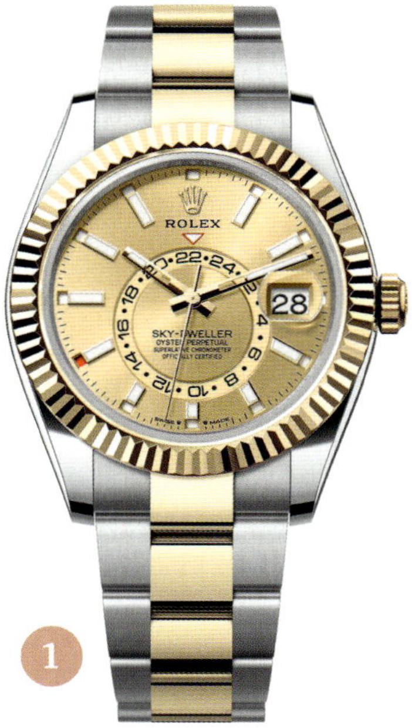

1

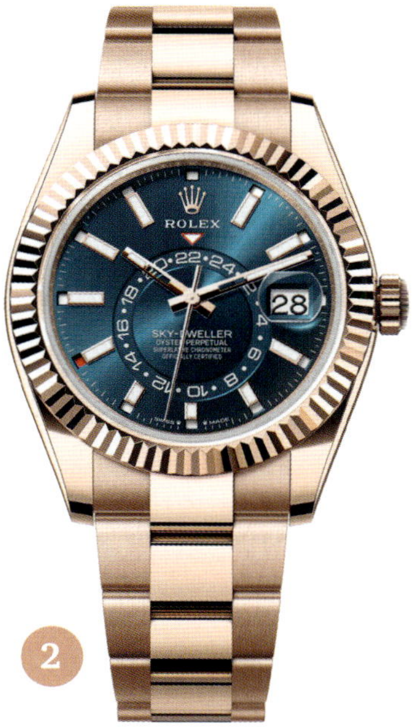

2

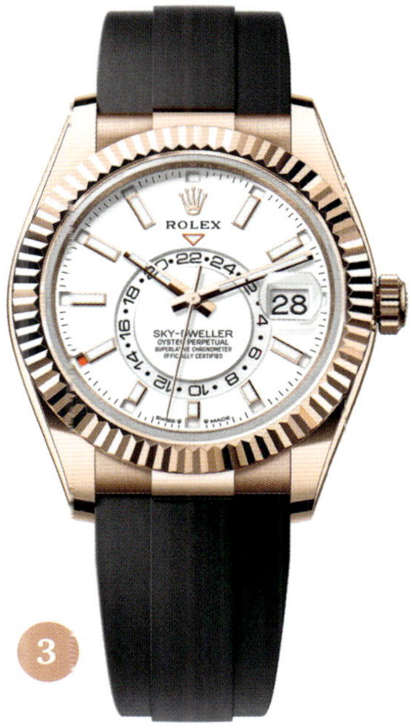

3

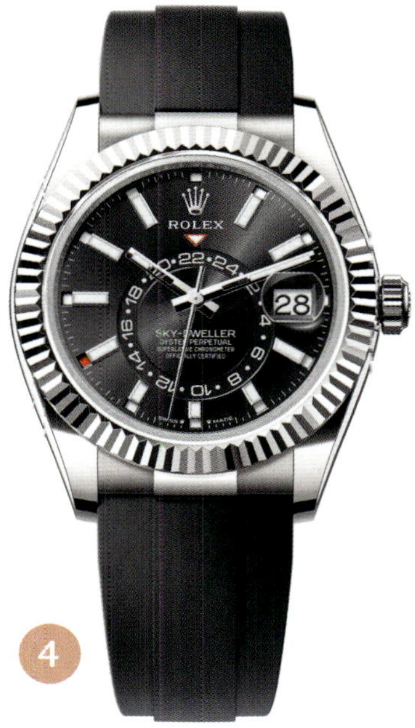

4

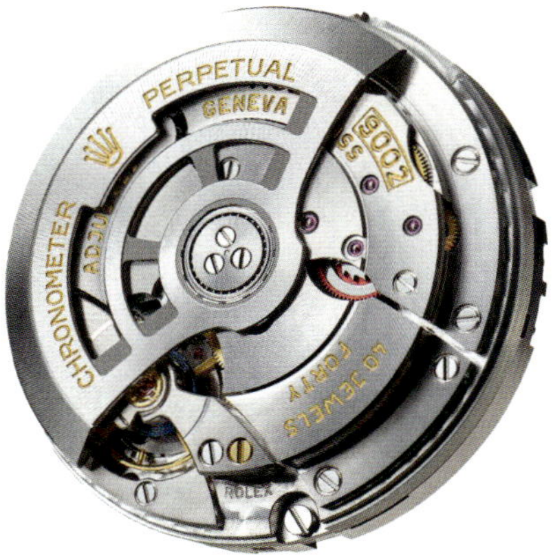

5

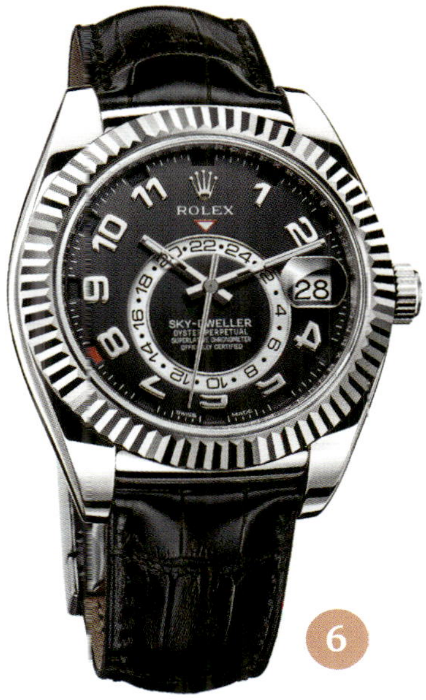

6

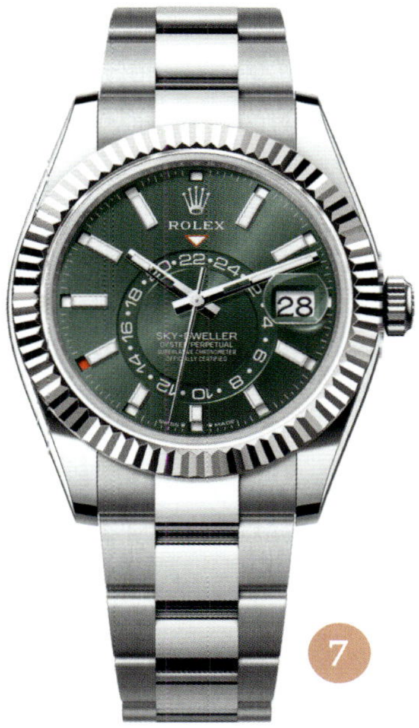

7

1. *Ref. 336933*
2. *Ref. 336935*
3. *Ref. 336235*
4. *Ref. 336239*
5. *Cal. 9002*
6. *Ref. 26139A, 2014*
7. *Ref. 336934*

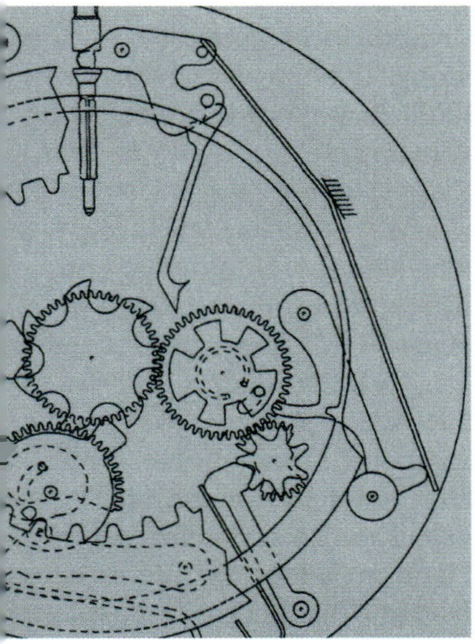

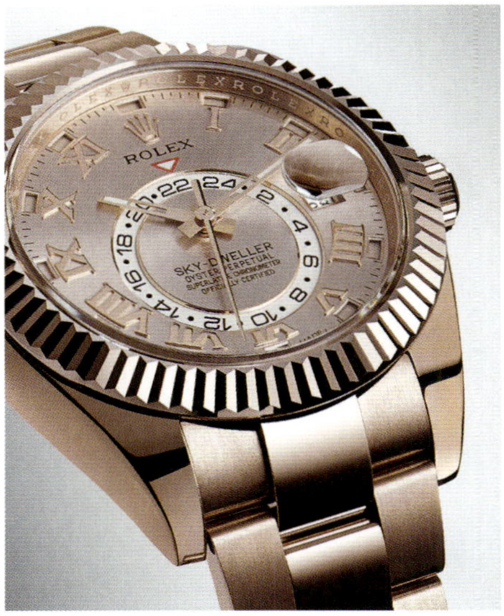

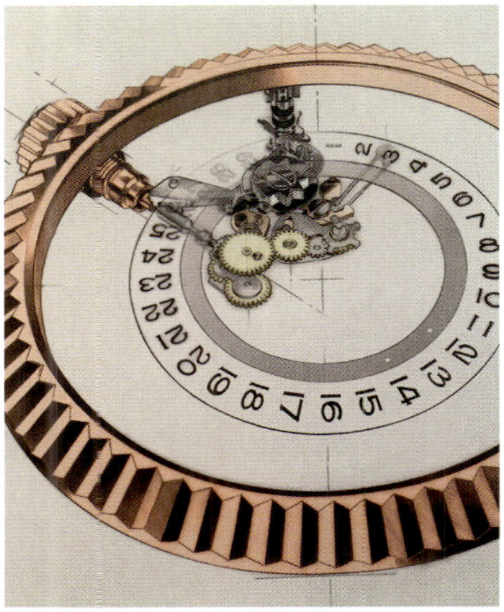

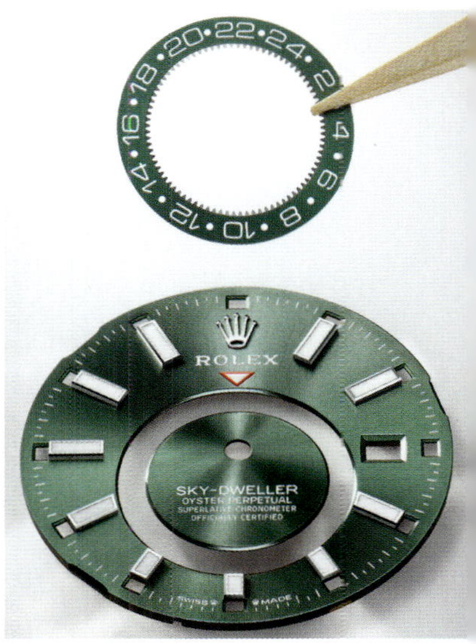

From left: Patented Saros annual calendar system | Oyster Perpetual Sky-Dweller, Everose gold, 2014 | Patented Ring Command system | Dial with 24-hour disc
Von links: Patentiertes „Saros"-Jahreskalender-System | Oyster Perpetual Sky-Dweller, Everose gold, 2014 | Patentiertes Ring-Command-System | Zifferblatt mit 24-Stunden-Scheibe

PROSPEX

PROFESSIONAL SPECIFICATIONS

In 1964, the Olympic Summer Games took place in Japan. Seiko acted as official timekeeper. The same year, the Tokyo watch manufactory founded in 1881 by Kintaro Hattori turned its attention to diver's watches. By designing a timepiece that could withstand depths of up to 150 metres, the Japanese watchmaker aimed to display its expertise to its European peers. The qualities of the sturdy deep-sea watch presented in 1965 included legibility even in moderate visibility conditions and high shock resistance. The 13.4-millimetre-high and 38-millimetre-wide steel case with rotating bezel protected a 12½-line automatic Calibre 6217 A movement. In 1966, the Reference 6217-8001 proved its mettle with the eighth Antarctic overwintering team. The following year Seiko presented the 6215-010, which was waterproof to 300 metres and also came with a proprietary automatic movement. Since then, diver's watches have been inseparably linked to the traditional Japanese brand. Even with the best of wills, the number of models developed and produced over the past six decades can no longer be counted. Multiple premières and top performances ensured attention. In this connection, the name Prospex springs to mind. The designation stands quite simply for "Professional Specifications". And they are satisfied by all watches specially designed by Seiko for underwater activities. A high level of waterproofness, optimum legibility, unidirectional diving-time bezel and user-friendly straps are a point of honour for Prospex timepieces. Seiko launches special editions on the market with a pleasing regularity.

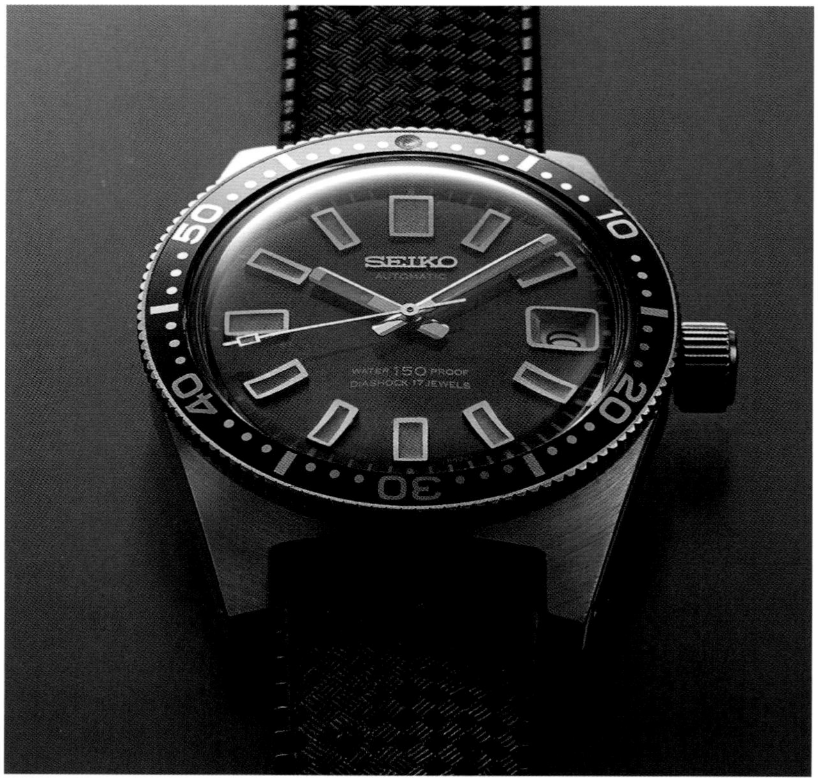

PROFESSIONELLE SPEZIFIKATIONEN

1964 gingen in Japan die Olympischen Sommerspiele über die Bühne. Dabei oblag Seiko die verantwortungsvolle Funktion eines offiziellen Zeitnehmers. In diesem Jahr wandte sich die 1881 von Kintaro Hattori in Tokio gegründete Uhrenmanufaktur erstmals auch dem Thema Taucherarmbanduhr zu. Mit der Kreation eines Zeitmessers, der Menschen bis zu 150 Meter unter den Meeresspiegel begleiten konnte, wollten die Uhrmacher im „Land der aufgehenden Sonne" den europäischen Kollegen ihre einschlägige Kompetenz demonstrieren. Zum Leistungsspektrum des 1965 vorgestellten Unterwasserboliden gehörten sichere Ablesbarkeit auch bei mäßigen Sichtverhältnissen und hohe Stoßfestigkeit. Das 38 Millimeter große und 13,4 Millimeter hohe Stahlgehäuse mit Drehlünette schützte ein 12½-liniges Automatikwerk vom Kaliber 6217 A. 1966 bewährte sich die Referenz 6217-8001 an den Handgelenken des „8. Überwinterungsteams" in der antarktischen Neumayer-Station III. Im folgenden Jahr präsentierte Seiko die bis 300 Meter wasserdichte 6215-010 ebenfalls mit hauseigenem Automatikwerk. Seitdem ist die Thematik Taucheruhren untrennbar verknüpft mit den japanischen Traditionalisten. Die Zahl der in beinahe 60 Jahren entwickelten und gefertigten Modelle lässt sich beim besten Willen nicht mehr zählen. Für Aufmerksamkeit sorgten zahlreiche Premieren und Bestleistungen. Ins Gedächtnis geprägt hat sich in diesem Zusammenhang der Name „Prospex". Und der steht ganz einfach für „professional specifications" oder auf gut Deutsch für „professionelle Spezifikationen". Und die erfüllen alle Armbanduhren, welche Seiko speziell für Unterwasseraktivitäten anbietet. Hohe Wasserdichtigkeit, optimale Ablesbarkeit, einseitig verstellbare Tauchzeitlünette und leicht handhabbare Armbänder sind bei den „Prospex"-Armbanduhren Ehrensache. Und mit schöner Regelmäßigkeit bringt Seiko Sondereditionen auf den Markt.

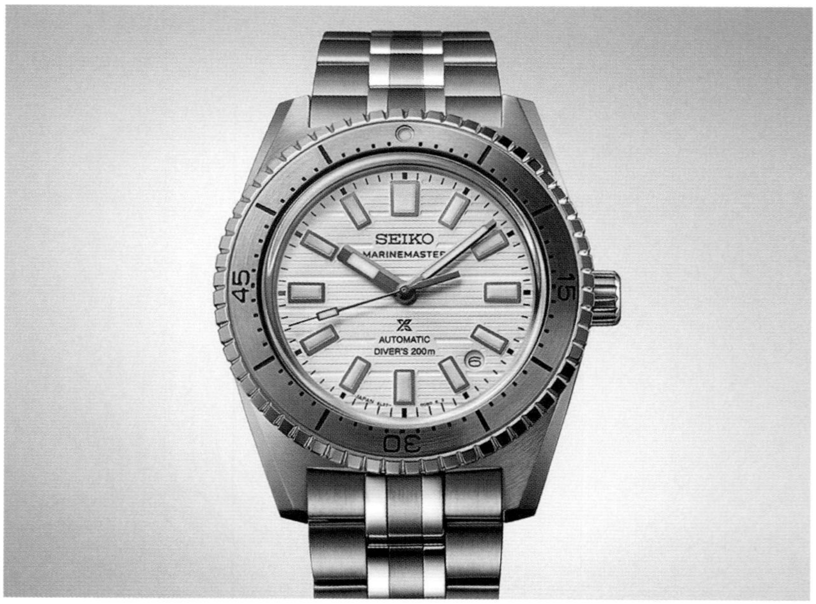

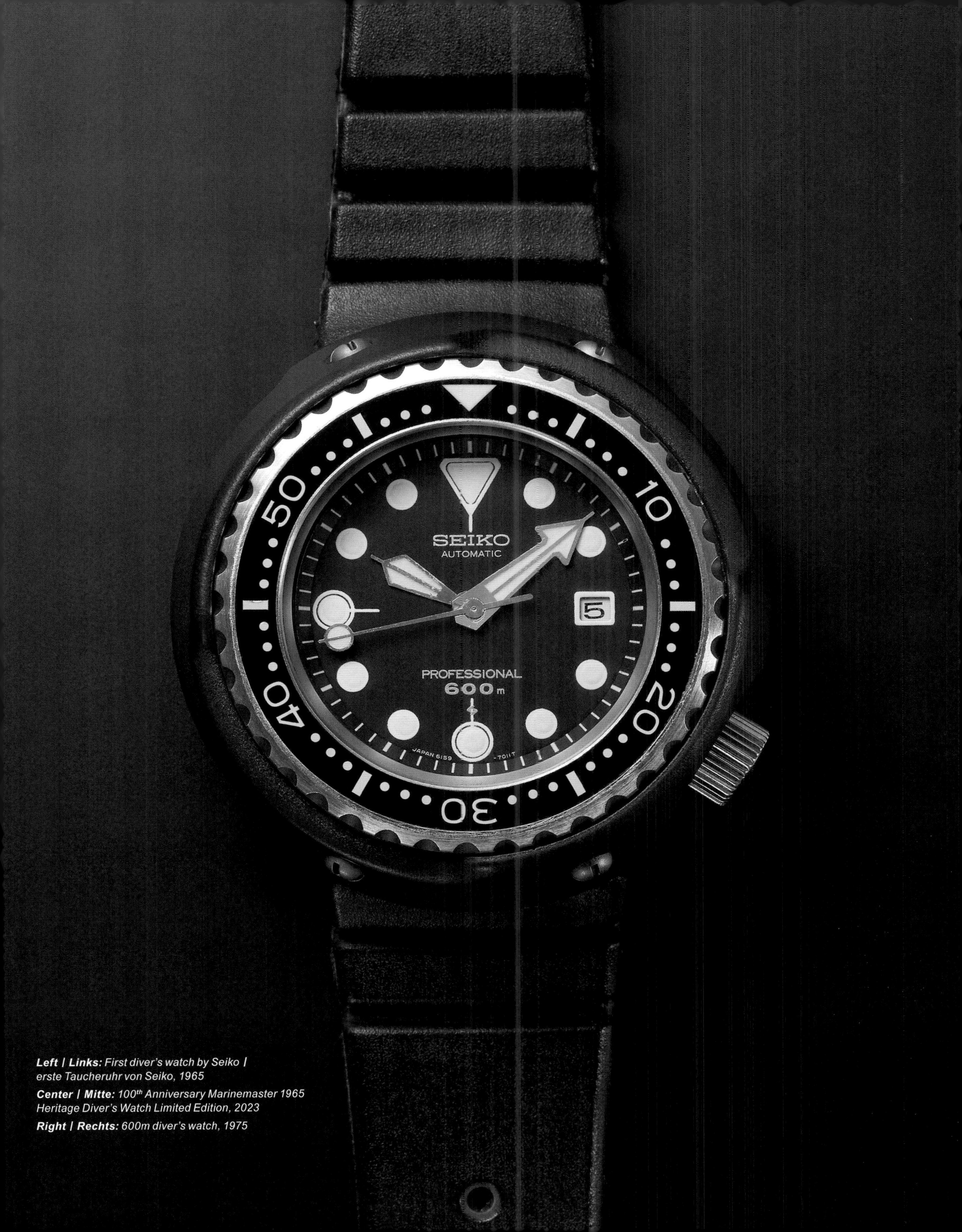

***Left | Links:** First diver's watch by Seiko | erste Taucheruhr von Seiko, 1965*

***Center | Mitte:** 100th Anniversary Marinemaster 1965 Heritage Diver's Watch Limited Edition, 2023*

***Right | Rechts:** 600m diver's watch, 1975*

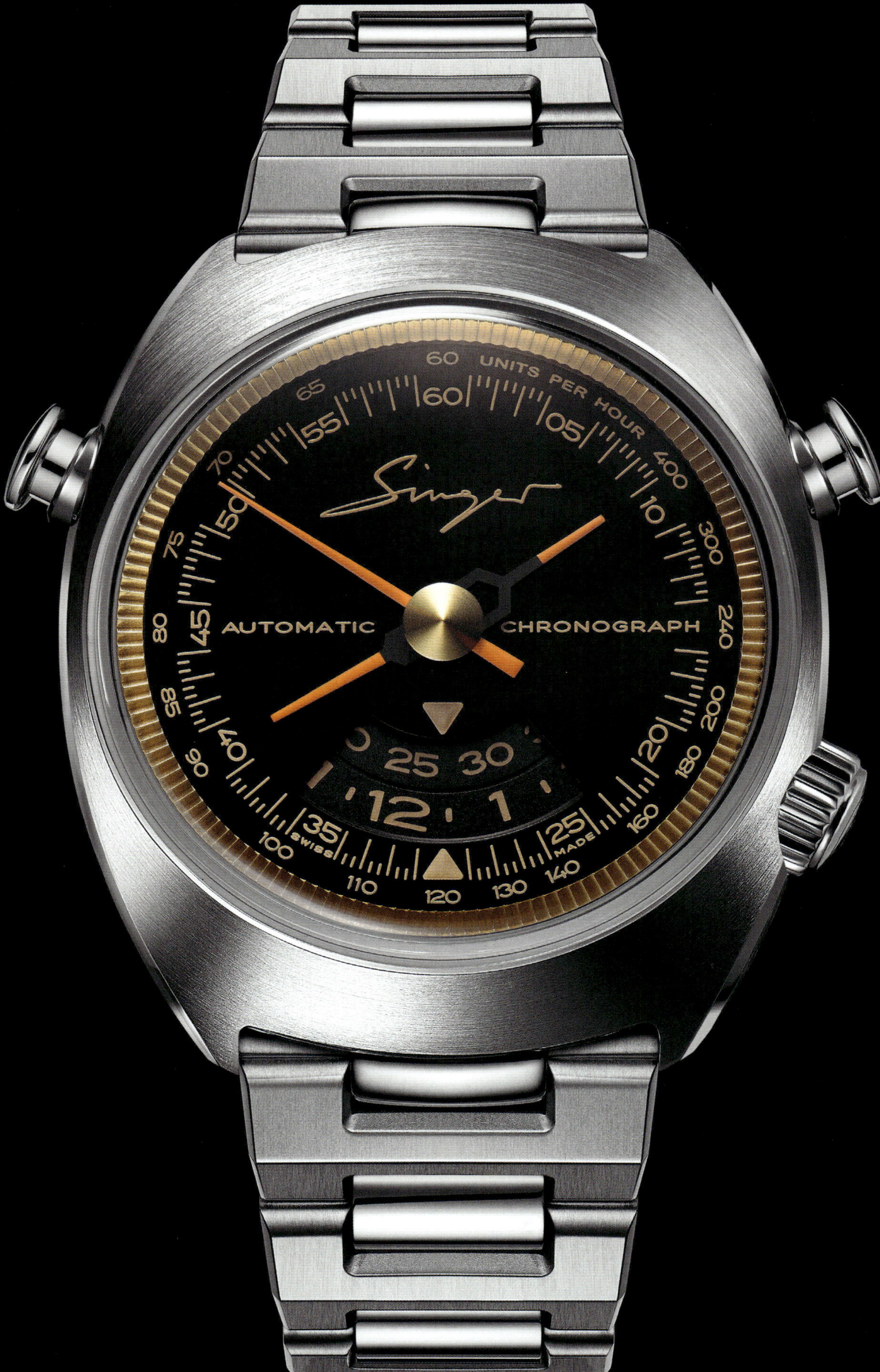

UNITS PER HOUR
Singer
AUTOMATIC
CHRONOGRAPH
SWISS
MADE

SINGER REIMAGINED 1969 CHRONOGRAPH

CHRONOGRAPH WITH A DIFFERENCE

The intellectual father of the extraordinary Calibre 6365 in the Singer Reimagined 1969 chronograph is Jean-Marc Wiederrecht. AgenHor, the company he founded in Geneva, engineered the movement called AgenGraphe. The product of a decade of development, it was presented together with the brand Singer Reimagined in 2017. The company is the watch subsidiary of an American firm specialising in the restoration of Porsche cars. The oscillating mass of the timepiece, composed of 491 components and 56 jewels, rotates under the dial, thus providing an unobstructed appreciation of the chronograph movement on the back with a special feature: three hands for the second chronograph, the 60-minute and – utterly unique – the 60-hour counter rotate at the centre of the dial. The digital totalisers for minutes and hours are characterised by accurately executed jumps. Springs return the chronograph hands to their vertical position. This snail-shaped cam and spring system required for the jumping of the totalisers ensures correct halting. The horizontal friction coupling is equally innovative. The reliability of the friction-driven function is due to a special coating on the toothless clutch wheels. The coarse material is akin to that of diamond nail files. Two wheels with specially shaped teeth ensure that the chronograph can only jump uncontrolled for a maximum of one second in the event of more serious bumps. The monobloc rotor winds the double mainspring barrels in both directions of rotation. A Magic Lever-type pawl system is responsible for polarising its rotational movements. Even after 60 hours with the chronograph switched on, the amplitude of the balance wheel oscillating at three hertz does not diminish by more than 20 degrees. The timepiece offers a power reserve of 72 hours. The innovative AgenPit regulator adjusts the effective length of the AtoKalpa balance spring. It is equipped with clamping rollers and a locking screw. These sophisticated inner workings are protected by a 40-millimetre stainless-steel case.

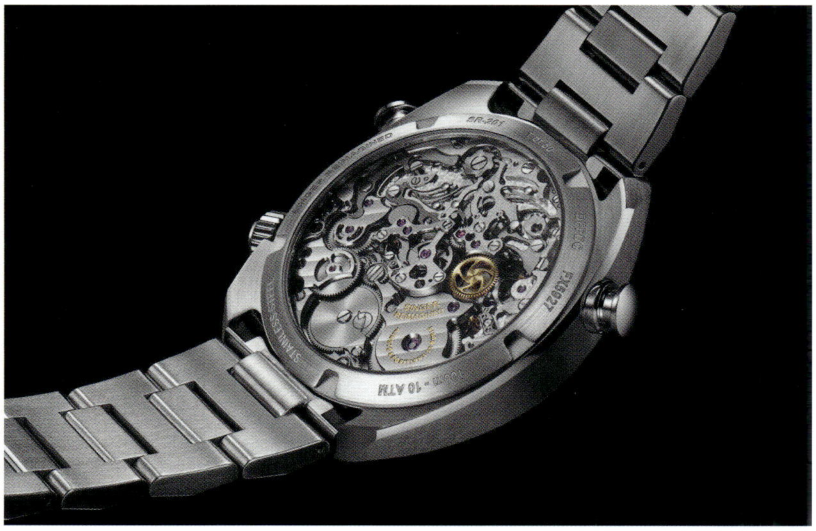

CHRONOGRAPH GANZ ANDERS

Geistiger Vater des außergewöhnlichen Kalibers 6365 im Singer Reimagined „1969"-Chronographen ist Jean-Marc Wiederrecht. AgenHor, die von ihm in Genf gegründete Werkeschmiede, hat die zeitschreibende Mechanik mit dem Beinamen „AgenGraphe" während zehn Jahren entwickelt und 2017 zusammen mit Singer Reimagined vorgestellt. Dabei handelt es sich um die Uhrentochter einer amerikanischen Firma, die sich auf die Restaurierung von Porsche-Fahrzeugen spezialisiert hat. Die Schwungmasse des aus 491 verschiedenen Teilen und dazu 56 funktionalen Steinen assemblierten Œuvres dreht unter dem Zifferblatt. Dieser Trick gewährleistet freie Sicht auf das rückwärtig angeordnete Chronographen-Schaltwerk mit einigen Besonderheiten: Drei Zeiger für den Sekunden-Chronographen, den 60-Minuten- und (absolut einzigartig) 60-Stunden-Zähler drehen im Zifferblattzentrum. Akkurate Sprünge zeichnen die digitalen Totalisatoren für Minuten und Stunden aus. Die Rückführung der Stoppzeiger-Zeiger in die Senkrechte übernehmen Spiralfedern. Jene Schneckenstufen, welche es für das Springen der Totalisatoren braucht, bewirken das korrekte Anhalten. Innovativ präsentiert sich auch die horizontale Friktionskupplung: Die zuverlässige Funktion per Reibung ist einer speziellen Beschichtung der zahnlosen Kupplungsräder zu verdanken. Das raue Material entspricht dem, was man von Diamantnagelfeilen kennt. Zwei Räder mit speziell geformten Zähnen sorgen dafür, dass der Chronograph bei härteren Stößen maximal eine Sekunde unkontrolliert springen kann. Der Rotor spannt das Zugfederduo in beiden Drehrichtungen. Für die Polarisierung seiner Drehbewegungen ist ein Exzenterwechsler vom Typ „Magic Lever" zuständig. Auch nach 60 Stunden mit eingeschaltetem Chronographen sinkt die Amplitude der mit drei Hertz oszillierenden Unruh nicht mehr als 20 Bogengrade ab. Insgesamt beträgt die Gangautonomie 72 Stunden. Ein neuartiger „AgenPit"-Rücker zur Veränderung der aktiven Länge der von „AtoKalpa" bezogenen Unruhspirale funktioniert mit Klemmrollen und einer fixierenden Schraube. Schutz für die komplexe zeitschreibende Mechanik bietet eine 40 Millimeter messende Schale aus Edelstahl.

< Cal. 6365

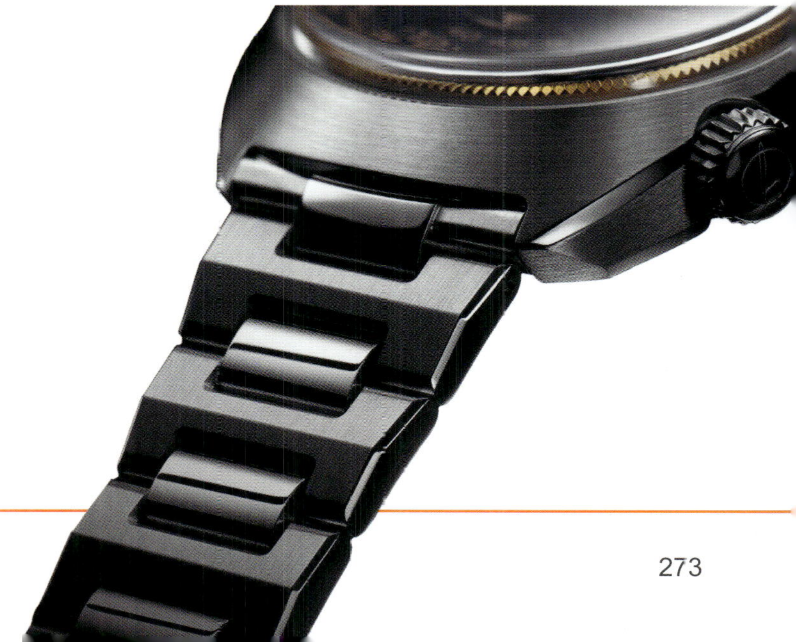

SINN

EZM

1997

MISSION TIMER

In 2021, Sinn Spezialuhren celebrated a round anniversary. Founded in 1961, Helmut Sinn's company looked back on six decades of watchmaking. The experienced pilot initially focused on sophisticated aviation clocks and striking pilot's chronographs. The watchmaker sold directly to his customers. After his 78th birthday, Helmut Sinn sold the company to Lothar Schmidt. On 1 September 1994, the latter, an engineer, rang in a new era for Sinn Spezialuhren GmbH. That the company name was more than justified is evidenced by the tool watches first presented in 1997. This chapter unfolds with the EZM 1 chronograph, its defining features a 40-millimetre titanium case and stalwart Lémania 5100 automatic calibre, which is no longer manufactured today. Target audience and customers included the maritime unit of the elite GSG 9 police special forces, the fire service, divers and professional pilots. Because it is a Sinn tradition never to leave anything to chance, many of the watches assembled in Frankfurt have to pass special certification processes. As in the case of the U2 model. It was tested in line with European diving equipment standards. Naturally, the 44-millimetre watch satisfies DIN 8310 technical requirements for waterproofness, being officially certified up to a diving depth of 2,000 metres. The watch, made of German submarine steel, must therefore withstand water pressure of 200 bar. Due to the application of the surface-hardening TEGIMENT technology, the case is particularly scratch-resistant. And of course, the diving-time rotating bezel can only be adjusted in one direction. Thanks to its special design, it is practically impossible to lose the hugely important bezel when diving. Driven by the Sellita SW200 calibre, the three luminescent hands can be admired through a scratch-resistant sapphire crystal.

EINSATZZEITMESSER

Bei der Sinn Spezialuhren GmbH gab es 2021 einen runden Geburtstag zu feiern. 60 Jahre wurde die 1961 von Helmut Sinn ins Leben gerufene deutsche Uhrenmarke. Anfangs konzentrierte sich der versierte Pilot auf die Produktion ausgereifter Navigations-Borduhren sowie markante Fliegerchronographen. Der Vertrieb erfolgte vom Hersteller ohne Umwege direkt zu den Kunden. Nach seinem 78. Geburtstag verkaufte Helmut Sinn die Firma an Lothar Schmidt. Mit dem Ingenieur begann am 1. September 1994 für die Sinn Spezialuhren GmbH eine neue Zeitrechnung. Dass der Begriff Spezialuhren dabei keineswegs aus der Luft gegriffen war, belegten die 1997 erstmals offerierten Einsatzzeitmesser. Den Beginn dieses Kapitels markierte der „EZM 1"-Chronograph mit 40 Millimeter starkem Titangehäuse und dem robusten, mittlerweile nicht mehr fabrizierten Automatikkaliber Lémania 5100. Adressaten und Nutzer waren die maritime Einheit der deutschen Bundespolizei GSG 9, die Feuerwehr, Taucher und auch Piloten im professionellen Einsatz.

Weil Sinn traditionsgemäß nichts dem Zufall überlässt, müssen viele der in Frankfurt assemblierten Armbanduhren spezielle Zertifizierungsprozesse durchlaufen. Das gilt auch für das Modell „U2". Seine Prüfung erfolgt in Anlehnung an die europäischen Tauchgerätenormen. Logischerweise erfüllt der 44 Millimeter große Zeitmesser die technischen Anforderungen der DIN 8310 für Wasserdichtigkeit. Offiziell bestätigt ist eine Wasserdichte bis zu 2.000 Meter Tauchtiefe. Ergo muss diese Armbanduhr aus deutschem U-Boot-Stahl einem Druck von 200 bar widerstehen. Durch die Anwendung der oberflächenhärtenden TEGIMENT-Technologie ist die Schale besonders kratzfest. Selbstverständlich lässt sich der Tauchzeitdrehring nur in einer Richtung verstellen. Durch die spezielle Konstruktion ist es praktisch unmöglich, die bei Tauchgängen immens wichtige Lünette zu verlieren. Auf das vom Automatikkaliber Sellita SW200 angetriebene Leuchtzeiger-Trio blickt man durch ein kratzfestes Saphirglas.

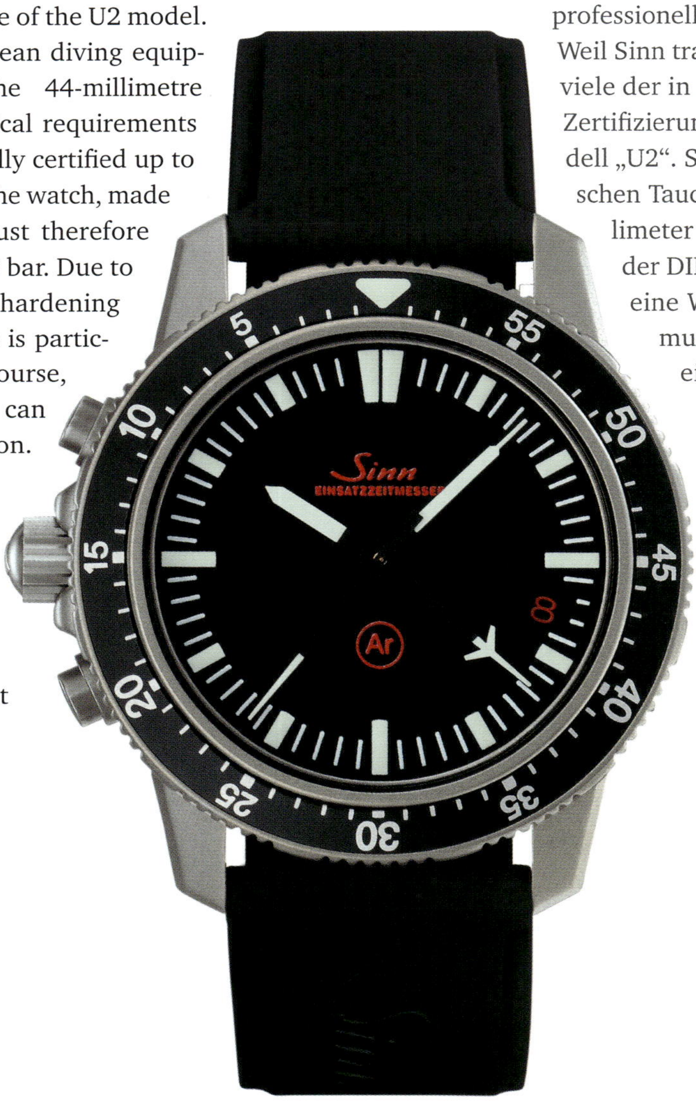

Left / Links: *EZM 1, 1997*
Right / Rechts: *EZM 5 "U2"*

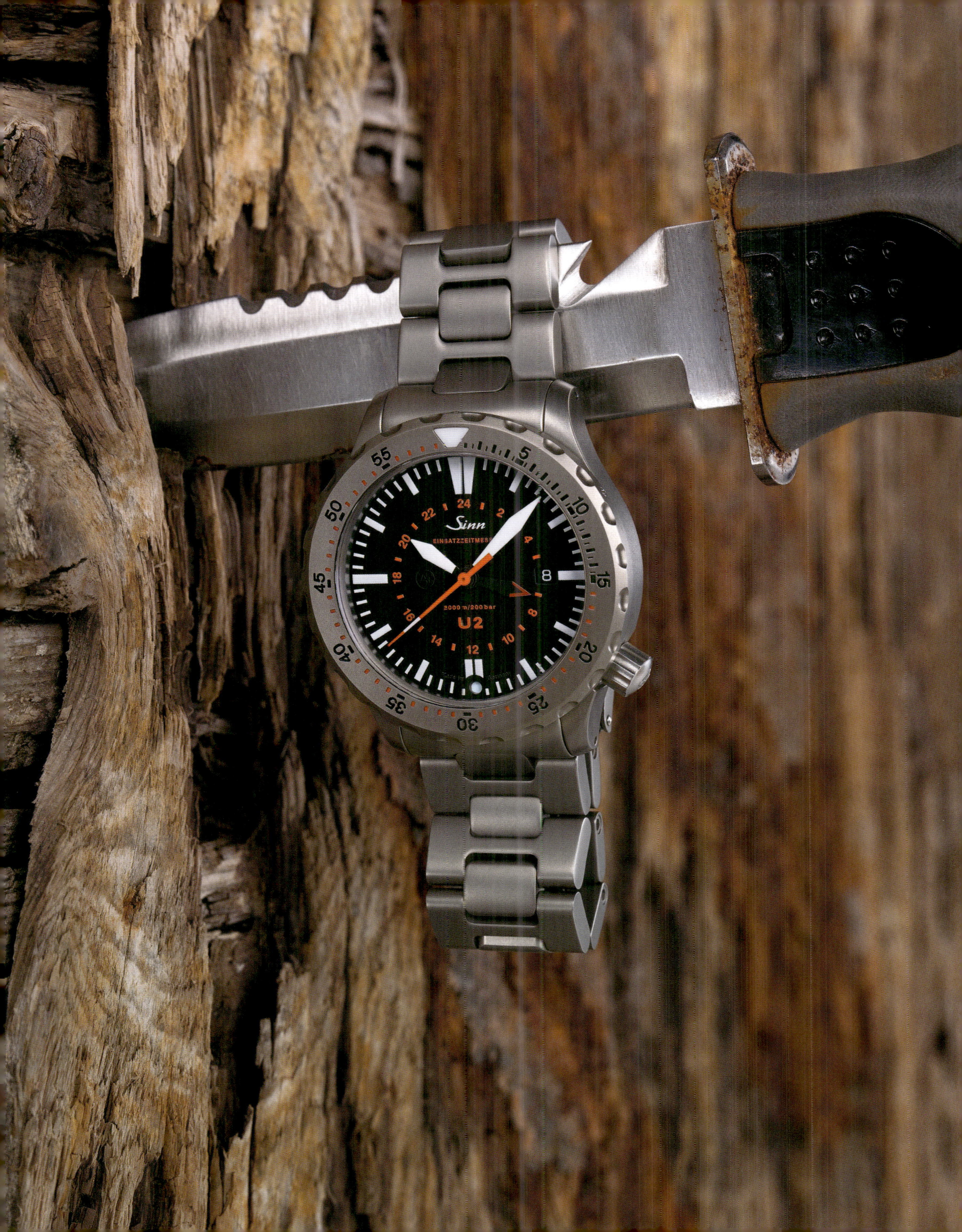
Sinn
2000 m/200 bar
U2

RIPPLES

THE RIPPLE EFFECT

The history of the small Geneva watch manufactory unfolds in 2002 with Peter Speake-Marin. Because the Englishman lacked the financial base for long-term development, Christelle Rosnoblet came on the scene in 2012. The watch lover identified the company's potential and set its course for the future. Since 2014, Le Cercle des Horlogers, a reputed fine watchmaking workshop specialised in the making, assembly and fine-tuning of timepieces, has crafted all Speake Marin calibres. This move achieved the sought-after independence from external suppliers and the quality of the movements to which the French company owner aspired. In 2017, the founder left the company. The mainstays of the collection include the Openworked with visible inner workings on display and the sporty and elegant Ripples, with which Speake Marin achieved an international breakthrough in 2022. The memorable name refers to the ripple structure of the dial. The highly unusual rotating second hand at 2 o'clock immediately catches the eye. It has become a sort of trademark and typical feature of all proprietary SMA automatic calibres with a micro-rotor. In its Ripples, Speake Marin uses the flat SMA03-T with a power reserve of 52 hours. For the case design Speake Marin was inspired by the City of London. The combination of design elements with curved and angular contours recalls London's financial district, where modern skyscrapers contrast with traces of traditional architecture. Seen in this light, the concept of squaring the circle is quite appropriate. Marking the celebration of the company's twentieth anniversary, the first limited edition of the unlimited Ripples Black Original was presented. The name Ripples Gold refers to the exactly 20 grams of yellow gold used by Speake Marin for each dial in the limited edition of 50. Other members of the collection include the Ripples Blue Jeans, the Ripples Metallic Green, Ripples Date and the 2024 Ripples Infinity Date and Ripples Dune Date. Further versions are envisaged. Annual production has increased from 100 to some 500 wristwatches. Significant growth that in no way detracts from the exclusive character of the Speake Marin brand.

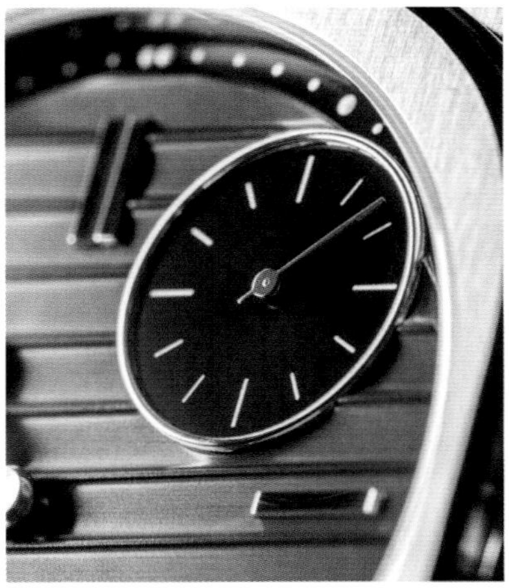

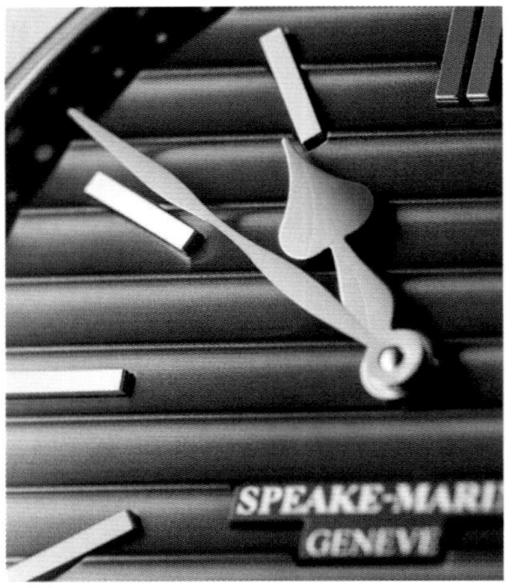

NOMEN EST OMEN

Die Geschichte der kleinen Genfer Uhrenmanufaktur begann 2002 mit Peter Speake-Marin. Weil dem gebürtigen Engländer für eine nachhaltige Entwicklung die Finanzkraft fehlte, trat 2012 Christelle Rosnoblet auf den Plan. Die Uhrenliebhaberin erkannte das Potenzial und brachte die Marke auf Zukunftskurs. Seit 2014 zeichnet „Le Cercle des Horlogers", ein anerkannter Mechanikspezialist für die Herstellung, Montage und Regulierung aller Speake Marin-Kaliber verantwortlich. Damit erreichte die Französin die intendierte Unabhängigkeit von externen Zulieferern und dazu den von ihr vorgegebenen Qualitätsstandard der verbauten Uhrwerke. 2017 schied der Gründer aus.

Zu den Säulen der Kollektion gehört einmal die „Openworked" mit vorne sichtbarer Zeitmechanik und zum anderen die sportlich-elegante „Ripples", mit der Speake Marin ab 2022 einen internationalen Durchbruch erzielte.

Der einprägsame Name resultiert aus der wellenförmigen Struktur des Zifferblatts. Sofort ins Auge sticht der höchst ungewöhnlich bei „2" rotierende Sekundenzeiger. Selbiger ist mittlerweile eine Art Markenzeichen und typisches Merkmal aller hauseigenen SMA-Automatikkaliber mit Mikrorotor. In der „Ripples" verbaut Speake Marin das flache SMA03-T, dessen Gangautonomie 52 Stunden beträgt. Beim Gehäusedesign ließ sich Speake Marin von der Londoner City inspirieren. Die Kombination aus runden und eckigen Gestaltungselementen erinnert an den Londoner Finanzdistrikt, wo moderne Wolkenkratzer im Kontrast stehen zu den Überresten traditionsreicher Architektur. So gesehen ist ein Vergleich mit der Quadratur des Kreises keineswegs aus der Luft gegriffen. Anlässlich der Feier des 20. Firmenjubiläums gesellte sich zur unlimitierten „Ripples Black Original" eine erste limitierte Edition. Der Name „Ripples Gold" resultiert bei den 50 Exemplaren aus der Tatsache, dass Speake Marin für jedes Zifferblatt exakt 20 Gramm Gelbgold verwendet. Weitere Familienmitglieder sind die „Ripples Blue Jeans", die „Ripples Metallic Green", die „Ripples Date" sowie die 2024 vorgestellten „Ripples Infinity Date" und „Ripples Dune Date". Zukünftige Kreationen sind in Planung. Mittlerweile ist die Jahresproduktion von 100 auf rund 500 Armbanduhren geklettert. Das ist ein beachtliches Wachstum, ändert jedoch nichts am exklusiven Charakter der Marke Speake Marin.

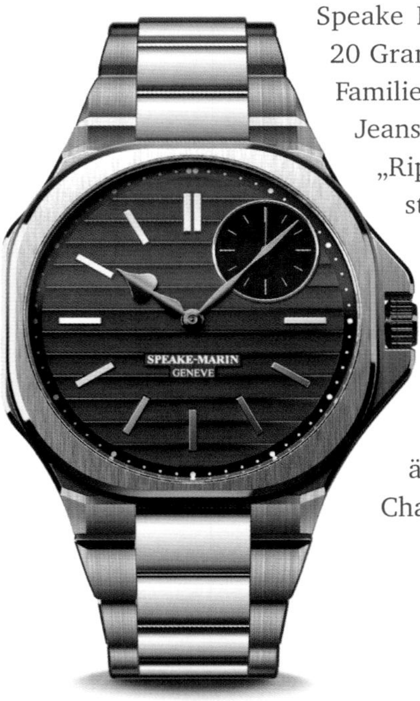

< Ripples Black, 2023

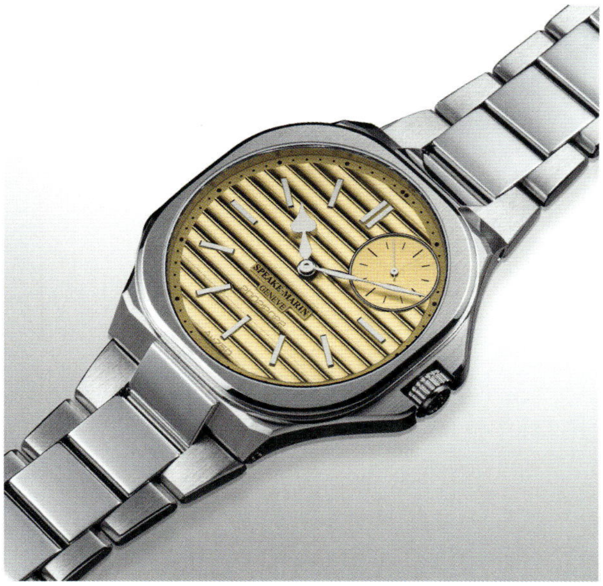

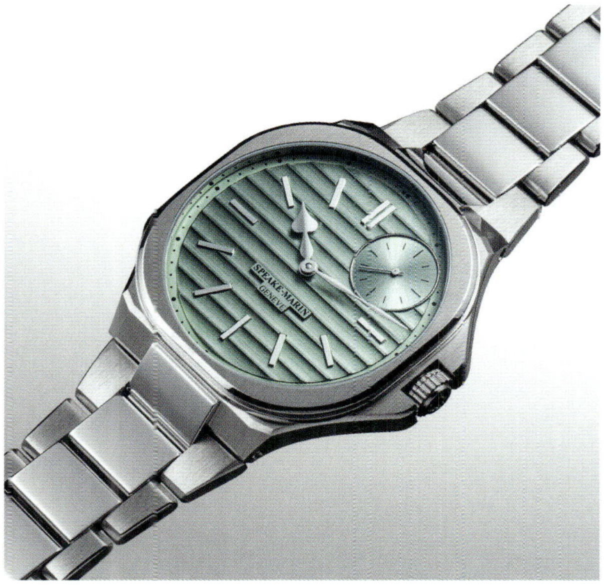

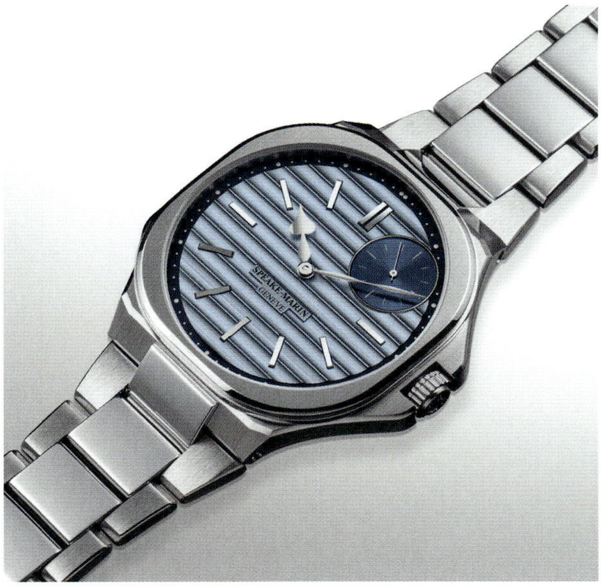

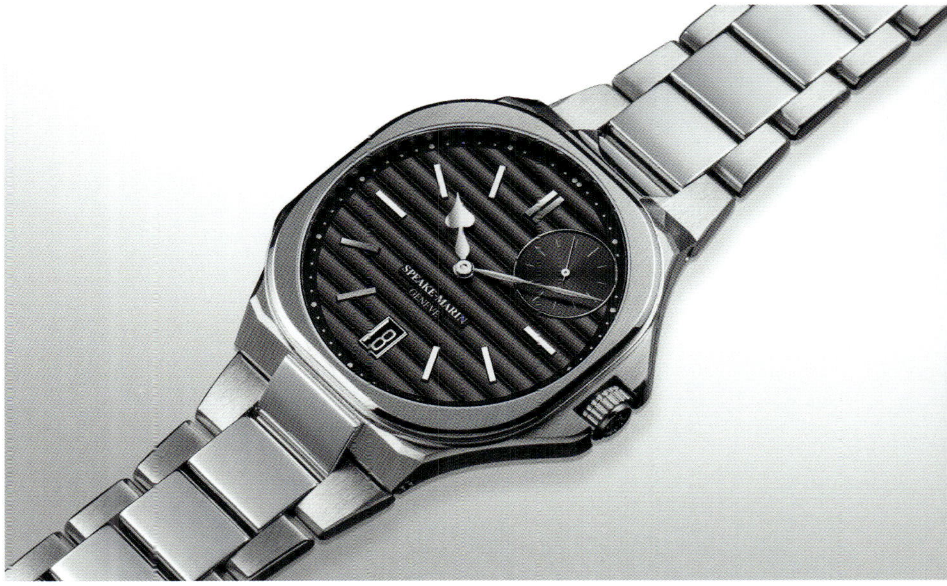

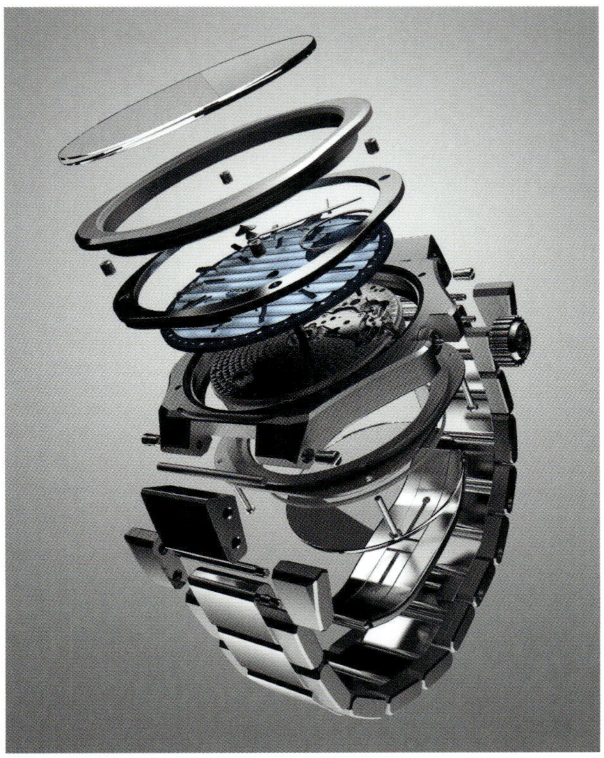

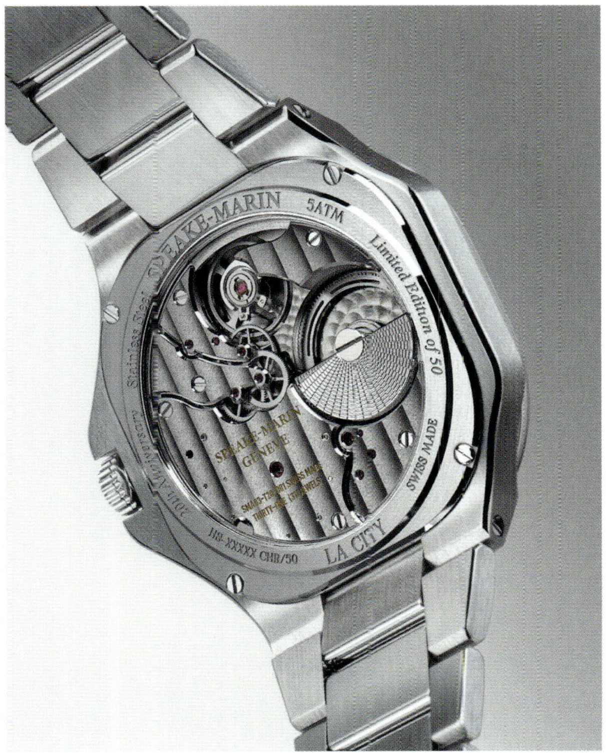

From top left, clockwise / Von oben links im Uhrzeigersinn:
Ripples Gold 20th Anniversary, 2023 / Ripples Metallic Green, 2023 / Ripples Date, 2023 / Ripples Gold 20th Anniversary, caseback, 2023 / Ripples Blue Jeans, exploded view / Explosionsdarstellung, 2023

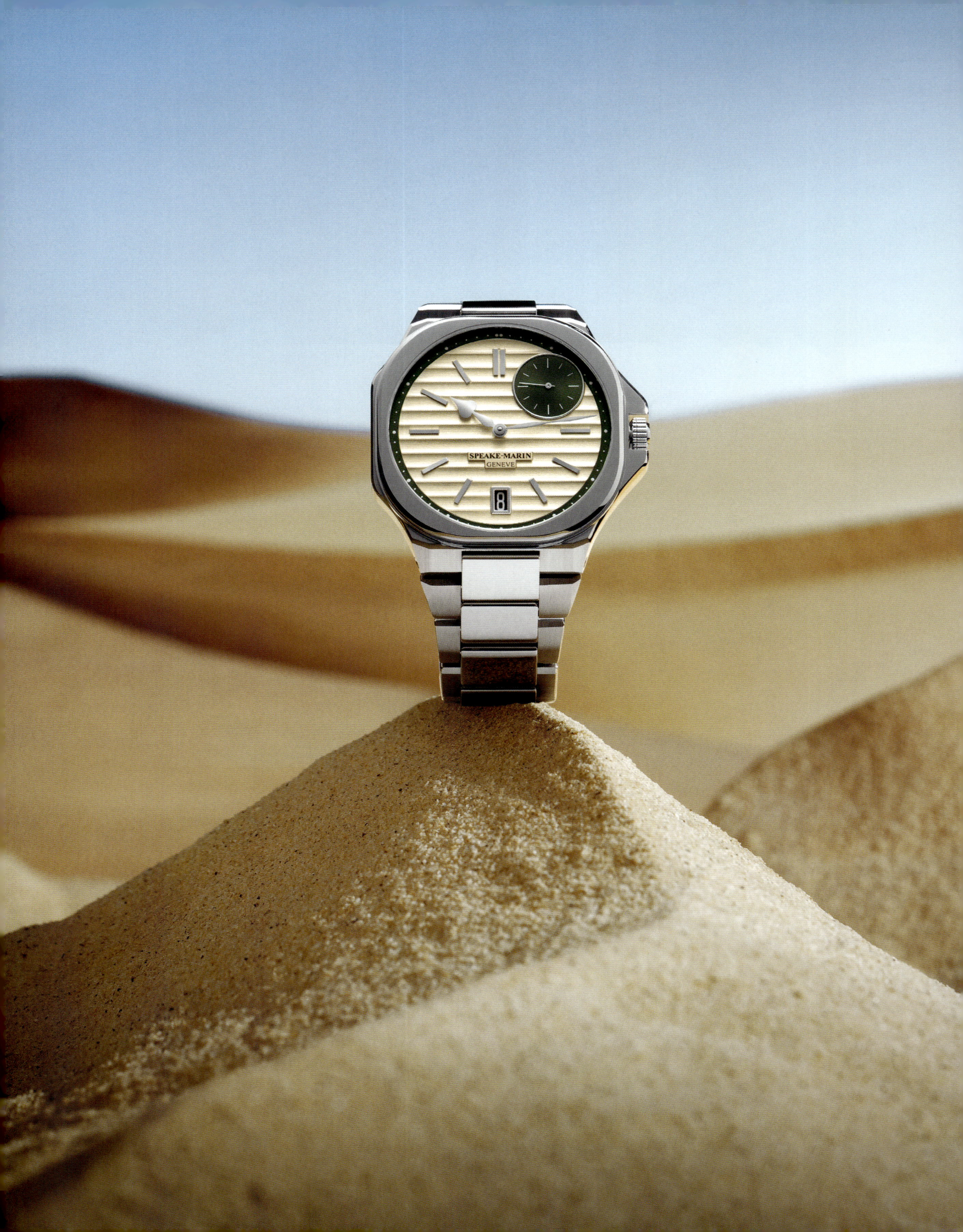
SPEAKE-MARIN
GENEVE

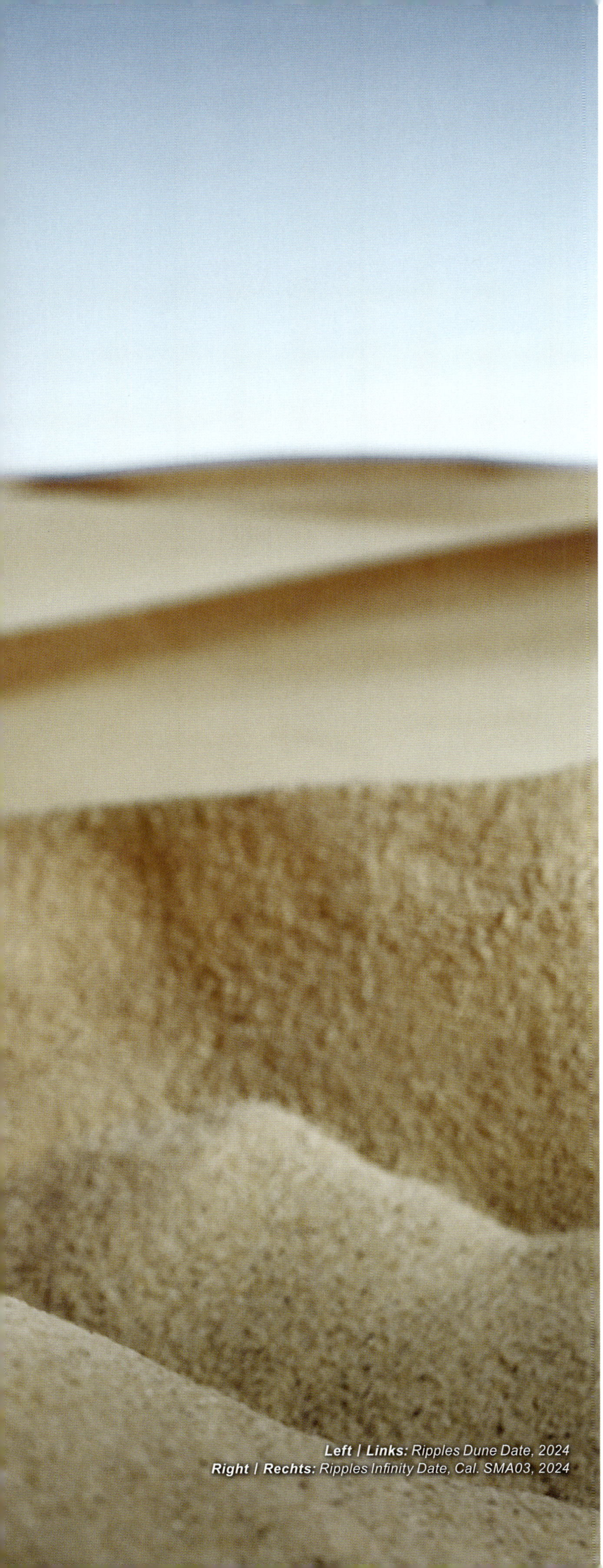

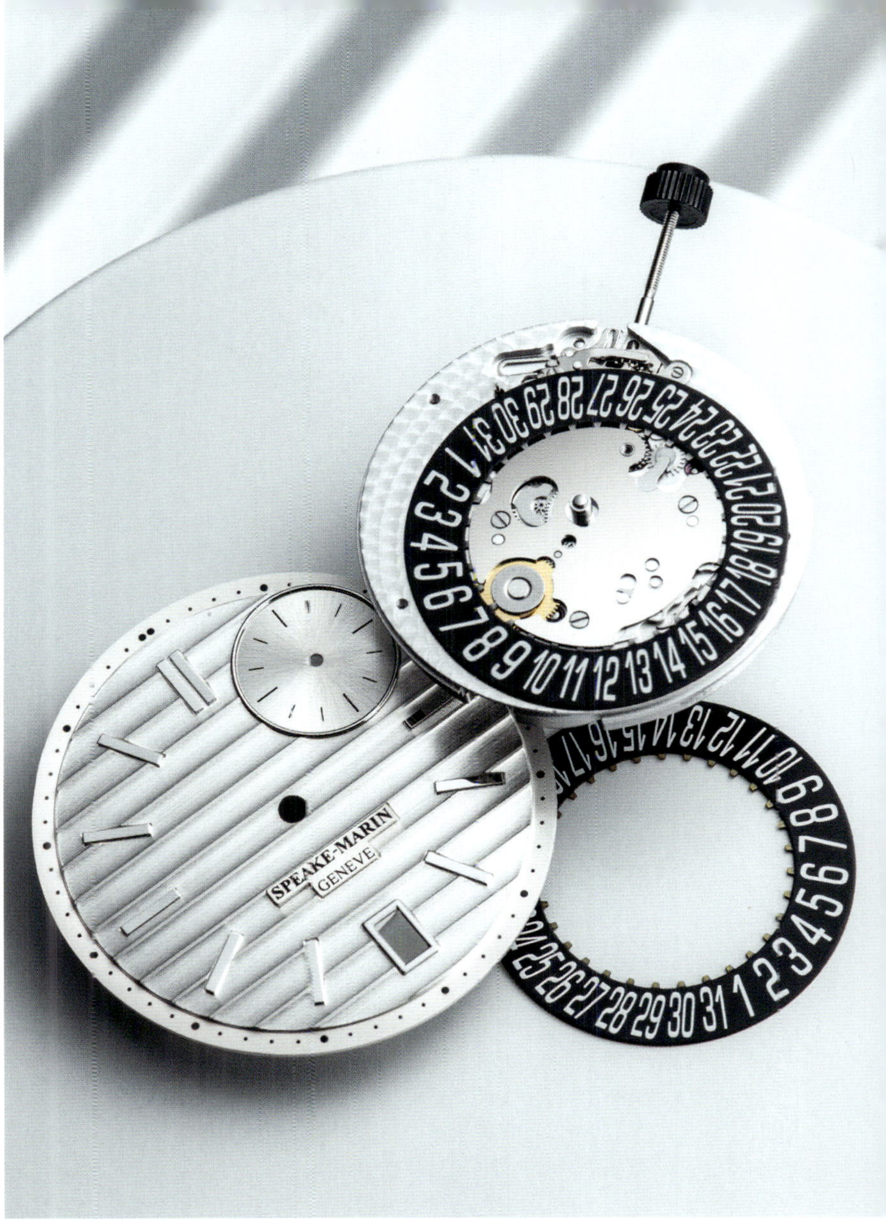

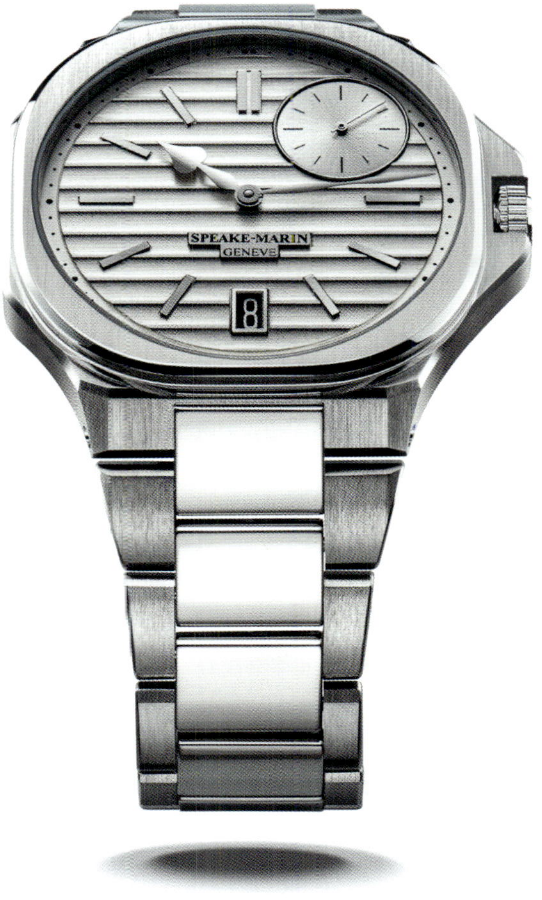

Left | Links: *Ripples Dune Date, 2024*
Right | Rechts: *Ripples Infinity Date, Cal. SMA03, 2024*

PILOT'S WATCHES

PILOT'S WATCHES FROM PFORZHEIM

Modern cockpits are deemed comfort zones in civil aviation. Navigation is via the Global Positioning System (GPS). Computers supply information on altitude, speed, range and fuel consumption. There is no mention any longer of deck watches. In the pre-electronic age, these precision tools were used by the military, research institutions, as well as in aviation and maritime navigation to monitor and compare time. In combat, 55-millimetre watches, strapped over the wearer's gear, ensured swift, reliable reading of the time accurate to the second. Not wanting to leave anything to chance, the relevant German offices and High Command divisions precisely defined the design, engineering and regulation specifications for instruments classed as Fl.23883. The 23 stood for navigation device. All movements destined for military purposes were fitted with an indirect-drive central second with a counterbalance and hack mechanism. Prior to delivery, all timepieces had to comply with the criteria for the issue of an official certificate of accuracy. There were no exceptions. One of only five manufacturers who supplied such deck watches was the Pforzheim watchmaker Walter Storz (Stowa). The 2812 calibre movements, decorated with fine Geneva stripes, were from Unitas in Switzerland. The historical models with a Type B dial are particularly rare. Their defining characteristic is a small sub-chapter ring with hour numerals in the centre of the dial. The archives show that probably only 42 of them were produced. Fans who appreciate this distinctive dial in its original form without a signature as well as the authentic provenance of their pilot's watch will find Stowa a good address. The Flieger Classic Baumuster B is available in different versions to suit your personal taste. Naturally, Stowa also supplies pilot's watches with Type A dials. In this case, the hour numerals are close to the minute track.

FLIEGERUHREN AUS PFORZHEIM

In der zivilen Luftfahrt gelten moderne Cockpits als Komfortzonen. Das Global Positioning System (GPS) übernimmt die Navigation. Computer liefern Informationen zu Flughöhe, Geschwindigkeit, Reichweite und Kraftstoffverbrauch. Niemand spricht mehr von Beobachtungsuhren. Derartige Präzisionsinstrumente nutzten Militär, Forschungseinrichtungen sowie Luft- und Seefahrt im vorelektronischen Zeitalter zum Überwachen und Vergleichen der Zeit. Bei Kampfeinsätzen gewährleisteten 55 Millimeter große, mit Hilfe langer Riemen über der Montur befestigte Armbanduhren rasches, sicheres und sekundengenaues Ablesen. Nachdem die einschlägigen deutschen Dienststellen und Oberkommandos nichts dem Zufall überlassen wollten, definierten sie die Konstruktion, Ausführung und Regulierung der mit Fl. 23883 klassifizierten Instrumente ganz genau. „23" stand dabei für Navigationsgerät. Alle für militärischen Zwecke verbauten Werke verfügten über einen zentral angeordneten und vom Uhrwerk indirekt angetriebenen Sekundenzeiger mit Gegengewicht sowie eine Unruh-Stoppvorrichtung. Vor ihrer Auslieferung mussten sämtliche Zeitmesser die Kriterien zur Ausfertigung eines offiziellen Gangscheins ohne Wenn und Aber erfüllen. Einer der nur fünf Fabrikanten, die derartige Beobachtungs-Armbanduhren lieferten, war der Pforzheimer Fabrikant Walter Storz (Stowa). Die mit feinem Genfer Streifenschliff dekorierten Rohwerke vom Kaliber 2812 stammten von Unitas aus der Schweiz. Besonders rar sind die historischen Modelle mit einem Zifferblatt des Baumusters B. Sein Kennzeichen ist ein kleiner, ins Zifferblattzentrum gerückter Stundenkreis. Aus den Archiven geht hervor, dass davon vermutlich nur 42 Exemplare entstanden sind. Fans, welche dieses markante Zifferblatt wie einst ohne jedwede Signatur schätzen und dazu eine authentische Provenienz ihrer Fliegerarmbanduhr, werden weiterhin bei Stowa fündig. Die „Flieger Klassik Baumuster B" gibt es dort gleich in verschiedenen Ausführungen, ganz nach persönlichem Gusto. Natürlich liefert Stowa auch Fliegeruhren mit Zifferblättern des Baumusters A. Hier befinden sich die Stundenziffern weiter außen neben der Minuterie.

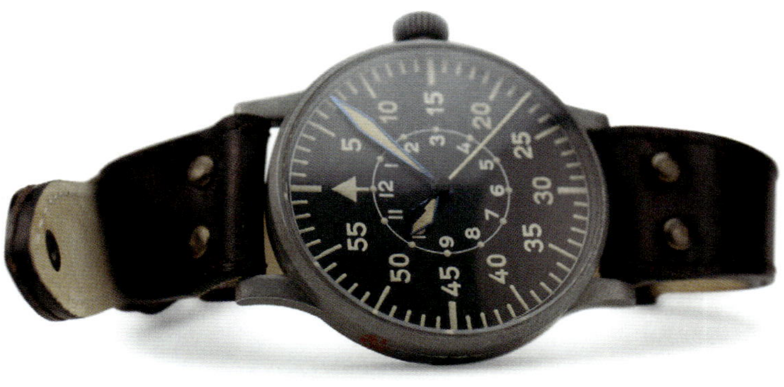

Type B dial | Baumuster B, 1942

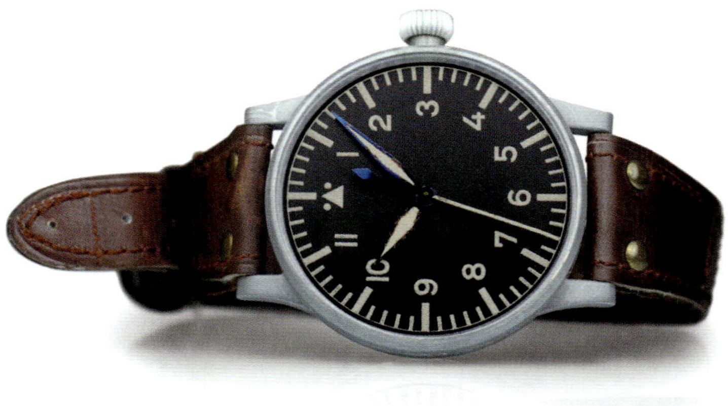

Type A dial | Baumuster A, 1940

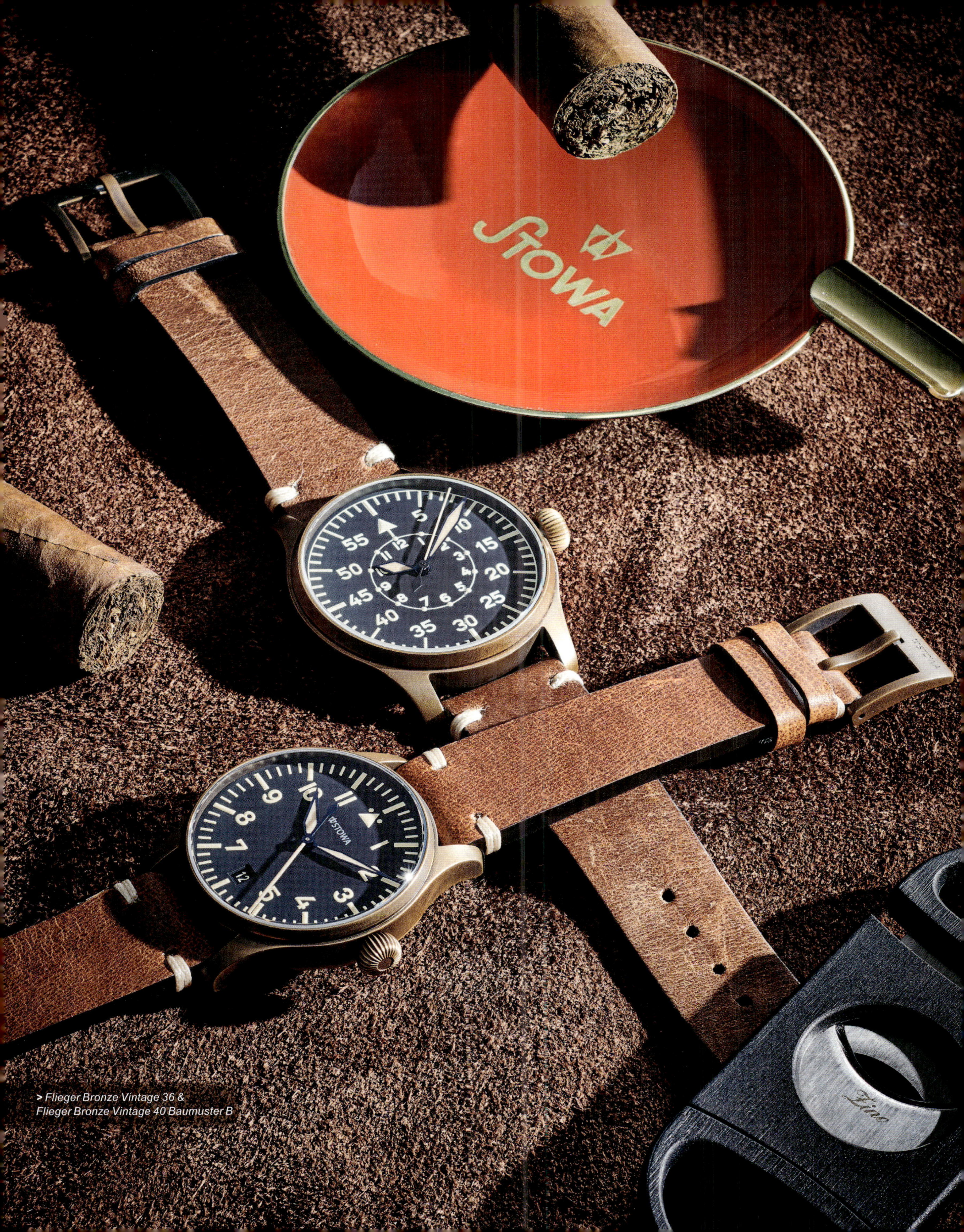

> *Flieger Bronze Vintage 36 &*
> *Flieger Bronze Vintage 40 Baumuster B*

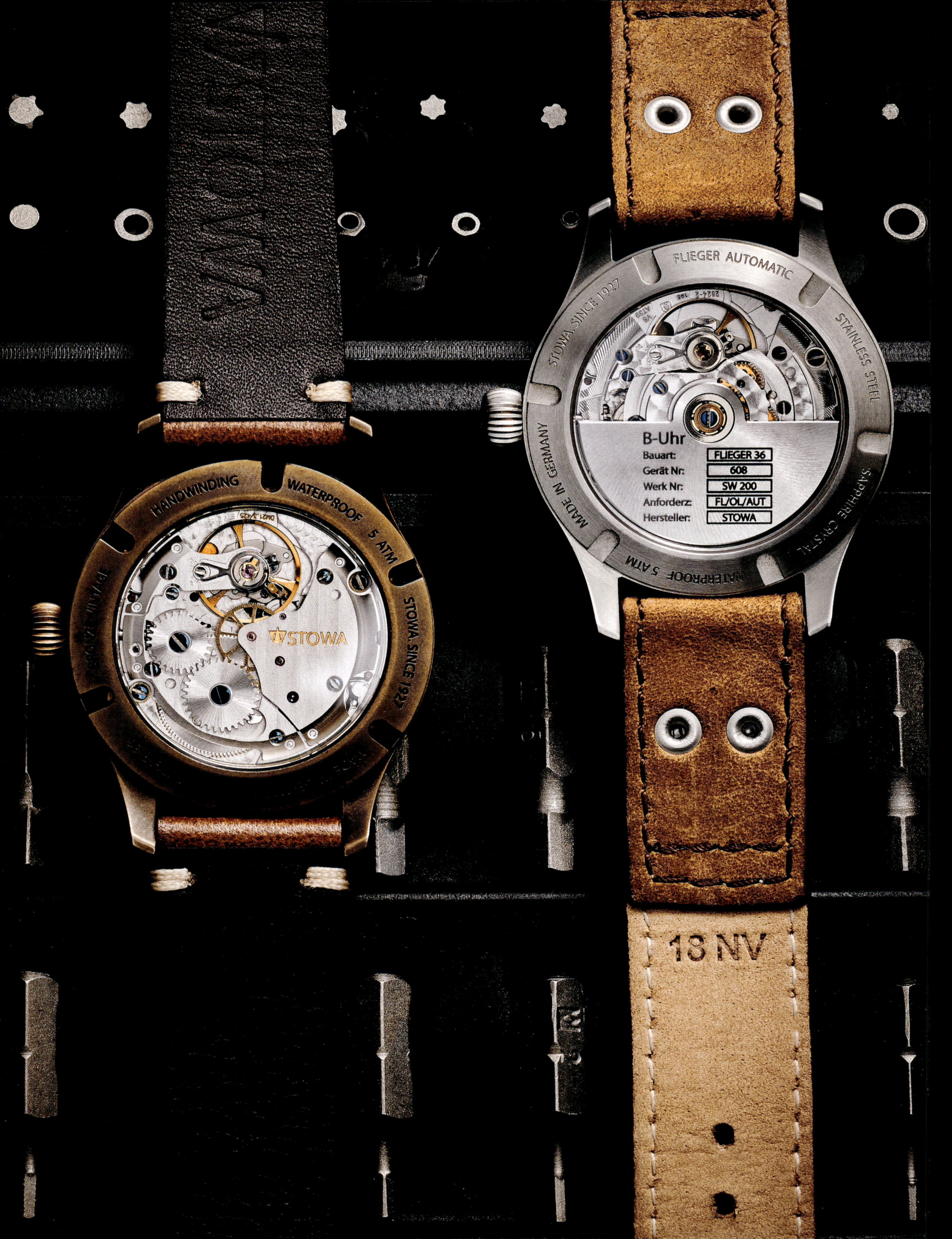
HANDWINDING
WATERPROOF 5 ATM
STOWA SINCE 1927
STOWA
FLIEGER AUTOMATIC
STAINLESS STEEL
SAPPHIRE CRYSTAL
WATERPROOF 5 ATM
MADE IN GERMANY
STOWA SINCE 1927
B-Uhr
Bauart: FLIEGER 36
Gerät Nr: 608
Werk Nr: SW 200
Anforderz: FL/OL/AUT
Hersteller: STOWA
18 NV

2015

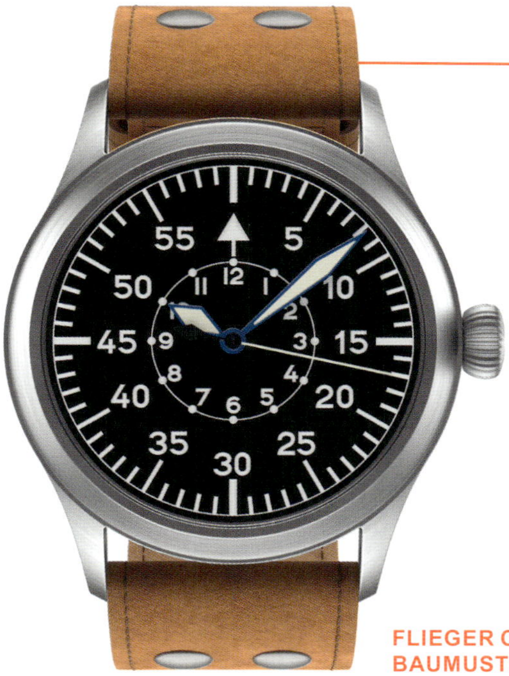

FLIEGER CLASSIC SPORT
BAUMUSTER B

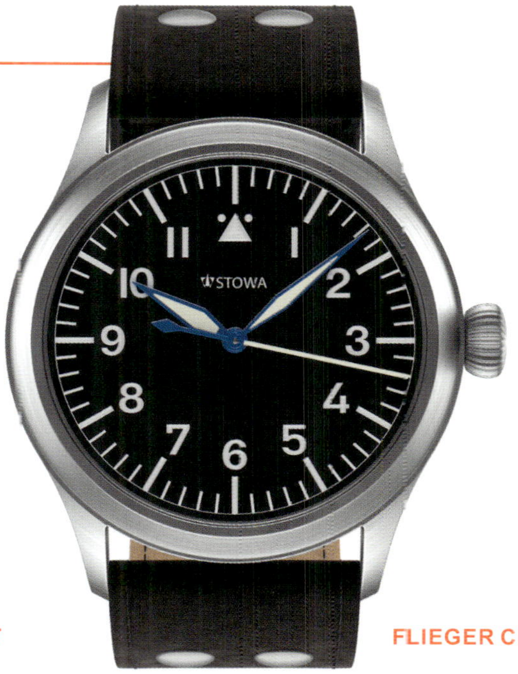

FLIEGER CLASSIC SPORT

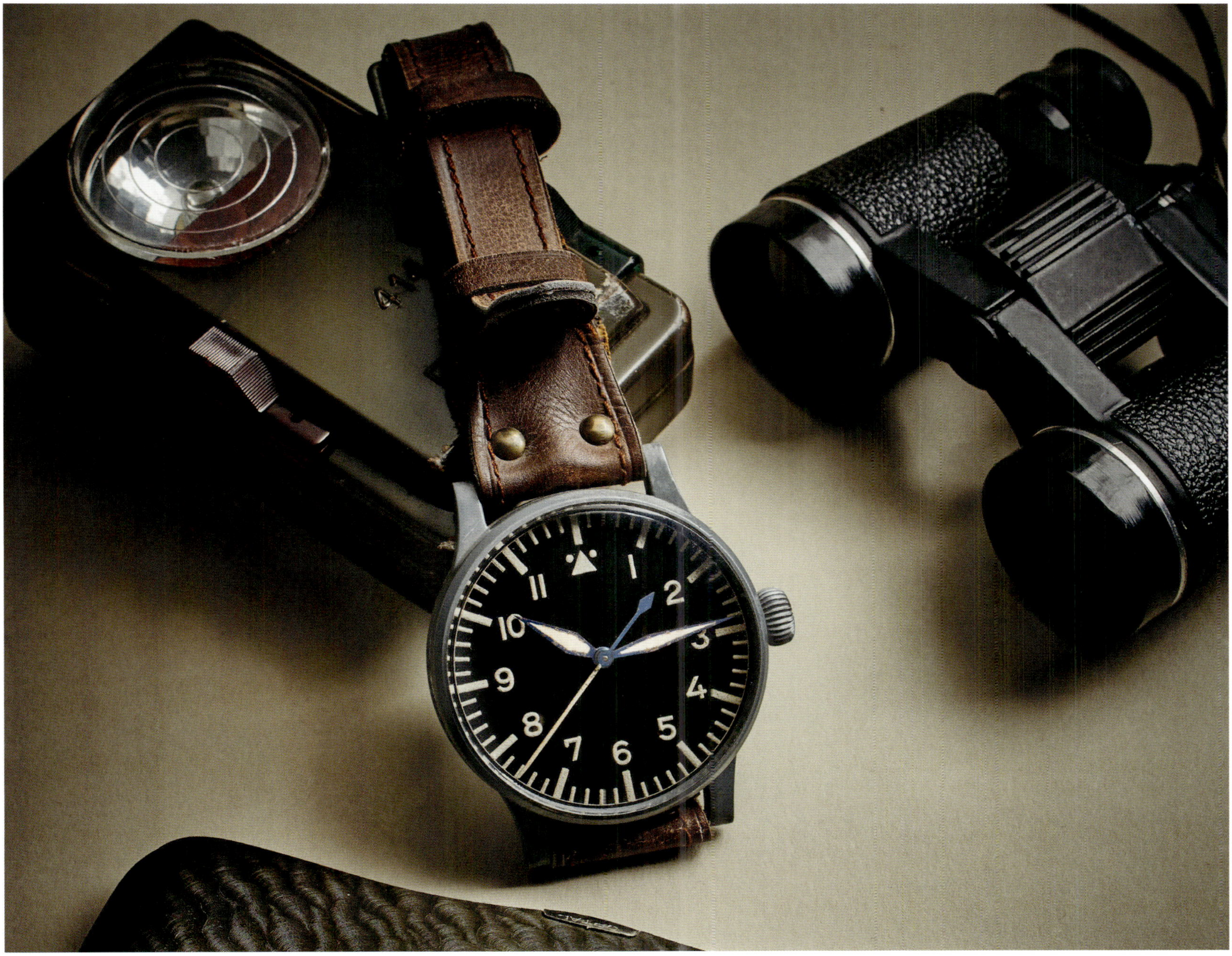

ARMIN STROM 2016

MIRRORED FORCE RESONANCE

BALANCED UNISON

Two are better than one. Metaphorically speaking, the old adage can be applied to mechanical watches. As is well known, the oscillating and escapement system consisting of the balance wheel, balance spiral, pallet anchor and escape wheel evenly divides the passing time into small, well-defined units. The gear train transmits them to the time display of whatever design. Modern wristwatches tick at a frequency of four hertz. The second hand therefore advances in increments of one eighth of a second. Deviations from the predefined oscillation duration will inevitably cause a watch to go wrong. In addition to mechanical faults, the gravity of the earth also influences the aimed-for precision – especially in a vertical position, if the oscillator's centre of gravity is not exactly in the centre of the balance wheel. Due to its position on the wrist and the latter's constant movement, a tourbillon has only limited effect. A precision-enhancing alternative such as Armin Strom's Resonance models proposes two oscillating balance wheels. The Mirrored Force Resonance with calibre ARF15 was presented 2016. In 2018 the Pure Resonance followed, equipped with the same hand-wound movement. New Resonance models – with calibre ARF21 – were released 2022 and 2023.

In this case, resonance refers to the rhythmic movement of two bodies in relation to one another, thus effecting their continuous mutual regulation. The lower regulator, which on the calibres ARF15 and ARF21 also controls the time display, can be loosely described as an exciter. Its counterpart, spurred by a special clutch spring, oscillates in the same direction, acting as a stabilising resonator. The watchmaker assembles the new calibre ARF21, which has a diameter of 37.2 millimetres and 276 components, in its Biel workshop. The timepiece offers a power reserve of 48 hours. Instead of the common 28,800 semi-oscillations per hour, this regulating duo executes a rarer 25,500 semi-oscillations every hour. This is the equivalent of a frequency of 3.5 hertz. The ingenious mechanism is protected by a 43-millimetre case that is waterproof to three bar.

IM EINKLANG GESCHWUNGEN

Doppelt genäht hält besser. Im übertragenen Sinn lässt sich die altbekannte Spruchweisheit auch auf mechanische Uhren anwenden. Bekanntlich unterteilt das Schwing- und Hemmungssystem bestehend aus Unruh, Unruhspirale, Anker und Ankerrad die gleichförmig verstreichende Zeit in kleine, wohl definierte Abschnitte. Das Räderwerk überträgt sie zur wie auch immer gearteten Zeitanzeige. Moderne Armbanduhren ticken mit vier Hertz. Der Sekundenzeiger bewegt sich also in Achtelsekundenschritten vorwärts. Abweichungen von der vordefinierten Schwingungsdauer lassen eine Uhr zwangsläufig falsch gehen. Neben mechanischen Fehlern wirkt insbesondere auch die Erdanziehungskraft der intendierten Präzision entgegen. Besagte Gravitation vor allem in senkrechter Position, wenn sich der Schwerpunkt des Oszillators nicht exakt im Zentrum der Unruhwelle befindet. Wegen permanenter Lagewechsel kann ein Tourbillon seine Wirkung am Handgelenk nur begrenzt entfalten. Als genauigkeitssteigernde Alternative bietet sich, wie von Armin Strom praktiziert, ein Paar schwingender Unruhen an. 2016 ging die „Mirrored Force Resonance" mit dem Kaliber ARF15 an den Start. 2018 folgte die mit dem gleichen Handaufzugswerk ausgestattete „Pure Resonance". 2022 und 2023 brachten neue Resonanz-Modelle, in denen das Kaliber ARF21 tickt.

Resonanz, zu Deutsch „Widerhallen", meint in diesem Fall das Mitschwingen eines Körpers mit einem anderen. Und dadurch auch die gegenseitige Synchronisation. Der untere Gangregler, welcher bei den Kalibern ARF15 und ARF21 auch die Zeitanzeige steuert, lässt sich salopp als „Erreger" bezeichnen, das durch eine spezielle Kupplungsfeder angeregte obere und in gleicher Richtung oszillierende Pendant als stabilisierender „Resonator". Das neue ARF21, Durchmesser 37,2 Millimeter, montieren die Uhrmacher in den Bieler Ateliers aus 276 Komponenten. Seine Gangautonomie beträgt 48 Stunden. In diesem Fall vollzieht das gangregelnde Duo stündlich nicht die verbreiteten 28.800, sondern eher seltene 25.200 Halbschwingungen. Umgerechnet entspricht das einer Frequenz von 3,5 Hertz. Die ausgeklügelte Mechanik schützt ein 43 Millimeter großes und bis drei bar wasserdichtes Gehäuse.

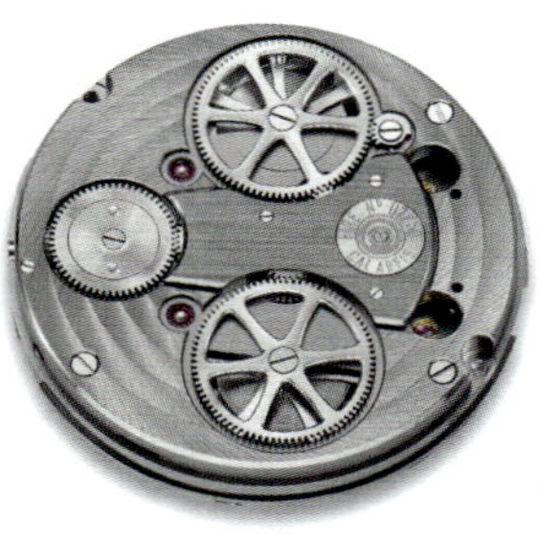

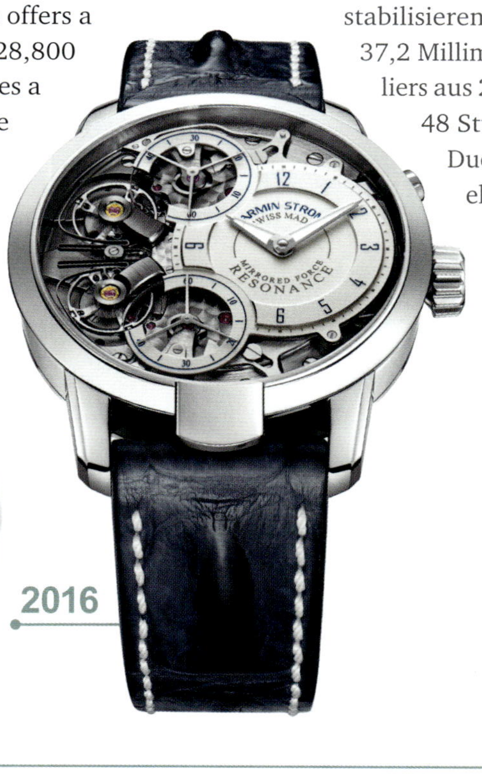

2016

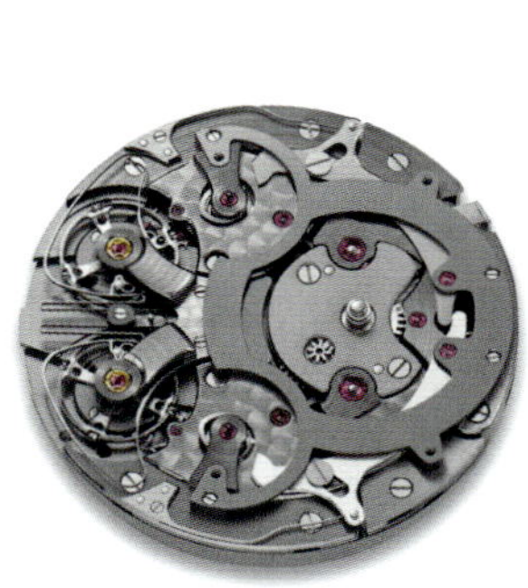

Cal. ARF15

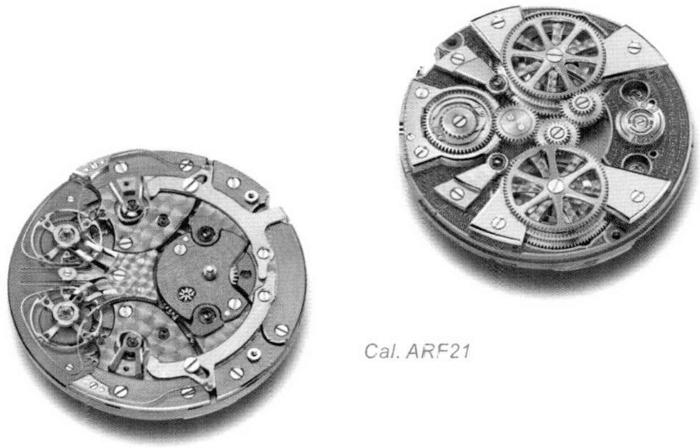

Cal. ARF21

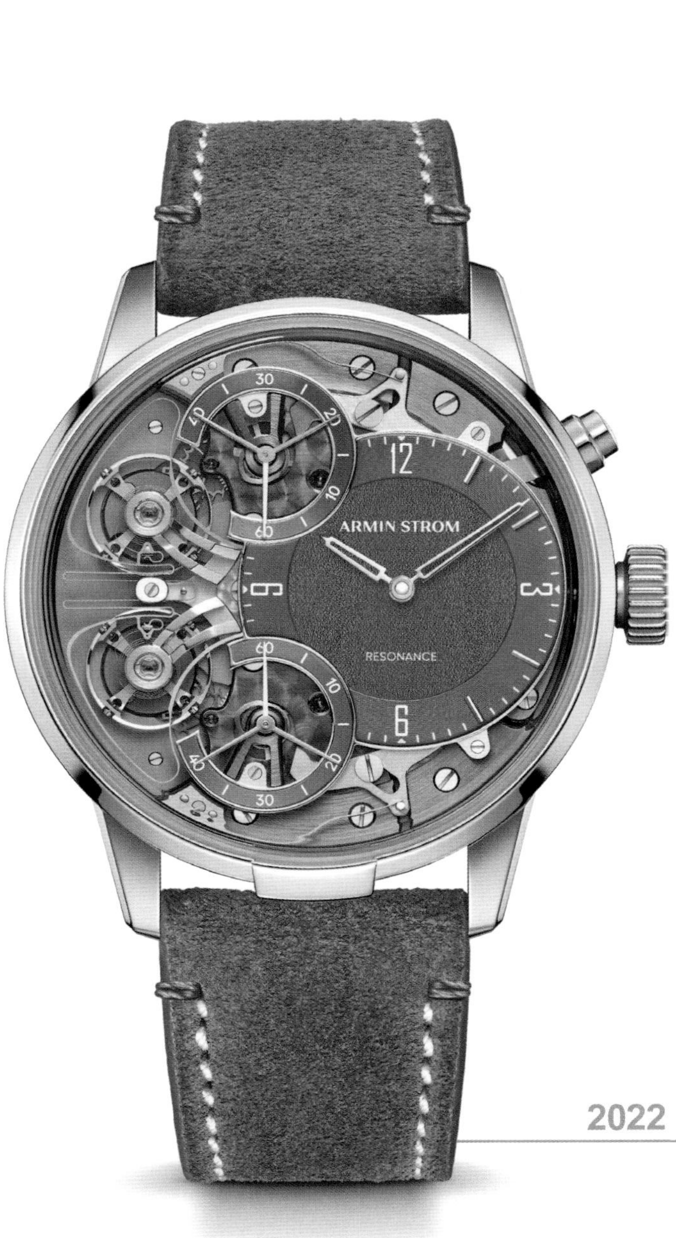

2022

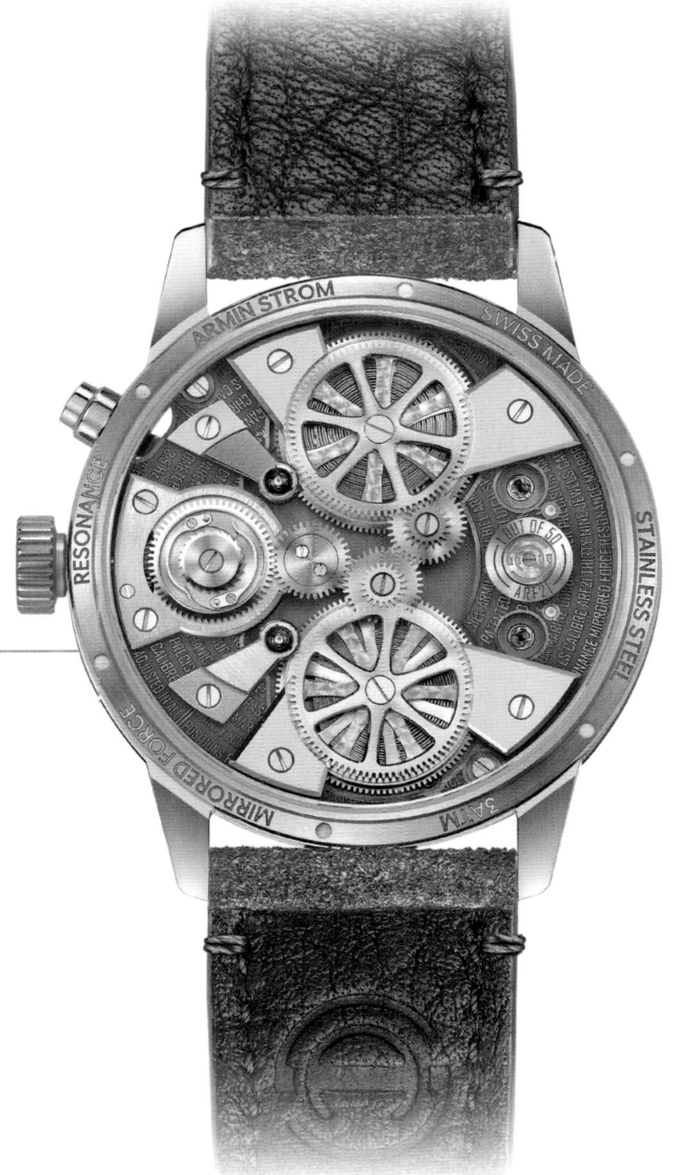

PURE RESONANCE,
MANUFACTURE EDITION GREEN

CARL SUCHY 2017

WALTZ

A WRISTFUL OF VIENNESE WALTZ

What do Emperor Franz Joseph I, his wife Elisabeth, better known as Sissi, and the psychologist Sigmund Freud have in common? It is easy! All three owned watches bearing the signature of Carl Suchy. The story unfolds in 1796, when the later watchmaker was born in Prague. Sometime in the early twentieth century, the traditional brand quietly bowed out. Its discreet renaissance took place in 2017. Despite the watchmaker's eminent past, Carl Suchy was only appreciated by connoisseurs with a deep affection for Viennese watchmaking. Its comeback therefore called for a flagship of outstanding character. Its distinctive design was inspired by Adolf Loos. Around 1900, this architect was one of Vienna's most influential but also most controversial personalities. What could be more fitting therefore than to pick up where Suchy left off.

Viennese Modernism with its minimalist, yet emotionally laden design is to the new Carl Suchy brand what the waltz is to the Austrian capital. They belong together. For this very reason, the outstanding and legally protected feature of the Carl Suchy Waltz is the seconds indicator designed as a waltzing disc. A wry nod to Vienna, where seconds never counted. And there is still no reason to call into question this traditional philosophy of life. In this spirit, our Waltz's allusion to the waltz is purely metaphorical. What matters is the spirited and personal lifestyle that the wristwatch expresses. Its joie de vivre is exuded by both versions of the Carl Suchy Waltz, the one closed, the other elaborately skeletonised. Both represent Viennese design, architecture and lifestyle, currently paired with Swiss watchmaking in a signature timepiece. Engineering excellence is evidenced by an ultra-thin automatic micro-rotor calibre by Vaucher, a member of Parmigiani Group. Long live the eternal waltz.

WIENER WALZER AM HANDGELENK

Was haben Kaiser Franz Joseph I., seine besser als „Sissi“ bekannte Gemahlin Elisabeth und der Psychologe Sigmund Freud gemeinsam? Ganz einfach! Alle drei blickten auf Uhren mit der Signatur Carl Suchy. Deren Geschichte reicht zurück bis 1796, als der spätere Uhrmacher in Prag das Licht der Welt erblickte. Irgendwann im frühen 20. Jahrhundert verschwand die Traditionsmarke still und leise von der Bildfläche. Die eher geräuschlose Renaissance erfolgte 2017. Trotz der bedeutenden Vergangenheit konnten da nur Insider mit ausgeprägtem Faible für Wiener Uhrmacherkunst etwas mit Carl Suchy anfangen. Für das Comeback brauchte es deshalb einen nicht alltäglichen Auftritt. Den prägte das von Adolf Loos inspirierte Design. Dieser Architekt gehörte um 1900 in Wien zu den einflussreichen, aber auch höchst umstrittenen Persönlichkeiten. So gesehen war es mehr als logisch, dort neu anzufangen, wo Suchy einst aufhörte.

Die Wiener Moderne mit sehr reduziertem und gleichzeitig hochemotionalem Design passt zur neuen Marke Carl Suchy wie der Walzer zur österreichischen Hauptstadt. Herausragendes und rechtlich geschütztes Merkmal der „Waltz“ ist genau deswegen die als Walzerscheibe gestaltete Indikation jener Sekunde, auf die es in Wien noch nie wirklich ankam. Und es gibt auch heute noch keinen Grund, an dieser tradierten Lebensphilosophie zu rütteln. In diesem Sinne geht es bei besagter „Waltz“ auch nur um den Walzer im übertragenen Sinn. Was zählt, ist ein beschwingter und persönlicher Lebensstil, den diese Armbanduhr am Handgelenk ausdrückt. Dabei spielt es keine Rolle, ob man auf eine Carl Suchy „Waltz“ in der geschlossenen oder in der aufwändig skelettierten Ausführung blickt. Beide repräsentieren Wiener Design, Architektur und Lebenskultur, aktuell gepaart mit Schweizer Uhrmacherkunst in einem charakteristischen Zeitmesser. Ausweis der Exzellenz ist ein ultraflaches Mikrorotor-Automatikkaliber von Vaucher, einem Mitglied der Parmigiani-Gruppe. Es lebe der ewige Walzer.

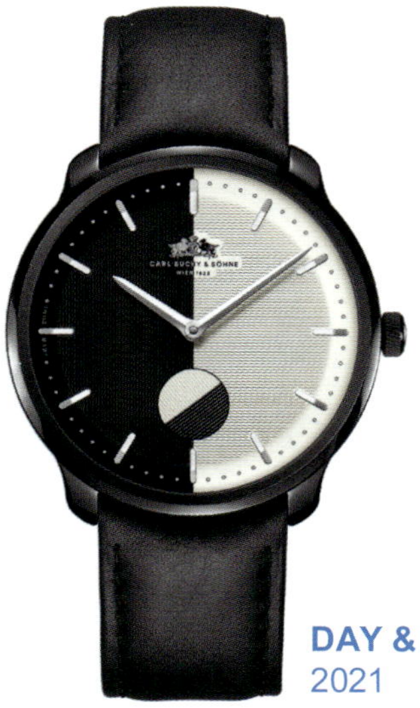

DAY & NIGHT, 2021

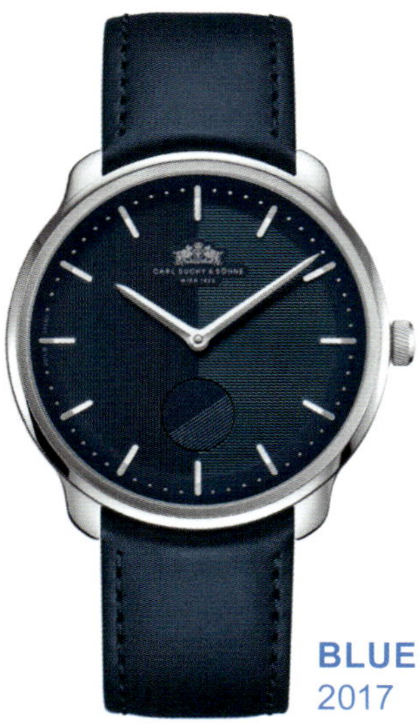

BLUE DANUBE, 2017

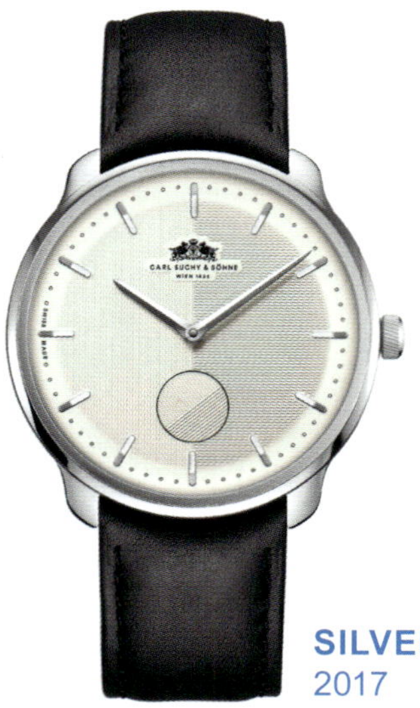

SILVER, 2017

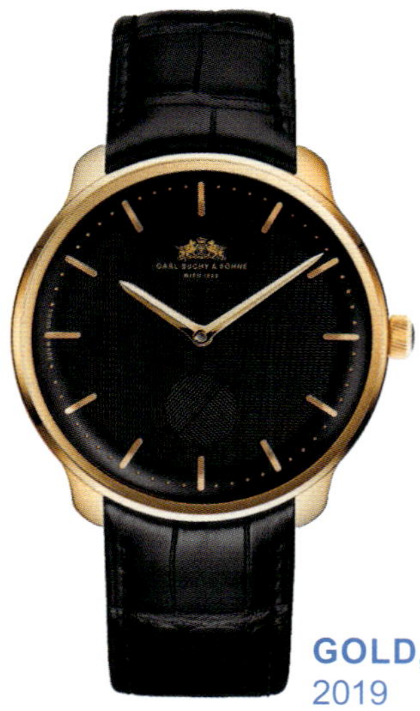

GOLD, 2019

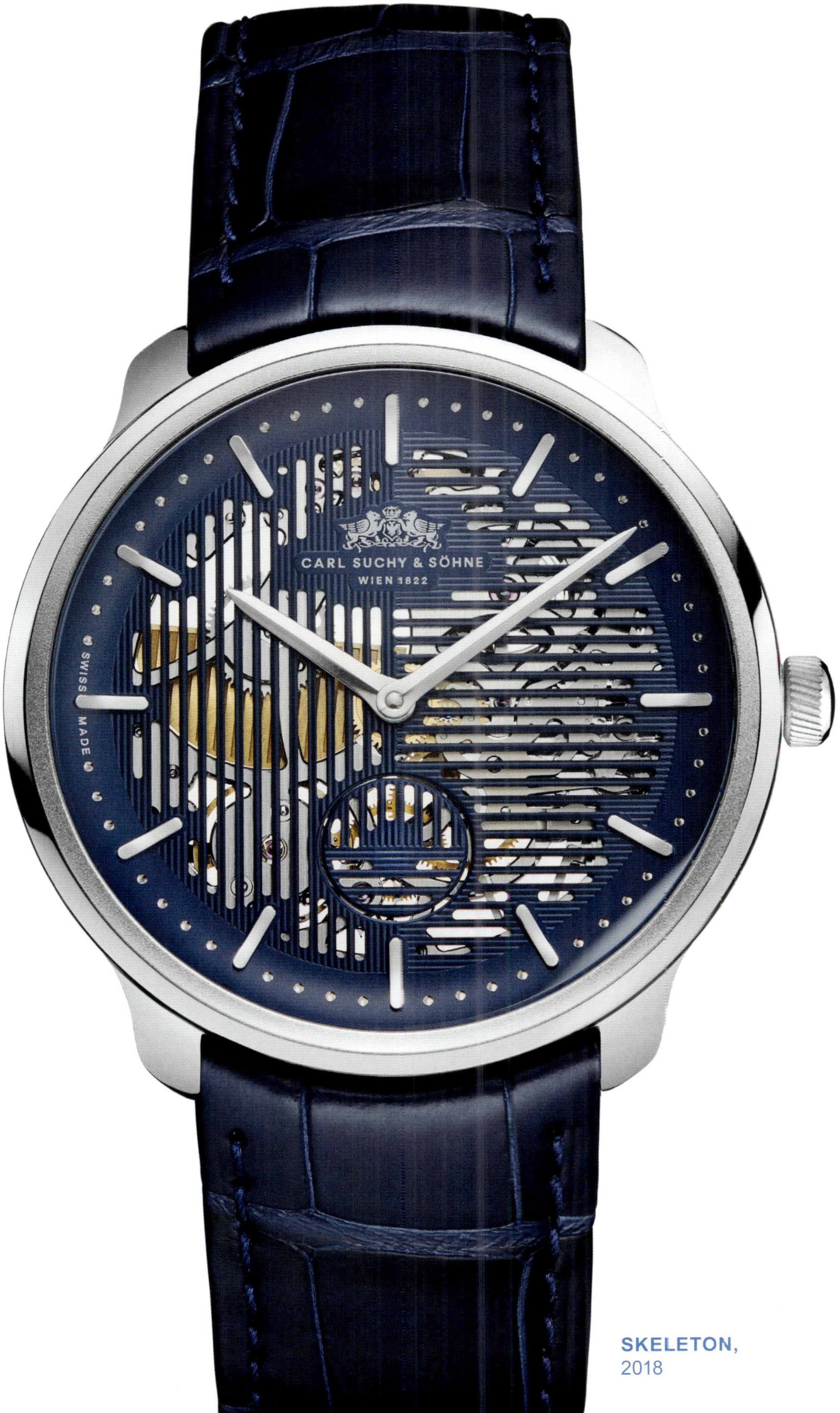

SKELETON,
2018

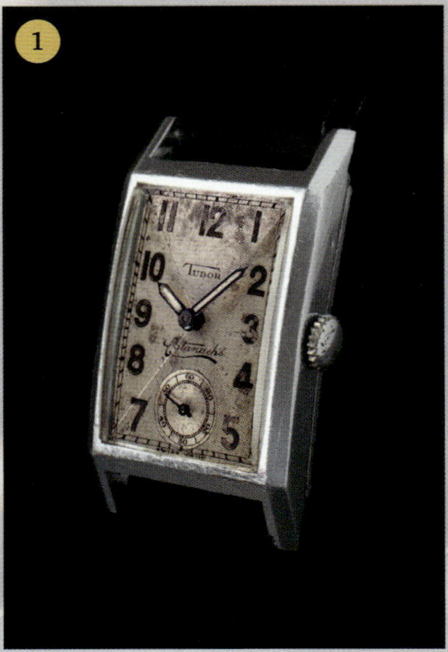

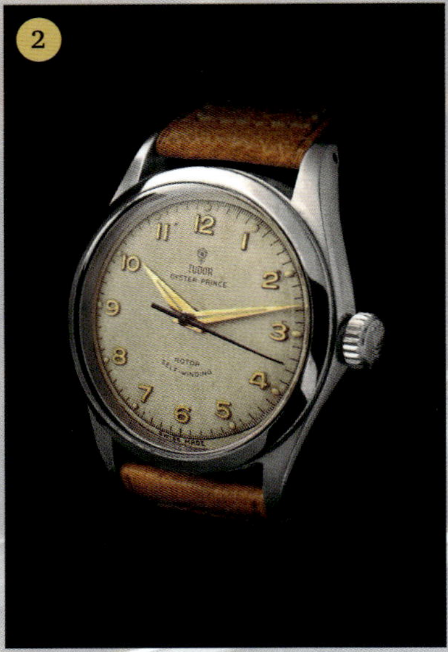

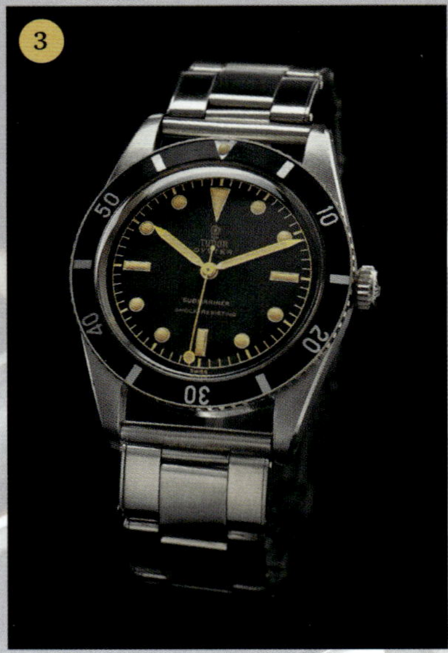

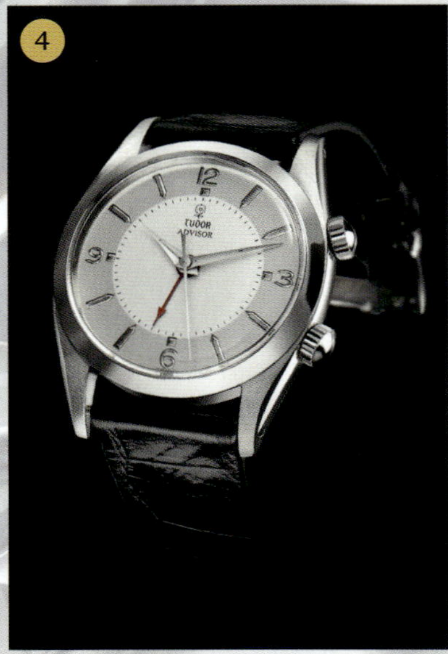

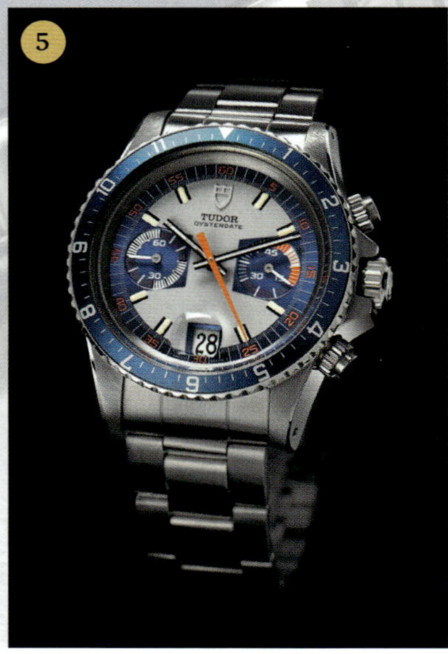

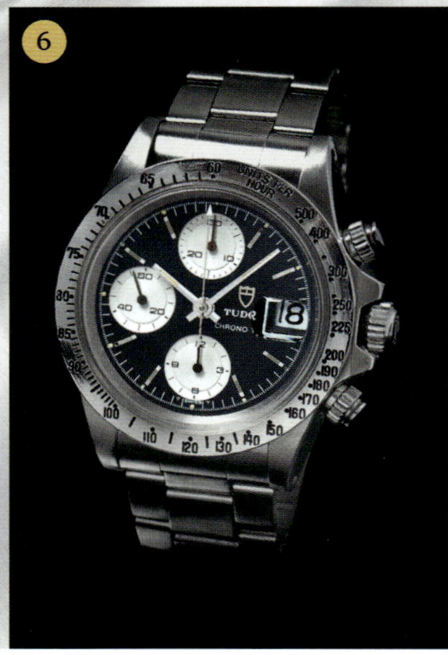

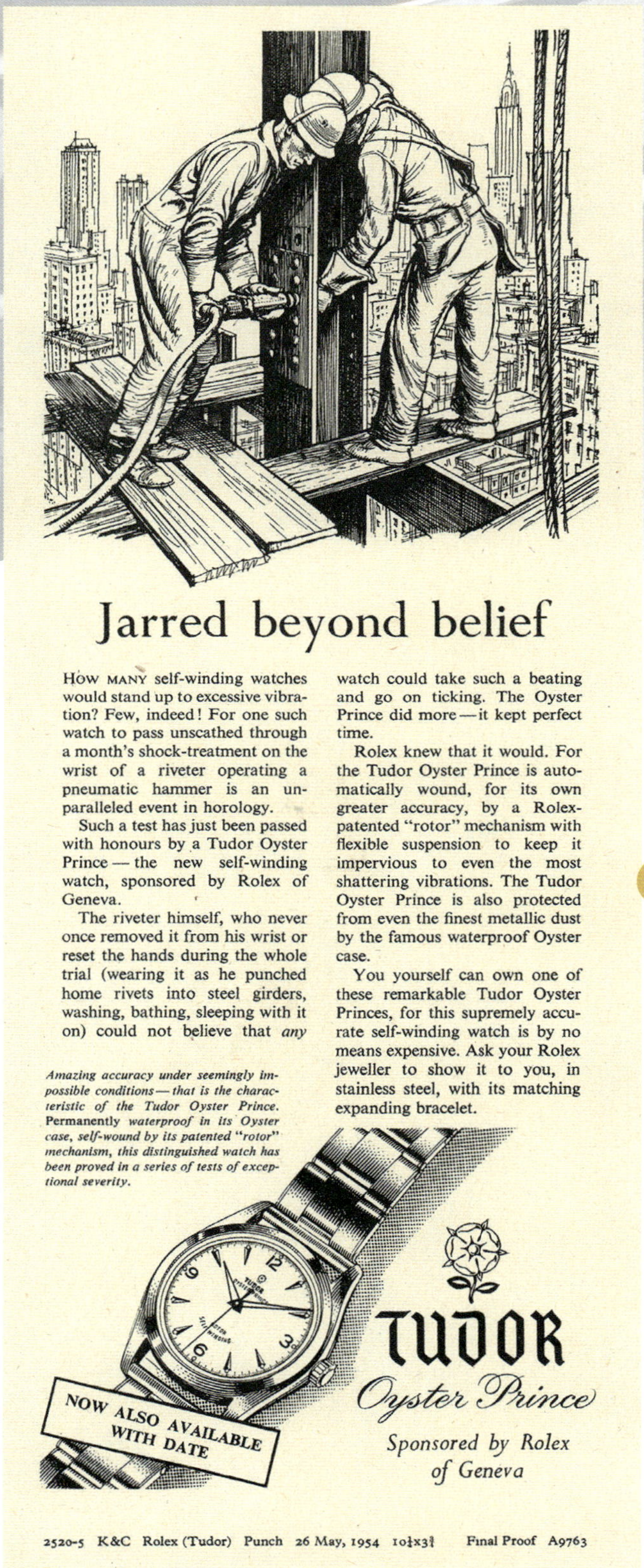

Jarred beyond belief

HOW MANY self-winding watches would stand up to excessive vibration? Few, indeed! For one such watch to pass unscathed through a month's shock-treatment on the wrist of a riveter operating a pneumatic hammer is an unparalleled event in horology.

Such a test has just been passed with honours by a Tudor Oyster Prince — the new self-winding watch, sponsored by Rolex of Geneva.

The riveter himself, who never once removed it from his wrist or reset the hands during the whole trial (wearing it as he punched home rivets into steel girders, washing, bathing, sleeping with it on) could not believe that *any* watch could take such a beating and go on ticking. The Oyster Prince did more — it kept perfect time.

Rolex knew that it would. For the Tudor Oyster Prince is automatically wound, for its own greater accuracy, by a Rolex-patented "rotor" mechanism with flexible suspension to keep it impervious to even the most shattering vibrations. The Tudor Oyster Prince is also protected from even the finest metallic dust by the famous waterproof Oyster case.

You yourself can own one of these remarkable Tudor Oyster Princes, for this supremely accurate self-winding watch is by no means expensive. Ask your Rolex jeweller to show it to you, in stainless steel, with its matching expanding bracelet.

Amazing accuracy under seemingly impossible conditions — that is the characteristic of the Tudor Oyster Prince. Permanently waterproof in its Oyster case, self-wound by its patented "rotor" mechanism, this distinguished watch has been proved in a series of tests of exceptional severity.

NOW ALSO AVAILABLE WITH DATE

TUDOR

Oyster Prince

Sponsored by Rolex of Geneva

2520-5 K&C Rolex (Tudor) Punch 26 May, 1954 $10\frac{1}{4}$x$3\frac{3}{8}$ Final Proof A9763

1) *Catanachs, 1932*
2) *Oyster Prince, 1953*
3) *Oyster Submariner, 1955*
4) *Advisor, 1957*
5) *Oysterdate Montecarlo, 1971*
6) *Oysterdate Big Block, 1976*
7) *Advertisements from the 1950s / Werbeanzeigen aus den 1950er-Jahren*

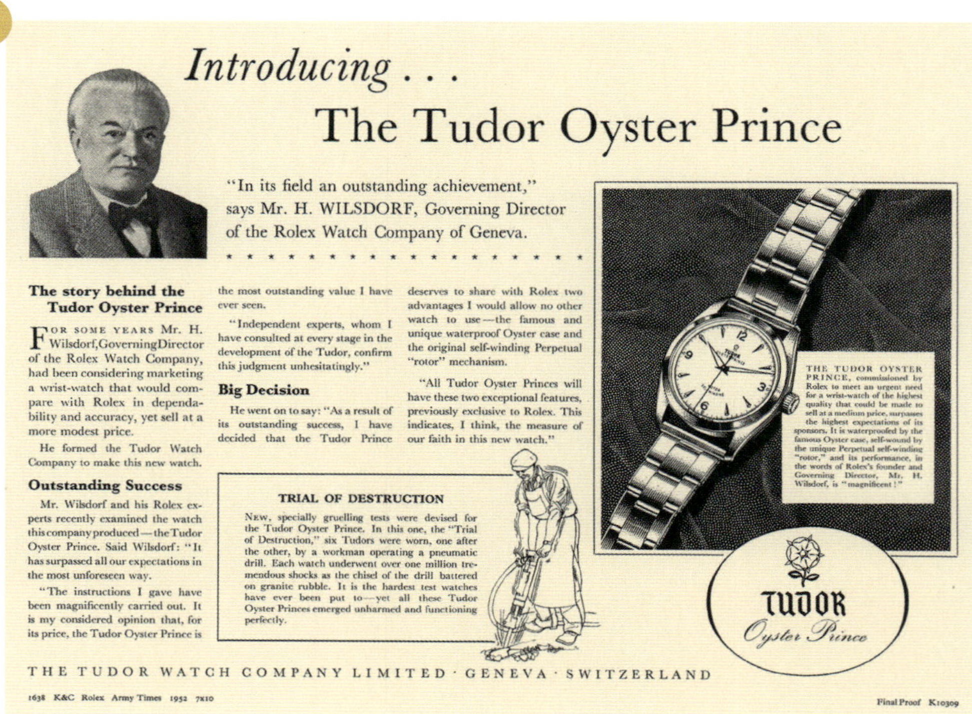

Introducing . . .

The Tudor Oyster Prince

"In its field an outstanding achievement," says Mr. H. WILSDORF, Governing Director of the Rolex Watch Company of Geneva.

The story behind the Tudor Oyster Prince

FOR SOME YEARS Mr. H. Wilsdorf, Governing Director of the Rolex Watch Company, had been considering marketing a wrist-watch that would compare with Rolex in dependability and accuracy, yet sell at a more modest price.

He formed the Tudor Watch Company to make this new watch.

Outstanding Success

Mr. Wilsdorf and his Rolex experts recently examined the watch this company produced — the Tudor Oyster Prince. Said Wilsdorf: "It has surpassed all our expectations in the most unforeseen way.

"The instructions I gave have been magnificently carried out. It is my considered opinion that, for its price, the Tudor Oyster Prince is the most outstanding value I have ever seen.

"Independent experts, whom I have consulted at every stage in the development of the Tudor, confirm this judgment unhesitatingly."

Big Decision

He went on to say: "As a result of its outstanding success, I have decided that the Tudor Prince deserves to share with Rolex two advantages I would allow no other watch to use — the famous and unique waterproof Oyster case and the original self-winding Perpetual "rotor" mechanism.

"All Tudor Oyster Princes will have these two exceptional features, previously exclusive to Rolex. This indicates, I think, the measure of our faith in this new watch."

TRIAL OF DESTRUCTION

NEW, specially gruelling tests were devised for the Tudor Oyster Prince. In this one, the "Trial of Destruction," six Tudors were worn, one after the other, by a workman operating a pneumatic drill. Each watch underwent over one million tremendous shocks as the chisel of the drill battered on granite rubble. It is the hardest test watches have ever been put to — yet all these Tudor Oyster Princes emerged unharmed and functioning perfectly.

THE TUDOR OYSTER PRINCE, commissioned by Rolex to meet an urgent need for a wrist-watch of the highest quality that could be made to sell at a medium price, surpasses the highest expectations of its sponsors. It is waterproofed by the famous Oyster case, self-wound by the unique Perpetual self-winding "rotor," and its performance, in the words of Rolex's founder and Governing Director, Mr. H. Wilsdorf, is "magnificent!"

TUDOR

Oyster Prince

THE TUDOR WATCH COMPANY LIMITED · GENEVA · SWITZERLAND

1638 K&C Rolex Army Times 1952 7x10 Final Proof K10309

2022

TUDOR

BLACK BAY 54

BACK TO THE ROOTS

With a nod to a venerable past, Tudor named the model it launched in 2022 Black Bay 54. Its curriculum vitae is closely related to that of the Rolex Submariner, which came onto the market in 1953. A year later – hence the number 54 in the model name – Tudor launched its Oyster Prince Submariner, Reference 7922. Apart from the lettering on the dial, there were no significant visual differences between the two watches for underwater enthusiasts. The Tudor timepiece with its three-part original Oyster case was more than a match for ten bar of water pressure. Screw-down case and crown came as standard for the Rolex subsidiary. Initially, the Reference 7922 winding and hand-setting crown had a modest diameter of just six millimetres. The French Navy divers involved in the development process found it too paltry, so the company switched to the patented eight-millimetre Twinlock crown, which Rolex also used for the Submariner References 6200 and 6538. In line with the Tudor philosophy cultivated by Rolex founder Hans Wilsdorf, timekeeping was left to a purchased automatic Calibre 390. History is now revived in the shape of the Black Bay 54. Harking back to the past, Tudor opted for a modest 37-millimetre case. The unguarded screw crown projects out from the steel case, which withstands pressure of up to 20 bar. Introduced in 1969 and now emblematic of the brand, the Snowflake hands rotate on the gently domed dot dial. For safety's sake, the diving-time bezel, now more finely fluted, can only be adjusted in a counterclockwise direction. In contrast to the past, Tudor now uses proprietary movements. The 70-hour power reserve of the chronometer-certified automatic Calibre MT5400 with amagnetic silicon balance coil ensures peace of mind.

ZURÜCK ZU DEN WURZELN

Mit Blick auf eine beachtliche Vergangenheit erfolgte bei Tudor die Namensgebung des 2022 lancierten Modells „Black Bay 54“. Das Curriculum Vitae steht in engem Zusammenhang mit dem der „Submariner“ aus dem Hause Rolex. Diese fand 1953 zum Uhrenfachhandel. Ein Jahr später, daher auch die Zahl 54 im Modellnamen, folgte Tudor mit der „Oyster Prince Submariner“, Referenz 7922. Bis auf die Beschriftung des Zifferblatts gab es zwischen den beiden Armbanduhren für Freunde des Unterwassersports keine nennenswerten optischen Unterschiede. Zehn bar Wasserdruck machten der Tudor-Kreation mit dreiteiliger Original „Oyster“-Schale nicht das Geringste aus. Schraubboden und -krone waren für die Rolex-Tochter eine Selbstverständlichkeit. Anfangs besaß die Aufzugs- und Zeigerstellkrone der Referenz 7922 nur sechs Millimeter Durchmesser. Weil das den Tauchern der am Entwicklungsprozess beteiligten französischen Marine zu wenig war, folgte ein Wechsel zu jener patentierten Acht-Millimeter-Twinlock-Krone, welche Rolex auch für die „Submariner“-Referenzen 6200 und 6538 verwendete. Gemäß der vom Rolex-Gründer Hans Wilsdorf postulierten Tudor-Philosophie bewahrte ein zugekauftes Automatikkaliber 390 die Zeit.
Diese Geschichte lebt nun wieder auf in besagter „Black Bay 54“. Als Reminiszenz hat Tudor moderate 37 Millimeter Gehäusedurchmesser übernommen. Ohne Flankenschutz ragt die Schraubkrone aus dem bis 20 bar druckdichten Stahlgehäuse. Vor dem leicht bombierten Dot-Zifferblatt drehen heutzutage die 1969 eingeführten und inzwischen charakteristischen „Snowflake“-Zeiger. Natürlich lässt sich die nun feiner geriffelte Tauchzeitlünette aus Sicherheitsgründen nur entgegen dem Uhrzeigersinn drehen. Im Gegensatz zu früher verbaut Tudor nun hauseigene Mechanik. Beruhigende 70 Stunden beträgt die Gangautonomie des chronometerzertifizierten Automatikkalibers MT5400 mit amagnetischer Silizium-Unruhspirale.

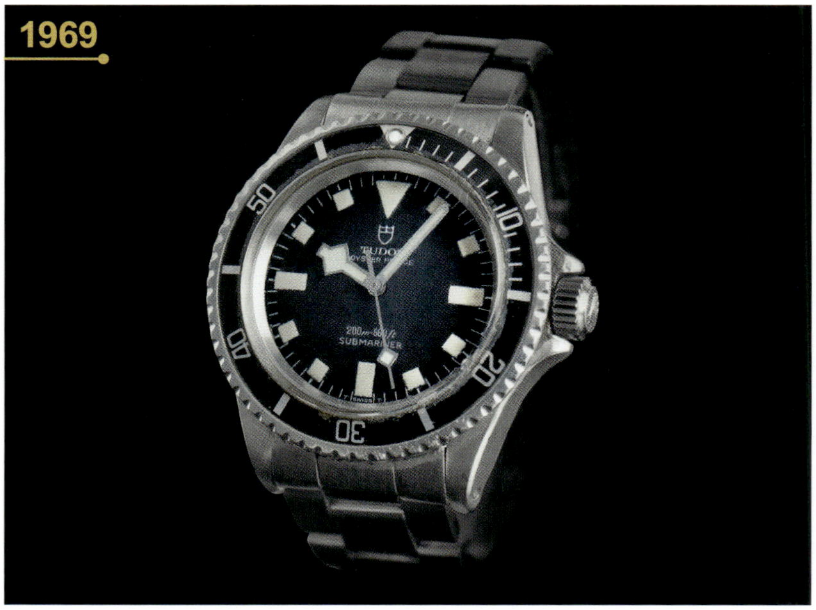

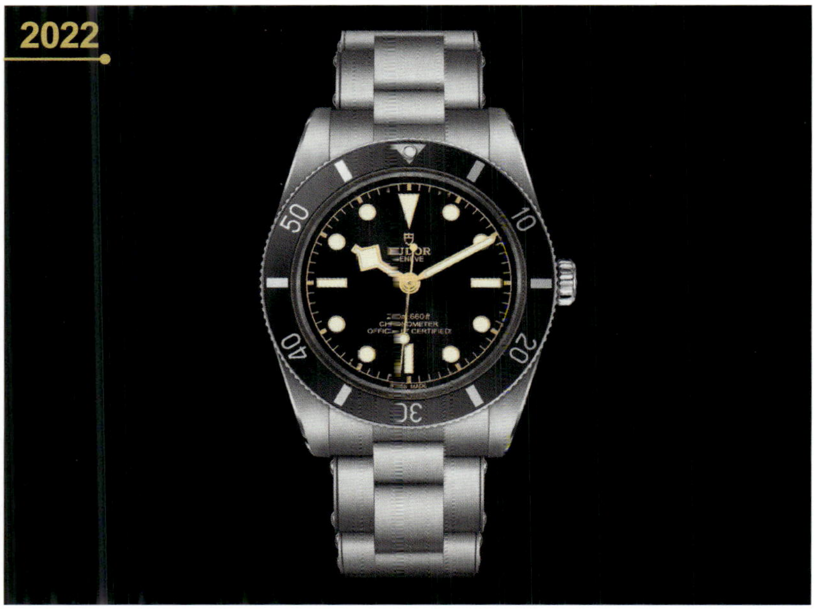

TUDOR 2021
PELAGOS FXD

MILITARY BACKGROUND

Tudor's inspiration for the Pelagos FXD with its lightweight, resilient and anti-allergic titanium case can be found in its well-filled archives, which includes wristwatches that were used for decades by professional US Navy divers. Back in the mid-1950s, various military institutions tested these heavy-duty timepieces from Geneva in real-life conditions. The results not only formed the subject of detailed reports, but also led to official recognition by the American Navy in 1958. Thus approved, generations of units purchased the Military Submariner – or Milsub for short – from 1962 into the late 1980s. These reliable tool watches were not only worn by members of the legendary Seals, but also by underwater demolition teams and experienced Navy diving instructors. The 2021 Pelagos FXD is modelled on the Tudor Reference 7016 from the late 1960s. Its 42-millimetre case is milled from one piece. To preclude any accidents in the midst of stressful situations, the strap push pins are fixed to the lugs. This of course means that you need special straps. The stainless-steel screw-down base withstands pressure of up to 20 bar or a water depth of 200 metres. Unlike in the past, the easy-grip titanium bezel with ceramic inlay can only be adjusted in a counterclockwise direction. The 60-minute track features noctilucent X1 SwissSuper-LumiNova. Tudor also uses the same luminous material on the hour indices and the snowflake-style hands. The exclusive MT5602 automatic movement with bidirectional ball-bearing rotor offers a power reserve of 70 hours and official COSC accuracy certification.

MILITÄRISCHER HINTERGRUND

Für die „Pelagos FXD" mit gleichermaßen leichtem, robustem und antiallergischem Titangehäuse recherchierte Tudor in seinen gut gefüllten Archiven. Dort fanden sich Armbanduhren, welche professionelle Taucher der US Navy über Jahrzehnte hinweg nutzten. Bereits Mitte der 1950er-Jahre hatten verschiedene Militärinstitutionen die belastbaren Zeitmesser aus Genf unter realistischen Bedingungen getestet. Die solcherart gewonnenen Erfahrungen mündeten nicht nur in ausführlichen Berichten, sondern führten 1958 zur offiziellen Anerkennung durch die amerikanische Marine. Danach stand dem Erwerb sogenannter „Military Submariners", kurz „Milsubs", für verschiedene Einheiten von 1962 bis in die späten 1980er-Jahre nichts im Wege. Auf die zuverlässigen Instrumente blickten neben Mitgliedern der legendären Seals auch Underwater Demolition Teams (UDTs) und erfahrene Tauchlehrer der Navy. Vorbild der „Pelagos FXD" von 2021 ist die Tudor-Referenz 7016 aus den späten 1960er-Jahren. Der Korpus ihrer 42 Millimeter großen Schale wird aus einem Stück gefräst. Damit bei stressigen Aktionen über und unter Wasser kein Malheur passieren kann, sind die Bandstege fest mit den Hörnern verbunden. Das verlangt logischerweise nach speziellen Armbändern. Aus Edelstahl besteht der Schraubboden des bis zu 20 bar Druck oder umgerechnet 200 Meter Tiefe wasserdichten Gehäuses. Im Gegensatz zu einst lässt sich die griffige Titanlünette mit Keramik-Inlay nur entgegen dem Uhrzeigersinn verstellen. Ausgelegt ist die 60-Minuten-Graduierung mit nachtleuchtendem X1 Swiss Super-LumiNova. Die gleiche Leuchtmasse verwendet Tudor auch für die applizierte Stundenindexierung und die Zeiger im „Snowflake"-Stil. MT5602 heißt das exklusive Automatikwerk mit beidseitig wirkendem Kugellagerrotor, rund 70 Stunden Gangautonomie und offizieller COSC-Genauigkeitszertifizierung.

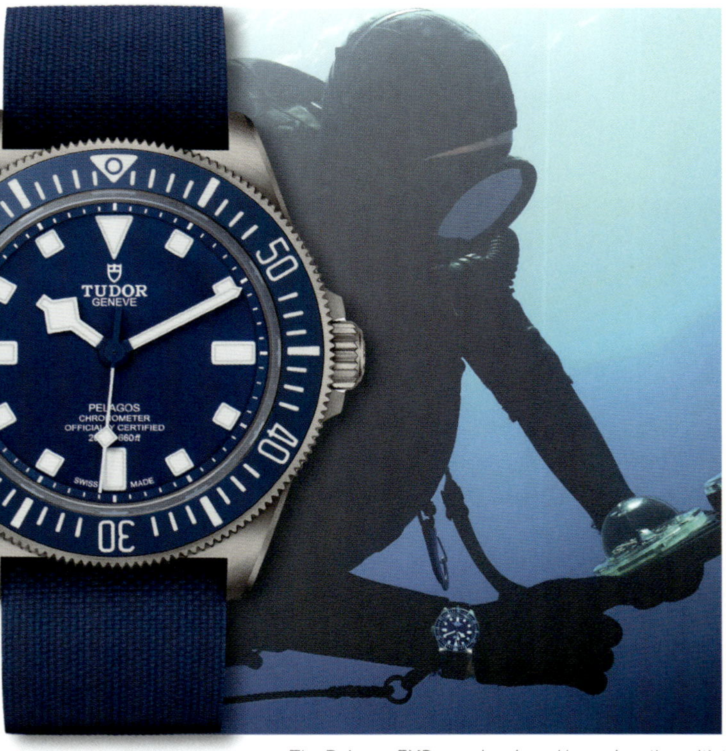

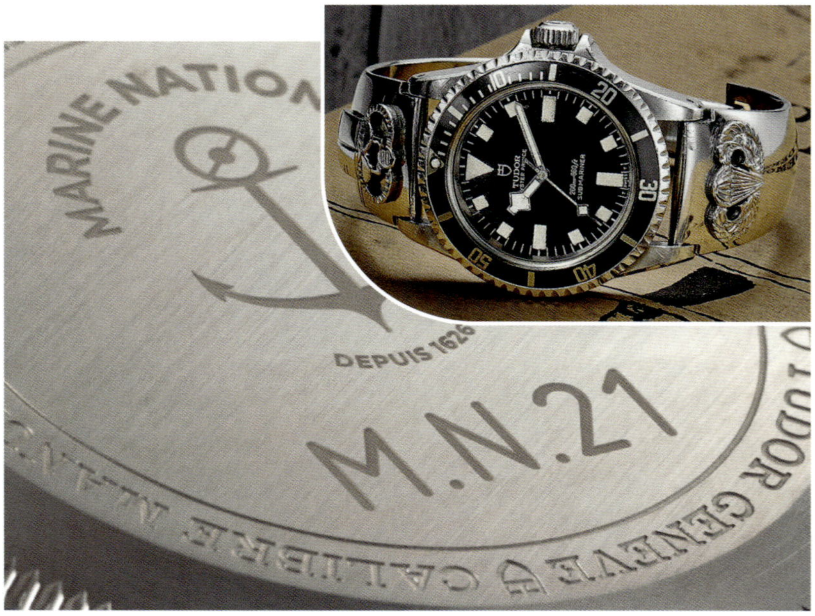

The Pelagos FXD was developed in conjunction with the French Navy / Die Pelagos FXD wurde in Zusammenarbeit mit der französischen Marine entwickelt.

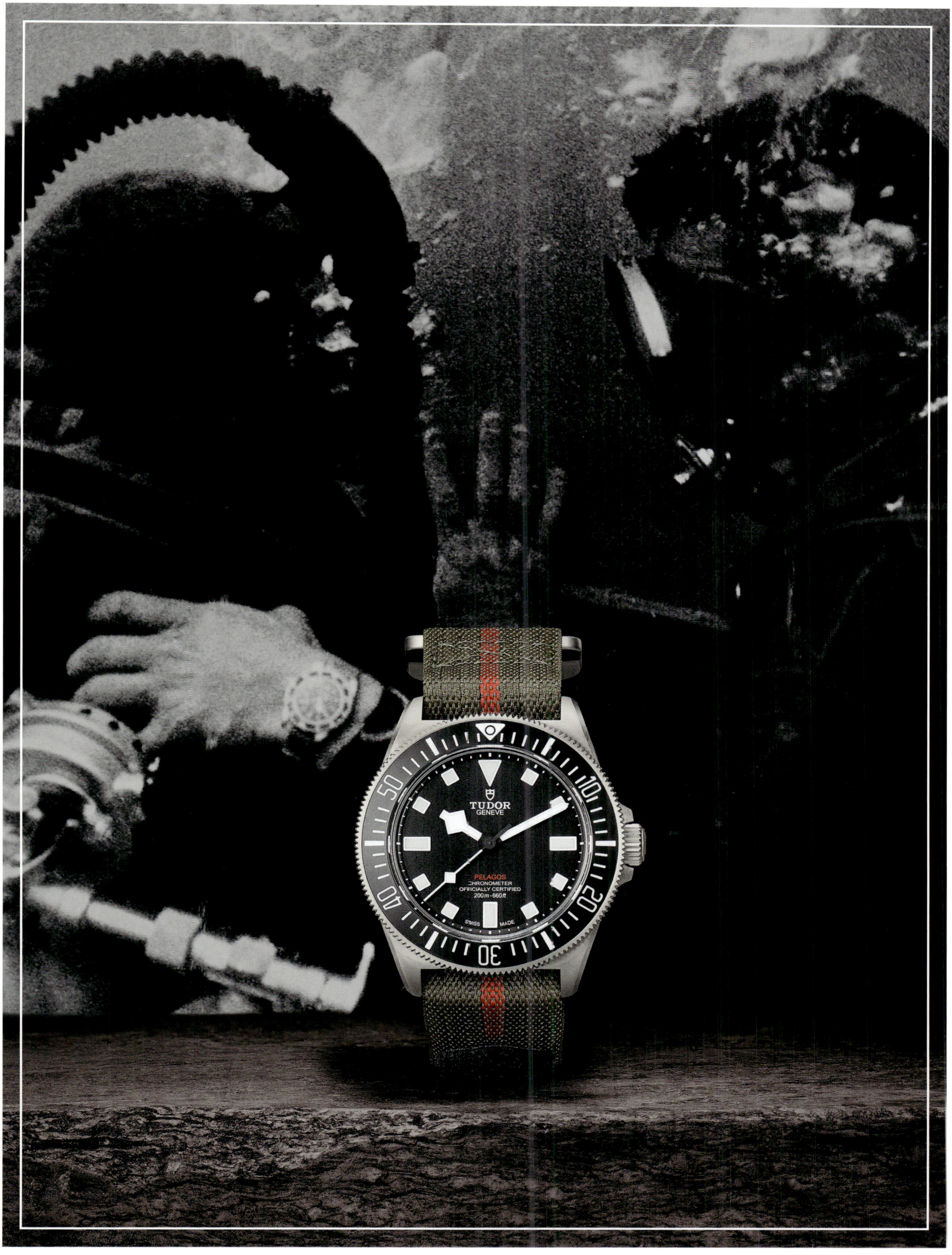
TUDOR
GENEVE
PELAGOS
CHRONOMETER
OFFICIALLY CERTIFIED
200m-660ft
SWISS MADE

TUTIMA GLASHÜTTE 2014

M2

GLASHÜTTE PILOT'S CHRONOGRAPH

Collectors regard the Tutima pilot's chronograph as a fully fledged icon. It was produced from 1941 to 1945. Believe it or not, it was powered by the equally legendary German manufactory Calibre Urofa 59. In the 1980s, the German Bundeswehr required an official pilot's chronograph. The military testing centre drafted an exhaustive detailed requirements specification for the purpose. Tutima submitted the winning offer and was awarded the contract in 1985. The automatic chronograph, which was also used by other NATO forces, is no longer driven by the long-discontinued Lémania 5100 calibre. The impressive features of this watch include ergonomic pushers and a highly legible 60-minute totaliser in the centre of the dial. The two icons enjoy almost museum status, which is why Tutima Glashütte seeks its inspiration for the future in the past. Launched in 2014, the M2 continues the success story of the 1985 pilot chronograph. As in the past, the large titanium pushers are ergonomically integrated into the 46-millimetre screw-down crown case of the same material. The solid base of the hermetically sealed 30-bar case with anti-reflective sapphire crystal is adorned with a silhouette of an aerobatic biplane. Super-LumiNova on the dial and hands is de rigueur. Tutima relies on proven technology from Switzerland for its movement. The automatic Calibre 521, which is protected from magnetic fields by a soft-iron inner case, is based on the Eta 7750 or Sellita clone SW500. Similar to the Lémania 5100, the highlight is the centrally positioned 60-minute counter and additional 24-hour hand. Straps are available in titanium or robust Kevlar weave.

1941

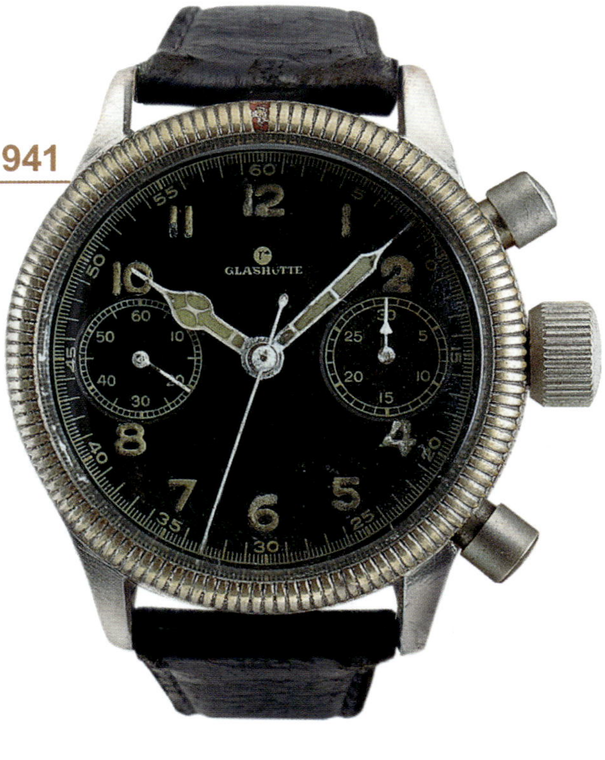

GLASHÜTTER FLIEGER-CHRONOGRAPH

In Sammlerkreisen gilt der Tutima-signierte Piloten-Chronograph als anerkannte Uhr-Ikone. Seine Produktion erstreckte sich von 1941 bis 1945. Die Zeit bewahrte und stoppte damals das nicht minder legendäre deutsche Manufakturkaliber Urofa 59. In den 1980er-Jahren benötigte die Deutsche Bundeswehr einen offiziellen Flieger-Chronographen. Hierfür legte die militärische Erprobungsstelle ein detailliertes und lückenlos zu erfüllendes Anforderungsprofil fest. Nach Prüfung der eingegangenen Offerten erhielt Tutima 1985 den Zuschlag. Im Automatik-Chronographen, den auch andere NATO-Streitkräfte nutzten, tickte das längst nicht mehr hergestellte Kaliber Lémania 5100. Zu den beeindruckenden Merkmalen dieser Armbanduhr gehörten ergonomische Bedientasten und ein optimal ablesbarer, weil in der Mitte des Zifferblatts drehender 60-Minuten-Totalisator.

Beide Ikonen besitzen inzwischen fast schon musealen Charakter. Gerade deswegen blickt Tutima Glashütte gerne zurück in die Zukunft. Die 2014 lancierte „M2"-Linie setzt jene Erfolgsgeschichte fort, welche der Piloten-Chronograph von 1985 geschrieben hatte. Wie einst sind die großflächigen Titandrücker ergonomisch ins 46 Millimeter große Schraubkronen-Gehäuse aus dem gleichen Werkstoff integriert. Der massive Boden der bis 30 bar druckdichten Schale mit entspiegeltem Saphirglas zeigt die Silhouette eines Kunstflug-Doppeldeckers. Quasi Pflicht ist Super-LumiNova-Leuchtausstattung für Zifferblatt und Zeiger. In Sachen Uhrwerk setzt Tutima auf Bewährtes aus der Schweiz. Das von einem Weicheisen-Innengehäuse gegen Magnetfelder geschützte Automatikkaliber 521 basiert auf dem Eta 7750 oder dem Sellita-Klon SW500. Der Clou besteht analog zum Lémania 5100 im zentral angeordneten 60-Minuten-Zähler sowie in einem zusätzlichen 24-Stunden-Zeiger. Armbänder gibt es wahlweise aus Titan oder aus robustem Kevlar-Gewebestoff.

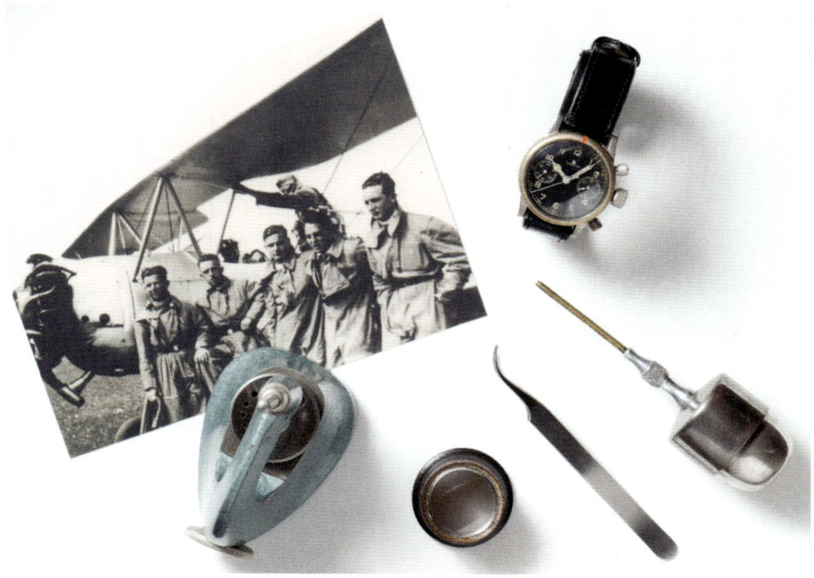

Vintage Chronograph

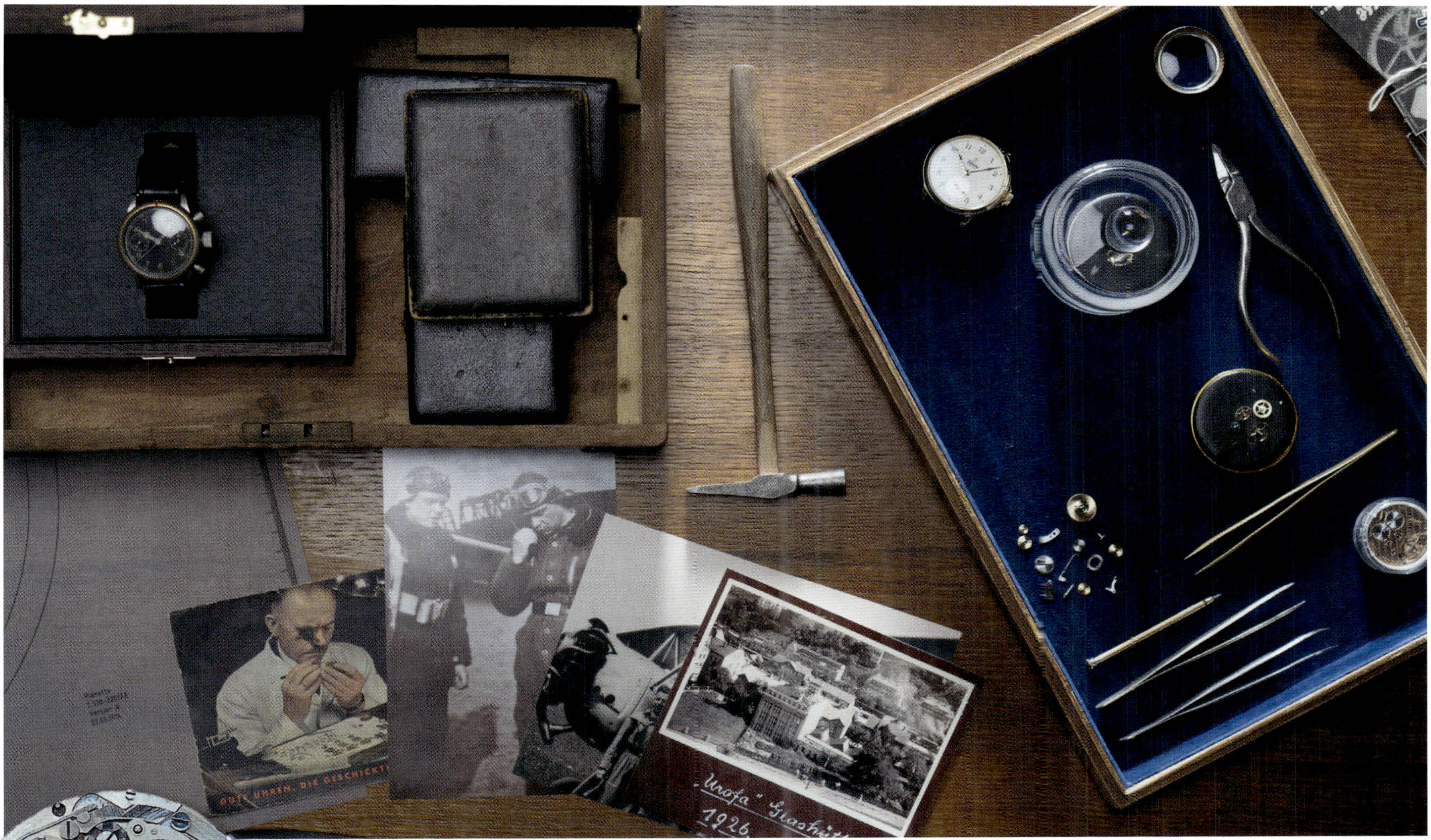

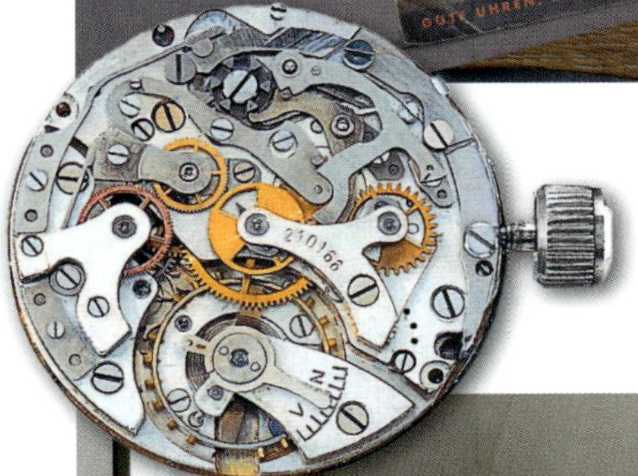

< Cal. Urofa 59

> Dr Ernst Kurtz, founder | Firmengründer

>> Uhren-Rohwerkefabrik AG (UROFA) & Uhrenfabrik AG (UFAG), Glashütte

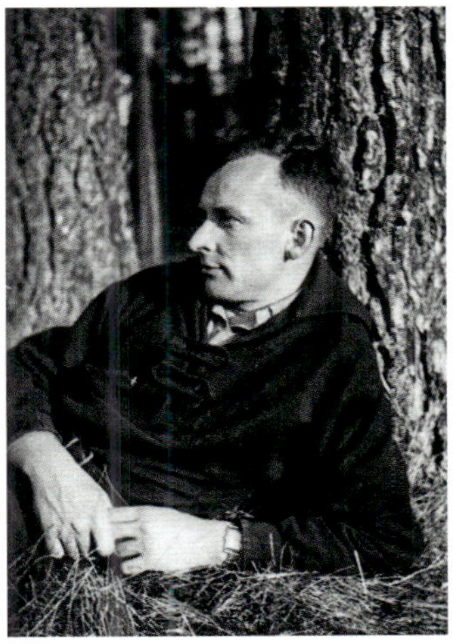

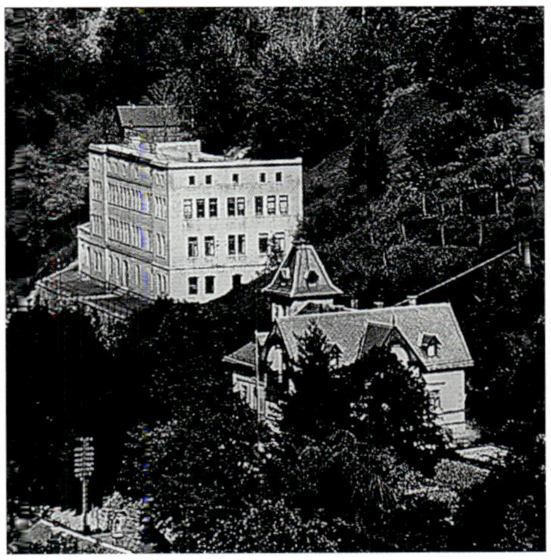

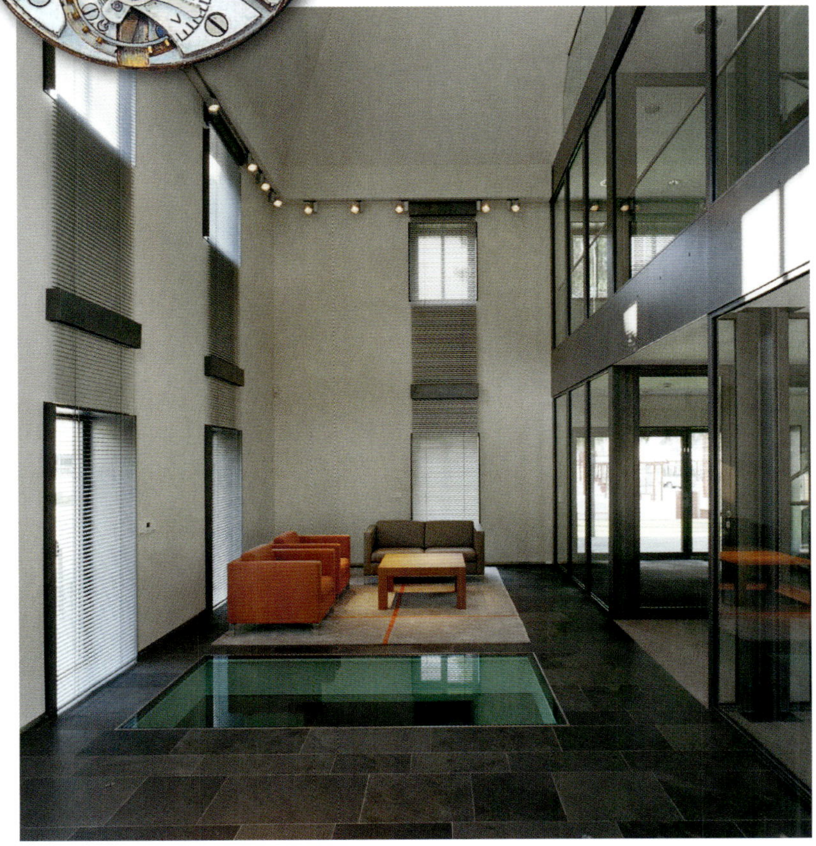

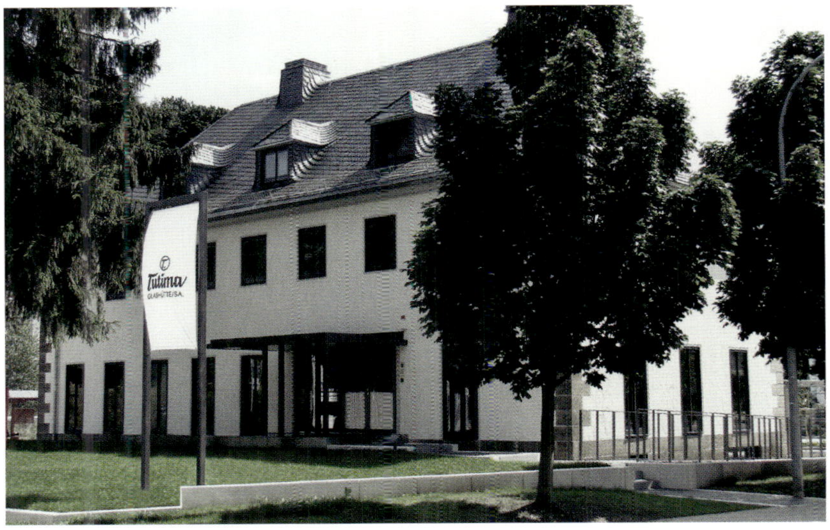

Company headquarters | Firmensitz in Glashütte

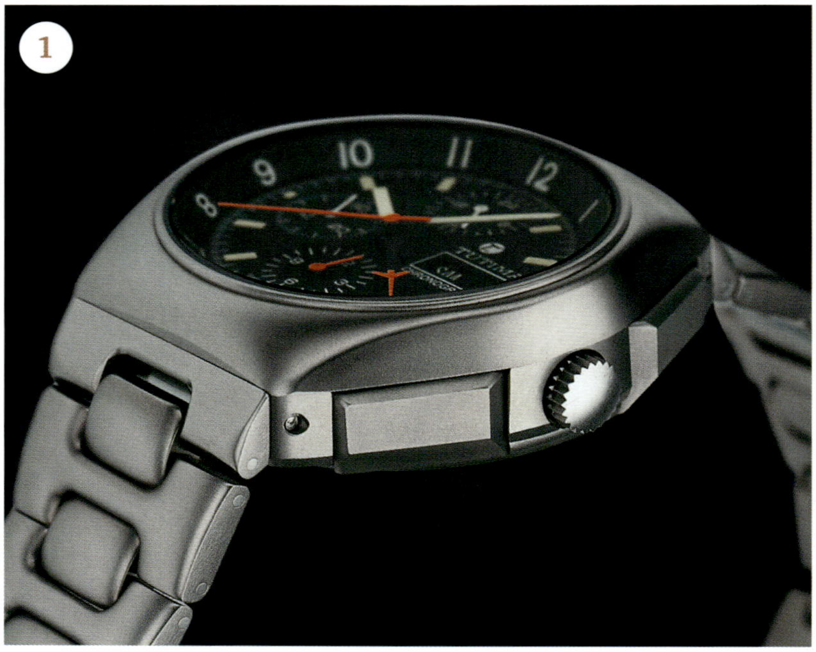

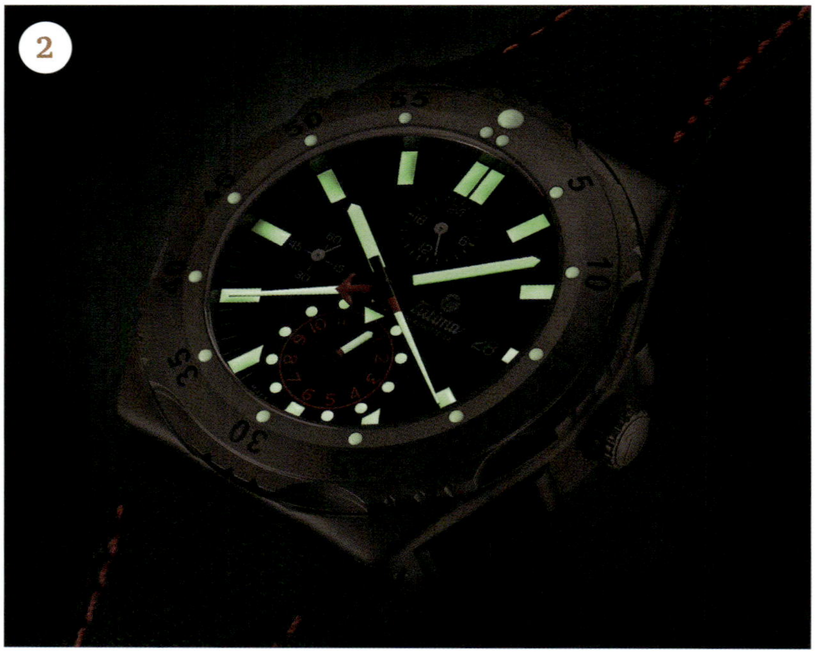

M2 REF. 6450-03

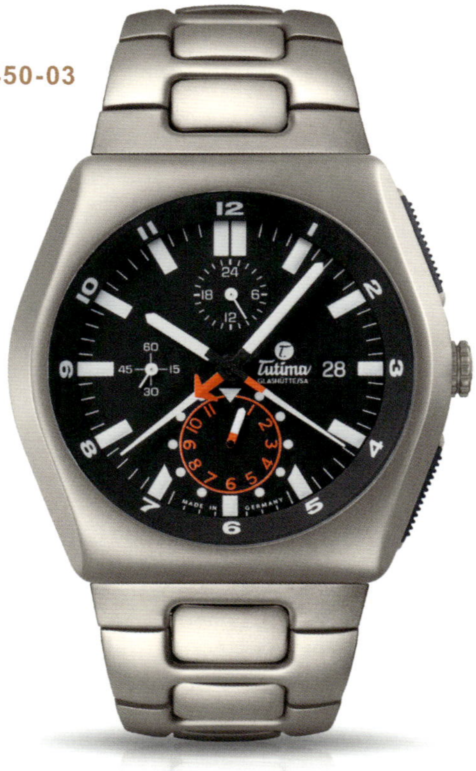

M2 REF. 6450-02

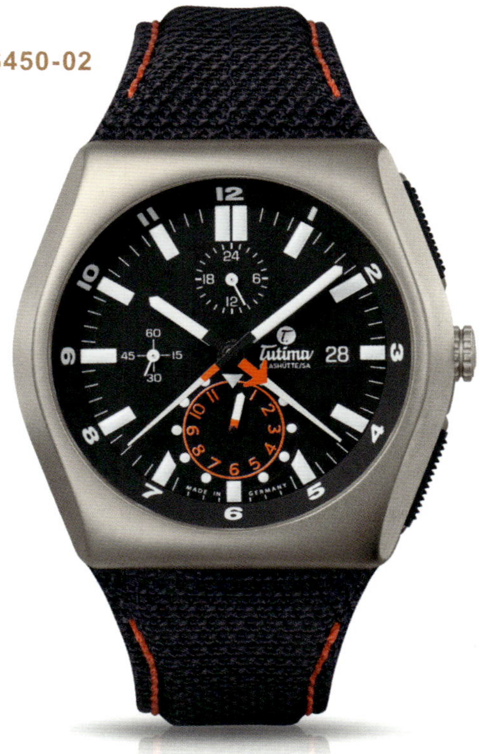

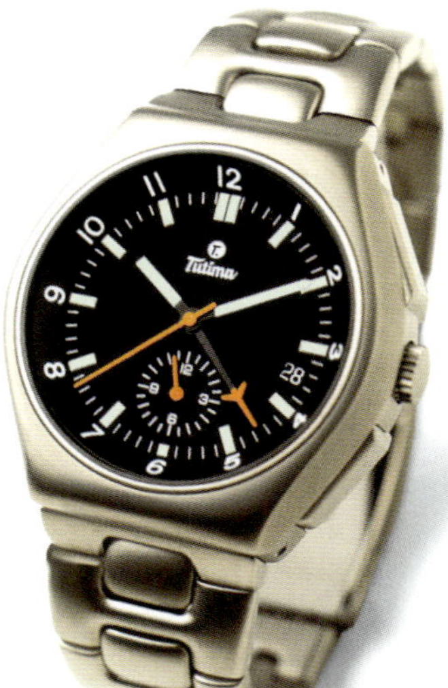

MILITARY
CHRONOGRAPH
REF. 760

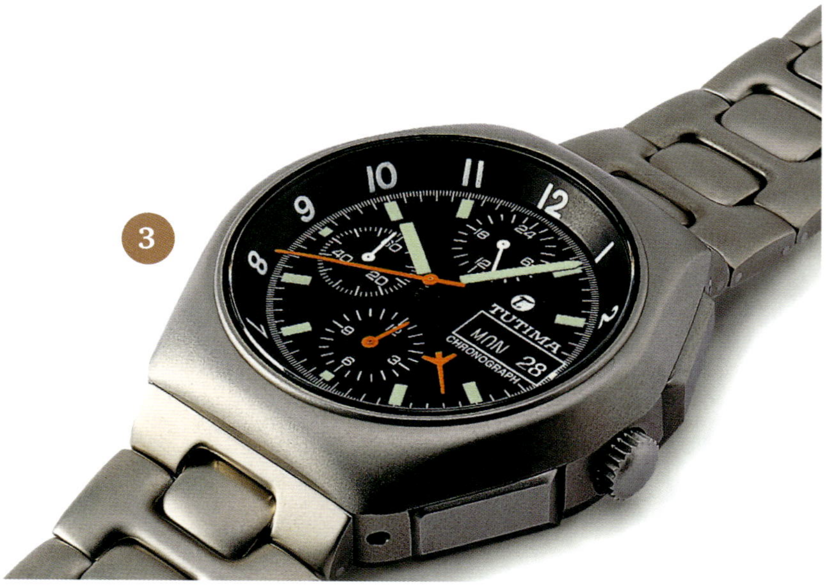

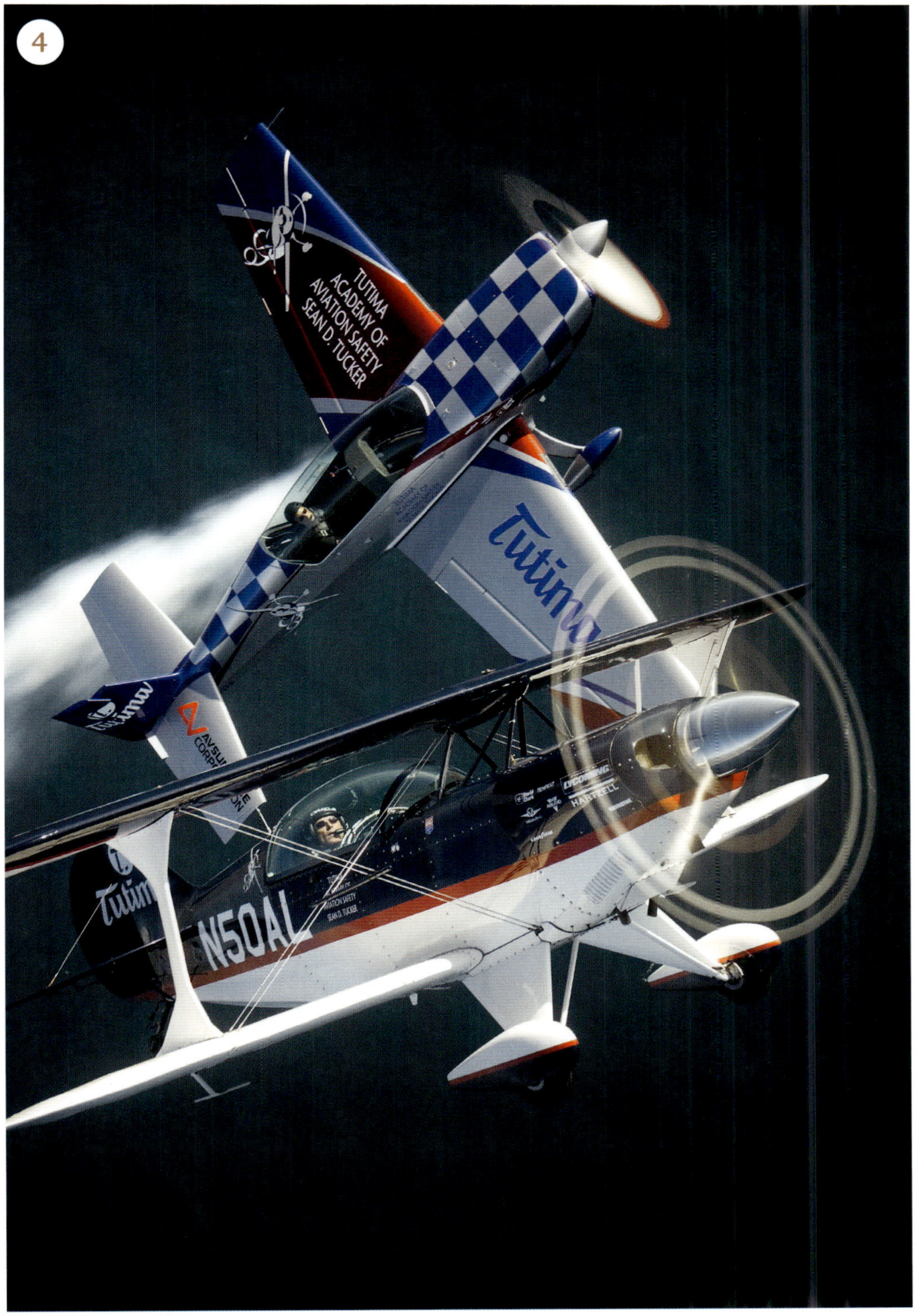

1 3 5 *Military Chronograph Ref. 760*

2 *Super-LumiNova on the dial and hands of the M2 | Super-LumiNova auf Ziffern und Zeigern der M2*

4 *In 2006, Tutima and American aerobatic flight legend Sean D. Tucker founded the Tutima Academy of Aviation Safety, a leading global school for precision flight training in California | Im Jahr 2006 gründeten Tutima und die amerikanische Kunstfluglegende Sean D. Tucker die Tutima Academy of Aviation Safety, eine der weltweit führenden Schulen für Präzisionsflug-Training.*

6 *The Cockpit of an AWACS aircraft | Im Cockpit eines AWACS-Flugzeugs*

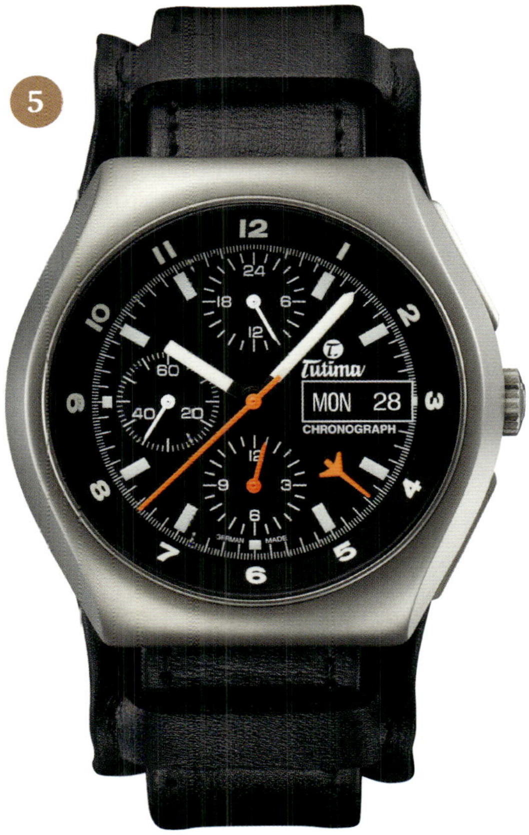

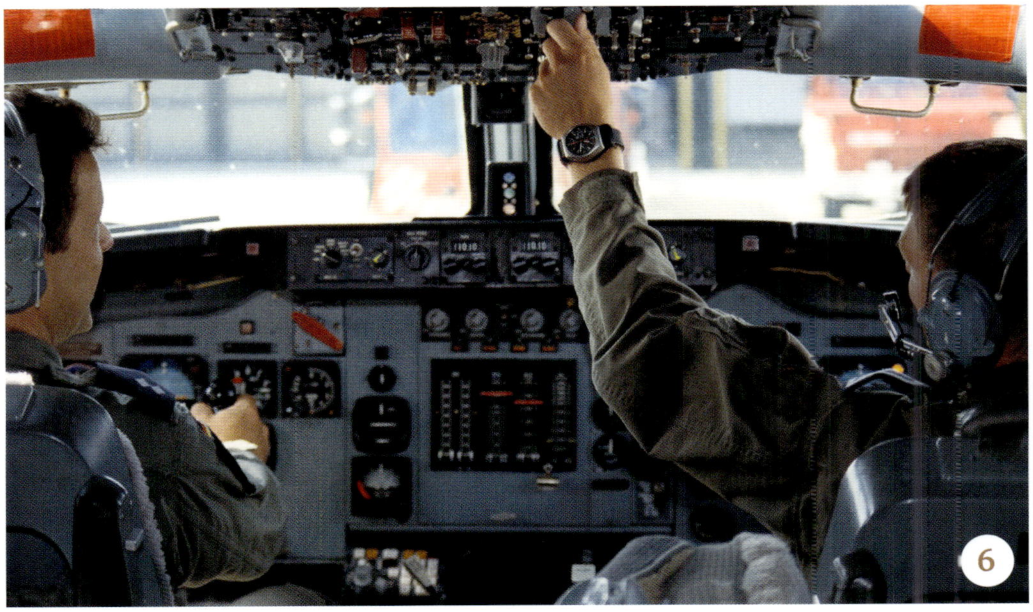

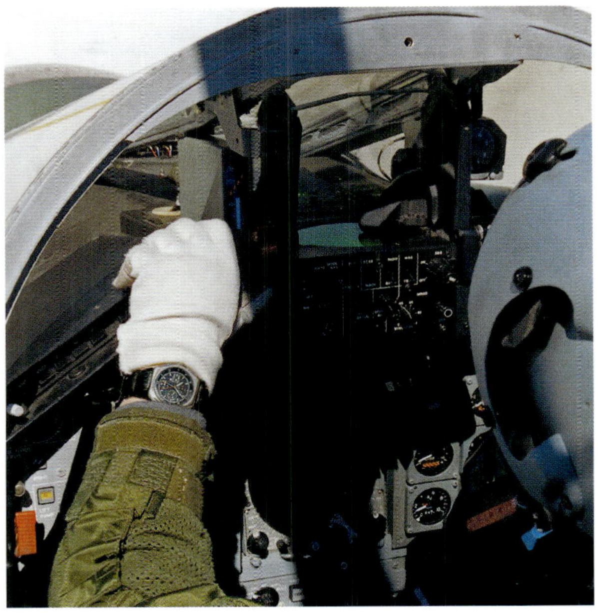

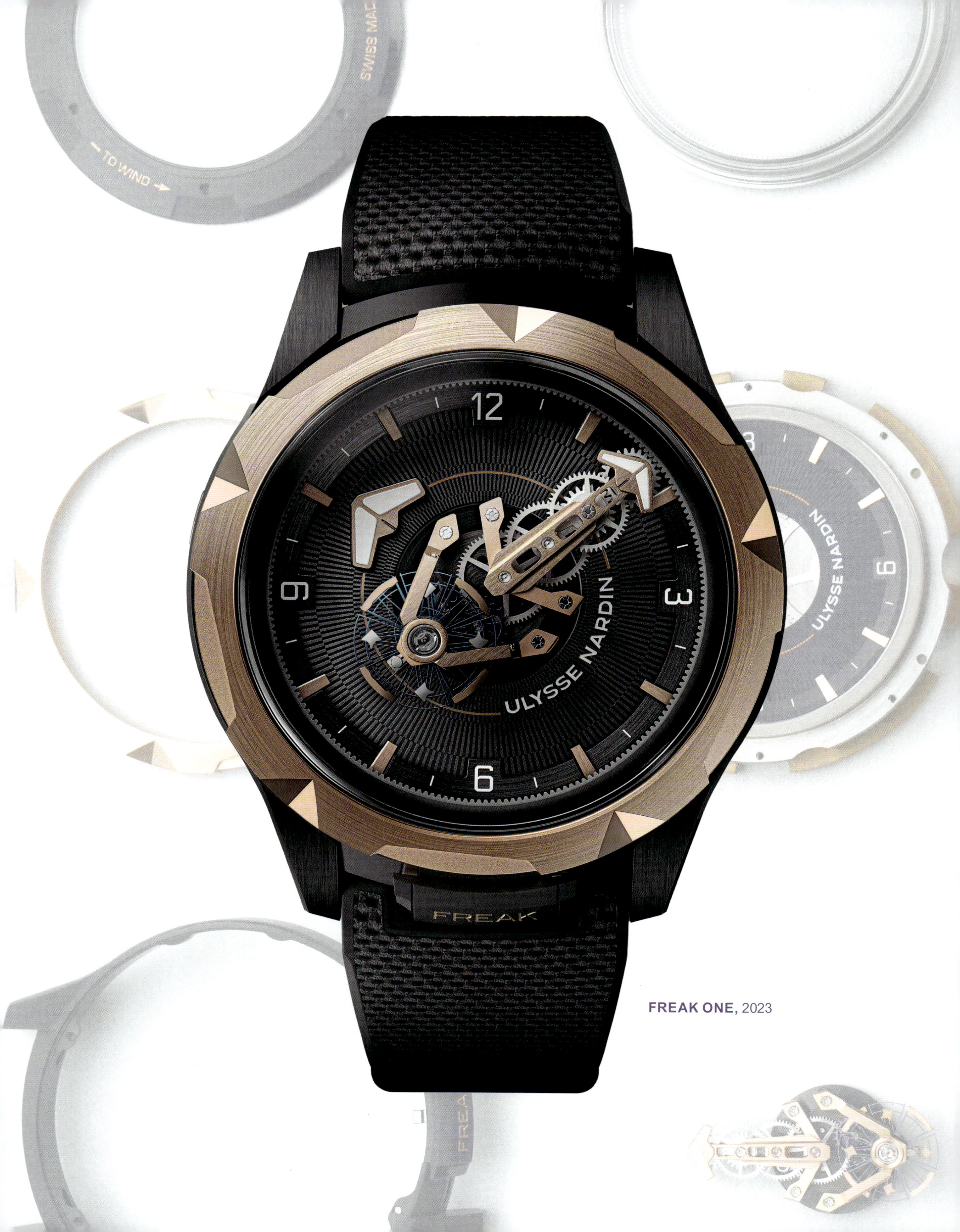

FREAK ONE, 2023

2001

ULYSSE NARDIN *FREAK*

A FREAKY PHENOMENON

The dictionary paraphrases the term "freak" variously. People of this ilk are those "who do not fit into normal civic life" and who are "fanatically enthusiastic" about something. Both descriptions somehow fit the majority owner of Ulysse Nardin, Rolf Schnyder, who died in 2011, and his then chief designer Ludwig Oechslin. It therefore comes as no surprise that their extraordinary engineering feat was memorably christened "Freak". After all, the engineering inside the striking case is anything but ordinary. And somehow, it does not quite fit in with timekeeping conventions. The entire mechanical process is displayed to the eyes of marvelling observers. Engineering nerds need to sacrifice some of their precious time to visually track in detail how the unusual and of course patent-pending technology works. They will seek in vain for a dial and hands. The movement, unique in its day, personally takes charge of indicating both hours and minutes. The oscillation and escapement system rotates 360 degrees per hour at a pace of four hertz, earning it the nickname "carousel". Along with the unusual tourbillon, the Freak offers another technical titbit in the shape of the patented Dual Direct escapement without a conventional lever. To minimise the mass of the motion components, the two escape and drive wheels are made of pure silicon, the material of computer chips, produced via deep reactive ion etching. Ulysse Nardin is therefore ascribed the undisputed role of silicon pioneer in the field of mechanical watchmaking. Based on this wristwatch, the now independent manufactory continues to focus on the subject up to the present day. True to the motto "No hands, no dial, no boundaries", various iterations of the Freak forge a bridge between the past and the future. The original model is by no means forgotten. Loyal to the core, the Freak One carries the torch. Cuvée 2024 is the Freak S Nomad.

VON FREAKS FÜR FREAKS

In Bezug auf Freaks hält der Duden gleich mehrere Umschreibungen bereit. Zu dieser Gattung Mensch gehören einmal solche, die „sich nicht in das normale bürgerliche Leben einfügen", und zum anderen zählen „fanatisch für etwas Begeisterte" dazu. Auf den 2011 verstorbenen Ulysse Nardin-Mehrheitseigentümer Rolf Schnyder sowie seinen damaligen „Chef-Konstrukteur" Ludwig Oechslin trifft beides irgendwie zu. Daher wunderte es 2001 nicht, dass ihr außergewöhnliches Mechanik-Opus den denkwürdigen Namen „Freak" erhielt.
Das, was sich im Inneren des markanten Gehäuses abspielt, ist nämlich alles andere als normal. Irgendwie passt es nicht so recht zu den Kriterien „bürgerlicher Zeitmessung". Das gesamte mechanische Geschehen spielt sich vor den Augen der staunenden Betrachter ab. Um die ungewöhnliche, natürlich zum Patent angemeldete Technik visuell detailliert verfolgen zu können, müssen Mechanikfreaks ein wenig ihrer kostbaren Zeit opfern. Zifferblatt und Zeiger zum Ablesen derselben werden sie vergeblich suchen. Das in seiner Art bis dahin einmalige Uhrwerk übernimmt die Indikation der Stunden und Minuten höchstpersönlich. Das Schwing- und Hemmungssystem, Frequenz vier Hertz, rotiert im Laufe einer Stunde um 360 Grad. Aus dieser Tatsache resultiert der Beiname „Karussell". Neben dem außergewöhnlichen Drehgang besitzt der „Freak" als weitere technische Finesse eine patentierte „Dual Direct Hemmung" ohne konventionellen Anker. Um die bewegten Massen möglichst gering zu halten, wurden die beiden Hemm- und Antriebsräder im Plasmaverfahren aus reinem Silizium, dem Werkstoff für Computerchips, geschnitten. Dadurch kommt Ulysse Nardin ohne Wenn und Aber die Rolle des Silizium-Pioniers auf dem Gebiet der mechanischen Uhrmacherei zu. Ausgehend von dieser Armbanduhr pflegt die mittlerweile eigenständige Manufaktur das Thema bis in die Gegenwart. Nach dem Motto „Keine Zeiger, kein Zifferblatt, keine Grenzen" bauen ganz unterschiedliche „Freak"-Modelle die Brücke zwischen Vergangenheit und Zukunft. Das Ursprungsmodell gerät dabei keineswegs in Vergessenheit. Es lebt in Gestalt des natürlich konsequent evaluierten „Freak One" weiter. Cuvée 2024 ist der „Freak S Nomad".

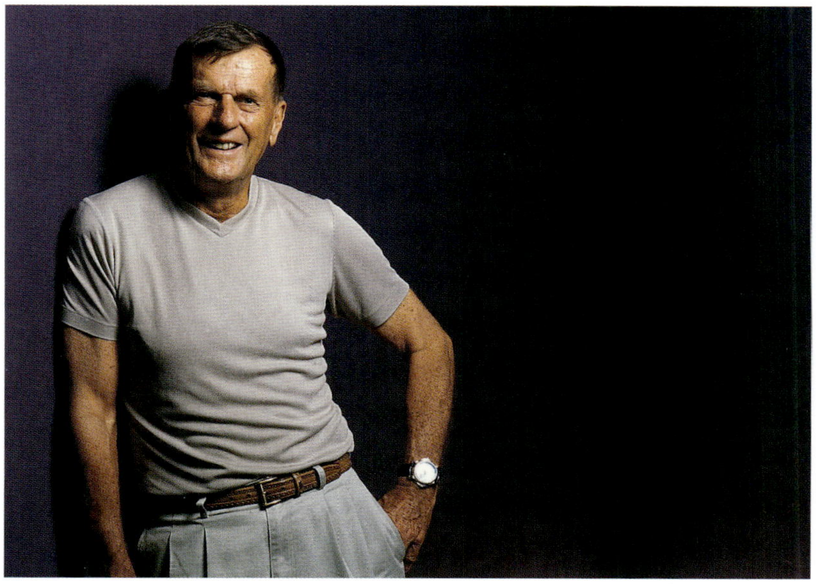

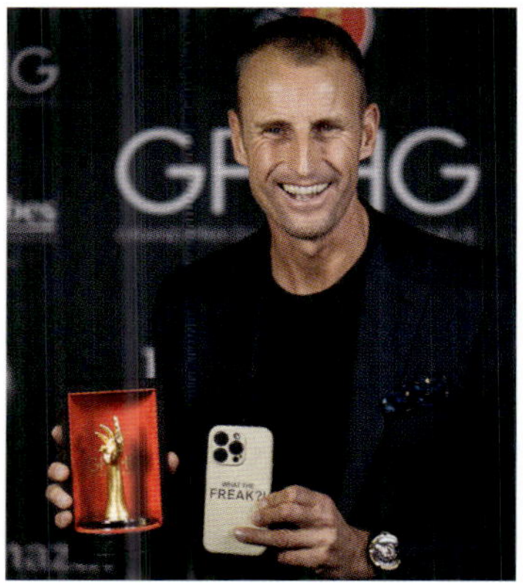

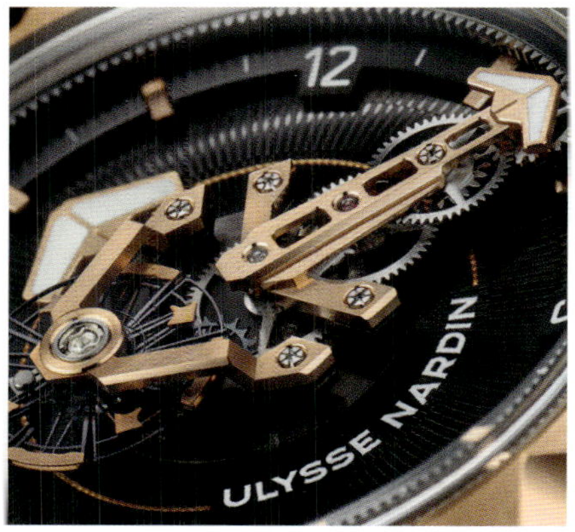

From left / von links: *Rolf Schnyder / Patrick Pruniaux, CEO of Ulysse Nardin at GPHG, where the Freak One was awarded the prize for Iconic Watch, 2023 / Patrick Pruniaux, Geschäftsführer von Ulysse Nardin beim Grand Prix d'Horlogerie de Genève, bei dem der Freak One den Preis in der Kategorie Iconic Watch gewonnen hat, 2023 / Detail of Freak One, 2023*

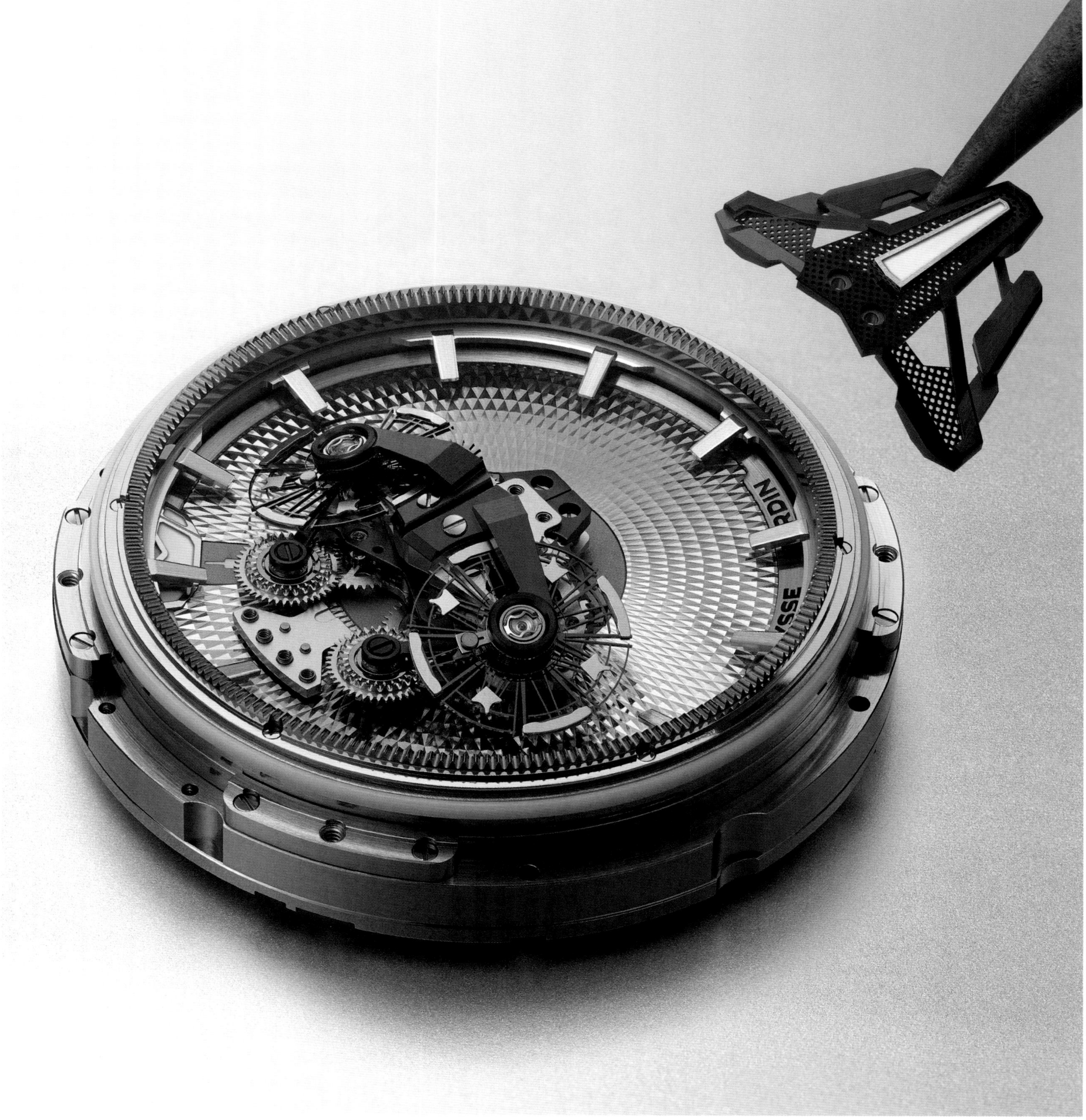

Assembly of Freak S Nomad | *Montage der Freak S Nomad*

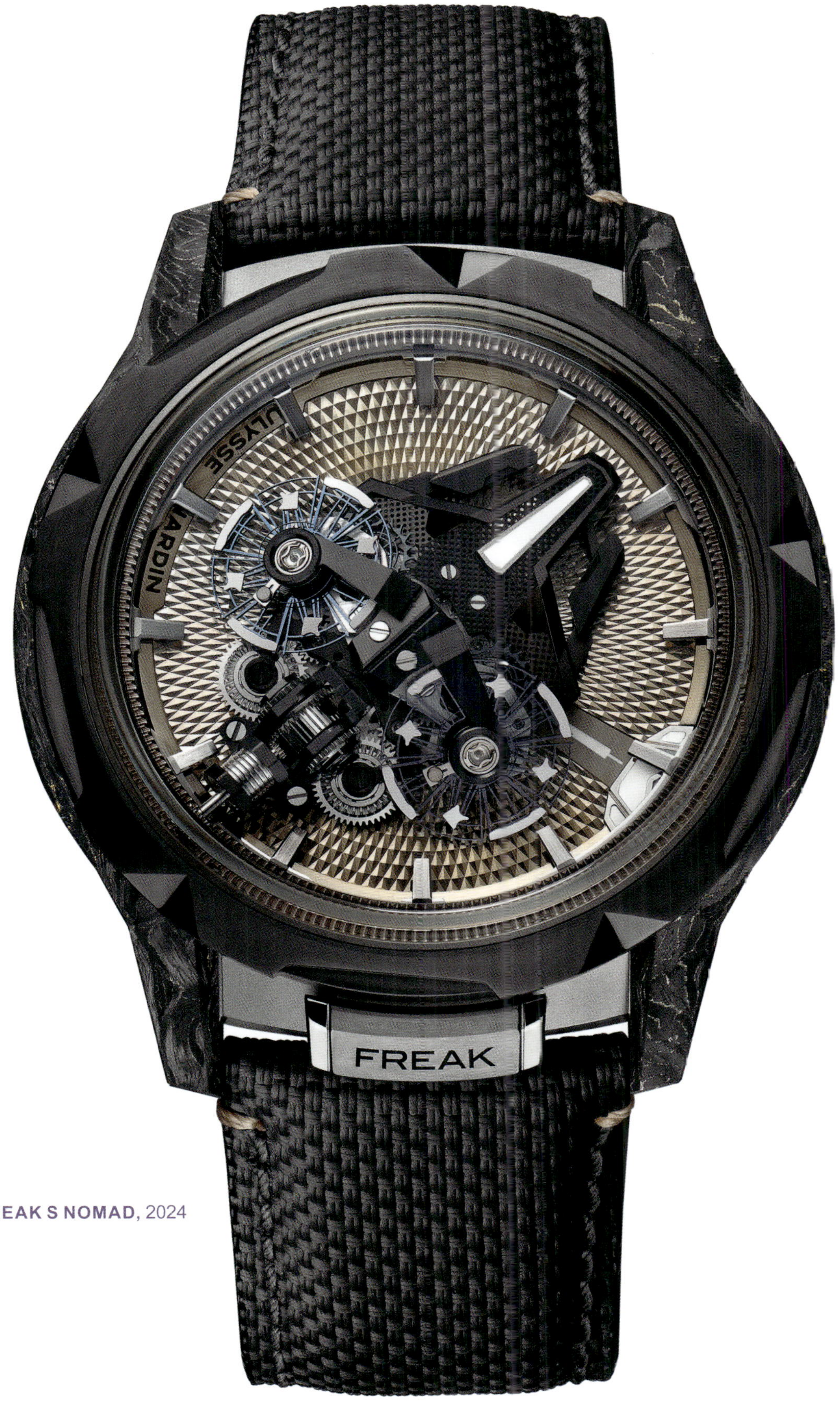

FREAK S NOMAD, 2024

VACHERON CONSTANTIN 1977

222 AND OVERSEAS

FROM 222 TO OVERSEAS

In 1976, Vacheron Constantin was in a quandary. The Geneva company was duty bound to respond to Audemars Piguet's Royal Oak and Patek Philippe's Nautilus. This situation gave birth to the 222, referencing the age of the traditional Geneva company. In his design and case concept the young German designer Jörg Hysek was of course influenced by the two iconic timepieces. The glass bezel was screwed to the case middle. A porthole-like look was thus created. Protection from water damage was additionally ensured by a double crown. A decade after its presentation with just around 500 models to its name, the Triple Two was superseded by two successor models 333 and Phidias, neither of which enjoyed much success either. At Vacheron Constantin, a certain reticence is encountered when mention is made of these models. Quite in contrast to the 222 and its triumphant comeback in 2022.

In the mid-1990s, Vacheron Constantin did not figure at all in the market for sporty, elegant luxury watches in stainless steel. The in-house design team applied itself to remedying this deficit. They used the brand's distinctive Maltese cross as a design element on the glass bezel. It is traditionally regarded as Vacheron Constantin's trademark. A total of 16 screws ensure the hermetic sealing of the stainless-steel case up to 15 bar pressure. Luxury sportiness is also exuded by the screw-down crown and flank guard, reinforced sapphire crystal, tritium-coated hour and minute markers and matching luminous hands. Marking the watchmaker's two hundred and fiftieth anniversary in 2005, the Overseas was the subject of a sensitive evolution. Its diameter was increased to 42 millimetres. The steel automatic model with central seconds and date window was powered by the Jaeger-LeCoultre 889 calibre, which was protected by a soft iron carriage. The third thoroughly revised edition of the Overseas generation made its début in 2016. The look is curvier, sleeker and more cuff-friendly. A constant is the increased protection against magnetic fields. This is once more supplied by soft iron casing. On the Overseas Automatic, resistance of up to 8,000 A/m or 100 Gauss is offered. A practical innovation is the patented interchangeability system for the bracelets and straps. The three bracelets supplied as standard can be easily exchanged to suit the occasion. In addition to the simple models in different sizes and materials, Vacheron Constantin also offers complications such as a chronograph, tourbillon and perpetual calendar.

VON DER „222" ZUR „OVERSEAS"

1976 blieb Vacheron Constantin fast keine andere Wahl. In Genf benötigte man dringend eine Antwort auf die Audemars Piguet „Royal Oak" und die Patek Philippe „Nautilus". So kam es zur „222", benannt nach dem damaligen Alter des Genfer Traditionsunternehmens. Beim Entwurf und der Gehäusekonzeption hatte sich der junge deutsche Designer Jörg Hysek natürlich ein wenig an den beiden großen Vorbildern orientiert. Ein Mittelteil trug den damit verschraubten Glasrand. Das Ensemble bewirkte wiederum eine Art Bullaugen-Look. Der Resistenz gegen schädliche Einflüsse des nassen Elements diente auch eine sogenannte „doppelte Krone". Zehn Jahre nach ihrer Vorstellung und nur rund 500 produzierten Exemplaren musste die „Triple Two" den ebenfalls nicht sonderlich erfolgreichen Nachfolgemodellen „333" und „Phidias" weichen. Über die beiden letztgenannten Armbanduhren spricht man bei Vacheron Constantin nicht besonders gern. Ganz im Gegensatz zur „222", die im Jahr 2022 ein erfolgreiches Comeback in der Uhrenszene feierte.

Doch Mitte der 1990er-Jahre war Vacheron Constantin am Markt der sportlich-eleganten Luxusuhren in Edelstahl gar nicht mehr vertreten. Zur Beseitigung dieses Missstands machte sich die interne Designabteilung ans Werk und nutzte als markantes Gestaltungselement für den Glasrand das Malteserkreuz. Traditionsgemäß diente es Vacheron Constantin als Markenzeichen. Für den hermetischen Verschluss der Edelstahlschale bis zu 15 bar Druck sorgten insgesamt 16 Schrauben. Weitere Merkmale luxuriöser Sportlichkeit: Schraubkrone mit Flankenschutz, verstärktes Saphirglas, mit Tritium beschichtete Stunden- und Minutenmarkierungen sowie entsprechende Leuchtzeiger.

Zum 250. Firmenjubiläum im Jahr 2005 unterzog Vacheron Constantin die „Overseas" behutsamer Evolution. Der Durchmesser wuchs auf 42 Millimeter. Im stählernen Automatikmodell mit Zentralsekunde und Fensterdatum tickte das von einem Weicheisenring umgebene Kaliber Jaeger-LeCoultre 889. Die dritte, nunmehr gründlich überarbeitete „Overseas"-Generation gab 2016 ihren Einstand.

Sie präsentiert sich merklich runder, geschmeidiger und manschettenfreundlicher. Geblieben ist ein erhöhter Schutz gegen Magnetfelder. Zu diesem Zweck umringt das Uhrwerk wiederum ein Weicheisengebilde. Der bewirkt bei der „Overseas Automatik" eine Resistenz bis 8.000 A/m oder 100 Gauß. Praktische Innovation verkörpert das patentierte Bandwechselsystem. Mit wenigen Handgriffen lassen sich drei serienmäßig mitgelieferte Armbänder nach Belieben gegeneinander austauschen. Neben den schlichten Ausführungen in unterschiedlichen Größen und Materialien offeriert Vacheron Constantin auch Komplikationen wie beispielsweise Chronograph, Tourbillon oder ewiger Kalender.

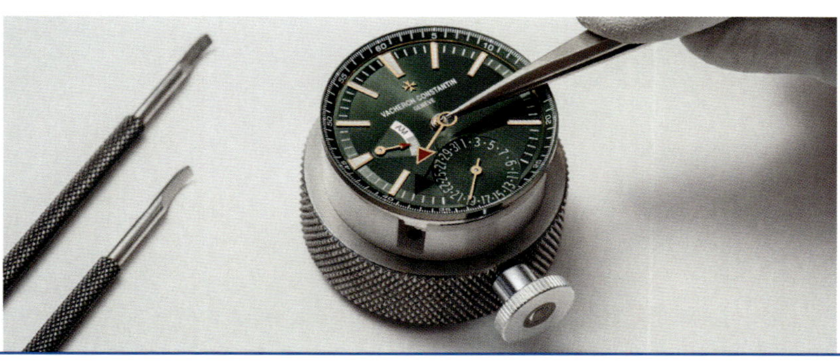

Overseas Ref. 7920V/210R-B965, Installation of watch hand | Montage der Zeiger

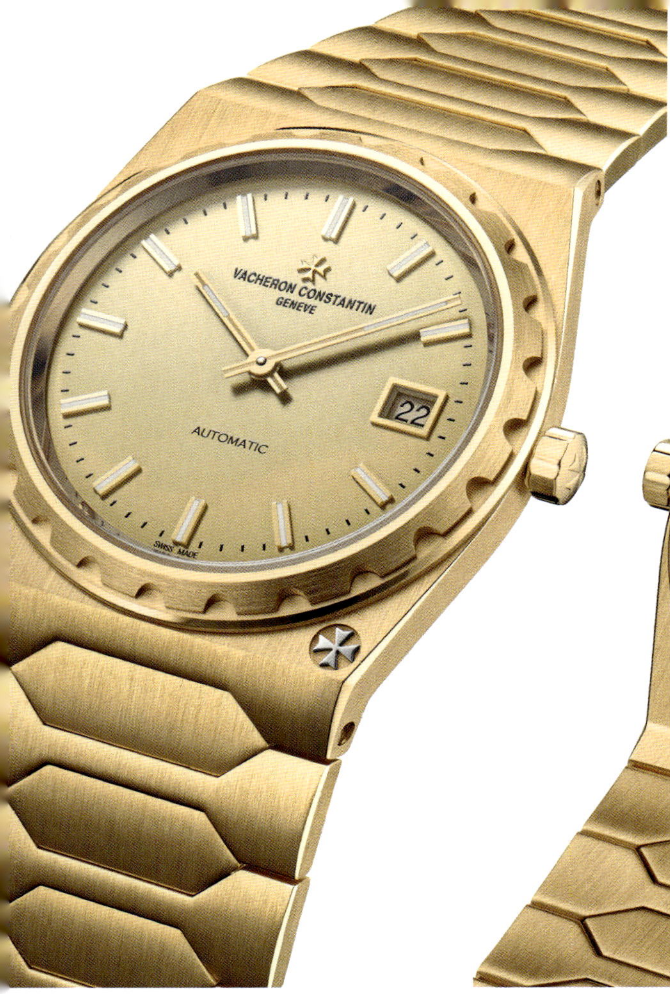

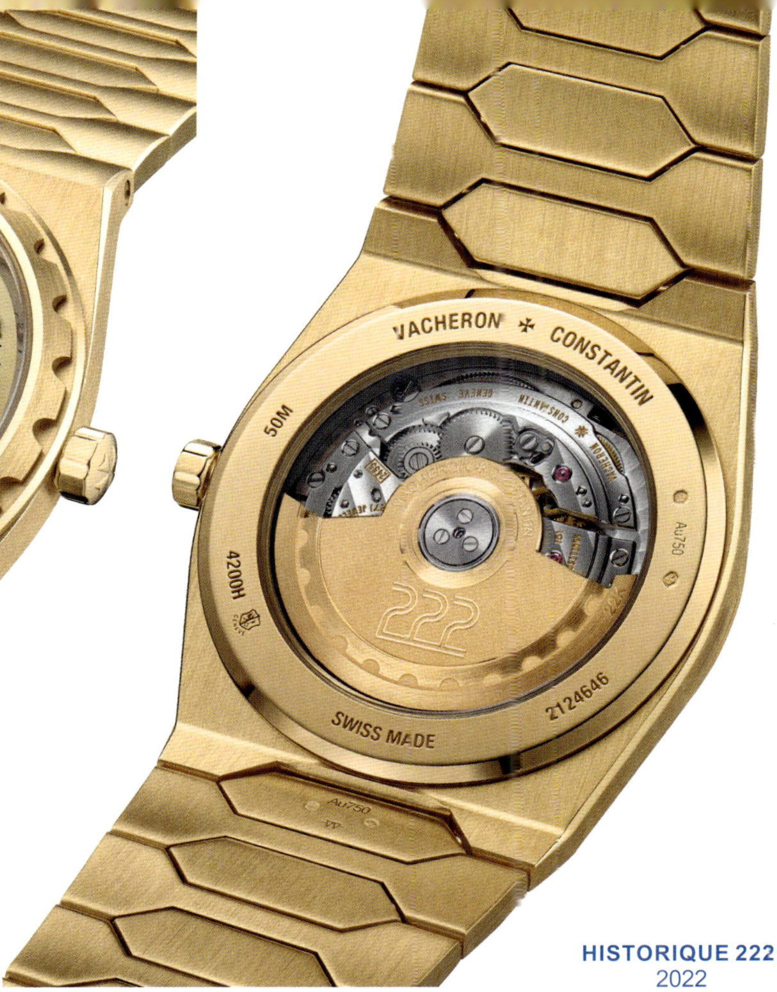

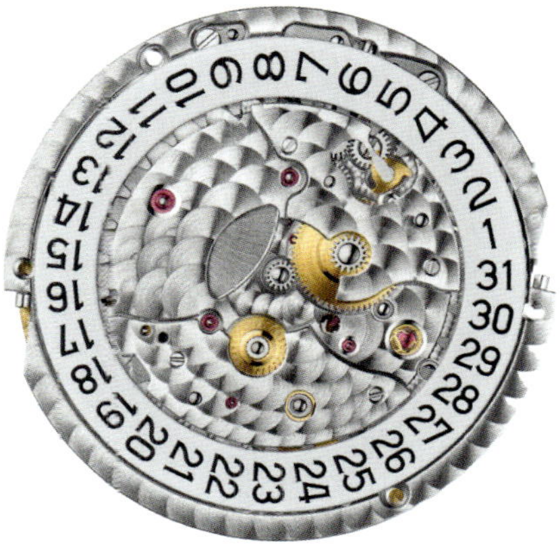

Cal. 2455

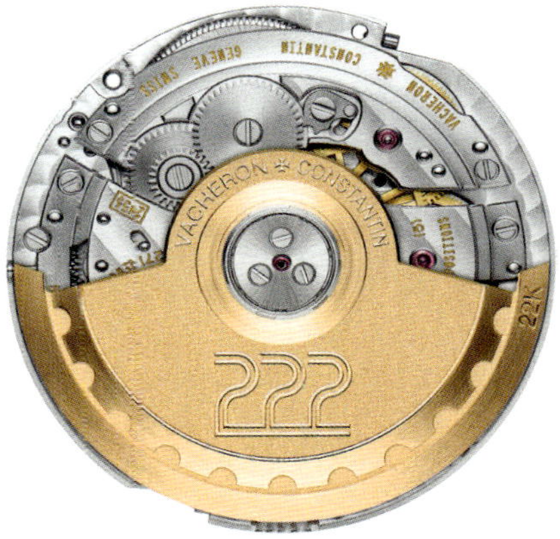

HISTORIQUE 222,
2022

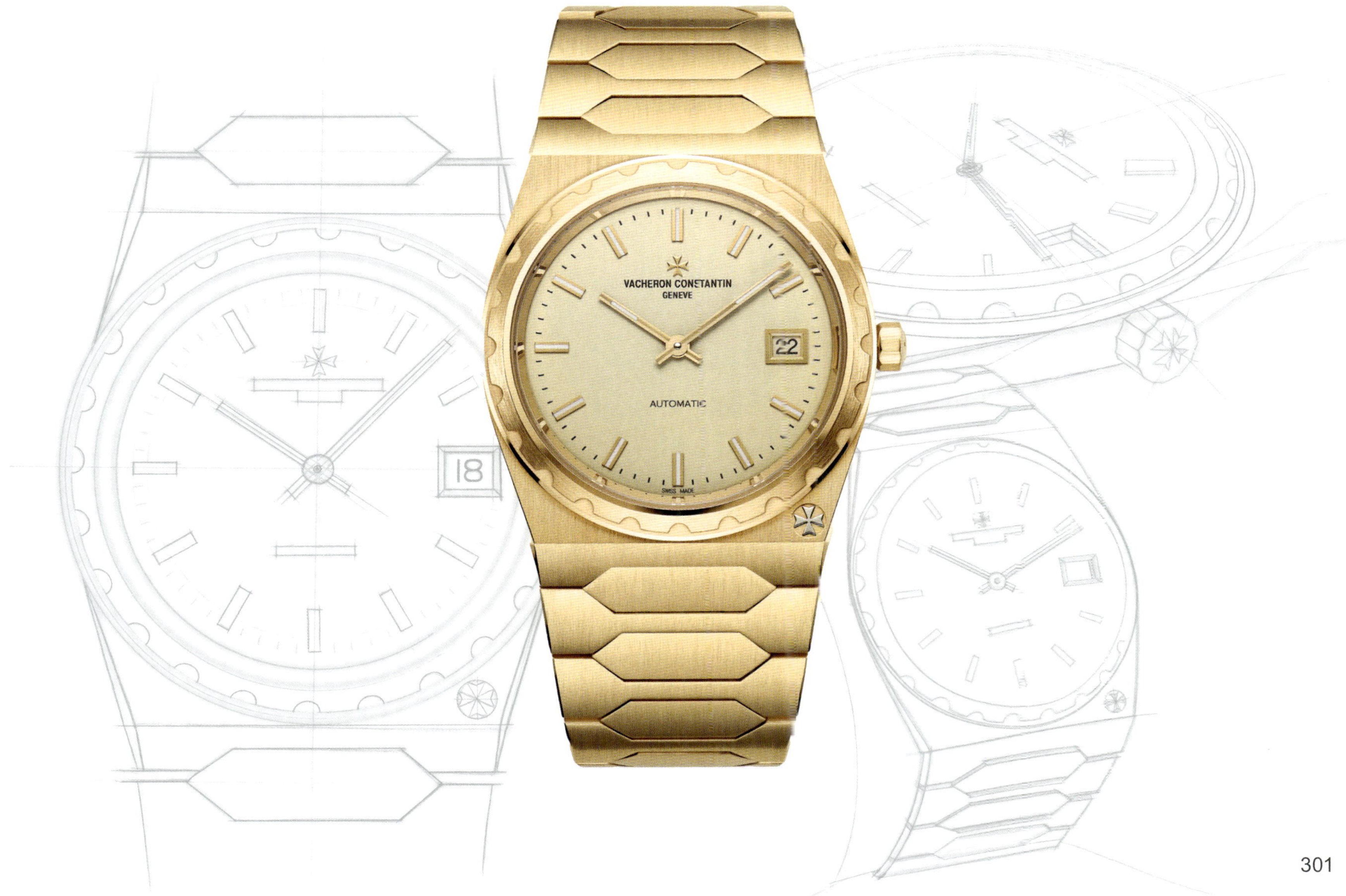

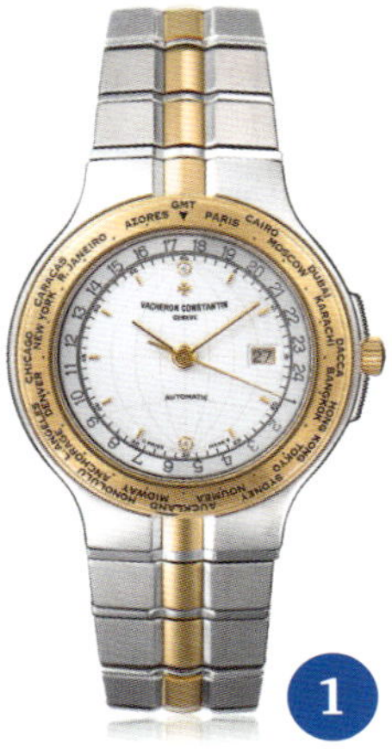

1

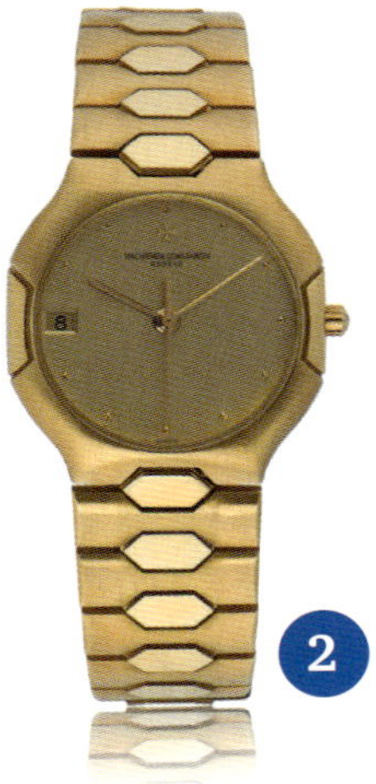

2

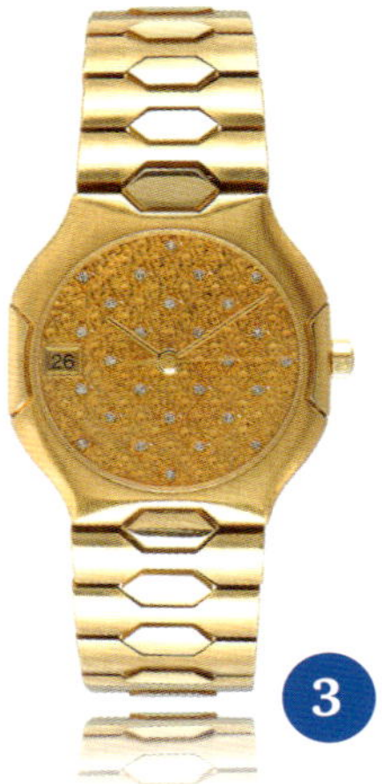

3

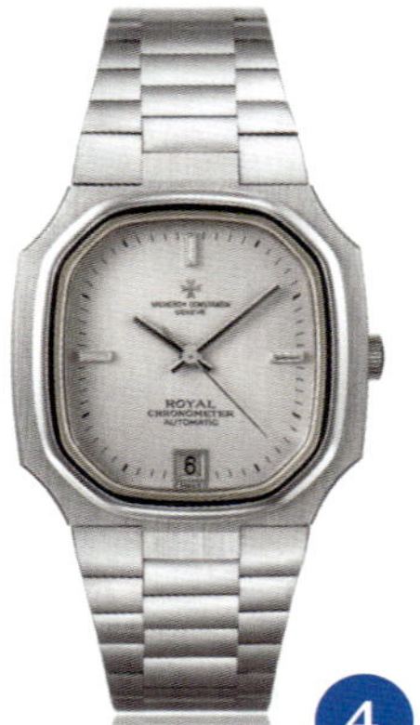

4

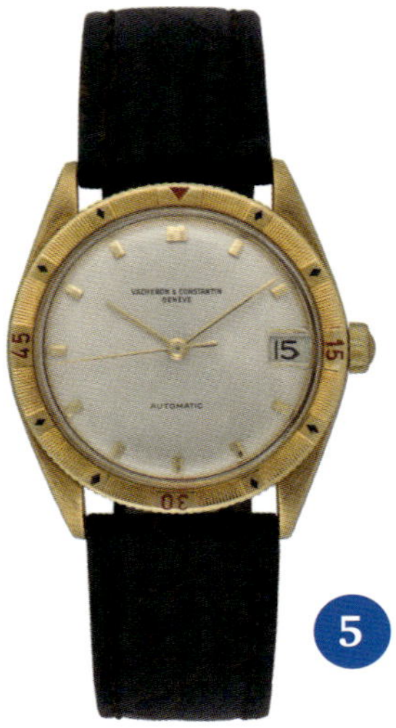

5

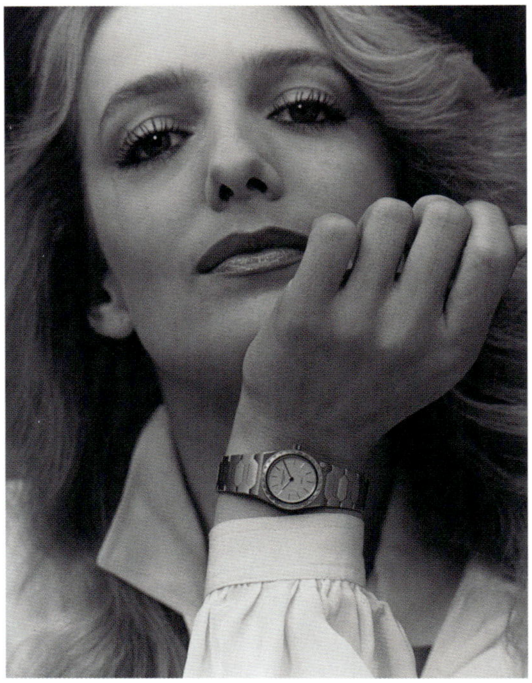

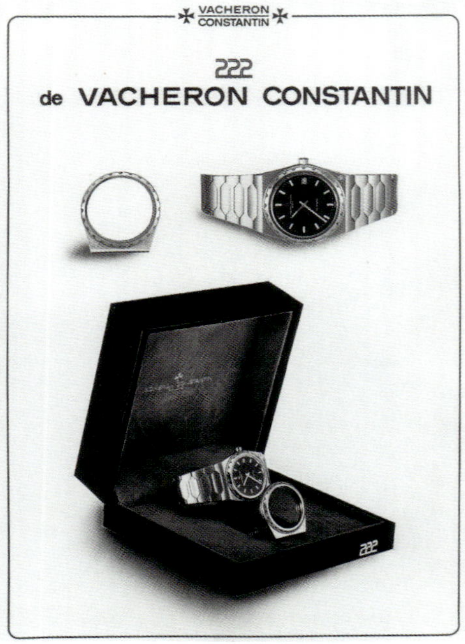

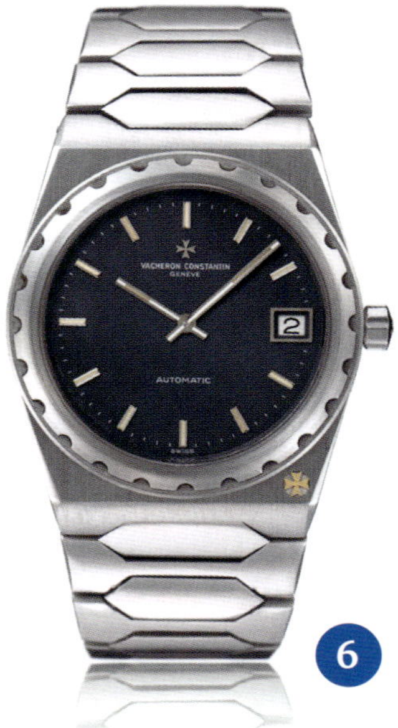

6

7

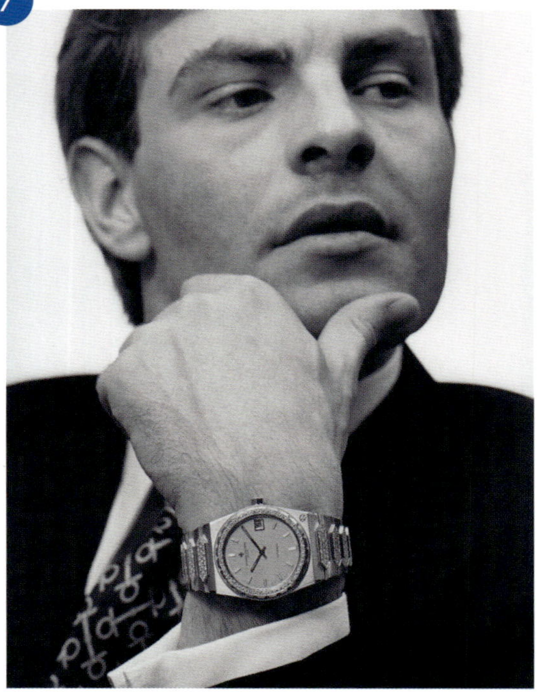

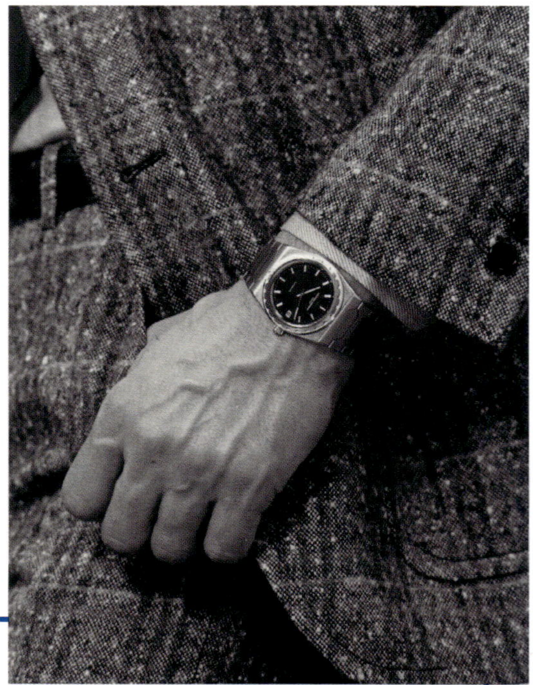

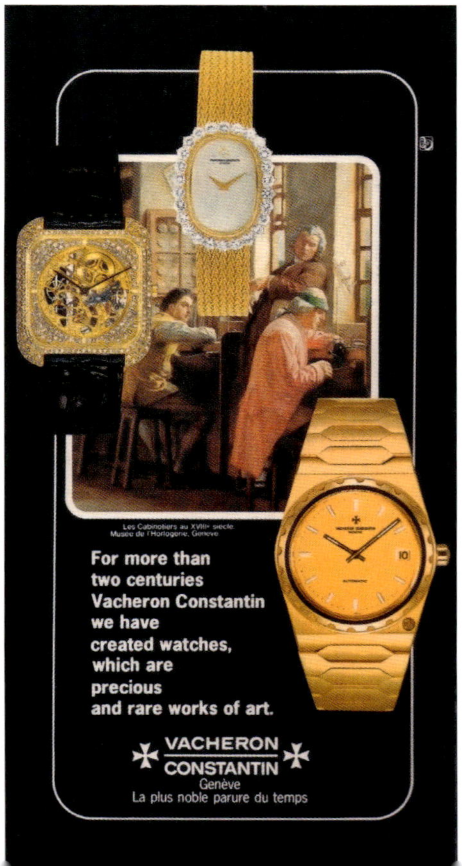

1) *Phidias Ref. 11754*
2) *333 Ref. 10218*
3) *333 Ref. 11503*
4) *Chronomètre Royal*
5) *Turnograph YG Ref. 6782*
6) *Régulateur I, 1988*
7) *Advertisements for the 222, 1977*

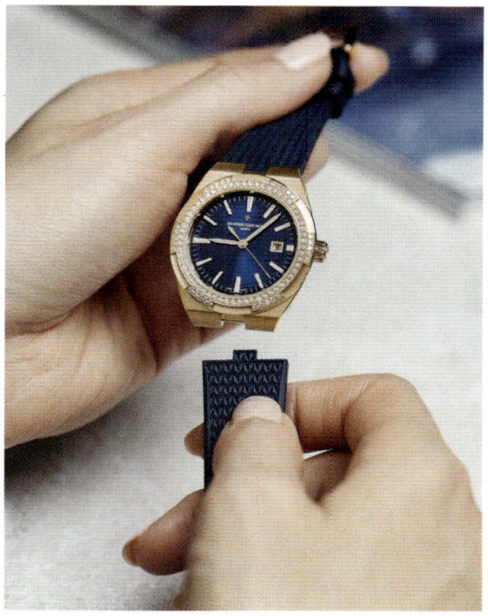

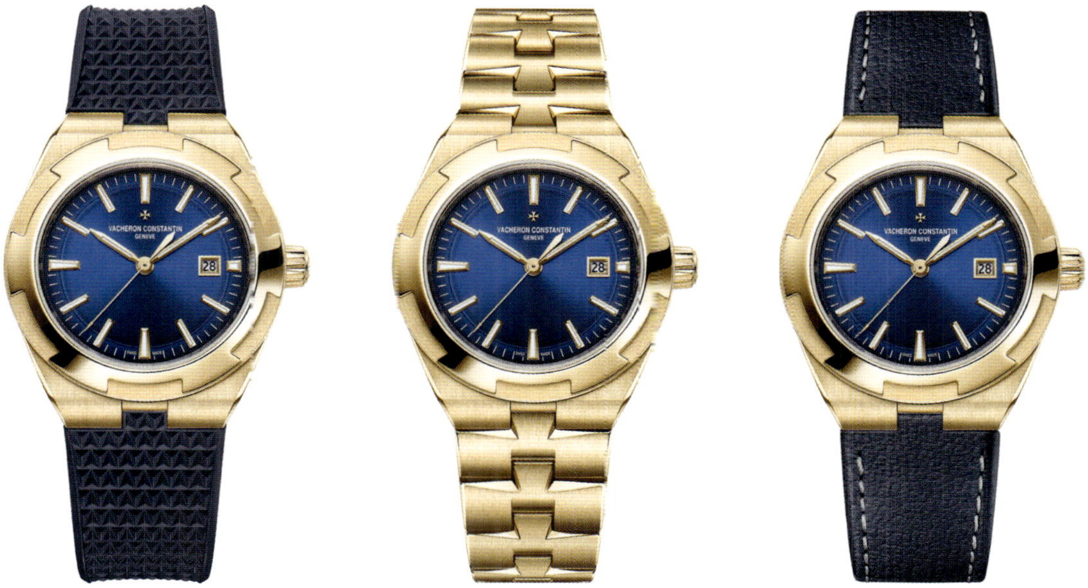

OVERSEAS REF. 4600V/200R-B979

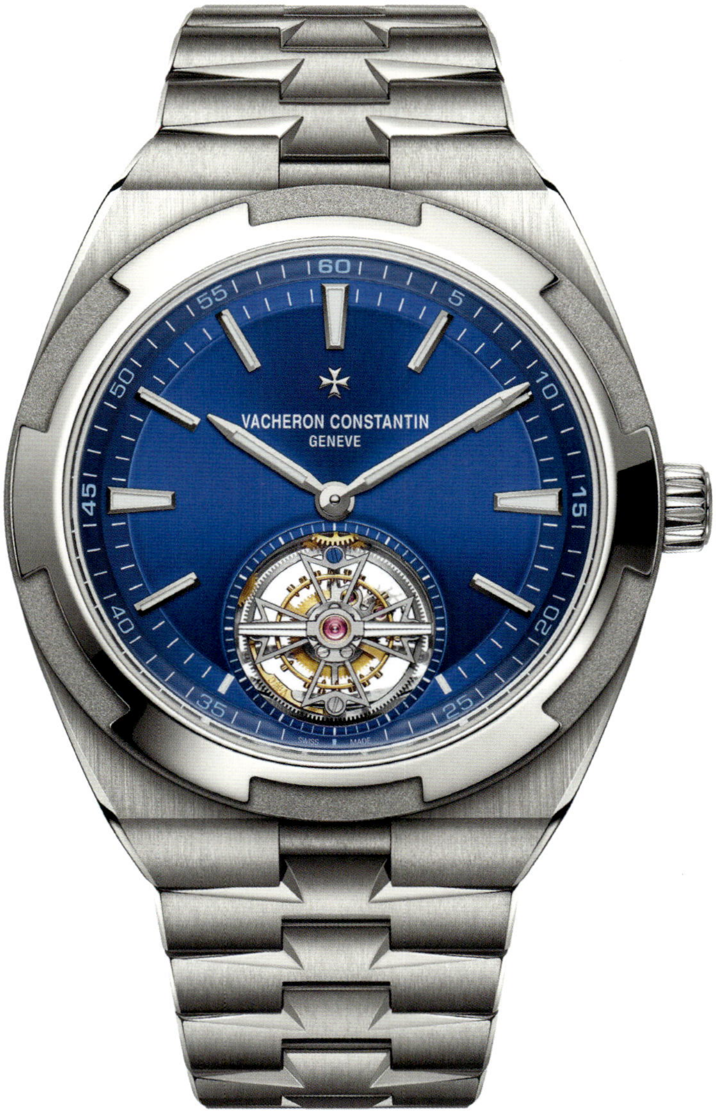

OVERSEAS TOURBILLON

OVERSEAS PERPETUAL CALENDAR
ULTRA-THIN SKELETON

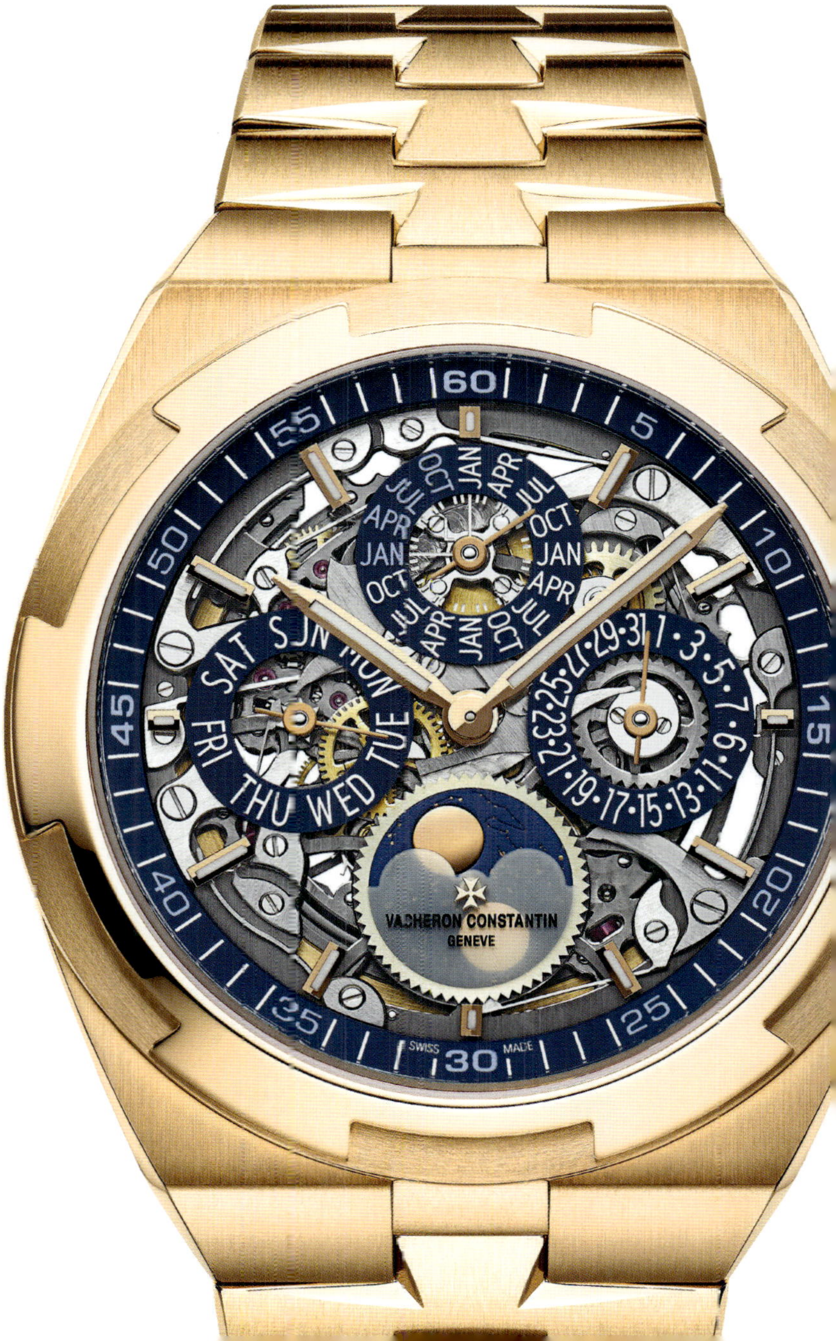

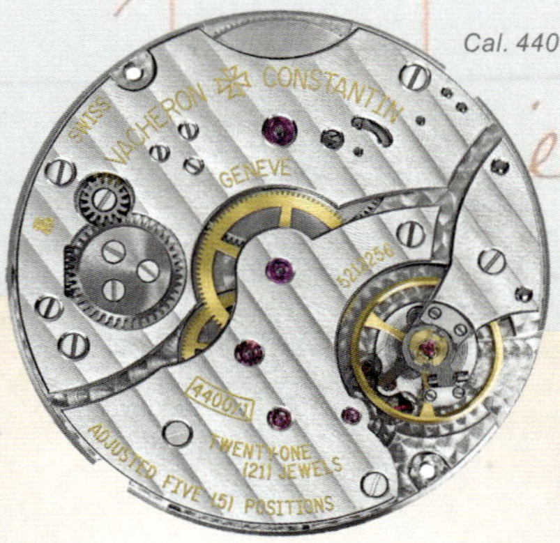

Cal. 4400AS

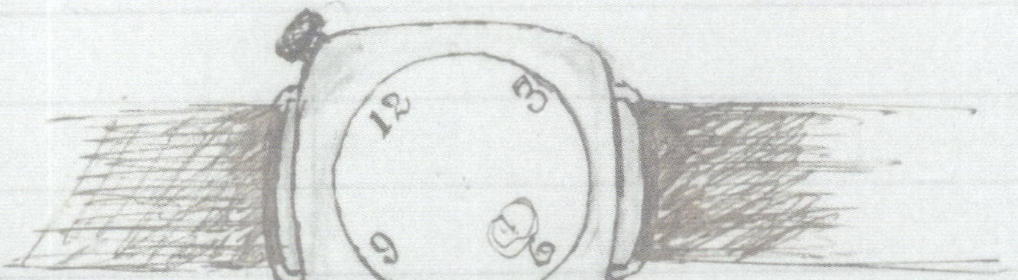

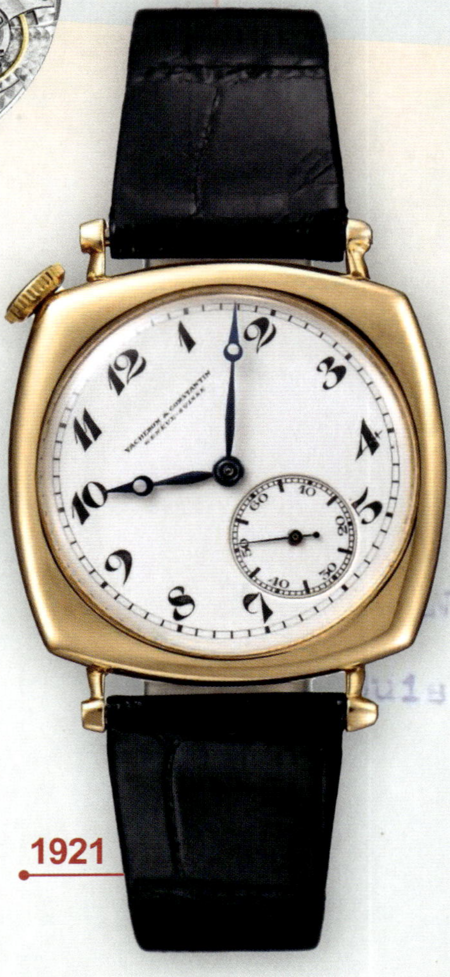

1921

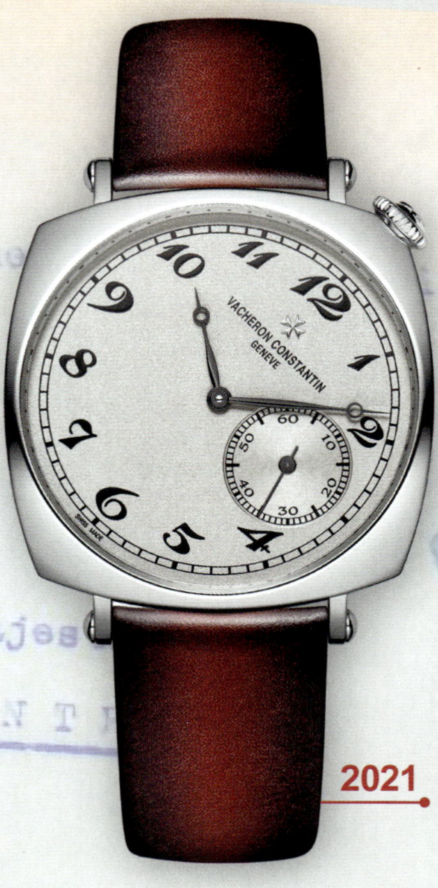

2021

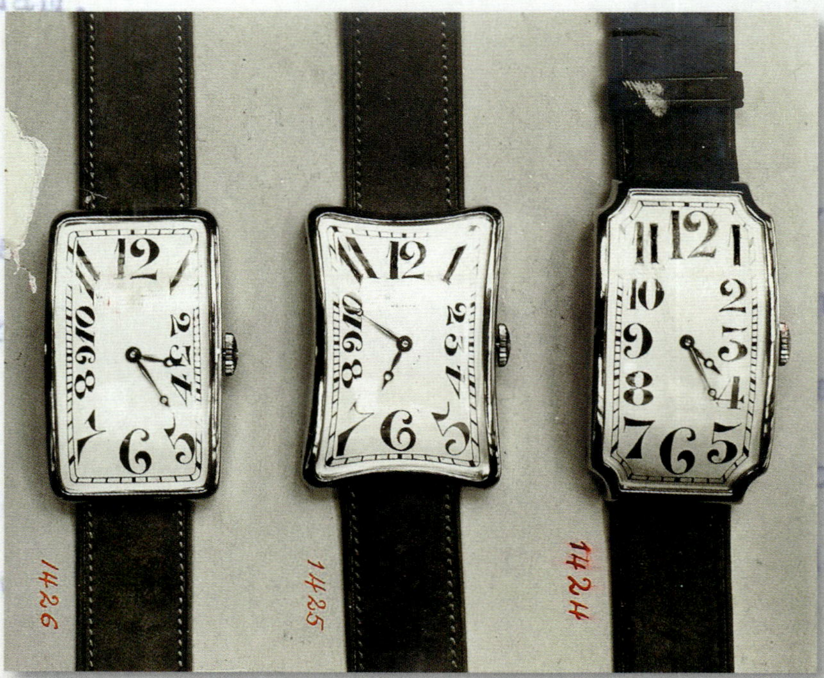

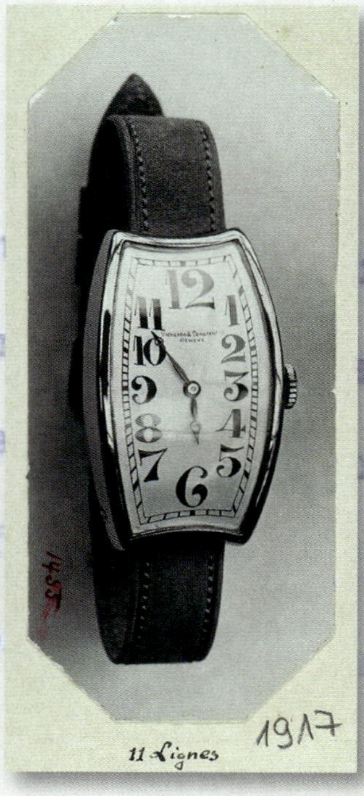

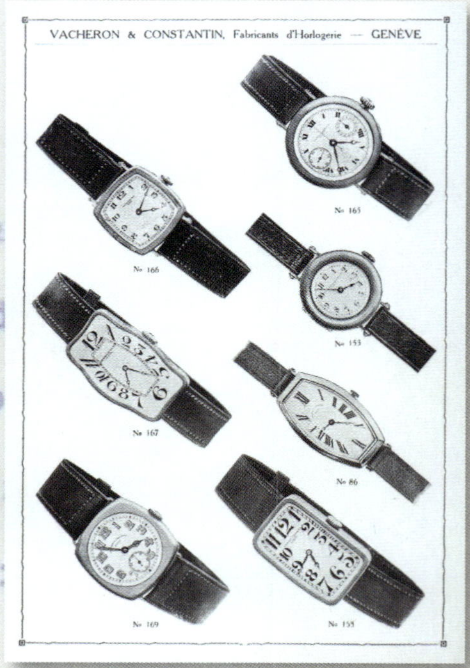

Product range | Sortiment, 1917

HISTORIQUES AMERICAN 1921

MADE FOR AMERICA

Vacheron Constantin's Historiques American 1921 quite literally introduced a new slant. The story unfolds as the hugely important US market called for timepieces of an exceptional nature. The Geneva-based company responded by creating an attention-grabber nicknamed the American, or Amerique in French. Its 31 by 31 millimetre cushion-shaped case was in complete harmony with the tastes of the early Roaring Twenties. Along with the lugs, the diagonal arrangement of the enamel dial with a small seconds at 6 o'clock defied conventions. Depending on the positioning of the Lépine R.A. 11'"62 Nouveau calibre, the winding and hand-setting crown was located in the top right or left corner of the solid gold case. The choice of positions was to accommodate right and left-handed wearers respectively. At your desk, this new slant enabled you to read off the time at a glance with no need to move your wrist. The small seconds was positioned at 6 o'clock, the crown directly opposite at 12 o'clock. It was also easy to read the time while driving as there was no need to take your hands off the steering wheel. One of the brand's customers was Samuel Parkes-Cadman. The American writer and clergyman fought against racial segregation and anti-Semitism. In 2009, the Geneva manufactory relaunched this iconic eye-catcher. It was primarily intended for right-handed wearers. The 40-millimetre case incorporated the then new hand-wound Calibre 4400. In 2021, a century after the icon was originally launched, Vacheron Constantin saw fit to devote some well-deserved attention to the American. The white gold case is sized 40 by 40 millimetres and is exactly 8.06 millimetres high. The centenary timepiece is driven by the hand-wound Calibre 4400 AS consisting of 127 components. A smaller version measures just 36.5 millimetres. At 7.41 millimetres, it is even slimmer. As you'd expect from Vacheron Constantin, both the movement and case bear the Geneva seal of quality.

ES WAR EINMAL FÜR AMERIKA

Schräg im wahrsten Wortsinn ist die „Historiques American 1921" von Vacheron Constantin. Einst hatte der immens wichtige US-amerikanische Markt nach einem außergewöhnlichen Zeitmesser fürs Handgelenk verlangt. Das Genfer Unternehmen handelte und kreierte das, was mit dem Beinamen „American" oder auf gut Französisch „Amerique" Aufmerksamkeit erregte. Ein kissenförmiges Gehäuse mit den Dimensionen 31 x 31 Millimeter entsprach voll und ganz dem Geschmack der frühen „Roaring Twenties". Neben den Bandanstößen setzte sich auch die diagonale Anordnung des Emailzifferblatts mit kleiner Sekunde bei „6" ganz deutlich vom Üblichen ab. Je nach Einbau des Lépine-Kalibers R.A. 11'"62 Nouveau fand sich die Aufzugs- und Zeigerstellkrone in der rechten oder linken oberen Ecke des massivgoldenen Gehäuses. Erstgenannte Positionierung war für Rechts-, die Alternative für Linkshänder gedacht. Auf diese Weise gestatteten die Modelle am Schreibtisch einen geraden Blick auf die Längsachse mit kleiner Sekunde bei „6" und der Krone exakt gegenüber bei „12". Auch in einem Automobil bot die „American" gute Ablesbarkeit, ohne die Hände vom Lenkrad zu nehmen. Zum Kundenkreis gehörte Samuel Parkes-Cadman. Der amerikanische Schriftsteller und Geistliche kämpfte gegen Rassentrennung und Antisemitismus.

2009 inszenierte die Genfer Manufaktur ein Comeback dieses tickenden Blickfängers, gedacht in erster Linie für Rechtshänder. Seine 40-Millimeter-Schale umfing das damals neue Handaufzugskaliber 4400. Rund 100 Jahre „American" waren 2021 für Vacheron Constantin Anlass genug, dieser augenfälligen Ikone erneut gebührende Aufmerksamkeit zu schenken. Am Handgelenk trägt das Weißgoldgehäuse in den Dimensionen 40 x 40 Millimeter exakt 8,06 Millimeter auf. Die Zeit bewahrt einmal mehr das aus 127 Komponenten assemblierte Handaufzugskaliber 4400 AS. Lediglich 36,5 Millimeter misst eine kleinere Version. Sie baut 7,41 Millimeter flach. Selbstverständlichkeit für Vacheron Constantin ist das Genfer Qualitätssiegel auf Uhrwerk und Gehäuse.

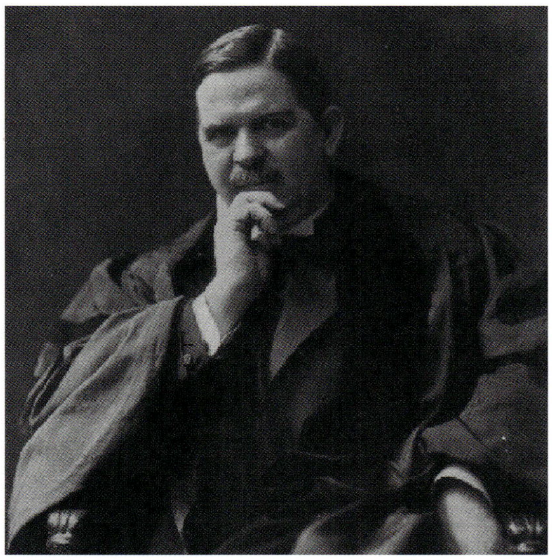

Samuel Parkes-Cadman

1919

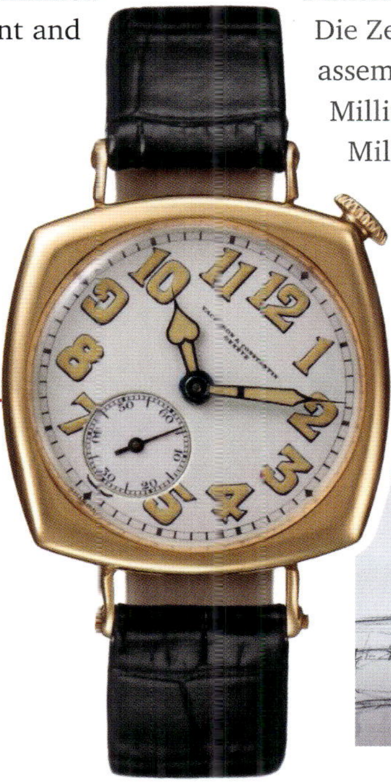

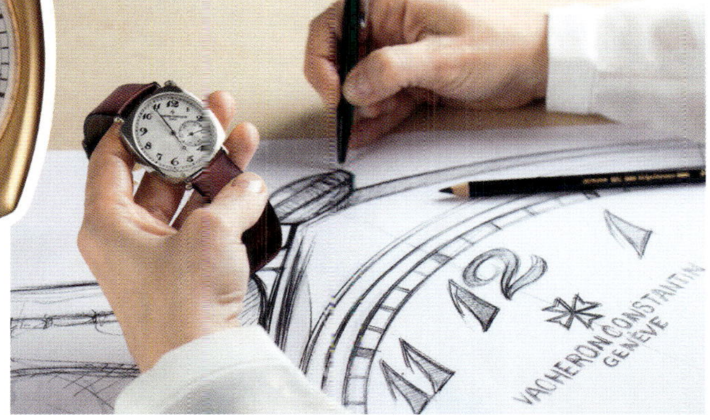

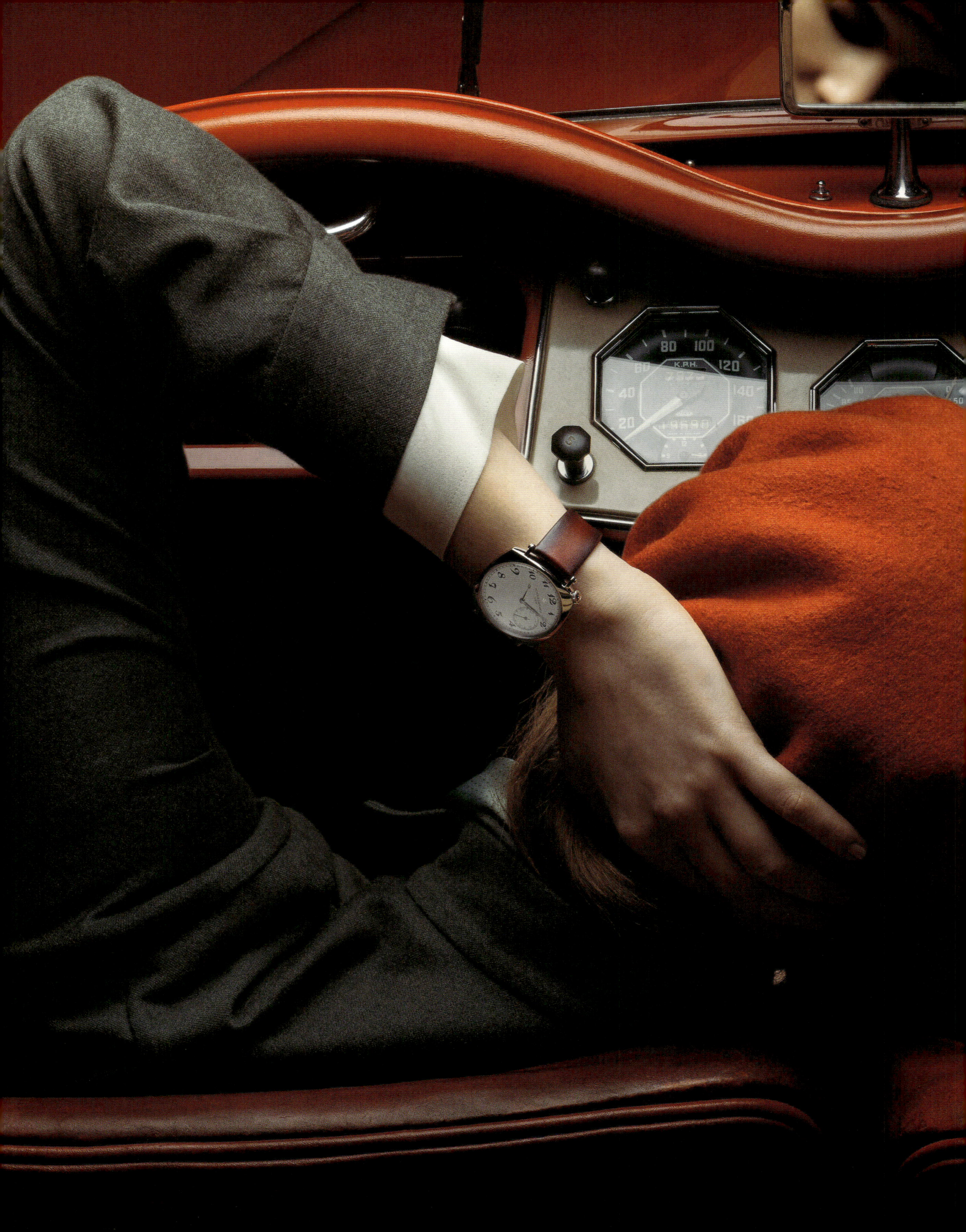
80
100
K.P.H.
120
40
140
20

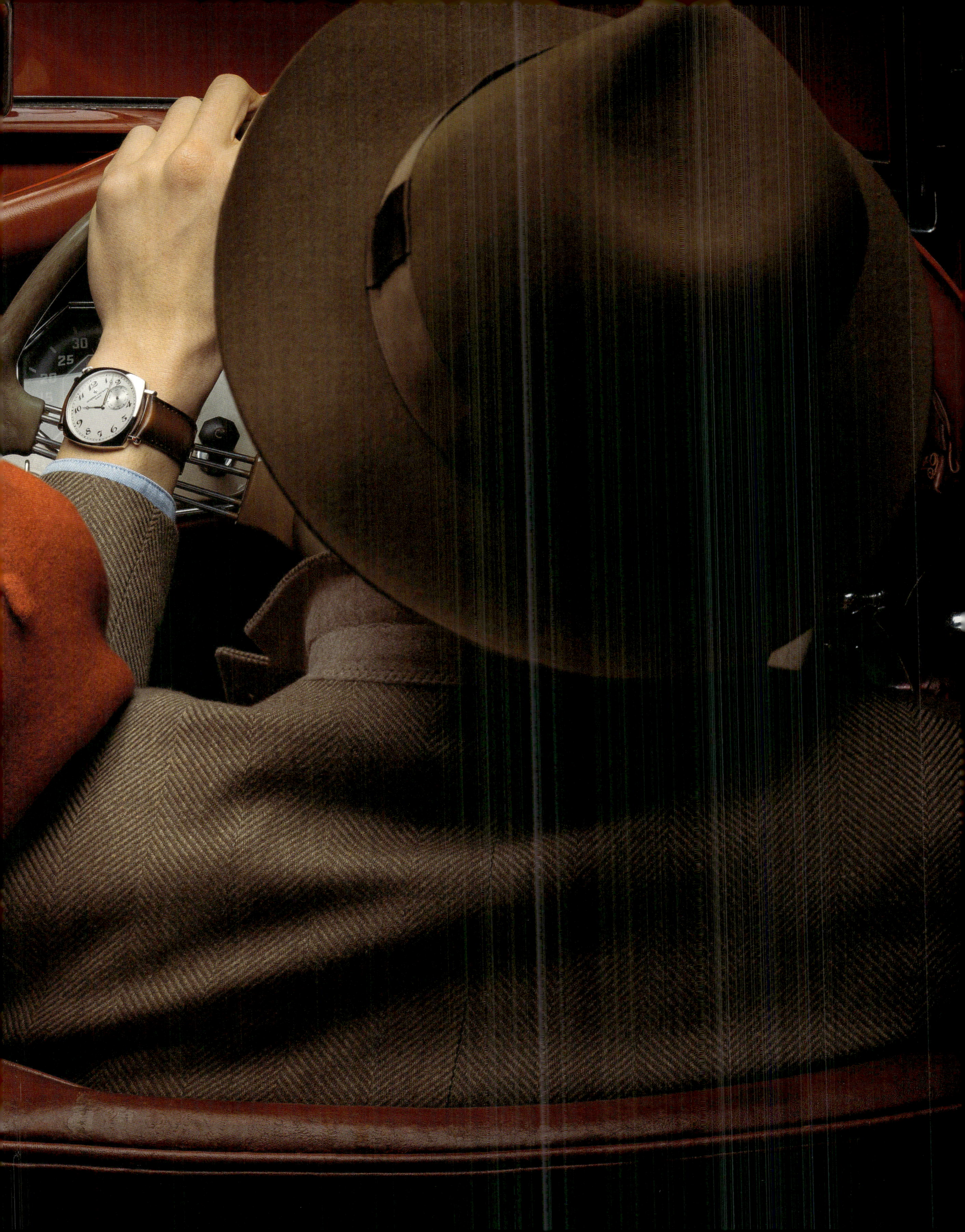

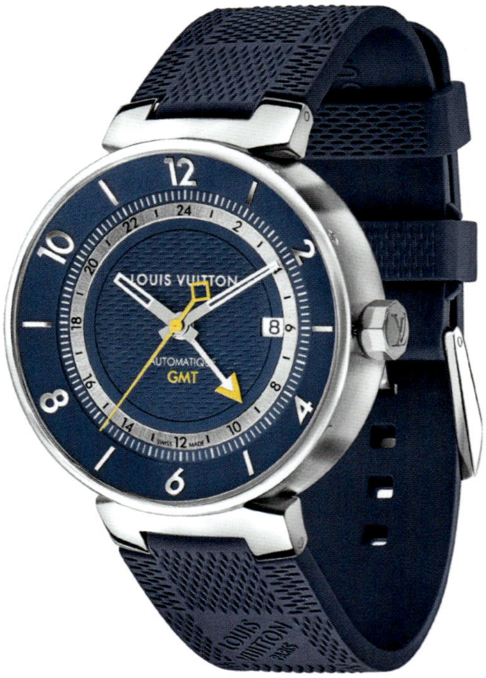

TAMBOUR MOON,
2021

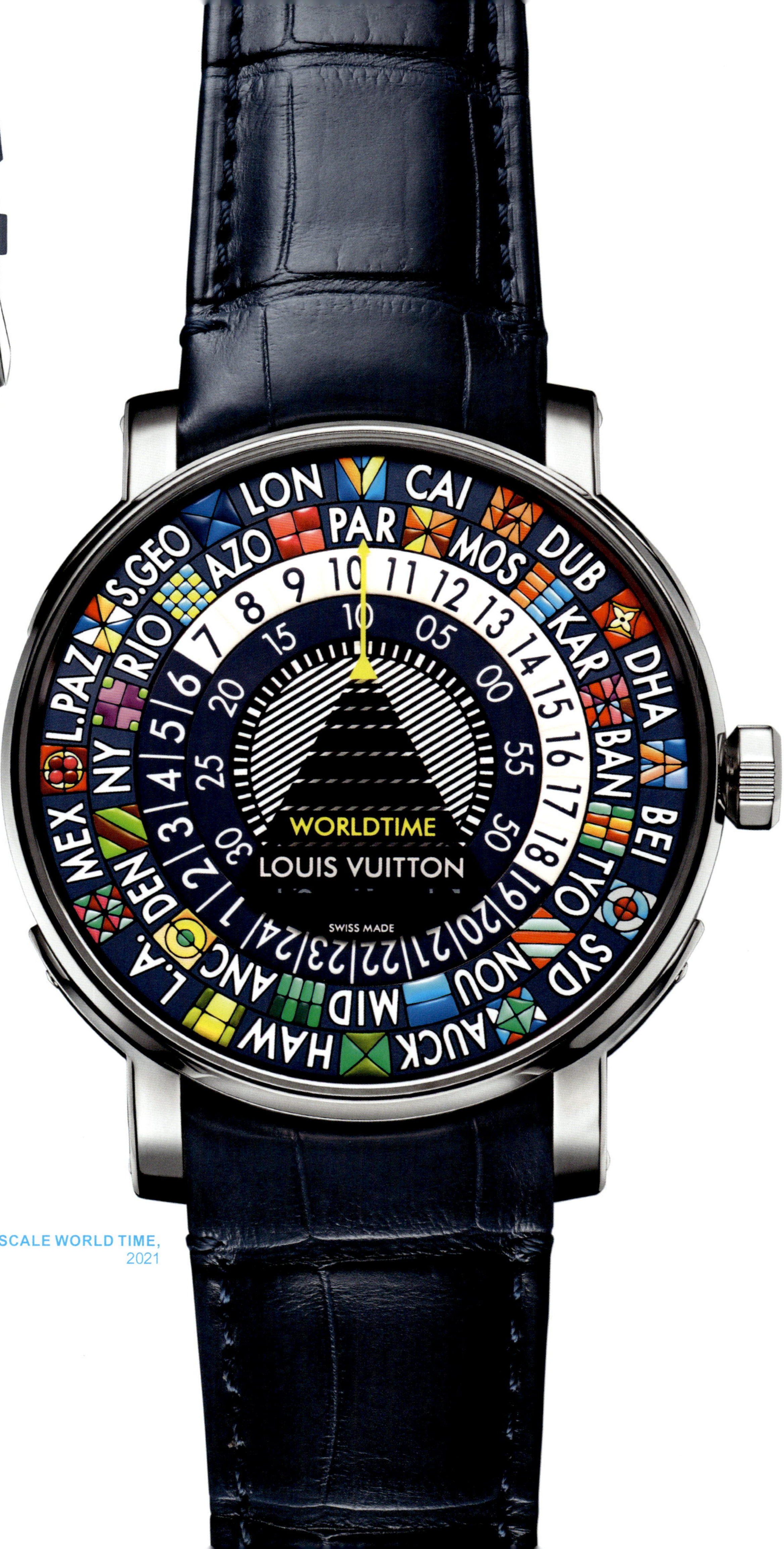

ESCALE WORLD TIME,
2021

2014

LOUIS VUITTON *TAMBOUR*

PARISIAN DRUM BEATS

Discerning globetrotters consider exclusive luggage and high-quality watches as must-haves. However, timepieces did not figure among the business challenges tackled by Monsieur Louis Vuitton and his descendants. LVMH, today owner of the trademark rights and international luxury multinational No. 1, takes a different view. Anyone in the travel accessory business should be equally acquainted with watchmaking and its shared focus on fine craftsmanship. Founded in 2002, the watch division of Louis Vuitton in La Chaux-de-Fonds aspired to do just that. The French luxury label's first excursion into the watch scene took place in 1989 with the travel watch, which was developed and produced by IWC. As Vuitton tradition dictates, the wristwatch with electronic inner workings and innovative time zone complication was sold exclusively through the company's worldwide network of boutiques, thus enabling full control over sales and prices.

By 2002, this timepiece was of course history. Henceforth, Louis Vuitton placed its trust in the power and magic of the drum. This was expressed by its new watch line Tambour, French for drum. With its rounded contours, it does recall that musical instrument which, if need be, can be very loud indeed. The first-generation chronographs, equipped with automatic El Primero calibre from the sister company Zenith, were followed by regular drum rolls introducing new and, in some cases, very complex Tambour iterations with a tourbillon or minute repeater. In 2011, Louis Vuitton acquired La Fabrique du Temps, a recognised specialist in mechanical complications, with which the brand had already worked very successfully. Tambour Bleu GMT, Tambour Evolution Chronograph GMT, Tambour Evolution Spin Time GMT and the colourful Escale Worldtime are just four examples of the rich portfolio that has evolved in the course of more than two decades. The latest member of the flagship Tambour watch line was launched in July 2023. The product designers treated the icon to a makeover for the occasion. The leaner flanks and an integrated link bracelet lend the watch, which is available in steel or gold, a sportier look to keep it in fine fettle for the next few years.

PARISER TROMMELWIRBEL

Anspruchsvolle Kosmopoliten betrachten edles Gepäck und hochwertige Uhren als beinahe unverzichtbar. Für Monsieur Louis Vuitton und seine familiären Nachkommen stellten Zeitmesser allerdings noch keine unternehmerische Herausforderung dar. LVMH, heutiger Inhaber der Markenrechte und internationaler Luxusmulti Nr. 1, denkt da ganz anders. Wer Reiseaccessoires herstellen kann, dem sollte das nicht minder handwerklich ausgerichtete Thema Armbanduhren keineswegs fremd sein. Diesem Anspruch trug die 2002 in La Chaux-de-Fonds gegründete Uhrendivision von Louis Vuitton Rechnung. Den ersten Ausflug in die Uhrenszene hatte das französische Luxuslabel indessen schon 1989 mit der von IWC entwickelten und gefertigten „Reiseuhr" unternommen. Der Verkauf dieser Armbanduhr mit elektronischem Innenleben und innovativem Zeitzonen-Dispositiv erfolgte, wie bei Louis Vuitton traditionsgemäß üblich, über das weltumspannende Netz eigener Boutiquen. Auf diese Weise lassen sich Vertrieb und Preise exakt kontrollieren. Von diesem Zeitmesser war 2002 logischerweise keine Rede mehr. Fortan setzte Louis Vuitton auf die Macht und Magie der Trommel. Selbige brachte die damals neu entwickelte Uhrenlinie „Tambour" zum Ausdruck. Deren Optik mit gerundeten Flanken erinnert in der Tat an das bei Bedarf sehr lautstarke Musikinstrument. Auf die Chronographen der ersten Generation, ausgestattet unter anderem mit dem Automatikkaliber „El Primero" der Schwester Zenith, folgten regelmäßige Trommelwirbel, begleitet von neuen, uhrmacherisch teilweise sehr komplexen „Tambour"-Modellen beispielsweise mit Tourbillon oder Minutenrepetition. 2011 erwarb Louis Vuitton mit „La Fabrique du Temps" einen ausgewiesenen Spezialisten für mechanische Komplikationen, mit der die Marke schon zuvor sehr gedeihlich zusammengearbeitet hatte. „Tambour Bleu GMT", „Tambour eVolution Chronograph GMT", „Tambour eVolution Spin Time GMT" oder die farbenfrohe „Escale Worldtime" sind nur einige Beispiele aus dem reichhaltigen Spektrum, welches im Laufe von mehr als 20 Jahren entstanden ist. Das neueste Mitglied der Flagship-Uhrenlinie „Tambour" ging im Juli 2023 an den Start. Im Vorfeld verpassten die Produktgestalter der Ikone eine modifizierte Optik. Weniger bauchige Flanken und ein integriertes Gliederband machen die in Stahl oder Gold erhältliche Armbanduhr sportlicher und damit auch fit für die nächsten Jahre.

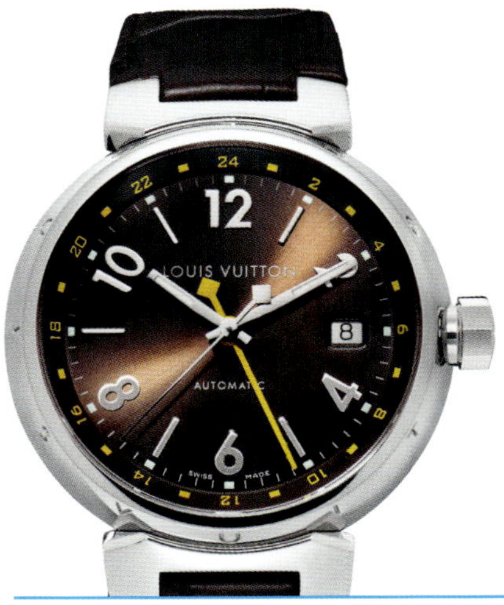

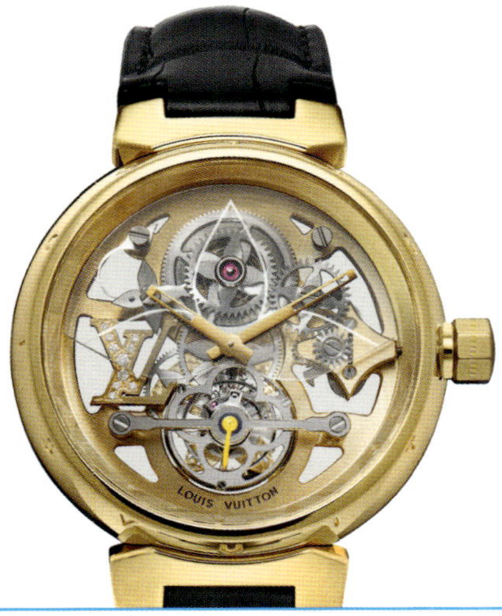

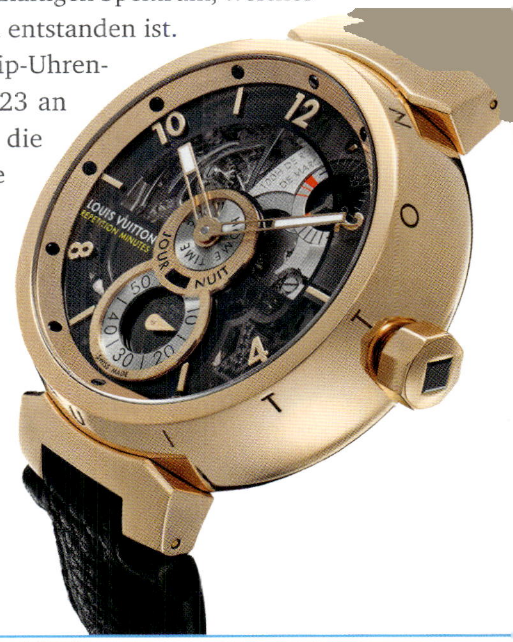

Left / Links: *Automatique GMT, 2002*
Center / Mitte: *Tourbillon Monogram, 2012*
Right / Rechts: *Répétition Minutes*

ZENITH 1969
DEFY

RESISTANCE FROM LE LOCLE

The name Defy is well chosen. In this spirit, some 50 years ago Zenith developed a watch called Defy A3642. Its début coincided with the presentation of the El Primero chronographs in 1969. In addition to its distinctive octagonal shape, the highly resistant case featured a fourteen-sided glass bezel. The style elements included substantial block-shaped indexes on the dial. Enthusiasm for the watch was muted to say the least. The edgy appearance of this chunky timepiece, which the French nicknamed *coffre-fort*, or vault in English, divided opinion. The "vault" was secured to the wrist with a steel rung-ladder bracelet designed by the experts Gay Frères. Over the years, Zenith continuously enhanced the Defy with technical advances. In the early 1980s, for instance, the company brought out a scratch-resistant sapphire crystal, also featured on the Revival model A3691. Zenith has remained loyal to its ladder bracelet. However, the folding clasp is now much more wrist-friendly. Another modern-day feature is the transparent caseback, which reveals the proprietary Elite 670 calibre with star-shaped ball-bearing rotor. The regulator oscillates at a pace of four hertz. A power reserve of 50 hours is guaranteed. With a nod to tradition, Zenith places the characteristic date window between 4 and 5 o'clock. Unlike in the past, non-radioactive Super-LumiNova coating is used today. With its 37-millimetre case diameter, this feisty icon cuts a good figure on male and female wrists. In spring 2024, the equally large diver's watch Defy Revival A3648 with a steel case was added to the collection.

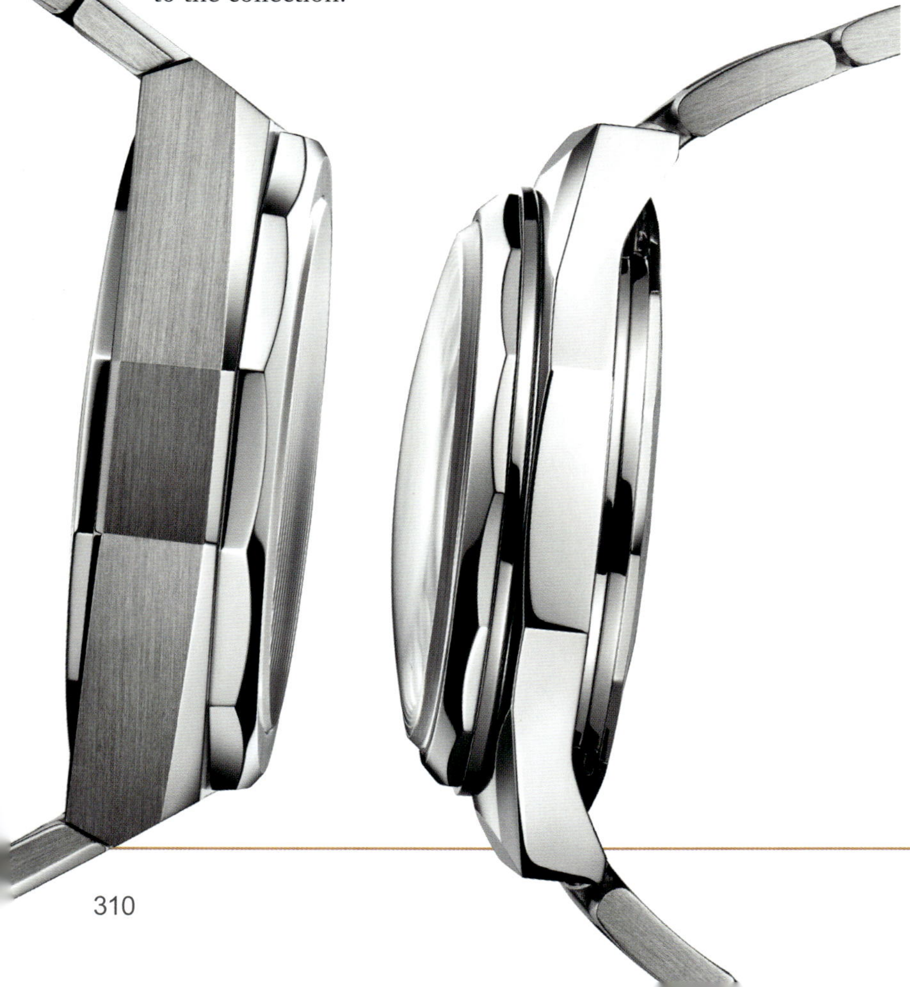

ROBUSTES AUS LE LOCLE

Auf gut Deutsch meint das englische Wort „Defy" nichts anderes als „widerstehen" oder „sich widersetzen". Unter diesem Vorzeichen entwickelte Zenith vor mehr als 50 Jahren eine Armbanduhr namens „Defy A3642". Das Debüt erfolgte zeitgleich zur Vorstellung der „El Primero"-Chronographen im Jahr 1969. Neben der markanten oktogonalen Gestalt besaß die in mehrfacher Hinsicht hoch belastbare Schale einen 14-eckigen Glasrand. Zu den gestalterischen Besonderheiten gehörten massive blockförmige Indexe am Zifferblatt. Von einhelliger Begeisterung konnte damals keine Rede sein. Der kantige Auftritt dieses Boliden mit dem französischen Spitznamen „coffre-fort" polarisierte mächtig. Am Handgelenk hielt diesen „Panzerschrank" ein stählernes Sprossendesign-Armband der einschlägig erfahrenen Gay Frères. Im Laufe der folgenden Jahre ließ Zenith der „Defy" immer wieder technische Fortschritte angedeihen. Im Zuge dessen brachten die frühen 1980er-Jahre ein kratzfestes Saphirglas, welches auch dem Revival-Modell „A3691" zu eigen ist. Auch am Leiterarmband hält Zenith fest. Allerdings verfügt es nun über eine deutlich ergonomischere Faltschließe. Ausdruck der Neuzeit ist ferner ein Sichtboden, durch den sich das hauseigene Automatikkaliber Elite 670 mit sternförmig ausgeführtem Kugellagerrotor zeigt. Sein Gangregler oszilliert mit vier Hertz. Rund 50 Stunden beträgt die Gangautonomie. Charakteristisch und gleichzeitig auch an die Tradition erinnernd, platziert Zenith das Fensterdatum zwischen „4" und „5". Im Gegensatz zu einst findet heutzutage die nicht radioaktive Leuchtmasse Super-LumiNova Verwendung. Dank 37 Millimeter Gehäusedurchmesser passt die ausdrucksstarke Ikone gleichermaßen gut an weibliche und männliche Handgelenke. Im Frühjahr 2024 fand die gleich große, bis zu 60 bar wasserdichte Taucherarmbanduhr „Defy Revival A3648" mit Stahlgehäuse in die Kollektion.

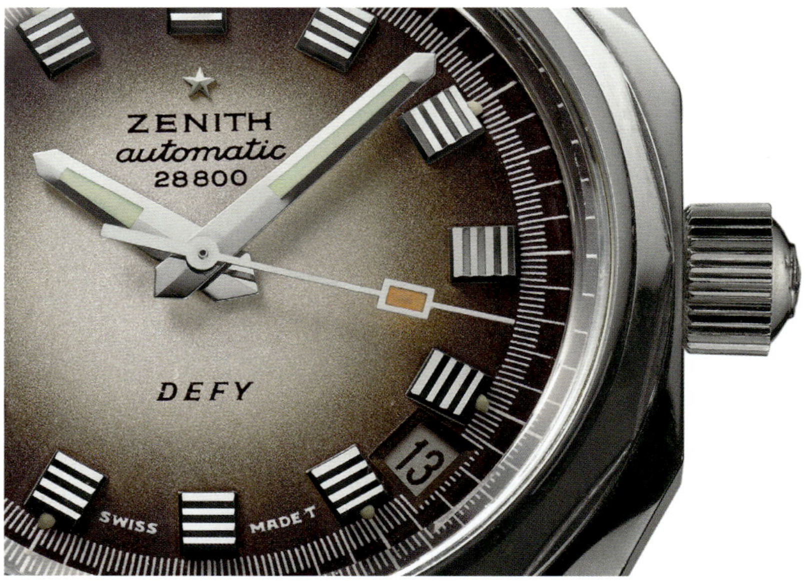

Revival A3642, 2022

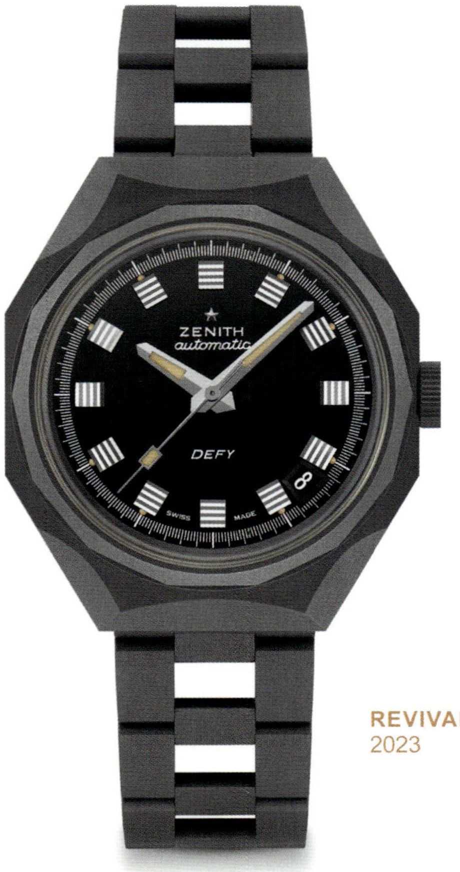

REVIVAL SHADOW, 2023

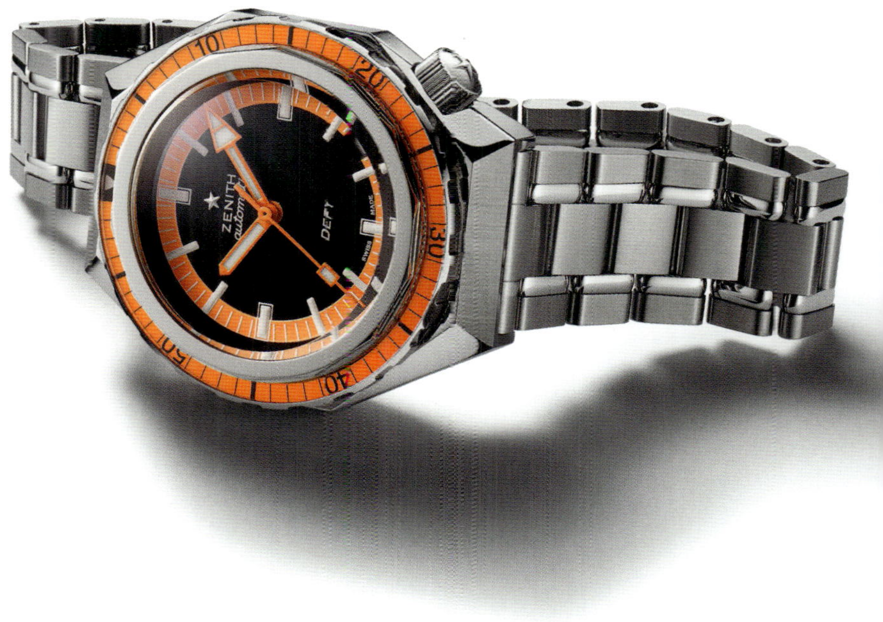

REVIVAL A3648, 2024

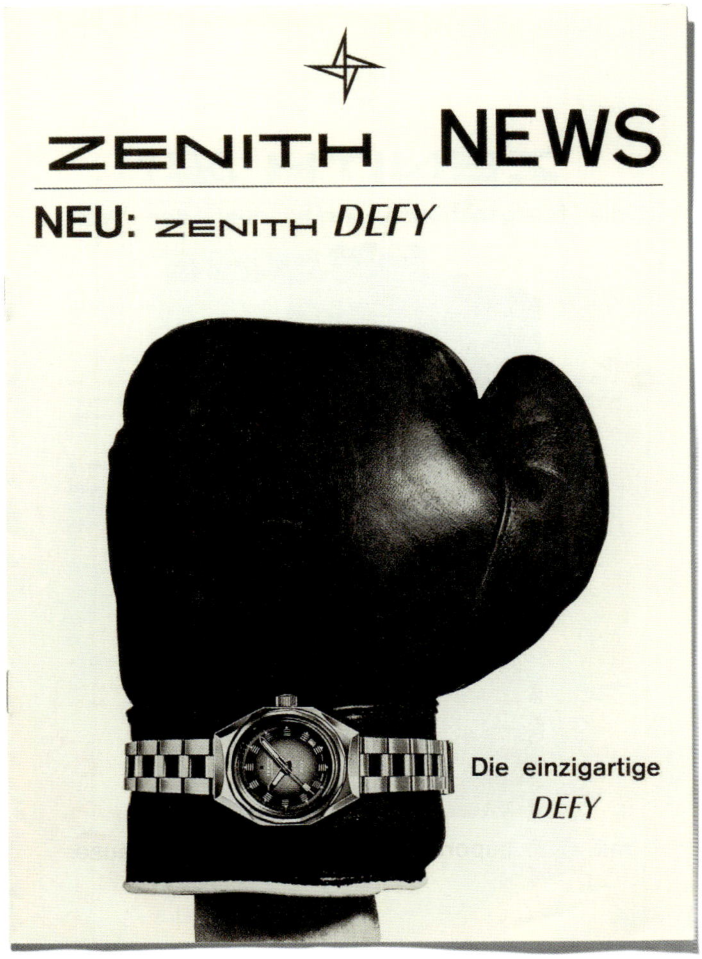

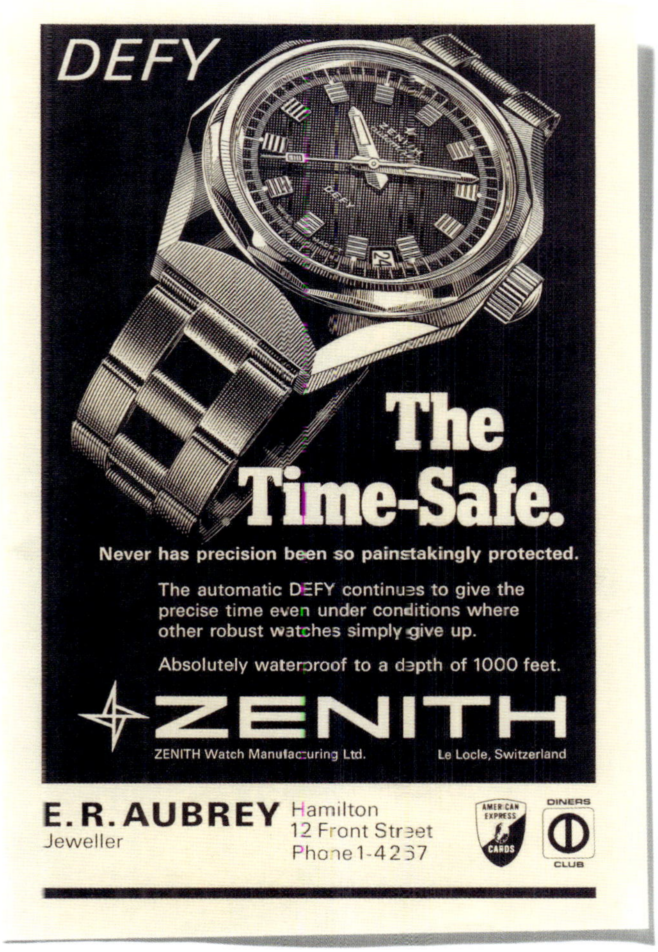

Advertisements / Werbeanzeigen, 1969

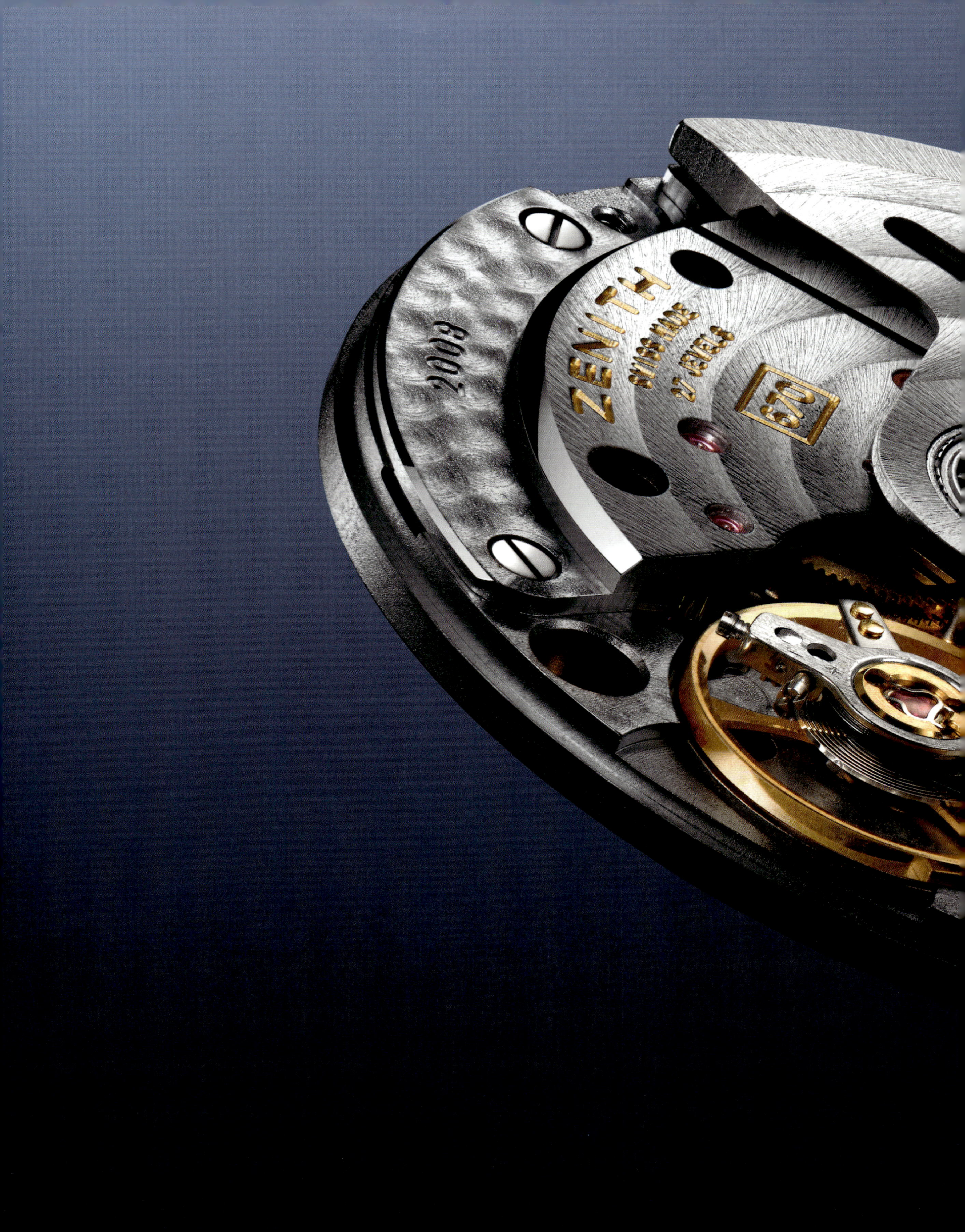
ZENITH
SWISS MADE
670

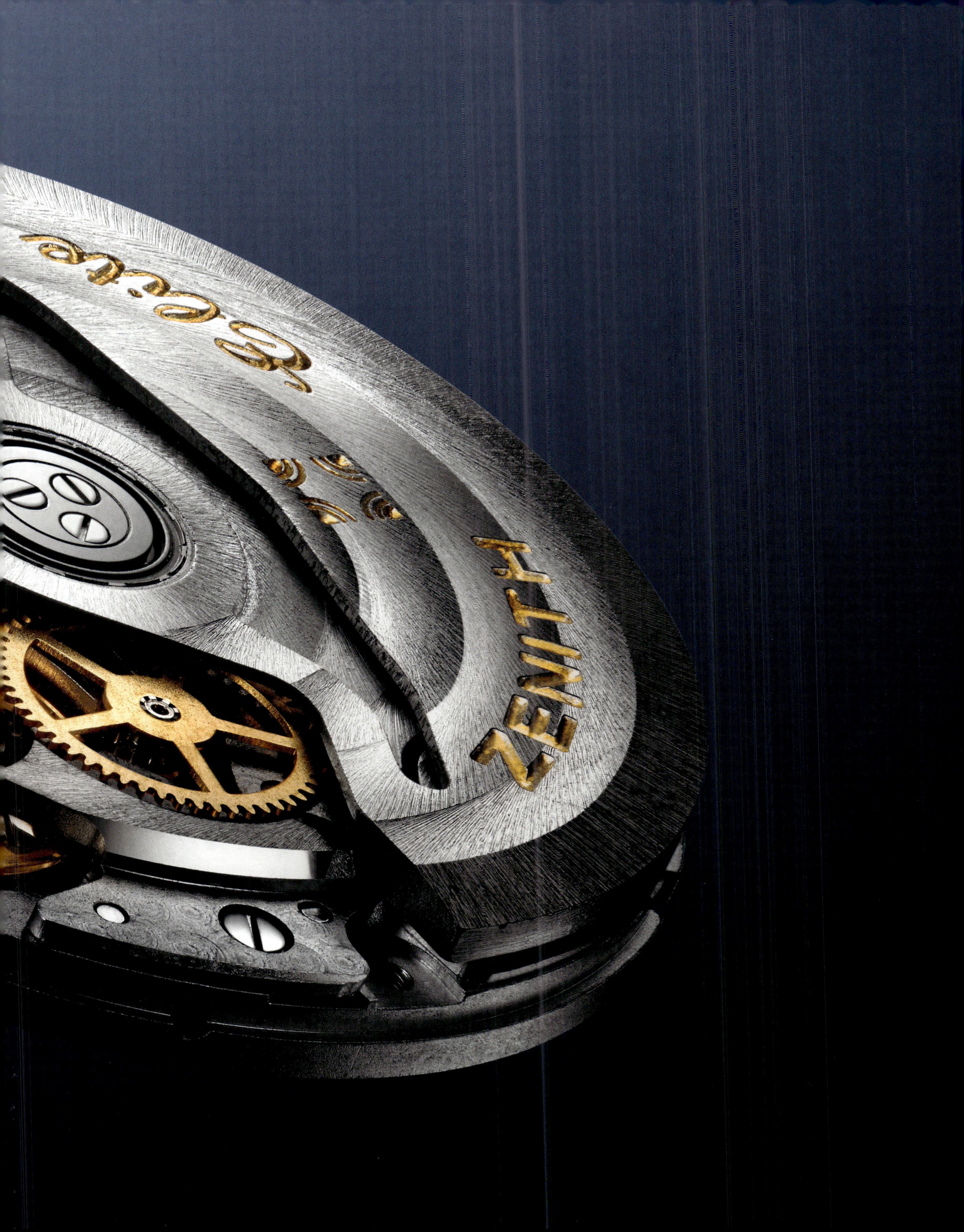
ZENITH

ZENITH
EL PRIMERO

1969

THE FIRST OF ITS ILK

Strictly speaking, the biography of the Zenith classic begins in 1965. This was the year in which the watchmakers of Martel Watch Co., bought in 1960, started to develop the world's first automatic chronograph with ball-bearing rotor, column wheel, horizontal clutch and five-hertz balance frequency. At its presentation to the press on 10 January 1969, El Primero's 29.33-millimetre diameter and height of just 6.5 millimetres ranked it as a superlative. The low volume was based in part on the development of a special automatic bridge, which the watchmakers fix above and next to the chronograph movement. Teamed with an ingenious reverser system to polarise the oscillating weight rotations, a power reserve of 50 hours was quickly built up. The 1969 model series was launched in the autumn and included the steel Reference AH 386 with a 30-minute and 12-hour counter and simple date display. A little later, the Espada iteration with a simple full calendar and moon phase display came onto the market. In the mid-1970s, Zenith Radio Corporation, the US owner of the watch manufacturer founded in 1865 by Georges Favre-Jacot, ordered the destruction of the remaining El Primero 3019 PHC and 3019 PHF calibres (354 parts) and all tools. The horological past was no longer of any significance. The way into an uncertain future was to be led by oscillating quartz crystals. Fortunately, Charles Vermot was not inclined to follow these instructions. The veteran watchmaker hid the treasures in the spacious attic of the factory building and bricked up its doors. Thanks to Ebel and later Rolex, the El Primero movement was revived in 1985. Initially, Zenith delivered the leftover stock. After that, production of the chronograph icon resumed afresh. Discontinuation is unthinkable given that the names Zenith and El Primero have become synonymous. The first automatic calibre with column wheel, horizontal clutch and five-hertz balance frequency has gained the status of an chronograph institution. Zenith has of course continued to improve and develop its models over the years. But even with the new calibre designation, the first of its ilk continues to thrive on its unchanged functionality and more than five decades of history.

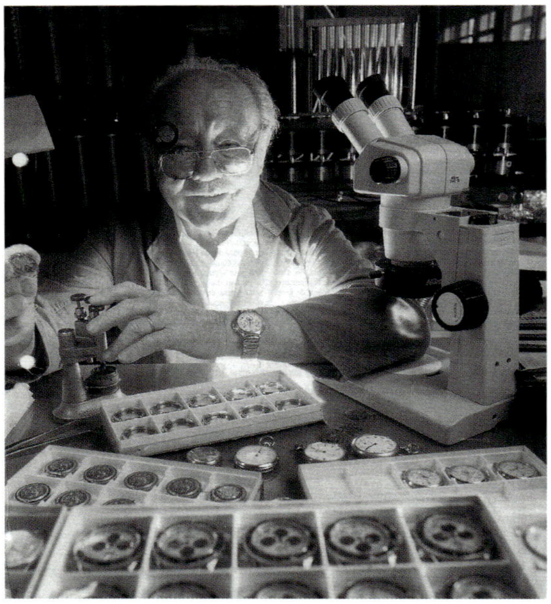

Charles Vermot

DER ERSTE SEINER ART

Genau genommen beginnt die Biographie des Zenith-Klassikers schlechthin im Jahr 1965. Da nämlich starteten die Uhrmacher der 1960 erworbenen Martel Watch Co. mit den Entwicklungsarbeiten zum weltweit ersten Automatik-Chronographen mit Kugellagerrotor, Schaltradsteuerung, Horizontalkupplung und fünf Hertz Unruhfrequenz. Als sich „El Primero" am 10. Januar 1969 der Presse präsentierte, tat er das mit seinen 29,33 Millimetern Durchmesser und nur 6,5 Millimetern Bauhöhe als Superlativ. Die Realisation des geringen Volumens bedingte u. a. die Entwicklung einer speziellen Automatikbrücke, welche die Uhrmacher über und neben dem Chronographen-Mechanismus fixierten. Zusammen mit einem durchdachten Wechsler zur Polarisierung der Rotordrehungen bauten sich zügig rund 50 Stunden Gangautonomie auf. Den 1969er-Modellreigen eröffneten im Herbst des Jahres unter anderem die Stahlreferenz AH 386 mit 30-Minuten-und 12-Stunden-Zähler sowie einfacher Datumsanzeige. Wenig später gelangte die Variante „Espada" mit einfachem Vollkalendarium und Mondphasenanzeige in die Geschäfte. Mitte der 1970er-Jahre ordnete die Zenith Radio Corporation als US-amerikanische Eigentümerin der 1865 von Georges Favre-Jacot ins Leben gerufenen Uhrenmanufaktur die Vernichtung der verbliebenen „El Primero"-Kaliber 3019 PHC und 3019 PHF (354 Werksteile) samt aller vorhanden Werkzeuge an. Die tickende Vergangenheit besaß keine Bedeutung mehr. Schwingende Quarze sollten den Weg in die ungewisse Zukunft weisen. Glücklicherweise mochte sich Charles Vermot den Weisungen nicht fügen. Der altgediente Uhrmacher versteckte die Schätze im weitläufigen Dachboden des Fabrikgebäudes und mauerte dessen Türen zu. Dank Ebel und später auch Rolex kam es ab 1985 zur Wiederbelebung des „El Primero"-Uhrwerks. Zunächst lieferte Zenith die verbliebene Ware. Dann startete die Produktion der chronographischen Ikone aufs Neue.

An ein Ende ist nicht zu denken, denn Zenith und „El Primero" sind mittlerweile so etwas wie Synonyme. Das erste Automatikkaliber mit Schaltradsteuerung, Horizontalkupplung und fünf Hertz Unruhfrequenz ist eine nicht mehr wegzudenkende chronographische Institution. Selbstredend hat Zenith im Laufe der Jahre Verbesserungen und Evolutionen vorgenommen. Aber auch mit neuer Kaliberbezeichnung lebt „Der Erste" von seiner gleich gebliebenen Funktionalität und der mehr als 50-jährigen Geschichte.

***From top left, clockwise:** Product brochure, 1969 | Attic of the Zenith factory building | Cal. 3019 PHC, 1969 | A7817 Espada | Catalogue, 1969–1970*

***Von oben links im Uhrzeigersinn:** Produktbroschüre, 1969 | Dachboden des Fabrikgebäudes | Kal. 3019 PHC, 1969 | A7817 „Espada" | Katalog, 1969–1970*

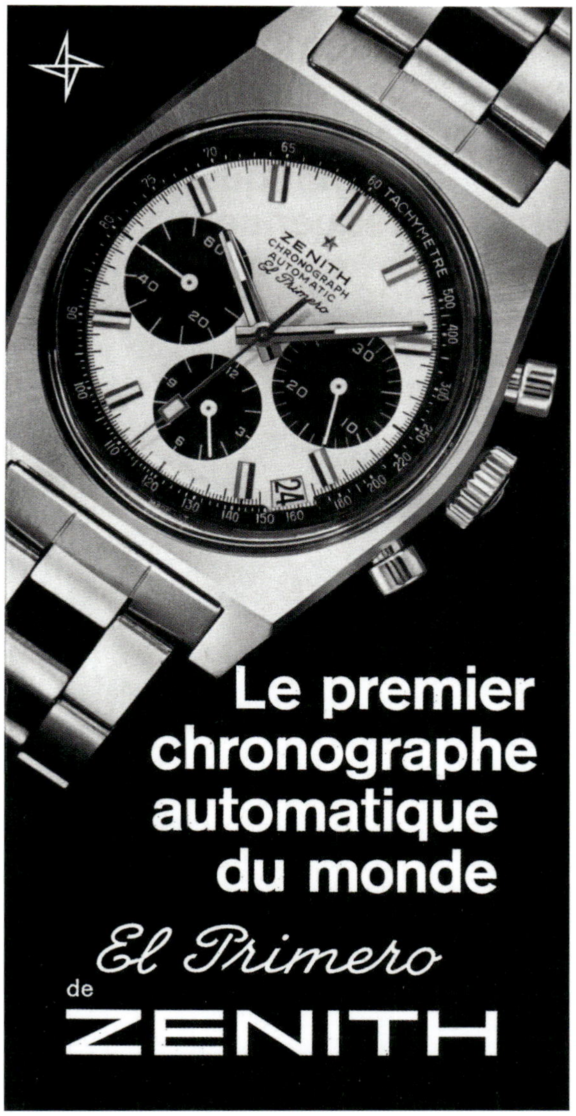
ZENITH
CHRONOGRAPH
AUTOMATIC
El Primero
Le premier
chronographe
automatique
du monde
El Primero
de
ZENITH

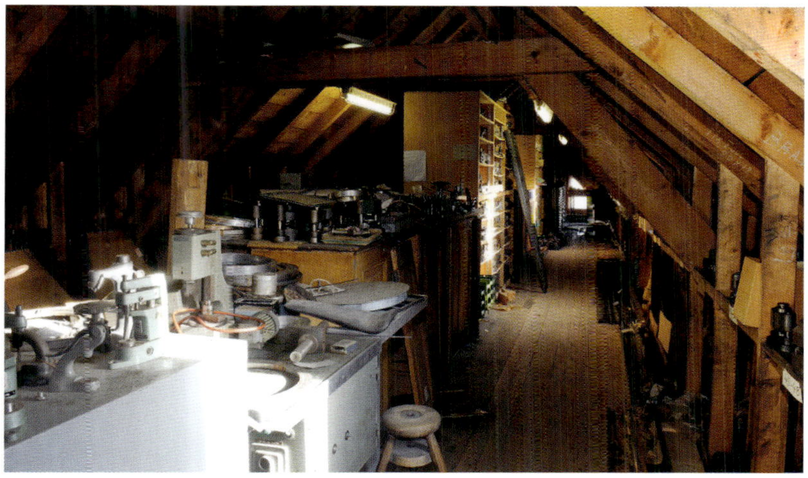

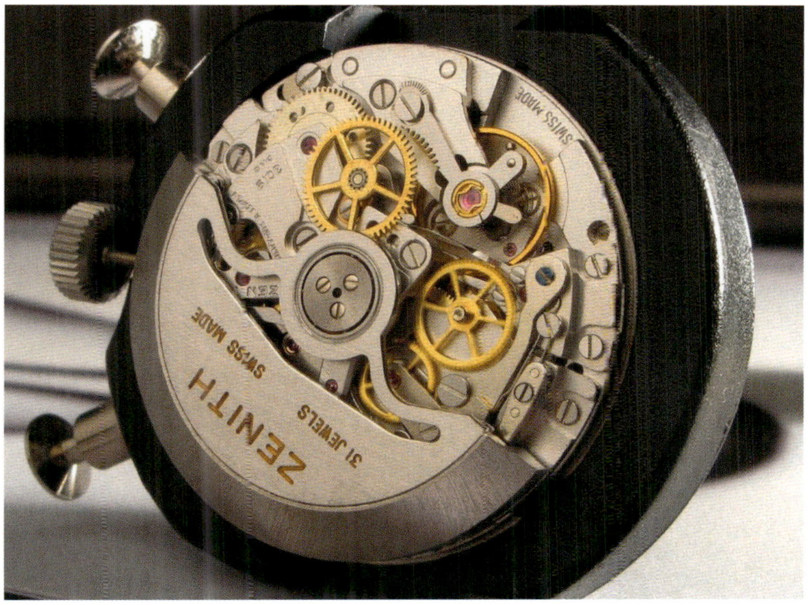
ZENITH
31 JEWELS
SWISS MADE

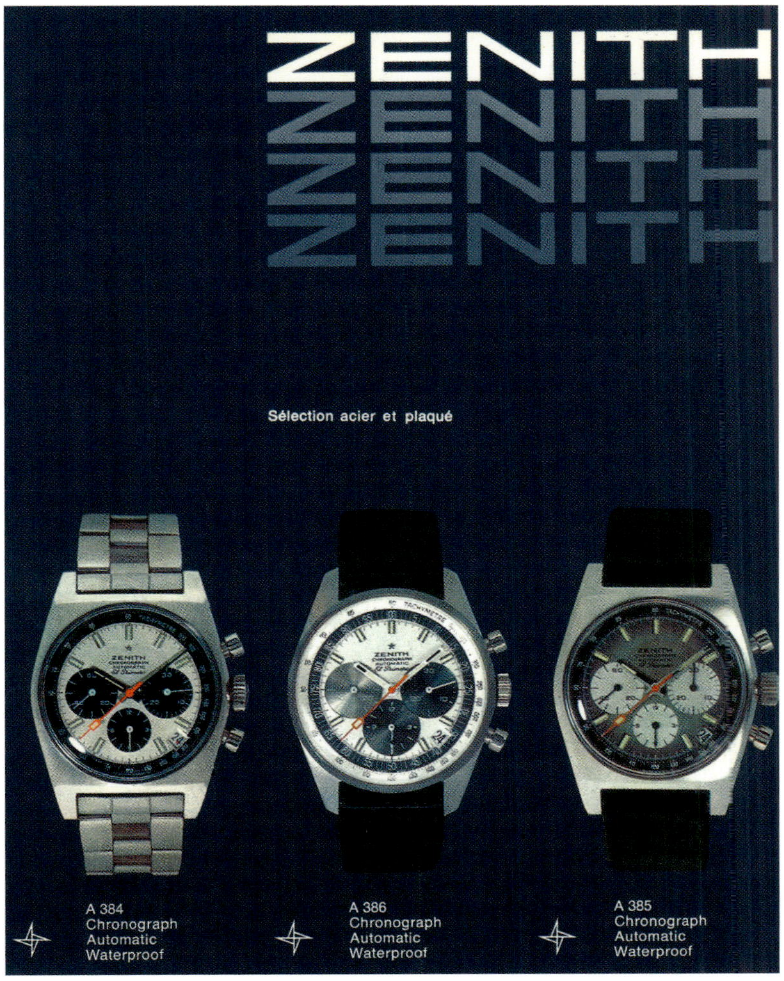
ZENITH
ZENITH
ZENITH
ZENITH
Sélection acier et plaqué
A 384
Chronograph
Automatic
Waterproof
A 386
Chronograph
Automatic
Waterproof
A 385
Chronograph
Automatic
Waterproof

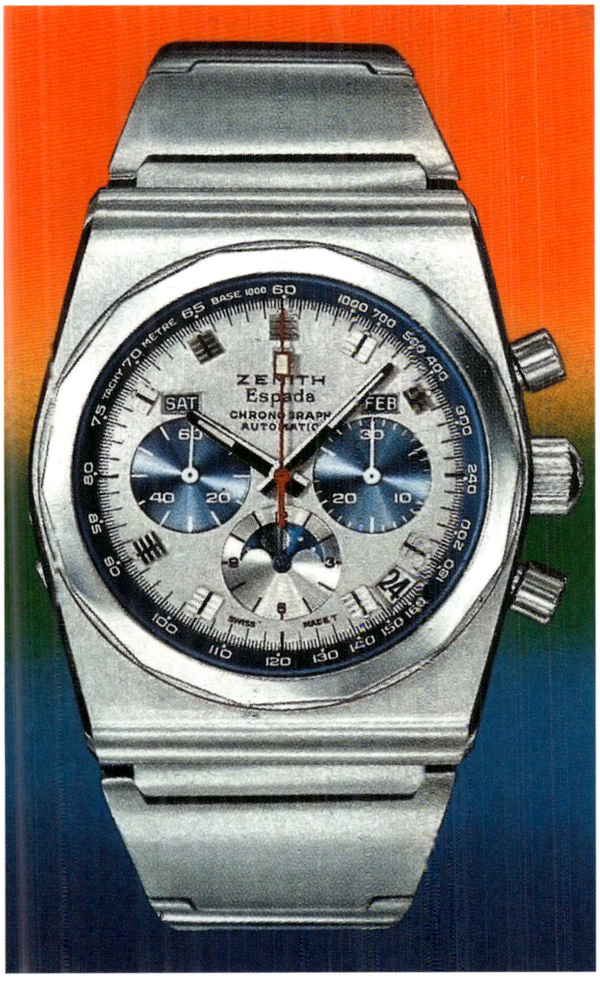
ZENITH
Espada
CHRONOGRAPH
AUTOMATIC
SAT
FEB

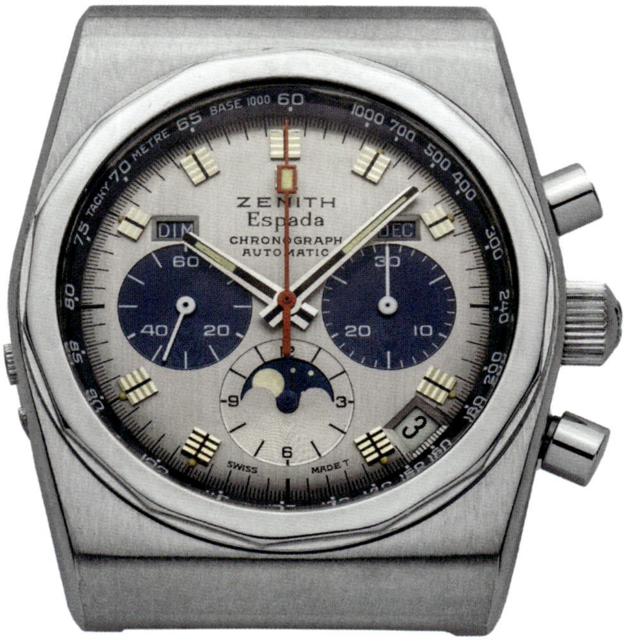

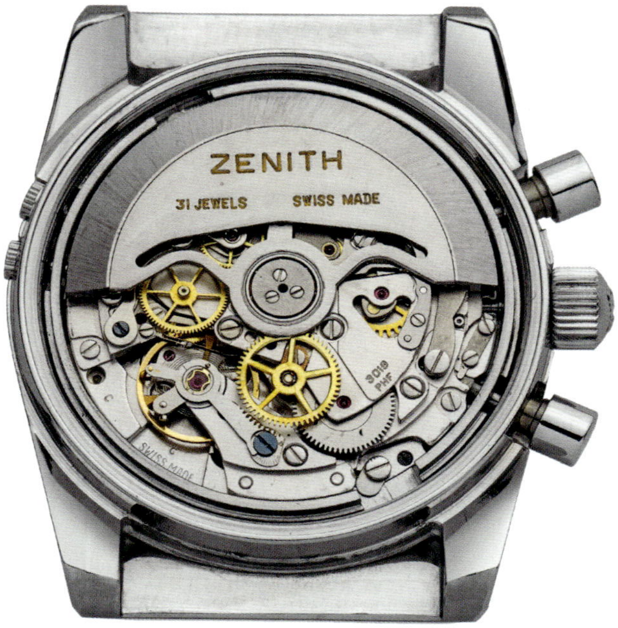

> A386, 1971
< A7817 "Espada" | Cal. 3019 PHF, 1972

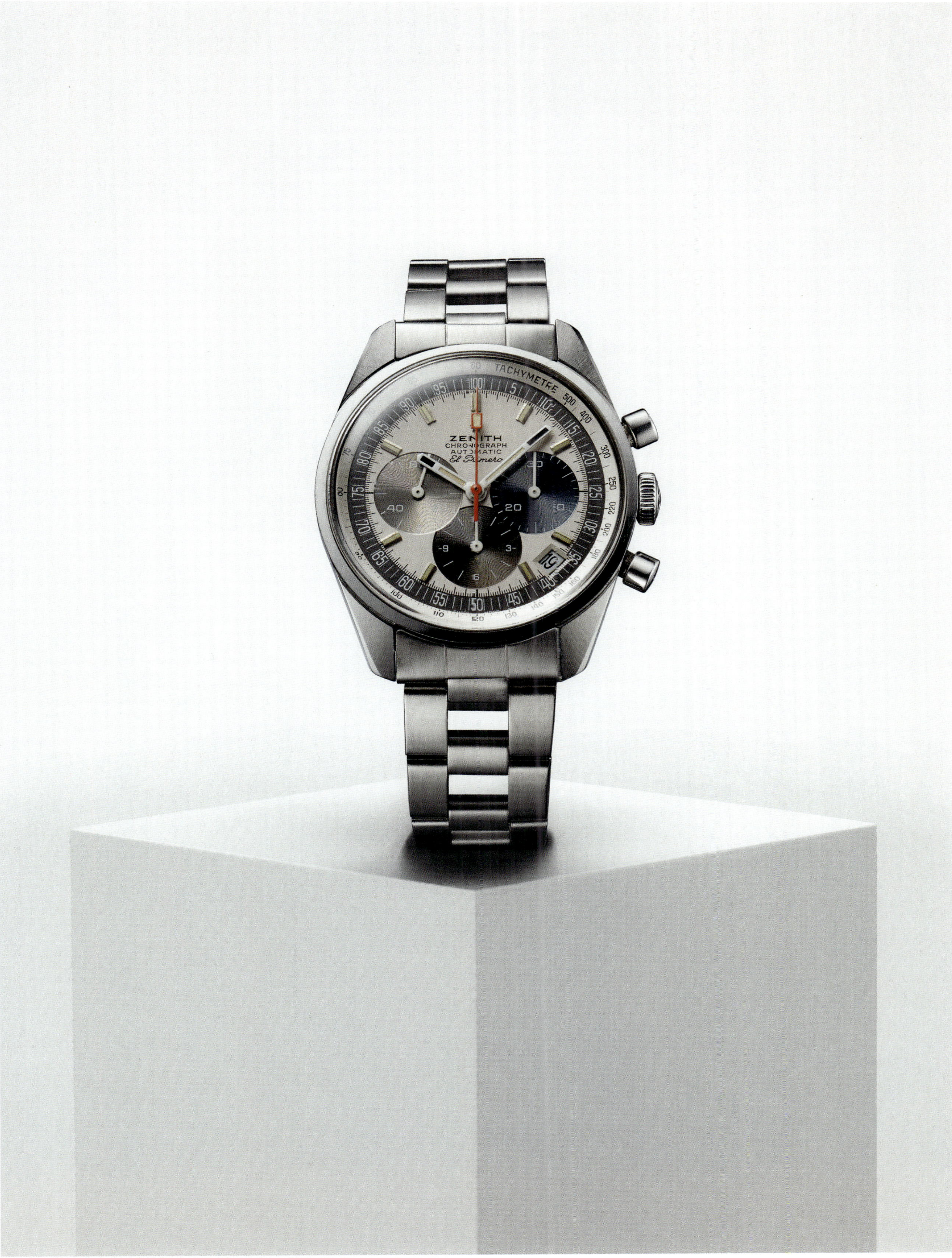
TACHYMETRE
ZENITH
CHRONOGRAPH
AUTOMATIC
El Primero

BIOGRAPHY BIOGRAFIE

GISBERT L. BRUNNER

Born in 1947, Gisbert L. Brunner has pursued his interest in wristwatches, pendulum clocks and other precision timepieces since 1964. During the 1970s quartz crisis, his love of mechanical watches, an ostensibly endangered species, grew. In the early 1980s his passion for collecting watches led him to pen his first articles. More than 25 books on the art of watchmaking were to follow. Brunner writes for publications such as *Chronos Japan*, *Handelszeitung*, *Prestige*, *Watch Around*, *0024 Horloges* and *ZEIT Magazin*. In 2018, together with his partner Wolfgang Winter, Brunner launched the popular internet platform *www.uhrenkosmos.com*.

Building on the success of *The Watch Book I* (2015) and *The Watch Book II* (2016), in June 2017 teNeues Verlag published *The Watch Book Rolex* in German, English and French. The third revised edition was published in 2023.

Recent works published by teNeues Verlag are *The Watch Book More Than Time I* and *II*, in 2020 and 2022, as well as *The Watch Book Oris* and the *Watchmaking History of Switzerland* and *The Watch Book Compendium* in 2023.

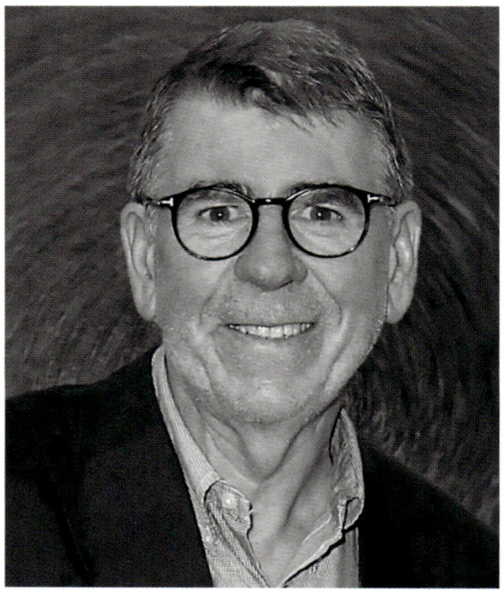

Gisbert L. Brunner

Gisbert L. Brunner, Jahrgang 1947, beschäftigt sich seit 1964 mit Armbanduhren, Pendeluhren und anderen Präzisionszeitmessern. Während der Quarzuhren-Krise in den 1970-Jahren wuchs seine Liebe zu den anscheinend aussterbenden mechanischen Zeitmessern. Ein leidenschaftliches Sammelhobby führte ab den frühen 1980er-Jahren zu ersten Zeitschriftenartikeln und inzwischen mehr als 25 Büchern über dieses Metier. Brunner ist u. a. für Publikationen wie *Chronos Japan*, *Handelszeitung Schweiz*, *Prestige*, *Watch Around*, *0024 Horloges* und *ZEIT Magazin* tätig.

2018 gründete Gisbert Brunner zusammen mit seinem Partner Wolfgang Winter die gut besuchte Internet-Plattform *www.uhrenkosmos.com*.

Nach den erfolgreichen Büchern *The Watch Book I* (2015) und *The Watch Book II* (2016) publizierte der teNeues Verlag im Juni 2017 *The Watch Book Rolex* in den Sprachen Deutsch, Englisch und Französisch. 2023 erschien die dritte überarbeitete Auflage. Weitere Werke im teNeues Verlag sind die 2020 und 2022 erschienenen *The Watch Book More Than Time I* und *II* (zu Deutsch „Mehr als die Zeit“) und die 2023 veröffentlichten *The Watch Book: Oris* und die *Schweizer Uhrengeschichte* und *The Watch Book Compendium.*

CREDITS BILDNACHWEISE

COVER

From left / Von links:

Audemars Piguet Royal Oak Ref. 16202ST.OO.1240ST.01 © Audemars Piguet | Chopard Happy Sport, 33mm Automatic © Chopard | Patek Philippe Nautilus Ref. 3800/1G © Patek Philippe
Cartier Tank Ref. WCL115A20 © Cartier | Rolex Cosmograph Daytona Ref. M126500LN © Rolex | Jaeger-LeCoultre Reverso Tribute Small Seconds © Jaeger-LeCoultre
A. Lange & Söhne Lange 1 Ref. 101.021 © Lange Uhren GmbH | Porsche Design 50Y Chronograph 1 – Limited Edition © Porsche Design

The images related to the wristwatches were made available by the companies' press offices:

Alpina (pp. 6/7) © Alpina | Audemars Piguet (pp. 8–15) © Audemars Piguet; photo: p. 10 (bottom left) Diode SA – Denis Hayoun; p. 12 (bottom left) archive Gisbert L. Brunner (GLB) | Baume & Mercier (pp. 16–19) © Baume & Mercier; photos: p. 17 (left) & p. 19 (bottom right) IMAGIE; p. 16 (top right & bottom) archive GLB | Blancpain (pp. 20/21) archive GLB | Bovet (pp. 22–25) © Bovet | Breguet (pp. 26–29) © Breguet | Breitling (pp. 30–39) © Breitling | Carl F. Bucherer (pp. 40–43) © Carl F. Bucherer; photos: pp. 42/43 WeArePepper, Switzerland | Bulgari (pp. 44–51) © Bulgari | Cartier (pp. 52–57) © Cartier | Chanel (pp. 58/59) © CHANEL | Chopard (pp. 60–73) © Chopard | Chronométrie Ferdinand Berthoud (pp. 74–75) © Chronométrie Ferdinand Berthoud; photos: p. 75 (top right, & bottom) Diode SA – Denis Hayoun, p. 75 (top left) F. Bertin | Chronoswiss (pp. 76–79) © Chronoswiss; photos: pp. 77/78 Anna Janson, p. 78 Koncina Media | Corum (pp. 80–83) © Corum; photos: p. 82 (middle & bottom right) IMAGIE | Cyrus (pp. 84/85) © Cyrus; photos: Marc Amiguet (3,5,7) | Doxa (pp. 86/87) © Doxa; photos: p. 87 (bottom left) Gustavo Kuri | Roger Dubuis (pp. 88–91) © Roger Dubuis; photo: p. 89 (middle right) Diode SA – Denis Hayoun | Philippe Dufour (pp. 92/93) © Philippe Dufour; photo: p. 92 Wollodja Jentsch | Ebel (pp. 94–97) © Ebel; photos: p. 96 Oliver Currat | Frederique Constant (pp. 98–101) © Frederique Constant | Girard-Perregaux (pp. 102–105) © Girard-Perregaux | Glashütte Original (pp. 106–107) © Glashütte Original; photo: p. 107 (bottom) René Gaens Fotografie | Grand Seiko (pp.108–111) © Grand Seiko | Greubel Forsey (pp. 112–113) © Greubel Forsey; photos: p. 112 Alex Teuscher, p. 113 Nicolas Boon | Grossmann (pp. 114/115) © Grossmann Uhren GmbH | Hanhart (pp. 116/117) © Hanhart | Heuer (pp. 118–127) © TAG Heuer | Hublot (pp. 128–133) © Hublot; photos: p. 131 (top) Laurent Xavier Moulin | IWC Schaffhausen (pp. 134–141) © IWC Schaffhausen | Jaeger-LeCoultre (pp. 4, 142–151) © Jaeger-LeCoultre; photos: p. 4 archive GLB, p. 146 (top) Oliver Currat, p. 148 (bottom) Claude Joray | Junghans (152–153) © Junghans | A. Lange & Söhne (pp. 3, 154–161) © Lange Uhren GmbH; photos: pp. 155 (top), 158/159 Erik Schimschar, pp. 156–159 Gieske Studios GmbH; p. 160 (bottom right) Philipp Bögle | Lang & Heyne (pp. 162/163) © Lang & Heyne | Longines (pp. 164/165) p. 164 © Uhrenkosmos; p. 165 © Longines | Maurice Lacroix (pp. 166–169) © Maurice Lacroix | MB&F (pp. 170/171) © MB&F | MeisterSinger (pp. 172–175) © MeisterSinger | Richard Mille (pp. 176–179) © Richard Mille; photo: p. 177 (top right) Paul Calver | Montblanc (pp. 180–183) © Montblanc | H. Moser & Cie. (pp. 184/185) © H. Moser & Cie. | Mühle-Glashütte (pp. 186/187) © Mühle-Glashütte; photo: p. 187 (middle right) Erik Schimschar | Franck Muller (pp. 188/189) © Franck Muller | Nomos Glashütte (pp. 190–195) © Nomos Glashütte | Omega (pp. 196–201) © Omega; archive GLB | Oris (pp. 202–207) © Oris; photo: pp. 207/208 Daan Verhoeven | Panerai (pp. 208–211) © Panerai; archive GLB | Parmigiani Fleurier (pp. 212–215) © Parmigiani Fleurier | Patek Philippe (pp. 216–233) © Patek Philippe; photos: p. 218 Claude Bossel, p. 224 (bottom left & middle) S.Duouchet, pp. 223, 224, 226, 227 Jean-Daniel Meyer | Pequignet (pp. 234–237) © Pequignet; photos: p. 235 (bottom right) Laurent Cheviet | Piaget (pp. 238–241) © Piaget | Porsche Design (pp. 242–245) © Porsche Design | Ressence (pp. 246/247) © Ressence | Rolex (pp. 1, 248–269) © Rolex; photos: p. 248 (top left), 269 (bottom left & second from right) archive GLB, pp. 248 (middle left), 249 (right), 250, 262, 263 (bottom right), 264, 269 (bottom right) Ulysse Fréchelin, p. 249 (left) Thomas Hensinger, p. 248 (bottom left) François Lacour, pp. 254, 255 (middle & bottom right), 259, 266, 267 JVA Studio, pp. 255 (bottom left), 260 (left top & bottom) archive GLB, p. 258 (left & bottom), 265 (top left) Jean-Daniel Meyer, p. 265 (bottom right) Duncan Cole, p. 268 (left) Joël von Allmen, p. 269 (#6) Claude Bossel, (bottom second from left) Alain Costa | Seiko (pp. 270/271) © Seiko | Singer Reimagined (pp. 272/273) © Singer Reimagined; photo: pp. 272 & 273 (left) Laurent Xavier Moulin | Sinn (pp. 274/275) © Sinn Spezialuhren GmbH | Speake Marin (pp. 276–279) © Speake Marin; photo: pp. 276 (left & middle), 277, 279 KUDAIBERGEN, p. 278 Alex Teuscher | Stowa (pp. 280–283) © Stowa | Armin Strom (pp. 284/285) © Armin Strom | Carl Suchy & Söhne (pp. 286/287) © Carl Suchy | Tudor (pp. 288–291) © Tudor | Tutima Glashütte (pp. 292–295) © Tutima Uhrenfabrik GmbH; photos: p. 293 (top) Borris Burba | Ulysse Nardin (pp. 296–299) © Ulysse Nardin; photo: p. 297 (middle) Miguel Bueno | Vacheron Constantin (pp. 300–307) © Vacheron Constantin; photos: p. 303 (top left) Charles-Elie Lathion, pp. 306/307 Alex Teuscher | Louis Vuitton (pp. 308/309) © Louis Vuitton | Zenith (pp. 310–316) © Zenith

Pictures from other sources are listed below:

p. 53 (left) & 56 (top) © Keystone-France/Gamma-Keystone/Getty Images
p. 54 (left) © AGIP / Bridgeman Images
p. 54 (second from right) © Monte Fresco/Mirrorpix/Getty Images
p. 54 (right) © Pierre Schermann/WWD/Penske Media/Getty Images
p. 56 (Jackie Onassis) © David Cairns/Express/Getty Images
p. 57 (top right) © Jean-Jacques LAPEYRONNIE/Gamma-Rapho/Getty Images
p. 57 (bottom right) Michel Dubois, Didier Boursin, Les Ateliers ABC © Cartier
p. 148 (left) © Hulton-Deutsch Collection/CORBIS/Corbis/Getty Images
p. 201 (Daniel Craig) © picture alliance / Everett Collection | ©MGM/Courtesy Everett Collection
p. 230 (left) © Allen J. Schaben / Los Angeles Times/Getty Images
p. 230 (right) © picture alliance / Everett Collection | Steven Ferdman/ Everett Collection
p. 243 (Tom Cruise) © picture alliance / ZUMAPRESS.com | Paramount Pictures
p. 252 (left) © picture alliance / Everett Collection | Courtesy Everett Collection
pp. 252 (second from left & right), 250 (bottom right) © Leonard Ortiz/ Digital First Media/Orange County Register/Getty Images
p. 260 (top middle) © ISC Images & Archives/Getty Images
p. 260 (top right) © Ron Galella/Ron Galella Collection/Getty Images
p. 260 (right middle) © Sotheby's
p. 265 (middle, second image from left) © picture-alliance/ dpa | Str
p. 265 (bottom middle) © picture alliance / AP Photo | KIN CHEUNG

IMPRINT IMPRESSUM

Third printing

Text by Gisbert L. Brunner
Translation into English by Fiona Haberland

Editorial Coordination by Nadine Weinhold, teNeues
Production by Alwine Krebber, teNeues
Design by Iris van Kempen
Color Separation by Robert Kuhlendahl, teNeues

Copyediting by Tillmann Courth (German), Daniel Ross (English)
Proofreading by Nadine Weinhold, teNeues

ISBN: 978-3-96171-610-4
Library of Congress Number: 2024938736

Printed in the Czech Republic by PBtisk a.s.

MIX
Paper | Supporting responsible forestry
FSC® C004378

Bibliographic information published by the Deutsche Nationalbibliothek: The Deutsche Nationalbibliothek lists this publication in the Deutsche Nationalbibliografie; detailed bibliographic data are available on the Internet at dnb.dnb.de.

Published by gestalten, Berlin 2026

Die Gestalten Verlag GmbH & Co. KG
Mariannenstrasse 9-10
10999 Berlin, Germany
hello@gestalten.com

Krefeld Office
Uerdinger Str. 265 / Villa Pattberg
47800 Krefeld, Germany
verlag@teneues.com

teNeues Press Department
press@gestalten.com

For more information and to order books, please visit
www.teneues.com and www.gestalten.com

https://instagram.com/teneuespublishing

www.teneues.com